Maryland Regulars in the War of 1812

Compiled by
Eric Eugene Johnson
President, Ohio Society, 2008–2011
Registrar General, General Society, 2017–2020

Foreword by
Christos Christou, Jr.
President, Maryland Society, 2012–2015
Registrar General, General Society, 2008–2017

HERITAGE BOOKS
2018

HERITAGE BOOKS
AN IMPRINT OF HERITAGE BOOKS, INC.

Books, CDs, and more—Worldwide

For our listing of thousands of titles see our website
at
www.HeritageBooks.com

Published 2018 by
HERITAGE BOOKS, INC.
Publishing Division
5810 Ruatan Street
Berwyn Heights, Md. 20740

Copyright © 2018 War of 1812 Society in the State of Maryland

If you find your ancestor listed in this book, it will likely make you eligible for membership in the War of 1812 Society (if you are male) or Daughters of 1812 (if you are female). Our Society accepts any male from birth up to any age. If interested in membership, please contact us at:

Christos Christou Jr.
President, Maryland Society, 2012-2015
303 Nicholson Rd
Essex MD 21221-6609
410-574-5467
cchristousoc@gmail.com

All rights reserved. No part of this book may be reproduced or transmitted in any form or by any means, electronic or mechanical, including photocopying, recording or by any information storage and retrieval system without written permission from the author, except for the inclusion of brief quotations in a review.

International Standard Book Number
Paperbound: 978-0-7884-5802-6

Dedication

Many thanks to Eric Johnson for his countless hours to compile this publication and to Christos Christou for his innumerable hours of editing and additions. These men have devoted many hours to ensuring the heroes of 1812 are remembered and that many join our society and learn about the deeds of their ancestors. This work is dedicated to our families and future generations to ensure the legacy of these men will always be preserved and revered.

(President Christou created this modified logo
to be used during the Bicentennial only)

Maryland Regulars in War of 1812

Table of Contents

FOREWORD I
ACKNOWLEDGEMENTS III
INTRODUCTION V
THE REGIMENTAL HISTORIES 1
    5ᵀᴴ Regiment of U.S. Infantry 1
    12ᴛʜ Regiment of U.S. Infantry 3
    14ᴛʜ Regiment of U.S. Infantry 4
    36ᴛʜ Regiment of U.S. Infantry 6
    38ᴛʜ Regiment of U.S. Infantry 7
    U.S. Sea Fencibles 8
    U.S. Volunteers 9
    2ɴᴅ Regiment of U.S. Artillery 10
    5ᴛʜ Military District 11
THE ROSTER 18
    A 18
    B 29
    C 47
    D 84
    E 106
    F 113
    G 131
    H 152
    I 187
    J 189
    K 202
    L 213
    M 232
    N 274
    O 282
    P 287
    Q 305
    R 305
    S 326
    T 367
    U 383
    V 384
    W 387
    Y 423
    Z 425
THE BIBLIOGRAPHY 427

By War of 1812 Soc. in MD

Foreword

The War of 1812 Society in the State of Maryland was founded by these soldiers themselves as they stood strong against the British attack in 1814 and stated: *"We will never disband"*. It is this spirit that still flows through our organization and continues today and is inspired by our Presidents of the various forms of our Society until its final formal incorporation in 1893. We thank these men for leading the society for the last 200 years.

Military Commanders (Veterans of 1812)

1st, Military Commander, Maj. Gen. Samuel Smith   1814-1815
2nd Military Commander, Maj. Gen. George H. Steuart 1815-1841

Presidents of Association of Defenders (Veterans of 1812)

1.	Maj. Gen. William McDonald	1841-1845
2.	Col. David Harris	1845-1847
3.	Maj. William Jackson	1847-1848
4.	Col. Joseph K. Stapleton	1848-1852
5.	Capt. Andrew E. Warner, Sr.	1852-1871
6.	Maj. Joshua Dryden	1871-1879
7.	Capt. John J. Danekar	1879-1882
8.	Col. Elijah Stansbury	1882-1884
9.	James C. Morford	1884-1888

Presidents of Association of Descendants

10.	Samuel A. Downs	1889-1890
11.	James Hyland	1890-1891

Presidents of War of 1812 Society
(Incorporated Nov. 2, 1893)

12.	Louis P. Griffith	1891-1894
13.	Hon. Edwin Warfield	1894-1898
14.	James Edward Carr, Jr.	1898-1901
15.	John Mason Dulany	1901-1903
16.	Albert Kimberly Hadel, M.D.	1903-1905
17.	Brig. Gen. Peter Leary, Jr., USA	1905-1911
18.	James D. Iglehart	1911-1914 (Centennial)
19.	Alfred D. Bernard	1914-1916
20.	Charles E. Sadtler, M.D.	1915-1917
21.	Alfred J. Carr (died in office)	1917-1/1923
22.	Thomas Maynadier	1/1923-1926
23.	Maj. Gen. Clinton Levering Riggs	1926-1929

Maryland Regulars in War of 1812

24. James Etchberger Hancock — 1929-1941
25. John Henry Orem, Jr. (died in office) — 1941-02/1942
26. John L. Sanford — 02/1942-1943
27. James Etchberger Hancock — 1943-1945
28. Lt. Col. George E. Ijams, ARNG Md. — 1945-1946
29. Hon. George Washington Williams — 1946-1948
30. William Henry Pitcher — 1948-1950
31. John A. Pentz — 1950-1953
32. Stanley Denmead Kolb — 1953-1956
33. Charles Francis Stein — 1956-1959
34. C. Elliott Baldwin — 1959-1962
35. Herbert Lee Trueheart — 1962-1964
36. Robert Emory Michel — 1964-1966
37. Gordon Malvern Fair Stick — 1966-1968
38. William Henry Lloyd (died in office) — 1968-1969
39. Lt. Col. Curtis Carroll Davis, USAR Ret. — 1970-1972
40. Dennis F. Blizzard — 1972-1974
41. Hon. Wilson King Barnes — 1974-1976
42. Edward Charles Beetem II — 1976-1978
43. S. Vannort Chapman — 1978-1980
44. Clement D. Erhardt, Jr. — 1980-1982
45. Richard Nicolai Hambleton — 1982-1984
46. Gordon Malvern Fair Stick, Jr. — 1984-1986
47. Col. Samuel A. Rittenhouse, USAR Ret. — 1986-1988
48. Wilson K. Barnes, Jr. — 1988-1990
49. PG Carl F. Bessent — 1990-1992
50. Brig. Gen. Edward G. Jones, Jr. Ret. — 1992-1994
51. PG Dandridge Brooke — 1994-1996
52. Lawrence Bailey Chambers — 1996-1998
53. PG Brig. Gen. M. Hall Worthington, Jr. Ret. — 1998-2000
54. Lt. Col. George E. Linthicum, III Ret. — 2000-2002
55. James Frederick Waesche — 2002-2004
56. Lee Crandall Park, M.D. — 2004-2006
57. Clarke Daniel Bowers — 2006-2008
58. Nelson M. Bolton — 2008-2010
59. Charles Pomeroy Ives, III — 2010-2012
60. Christos Christou, Jr. — 2012-2015 (Bicentennial)
61. Joseph Patrick Warner — 2015-2016
62. Cato Douglas Glover Carpenter — 2016-2018

By War of 1812 Soc. in MD

Acknowledgements

Eric Johnson, member of the Ohio Society, has generously given of his time and resources to research the soldiers who served in the War of 1812 in the regular army. Eric's hard work and hours of typing has produced the largest publication of 1812 regular soldiers in Maryland. This work is **not** a complete listing of all soldiers who served in regular service and does not include any of the tens of thousands that served in the militia. That is not the purpose of this work. This work is based upon the Army's Register of Enlistments which is a consolidation of the army's enlistment records except for the period 1798 through 1815. Few of the enlistment records survived during this period so the army used all of its known documents to reconstruct the enlistments. This period which covers the War of 1812 is not complete and there are missing men, many missing men, who served in the army. The Register of Enlistment was used to find the Marylanders and then each researched to see if they had a military land bounty and a pension.

Eric has served in many roles in the War of 1812 Society and published several books, articles, newsletters, etc. In the General Society, he is the current Registrar General 2017-2020, he has been Archivist General 2014-2017, Historian General 2011-2014, Assistant Archivist General 2011-2014, Vice President of Publications 2005-2011. In Ohio, he has served as State President 2008-2011 and current Registrar-Genealogist, Editor 2006-, and Vice President 2005-2008.

Christos Christou, President of the Maryland Society 2012-2015, edited and formatted the final work, made corrections and added information where necessary to support this great publication. He has served in many roles in the War of 1812 Society and published several books and articles, newsletters, etc. for numerous hereditary societies. In the General Society he served nine years as the national Registrar General 2008-2017. He was State President of the Maryland Society during the great Bicentennial 2012-2015 (the state bylaws were changed to allow him to serve three consecutive years), Committee Chairman and member of numerous committees, Vice President, and after his President service as the current Corresponding Secretary 2015-2018. Christos is an expert genealogist and published many genealogical works – books, articles, and newsletters and helped hundreds of people join various hereditary societies.

These men have devoted countless hours to ensuring the heroes of 1812 are remembered and that many join our society and learn about the deeds of their ancestors. This work is dedicated to our families and future generations to ensure the legacy of these men will always be preserved and revered.

Maryland Regulars in War of 1812

By War of 1812 Soc. in MD

Introduction

Marylanders take great pride in the War of 1812 and particularly with its veterans who saved the City of Baltimore in 1814 during the Battles of Fort McHenry and North Point. This state had an effective militia organization that was second to none in this nation.

Most of the state militias which served during this war, never saw combat, refused to cross the borders into Canada, and would only serve for three months. The U.S. Army had to deal with this and these militia units were used to man minor forts, guarding British prisoners of war, running supplies between supply depots and the front lines. Besides Maryland, the militia forces of Kentucky, Louisiana, Ohio, and Tennessee were not afraid of a fight. Other states did have some militia companies which also proved their worth to the war effort.

A very big misconception is that the militia forces of the United States won the War of 1812. In reality, it was the U.S. Army and its' U.S. Voluntary Corps which fought the British to a stand-still. The U.S. Volunteers were raised by the army. The president of the United States awarded volunteer officer commissions to qualified men as colonels and captains. These men raised the regiments and independent companies which made up the U.S. Volunteers.

The U.S. Volunteers served in one year enlistments. The force was renewed for another year and in the third year, they were incorporated into the U.S. Army as the 45th through 48th Regiments of U.S. Infantry. So successful was this branch of the U.S. Army that the U.S. Volunteers became the primary fighting force during the Mexican-American War, the Civil War, the Spanish-American War and the Philippine War. The U.S. Volunteers would be replaced in 1903 by the National Guard.

Researching your militia veteran who served in the War of 1812 is simple. Most served for only a few days during an emergency or signed up for only three months or six months. They served in a single regiment and usually had only one company commander during their tour of duty. Their *Consolidated Military Service Records* will usually contain excerpts from a single muster roll and from a single pay roll report. If the veterans were alive during the 1850's, they qualified for land bounties and if alive during the 1870's, they qualified for service pensions. If they were wounded or had died during the war, then they or their heirs qualified for military pensions.

On the other hand, researching your U.S. Army veteran is far more detailed than your militia veteran. Many officers and enlisted men who served during the war had joined the army prior to 1800 and many of them who were commissioned or enlisted during the war, served through the Mexican-American War and into the early part of the Civil War.

Prior to 1808, the army consisted only of the 1st and 2nd Regiments of Infantry and the Artillery Regiment along with the various departments of the army. These three regiments recruited nation-wide.

Maryland Regulars in War of 1812

In 1808, with the renewed threat of war with Great Britain, the U.S. Congress created the 3rd through 7th Regiments of Infantry, the Rifle Regiment, the Light Dragoons Regiment and the Light Artillery Regiment. The infantry regiments were raised in designated regions within the nation, while the other regiments were raised nationally with each region raising a given number of companies for each regiment.

The 3rd Regiment of Infantry was raised in the Carolinas and Georgia. The 4th Infantry was raised in New England. The 5th Infantry was raised in Maryland, District of Columbia, Virginia and Pennsylvania. The 6th Infantry was raised in New York. The 7th Infantry was raised in Kentucky and Tennessee. These regiments also recruited in the states surrounding their designated recruiting areas.

After the Battle of Tippecanoe in the fall of 1811, Congress added another ten infantry regiments, two artillery regiments and a light dragoons regiment to the U.S. Army on 11 January 1812. Maryland was assigned the raising of the 14th Regiment of Infantry while Virginia raised the 12th Regiment of Infantry.

Under the Act of 26 June 1812, eight additional infantry regiments were created. The 20th Regiment of U.S. Infantry was assigned to Virginia. This regiment was raised primarily from what is now West Virginia and western Virginia while the 12th Infantry continued to be raised in central and eastern Virginia. The 12th Infantry also recruited in the panhandle counties of Maryland.

On 29 January 1813, Congress created twenty more infantry regiments but converted the last regiment into ten companies of U.S. Rangers for duty in Ohio, Tennessee, and in the Territories of Indiana, Illinois and Missouri. The 35th Regiment of U.S. Infantry was assigned to Virginia while the 36th Regiment of U.S. Infantry was assigned to the District of Columbia and the 38th Regiment of U.S. Infantry was assigned to Maryland.

The army created nine military districts on 19 March 1813 before the new regiments became operational. Virginia (including West Virginia), the District of Columbia and Maryland became the 5th Military District. Each military district was commanded by either a brigadier general or a general major from the U.S. Army.

The military districts continued to support and to recruit new men for the local regiments and companies that were sent to the three armies fighting the British. These armies were the Army of the North, the Army of the Northwest, and the Army of the Southwest. The newer regiments remained in the military district and along with the militia regiments, they were used to defend the districts.

The 35th Infantry was raised in southeastern Virginia. The 36th Infantry was raised in the District of Columbia, western Maryland and eastern Virginia while the 38th Infantry was raised in southern Maryland. The 5th Infantry was assigned to the 4th Military District and continued to recruit men primarily in Pennsylvania. However, the regiment maintained a recruiting station in Baltimore until the end of the war.

By War of 1812 Soc. in MD

The 48th Regiment of U.S. Infantry was assigned to Vermont and on 12 May 1814 the regiment was renumbered as the 26th Infantry. The original 26th Infantry, which was raised in Ohio, had been disbanded and its men merged into the 19th Regiment of U.S. Infantry. The new 26th Infantry had recruiting stations in New York City, Philadelphia and Baltimore.

The 5th Military District raised numerous companies of artillery, rifles and light dragoons for the U.S. Army. The district also raised companies of U.S. Volunteers and two Sea Fencibles companies. Sea Fencibles were army companies made up of army officers and enlisted seaman. They were primarily used as heavy artillery units in the major forts on the eastern seaboard.

On 17 May 1815 the army was downsized to its pre-war strength. All of the regiments were merged into eight infantry regiments, light artillery regiment, a corps of artillery and a rifle regiment. The enlisted men who had enlistments of twelve months, eighteen months and the term of "during the war" were dismissed from the service and the men who had enlisted for five years were assigned to these new regiments.

The three Virginia regiments merged into the new 8th Regiment of U.S. Infantry while the regiments from the District of Columbia and Maryland merged into the new 4th Regiment of U. S. Infantry. The 5th Regiment of U.S. Infantry merged into the new 3rd Regiment of U.S. Infantry.

Regiment	Military District	Headquarters
5th	9th Military District	Harrisburg, PA
12th	9th Military District	Staunton, VA
14th	9th Military District	Baltimore, MD
20th	5th Military District	Fredericksburg, VA
35th	5th Military District	Petersburg, VA
36th	5th Military District	Georgetown, DC Richmond, VA
38th	5th Military District	Baltimore, MD
1st Rifle	6th & 9th Military Districts	Shepherdstown, VA Savannah, GA
3rd Rifle	5th & 6th Military Districts	Charlotte, NC Bath Court House, VA Gallatin, TN

Maryland Regulars in War of 1812

The Army's Register of Enlistments pertaining to the War of 1812

The book, *Register of Enlistments in the U.S. Army, 1798-1914,* is an extract of personnel information for the land forces of the United States obtained from documents generated by the U.S. Army. This book can be found at the National Archives in Washington, D.C. and on-line through Ancestry.com.

The first section of this book covers the period from 1798 to 17 May 1815. This section of the book is also called the *Records of the Men Enlisted in the U.S. Army Prior to the Peace Establishment, May 17, 1815.* The majority of the 7,360 pages in this section include the enlistments and commissions of men who served during the War of 1812.

The Army branches included in this section are the infantry, light dragoons, artillery, rifles and rangers. Men who served in the various army departments are also listed. These departments are the adjutant general's, inspector general's, judge advocate general's, quartermaster's, subsistence, medical, pay, corps of engineers, topographical engineers, and ordnance.

Officers and enlisted men who served on the staffs of the regiments, brigades and division plus the auxiliary units (U.S. Volunteers, Sea Fencibles, Canadian Volunteers and West Point cadets) are also listed. There are entries for militiamen and British prisoners of war if they were mentioned in any army document. Finally, the men from the U.S. Marines can also be found in this register. If your ancestor served in the War of 1812, it may be beneficial to search these pages even though your ancestor did not serve in the U.S. Army.

The U.S. Army required a number of different types of reports to be completed in a timely manner. Certain reports were required daily, weekly, monthly, quarterly, or semi-annually. Each regiment maintained their own 'recruiting service' and as such, produced weekly recruiting reports. These reports were consolidated by each recruiting district into monthly and quarterly reports. Besides the recruiting reports, there were initial clothing and equipment reports, needed to be completed for the new soldiers.

A Muster Roll containing the names and ranks of all of the men in each company was required every two months. These reports were completed on the last day of February, April, June, August, October and December. Another type of muster roll was required each month showing only the number of men by each rank who were fit for duty.

This report also included the names of the men who were not fit for duty, where they were located and why they were not fit for duty. In most cases, these men were listed by their surnames and the first initial of their given name.

An inspection report was required quarterly. This report listed each man in a company plus what type of small arms that man had been issued. Both the muster roll and the inspection report were required as a semi-annual report covering a six-month period ending each June 30th and December 31st.

While in camp, regiments were required to produce Morning Reports. These reports listed the number of men by rank who were fit for duty that day, who were on sick call, who were in hospitals, who recently had died, or who were 'on command.' A soldier 'on command' was performing duties away from his company.

Each company maintained a Descriptive Roll listing every man in the company, his age, height, weight, color of eyes and complexion, where born, where enlisted, the name of the officer who enlisted them, enlistment date, length of enlistment, and bonus amounts that were paid and what was due. These reports could also include a list of clothing and other supplies that were issued to the soldiers.

A Clothing Report was required when new uniforms or replacement uniforms, new equipment or replacement equipment were issued to the men. Paydays in the army occurred every two months at the end of odd numbered months. Paydays produced payroll reports.

The last required muster was the Muster Out Report, which ended the military service of an individual. Normally all of the men within a regiment who were being discharged were organized into a 'muster out' detachment. The commanders of these 'muster out' detachments were normally officers who were also being discharged, usually lieutenants or ensigns.

Company, regiments and brigades were required to maintain an Orders (or Orderly) Book and a Record Book. All up and down the chain of command clerks copied reports and letters and then posted them in these books. Depots, forts, recruiting facilities, hospitals and other military installations also were required to maintain an orderly book and a record book. General orders and special orders were written down, copied many times and distributed to all concerned parties.

Regiments were also required to keep copies of the Description Rolls, all correspondences, and all monthly returns. Other types of reports include a soldier's transfer papers from one company or regiment to another, court martial proceedings, prisoner of war listings, discharged certificates and certificates of death.

As stated, the *Register of Enlistments in the U.S. Army, 1798-1914* contains personnel information extracted from army reports. This information includes the name and rank of each soldier or marine, his regiment, his company commander's name, his regimental commander's name, a physical description of the soldier plus his peacetime occupation, where he was born (county and state, or country), where he enlisted and the period of enlistment, and any additional remarks. All of the columns may not be filled in.

List of abbreviations found in the *Register*:

MoRet or MRet - Monthly Returns
MR - Morning Reports or Muster Rolls
RR - Recruiting Returns
DR - Descriptive Rolls
GO - General Orders
SO – Special Orders
IR - Inspection Returns
CoBook - Company book
SAMR – Semi-Annual Muster Rolls
SAIR – Semi-Annual Inspection Returns
PR - Payroll Reports
OB - Orders Book or Orderly Book

Bibliography

Register of Enlistments in the U.S. Army, 1798-1914; (National Archives Microfilm Publication M233, 81 rolls); Records of the Adjutant General's Office, 1780's-1917, Record Group 94; National Archives, Washington, D.C.

Adjutant and Inspector General's Office, *Military Laws and Rules and Regulations for the Army of the United States*, September 1816, (E. De Krafft, Printer: Washington, DC).

By War of 1812 Soc. in MD

The Plan of Action

The U.S. Army kept far better records than the state and territorial militias. The army, however, did not annotate on the enlistment records the residences of its soldiers, only where they were born and where they enlisted. The majority of these records are incomplete so there will be errors in identifying men who were citizens of Maryland during the War of 1812 and who were commissioned or enlisted in the U.S. Army. This book does not include the men who served in the army from other states and who moved to Maryland after the war.

I established a 4-part criterion in order to identify Marylanders who served in the U.S. Army between 1798 and 1820:

1) If a soldier was a member of the 14th or 36th Regiments of U.S. Infantry, Captain Nathan Towson's artillery company, the Baltimore Volunteers and the two U.S. Sea Fencibles companies (raised by Maryland) regardless of the birth place or the enlistment place of the soldier.
2) If a soldier was born in Maryland (enlistment records)
3) If a soldier enlisted in Maryland (enlistment records) or
4) If the soldier was married in Maryland during the war (pension indexes)

It is to be noted that the men who enlisted for five years in 1808 were still serving during the war and that the men who had enlisted for five year during the war could still be serving up to 1820. Many men made a career in the army serving before and after these years.

The 14th Regiment of U.S. Infantry, Captain Towson's artillery company and the Baltimore Volunteers continued to recruit men after being sent to New York State early in the war. These new men were from New York or New England.

Besides determining the criteria of those soldiers who were Marylanders, I needed to identify the source material that I would be using for this book. I first created a database and entered the names of the army officers from Maryland who were serving in the army between 1798 and 1815. This list was obtained from the *List of Officers of the Army of the United States from 1779-1900*.

I then used the *U.S. Army, Register of Enlistments, 1798-1914*, the *War of 1812 Pension Application Files Index, 1812-1815*, and the *U.S. War Bounty Land Warrants, 1789-1858* files. The *Register of Enlistments* is an extract of personnel information from the land forces of the United States obtained from documents generated by the U.S. Army. These documents include enlistment records, recruiting reports, discharge records, commissioning records, muster rosters, inspection reports, morning reports, court-martial records, etc.

The first section of *Register of Enlistments* covers the period from 1798 to 17 May 1815. This section of the book is also called the *Records of the Men Enlisted in the U.S. Army Prior to the Peace Establishment, May 17, 1815*. It contains 7,360 pages with approximately 10 to 20 men listed on each page. The majority of these pages were generated during the War of 1812. I went through this section reading each name and determining each man's residence.

I then tackled the 76,126 records of the *War of 1812 Pension Application Files Index, 1812-1815* file. As I did with the *Register of Enlistments*, I entered into my database the Marylanders who I found in this index and copied their personal information.

Finally, I reviewed the *U.S. War Bounty Land Warrants, 1789-1858* file. This file contained 27,433 warrants that were issued to soldiers under the Congressional Acts of 1812, 1814 and 1842. The majority of these warrants were issued to soldiers between 1817 and 1819. I did use the *U.S. General Land Office Records, 1796-1907* file in my research to identify the officers who received land bounties in 1850 or 1855.

Conclusion

As stated, the purpose of this book is to identify Maryland residents who served in the U.S. Army during the War of 1812 and not to create a detailed service record for each individual man. There are mistakes! Due to misspelling of surnames, missing records, and other factors, some men are missing, some men are listed twice, and some men are not properly identified. Overall, this book should be used to identify soldiers and to point out what records are available for each individual.

It is entirely safe to say that Marylanders served in nearly every U.S. Army regiment during the war and that they also fought in every major battle in both the United States and in Canada. Marylanders, by birth, had migrated to all parts of the United States by the start of the war. Marylanders, who were seamen, joined the army at every major U.S. seaport after the British blockade became too dangerous to sail into the open waters around our coastline.

There are a total of 5,452 men listed in this book. Twenty-seven men have been identified as being African-Americans; 545 men became prisoners of war; 541 men died during the war either in battle, from wounds, from diseases, or from injuries.

There were ten Marylanders who graduated from the U.S. Military Academy at West Point, NY, and who served in this war. There is also one woman, Mary Harker, who may have been a washerwoman in the 14th Infantry.

I hope this book will honor the memory of those Marylanders who fought in this war while serving as members of the U.S. Army.

Eric E. Johnson   1 January 2015

The Regimental Histories
5th Regiment of U.S. Infantry

The 5th Regiment of U.S. Infantry was organized under the Congressional Act of 12 April 1808. The regiment consisted of ten companies, with a headquarters staff, totally 849 men. The regiment was restructured under the act of 26 June 1812 increasing the number of privates to 220. This gave the regiment the manpower strength of 1,070 officers and men.

The headquarters and the main recruiting rendezvous was established at Harrisburg, Pennsylvania, and once the regiment was raised, it was assigned to the Army of the North in New York State. The regiment recruited in the middle Atlantic states, which included, Pennsylvania, southern New Jersey, Delaware, Maryland, the District of Columbia, and Virginia. West Virginia was a part of Virginia during this war and the regiment did recruit in this area of Virginia.

The enlistment period was set at five years for the enlisted men. On 12 December 1812, 'during the war' enlistments were approved for this regiment. This simply met that these men would enlist for the duration of the war and then they would be discharged at the end of the conflict.

Each army regiment formed its own recruiting service during the war. Qualified men were transferred from other regiments, from the militia or from civilian life to become officers. Once commissioned, these officers were assigned to a recruiting station within the middle Atlantic states and they began the process of recruiting the enlisted men.

Weekly recruiting reports from the recruiting stations were consolidated at the recruiting rendezvous at Harrisburg and when the total number of recruits reached 100, they were ordered to this city. The recruits plus selective officers from the recruiting service were then formed into a new company, and the company then march off to New York State. It was rare to use new recruits to backfill companies in the field. Once companies dwindled in size in the field, they were combined into new companies and the excess officers were then sent back to the recruiting service.

William D. Beall of Maryland was the regiment's first colonel. He served from 24 Apr 1812 to 15 Aug 1812. Daniel Bissell of Connecticut was the regiment's second colonel, he served from 15 Aug 1812 to 9 Mar 1814 then he was promoted to brigadier general. The regiments final colonel was John Bowyer of Virginia. He served from 13 Mar 1814 to 17 Mar 1815.

Maryland Regulars in War of 1812

Under the act of 3 March 1813 a second major's position was authorized along with ten 3rd lieutenant's positions and ten additional sergeants positions. This act set the new manpower strength for the regiment at 1,091 men. The additional manpower was used to supplement the regiment's recruiting efforts.

The regiment was assigned to the 4th Military District on 19 March 1813 which included Pennsylvania, southern New Jersey and Delaware. The regiment continued to operate a recruiting station in Baltimore until the end of the war.

Under the act of 3 March 1815 selected officers and the enlisted men who had enlisted for five years were assigned to the new 3rd Regiment of U.S. Infantry. The other officers and the men who had 'during the war' enlistments were discharged. 298 Marylanders have been identified as serving in this regiment.

Battles, Actions, Etc.

6 Jun 1813	Stony Point, UC
26 Oct 1813	Chateaugay, UC
1-2 Nov 1813	French Creek, NY
30 Mar 1814	La Cole Mill, LC
15 Oct 1814	Chippewa, UC
19 Oct 1814	Cook's Mill, UC (also called Lyon's Creek)

By War of 1812 Soc. in MD

12th Regiment of U.S. Infantry

The 12th Regiment of U.S. Infantry was organized under the Congressional Act of 11 January 1812 as part of a ten-regiment buildup of the army prior to the beginning of the War of 1812. The regiment was authorized for a maximum period of five years. The regiment was structured into two battalions of nine 110-man companies. A lieutenant colonel commanded each of the battalions with a colonel serving as the regimental commander.

The 12th Infantry was reorganized under the act of 26 July 1812 and restructured into ten 102-man companies. One lieutenant colonel's position was eliminated. The regiment's authorized manning was set at 1,070 men. Enlistments were authorized for five years and 18-month enlistments were added on 8 April 1812. On 12 December 1812 'during the war' enlistments were approved for this regiment.

The headquarters and the main recruiting rendezvous was established at Staunton, Virginia, and once the regiment was raised, it was assigned to the Army of the North in New York State. The regiment recruited in Virginia, what is now West Virginia, and in the panhandle counties of Maryland.

Three Virginians served as colonels of the 12th Infantry: Thomas Parker commanded from 12 March 1812 to 12 March 1813 when he was promoted to brigadier general; Isaac A. Coles commanded from 12 March 1813 to 24 January 1815; and Robert Pegram commanded from 21 December 1814 to 17 May 1815.

Under the Act of 3 March 1813 a second major's position was authorized along with ten 3rd lieutenant's positions and ten additional sergeants positions. This act set the new manpower strength for the regiment at 1,091 men. The additional manpower was used to supplement the regiment's recruiting efforts. The regiment was assigned to the 5th Military District on 19 March 1813.

Under the act of 3 March 1815 selected officers and the enlisted men who had enlisted for five years were assigned to the new 8th Regiment of U.S. Infantry. The other officers and the men who had 'during the war' enlistments were discharged. 122 Marylanders have been identified as serving in this regiment.

Battles, Actions, Etc.
28 Nov 1812	Black Rock, NY
11 Nov 1813	Crystler's Field, UC
30 Mar 1814	La Cole Mill, LC
28 Jun 1814	Odelltown, NY
24 Aug 1814	Bladensburg, MD
13 Sep 1814	Ft. McHenry, MD

Maryland Regulars in War of 1812

14th Regiment of U.S. Infantry

The 14th Regiment of U.S. Infantry was organized under the Congressional Act of 11 January 1812 as part of a ten-regiment buildup of the army prior to the beginning of the War of 1812. The regiment was authorized for a maximum period of five years. The regiment was structured into two battalions of nine 110-man companies. A lieutenant colonel commanded each of the battalions with a colonel serving as the regimental commander.

The 14th Infantry was reorganized under the act of 26 July 1812 and restructured into ten 102-man companies. One lieutenant colonel's position was eliminated. The regiment's authorized manning was set at 1,070 men. Enlistments were authorized for five years and 18-month enlistments were added on 8 April 1812. On 12 December 1812 'during the war' enlistments were approved for this regiment.

The headquarters and the main recruiting rendezvous was established at Baltimore Maryland, and once the regiment was raised, it was assigned to the Army of the North in New York State. The regiment recruited in Maryland, the District of Columbia, and parts of Pennsylvania and Delaware.

Two Marylanders served as colonels of the 14th Infantry. William H. Winder served from 6 July 1812 to 12 March 1813 when he was promoted to brigadier general. Charles G. Boerstler served from 20 Jun 1813 to 17 May 1815.

Under the Act of 3 March 1813 a second major's position was authorized along with ten 3rd lieutenant's positions and ten additional sergeants positions. This act set the new manpower strength for the regiment at 1,091 men. The additional manpower was used to supplement the regiment's recruiting efforts. The regiment was assigned to the 5th Military District on 19 March 1813.

Under the act of 3 March 1815 selected officers and the enlisted men who had enlisted for five years were assigned to the new 4th Regiment of U.S. Infantry. The other officers and the men who had 'during the war' enlistments were discharged.

By War of 1812 Soc. in MD

A total of 1,351 men were commissioned or were recruited to serve in this regiment. Of these men, 463 became prisoners of war and 220 men were killed or had died while serving.

<u>Battles, Actions, Etc.</u>
12 Jun 1813	Stony Creek, UC
21 Nov 1812	Bombardment of Ft. Niagara, NY
28 Nov 1812	Black Rock, NY
27 May 1813	Capture of Ft. George, UC
24 Jun 1813	Beaver Dams, UC
11 Nov 1813	Crystler's Field, UC
30 Mar 1814	La Cole Mill, LC
13 Sep 1814	Ft. McHenry, MD
15 Oct 1814	Chippewa, UC
19 Oct 1814	Cook's Mill, UC (also called Lyon's Creek)

Maryland Regulars in War of 1812

36th Regiment of U.S. Infantry

The 36th Regiment of U.S. Infantry was organized under the Congressional Act of 29 January 1813 as part of a twenty-regiment buildup of the U.S. Army. Initially these newer infantry regiments were to be raised for only one year with one-year enlistments but under the act of 5 July 1813 fourteen regiments, including the 36th Infantry, were switched to 'five years or during the war' reenlistments, and they were continued until the end of the war through the act of 28 January 1814.

The headquarters of the regiment was established at Georgetown, District of Columbia, and two recruiting rendezvous were established at Richmond (Virginia) and at Georgetown. The regiment recruit in northern Maryland, the District of Columbia, and in eastern Virginia. The 36th Infantry was assigned to the defense of the 5th Military District.

Henry Carberry of Maryland was the only colonel of this regiment. He served from 22 March 1813 to 4 March 1815.

Under the Act of 3 March 1813 a second major's position was authorized along with ten 3rd lieutenant's positions and ten additional sergeants positions. This act set the new manpower strength for the regiment at 1,091 men. The additional manpower was used to supplement the regiment's recruiting efforts. The regiment was assigned to the 5th Military District on 19 March 1813.

Under the act of 3 March 1815 selected officers and the enlisted men who had enlisted for five years were assigned to the new 4th Regiment of U.S. Infantry. The other officers and the men who had 'during the war' enlistments were discharged.

The regiment fell short on obtaining its full intended strength of 1,091 men. A total of 856 men were commissioned as officers and who enlisted as soldiers. Very few men who had one-year enlistments re-enlisted. The regiment's recruiting service was trying to recruit to full strength and at the same time replace the men who were being discharged after serving only one year.

Battles, Actions, Etc.
24 Aug 1814	Bladensburg, MD
13 Sep 1814	Ft. McHenry, MD

By War of 1812 Soc. in MD

38th Regiment of U.S. Infantry

The 38th Regiment of U.S. Infantry was organized under the Congressional Act of 29 January 1813 as a one-year regiment with one year enlistments. Under the act of 5 July 1813, the 38th Infantry was switched to 'five years or during the war' reenlistments, and then continued until the end of the war through the act of 28 January 1814.

The headquarters of the regiment was established at Baltimore, Maryland, and two recruiting rendezvous were established at Baltimore and Craney Island, Virginia. The regiment recruit in eastern Maryland, in southeastern Virginia. The 38th Infantry was assigned to the defense of the 5th Military District.

Peter Little of Maryland was the only colonel of this regiment. He served from 19 May 1813 to 4 March 1815.

Under the Act of 3 March 1813 a second major's position was authorized along with ten 3rd lieutenant's positions and ten additional sergeants positions. This act set the new manpower strength for the regiment at 1,091 men. The additional manpower was used to supplement the regiment's recruiting efforts. The regiment was assigned to the 5th Military District on 19 March 1813.

Under the act of 3 March 1815 selected officers and the enlisted men who had enlisted for five years were assigned to the new 4th Regiment of U.S. Infantry. The other officers and the men who had 'during the war' enlistments were discharged.

The regiment fell short on obtaining its full intended strength of 1,091 men. A total of 871 men were commissioned as officers and who enlisted as soldiers. Very few men who had one-year enlistments re-enlisted. The regiment's recruiting service was trying to recruit to full strength and at the same time replace the men who were being discharged after serving only one year.

Battles, Actions, Etc.
24 Aug 1814	Bladensburg, MD
13 Sep 1814	Ft. McHenry, MD

Maryland Regulars in War of 1812

U.S. Sea Fencibles
The Baltimore companies

Ten companies of sea fencibles were created under the act of 26 July 1813 to serve for one year guarding the ports and harbors of the United States. These companies could be used on both land or on water in order to perform their missions. Most of these companies were used as heavy artillery units in the major forts on the eastern seaboard.

These companies were unique in that the officers were army officers and naval enlisted personnel. The companies were made up of a captain, a 1st lieutenant, a 2nd lieutenant and a 3rd lieutenant for the officer's ranks. The enlisted men consisted of the warrant officer's rank of a boatswain, six gunners and six quarter gunners along with 90 seamen.

Maryland raised two of the ten authorized sea fencibles companies. M. Simmons Bunbury was commissioned as a captain on 1 October 1813 and served until discharged on 15 June 1815. John Gill was commissioned as a captain on 24 November 1813 and he served until 17 March 1814 when his commission was negated by the U.S. Senate. William H. Addison was then commissioned as a captain to replace Gill. He served from 27 April 1814 until his death on 9 December 1814. All three men were Marylanders. Both companies served at Fort McHenry.

The Sea Fencibles were retained in service until 15 June 1815 when discharged under the act of 3 March 1815. No field officers were authorized or appointed. Each companies served under the commandant of the fort in which they had been assigned. A total of 243 men served in the two companies from Maryland.

<u>Battles, Actions, Etc.</u>
13 Sep 1814Fort McHenry, MD

By War of 1812 Soc. in MD

U.S. Volunteers
The Baltimore Volunteers

Congress created the U.S. Voluntary Corps under the Act of 6 February 1812. This corps was made up of militiamen who would serve a one-year enlistment with the U.S. Army.

The President had the power to accept into the U.S. Voluntary Corps up to 50,000 militiamen in companies of artillery, cavalry and infantry. Those militiamen in the cavalry had to supplied their own horses. Besides accepting militia companies into the corps, the President could also accept militia battalions, regiments, brigades or divisions.

Some of the companies were attached to existing army regiments and many of these militiamen fought shoulder to shoulder along with their regular army counterparts. The volunteers were covered by the same rules and regulations as the regular army and these men were entitled to the same pension programs as regular soldiers.

The Act of 24 February 1814 authorized the President to call into service members of the U.S. Volunteer Corps for five years or during the war enlistments. This act created the last four infantry regiments, which were directly made up of militiamen and not men who were recruited from civilian status.

The *Baltimore Volunteers* was formed during the summer of 1812 to serve with the U.S. Army as part of the U.S. Voluntary Corps. The company was organized under the command of Captain Stephen H. Moore in August 1812 as a light infantry company made up of 111 men.

Once the *Baltimore Volunteers* was accepted into federal service in early September 1812 they were ordered to report to the Niagara Theater in New York to begin operations with the Army of the North. The *Baltimore Volunteers* were assigned to the battalion commanded by Major Darby Noon of Lieutenant Colonel Francis McClure's regiment of the U.S. Voluntary Corps. The company was discharged on 7 September 1813.

Battles, Actions, Etc.
28 Nov 1812	Black Rock, NY
27 Apr 1813	York, UC
27 May 1813	Fort George, UC

Maryland Regulars in War of 1812

2nd Regiment of U.S. Artillery
Captain Nathan Towson's Company

The 2nd Regiment of U.S. Artillery was organized under the act of 11 January 1812, to consist of two battalions, each with ten companies. Enlistments were set at five years upon creation of this regiment. On 8 Apr 1812, eighteen-month enlistments were authorized and on 12 December 1812 "during the war" enlistments were authorized.

George Izard of South Carolina was the regiment's first colonel, serving from 12 March 1812 until he was promoted to brigadier general on 12 March 1813. Winfield Scott of Virginia became the regiment's second colonel. He served from 12 March 1813 until he was promoted to brigadier general on 9 March 1814.

Nathan Towson of Maryland was commissioned as a captain on 12 March 1812 and he was ordered to raise an artillery company in his state. Once raised, the company reported to the left division of the Army of the North along the Niagara River in New York State.

Captain Towson led a party of army volunteers on the night of 10 October 1812 and captured the British Brig *Caledonia* as it was anchored near Fort Erie, Upper Canada. Lieutenant Jesse D. Elliot, USN, led a second party, made up of volunteer sailors, and captured the former U.S. Army Brig *Adams* (renamed the HMS Brig *Detroit*). For his actions, Towson was promoted to brevet major.

The company participated in all of the major battles along with Niagara River during the War of 1812. During the Battles of Chippewa and Lundy's Lane, the company operated two 6-pound field pieces and a 5 ½-pound howitzer.

On 12 May 1814, under the act of 30 March 1814, the regiment was consolidated with the 1st and the 3rd Regiments of Artillery to form the Corps of Artillery. There were a total of 113 men who were commissioned or recruited for this company.

Battles, Actions, Etc. (Towson's Company)
10 Oct 1812	Capture of the British brigs *Detroit* and *Caledonia*
13 Oct 1812	Queenston Heights, UC
27 May 1813	Fort George, UC
12 Jun 1813	Stoney Creek, UC
5 Jul 1814	Chippewa, UC
25 Jul 1814	Lundy's Lane, UC (also called Bridgewater)
13-15 Aug 1814	Fort Erie, UC

By War of 1812 Soc. in MD

5th Military District

In June 1810 the army divided the country into two military sections called the Northern Department and the Southern Department. On 19 March 1813, Secretary of War James Armstrong re-divided the country into nine military districts in which the army was to operate.[1] A general officer was in charge of each military district and he had under his control various infantry regiments and supporting artillery, dragoons, rifle and rangers units as needed. The various state and territorial militia divisions would also be a part of these districts. The purpose of these districts was to have an organized defense for each area of the country.

On 2 July 1814 the secretary of war created a tenth district centering on Washington, D.C., and the surrounding area. This district would be short-lived and it would be eliminated in January 1815.

Military Districts	States and Territories Assignment
1	Massachusetts, District of Maine, New Hampshire
2	Rhode Island, Connecticut
3	Lower New York, eastern New Jersey
4	Western New Jersey, Pennsylvania, Delaware
5	Maryland, Virginia, District of Columbia
6	North Carolina, South Carolina, George
7	Louisiana, Tennessee, Territory of Mississippi
8	Kentucky, Ohio, Territories of Michigan, Indiana, Illinois and Missouri
9	Upper New York, Vermont
10	District of Columbia, Maryland and Virginia east of the Rappahannock River

The military districts were abolished on 17 May 1815 and they were replaced by nine military departments divided between the North Division and the South Division.

---

[1] **American State Papers**, Documents, Legislative and Executive, of the Congress of the United States, 1789-1819, (Washington, D.C.: Gales and Seaton, 1832), Military Affairs, Military Districts 1 May 1813, page 432.

Maryland Regulars in War of 1812

Regimental Assignments 1813 [2]

5th Military District
- 2 companies of light artillery
- 1 company of light dragoons
- 6 companies of 2nd artillery
- 12th Regiment of U.S. Infantry
- 14th Regiment of U.S. Infantry
- 20th Regiment of U.S. Infantry

Regimental Assignments as of 1 July 1814

5th Military District
- Artillery
- 20th Regiment of U.S. Infantry
- 35th Regiment of U.S. Infantry
- 36th Regiment of U.S. Infantry
- 38th Regiment of U.S. Infantry
- Sea Fencibles

A report of the Army, its strength and distribution; previous to the 1st of July 1814 [3]

Effectives – men fit for duty
Aggregate – total number of men

5th Military District

Norfolk, VA	Effectives	Aggregate
Artillery	210	224
20th, 35th & 38th (1st Battalion)	873	912

Baltimore, MD	Effectives	Aggregate
Artillery	65	111
38th (2nd Battalion)	300	316
Sea Fencibles	167	173

Annapolis, MD	Effectives	Aggregate
Artillery	40	40

Fort Washington, MD	Effectives	Aggregate
Artillery	79	82

St. Mary's, MD	Effectives	Aggregate
36th	320	350

---

[2] *Ibid., Rules and Regulations of the Army for 1813, page 433.*

[3] *Ibid., Capture of the City of Washington, page 535.*

By War of 1812 Soc. in MD

9th Military District	Effectives	Aggregate
12th Infantry	482	752
14th Infantry	137	262

Recruiting reports for 30 Sep 1814 [4]

Recruiting HQs	**Recruits**
12th – Staunton, VA	159
14th – Baltimore	171
20th – Fredericksburg, VA	163
35th – Petersburg, VA	522
36th – Georgetown, DC	
	Richmond, VA 75
38th – Baltimore	225
1st Rifles – Shepherdstown, VA	
	Savannnah, GA 18
3rd Rifles – Charlotte, NC	
	Bath Court House, VA
	Gallatin, TN 187

Recruits between Jan and Oct 1814 – 2 Nov 1814 [5]

Regiment	**Recruits**
12th Infantry	194
14th Infantry	180
20th Infantry	170
35th Infantry	565
36th Infantry	289
38th Infantry	692 (Craney Island, VA and Baltimore)

Abbreviations

This chapter will explain the data fields, abbreviations, terms and phrases used in creating the 'service record' of each officer and enlisted personnel listed in the following chapter.

---

[4] ***Ibid.***, *Military Affairs, pp. 512-513.*

[5] ***Ibid.***, *Return of Enlistments, pp. 521-522.*

Maryland Regulars in War of 1812

Data Field	Explanation
Rank	The highest known military rank is listed for each soldier. The ranks of third lieutenant and ensign (also cornet) were discontinued by the army after the War of 1812.

Officer ranks:

Major General	Captain
Brigadier General	First Lieutenant
Colonel	Second Lieutenant
Lieutenant Colonel	Third Lieutenant
Major	Ensign or Cornet

The lowest officer's rank in the infantry, rifles, and artillery was an ensign while a cornet was the lowest officer's rank in the light dragoons.

Enlisted ranks:

Sergeant Major
Quartermaster Sergeant
Fife Major
Sergeant
Corporal
Private
Recruit

> Equal in rank to privates were: musicians, artificers, saddlers, farriers, blacksmiths and drivers of artillery. Musicians were found in all branches of the army while artificers were used in the artillery instead of privates. The other ranks were used in the artillery and light dragoons regiments.

Regiment — The names of the regiments are abbreviated:

5th Infantry = 5th Regiment of U.S. Infantry
12th Infantry = 12th Regiment of U.S. Infantry
14th Infantry = 14th Regiment of U.S. Infantry
35th Infantry = 35th Regiment of U.S. Infantry
36th Infantry = 36th Regiment of U.S. Infantry
38th Infantry = 38th Regiment of U.S. Infantry

Other regiment(s) — If a soldier was transferred to another regiment or regiments, the name of the additional regiment or regiments are listed in this field.

Company — The company's commander name

By War of 1812 Soc. in MD

Prior to 1816 the army identified a company by the company commander's name. In 1816 the army started to identify a company by using alpha characters: A through K (J not used).

Age	Age of the soldier at the time of his enlistment or commissioning
Height	Height of the soldier in feet and inches.
Birth Place	The birth place of a soldier by state or country, county and city.
Trade	Civilian trade of a soldier at the time of enlistment.
Enlistment date	Date that an enlisted man entered service.
Enlistment Place	The place of enlistment by state, county and city.
Enlistment Period	There were four enlistment periods which a soldier could select. All soldiers could choose to re-enlist once their initial enlistment period ended. Those soldiers who enlisted "during the war" were discharged at the end of the War of 1812. 18 Mos = 18 months 1 Yr = 1 year 5 Yrs = 5 years War = "during the war" All officers had "during the war' enlistments. In the spring of 1815, officers were selected to be retained for peacetime duty while the remaining officers were released from duty.
Pension	I-9999 – Invalid IC-9999 – Invalid's Certificate IF-9999 – Invalid's File IO-9999 – Invalid's Original MC-9999 – Minor's Certificate MO-9999 – Minor's Original SC-9999 – Survivor's Certificate SF-9999 – Survivor's File SO-9999 – Survivor's Original WC-9999 – Widow's Certificate

Maryland Regulars in War of 1812

WF-9999 – Widow's File
WO-9999 – Widow's Original

Bounty Number   BLW 123456-160-12

BLW 999999 = Bounty land warrant number
-160- = number of acres issued for the warrant
-12 or -14 or -42 or -50 or -55 = Years of the Land Bounty Acts

Comments   Any additional comments for a soldier.

Abbreviations

Standard U.S. Postal Service abbreviations are used for U.S. states plus
LC = Lower Canada (now Quebec)
UC = Upper Canada (now Ontario)
NS = Nova Scotia

Alexandria, VA, was a part of the District of Columbia during the War of 1812 so this city is listed as "Alexandria, DC."

Terms and phases

Discharged by writ of habeas corpus
If an under aged soldier had enlisted without his father's permission, the father could go to the county courthouse and obtain a "writ of habeas corpus" and the army would then discharge the soldier.

Heirs obtained half pay for five years in lieu of military bounty land
Early in the war, the heirs of soldiers who had died (or were killed) during the war and who had enlisted for one-year or 18-months, could elect to receive the soldiers' half-month pay for five years or receive 160 acres of land. This was changed so that the heirs only received the half pay. The heirs of soldiers who enlisted for five years or during the war always received either 160 or 320 acres of land. 320 acres of land was issued to soldiers who enlisted or re-enlisted after 1 February 1814.

Land bounty to "name of heirs" heirs at law of "name of soldier"
This phase lists the name of the heir or heirs (by law) who received the land bounty of a deceased soldier.

On aboard the US "name of ship" or squadron
This phase identifies the ship in which the soldier served as an officer, sailor or marine on the Great Lakes or Lake Champlain.

Surgeon's Certificate

An army surgeon could issue a "Surgeon's Certificate of Disability" to a wounded, sick or injured soldier and this would release the soldier from the service. The soldier's enlistment would end and he was still entitled to all his back pay, bonuses and land bounties.

Brevet

Officers could be promoted to the next highest rank by brevet, that is, they would wear the insignia of the next highest officer's rank without an increase in pay.

Deserted

This term denotes soldiers who left their regiments and who did not return to duty. When found, some men were executed for this crime, while others returned to duty and received either a punishment or extended enlistment period. In early 1815, many soldiers simply left their regiments without permission and returned home. These men were not entitled to their enlistment benefits.

Prisoners of War

Prisoner of War (Halifax, NS)

An American soldier who was held at the British prisoner of war camp at Halifax, Nova Scotia.

Prisoner of War (Montreal, LC)

An American soldier who was held at the British prisoner of war camp at Montreal, Quebec.

Prisoner of War (Quebec, LC)

An American soldier who was held at the British prisoner of war camp at Quebec City, Quebec.

The Roster
A

Abaits, J. - Private - 38th Infantry - Company: Shepperd Leakin - Enlistment date: 28 Mar 1814.

Abbot, Nehemiah - Private - 36th Infantry - Age: 24 - Height: 5' 3" - Born: Manchester, Essex Cty, MA - Enlistment date: 29 Apr 1813 - Place: Baltimore - Period: 1 Yr.

Abbott, James - Corporal - 38th Infantry - Company: Isaac Aldridge - Age: 34 - Height: 5' 9" - Born: Rockingham County, NH - Trade: Seaman - Enlistment date: 5 Sep 1814 - Place: Baltimore - Period: War - BLW 13461-160-12 - Discharged at Baltimore on 6 Apr 1814.

Abbott, John - Private - 14th Infantry - Company: Joseph Marshall - Age: 28 - Height: 5' 4 1/2" - Born: Dorchester Cty, MD - Trade: Seaman - Enlistment date: 22 Jun 1814 - Place: Baltimore - Period: War - BLW 14442-160-12 - Discharged at Greenbush, NY, on 16 May 1814.

Abbott, William - Private - 29th Infantry - Company: John Rochester - Other regiment: 6th Infantry (New) - Age: 21 - Height: 5' 5" - Born: Salisbury, MD - Trade: Laborer - Enlistment date: 11 Mar 1814 - Place: Sackets Harbor, NY - Period: 5 Yrs - Discharged at Greenbush, NY, on 15 Jul 1815, for disability.

Abercrombie, James - Private - 38th Infantry - Company: John Rothrock - Enlistment date: 2 Jun 1813 - Period: War - Pension: WO-40106 - Died at Baltimore on 31 Jul 1814.

Abert, John J. - Colonel - Topographical Department - Born: Maryland - BLW 414-160-50 - Graduated from the US Military Academy on 1 Apr 1811 and resigned; commissioned as a brevet major on 22 Nov 1814; promoted to brevet lieutenant colonel on 22 Nov 1814; promoted to colonel on 7 Jan 1838; retired on 9 Sep 1861; died on 27 Jan 1863.

Abert, Michael - Private - 14th Infantry - Prisoner of War (Halifax), discharged on 3 Feb 1814.

Achcum, James - 2nd Light Dragoons - Company: John Burd - Age: 21 - Height: 5' 6 1/2" - Born: Maryland - Trade: Shoemaker - Enlistment date: 11 Jun 1812 - Period: 18 Mos - Proved to be a minor.

Acker, John - Private - 2nd Infantry - Company: William Boots - Age: 33 - Height: 5' 9 1/2" - Born: Middlesex Cty, NJ - Trade: Farmer - Enlistment date: 16 Feb 1807 - Place: Baltimore - Period: 5 Yrs - Re-enlisted on 28 Feb 1812 for five years at Fort Stoddard; discharged on 2 Sep 1815 at Pass Christian, supernumerary.

Ackle, John P. - Private - 38th Infantry - Age: 32 - Height: 5' 7 1/2" - Born: Germany - Trade: Wheelwright - Enlistment date: 7 Dec 1814 - Place: Baltimore - Period: War.

Ackley, David - Private - 2nd Artillery - Company: Frederick Evans - Age: 21 - Height: 5' 7" - Born: Pennsylvania - Enlistment date: 11 Jan 1813 - Place: Annapolis - Period: 5 Yrs - BLW 14489-160-12 - Discharged at Fort McHenry on 11 Jan 1818.

By War of 1812 Soc. in MD

Ackley, William - Private - 14th Infantry - Age: 40 - Height: 5' 5 1/2" - Born: Boston - Trade: Sailor - Enlistment date: 20 Apr 1812 - Place: Baltimore - Period: 5 Yrs - Prisoner of War (Halifax), captured at Beaver Dams on 24 Jun 1813, exchanged on 15 Apr 1814; discharged on 20 Apr 1815 at Greenbush, NY, for wounds=.

Ackman, William - Private - 38th Infantry - Company: John Buck - Other regiment: 4th Infantry (New) - Age: 18 - Height: 5' - Born: Princess Anne's or Somerset Cty, MD - Trade: Chair maker - Enlistment date: 14 May 1814 - Place: Baltimore - Period: 5 Yrs - BLW 22038-160-12 - Discharged on 13 May 1819.

Adams Jr., William - Private - 21st Infantry - Age: 28 - Height: 5' 11" - Born: Easton, MD - Enlistment date: 17 Jan 1814 - Place: Portsmouth, NH.

Adams, Clement - 14th Infantry - Company: Reuben Gilder - Other regiment: 4th Infantry (New) - Age: 22 - Height: 5' 9" - Born: Maryland - Trade: Cordwainer - Enlistment date: 19 Apr 1813 - Place: Maryland - Period: 5 Yrs - Deserted on 15 Jul 1815.

Adams, Daniel - Private - 14th Infantry - Company: Thomas Montgomery - Enlistment date: 11 Mar 1813 - Period: 5 Yrs - Deserted on 12 Jul 1813, probably at Lewiston, NY.

Adams, George - Corporal - 5th Infantry - Company: William Henshaw - Other regiment: 3rd Infantry (New) - Age: 22 - Height: 5' 7" - Born: Derby, Chester Cty, PA - Trade: Miller - Enlistment date: 28 May 1812 - Place: Baltimore - Period: 5 Yrs - Deserted on 6 Feb 1816.

Adams, John - Private - 14th Infantry - Company: James Britton - Age: 28 - Height: 5' 5" - Born: Sussex Cty, DE - Trade: Laborer - Enlistment date: 31 Aug 1812 - Place: Sussex Cty, DE - Period: 18 Mos - Discharged on 27 Feb 1814.

Adams, Joseph - Private - 5th Infantry - Age: 29 - Height: 5' 6" - Born: Philadelphia - Trade: Shoemaker - Enlistment date: 2 Mar 1813 - Place: Baltimore - Period: 5 Yrs - Deserted.

Adams, Moses - Private - 36th Infantry - Age: 41 - Height: 5' 7" - Born: St. Mary's, MD - Enlistment date: 22 May 1813 - Place: Baltimore - Period: 1 Yr.

Adams, Robert - Private - 38th Infantry - Company: James Haslett - Enlistment date: 4 Jun 1813 - Period: 1 Yr - Died or discharged on 15 Jul 1813.

Adams, Robert - Private - 14th Infantry - Age: 42 - Height: 5' 4 1/2" - Born: Loudoun Cty, VA - Enlistment date: 9 Jun 1814 - Place: Alexandria, DC - Period: 5 Yrs.

Adams, Robert - Private - 14th Infantry - Company: Thomas Montgomery - Enlistment date: 2 Dec 1812 - Period: 18 Mos - Discharged on 2 Jun 1814.

Adams, Walker - Private - 36th Infantry - Age: 22 - Height: 6' 1 1/2" - Born: Harford Cty, MD - Enlistment date: 20 Jun 1813 - Place: Havre de Grace, MD.

Adams, William - Private - 38th Infantry - Enlistment date: 22 Feb 1815 - Period: 5 Yrs.

Maryland Regulars in War of 1812

Adams, William H. - Private - 38th Infantry - Age: 22 - Height: 5' 8" - Born: Madison, VA - Trade: Carpenter - Enlistment date: 20 Feb 1815 - Period: 5 Yrs.

Addams, William - Private - 38th Infantry - Company: James Smith - Age: 22 - Height: 5' 11 1/2" - Born: King & Queen Cty, VA - Trade: Farmer - Enlistment date: 8 Mar 1814 - Place: Norfolk, VA - Period: War - Discharged at Craney Island, VA, on 15 Mar 1815.

Addison, William H. - Captain - US Sea Fencibles - Other regiment: ex 38th Infantry - Born: Maryland - Pension: Wife Anna, Old War WF-15507, Navy WF-1040 Rejected; heirs received half pay for five years in lieu of military bounty land - Commissioned as an ensign, 38th Infantry, on 20 Sep 1813; commissioned as a captain, Sea Fencibles, on 27 Apr 1814; died on 9 Dec 1814.

Ader, Joseph - Fifer - 17th Infantry - Company: Harris Hickman - Age: 42 - Height: 5' 10" - Born: Maryland - Trade: Laborer - Enlistment date: 24 Feb 1814 - Period: War - BLW 6170-160-12 - Discharged at Chillicothe, OH, on 9 Jun 1815.

Agane, John - Private - 38th Infantry - Company: Isaac Aldridge - Age: 28 - Height: 5' 5" - Born: Ireland - Trade: Laborer - Enlistment date: 24 Sep 1814 - Place: Baltimore - Period: War - Discharged at Baltimore on 6 Apr 1815.

Aggin, James - Private - 7th Infantry - Age: 21 - Height: 5' 6 1/2" - Born: Caroline Cty, MD - Trade: Farmer - Enlistment date: 13 Dec 1808 - Place: Tennessee.

Aggin, Washington - Private - 7th Infantry - Company: Thomas Van Dyke - Age: 31 - Height: 5' 7 1/2" - Born: Caroline Cty, MD - Trade: Shoemaker - Enlistment date: 13 Dec 1808 - Place: Tennessee.

Agnew, Patrick - Private - 14th Infantry - Company: Thomas Montgomery - Age: 30 - Height: 5' 5" - Born: Montgomery Cty, MD - Trade: Laborer - Enlistment date: 27 Apr 1812 - Place: Baltimore - Period: 5 Yrs - BLW 11825-160-12 - Discharged at Greenbush, NY, on 28 Feb 1815 rupture.

Ahrens, John D. - Private - US Volunteers - Company: Stephen Moore - Enlistment date: 8 Sep 1812 - Period: 1 Yr.

Aiken, John - Private - Light Dragoons - Age: 26 - Height: 5' 10" - Born: Delaware - Trade: Blacksmith - Enlistment date: 19 Feb 1813 - Place: Elkton, MD - Period: War.

Aikman, Thomas - Private - 36th Infantry - Company: Joseph Hook - Age: 22 - Height: 5' 7 1/2" - Born: Somerset Cty, MD - Trade: Blacksmith - Enlistment date: 16 Aug 1813 - Place: Snow Hill, MD - Period: 1 Yr - Pension: Land bounty to William Aikman, brother & other heirs at law of Thomas Aikman - BLW 23224-160-12 - Re-enlisted at Sandy Point, VA, on 18 Mar 1814 for the war; died on 1 Mar 1815.

Ailer, John - Private - 38th Infantry - Company: Shepperd Leakin - Age: 32 - Height: 5' 7 1/2" - Born: Carlisle, PA or Queen Anne's Cty, MD - Trade: Plasterer - Enlistment date: 9 Jun 1814 - Period: War - BLW 3127-160-12 - Discharged at Fort McHenry on 28 Mar 1815.

Aimes, Amos - Musician - 5th Infantry - Company: Leroy Opie - Age: 14 -

By War of 1812 Soc. in MD

Height: 4' 10" - Born: West Chester, MD - Trade: Laborer - Enlistment date: 30 Apr 1812 - Place: Baltimore - Period: 5 Yrs.

Alberry, John - Private - 19th Infantry - Age: 19 - Height: 6' 2 1/2" - Born: Allegany Cty, MD - Enlistment date: 22 Jul 1812 - Place: Zanesville, OH - Period: 18 Mos.

Albert, Peter - Private - 38th Infantry - Company: Thomas Sangster - Age: 35 - Height: 5' 8" - Born: England - Trade: Wool stapler - Enlistment date: 2 Nov 1814 - Place: Baltimore - Period: 5 Yrs - Discharged on 10 May 1815.

Albright, Godfrey - Private - 2nd Infantry - Company: Peter Schuyler - Age: 25 - Height: 5' 8 1/2" - Born: Lancaster, PA - Trade: Hatter - Enlistment date: 17 Feb 1808 - Place: Frederick, MD - Period: 5 Yrs - Discharged at New Orleans on 16 Feb 1813.

Alcherson, James - Private - 38th Infantry - Company: John Rothrock - Enlistment date: 4 Aug 1813 - Period: 1 Yr - Discharged at Craney Island, VA, on 4 Aug 1814.

Aldridge, Isaac - Captain - 38th Infantry - Commissioned as a captain on 20 May 1813; discharged on 15 Jun 1815.

Alexander, John - Private - 6th Infantry - Other regiment: 2nd Infantry (New) - Age: 19 - Height: 5' 11" - Born: Baltimore or Cecil County, MD - Trade: Shoemaker - Enlistment date: 15 Apr 1813 - Place: Baltimore - Period: 5 Yrs - Deserted on 1 Apr 1815.

Alexander, Michael - Private - 12th Infantry - Age: 39 - Height: 5' 9 1/4" - Born: Cecil Cty, MD - Trade: Nailor - Enlistment date: 2 Jan 1814 - Place: Staunton, VA - Period: War - Discharged at Winchester, VA, on 18 Mar 1815.

Alford, Beniah - Private - 38th Infantry - Company: Anthony Miltenberger - Enlistment date: 18 Oct 1813 - Period: 1 Yr.

Alford, Jacob - Musician - US Sea Fencibles - Company: William Addison - BLW 54798-160-55.

Alford, Joseph - Private - Light Artillery - Company: Luther Leonard - Age: 34 - Height: 5' 10" - Born: Montgomery Cty, MD - Trade: Mason - Enlistment date: 8 Mar 1812 - Place: Washington, DC - Period: 5 Yrs - Discharged on 8 Mar 1817.

Alford, Thomas - Seaman - US Sea Fencibles - Company: John Gill - Age: 21 - Height: 5' 7" - Born: Baltimore - Enlistment date: 8 Jan 1814.

Algier, George - Corporal - 12th Infantry - Company: Andrew Madison - Other regiment: 8th Infantry (New) - Age: 22 - Height: 5' 1 1/2" - Born: Charles Cty, MD - Trade: Shoemaker - Enlistment date: 14 Sep 1812 - Place: Wheeling, VA - Period: 5 Yrs - BLW 13014-160-12 - Discharged at Fort Osage, MO, on 14 Sep 1817.

All, Micajah - Private - 20th Infantry - Company: John Stanard - Other regiment: 8th Infantry (New) - Age: 33 - Height: 5' 10" - Born: Harford Cty, MD - Trade: Brick layer - Enlistment date: 14 May 1812 - Place: Suffolk C.H, VA - Period: 5 Yrs - Discharged on 14 May 1817.

Allabaugh, Adam - Corporal - 1st Rifles - Company: Daniel Appling - Age: 20 - Height: 5' 8 1/2" - Born: Maryland - Trade: Laborer - Enlistment date:

30 Nov 1808 - Place: Hagerstown, MD - Period: 5 Yrs - Discharged at Sackets Harbor, NY, on 30 Nov 1813.

Allen, James - Private - 14th Infantry - Company: Reuben Gilder - Age: 21 - Height: 5' 6" - Born: Baltimore - Trade: Farmer - Enlistment date: 15 Mar 1813 - Place: Baltimore - Period: War - Pension: SO-7660, SC-4470; soldier died on 29 Dec 1872 - BLW 21883-160-12 - Discharged on 4 May 1815.

Allen, James - Private - 14th Infantry - Company: David Cummings - Age: 26 - Height: 5' 10" - Born: Harford Cty, MD - Trade: Carpenter - Enlistment date: 8 Oct 1814 - Place: Baltimore - Period: War - BLW 14227-160-12 - Discharged at Baltimore on 16 Mar 1815.

Allen, John - Private - 36th Infantry - Company: Samuel Raisin - Age: 19 - Height: 5' 6 1/2" - Born: Annapolis - Trade: Farmer - Enlistment date: 31 Mar 1814 - Place: Westmoreland or Sandy Point, Kinsale, VA - Period: War - Died on 4 Feb 1815.

Allen, Owen - Private - 2nd Artillery - Company: Frederick Evans - Age: 25 - Height: 6' - Born: Maryland - Trade: Farmer - Enlistment date: 6 Feb 1814 - Place: Baltimore - Period: War - Discharged on 24 Mar 1815.

Allen, Robert - Private - 14th Infantry - Company: Reuben Gilder - Other regiment: 4th Infantry (New) - Age: 26 - Height: 5' - Born: Chester Cty, PA - Trade: Laborer - Enlistment date: 5 Sep 1812 - Place: Pennsylvania - Period: 5 Yrs - Discharged on 4 Sep 1817.

Allen, William - Private - 38th Infantry - Company: John Rothrock - Age: 35 - Height: 5' 8" - Born: Baltimore - Trade: Farmer - Enlistment date: 14 Apr 1814 - Place: Craney Island, VA - Period: War - BLW 16828-160-12 - Discharged at Craney Island, VA, on 15 Mar 1815.

Allison, Daniel - Private - 38th Infantry - Company: James Hook - Age: 39 - Height: 6' 1" - Born: Montgomery Cty, MD - Trade: Carpenter - Enlistment date: 28 Feb 1814 - Period: War.

Allison, James - Private - 2nd Artillery - Company: Frederick Evans - Other regiment: Corps of Artillery - Age: 40 - Height: 5' 10" - Born: Delaware - Trade: Plasterer - Enlistment date: 11 Jul 1812 - Place: Annapolis - Period: 5 Yrs - BLW 14492-160-12 - Discharged at Fort McHenry on 25 Aug 1817.

Allison, John - Private - 12th Infantry - Company: Thomas Post - Age: 30 - Height: 5' 10" - Born: Montgomery Cty, MD - Trade: Cooper - Enlistment date: 10 Oct 1813 - Place: Fredericksburg or Staunton, VA - Period: War - Discharged at Buffalo on 31 May 1815.

Alspaugh, Frederick - Private - 1st Infantry - Company: John Symmes - Other regiment: 3rd Infantry (New) - Age: 20 - Height: 5' 10" - Born: Baltimore - Trade: Rope maker - Enlistment date: 15 Apr 1807 - Place: Philadelphia - Period: 5 Yrs - Re-enlisted at Belle Fontaine, MO, on 27 May 1812 ; present during the Battle of Chippewa, UC; discharged on 26 May 1817.

Alter, Adam - Private - 38th Infantry - Company: Charles Stansbury - Age: 40 - Height: 5' 9 1/2" - Born: Pennsylvania - Trade: Carpenter - Enlistment date: 15 Nov 1814 - Place: Baltimore - Period: War - BLW 20306-160-12 - Discharged at Baltimore on 6 Apr 1815.

Altoo, Samuel - Private - 38th Infantry - Company: Thomas Sangster - Other

By War of 1812 Soc. in MD

regiment: 4th Infantry (New) - Age: 25 - Height: 5' 5" - Born: Frederick, MD - Trade: Wagoner - Enlistment date: 1 Feb 1815 - Period: 5 Yrs - Discharged at Annapolis on 24 Jul 1815, supernumerary.

Alvers, Peter - Private - 38th Infantry - Company: John Rothrock - BLW 10019-160-12.

Alvey, William - Private - 1st Infantry - Company: Hugh Moore - Other regiment: 3rd Infantry (New) - Age: 42 - Height: 5' 10" - Born: Maryland - Trade: Laborer - Enlistment date: 17 Dec 1810 - Place: Fort Wayne, IN - Period: 5 Yrs - Discharged on 17 Dec 1815.

Ambrose, Matthias - Private - 43rd Infantry - Other regiment: Corps of Artillery - Age: 27 - Height: 5' 8" - Born: Frederick Cty, MD - Trade: Currier - Enlistment date: 21 Sep 1814 - Period: 5 Yrs - Deserted on 1 Jan 1816.

Amick, John - Private - 5th Infantry - Company: William Oliver - Other regiment: 3rd Infantry (New) - Age: 23 - Height: 5' 8" - Born: Baltimore - Trade: Shoemaker - Enlistment date: 1 Sep 1814 - Place: Gettysburg, PA - Period: 5 Yrs - BLW 19294-160-12 - Discharged at Fort Mackinaw, MI, on 31 Aug 1818.

Amos, Scott - Private - 38th Infantry - Company: John Mowton - Age: 19 - Height: 5' 11" - Born: Harford Cty, MD - Trade: Farmer - Enlistment date: 23 Mar 1814 - Place: Craney Island, VA - Period: War - BLW 248-160-12 - Discharged at Craney Island, VA, on 15 Mar 1615.

Anders, Samuel - Corporal - 2nd Artillery - Company: George Richards - Age: 30 - Height: 6' - Born: Hagerstown - Trade: Baker - Enlistment date: 20 Feb 1810 - Place: New York - Period: 5 Yrs - Discharged on 21 Feb 1815.

Anderson, Andrew - 14th Infantry - Enlistment date: 13 Jan 1813 - Period: 1 Yr - Prisoner of War (Halifax), sent to England; discharged on 31 Mar 1815.

Anderson, Enoch - Private - 17th Infantry - Company: William Adair - Age: 38 - Height: 5' 5" - Born: Maryland - Trade: Rope maker - Enlistment date: 7 Feb 1814 - Place: Louisville, KY - Period: War - BLW 8712-160-12 - Discharged at Chillicothe, OH, on 4 Jun 1814.

Anderson, George - Private - 38th Infantry - Company: John Mowton - BLW 2772-160-12.

Anderson, James - Artificer - 2nd Artillery - Company: James Barker - Age: 39 - Height: 5' 7 1/2" - Born: Maryland - Trade: Carpenter - Enlistment date: 13 Jun 1812 - Place: Philadelphia - Period: 5 Yrs.

Anderson, James H. - Private - 14th Infantry - Company: James McDonald - Other regiment: 4th Infantry (New) - Age: 30 - Height: 5' 10" - Born: Ireland - Trade: Carpenter - Enlistment date: 1 Feb 1814 - Period: 5 Yrs - Discharged on 10 Feb 1819.

Anderson, John - Private - 36th Infantry - Age: 43 - Height: 5' 5" - Born: County Down, Ireland - Enlistment date: 16 Jul 1813 - Place: Waynesburg, PA - Period: 1 Yr.

Anderson, Mathias - Private - 2nd Artillery - Company: Nathan Towson - Pension: Old War IF-2911 - BLW 7992-160-12.

Anderson, Michael - Private - 38th Infantry - Company: Shepperd Leakin - Age: 25 - Height: 5' 7" - Born: Baltimore - Trade: Carpenter - Enlistment date: 28 Mar 1814 - Place: Baltimore - Period: War - Discharged at Baltimore on 30 Mar 1815.

Anderson, Stephen - Private - 36th Infantry - Age: 18 - Height: 5' - Born: Dorchester Cty, MD - Enlistment date: 25 Jun 1813 - Place: Baltimore - Period: 1 Yr.

Anderson, William - Private - 38th Infantry - Company: James Hook - Age: 22 - Height: 5' 6" - Born: Talbot Cty or Baltimore - Trade: Tailor - Enlistment date: 28 May 1814 - Place: Baltimore - Period: War - BLW 27356-160-42 - Discharged on 31 Mar 1815.

Andrew, Edward - Private - 14th Infantry - Enlistment date: 26 Jan 1813 - Period: 18 Mos - Prisoner of War, sent to England; discharged on 31 Mar 1815.

Andrew, Philip - Private - 14th Infantry - Company: Thomas Montgomery - Enlistment date: 21 May 1812 - Period: 5 Yrs - Pension: Land bounty to Michael Andrew, brother and other heirs at law of Philip Andrew - BLW 20267-160-12 - Died on 13 Jan 1813.

Andrews, David - Private - 14th Infantry - Enlistment date: 15 Aug 1812 - Period: 5 Yrs - Prisoner of War (Halifax), sent to England; deserted at Fort McHenry on 25 Apr 1815.

Andrews, John - Private - 14th Infantry - Company: Reuben Gilder - Other regiment: 4th Infantry (New) - Age: 17 - Height: 5' 5" - Born: Annapolis - Trade: Laborer - Enlistment date: 7 Dec 1812 - Place: Maryland - Period: 5 Yrs.

Andrews, Luke - Private - 14th Infantry - Company: Reuben Gilder - Age: 24 - Height: 5' 5 1/2" - Born: Sussex Cty, DE - Trade: Farmer - Enlistment date: 14 Nov 1812 - Period: 18 Mos - Pension: Land bounty to John Andre, brother and only heir at law of Luke Andre - BLW 27782-160-42 - Re-enlisted on 1 Mar 1814 for the war; died on 24 Apr 1814 at Plattsburgh, NY, from sickness.

Andrews, Nathaniel - Private - US Volunteers - Company: Stephen Moore - Enlistment date: 8 Sep 1812 - Period: 1 Yr.

Andrews, Reuben G. - Private - 3rd Rifles - Company: Edward Carrington - Other regiment: Rifles - Age: 29 - Height: 5' 11" - Born: Maryland - Trade: Shoemaker - Enlistment date: 9 May 1814 - Place: Beverly, VA - Period: 5 Yrs.

Andrews, Stephen - Private - 36th Infantry - Company: Joseph Hook - Enlistment date: 25 Jun 1813 - Pension: Wife Sarah A., WO-26386, WC-24208; married on 15 Jul 1860, Otter, IN; soldier died on 8 Aug 1868, Otter, IN; first wife's surname was Scudder - BLW 22963-160-50 - Discharged on 23 Jun 1814.

Angel, David - Private - 22nd Infantry - Company: William Oliver - Age: 23 - Height: 5' 8" - Born: Frederick Cty, MD - Trade: Hatter - Enlistment date: 25 Oct 1813 - Place: Hanover, York Cty, PA - Period: 5 Yrs.

Angel, Thomas - Private - 36th Infantry - Company: Joseph Hook - Age: 27 - Height: 5' 7 1/4" - Born: Baltimore Cty - Trade: Cooper - Enlistment date: 1 Jun 1814 - Place: Georgetown, DC - Period: War - BLW 198-160-12 -

By War of 1812 Soc. in MD

Discharged at Fort Covington, MD, on 30 Mar 1815.

Angst, George - Private - 38th Infantry - Company: Isaac Aldridge - BLW 20566-160-12.

Annis, Tegar - Private - 36th Infantry - Company: Thomas Carbery - Other regiment: Ordnance Department - Age: 25 - Height: 5' 9" - Born: Accomack Cty, VA - Trade: Farmer - Enlistment date: 27 Oct 1814 - Place: Georgetown, DC - Period: 5 Yrs - BLW 22902-160-12 - Transferred to Ordnance Department on 21 Jun 1815; discharged at Greenleaf Point, DC, on 27 Oct 1819.

Ansel, Philip - Private - 14th Infantry - Prisoner of War (Halifax), discharged on 3 Feb 1814.

Anson, John - Private - 38th Infantry - Born: Charleston, SC - Enlistment date: 10 Jun 1814 - Place: Norfolk, VA - Period: War.

Anthony, Hardeman - Private - 38th Infantry - Company: John Rothrock - BLW 20660-160-12.

Archer, John - Private - 14th Infantry - Company: Reuben Gilder - Other regiment: 4th Infantry (New) - Age: 25 - Height: 5' 8 1/2" - Born: Warm Springs, VA - Trade: Laborer - Enlistment date: 14 Feb 1814 - Place: Maryland - Period: 5 Yrs - Discharged on 14 Feb 1819.

Archer, William T. - Private - 38th Infantry - Company: John Rothrock - Age: 26 - Height: 5' 7" - Born: York, VA - Trade: Clerk - Enlistment date: 25 Apr 1814 - Place: Richmond, VA - Period: War - BLW 3695-160-12 - Discharged at Craney Island, VA, on 15 Mar 1815.

Archibald, James J. - Private - 12th Infantry - Company: Thomas Sangster - Age: 21 - Height: 5' 6" - Born: Frederick, VA or MD - Trade: Farmer - Enlistment date: 7 May 1814 - Period: War - BLW 5282-160-12 - Discharged at Fort Covington, MD, on 30 Mar 1815.

Archy, Peter - Private - 2nd Artillery - Company: Frederick Evans - Other regiment: Corps of Artillery - Age: 30 - Height: 5' 8 1/2" - Born: Maryland - Trade: Laborer - Enlistment date: 1 Feb 1812 - Place: Annapolis - Period: 5 Yrs - Discharged on 30 Jun 1817.

Ardery, William - Private - 36th Infantry - Company: Thomas Carbery - Age: 22 - Height: 5' 8" - Born: South Carolina - Trade: Blacksmith - Enlistment date: 16 Aug 1814 - Place: Washington, DC - Period: War - BLW 15447-160-12 - Discharged at Baltimore on 31 Mar 1815.

Arell, Richard - Captain - 14th Infantry - Age: 24 - Born: Pennsylvania - BLW 358-160-50 - Commissioned as a 1st lieutenant on 12 Mar 1812; promoted to captain on 14 Nov 1813; Prisoner of War (Dartmouth), released on 31 May 1814; discharged on 15 Jun 1815.

Arkwright, Isaac - Private - 2nd Artillery - Company: Stanton Sholes - Age: 24 - Height: 5' 8" - Born: Harford Cty, MD - Trade: Farmer - Enlistment date: 15 Feb 1813 - Place: Greensburg, Beaver Cty, PA - Period: 5 Yrs - Pension: Old War IF-12847 - Discharged on 16 Oct 1815 on Surgeon's Certificate.

Armatage, John - Private - 36th Infantry - Age: 40 - Height: 5' 7" - Born: New York - Enlistment date: 21 Apr 1813 - Place: Baltimore.

Maryland Regulars in War of 1812

Armitage, Joseph - Private - US Volunteers - Company: Stephen Moore - Enlistment date: 8 Sep 1812 - Period: 1 Yr - Deserted on 13 Dec 1812.

Armour, John - Private - 38th Infantry - Company: Isaac Aldridge - Enlistment date: 5 Jun 1813 - Period: 1 Yr - Discharged at Craney Island, VA, on 5 Jun 1814.

Armstrong, Archibald - Corporal - 2nd Artillery - Company: Daniel Cushing - Enlistment date: 16 Dec 1812 - Period: 5 Yrs - Pension: Wife Nancy Dayton, former widow of Archibald Armstrong, Old War Widow File 9831; land bounty to David Armstrong, son and other heirs at law of Archibald Armstrong - BLW 26335-160-12 - Transferred to Captain Cushing's Company, 2nd Artillery; died on 3 Dec 1813.

Armstrong, Daniel - Sergeant - 14th Infantry - Company: David Cummings - Age: 22 - Height: 5' 11" - Born: Ireland - Trade: Cooper - Enlistment date: 27 Mar 1814 - Place: Baltimore or Towsontown, MD - Period: War - Discharged on 18 Mar 1815.

Armstrong, David - Private - US Volunteers - Company: Stephen Moore - Enlistment date: 8 Sep 1812 - Period: 1 Yr.

Armstrong, Logan - Private - 36th Infantry - Company: Thomas Carbery - Age: 18 - Height: 5' 6" - Born: Washington, DC - Trade: Barber - Enlistment date: 16 Jun 1814 - Place: Washington, DC - Period: War - BLW 11850-160-12 - Discharged at Baltimore on 31 Mar 1815.

Armstrong, Patrick - 36th Infantry - Company: Joseph Hook - Enlistment date: 7 Jul 1813 - Period: 1 Yr - Transferred to the U.S. Navy on 26 Oct 1813.

Armstrong, Peter - Sergeant Major - 36th Infantry - Age: 23 - Height: 5' 3 1/2" - Born: Switzerland - Trade: Sailor - Enlistment date: 23 Jul 1814 - Place: Marlboro - Period: War - BLW 12412-160-12 - Discharged at Baltimore on 31 Mar 1815.

Armstrong, Thomas - Private - 5th Infantry - Company: George Brooks - Age: 34 - Height: 5' 7" - Born: Harford Cty, MD - Trade: Carpenter - Enlistment date: 1 May 1813 - Place: Baltimore - Period: 5 Yrs - Deserted at Philadelphia on 15 Sep 1815.

Armstrong, Thomas - Private - 38th Infantry - Age: 30 - Height: 5' 6" - Born: Harford Cty, MD - Trade: Carpenter - Enlistment date: 13 Aug 1814 - Place: Baltimore - Period: War.

Armstrong, William - Private - 14th Infantry - Company: William McIlvane - Other regiment: 4th Infantry (New) - Enlistment date: 22 Sep 1812 - Period: 5 Yrs - Prisoner of War (Halifax), captured at Beaver Dams on 24 Jun 1813, exchanged on 15 Apr 1814; deserted at Fort Severn, MD, on 14 Jul 1815.

Arnold, John - Private - 2nd Artillery - Company: Daniel Cushing - Age: 26 - Height: 5' 8" - Born: Allegany Cty, MD - Enlistment date: 6 Jul 1812 - Place: St. Clairsville, OH - Period: 5 Yrs.

Arnold, Samuel - Private - 17th Infantry - Company: Harris Hickman - Age: 35 - Height: 5' 6" - Born: Westminster, MD - Trade: Farmer - Enlistment date: 17 Oct 1813 - Place: Cadiz, Harrison County, OH - Period: War - BLW 6536-160-12 - Discharged at Chillicothe, OH, on 9 Jun 1815.

Arrinder, Isaac - Private - 18th Infantry - Company: Benjamin Elmore - Other regiment: 4th Infantry (New) - Age: 40 - Height: 5' 8" - Born: Georgetown, MD or Surry County, NC - Trade: Farmer - Enlistment date: 25 Oct 1814 - Place: Edgefield, Newbury C.H. - Period: 5 Yrs - Deserted on 10 Nov 1815.

Arthur, James - Private - 14th Infantry - Company: Henry Fleming - Enlistment date: 15 Feb 1813 - Period: 5 Yrs - Pension: Land bounty to Daniel Arthur, brother and other heirs at law of James Arthur - BLW 22563-160-12 - Prisoner of War, captured at Beaver Dams, died at Halifax, NS.

Arthurs Jr., James - Private - Light Dragoons - Age: 25 - Height: 5' 5" - Born: Maryland - Trade: Clerk - Enlistment date: 17 Feb 1813 - Place: Baltimore - Period: 5 Yrs.

Artis, John - Private - 14th Infantry - Company: Clement Sullivan - Age: 44 - Height: 5' 6" - Trade: Farmer - Enlistment date: 25 May 1812 - Period: 18 Mos - Prisoner of War (Halifax), captured at Beaver Dams on 24 Jun 1813, exchanged on 15 Apr 1814; discharged on 7 Aug 1814.

Aseburg, Philip - Private - 14th Infantry - Prisoner of War, exchanged on 15 Apr 1814.

Ashcombe, Samuel H. - Private - Artillery - Age: 25 - Height: 5' 7" - Born: St. Mary's, MD - Trade: Sailor - Enlistment date: 20 Jul 1813 - Place: Norfolk, VA - Period: 5 Yrs.

Asher, Frederick - Private - 14th Infantry - Prisoner of War (Halifax); died at Halifax on 7 Oct 1813.

Asherlon, John - Private - 14th Infantry - Company: David Cummings - Age: 25 - Height: 5' 10" - Born: England - Trade: Fuller - Enlistment date: 23 Mar 1814 - Place: Towsontown, MD - Period: War - BLW 20373-160-12 - Discharged on 16 Mar 1815.

Askew, Charles - Seaman - US Sea Fencibles - Company: John Gill - Age: 29 - Height: 5' 11" - Born: Baltimore City - Enlistment date: 3 Jan 1814.

Askew, William - Private - 2nd Artillery - Company: Samuel Archer - Age: 34 - Height: 5' 10" - Born: Baltimore - Trade: Trader - Enlistment date: 25 Jun 1812 - Period: 5 Yrs - Discharged at Sackets Harbor, NY, on 19 Jul 1814, for debility.

Atcherson, James - Private - 38th Infantry - Enlistment date: 4 Aug 1813 - Period: 1 Yr.

Athey, Benjamin - Private - 5th Infantry - Company: George Brooks - Other regiment: 3rd Infantry (New) - Age: 25 - Height: 6' 1/2" - Born: Prince George's Cty, MD - Trade: Carpenter - Enlistment date: 16 Jun 1812 - Place: Hagerstown, MD - Period: 5 Yrs.

Athey, Joseph - Private - 36th Infantry - Company: Thomas Carbery - Age: 20 - Height: 5' 10" - Born: Charles Cty, MD - Trade: Laborer - Enlistment date: 16 Jun 1814 - Place: Georgetown, DC - Period: War - BLW 79-160-12 - Discharged at Baltimore on 31 Mar 1815.

Atkins, James - Private - 36th Infantry - Age: 40 - Height: 5' 8".

Atkins, John - Private - 14th Infantry - Company: Reuben Gilder - Enlistment date: 18 Feb 1813 - Period: War - Pension: Land bounty to William Atkins, brother and other heirs at law of John Atkins - BLW 20522-160-12 - Died at Plattsburgh, NY, on 31 Mar 1814.

Maryland Regulars in War of 1812

Atkins, John C. - Private - 14th Infantry - Age: 18 - Height: 5' 3" - Born: Baltimore - Trade: Bookbinder - Enlistment date: 17 Aug 1814 - Period: War - Pension: Land bounty to William Atkins, brother and other heirs at law of John Atkins - BLW 20522-160-12 - Died at Baltimore in Oct 1814.

Atkins, Samuel - Private - 12th Infantry - Company: Thomas Post - Age: 20 - Height: 5' 4" - Born: Frederick, MD - Trade: Shoemaker - Enlistment date: 5 Mar 1814 - Place: Charles Town, VA - Period: War.

Attire, Peter - Recruit - 26th Infantry - Company: William Bezeau - Race: 19 - Age: 19 - Height: 5' 6" - Born: Maryland - Trade: Farmer - Enlistment date: 2 Jan 1815 - Place: Philadelphia - Period: War - Listed on the Descriptive Roll of Colored Men, 1 Apr 1815; deserted on 18 Jan 1815.

Atwell, James - Private - 38th Infantry - Company: James Haslett - Enlistment date: 3 Jul 1813 - Period: 1 Yr - Pension: Heirs received half pay for five years in lieu of land bounty - Died on 29 Mar 1814.

Atwood, John - Private - 36th Infantry - Age: 28 - Height: 5' 6" - Born: England - Enlistment date: 15 Sep 1813 - Place: Baltimore.

Augh, Lewis - Private - 14th Infantry - Company: Reuben Gilder - Age: 25 - Height: 5' 7" - Born: Adams Cty, PA or Germany - Trade: Laborer - Enlistment date: 16 Apr 1813 - Place: Baltimore - Period: War - BLW 11828-160-12 - Discharged at Greenbush, NY, on 4 May 1815.

Augst, George - Private - 38th Infantry - Company: Isaac Aldridge - Age: 29 - Height: 5' 9" - Born: Dauphin Cty, PA - Trade: Blacksmith - Enlistment date: 25 Nov 1814 - Place: Baltimore - Period: War - Discharged at Baltimore on 8 Aug 1815.

Auld, James - Sergeant - US Volunteers - Company: Stephen Moore - Enlistment date: 8 Sep 1812 - Period: 1 Yr - Died on 3 Jan 1813.

Aulguire, Henry - Private - 38th Infantry - Other regiment: 4th Infantry (New) - Age: 32 - Height: 5' 4" - Born: Holland - Trade: Tailor - Enlistment date: 10 Oct 1814 - Place: Baltimore - Period: 5 Yrs - Died at Mobile, AL, on 1 May 1819.

Aull, James - Seaman - US Sea Fencibles - Company: John Gill - Age: 21 - Height: 5' 6 1/2" - Born: Washington, DC - Enlistment date: 3 Jan 1814 - Under aged.

Austin, Benjamin - Private - 14th Infantry - Company: Reuben Gilder - Age: 26 - Height: 5' 10" - Born: England - Trade: Millwright - Enlistment date: 15 Jan 1813 - Period: 5 Yrs - Died at Cheektowaga Hospital, NY, on 18 Jan 1815 from dysentery.

Austin, John - Private - 36th Infantry - Age: 24 - Height: 5' 8" - Born: New York City - Enlistment date: 9 Jul 1814 - Place: Norfolk, VA.

Austin, Lawless - Private - 36th Infantry - Other regiment: 26th Infantry - Age: 25 - Height: 5' 5" - Born: Baltimore - Trade: Carpenter - Enlistment date: 10 May 1813 - Place: Baltimore - Period: 18 Mos - Re-enlisted on 31 Oct 1814 in the 26th Infantry; discharged on 23 Mar 1815 at Philadelphia, PA.

Ausunt, William - Private - 14th Infantry - Died at Cheektowaga Hospital, NY, on 10 Jan 1815 from dysentery.

Avis, Jarvis - Private - 14th Infantry - Company: David Cummings -

By War of 1812 Soc. in MD

Enlistment date: 25 May 1812 - Period: 5 Yrs - Prisoner of War, sent to England, captured at Beaver Dams on 24 Jun 1813; deserted on 18 Apr 1815 from Fort McHenry.

Awalt, Baltzer - Private - 38th Infantry - Company: John Buck - Age: 19 - Height: 5' 8" - Born: Baltimore - Trade: Cooper - Enlistment date: 2 Feb 1814 - Place: Baltimore - Period: War - BLW 18804-160-12 - Discharged on 30 Mar 1815.

Ayres, John - Private - 36th Infantry - Company: Samuel Raisin - Age: 22 - Height: 5' 7" - Born: Elizabeth City, VA - Trade: Brick layer - Enlistment date: 28 Jul 1814 - Place: Richmond or Manchester - Period: War.

B

Bageley, Abraham - Musician - 19th Infantry - Company: William Gill - Age: 16 - Height: 5' 2" - Born: Maryland - Trade: Farmer - Enlistment date: 2 May 1814 - Period: War - Discharged at Chillicothe, OH, on 5 Jun 1815.

Bailey, Alexander - Private - 14th Infantry - Company: Joseph Marshall - Age: 18 - Height: 5' 4" - Born: Baltimore - Trade: Laborer - Enlistment date: 17 Aug 1814 - Place: Elkton, MD - Period: War - BLW 24570-160-12 - Discharged on 3 May 1815.

Bailey, Esma - Seaman - US Sea Fencibles - Company: Simmones Bunbury.

Bailey, Isaac - Private - 38th Infantry - Company: Shepperd Leakin - BLW 20697-160-12.

Bailey, James - Matross - 2nd Artillery - Company: Nathan Towson - BLW 13281-160-12.

Bailey, Joshua - Private - 19th Infantry - Company: Joel Collins - Age: 19 - Height: 5' 7" - Born: Maryland - Trade: Carpenter - Enlistment date: 28 Jun 1813 - Place: Detroit - Period: War - BLW 27712-160-42 - Discharged on 19 Jul 1815.

Baird, Jacob - Private - 19th Infantry - Company: George Kesling - Age: 18 - Height: 5' 8" - Born: Baltimore Cty - Trade: Farmer - Enlistment date: 17 Jan 1815 - Place: Lebanon, Warren Cty, OH - Period: War.

Baird, William - Corporal - 2nd Artillery - Company: Nathan Towson - Pension: Land bounty to William Baird, son and other heirs at law of William Baird - BLW 18182-160-12.

Baker, Asyah H. - Seaman - US Sea Fencibles - Company: John Gill - Age: 24 - Height: 5' 7 1/2" - Born: Baltimore City - Enlistment date: 8 Jan 1814.

Baker, Christian - Private - US Volunteers - Company: Stephen Moore - Enlistment date: 8 Sep 1812 - Period: 1 Yr.

Baker, Henry - Private - 38th Infantry - Company: Shepperd Leakin - BLW 11836-160-12.

Baker, Henry - Private - US Volunteers - Company: Stephen Moore - Enlistment date: 8 Sep 1812 - Period: 1 Yr.

Baker, John - Private - 2nd Rifles - Company: John O'Fallon - Age: 18 - Height: 5' 1 1/2" - Born: Worcester Cty, MD - Trade: Shoemaker - Enlistment date: 16 Apr 1814 - Place: Lexington, KY - Period: 5 Yrs - Pension: Land bounty to John Baker, father & only heir at law of John Baker,

Maryland Regulars in War of 1812

Jr. - BLW 26211-160-12 - Died at Amherstburg, UC, on 27 Feb 1815.

Baker, Theophilus - Sergeant - 2nd Artillery - Company: Nathan Towson - Pension: Land bounty to George Baker, brother and other heirs at law of Theophilus Baker - BLW 17865-160-12.

Baker, William - Private - 14th Infantry - Company: Thomas Montgomery - Pension: Land bounty to Jess Baker, brother and other heirs at law of William Baker - BLW 12392-160-12.

Baker, William - Private - 14th Infantry - Pension: Land bounty to Mary Ann Baker, only heir at law of William Baker - BLW 26961-160-12 Cancelled.

Balch, Hezekiah T. - Ensign - 14th Infantry - Company: William McIlvane - Born: New Hampshire - BLW 119-160-12 - Commissioned as an ensign on 21 May 1814; dismissed on 14 Jul 1814.

Ball, Adam A. - Corporal - 19th Infantry - Company: David Holt - Other regiment: 17th Infantry - Age: 20 - Height: 5' 5 1/2" - Born: Maryland - Enlistment date: 7 Apr 1814 - Place: Marietta, OH - Period: War - Pension: Land bounty to Dinah Bryam, only heir at law of Adam Ball - BLW 24687-160-12 - Died at Williamsville, NY, on 7 Jan 1815; buried in the War of 1812 Cemetery in Cheektowaga, Erie Cty, NY.

Ballard, James Hudson - Captain - 36th Infantry - Other regiment: 4th Rifles - Born: Maryland - Commissioned as a 2nd lieutenant, 36th Infantry, on 30 Apr 1813; promoted to 1st lieutenant, 4th Rifles, on 17 Mar 1814; transferred to Rifles on 17 May 1815; promoted to captain on 22 Apr 1817; transferred to 2nd Infantry on 1 Jun 1821; transferred to 4th Artillery on 10 Oct 1822; died on 15 Jan 1823.

Ballard, Martin - Private - 14th Infantry - Prisoner of War, sent to England, captured at Beaver Dams on 24 Jun 1813.

Banes, William - Private - 38th Infantry - Company: James Smith - Pension: Land bounty to Bethshaba and Polly Banes, only heirs at law of William Banes - BLW 26901-160-12.

Banks, Everhard - Private - 14th Infantry - Company: Reuben Gilder - BLW 8074-160-12.

Banner, John - Private - 38th Infantry - Company: John Brookes - BLW 5232-160-12.

Barber, Amon - Private - 4th Rifles - Company: Joseph Kean - Other regiment: Rifles - Age: 37 - Height: 5' 7" - Born: Montgomery Cty, MD - Trade: Shoemaker - Enlistment date: 31 May 1814 - Place: Uniontown, PA - Period: 5 Yrs - BLW 25149-160-12 - Discharged at Fort Crawford, IL, on 31 May 1819.

Barbine, Charles - Seaman - US Sea Fencibles - Company: Simmones Bunbury.

Bark, James - Private - 14th Infantry - Pension: Placed on the pension rolls on 17 May 1815.

Barker, Archibald - Private - 38th Infantry - Company: James Smith - Age: 18 - Height: 5' 4" - Born: Camden Cty, NC - Trade: Farmer - Enlistment date: 28 Jul 1814 - Place: Suffolk Cty or Norfolk, VA - Period: War - BLW 3129-160-12 - Discharged at Craney Island, VA, on 15 Mar 1815.

By War of 1812 Soc. in MD

Barkley, Samuel - Second Lieutenant - 38th Infantry - Company: James Hook - BLW 8895-160-50 - Commissioned as a 3rd lieutenant on 20 May 1813; promoted to 2nd lieutenant on 1 May 1814; discharged on 15 Jun 1815.

Barks, John - Private - 14th Infantry - Prisoner of War (Halifax), discharged on 3 Feb 1814.

Barlow, Henry - Private - 7th Infantry - Company: James Doherty - Age: 33 - Height: 5' 2" - Born: Maryland - Trade: Farmer - Enlistment date: 28 May 1813 - Place: Tennessee - Period: War - Discharged at New Orleans on 8 Apr 1815.

Barlow, John - Private - 14th Infantry - Prisoner of War (Halifax), discharged on 27 Aug 1813.

Barnard, Isaac D. - Major - 14th Infantry - Born: Pennsylvania - Place: Pennsylvania - Commissioned as a captain on 12 Mar 1812; promoted to major on 26 Jun 1813; discharged on 15 Jun 1815; died on 19 Feb 1834.

Barnes, Asel - Musician - 14th Infantry - Company: Henry Grindage - Age: 18 - Height: 5' 10" - Born: Harford Cty, MD - Enlistment date: 12 Jun 1812 - Period: 18 Mos.

Barnes, Jonathan - Sergeant - 38th Infantry - Company: John Brookes - Enlistment date: 23 Feb 1814 - Pension: Wife Charlotte Moore, WO-31789, WC-24761; married on 27 Mar 1815, Craney Island, VA; soldier died on 26 Oct 1860 at sea - BLW 10397-160-12 - Discharged on 15 Mar 1815.

Barnes, Robert - Private - 36th Infantry - Company: Joseph Hook - Pension: Wife Elizabeth A., Old War WI-44311.

Barnes, Samuel - Private - 14th Infantry - Pension: Placed on the pension roll on 15 Dec 1815.

Barnes, Samuel - Private - 14th Infantry - Company: Richard Arell - Prisoner of War (Halifax), discharged on 3 Feb 1814.

Barnes, Truman - Sergeant - 36th Infantry - Company: Joseph Hook - BLW 12680-160-12.

Barnett, John - Private - 36th Infantry - Company: Joseph Hook - Pension: No pension claim - BLW 9121-160-12.

Barnhart, Henry - Seaman - US Sea Fencibles - Company: William Addison.

Barnhouse, Joseph – Private - Light Dragoons 1st Regt. – Born St. Mary's Co, Maryland – Age 25 – Enlisted: 18 Jun 1812 for 5 years -Wife: married Margaret Jane (White) 19 Nov 1849 widow of his brother Corporal Richard Barnhouse who died 24 Dec 1845 Loudoun Co VA - Company: Nelson Luckett – Served in Lower Canada, at Cristers Fields Nov 1813 wounded from musketball in left knee fracturing bone and at Battle of Lundy's Lane received head wound left side fracturing skull. Discharged Nov 1815 BLW 27690-160-12, Old War IF -12241,

Barnover, Adam - Private - 2nd Infantry - Company: Reuben Chamberlain - Age: 31 - Height: 5' 10" - Born: Baltimore - Trade: Carpenter - Enlistment date: 29 Feb 1808 - Place: Fredericksburg, VA - Period: 5 Yrs.

Barr, Daniel - Private - US Volunteers - Company: Stephen Moore - Enlistment date: 8 Sep 1812 - Period: 1 Yr - Deserted on 10 Dec 1812.

Barr, James - Private - 14th Infantry - Company: Thomas Montgomery - BLW

Maryland Regulars in War of 1812

6446-160-12.

Barrack, William - Private - 38th Infantry - Company: Isaac Aldridge - BLW 3644-160-12.

Barrackman, William - Private - 2nd Artillery - Company: Nathan Towson - Pension: Placed on the pension rolls.

Barrett, Alexander - Private - 36th Infantry - Company: Thomas Carbery - BLW 881-160-12.

Barrett, Alexander - Private - 36th Infantry - Pension: Placed on the pension roll on 12 May 1816.

Barrett, Alexander - Private - 38th Infantry - Company: James Smith - Age: 40 - Height: 5' 6" - Born: Norfolk, VA - Trade: Tailor - Enlistment date: 3 Mar 1814 - Place: Norfolk, VA - Period: War - BLW 1666-160-12 - Discharged at Craney Island, VA, on 15 Mar 1815.

Barrett, John M. - First Lieutenant - 38th Infantry - Born: Maryland - Commissioned as a 2nd lieutenant on 20 May 1813; promoted to 1st lieutenant on 15 Aug 1813; discharged on 15 Jun 1815; died on 16 Oct 1819.

Barrett, Patrick - Private - 38th Infantry - Company: Isaac Aldridge - BLW 13402-160-12.

Barrington, John - Private - 14th Infantry - Prisoner of War (Halifax); died at Halifax on 3 Sep 1813.

Barrow, Andrew D. - Sergeant - 14th Infantry - Company: Reuben Gilder - Period: 18 Mos - Discharged on 22 Feb 1814.

Barry, Garrett - Surgeon's Mate - 38th Infantry - Born: Maryland - Commissioned as a surgeon's mate, 38th Infantry, on 8 Apr 1814; died on 22 Apr 1815.

Bartlett, James - Private - 14th Infantry - Prisoner of War (Halifax), discharged on 3 Feb 1814.

Barton, Thomas - Second Lieutenant - 14th Infantry - Commissioned as an ensign, 14th Infantry, on 13 Mar 1813; promoted to 2nd lieutenant on 14 Oct 1813; discontinued on 1 Jul 1814.

Bartow, Peter - Private - 38th Infantry - Company: Charles Stansbury - BLW 12413-160-12.

Bash, Jacob - Private - 28th Infantry - Company: George Stockton - Age: 21 - Height: 5' 9" - Born: Maryland or Cleveland, OH - Trade: Carpenter - Enlistment date: 16 Oct 1814 - Place: Cleveland, OH - Period: War - Discharged at Detroit on 30 Jul 1815.

Bassord, Elijah - Private - 36th Infantry - Company: Thomas Carbery - BLW 3511-160-12.

Bassord, Ezekiel - Private - 36th Infantry - Company: Thomas Carbery - BLW 3510-160-12.

Batchelor, Joseph - Private - 14th Infantry - Company: Reuben Gilder - Pension: Land bounty to Nathaniel Batchelor, son and other heirs at law of Joseph Batchelor - BLW 25973-160-12.

Bateman, Lemuel - Third Lieutenant - 38th Infantry - Commissioned as an ensign on 20 May 1814; promoted to 3rd lieutenant on 1 Oct 1814; resigned on 19 Oct 1814.

By War of 1812 Soc. in MD

Bateman, Nicholas - Second Lieutenant - 36th Infantry - Company: Joseph Hook - Pension: Land bounty to Constance Bateman, widow of Nicholas Bateman - BLW 4779-160-50 - Served as a private and sergeant, 36th Infantry; commissioned as an ensign on 3 May 1814; promoted to 2nd lieutenant on 1 Jan 1815; discharged on 15 Jun 1815.

Bates, James - Private - 38th Infantry - Company: John Mowton - BLW 24242-160-12.

Bates, William H. - First Lieutenant - 38th Infantry - Pension: Land bounty to Eliza Bates, widow of William H. Bates - BLW 10999-160-50 - Commissioned as a 2nd lieutenant on 20 May 1813; promoted to 1st lieutenant on 15 Aug 1813; discharged on 15 Jun 1815.

Battest, John - Private - 36th Infantry - Company: Joseph Hook - BLW 7994-160-12.

Baubalier, Henry - Private - 38th Infantry - Company: John Rothrock - BLW 16412-160-12.

Baxter, George - Private - 36th Infantry - Company: Joseph Hook - BLW 24598-160-12.

Bay, John - Private - 14th Infantry - Prisoner of War (Halifax), discharged on 3 Feb 1814.

Bayard, James A. - Private - 14th Infantry - Company: Joseph Marshall - BLW 17091-160-12.

Bayce, Nelson - Private - 14th Infantry - Prisoner of War (Halifax), discharged on 3 Feb 1814.

Baynes, Thomas - Private - 14th Infantry - Prisoner of War (Halifax), discharged on 3 Feb 1814.

Beachman, Stephen - Private - 38th Infantry - Company: John Buck - BLW 6449-160-12.

Beales, John - Private - 14th Infantry - Prisoner of War (Halifax), captured at Beaver Dams on 24 Jun 1813, discharged on 9 Nov 1813.

Beall, George M. - Third Lieutenant - 17th Infantry - Company: David Holt - Born: Maryland - Commissioned as an ensign on 6 Apr 1813; promoted to 3rd Lieutenant on 19 Jan 1814; resigned on 8 Nov 1814.

Beall, John Hamilton - Surgeon's Mate - 36th Infantry - Born: Maryland - Commissioned as a surgeon's mate, 36th Infantry, on 31 Jul 1813; resigned on 24 Mar 1814.

Beall, Lloyd - Captain - 1st Artillery - Born: Maryland - Served during the Revolutionary War; commissioned as a captain, 9th Infantry, on 8 Jan 1799; discharged on 15 Jun 1800; commissioned as a captain, 2nd Artillery and Engineers, on 16 Feb 1801; transferred to 1st Artillery on 1 Apr 1802; transferred to Corps of Artillery on 12 May 1814; promoted to brevet major on 10 Jul 1812; discharged on 15 Jun 1815; died on 5 Oct 1817.

Beall, Robert - First Lieutenant - 14th Infantry - Other regiment: Corps of Artillery - Born: Maryland - Commissioned as a 3rd lieutenant, 14th Infantry, on 30 Mar 1813; promoted to 2nd lieutenant on 14 Nov 1813; transferred to Corps of Artillery on 17 May 1815; promoted to 1st lieutenant on 20 Mar 1818; cashiered on 4 Dec 1819.

Maryland Regulars in War of 1812

Beall, Thomas J. - Captain - 1st Artillery - Other regiment: 4th Rifles - Born: District of Columbia - Graduated from the US Military Academy on 6 Mar 1806; commissioned as a 2nd lieutenant, Artillery, on 1 Mar 1811; promoted to 1st lieutenant on 3 Mar 1813; promoted to captain, 4th Rifles, on 17 Mar 1814; discharged on 15 Jun 1815; reinstated on 1 Jan 1816; died on 26 Oct 1832.

Beall, Walter T. G. - Ensign - 36th Infantry - Born: Maryland - Commissioned as an ensign on 30 Apr 1813; resigned on 8 Nov 1813.

Beall, William Dent. - Colonel - 3rd Infantry - Other regiment: 5th Infantry - Born: Maryland - Served during the Revolutionary War; commissioned as a major, 9th Infantry, on 8 Jan 1799; promoted to lieutenant colonel, 5th Infantry, on 12 Dec 1808; promoted to colonel, 3rd Infantry, on 30 Nov 1810; transferred to 5th Infantry on 24 Apr 1812; resigned on 15 Aug 1812; died on 24 Sep 1829.

Beard, Anthony - Private - 5th Infantry - Other regiment: 3rd Infantry (New) - Age: 27 - Height: 5' 10 1/2" - Born: Rockingham Cty, VA - Trade: Farmer - Enlistment date: 18 Feb 1813 - Place: Maryland - Period: 5 Yrs - Discharged at Detroit on 17 Sep 1815.

Beard, Jacob - Private - 19th Infantry - Age: 18 - Height: 5' 8" - Born: Maryland - Trade: Farmer - Enlistment date: 17 Jan 1815 - Period: War - BLW 463-320-14 - Discharged at Zanesville, OH, on 27 Mar 1815.

Beard, John - Private - 14th Infantry - Prisoner of War, sent to England, captured at Beaver Dams on 24 Jun 1813.

Beard, Richard - Private - 14th Infantry - Prisoner of War (Halifax), captured at Beaver Dams on 24 Jun 1813, discharged on 9 Nov 1813.

Beard, William C. - Captain - 1st Rifles - Born: Maryland - Pension: Land bounty to Matilda H. Beard, widow of William C. Beard - BLW 8897-160-50 - Commissioned as a 2nd lieutenant, Rifles, on 19 May 1812; promoted to 1st lieutenant on 27 Sep 1812; promoted to captain on 20 Aug 1814; discharged on 15 Jun 1815; died on 18 Apr 1837.

Beatty, Abraham - Private - 14th Infantry - Company: Reuben Gilder - Age: 37 - Height: 5' 8 1/2" - Born: Chester Cty, PA - Trade: Mason - Enlistment date: 10 Mar 1814 - Period: War - BLW 26710-160-12 - Discharged at Greenbush, NY, on 4 May 1814.

Beatty, George - Private - 38th Infantry - Company: John Rothrock - BLW 4749-160-12.

Beatty, John - Private - 14th Infantry - Company: Joseph Marshall - BLW 14898-160-12.

Bebee, Edward - Seaman - US Sea Fencibles - Company: William Addison.

Bechm, Joseph - Private - 14th Infantry - Prisoner of War (Halifax), captured at Beaver Dams on 24 Jun 1813, exchanged on 15 Apr 1814.

Beck, Anthony - Private - 28th Infantry - Company: Joseph Belt - Age: 40 - Height: 5' 10" - Born: Maryland - Trade: Carpenter - Enlistment date: 24 May 1814 - Place: Flemingsburg, KY - Period: War - BLW 7915-160-12 - Discharged at Olympian Springs, KY, on 31 Mar 1815.

Beckett, John - First Lieutenant - 14th Infantry - Pension: Land bounty to

By War of 1812 Soc. in MD

Susan N. Beckett, widow of John Beckett - BLW 21922-160-50 - Commissioned as a 2nd lieutenant on 12 Mar 1812; promoted to 1st lieutenant on 13 Mar 1813; discharged on 15 Jun 1815; died on 9 Jun 1850.

Beckwith, Jennings - Second Lieutenant - 14th Infantry - Born: Maryland - BLW 7557-160-50 - Commissioned as an ensign on 15 Apr 1814; promoted to 3rd Lieutenant on 1 May 1814; promoted to 2nd Lieutenant on 1 Dec 1814; discharged on 15 Jun 1815.

Bell, Bazel - Private - 14th Infantry - Company: Reuben Gilder - Other regiment: 4th Infantry (New) - Age: 30 - Height: 5' 2" - Born: New Town, Somerset Cty, MD - Trade: Farmer - Enlistment date: 1 Sep 1812 - Place: Maryland - Period: 5 Yrs - Discharged on 1 May 1815 after furnishing a substitute.

Bell, George - Private - 36th Infantry - Company: Thomas Carbery - BLW 18652-160-12.

Bell, John - Private - 14th Infantry - Pension: Heirs received half pay for five years in lieu of land bounty - Died on 24 Jun 1813.

Belott, William - Seaman - US Sea Fencibles - Company: William Addison.

Belt, Basil - Private - 5th Infantry - Company: William Bird - Age: 31 - Height: 5' 11" - Born: Baltimore - Trade: House joiner - Enlistment date: 28 Mar 1813 - Place: Baltimore - Period: 5 Yrs - Deserted on 15 Jan 1814.

Belton, Francis Smith - Captain - 1st Light Dragoons - Born: Maryland - BLW 153-181-50 - Commissioned as a 2nd lieutenant, 1st Light Dragoons, on 27 Mar 1812; promoted to 1st Lieutenant on 20 Jan 1813; served as a major, assistant adjutant general between 18 Oct 1814 and 15 Jun 1815; retired as a brigadier general on 28 Aug 1861; died on 10 Sep 1861.

Belville, Samuel - Private - 14th Infantry - Company: Joseph Marshall - BLW 7126-160-12.

Bennet, James - Private - 38th Infantry - Company: John Brookes - BLW 19983-160-12.

Bennett Sr., Wooten - Private - 36th Infantry - Company: Thomas Carbery - Enlistment date: 4 Jan 1814 - Period: 1 Yr - Discharged on 3 Jan 1815.

Bennett, Benjamin - Corporal - 17th Infantry - Company: John Chunn - Age: 19 - Height: 5' 5 1/2" - Born: Baltimore Cty - Trade: Wheelwright - Enlistment date: 16 Aug 1813 - Place: Newport, KY - Period: War - BLW 9362-160-12 - Discharged at Chillicothe, OH, on 7 Jun 1815.

Bennett, Edward - Private - 36th Infantry - BLW 12730-160-12.

Bennett, Freeman - Seaman - US Sea Fencibles - Company: Simmones Bunbury.

Bennett, John B. - Private - US Volunteers - Company: Stephen Moore - Enlistment date: 8 Sep 1812 - Period: 1 Yr - Deserted on 5 Dec 1812.

Benson, William B. - Sergeant - 2nd Light Dragoons - Company: William Littlejohn - Other regiment: Corps of Artillery - Age: 25 - Height: 5' 8" - Born: Talbot Cty, MD - Trade: Carpenter - Enlistment date: 16 Feb 1814 - Place: Centreville, MD - Period: 5 Yrs - BLW 20770-160-12 - Discharged at Fort McHenry on 16 Feb 1819.

Bentalou, Paul - Colonel - Quartermaster Department - Served during the

Revolutionary War in the German Regiment and Pulaski Legion; commissioned as a major, deputy quartermaster general, on 29 Jun 1813; promoted to colonel, quartermaster general, on 17 Aug 1814; discharged on 15 Jun 1815; died on 14 Apr 1826.

Bernard, Hezekiah - Private - 38th Infantry - Pension: Old War WI-12381 Rejected.

Berry, Bayne L. - Private - 14th Infantry - Company: Reuben Gilder - Age: 24 - Height: 5' 4" - Born: Port Tobacco, MD - Trade: Farmer - Enlistment date: 31 Dec 1812 - Place: Alexandria, DC - Period: 5 Yrs - Wounded at Cook's Mills, UC; discharged on 19 Oct 1814 at Greenbush, NY, because of wound to right thigh.

Berry, Bayne S. - Private - 14th Infantry - Company: Reuben Gilder - Pension: Placed on the pension roll on 14 Aug 1815 - BLW 134-160-12 - Died on 24 Sep 1821.

Berry, Hezekiah - Private - 36th Infantry - Company: Henry Neale - BLW 4803-160-12.

Bert, Peter O. - Private - 38th Infantry - Company: Alexander Stuart - Pension: Land bounty to Charles H. O. Bert, brother and only heir at law of Peter O. Bert - BLW 22947-160-12.

Beverlo, John - Private - 14th Infantry - Company: Joseph Marshall - BLW 15181-160-12.

Bhare, Charles - Seaman - US Sea Fencibles - Company: Simmones Bunbury.

Bickors, John - Private - 38th Infantry - Company: John Mowton - BLW 18253-160-12.

Biddle, Richard - Private - 1st Rifles - Company: Joshua Hamilton - Other regiment: Rifles - Age: 38 - Height: 5' 8 1/4" - Born: Maryland - Trade: Cabinet maker - Enlistment date: 17 Aug 1813 - Place: Knoxville, TN - Period: War - Discharged at Conjocta Creek, NY, on 11 Jun 1815.

Bignell, Nathaniel B. - Corporal - 14th Infantry - Company: Richard Arell - Pension: Land bounty to Julia Ann Bignell, daughter and other heirs at law of Nathaniel B. Bignell - BLW 25267-160-12.

Bingham, Thomas - Private - 14th Infantry - Company: Joseph Marshall - BLW 4852-160-12.

Bird, Allen - Private - 14th Infantry - Company: David Cummings - Age: 32 - Height: 5' 4 1/2" - Born: Essex Cty, MA - Trade: Stone mason - Enlistment date: 28 Sep 1814 - Place: Baltimore - Period: War - Discharged on 18 Mar 1815.

Bird, Oliver - Private - 14th Infantry - BLW 14239-160-12.

Bird, William C. - First Lieutenant - 5th Infantry - Born: Pennsylvania - Commissioned as a 2nd lieutenant on 3 Jan 1812; promoted to 1st lieutenant on 1 Sep 1812; resigned on 1 Aug 1814.

Birely, John - Third Lieutenant - 14th Infantry - Pension: Wife Mary Beideman, WO-15709, WC-27364; married on 7 Jun 1817, Philadelphia, PA; soldier died about 1881 - BLW 10535-160-55 - Commissioned as a 3rd lieutenant on 3 Jul 1814; discharged on 15 Jun 1815.

Bishop, Abraham - Private - 2nd Light Dragoons - Company: Samuel Hopkins

By War of 1812 Soc. in MD

- Age: 21 - Height: 5' 5 1/2" - Born: Baltimore City - Trade: Hatter - Enlistment date: 13 Jul 1812 - Place: Frankford, KY - Period: 5 Yrs.

Bishop, Henry - Private - US Volunteers - Company: Stephen Moore - Enlistment date: 8 Sep 1812 - Period: 1 Yr - Pension: Land bounty to James Legg, John Henry Legg, William E. Legg, and Charles M. Legg, the nephews and only heirs at law of Henry Bishop - BLW 27388-160-12 - Died on 6 Aug 1813.

Bishop, Jesse - Private - 14th Infantry - Prisoner of War (Halifax).

Bishop, Nicholas - Sergeant - 38th Infantry - Company: James Smith - BLW 11813-160-12.

Bishop, William - Private - 5th Infantry - Company: Richard Bell - Age: 25 - Height: 5' 5" - Born: New Haven, CT - Trade: Hatter - Enlistment date: 3 Apr 1813 - Place: Baltimore - Period: 5 Yrs - BLW 904-160-12 - Discharged at Greenbush, NY, on 15 Apr 1815, on Surgeon's Certificate.

Black, John - Private - 14th Infantry - Prisoner of War, sent to England, captured at Beaver Dams on 24 Jun 1813.

Black, Samuel - Private - 38th Infantry - Company: James Hook - BLW 25184-160-12.

Black, William - Private - Light Artillery - Age: 21 - Height: 5' 2 1/2" - Born: Baltimore - Trade: Mariner - Enlistment date: 11 Jul 1814 - Place: Boston - Period: War.

Black, William - Private - 38th Infantry - Company: John Brookes - BLW 18914-160-12.

Blackwood, Hugh - Private - 2nd Artillery - Company: Nathan Towson - Pension: Land bounty to Delilah Cornel, sister and other heirs at law of Hugh Blackwood - BLW 18766-160-12.

Blain, Alexander - Private - 26th Infantry - Company: William Bezeau - Age: 28 - Height: 5' 10" - Born: Pennsylvania - Enlistment date: 18 May 1813 - Place: Annapolis.

Blaze, William - Private - 12th Infantry - Company: James Dorman - Other regiment: 8th Infantry (New) - Age: 31 - Height: 5' 10" - Born: Frederick Cty, MD - Trade: Farmer - Enlistment date: 29 Dec 1814 - Place: Fincastle - Period: 5 Yrs - BLW 313-320-12 - Discharged at Camp Jefferson, AL, on 1 Mar 1819.

Bloice, David - Private - 14th Infantry - Company: Henry Grindage - Pension: Land bounty to Rebecca Bloice, sister and other heirs at law of David Bloice - BLW 22887-160-12.

Bloise, Zachariah - Private - 14th Infantry - Company: Henry Grindage - Enlistment date: 10 Sep 1812 - Period: 5 Yrs - Pension: Land bounty to Morecai Bloice, son and other heirs at law of Zachariah Bloice - BLW 22886-160-12 - Prisoner of War (Halifax), captured at Beaver Dams on 24 Jun 1813, exchanged on 15 Apr 1814.

Bloodworth, William C. - Private - 5th Infantry - Company: William Henshaw - Age: 22 - Height: 5' 8" - Born: Princeton, Baltimore Cty - Trade: Merchant - Enlistment date: 25 Oct 1813 - Place: Baltimore - Period: 5 Yrs.

Blue, Henry - Private - 14th Infantry - Prisoner of War (Halifax), discharged

Maryland Regulars in War of 1812

on 9 Nov 1813.

Blunt, Joseph - Seaman - US Sea Fencibles - Company: Simmones Bunbury.

Boerstler, Charles G. - Colonel - 14th Infantry - Born: Maryland - Place: Maryland - Commissioned as a lieutenant colonel, 14th Infantry, on 12 Mar 1812; promoted to colonel on 20 Jun 1813; discharged on 15 Jun 1815.

Boestaller, Andrew - Private - 38th Infantry - Company: Shepperd Leakin - Age: 25 - Height: 5' 11" - Born: Hagerstown, MD - Trade: Farmer - Enlistment date: 8 Feb 1814 - Place: Hagerstown, MD - Period: 5 Yrs.

Bogert, Paul C. F. - Private - 38th Infantry - Company: John Mowton - BLW 10289-160-12.

Bohem, Joseph - Musician - 14th Infantry - Company: William McIlvane - Pension: Old War IF-3030.

Boice, Elijah - Private - 38th Infantry - Company: John Brookes - BLW 27307-160-42.

Bolar, Michael - Private - 38th Infantry - Company: John Rothrock - BLW 15177-160-12.

Boldine, Alexander - Private - 2nd Light Dragoons - Company: Samuel Hopkins - Age: 17 - Height: 5' 6" - Born: Baltimore City - Enlistment date: 29 Mar 1814 - Place: Philadelphia - Period: 5 Yrs - Discharged on 6 Sep 1815, invalid.

Bolster, William - Private - Corps of Artillery - Company: Evans Humphrey - Age: 37 - Height: 5' 7 1/2" - Born: Harford Cty, MD - Trade: Farmer - Enlistment date: 6 Aug 1814 - Period: War - BLW 9852-160-12 - Discharged on 9 Apr 1815.

Bolton, William - Corporal - 2nd Artillery - Company: Isaac Roach - Age: 22 - Height: 6' 1/4" - Born: Hancock, MD - Trade: Blacksmith - Enlistment date: 12 Aug 1814 - Place: Annapolis - Period: 5 Yrs - Deserted on 20 Sep 1817.

Bolton, William - Private - 2nd Artillery - Company: Nathan Towson - Age: 21 - Height: 5' 8" - Born: New Jersey - Trade: Laborer - Enlistment date: 19 Jan 1813 - Place: Lancaster - Period: 5 Yrs.

Bolvin, Thomas - Private - 38th Infantry - Company: John Brookes - BLW 3909-160-12.

Boman, William - Private - 14th Infantry - Company: Samuel Lane - Age: 22 - Height: 6' 1" - Born: Philadelphia - Enlistment date: 12 Aug 1812 - Period: 5 Yrs.

Bond, Charles - Sergeant - 14th Infantry - BLW 7683-160-12.

Boner, Arthur - Private - 17th Infantry - Company: Caleb Holder - Other regiment: 3rd Infantry (New) - Age: 57 - Height: 5' 11" - Born: Harford Cty, MD - Trade: Farmer - Enlistment date: 10 Sep 1813 - Place: Cincinnati - Period: 5 Yrs - Discharged at Newport, KY, on 9 Nov 1815 of old age.

Bongers, Peter C. - Seaman - US Sea Fencibles - Company: William Addison.

Bonner, Arthur - Private - 17th Infantry - Company: Caleb Holder - Other regiment: 3rd Infantry (New) - Age: 57 - Height: 5' 11" - Born: Harford Cty, MD - Trade: Farmer - Enlistment date: 10 Sep 1813 - Place: Cincinnati - Period: 18 Mos - Discharged at Newport, KY, on 9 Nov 1815 of old age.

By War of 1812 Soc. in MD

Boon, William - Private - 1st Infantry - Company: Moses Hook - Age: 25 - Height: 5' 9" - Born: Maryland - Trade: Sailor - Enlistment date: 25 Nov 1806 - Place: New Orleans - Period: 5 Yrs - Discharged at Fort Stoddard, AL, on 26 Nov 1811.

Boon, William C. - Sergeant - 44th Infantry - Age: 30 - Height: 5' 11 3/4" - Born: Caroline Cty, MD - Enlistment date: 19 Oct 1813 - Period: War.

Booth, George - Private - 14th Infantry - Prisoner of War (Halifax), discharged on 9 Nov 1813.

Booth, John - Private - 14th Infantry - Prisoner of War (Halifax), discharged on 3 Feb 1814.

Boran, John - Private - 5th Infantry - Pension: Heirs received half pay for five years in lieu of land bounty - Died in 1813.

Borgea, John - Private - 14th Infantry - Prisoner of War (Halifax), discharged from Halifax on 3 Feb 1814.

Bosely, Thomas - Private - 14th Infantry - Prisoner of War (Halifax), discharged on 3 Feb 1814.

Bosford, Thomas - Corporal - 36th Infantry - Company: Samuel Raisin - BLW 13806-160-12.

Bosley, Thomas - Servant - US Sea Fencibles - Company: Simmones Bunbury - Discharged at Fort McHenry on 24 Mar 1815.

Bosley, William - Private - 26th Infantry - Company: William Bezeau - Age: 32 - Height: 5' 9" - Born: Baltimore City - Enlistment date: 29 Sep 1814 - Place: Baltimore.

Bosley, William P. B. - Private - 38th Infantry - Company: Isaac Aldridge - Age: 23 - Height: 5' 9" - Born: Baltimore - Trade: Farmer - Enlistment date: 31 Dec 1814 - Place: Baltimore - Period: War - BLW 577-320-12 - Discharged on 6 Apr 1815.

Boswell, William - Private - 38th Infantry - Company: Charles Stansbury - Age: 32 - Height: 5' 3" - Born: Baltimore - Trade: Wheelwright - Enlistment date: 25 Nov 1814 - Place: Baltimore - Period: War - BLW 15056-160-12 - Discharged on 5 Apr 1815.

Botts, John - Private - 2nd Artillery - Company: Joseph Philips - Other regiment: Corps of Artillery - Age: 24 - Height: 5' 7 1/2" - Born: Maryland - Trade: Laborer - Enlistment date: 16 Jun 1812 - Period: 18 Mos - BLW 22188-160-12 - Re-enlisted at Fort Massac, IL, on 1 May 1814 for five years discharged on 1 May 1819.

Boulder, James - Private - 14th Infantry - Pension: Heirs received half pay for five years in lieu of land bounty - Died on 19 Oct 1814.

Bowan, Benjamin - Private - 28th Infantry - Pension: Heirs received half pay for five years in lieu of land bounty - Died in Jun 1814.

Bowden, James - Private - 14th Infantry - BLW 6221-160-12.

Bowen, Artemus - Sergeant - 14th Infantry - Company: David Cummings - Enlistment date: 20 Mar 1813 - Period: 5 Yrs - Prisoner of War (Halifax), sent to England, discharged on 3 Apr 1815.

Bowen, Peter - Private - 14th Infantry - Company: Joseph Marshall.

Bowen, Thomas - Private - 2nd Artillery - Company: Nathan Towson - Age:

Maryland Regulars in War of 1812

24 - Prisoner of War, captured during the Battle of Stoney Creek.

Bowen, Thomas - Private - 14th Infantry - Company: Joseph Marshall - Pension: Land bounty to Peter Bowen, son and heir at law of Thomas Bowen - BLW 26505-160-12.

Bowen, William - Private - 14th Infantry - Company: Clement Sullivan - Enlistment date: 22 May 1812 - Period: 18 Mos - Pension: Old War IF-10094 - Prisoner of War, sent to England, captured at Beaver Dams on 24 Jun 1813; discharged on 31 Mar 1815.

Bowermaster, Henry - Private - 2nd Artillery - Company: Nathan Towson - Age: 18 - Prisoner of War, captured during the Battle of Stoney Creek.

Bowers, Daniel - Private - US Volunteers - Company: Stephen Moore - Enlistment date: 8 Sep 1812 - Period: 1 Yr.

Bowers, John - Private - 38th Infantry - Company: Charles Stansbury - BLW 5447-160-12.

Bowers, William - Private - 2nd Light Dragoons - Company: William Littlejohn - Age: 38 - Height: 5' 10 1.2" - Born: Kent Cty, MD - Trade: Farmer - Enlistment date: 7 Nov 1813 - Place: Alexandria, DC - Period: War.

Bowie, Benjamin - Sergeant - 38th Infantry - Company: John Buck - Pension: Old War WI-12574 Rejected.

Bowstred, Samuel - Private - US Volunteers - Company: Stephen Moore - Enlistment date: 8 Sep 1812 - Period: 1 Yr - Died at Buffalo on 6 Dec 1812.

Boyce, Joseph - Drummer - 14th Infantry - Prisoner of War (Halifax), discharged from Halifax on 3 Feb 1814.

Boyd, Francis - Private - 2nd Rifles - Other regiment: Rifles - Age: 27 - Height: 5' 8 3/4" - Born: Cecil Cty, MD - Trade: Saddler - Enlistment date: 25 May 1814 - Place: Chillicothe, OH - Period: 5 Yrs - Discharged at Detroit on 22 Sep 1815 for inability, abscess in jaw.

Boyd, Francis - Private - 19th Infantry - Age: 27 - Height: 5' 10 1/2" - Born: Maryland - Enlistment date: 31 Mar 1814 - Place: Lebanon, Warren Cty, OH - Period: 18 Mos.

Boyd, Samuel - Private - 14th Infantry - Company: Clement Sullivan - Age: 52 - Pension: Old War IF-3017 - Prisoner of War, captured during the Battle of Stoney Creek.

Boyd, Walter - Private - 5th Infantry - Company: Peter Mulhenberg - Age: 25 - Height: 5' 5" - Born: Maryland - Trade: Laborer - Enlistment date: 26 Feb 1812 - Place: Hudson - Period: 5 Yrs - Discharged on 9 May 1816.

Boyd, William - Private - 2nd Artillery - Company: Nathan Towson - Pension: Land bounty to Elizabeth Boyd, sister and only heir at law of William Boyd - BLW 21735-160-12 - Killed during the Battle of Chippewa.

Boyer, Benjamin - Private - 14th Infantry - Company: Joseph Marshall.

Boyle, Edward J. - Private - US Volunteers - Company: Stephen Moore - Enlistment date: 8 Sep 1812 - Period: 1 Yr.

Boyle, Richard - Private - 14th Infantry - Company: William McIlvane.

Bradford, James - Sergeant - 12th Infantry - Company: James Paxton - Pension: Land bounty to Joseph Bradford, father & only heir at law of James Bradford - BLW 24548-160-12 - Died on 23 Feb 1814.

By War of 1812 Soc. in MD

Bradford, William - Seaman - US Sea Fencibles - Company: John Gill - Age: 33 - Height: 5' 9" - Born: Salem, Salem Cty, NJ - Enlistment date: 1 Jan 1814.

Brady, William - Private - 14th Infantry - Company: Reuben Gilder - Period: 18 Mos - Discharged at Plattsburgh, NY, on 4 Jul 1814.

Brady, William - Private - 23rd Infantry - Age: 21 - Height: 5' 7" - Born: Hillsborough, MD - Trade: Tailor - Enlistment date: 14 Jul 1814 - Place: Albany, NY - Period: War - Discharged on 1 Jan 1818 or 1 Jul 1814.

Brady, William - Private - 24th Infantry - Company: James Campbell - Age: 38 - Height: 5' 8" - Born: Maryland - Trade: Farmer - Enlistment date: 2 Jul 1814 - Place: Sumner County, TN - Period: War - BLW 12159-160-12 - Discharged at Camp Mandeville, AL, on 22 Mar 1815.

Branard, George - Private - 39th Infantry - Age: 25 - Height: 6' 1" - Born: Frederick Cty, MD - Trade: Farmer - Enlistment date: 28 Oct 1814 - Place: Knoxville, TN - Period: 5 Yrs - Died on 17 Feb 1815.

Branch, Alexander - Private - 38th Infantry - Enlistment date: 26 Mar 1814 - Period: War.

Branch, Reaves A. - Private - 38th Infantry - Company: John Rothrock - BLW 12331-160-12.

Brand, George - Private - 7th Infantry - Company: Carey Nicholas - Age: 26 - Height: 5' 7" - Born: Maryland - Trade: Shoemaker - Enlistment date: 21 Dec 1813 - Place: Nashville, TN - Period: War - Discharged at New Orleans on 8 Apr 1815.

Brashears, George T. - Private - 36th Infantry - Company: Samuel Raisin - BLW 9159-160-12.

Brasheart, Osborn D. - Musician - 14th Infantry - Company: Kenneth McKenzie - Pension: Land bounty to Cecilia Brasheart, sister and other heirs at law of Osborn D. Brasheart - BLW 10884-160-12.

Brazier, Thomas - Private - 14th Infantry - Company: Joseph Marshall - Pension: Land bounty to William Brazier, brother and other heirs at law of Thomas Brazier - BLW 21483-160-12 - Died at Cheektowaga Hospital, NY, on 10 Jan 1815.

Breeding, James - Private - 24th Infantry - Company: James Campbell - Age: 39 - Height: 5' 10" - Born: Maryland or North Carolina - Trade: Farmer - Enlistment date: 24 Jun 1814 - Place: Jackson County, TN - Period: War - BLW 10450-160-12 - Discharged at Camp Russell, AL, on 6 Apr 1815.

Brener, William - Private - 38th Infantry - Age: 23 - Height: 5' 4" - Born: North Hampton, VA - Enlistment date: 14 Jul 1814 - Place: Suffolk, VA.

Brent, Robert C. - First Lieutenant - Corps of Artillery - Born: Virginia - Graduated from the US Military Academy on 2 Mar 1815; commissioned as a 3rd lieutenant, Corps of Artillery, on 2 Mar 1815; promoted to 2nd lieutenant on 13 May 1817; promoted to 1st lieutenant, 1st Artillery, on 1 Jun 1821; transferred to 1st Infantry on 16 Nov 1821; transferred to 1st Artillery on 21 Dec 1822; resigned on 1 Nov 1823; died on 15 May 1837.

Brewer, Zachariah - Corporal - 14th Infantry - Pension: Wife Rebecca, Old War WF-12364.

Maryland Regulars in War of 1812

Brewington, William - Private - 38th Infantry - Company: John Brookes - Age: 20 - Height: 5' 10 1/2" - Born: Somerset Cty, MD - Trade: Waterman - Enlistment date: 10 Mar 1814 - Place: Norfolk, VA - Period: War - Deserted on 2 Mar 1815.

Brice, William - Private - 2nd Light Dragoons - Company: George Haig - Other regiment: Corps of Artillery - Age: 30 - Height: 5' 9" - Born: Bladensburg, MD - Trade: Farmer - Enlistment date: 20 Jun 1812 - Place: Baltimore - Period: 5 Yrs.

Briers, John R. - Private - US Volunteers - Company: Stephen Moore - Enlistment date: 8 Sep 1812 - Period: 1 Yr - Died at Buffalo on 15 Jan 1813.

Brightwell, Barnet - Private - 38th Infantry - Company: John Brookes - Enlistment date: 30 Aug 1814 - Pension: Wife Sally Crittenden, WO-31169, WC-28058; married on 17 Feb 1821, Henrico Cty, VA; soldier died in 1840, Henrico Cty, VA - BLW 10154-160-12; BLW 109515-160-55 - Discharged on 22 Feb 1815.

Briley, John - Private - 38th Infantry - Company: John Rothrock - BLW 16621-160-12.

Brinklinger, Andrew - Private - 17th Infantry - Company: William Adair - Age: 21 - Height: 5' 8" - Born: Maryland - Trade: Farmer - Enlistment date: 20 Feb 1814 - Place: Louisville, KY - Period: War - BLW 14041-160-12 - Discharged at Chillicothe, OH, on 4 Jun 1814.

Brinkman, John - Seaman - US Sea Fencibles - Company: Simmones Bunbury.

Brison, James - Private - 14th Infantry - Prisoner of War (Halifax), captured at Beaver Dams on 24 Jun 1813, discharged on 9 Nov 1813.

Brittain, William - Private - 38th Infantry - Company: Shepperd Leakin - Age: 31 - Height: 5' 8 1/2" - Born: Baltimore - Trade: Brick layer - Enlistment date: 2 Mar 1814 - Period: War - BLW 11103-160-12 - Discharged at Fort McHenry on 28 Mar 1815.

Brittenham, Nehemiah - Private - 14th Infantry - Company: Joseph Marshall.

Britton, James - Captain - 14th Infantry - Commissioned as a captain on 12 Mar 1812; resigned on 1 Oct 1813.

Britton, John H. - Private - 14th Infantry - Company: Robert Kent - Pension: Land bounty to Elizabeth Lang, sister and other heirs at law of John H. Britton - BLW 12784-160-12.

Britton, William - Private - 36th Infantry - Company: Thomas Carbery - Enlistment date: 18 Jan 1814 - Period: 1 Yr - Discharged on 17 Jan 1815.

Britton, William - Corporal - 38th Infantry - Company: Shepperd Leakin - Pension: Land bounty to Mary Nelson, sister and other heirs at law of William Britton - BLW 11103-160-12.

Brook, William - Seaman - US Sea Fencibles - Company: Simmones Bunbury - Age: 45 - Height: 5' 7 1/2" - Born: Annapolis - Enlistment date: 11 Mar 1814 - Period: 1 Yr.

Brooke, Edmund - First Lieutenant - US Marine Corps - Born: Maryland - BLW 825-160-50 - Graduated from the US Military Academy on 13 Apr 1814; commissioned as a 2nd lieutenant, U.S. Marine Corps, on 15 Apr

By War of 1812 Soc. in MD

1814; participated in the Battle of Bladensburg, MD; promoted to 1st lieutenant on 18 Jun 1815; resigned on 3 Feb 1817; died in 1855.

Brookes, John - Captain - 38th Infantry - Pension: Wife Ester Jane Fowie (3rd wife), WO-14447, WC-8382; first wife Sarah Taliaferro, second wife Ellen H. Waring; married on 23 Apr 1846, Alexandria, Alexandria Cty, DC; soldier died on 13 Jul 1858 in Prince George's Cty, MD - BLW 22680-160-50 - Commissioned as a captain on 20 May 1813; discharged on 15 Jun 1815.

Brookes, Thomas - Private - 38th Infantry - Company: James Smith - BLW 10534-160-12.

Brooks, Jesse - Private - 38th Infantry - Company: John Rothrock - BLW 11140-160-12.

Brooks, Thomas - Private - 38th Infantry - Company: Charles Stansbury - BLW 17527-160-12.

Broom, Thomas - Quarter Gunner - US Sea Fencibles - Company: Simmones Bunbury.

Broomhall, Thomas - Private - 14th Infantry - Company: Robert Kent - Pension: Land bounty to John Broomhall, son and other heirs at law of Thomas Broomhall - BLW 20442-160-12.

Broudent, Moses - Private - 14th Infantry - Prisoner of War (Halifax), captured at Beaver Dams on 24 Jun 1813, discharged from Halifax on 9 Nov 1813.

Broughton, George - Private - 38th Infantry - Company: Shepperd Leakin - BLW 14932-160-12.

Brown, Abel - Private - 14th Infantry - Company: Reuben Gilder - Age: 42 - Height: 5' 9" - Born: Maryland - Enlistment date: 2 Dec 1812 - Period: 18 Mos - Discharged at Burlington, VT, on 29 Mar 1814, old age and inability.

Brown, Andrew - Private - 36th Infantry - Age: 32 - Height: 5' 7" - Born: Ireland - Enlistment date: 23 Jun 1813 - Place: Baltimore.

Brown, Aquila - Sergeant - 2nd Light Dragoons - Company: Henry Hall - Age: 21 - Height: 6' 1" - Born: Harford Cty, MD - Trade: Laborer - Enlistment date: 11 Apr 1812 - Place: Havre de Grace, MD - Period: 5 Yrs - Discharged on 11 Aug 1815.

Brown, Bazil - Corporal - 38th Infantry - Company: Shepperd Leakin - BLW 17028-160-12.

Brown, Charles W. - Corporal - 14th Infantry - Pension: Wife Rachel C., Old War WF-12660 Rejected.

Brown, Christian - Private - 38th Infantry - Company: James Hook - Pension: Old War IF-15001 - BLW 22152-160-12.

Brown, David - Private - 38th Infantry - Company: James Hook - BLW 24403-160-12.

Brown, James - Servant - US Sea Fencibles - Company: Simmones Bunbury - Enlistment date: 1 Nov 1814 - Period: 1 Yr - Discharged at Fort McHenry on 24 Mar 1815.

Brown, James - Private - 14th Infantry - Company: Kenneth McKenzie - BLW 3234-160-12 - Prisoner of War (Halifax), discharged from Halifax on 3 Feb 1814.

Brown, John - Seaman - US Sea Fencibles - Company: Simmones Bunbury.

Maryland Regulars in War of 1812

Brown, John - Seaman - US Sea Fencibles - Company: Simmones Bunbury.

Brown, John - Private - 38th Infantry - Company: John Mowton - BLW 11251-160-12.

Brown, John - Private - 14th Infantry - Prisoner of War (Halifax), discharged from Halifax on 27 Aug 1813.

Brown, Michael - Private - 14th Infantry - Prisoner of War (Halifax), sent to England.

Brown, Perry - Private - 14th Infantry - Prisoner of War (Halifax), discharged from Halifax on 3 Feb 1814.

Brown, Thomas - Recruit - 26th Infantry - Company: William Bezeau - Race: 24 - Age: 24 - Height: 5' 8" - Born: Maryland - Trade: Laborer - Enlistment date: 8 Oct 1814 - Place: Philadelphia - Period: War - BLW 25756-160-12 - Listed as "Col'd" in his service record; discharged on 23 Mar 1815 at Philadelphia.

Brown, William - Private - 38th Infantry - Company: John Brookes - Age: 33 - Height: 5' 8 1/4" - Born: Anne Arundel Cty, MD - Trade: Stone paver - Enlistment date: 7 Arp 1814 - Place: Richmond, VA - Period: War - BLW 22597-160-12 - Discharged at Craney Island, VA, on 15 Mar 1815.

Brown, William - Private - 36th Infantry - Company: Joseph Merrick - Age: 27 - Height: 5' 7" - Born: Charles City, VA - Enlistment date: 6 May 1813 - Place: Baltimore - Transferred to the U.S. Navy on 26 Oct 1813.

Brown, William - Private - 2nd Infantry - Company: William Boots - Other regiment: 1st Infantry (New) - Age: 18 - Height: 5' 8" - Born: Anne Arundel Cty, MD - Trade: Planter - Enlistment date: 24 Oct 1800 - Place: Washington, DC - Period: 5 Yrs - Re-enlisted on 21 Oct 1805 and on 1 Nov 1810; discharged on 27 Nov 1815 at New Orleans, LA.

Brown, William - Private - 38th Infantry - Company: Shepperd Leakin - Discharged on 19 Apr 1814.

Brown, William - Private - 38th Infantry - Company: John Rothrock - Age: 26 - Height: 5' 8" - Born: Pennsylvania - Trade: Sailor - Enlistment date: 2 Mar 1814 - Place: Norfolk, VA - Period: War - BLW 24740-160-12 - Discharged at Craney Island, VA, on 15 Mar 1815.

Brown, William - Private - 14th Infantry - Company: Thomas Montgomery - Enlistment date: 10 Mar 1813 - BLW 5259-160-12 - Discharged at Plattsburgh, NY, on 25 May 1814 on Surgeon's Certificate.

Brown, William - Private - 41st Infantry - Company: James McCulten - Age: 35 - Height: 5' 9" - Born: Baltimore - Trade: Seaman - Enlistment date: 9 Aug 1814 - Place: New York City - Period: 5 Yrs - Discharged on 13 Jul 1815 on Surgeon's Certificate.

Brown, William - Private - 38th Infantry - Company: John Mowton - Age: 36 - Height: 5' 2 1/2" - Born: Ireland - Trade: Sailor - Enlistment date: 2 Apr 1814 - Place: Suffolk, VA - Period: War.

Brown, William - Private - 36th Infantry - Age: 25 - Height: 5' 10" - Born: Frederick Cty, MD - Enlistment date: 17 Feb 1814 - Place: Georgetown, DC.

Brown, William - Private - 2nd Light Dragoons - Company: John Burd - Age: 25 - Height: 5' 6" - Born: England - Enlistment date: 20 Jan 1814 - Place:

By War of 1812 Soc. in MD

Maryland - Period: 5 Yrs.

Brown, William - Private - 38th Infantry - Company: Isaac Aldridge - Enlistment date: 22 Jun 1813 - Period: 1 Yr - Died on 21 Feb 1814.

Brown, William - Private - 2nd Light Dragoons - Company: Stephen Proctor - Age: 39 - Height: 5' 6" - Born: Kilkenny, Ireland - Trade: Brick layer - Enlistment date: 1 Jan 1812 - Place: Baltimore - Period: 5 Yrs.

Brown, Zachariah - Corporal - 14th Infantry - Company: Thomas Montgomery - Enlistment date: 6 Apr 1813 - Period: 5 Yrs - Wounded at La Cole Mill, LC,; died on 13 May 1814 at Plattsburgh, NY.

Bruffett, Daniel - Sergeant - 14th Infantry - Company: Thomas Kearney - Pension: Land bounty to William Bruffett, brother and only heir at law of Daniel Bruffett - BLW 26707-160-12.

Brunan, John - Private - 14th Infantry - Company: Joseph Marshall.

Bruner, William - Sergeant - 2nd Rifles - Company: Batteal Harrison - Age: 22 - Height: 5' 10" - Born: Maryland - Trade: Farmer - Enlistment date: 24 Jun 1814 - Place: Cincinnati - Period: War - Discharged on 29 Jun 1815.

Brusus, Jacob - Private - 14th Infantry - Company: William McIlvane - Pension: Placed on the pension rolls on 6 Feb 1816.

Buck, John - Captain - 38th Infantry - Born: Maryland - Commissioned as a captain on 20 May 1813; discharged on 15 Jun 1815.

Bulgar, William - Private - 38th Infantry - Company: James Hook - Age: 20 - Height: 5' 10" - Born: Shenandoah Cty, VA - Trade: Farmer - Enlistment date: 10 Mar 1814 - Place: Norfolk, VA - Period: War - BLW 4346-160-12 - Discharged at Fort Covington, MD, on 31 Mar 1815.

Bumigan, Nicholas - Private - 38th Infantry - Company: Shepperd Leakin - BLW 16022-160-12.

Bunbury, M. Simmones - Captain - US Sea Fencibles - Commissioned as a captain on 1 Oct 1813; discharged on 15 Jun 1815.

Burch, Theophilus Y. - Private - 14th Infantry - Pension: Heirs received half pay for five years in lieu of land bounty.

Burch, William - Private - 36th Infantry - Age: 21 - Height: 5' 3 1/4" - Born: Queen Anne's Cty, MD - Enlistment date: 30 May 1813 - Place: Centreville, MD.

Burgess, John M. - Second Lieutenant - 36th Infantry - Other regiment: 26th Infantry - Commissioned as a 3rd lieutenant, 36th Infantry, on 30 Apr 1813; transferred to 26th Infantry on 10 Aug 1814; promoted to 2nd lieutenant on 1 May 1814; discharged on 15 Jun 1815.

Burk, James - Private - 14th Infantry - Company: Reuben Gilder - Pension: Old War IF-15019.

Burk, William - Private - 12th Infantry - Company: Thomas Sangster - Age: 21 - Height: 5' 2 1/2" - Born: Baltimore - Trade: Blacksmith - Enlistment date: 8 Apr 1814 - Place: Clarksburg, VA - Period: 5 Yrs.

Burk, William - Private - 6th Infantry - Company: Alexander Thompson - Other regiment: 2nd Infantry (New) - Age: 21 - Height: 5' 7 1/2" - Born: Carlisle, PA - Trade: Farmer - Enlistment date: 4 Feb 1812 - Place: Hagerstown, MD - Period: 5 Yrs - Discharged on 4 Feb 1817.

Burk, William - Private - 4th Infantry - Other regiment: 2nd Infantry (New) - Age: 21 - Height: 5' 2 1/2" - Born: Maryland - Trade: Blacksmith - Enlistment date: 8 Apr 1814 - Period: 5 Yrs - BLW 13000-160-12.

Burke, John B. - Private - 14th Infantry - Company: Reuben Gilder - Pension: Land bounty to Ann Burke, daughter and heir at law of John B. Burke - BLW 21275-160-12.

Burke, John T. - Private - 38th Infantry - Company: John Mowton - BLW 2696-160-12.

Burlew, Daniel - Private - 2nd Rifles - Company: Batteal Harrison - Age: 33 - Height: 5' 9" - Born: Baltimore Cty - Trade: Carpenter - Enlistment date: 22 May 1814 - Place: Cincinnati - Period: War - BLW 12165-160-12 - Discharged at Detroit on 30 Jul 1815.

Burlue, Gilbert - Private - 14th Infantry - Prisoner of War (Halifax), sent to England.

Burnell, John - Private - 14th Infantry - Prisoner of War (Halifax), captured at Beaver Dams on 24 Jun 1813, discharged from Halifax on 3 Feb 1813.

Burnett, Archibald - Recruit - 26th Infantry - Company: William Bezeau - Age: 21 - Height: 5' 9" - Born: Maryland or Dauphin Cty, PA - Trade: Tailor - Enlistment date: 19 Nov 1814 - Place: Baltimore - Period: War - BLW 7350-160-12 - Discharged at Philadelphia on 23 Mar 1815.

Burnett, John - Private - 36th Infantry - Company: Joseph Hook - BLW 4488-160-12.

Burnham, George - Private - 38th Infantry - Company: James Smith - BLW 7223-160-12.

Burns, Andrew - Private - 36th Infantry - Company: Joseph Hook - Age: 32 - Height: 5' 11" - Born: Fredericktown, MD - Enlistment date: 14 May 1813 - Place: Westminster, MD - Period: 1 Yr.

Burns, John - Private - 38th Infantry - Pension: Heirs received half pay for five years in lieu of land bounty - Died on 4 May 1814.

Burns, Truman - Sergeant - 36th Infantry - Company: Joseph Hook - Pension: No pension claim - BLW 12680-160-12 & 78-160-50.

Burns, William - Corporal - 38th Infantry - Company: Isaac Aldridge - Age: 18 - Height: 5' 4 1/2" - Born: Dorchester Cty, MD - Trade: Laborer - Enlistment date: 17 Nov 1814 - Place: Baltimore - Period: War - BLW 14099-160-12 - Discharged at Baltimore on 6 Apr 1815.

Burny, James - Private - 38th Infantry - Company: Shepperd Leakin - BLW 6921-160-12.

Burton, Joshua - Private - 38th Infantry - Company: John Buck - BLW 13234-160-12.

Burton, William - Private - 36th Infantry - Age: 23 - Height: 5' 6" - Born: Worcester Cty, MD - Enlistment date: 28 Jul 1813 - Place: Snow Hill, MD - Period: 1 Yr.

Busey, Jesse - Private - 2nd Artillery - Company: Daniel Cushing - Other regiment: Corps of Artillery - Age: 18 - Height: 5' 4 1/2" - Born: Prince George's Cty, MD - Trade: Hatter - Enlistment date: 1 Jan 1813 - Place: Lexington, KY - Period: 5 Yrs - Discharged on 29 Jan 1816.

By War of 1812 Soc. in MD

Bush, Abraham - Private - 1st Artillery - Age: 48 - Born: Maryland - Prisoner of War (Quebec), died at Quebec on 22 Mar 1814.

Bussel, George - Gunner - US Sea Fencibles - Company: Simmones Bunbury.

Bussel, John - Private - 36th Infantry - BLW 6050-160-12.

Bussell, Charles - Corporal - 36th Infantry - Company: Charles Randolph - Enlistment date: 14 Jan 1815 - Pension: Wife Lucy Ann Wine, SC-4965, WO-13796, WC-10506; married on 18 Jan 1867, Mountain View, VA; soldier died on 16 Nov 1876, Mountain View, VA; first wife was Mary Black - BLW 940-320-12 - Discharged on 13 Jan 1820.

Butcher, William - Private - 38th Infantry - Company: Shepperd Leakin - Age: 35 - Height: 6' 1 1/2" - Born: Anne Arundel Cty, MD - Trade: Carpenter - Enlistment date: 16 Sep 1814 - Place: Baltimore - Period: War - BLW 14123-160-12 - Discharged at Fort McHenry on 28 Mar 1815.

Butcher, William - Private - 14th Infantry - Company: Thomas Montgomery - Age: 27 - Height: 6' 1" - Born: Anne Arundel Cty, MD - Enlistment date: 27 Jun 1812 - Place: F. County - Period: 18 Mos - Pension: Old War IF-15016 - Discharged on 27 Dec 1813.

Butler, William - Private - 2nd Light Dragoons - Company: William Littlejohn - Age: 15 - Height: 5' 1" - Born: Pennsylvania - Enlistment date: 4 Dec 1813 - Place: Baltimore - Period: 5 Yrs - Discharged on 6 Sep 1815 on Surgeon's Certificate.

Butler, William - Sergeant - 14th Infantry - Company: Reuben Gilder - Re-enlisted on 10 May 1814 in the 1st Rifles.

Butts, Noah - Private - 2nd Artillery - Company: Nathan Towson - Pension: Old War IF-45888 - BLW 7298-160-12.

Byard, Richard - Private - 36th Infantry - Company: Joseph Hook - BLW 13232-160-12.

Byrd, Francis Otway - Second Lieutenant - 2nd Artillery - Company: Nathan Towson - Pension: Wife Elizabeth R. Pleasants, WO-18433, WC-17300; married on 9 Jun 1817, Philadelphia, PA; soldier died on 2 May 1860, Baltimore, MD - BLW 10552-160-50 - Commissioned on 6 Jul 1812; discharged on 23 March 1813.

Byrum, William - Private - 19th Infantry - Age: 20 - Height: 5' 9 1/2" - Born: Maryland - Enlistment date: 7 Apr 1814 - Place: Marietta, OH.

C

Cable, Henry - Private - 38th Infantry - Company: James Hook - Age: 57 - Height: 5' 4" - Born: Frederick, MD - Trade: Brick layer - Enlistment date: 28 Feb 1814 - Place: Alexandria, DC - Period: War - Pension: Placed on the pension rolls on 21 Oct 1816 - BLW 7097-160-12 - Discharged at Fort Covington, MD, on 31 Mar 1815, for disability; died on 3 Dec 1817.

Caccaher, Stephen - Private - 14th Infantry - Prisoner of War (Halifax), captured at Beaver Dams on 24 Jun 1813, discharged from Halifax on 3 Feb 1813.

Caddern, Peter - Private - 42nd Infantry - Company: Edward Mendenhall - Age: 24 - Height: 5' 3 1/2" - Born: Maryland - Enlistment date: 20 May 1814

Maryland Regulars in War of 1812

- Period: War - Discharged at Philadelphia on 19 May 1815.

Cadell, William - Corporal - Corps of Artillery - Pension: Heirs received half pay for five years in lieu of land bounty - Died in Dec 1814.

Caent, Levie - Private - 14th Infantry - Prisoner of War (Halifax), captured at Beaver Dams on 24 Jun 1813, discharged from Halifax on 3 Feb 1813.

Cafferty, William - Private - 14th Infantry - Company: William McIlvane - Enlistment date: 13 Jul 1812 - Period: 5 Yrs - Prisoner of War (Halifax), captured at Beaver Dams on 24 Jun 1813, exchanged on 15 Apr 1814.

Caffrey, John R. - Seaman - US Sea Fencibles - Company: John Gill - Enlistment date: 26 Dec 1813.

Caffry, John R. - Private - US Volunteers - Company: Stephen Moore - Enlistment date: 8 Sep 1812 - Period: 1 Yr.

Cage, Thomas S. - Sergeant - 38th Infantry - Company: John Brookes - Age: 28 - Height: 5' 8" - Born: Prince George's Cty, MD - Trade: Farmer - Enlistment date: 28 Apr 1814 - Place: Craney Island, VA - Period: War - Pension: Wife Susanna Lawson, SO-19194, SC-12804; married on 26 Jan 1817, Prince George's Cty, MD - BLW 496-160-12 - Discharged at Craney Island, VA, on 15 Mar 1815.

Caheve, Patrick - Private - 14th Infantry - Prisoner of War (Halifax), sent to England.

Cahoon, John - Private - 2nd Artillery - Company: Frederick Evans - Other regiment: Ordnance Department - Age: 21 - Height: 5' 9" - Born: Philadelphia - Trade: Cooper - Enlistment date: 28 Nov 1811 - Place: Greenleaf Point, DC - Period: 5 Yrs - Transferred to Ordnance Department on 16 Nov 1815.

Cain, Jacob - Private - 38th Infantry - Company: Anthony Miltenberger - Enlistment date: 6 Jun 1813 - Period: 1 Yr.

Cain, John - Private - 14th Infantry - Company: Kenneth McKenzie - Age: 29 - Height: 5' 9" - Trade: Farmer - Enlistment date: 18 Dec 1812 - Period: 18 Mos - Prisoner of War (Halifax), captured at Beaver Dams on 24 Jun 1813; discharged on 7 Aug 1814.

Cain, William - Private - 38th Infantry - Company: Shepperd Leakin - Age: 25 - Height: 5' 8" - Born: Harford Cty, MD - Trade: Farmer - Enlistment date: 14 Feb 1814 - Period: War - Discharged at Fort McHenry on 28 Mar 1815.

Cain, William - Private - 14th Infantry - Age: 52 - Height: 5' 8" - Born: Lancaster, PA - Trade: Distiller - Enlistment date: 18 Aug 1813 - Place: Hagerstown, MD - Period: War - Discharged at Greenbush, NY, on 15 Mar 1815.

Cairn, Hugh - Private - 5th Infantry - Company: William Bird - Age: 43 - Height: 5' 4 1/2" - Born: Baltimore - Enlistment date: 15 Apr 1813 - Period: 5 Yrs.

Caldwell, Benjamin W. - Private - 5th Infantry - Company: James Dorman - Other regiment: 3rd Infantry (New) - Age: 29 - Height: 6' - Born: Baltimore - Trade: Shoemaker - Enlistment date: 5 Jun 1813 - Place: Baltimore - Period: 5 Yrs.

By War of 1812 Soc. in MD

Caldwell, Isaac - Corporal - 2nd Artillery - Company: Nathan Towson - BLW 14674-160-12.

Caldwell, John - Corporal - 14th Infantry - Company: Thomas Montgomery - Prisoner of War, exchanged on 15 Apr 1814.

Caldwell, Philemon - Private - 14th Infantry - Company: Thomas Montgomery - Other regiment: 4th Infantry (New) - Age: 19 - Height: 5' 6" - Born: Dorchester Cty, MD - Trade: Laborer - Enlistment date: 16 Jun 1812 - Place: Baltimore - Period: 5 Yrs - Deserted at Fort Severn, MD, on 17 Jul 1815.

Caldwell, Thomas - Private - 3rd Infantry - Company: William Johnson - Other regiment: 1st Infantry (New) - Age: 30 - Height: 6' 2" - Born: Maryland - Trade: Shoemaker - Enlistment date: 26 Aug 1811 - Place: Washington, MS - Period: 5 Yrs - Discharged on 25 Aug 1816.

Caldwell, Thomas - Corporal - 42nd Infantry - Company: George Barker - Age: 24 - Height: 5' 11" - Born: Maryland - Trade: Laborer - Enlistment date: 6 Aug 1814 - Place: Wilmington, DE - Period: 5 Yrs - Deserted on 22 Mar 1815.

Caley, James - Private - 2nd Artillery - Company: Nathan Towson - Other regiment: Corps of Artillery - Enlistment date: 29 Jun 1812 - Period: 5 Yrs - Discharged at Fort Moultrie, SC, on 5 Jan 1815 for sore legs.

Calhoun Jr., James - Deputy Commissary of Purchasing - Purchasing Department - Commissioned as the deputy commissary of purchasing on 29 Jun 1813; discharged on 15 Jun 1815.

Calhoun, Andrew - Private - 4th Rifles - Company: Matthew Magee - Age: 27 - Height: 5' 9" - Born: Barron, MD - Trade: Miller - Enlistment date: 2 Sep 1814 - Place: Bedford, PA - Period: 5 Yrs - Discharged at Fort Smith on 1 Apr 1819 and then re-enlisted.

Callaghan, Samuel - Musician - 1st Infantry - Company: John McClary - Other regiment: 2nd Infantry and 1st Infantry (New) - Height: 5' 5" - Born: Annapolis - Trade: Soldier - Enlistment date: 1 Jul 1802 - Period: 5 Yrs - Re-enlisted on 3 Jul 1807 and again on 12 Aug 1812; discharged on 9 Feb 1817 at Baton Rouge, LA, and re-enlisted.

Callahan, Jeremiah - Private - 32nd Infantry - Company: John Steele - Age: 31 - Height: 5' 10" - Born: Maryland or Delaware - Trade: Farmer - Enlistment date: 8 Jun 1813 - Period: 1 Yr - BLW 18886-160-12 - Re-enlisted at Utrecht, NY, on 24 Feb 1814 for the war; discharged on 11 May 1815 at New York, NY.

Calvin, John - Private - 38th Infantry - Company: John Hook - Discharged on 31 Dec 1814.

Cameron, Simeon - Private - 38th Infantry - Company: John Buck - BLW 17550-160-12.

Cameron, Thomas - Private - 5th Infantry - Company: George Brooks - Other regiment: 3rd Infantry (New) - Age: 22 - Height: 5' 7" - Born: Scotland - Trade: Farmer - Enlistment date: 6 Mar 1813 - Place: Hagerstown, MD - Period: 5 Yrs - Discharged in Sep or Oct 1815 after furnishing a substitute.

Camp Jr., William - Ensign - 38th Infantry - Company: Anthony Miltenberger

Maryland Regulars in War of 1812

- Born: Maryland - BLW 7414-160-50 - Commissioned as an ensign on 20 May 1813; resigned on 1 May 1814.

Camp, Jacob - Sergeant - 26th Infantry - Company: William Bezeau - Age: 22 - Height: 5' 7" - Born: Johnstown, PA - Trade: Shoemaker - Enlistment date: 20 Sep 1814 - Place: Baltimore - Period: War - BLW 311-160-12 - Discharged in Jan 1815.

Camp, John - Private - 17th Infantry - Company: John Chunn - Age: 26 - Height: 5' 8 3/4" - Born: Maryland - Trade: Clock maker - Enlistment date: 15 Sep 1814 - Period: War - BLW 24828-160-12 - Discharged at Chillicothe, OH, on 11 Jul 1815.

Campbell, Andrew - Private - 38th Infantry - Company: Shepperd Leakin - Age: 38 - Height: 5' 3" - Born: Ireland - Trade: Nailor - Enlistment date: 7 Apr 1814 - Period: War - BLW 2387-160-12 - Discharged at Fort McHenry on 28 Mar 1815.

Campbell, Charles - Private - 32nd Infantry - Company: John Steele - Age: 20 - Height: 5' 7" - Born: Maryland - Trade: Farmer - Enlistment date: 19 Jul 1813 - Period: 1 Yr - BLW 15803-160-12 - Re-enlisted.

Campbell, Daniel - Private - 14th Infantry - Company: Reuben Gilder - Discharged at Plattsburgh, NY, on 12 Jul 1814 for old age.

Campbell, Francis - Private - 38th Infantry - Company: Shepperd Leakin - Enlistment date: 28 May 1813 - Period: 1 Yr - BLW 20327-160-12.

Campbell, Frederick - Private - 14th Infantry - Prisoner of War (Halifax), discharged from Halifax on 3 Feb 1814.

Campbell, George - Private - 38th Infantry - Company: John Mowton - Age: 34 - Height: 5' 9" - Born: Nova Scotia - Trade: Clerk - Enlistment date: 14 Apr 1814 - Place: Craney Island, VA - Period: War - BLW 11909-160-12 - Discharged at Craney Island, VA, on 15 Mar 1815.

Campbell, Henry M. - Second Lieutenant - 2nd Artillery - Company: Nathan Towson - Other regiment: Corps of Artillery - Born: Pennsylvania - Commissioned as a 3rd lieutenant, 2nd Artillery, on 19 Jul 1813; promoted to 2nd lieutenant on 12 Oct 1813; transferred to Corps of Artillery on 12 May 1814; resigned on 1 May 1817; died on 12 May 1824.

Campbell, John - Corporal - 1st Infantry - Company: Eli Clemson - Age: 28 - Height: 5' 9" - Born: Maryland - Trade: Farmer - Enlistment date: 24 Mar 1814 - Place: Newport, KY or Cincinnati, OH - Period: War.

Campbell, John - Private - 38th Infantry - Company: James Hook - Enlistment date: 27 Jan 1814 - Period: War - Discharged on 27 Jan 1815.

Campbell, Mathew - Private - 14th Infantry - Prisoner of War (Halifax), sent to England.

Campbell, Robert - Private - 38th Infantry - Age: 21 - Height: 6' - Born: Somerset Cty, MD - Trade: Ship's carpenter - Enlistment date: 28 Oct 1814 - Place: Newtown, MD - Period: War.

Campbell, Thomas - Private - 36th Infantry - Age: 28 - Height: 5' 4" - Born: Prince George's Cty, MD - Enlistment date: 9 Aug 1813 - Place: Waynesburg, PA - Period: 1 Yr.

Campbell, William L. - Private - 38th Infantry - Age: 31 - Height: 5' 7" - Born:

By War of 1812 Soc. in MD

Surry County, VA - Trade: Clerk - Enlistment date: 16 Jul 1814 - Period: War.

Camper, John - Private - 2nd Light Dragoons - Company: William Littlejohn - Age: 18 - Height: 5' 8 1/2" - Born: Baltimore Cty - Trade: Shoemaker - Enlistment date: 14 Dec 1813 - Place: Baltimore - Period: 5 Yrs.

Camper, Jonathan - Private - US Volunteers - Company: Stephen Moore - Enlistment date: 8 Sep 1812 - Period: 1 Yr.

Camper, Jonathan - Private - 2nd Light Dragoons - Company: William Littlejohn - Other regiment: Corps of Artillery - Age: 35 - Height: 5' 9" - Born: Baltimore Cty - Trade: Farmer - Enlistment date: 8 Dec 1813 - Place: Baltimore - Period: 5 Yrs - Died on 17 Mar 1816.

Canaan, John - Private - 38th Infantry - Company: John Rothrock - Enlistment date: 11 Jul 1813 - Period: 1 Yr.

Canahan, Edmund - Private - 38th Infantry - Company: John Rothrock - Age: 18 - Height: 5' 5" - Born: Philadelphia - Trade: Printer - Enlistment date: 1 Sep 1813 - Place: Craney Island, VA - Period: War.

Cane, Thomas - Private - 14th Infantry - Enlistment date: 12 Dec 1812 - Period: 18 Mos - Prisoner of War (Halifax), sent to England; discharged on 31 Mar 1815.

Cannon, Dominick - Private - 14th Infantry - Company: David Cummings - Age: 30 - Height: 5' 7 1/2" - Born: Ireland - Trade: Laborer - Enlistment date: 19 Sep 1812 - Place: Baltimore - Period: 5 Yrs - Pension: Old War IF-45966 - Prisoner of War, sent to England; discharged at Baltimore on 20 Apr 1815 due to wound in left shoulder received at York, UC, on 27 Apr 1813.

Cannon, John - Private - 36th Infantry - Company: Thomas Carbery - Age: 28 - Height: 5' 11 1/2" - Born: Trenton, NJ - Trade: Blacksmith - Enlistment date: 6 Apr 1814 - Place: Georgetown, DC - Period: War - BLW 10288-160-12 - Discharged at Baltimore on 31 Mar 1815.

Cannon, Thomas - Private - 38th Infantry - Company: John Brookes - Age: 19 - Height: 5' 5" - Born: Somerset Cty, MD - Trade: Farmer - Enlistment date: 30 Jul 1813 - Period: 1 Yr - BLW 10683-160-12 - Re-enlisted at Craney Island, VA, on 15 Apr 1814 for during the war; discharged on 15 Mar 1815 at Craney Island, VA.

Cannon, William - Private - 14th Infantry - Company: Reuben Gilder - Other regiment: 4th Infantry (New) - Age: 62 - Height: 5' 10" - Born: Chester, Chester Cty, PA - Trade: Laborer - Enlistment date: 18 May 1813 - Place: Baltimore - Period: 5 Yrs - Deserted on 16 Jul 1815.

Cantwell, John - Private - 1st Infantry - Company: Daniel Bissell - Other regiment: 3rd Infantry (New) - Height: 5' 8" - Born: Maryland - Enlistment date: 25 Mar 1805 - Period: 5 Yrs - Re-enlisted at Belle Fontaine, MO, on 22 May 1810 for five years; discharged on 22 May 1815.

Cantwell, Richard - Private - 16th Infantry - Age: 27 - Height: 5' 7" - Born: Maryland - Trade: Farmer - Enlistment date: 23 Oct 1812 - Place: Columbia, MD - Period: 5 Yrs.

Carberry, Henry - Colonel - 36th Infantry - Born: Maryland - Place: Maryland - Served as a captain during the Revolutionary War; served as a captain of levies under Major General St. Clair; commissioned as a colonel, 36th Infantry, on 22 Mar 1813; resigned on 4 Mar 1815; died on 26 May 1822.

Carberry, Thomas - Captain - 36th Infantry - BLW 11236-160-50 - Commissioned as a captain on 30 Apr 1815; discharged on 15 Jun 1815.

Carey, Dennis - Seaman - US Sea Fencibles - Company: John Gill - Age: 40 - Height: 5' 7" - Born: Derry, County Derry, Ireland - Enlistment date: 6 Jan 1814.

Carey, James - Private - 14th Infantry - Company: Samuel Lane - Age: 27 - Height: 5' 10" - Born: Virginia - Enlistment date: 31 Jul 1812 - Place: Carlisle, PA - Period: 5 Yrs - Prisoner of War (Halifax), sent to England.

Carey, John - Private - 38th Infantry - Company: Charles Stansbury - BLW 24265-160-12.

Carico, Joseph - Private - 3rd Rifles - Company: Edward Carrington - Age: 30 - Height: 5' 11" - Born: Charles Cty, MD - Trade: Farmer - Enlistment date: 9 May 1814 - Place: Morgantown, VA - Period: War - Discharged at Carlisle Barracks, PA, on 26 Feb 1815.

Carlin, Cornelius - Private - 38th Infantry - Company: John Buck - Age: 26 - Height: 5' 9" - Born: Montgomery Cty, MD - Trade: Blacksmith - Enlistment date: 12 Apr 1814 - Place: Annapolis - Period: War - Pension: No pension claim - BLW 236-160-12 - Discharged at Baltimore on 30 Mar 1815.

Carlisle, Howard W. - Private - 2nd Rifles - Company: John O'Fallon - Age: 21 - Height: 5' 8" - Born: St. Mary's Cty, MD - Trade: Laborer - Enlistment date: 21 May 1814 - Place: Georgetown, KY - Period: War - BLW 7524-160-12 - Discharged at Detroit on 30 Jun 1815.

Carmean, Jacob - Private - 21st Infantry - Company: Charles Proctor - Other regiment: 11th Infantry - Height: 5' 8" - Born: Denton, MD - Trade: Mariner - Enlistment date: 12 Aug 1812 - Period: 18 Mos - Re-enlisted on 27 Feb 1814 in 11th Infantry at Burlington, VT, for the war.

Carmickle, George - Private - 36th Infantry - Age: 35 - Height: 5' 8" - Born: Virginia - Enlistment date: 28 Aug 1813 - Place: Baltimore - Period: 1 Yr.

Carmickle, James - Private - 1st Infantry - Company: Simon Owens - Age: 19 - Height: 5' 7 3/4" - Born: Frederick, MD - Trade: Laborer - Enlistment date: 2 Aug 1809 - Place: Winchester - Period: 5 Yrs - Died at Fort Clark, IL, on 30 Jan 1814.

Carmody, Darby - Private - 14th Infantry - Enlistment date: 23 Nov 1812 - Period: 18 Mos - Prisoner of War, sent to England; discharged on 31 Mar 1815.

Carnacombe, George - Private - 5th Infantry - Company: James Dorman - Age: 21 - Height: 5' 5" - Born: Harristown, Washington Cty, MD - Trade: Farmer - Enlistment date: 7 Jan 1812 - Place: Hagerstown, MD - Period: 5 Yrs - Died at Buffalo on 13 Nov or 22 Dec 1814.

Carnahan, George - Private - 38th Infantry - Age: 23 - Height: 5' 7" - Born: Loudoun Cty, VA - Trade: Farmer - Enlistment date: 27 Jun 1814 - Place: Norfolk, VA - Period: 5 Yrs.

Carnahan, James - Private - 14th Infantry - Company: Joseph Marshall - BLW 11837-160-12.

By War of 1812 Soc. in MD

Carney, Edward - Private - Light Dragoons - Age: 28 - Height: 5' 7" - Born: County Derry, Ireland - Trade: Wheelwright - Enlistment date: 9 Apr 1814 - Place: Baltimore - Period: 5 Yrs - Deserted at Baltimore.

Carpenter, Elijah - Private - 38th Infantry - Company: James Smith - Age: 21 - Height: 5' 10" - Born: Chesterfield, VA - Trade: Farmer - Enlistment date: 9 Mar 1814 - Place: Norfolk, VA - Period: War - BLW 19515-160-12 - Discharged at Craney Island, VA, on 15 Mar 1815.

Carpenter, John - Private - 14th Infantry - Company: Samuel Lane - Other regiment: 4th Infantry (New) - Age: 16 - Height: 5' 4" - Born: Frederick, MD - Trade: Farmer - Enlistment date: 28 May 1812 - Place: Cumberland, MD - Period: 5 Yrs - Prisoner of War (Halifax), captured at Beaver Dams on 24 Jun 1813, discharged from Halifax on 3 Feb 1813; deserted at Fort Severn, MD, on 16 Jul 1815.

Carr, George - Seaman - US Sea Fencibles - Company: John Gill - Age: 29 - Height: 5' 6 1/2" - Born: Baltimore City - Enlistment date: 5 Jan 1814.

Carr, James - Private - 14th Infantry - Other regiment: 4th Infantry (New) - Age: 27 - Height: 5' 5" - Born: Baltimore City - Trade: Laborer - Enlistment date: 3 Jun 1812 - Place: Baltimore - Period: 5 Yrs - Prisoner of War (Halifax), exchanged on 15 Apr 1814; deserted at Fort Severn, MD, on 16 Jul 1815.

Carr, John - Third Lieutenant - 14th Infantry - Commissioned as an ensign on 15 Apr 1814; promoted to 3rd lieutenant on 2 May 1814; discharged on 15 Jun 1814.

Carr, John - Private - 14th Infantry - Enlistment date: 9 Nov 1812 - Period: 18 Mos - Prisoner of War (Halifax), sent to England; discharged on 31 Mar 1815.

Carr, William - Private - 14th Infantry - Other regiment: 4th Infantry (New) - Age: 30 - Height: 5' 9 3/4" - Born: Wilmington, DE - Trade: Laborer - Enlistment date: 9 Sep 1812 - Place: Chester Cty, PA - Period: 5 Yrs - Prisoner of War (Halifax), exchanged on 15 Apr 1814; deserted at Fort Severn, MD, on 12 Jul 1815.

Carrick, Obadiah - Corporal - 5th Infantry - Company: James Dorman - Other regiment: 3rd Infantry (New) - Age: 23 - Height: 5' 8" - Born: Cumberland, MD - Trade: Nailor - Enlistment date: 2 Apr 1813 - Place: Baltimore - Period: 5 Yrs - Discharged on 1 Apr 1818.

Carrington, Elihu - Private - 36th Infantry - Company: Joseph Hook - Age: 38 - Height: 5' 6" - Born: New Haven, CT - Trade: Chair maker - Enlistment date: 3 Jul 1814 - Place: Georgetown, DC - Period: War - BLW 10239-160-12 - Discharged at Fort Covington, MD, on 30 Mar 1815.

Carroll, Daniel - Private - 20th Infantry - Age: 19 - Height: 5' 7" - Born: Harford Cty, MD - Trade: Brick layer - Enlistment date: 16 Jan 1815 - Place: Alexandria, DC - Period: War - BLW 778-320-12 - Discharged at Fredericksburg, VA, on 15 Mar 1815.

Carroll, Isaac - Private - 14th Infantry - Company: William McIlvane - Enlistment date: 27 Aug 1812 - Period: 5 Yrs - Pension: Placed on the pension rolls on 15 Aug 1816 - Prisoner of War, exchanged on 15 Apr 1814;

discharged on 3 Aug 1814.

Carroll, John - Private - 36th Infantry - Age: 23 - Height: 5' 8" - Born: Harford Cty, MD - Enlistment date: 23 Jun 1812 - Place: Havre de Grace, MD.

Carroll, John - Private - 36th Infantry - Age: 24 - Height: 5' 8" - Born: Harford Cty, MD - Enlistment date: 1 Mar 1814 - Place: Annapolis.

Carroll, Martin - Private - 36th Infantry - Company: Samuel Raisin - BLW 14312-160-12.

Carroll, Martin - Private - 36th Infantry - Company: Samuel Raisin - Age: 28 - Height: 5' 6" - Born: Ireland - Trade: Farmer - Enlistment date: 30 Oct 1814 - Place: Fredericktown, MD - Period: War - BLW 13687-160-12 - Discharged at Washington, DC, on 20 Mar 1815.

Carroll, Michael - Private - 38th Infantry - Company: John Buck - Enlistment date: 19 Jun 1813 - Period: 1 Yr - Discharged on 19 Jul 1813.

Carroll, Michael - Private - 38th Infantry - Company: Anthony Miltenberger.

Carson, Robert - First Lieutenant - 5th Infantry - Born: Maryland - Commissioned as an ensign on 3 May 1808; promoted to 2nd lieutenant on 16 Apr 1809; promoted to 1st lieutenant on 1 Sep 1811; resigned on 1 Sep 1812.

Carter, Elias - Private - 2nd Infantry - Age: 28 - Height: 5' 7" - Born: Charles Cty, MD - Trade: Laborer - Enlistment date: 28 Nov 1814 - Place: Washington, DC - Period: 5 Yrs.

Carter, George - Private - 2nd Infantry - Company: Perrin Willis - Other regiment: 8th Infantry (New) - Age: 22 - Height: 5' 6" - Born: St. Mary's Cty, MD - Trade: Sailor - Enlistment date: 17 Nov 1814 - Place: Washington, DC - Period: 5 Yrs - BLW 18580-160-12 & 24276-160-12 - Discharged at Camp Young, AL, in Nov 1819.

Carter, John - Private - 2nd Infantry - Company: Perrin Willis - Age: 24 - Height: 5' 6 1/2" - Born: St. Mary's Cty, MD - Trade: Sailor - Enlistment date: 8 Dec 1814 - Place: Washington, DC - Period: 5 Yrs - BLW 7020-160-12 - Discharged at Camp Young, AL, on 8 Dec 1819.

Carter, John - Private - 5th Infantry - Company: Richard Bell - Age: 40 - Height: 5' 11" - Born: New York or Ireland - Trade: Farmer - Enlistment date: 19 May 1813 - Place: Towsontown, MD - Period: 5 Yrs - BLW 325-160-12 - Discharged on 21 Jun 1815, unfit for service.

Carter, John - Private - 36th Infantry - Company: Samuel Raisin - Age: 25 - Height: 5' 8" - Born: Virginia - Trade: Farmer - Enlistment date: 19 Sep 1814 - Place: Norfolk, VA - Period: War - Died on 2 Mar 1815.

Carter, Joseph - Private - 12th Infantry - Company: Thomas Sangster - Age: 21 - Height: 5' 7 1/2" - Born: Baltimore - Trade: Carpenter - Enlistment date: 13 Jun 1812 - Place: Baltimore - Period: 18 Mos - Prisoner of War (Halifax), exchanged on 15 Apr 1814.

Carter, Nathaniel - Private - 14th Infantry - Prisoner of War (Halifax), died at Halifax.

Carter, Thomas - Private - 36th Infantry - Age: 27 - Height: 5' 7" - Born: Harford Cty, MD - Enlistment date: 5 Jun 1813 - Place: Annapolis - Period: 1 Yr.

By War of 1812 Soc. in MD

Carter, William - Private - 36th Infantry - Age: 40 - Height: 5' 5" - Born: Delaware - Enlistment date: 30 Apr 1813 - Place: Baltimore - Period: 1 Yr.

Cartwright, Henry G. - Private - 38th Infantry - Company: James Smith - Age: 29 - Height: 6' 1/2" - Born: St. Mary's Cty, MD - Trade: Farmer - Enlistment date: 6 Apr 1814 - Place: Richmond, VA - Period: War - BLW 14698-160-12 - Discharged at Craney Island, VA, on 15 Mar 1815.

Cartwright, Malica - Private - 38th Infantry - Age: 25 - Height: 5' 11" - Born: Passaquotank, NC - Enlistment date: 22 Jun 1814 - Place: Norfolk, VA - Period: War - Deserted.

Cary, Dennis - Seaman - US Sea Fencibles - Company: William Addison.

Cary, James - Private - 14th Infantry - Company: David Cummings - Enlistment date: 29 May 1812 - Period: 5 Yrs - Pension: Land bounty to Moses Cary & other heirs at law of James Cary - BLW 15767-160-12 - Prisoner of War, sent to England; died at Sacket's Harbor, NY, in Oct 1814.

Case, Thomas S. - Corporal - 38th Infantry - Company: John Brookes - Enlistment date: 26 Aug 1813 - Period: 1 Yr.

Casey, John M. - Private - 36th Infantry - Company: Thomas Carbery - Age: 25 - Height: 5' 9 3/4" - Born: Charles Cty, MD - Trade: Ship's carpenter - Enlistment date: 24 Feb 1814 - Place: Leonardstown, MD - Period: War - BLW 10346-160-12 - Discharged at Baltimore on 31 Mar 1815.

Casey, Zachariah - Private - 14th Infantry - Company: Joseph Marshall - Age: 23 - Height: 5' 5" - Born: Washington Cty, MD - Trade: Farmer - Enlistment date: 30 May 1814 - Place: Sharpsburg or Boonsberry, MD - Period: War - BLW 4237-160-12 - Discharged at Greenbush, NY, on 16 May 1815.

Cashman, William - Private - 14th Infantry - Prisoner of War (Halifax), sent to England.

Caspost, Jeremiah - Seaman - US Sea Fencibles - Company: John Gill - Age: 47 - Height: 5' 6" - Born: Philadelphia - Enlistment date: 3 Jan 1814.

Cass, Philip - Private - 38th Infantry - Company: John Brookes - Enlistment date: 18 Sep 1813 - Period: 1 Yr.

Cassaday, Michael - Private - 38th Infantry - Company: Sheppard Leakin - Enlistment date: 21 Jun 1813 - Period: 1 Yr - Died on 20 or 26 Mar 1814.

Castell, Jeremiah - Private - 14th Infantry - Age: 24 - Height: 6' 1" - Born: Allegany Cty, MD - Enlistment date: 21 May 1812 - Period: 5 Yrs - Died at Carlisle Barracks, PA, on 31 Jul or 31 Aug 1812.

Caster, James M. - Private - 20th Infantry - Company: John Stanard - Age: 22 - Height: 5' 5 1/2" - Born: Maryland - Enlistment date: 12 Aug 1812 - Period: 5 Yrs - Pension: Land bounty to Elizabeth Meyers & other heirs at law of James M. Caster - BLW 26642-160-12 - Died at Buffalo on 1 Dec 1812.

Castle, George V. - Private - 36th Infantry - Company: Joseph Hook - Enlistment date: 15 Sep 1813 - Pension: Wife Catharine Horine, WO-2252, WC-1203; married on 4 Dec 1905, Frederick Cty, MD; soldier died on 7 Jul 1850, Butler Cty, OH - BLW 26971-160-50 - Discharged on 15 Sep 1814.

Caswell, William - Musician - 2nd Artillery - Company: Alexander Williams - Other regiment: 22nd and 2nd Infantry (New) - Age: 14 - Height: 4' 7" - Born: Maryland - Trade: Laborer - Enlistment date: 7 Apr 1813 - Place:

Philadelphia - Period: 5 Yrs - Discharged at Sackets Harbor, NY, on 1 Jan 1818 and re-enlisted.

Cater, Benjamin - Corporal - 10th Infantry - Company: Philip Brittain - Age: 21 - Height: 5' 7" - Born: Prince George's Cty, MD - Trade: Brick layer - Enlistment date: 13 Mar 1814 - Place: Washington, DC - Period: War - BLW 5870-160-12 - Discharged at Buffalo on 31 May 1815.

Cather, Samuel - Private - 26th Infantry - Company: William Bezeau - Age: 32 - Height: 6' 1" - Born: Maryland - Trade: Carpenter - Enlistment date: 5 Dec 1814 - Place: Philadelphia - Period: War - Deserted on 5 Dec 1814.

Cather, William - Private - 36th Infantry - Age: 35 - Height: 5' 8" - Born: Cecil Cty, MD - Enlistment date: 22 Jun 1813 - Place: Baltimore - Period: 1 Yr.

Cato, John - Private - 38th Infantry - Company: John Rothrock - Enlistment date: 6 Sep 1813 - Period: 1 Yr.

Caulder, William - Corporal - 38th Infantry - Company: James Smith - Age: 35 - Height: 5' 5" - Born: Annapolis - Trade: Tailor - Enlistment date: 19 Feb 1814 - Place: Craney Island, VA - Period: War - BLW 3004-160-12 - Discharged at Craney Island, VA, on 15 Mar 1815.

Caulk, Peter - Private - 1st Rifles - Company: Abraham Massias - Other regiment: Rifles - Age: 37 - Height: 5' 10 1/2" - Born: Caroline Cty, MD - Trade: Shoemaker - Enlistment date: 3 Jul 1813 - Place: Abbeville Court House, SC - Period: 5 Yrs - BLW 18927-160-12 - Discharged at Fort Crawford, IL, on 3 Jul 1818.

Causer, Henry - Private - 14th Infantry - Company: Reuben Gilder - Age: 28 - Height: 5' 8" - Born: Bellevdeer, County Armagh, Ireland - Trade: Weaver - Enlistment date: 15 Apr 1814 - Place: Towsontown or Baltimore, MD - Period: War - BLW 1768-160-12 - Discharged at Greenbush, NY, on 4 May 1815.

Cavatt, Thomas - Sergeant - 14th Infantry - Company: David Cummings - Other regiment: 4th Infantry (New) - Enlistment date: 16 Apr 1812 - Place: Annapolis - Period: 5 Yrs - Discharged on 16 Apr 1817.

Cawford, James - Private - 36th Infantry - Age: 25 - Height: 5' 4" - Born: Halifax, England - Enlistment date: 16 May 1813 - Place: Havre de Grace, MD.

Cawood, Benjamin - Private - 12th Infantry - Company: Thomas Sangster - Age: 29 - Height: 5' 8" - Born: Prince George's or Charles County, MD - Trade: Cooper - Enlistment date: 16 Jun 1812 - Place: Alexandria, DC - Period: 18 Mos - Re-enlisted at Alexandria, DC, on 8 Nov 1814 for the war; discharged on 15 Mar 1815 at Fredericksburg.

Cease, John - Private - 12th Infantry - Company: Andrew Madison - Age: 38 - Height: 5' 6 1/2" - Born: Frederick Cty, MD - Trade: Laborer - Enlistment date: 19 May 1812 - Place: M Town - Period: 5 Yrs - Prisoner of War (Halifax); died on 1 Dec 1814 at Buffalo, NY, from an accident.

Cecil, Henry B. - Sergeant - 19th Infantry - Company: Martin Hawkins - Other regiment: 17th Infantry - Age: 31 - Height: 5' 10" - Born: Fredericktown, MD - Trade: Laborer - Enlistment date: 4 Apr 1814 - Place: Detroit - Period: War - BLW 5718-160-12 - Discharged at Chillicothe, OH, on 9 Jun 1815.

By War of 1812 Soc. in MD

Chadborn, John - Private - 2nd Light Dragoons - Company: William Littlejohn - Age: 18 - Height: 5' 9" - Born: Baltimore - Enlistment date: 18 Jan 1814 - Period: 5 Yrs - Deserted at Baltimore on 7 Dec or 7 Sep 1814.

Chamberlain, Barney - Private - 36th Infantry - Age: 19 - Height: 5' 6" - Born: Havre de Grace, MD - Enlistment date: 1 Mar 1814 - Place: Annapolis.

Chambers, Henry - Sergeant - 14th Infantry - Company: David Cummings - Age: 27 - Height: 5' 10" - Born: Virginia - Enlistment date: 18 May 1812 - Place: Chambersburg, PA - Period: 5 Yrs - BLW 24843-160-12 - Prisoner of War, sent to England; discharged on 20 Apr 1815 on Surgeon's Certificate.

Chambers, James - Private - 36th Infantry - Company: Joseph Hook - Age: 38 - Height: 5' 6" - Born: Lancaster, PA - Trade: Farmer - Enlistment date: 21 Jul 1814 - Place: Georgetown, DC - Period: War - BLW 100-160-12 - Discharged at Fort Covington, MD, on 30 Mar 1815.

Chambers, John - Private - 36th Infantry - Enlistment date: 11 Dec 1814 - Pension: Wife Ellen Marie O. Boyle, WO-16724, WC-15506; married on 4 Jun 1829, Fredericktown, MD; soldier died on 14 Mar 1850, Harpers Ferry, VA; first wife was Ann Jeffries - BLW 1034-320-14 - Discharged on 13 Mar 1815.

Chambers, Samuel - Private - 16th Infantry - Other regiment: 2nd Infantry (New) - Age: 28 - Height: 5' 7 1/2" - Born: Baltimore - Trade: Clerk - Enlistment date: 10 Feb 1813 - Place: Philadelphia - Period: 5 Yrs - Pension: Land bounty to Ann Martin, sister & other heirs at law of Samuel Chambers - BLW 25328-160-12 - On board the fleet on Lake Ontario; prisoner of war at Halifax; died on 27 Jan 1817 at Sackets Harbor, NY.

Chance, Clement - Private - 36th Infantry - Age: 17 - Height: 5' 6 1/4" - Born: Caroline Cty, MD - Enlistment date: 11 Aug 1813 - Place: Greensboro, MD - Period: 1 Yr.

Chance, Henry - Private - 2nd Infantry - Company: Thomas Swaine - Height: 5' 9 1/2" - Born: Caroline Cty, MD - Trade: Farmer - Enlistment date: 1804 - Re-enlisted on 30 Mar 1809; discharged on 30 Mar 1814 at Mobile Point, AL.

Chance, John - Private - 36th Infantry - Company: Joseph Hook - Age: 21 - Height: 5' 7" - Born: Delaware - Enlistment date: 1 Jun 1813 - Place: Baltimore - Period: 1 Yr - Pension: No pension claim - BLW 9517-40-50 Cancelled; BLW 166-160-50 - Also served in Captain Peyton's Company, DC Militia.

Chance, Peter - Recruit - 26th Infantry - Company: William Bezeau - Race: 27 - Age: 27 - Height: 5' 5" - Born: Maryland - Trade: Laborer - Enlistment date: 7 Nov 1814 - Place: Philadelphia - Period: 5 Yrs - Listed as a "Colored man" in his service record; discharged on 18 May 1815.

Chandler, George - Private - 38th Infantry - Company: John Buck - Age: 37 - Height: 5' 6" - Born: Somerset Cty, MD - Trade: Shipwright - Enlistment date: 21 May 1814 - Place: Baltimore - Period: War - BLW 11179-160-12 - Discharged at Baltimore on 30 Mar 1815.

Chaney, John - Private - 38th Infantry - Company: James Hook - Age: 27 -

Height: 5' 8 1/2" - Born: Calvert Cty, MD - Trade: Farmer - Enlistment date: 14 Apr 1814 - Period: War - Discharged at Fort Covington, MD, on 31 Mar 1815.

Chaney, Samuel - Private - 28th Infantry - Age: 26 - Height: 5' 8" - Born: Maryland - Enlistment date: 11 Oct 1814 - Place: Flemingsburg, KY.

Chaney, Samuel - Private - 38th Infantry - Company: James Haslett - Enlistment date: 11 Aug 1813 - Period: 1 Yr.

Chaney, Samuel - Private - 38th Infantry - Company: Shepperd Leakin - Enlistment date: 13 Jun 1813 - Period: 1 Yr - Discharged on 19 Apr 1814.

Chapman, Eliphalet - Private - 14th Infantry - Company: William McIlvane - Other regiment: 4th Infantry (New) - Enlistment date: 18 Mar 1813 - Period: 5 Yrs - Prisoner of War (Halifax), discharged from Halifax on 3 Feb 1814; deserted at Fort Severn, MD, on 18 Jul 1815.

Chapman, John - Private - 7th Infantry - Company: Gilbert Russell - Other regiment: 1st Infantry (New) - Age: 22 - Height: 5' 8" - Born: Maryland - Trade: Barber - Enlistment date: 11 Aug 1808 - Period: 5 Yrs - Re-enlisted on 22 Aug 1813; discharged on 28 May 1818.

Chapman, Jonathan - Private - US Volunteers - Company: Stephen Moore - Enlistment date: 8 Sep 1812 - Period: 1 Yr.

Chapman, Thomas - Private - 28th Infantry - Company: Thomas Edmondson - Age: 21 - Height: 5' 11" - Born: Maryland - Trade: Farmer - Enlistment date: 24 Mar 1814 - Place: Kentucky - Period: War - BLW 17459-160-12 - Discharged at Detroit on 30 Jun 1815.

Chapman, William - Private - 36th Infantry - Transferred to the U.S. Navy on 26 Oct 1813.

Chapman, Zachariah - Private - 28th Infantry - Company: George Stockton - Age: 26 - Height: 5' 7 1/2" - Born: Maryland - Trade: Cooper - Enlistment date: 7 Feb 1814 - Place: Flemingsburg, KY - Period: War - BLW 7053-160-12 - Discharged at Detroit on 30 Jun 1815.

Charlton, Armstrong - Private - Light Dragoons - Other regiment: Corps of Artillery - Age: 34 - Height: 5' 7 1/2" - Trade: Printer - Enlistment date: 9 Jun 1812 - Place: Baltimore - Period: 5 Yrs - Discharged on 21 Jul 1815, invalid.

Chase, Joshua - Black servant - Unknown - Race: 21 - Age: 21 - Born: Baltimore - Prisoner of War (Quebec), captured during the Battle of Stoney Creek.

Chatham, Aquila - Private - 8th Infantry - Company: Matthew Keith - Age: 24 - Height: 6' 2" - Born: Maryland - Trade: Farmer - Enlistment date: 29 May 1813 - Place: Eatonton - Period: War - Discharged at Camp Flournoy, GA, on 2 Mar 1815.

Chatton, John - Musician - 38th Infantry - Company: Shepperd Leakin - Enlistment date: 25 Jun 1813 - Period: 1 Yr.

Chauncey, John - First Lieutenant - 36th Infantry - Born: Maryland - Commissioned as a 1st lieutenant on 30 Apr 1813; discharged on 15 Jun 1815.

Cheatham, John - Private - 38th Infantry - Company: James Smith - Age: 22 -

Height: 5' 6" - Born: Currituck, NC - Trade: Shoemaker - Enlistment date: 9 Mar 1814 - Place: Norfolk, VA - Period: War - BLW 5042-160-12 - Discharged at Craney Island, VA, on 15 Mar 1815.

Cherry, Javan - Private - 38th Infantry - Company: John Brookes - Age: 25 - Height: 5' 7" - Born: Norfolk, VA - Trade: Farmer - Enlistment date: 30 Aug 1814 - Place: Norfolk, VA - Period: War - Pension: Land bounty to Ezekiel Cherry and other heirs at law of Javan Cherry - BLW 27688-160-42 - Discharged at Craney Island, VA, on 15 Mar 1815.

Chesney, Thomas E. - Private - 14th Infantry - Company: Thomas Montgomery - Age: 55 - Height: 5' 5" - Born: Harford Cty, MD - Trade: Gunsmith - Enlistment date: 16 Jun 1812 - Place: Baltimore - Period: 5 Yrs - Pension: Old War IF-12459 - BLW 436-160-12 - Discharged at Greenbush, NY, on 4 May 1815, rupture.

Chick, Francis - Private - 36th Infantry - Age: 21 - Height: 5' 11" - Born: Dumfries, Prince Williams Cty, VA - Enlistment date: 2 Sep 1814 - Place: Fredericksburg, VA.

Chick, James - Sergeant - 36th Infantry - Age: 18 - Height: 5' 8" - Born: Dumfries, Prince Williams Cty, VA - Trade: Teacher - Enlistment date: 2 Sep 1814 - Period: War - BLW 24926-160-12 - Discharged at Georgetown, DC, on 13 Mar 1815.

Chick, William - Private - 38th Infantry - Company: John Rothrock - Age: 21 - Height: 5' 5" - Born: Campbell, VA - Trade: Farmer - Enlistment date: 27 Mar 1814 - Place: Richmond, VA - Period: War - BLW 13051-160-12 - Discharged at Craney Island, VA, on 15 Mar 1815.

Childers, Joseph - Private - 14th Infantry - Prisoner of War (Halifax), sent to England.

Childes, James - Seaman - US Sea Fencibles - Company: John Gill - Age: 39 - Height: 5' 9" - Born: Baltimore - Enlistment date: 29 Dec 1813.

Chipley, Edward - Private - 7th Infantry - Company: Carey Nicholas - Age: 24 - Height: 5' 10" - Born: Maryland - Trade: Shoemaker - Enlistment date: 2 Mar 1809 - Period: 5 Yrs - Re-enlisted at Fort Claiborne on 1 Jun 1813 for the war.

Chipley, Robert - Private - 2nd Rifles - Other regiment: 24th and 7th Infantry (New) - Age: 39 - Height: 5' 11" - Born: Maryland - Trade: Shoemaker - Enlistment date: 14 Nov 1814 - Period: 5 Yrs - Pension: Land bounty to John Chipley, son & other heirs at law of Robert Chipley - BLW 26169-160-12 - Transferred from 24th Infantry; died on 21 Jul 1819, probably at Fort Scott, GA.

Chockley, David - Private - 36th Infantry - Age: 27 - Height: 5' 6" - Born: Virginia - Enlistment date: 10 Jul 1814 - Place: Norfolk, VA.

Chrise, John - Private - 5th Infantry - Company: James Dorman - Other regiment: 3rd Infantry (New) - Age: 23 - Height: 5' 8 1/2" - Born: Maryland - Trade: Blacksmith - Enlistment date: 5 Jun 1812 - Place: Creagerstown, MD - Period: 5 Yrs - Pension: Land bounty to Henry Chrise, father & only heir at law of John Chrise - BLW 10617-160-12 - Prisoner of War (Halifax), exchanged on 15 Apr 1814; Drowned at Pittsburgh, PA, on 24 Jul 1815.

Maryland Regulars in War of 1812

Chrisman, Powel - Private - 14th Infantry - Company: David Cummings - Age: 35 - Height: 5' 7" - Born: Washington Cty, MD - Enlistment date: 21 May 1812 - Place: Cumberland, MD - Period: 5 Yrs - Pension: Land bounty to Powel Chrisman, son and only heir at law of Powel Chrisman - BLW 27243-160-42 - Killed at Beaver Dams, UC, on 24 Jun 1813.

Christfield, Absalom - Private - 5th Infantry - Other regiment: 3rd Infantry (New) - Age: 25 - Height: 6' 2" - Born: Kent Cty, MD - Trade: Chair maker - Enlistment date: 18 Aug 1814 - Place: Harrisburg, PA - Period: 5 Yrs - Died at Detroit on 30 Sep 1817 from consumption.

Christie, James - First Lieutenant - 14th Infantry - Commissioned as a 2nd lieutenant on 12 Mar 1812; promoted to 1st lieutenant on 1 Oct 1813; dismissed on 2 Aug 1814.

Christopher, Samuel - Private - 14th Infantry - Other regiment: 4th Infantry (New) - Age: 19 - Height: 5' 6 1/2" - Born: Worcester Cty, MD - Trade: Farmer - Enlistment date: 27 Apr 1812 - Place: Salisbury, MD - Period: 5 Yrs - Prisoner of War (Halifax), captured at Beaver Dams on 24 Jun 1813, exchanged on 15 Apr 1814; discharged on 27 Aug 1817.

Chunn, Zachariah - Sergeant - 14th Infantry - Company: Reuben Gilder - Age: 29 - Height: 6' 2" - Born: Maryland - Trade: Merchant - Enlistment date: 1 May 1813 - Period: 5 Yrs - Died at Burlington, VT, in Apr 1815.

Church, Thomas - Private - 14th Infantry - Company: Thomas Montgomery - Height: 5' 6" - Born: Hopkinton, RI - Trade: Sailor - Enlistment date: 10 Nov 1812 - Period: 18 Mos - Re-enlisted at Plattsburgh, NY, on 28 Feb 1814 for the war.

Church, Thomas - Private - 36th Infantry - Company: Thomas Carbery - Age: 18 - Height: 5' 7" - Born: Virginia - Trade: Baker - Enlistment date: 16 May 1814 - Place: Georgetown, DC - Period: War - Discharged on 13 Mar 1815.

Churchill, Israel - Private - 38th Infantry - Company: James Haslett - Enlistment date: 13 Aug 1813 - Period: 1 Yr.

Churchman, Ashel - Private - 14th Infantry - Age: 34 - Height: 5' 11" - Born: Pennsylvania - Trade: Miller - Enlistment date: 3 Aug 1814 - Period: War.

Ciders, Solomon (Sides) - Private - 35th Infantry - Company: Benjamin Hardaway - Age: 32 - Height: 5' 9" - Born: Washington Cty, VA or MD - Trade: Farmer - Enlistment date: 8 Apr 1814 - Place: Norfolk, VA - Period: War - Discharged on 15 Mar 1815.

Clabough, James (or Jacob) - Private - 4th Rifles - Company: Matthew Magee - Age: 38 - Height: 5' 8" - Born: Hagerstown, MD - Trade: Carpenter - Enlistment date: 10 Feb 1815 - Period: War - Discharged at Carlisle Barracks, PA, on 25 Mar 1815.

Clapsaddle, Paul - Private - 14th Infantry - Company: Samuel Lane - Other regiment: 4th Infantry (New) - Age: 21 - Height: 5' 11" - Born: Woodsboro, MD - Trade: Blacksmith - Enlistment date: 6 Jun 1812 - Place: Westminster, MD - Period: 5 Yrs - Deserted at Fort Severn, MD, on 15 Jul 1815.

Clarage, Levin - Private - 36th Infantry - Company: Thomas Carbery - Enlistment date: 8 Oct 1813 - Period: 1 Yr - Discharged on 7 Oct 1814.

Clark, Abraham - Second Lieutenant - 14th Infantry - Age: 25 - Pension: Wife

By War of 1812 Soc. in MD

Maria, WO-15768, WC-7243; married on 14 Oct 1818, Bladensburg, MD; soldier died in 1839 - BLW 10217-160-50 - Commissioned as an ensign on 12 Mar 1812; promoted to 3rd lieutenant on 13 Mar 1813; promoted to 2nd lieutenant on 1 Oct 1813; Prisoner of War (Dartmouth), released on 31 May 1814; discharged on 15 Jun 1815; died on 4 Sep 1839.

- Clark, Charles - Private - 12th Infantry - Company: James Paxton - Other regiment: 8th Infantry (New) - Age: 28 - Height: 5' 10" - Born: Maryland - Trade: Shoemaker - Enlistment date: 18 Jun 1812 - Period: 5 Yrs - Discharged on 18 Jun 1817.
- Clark, Charles G. - Private - 14th Infantry - Company: James McDonald - Age: 30 - Height: 5' 9" - Born: Connecticut - Trade: Wheelwright - Enlistment date: 15 Jan 1815 - Period: 5 Yrs - Died at Burlington, VT, on 7 Oct 1814.
- Clark, David - Private - 36th Infantry - Age: 20 - Height: 5' 8" - Born: Allegany Cty, MD - Trade: Farmer - Enlistment date: 18 Mar 1814 - Place: Sandy Point, VA - Period: War.
- Clark, Hugh - Sergeant - 14th Infantry - Company: Reuben Gilder - Age: 23 - Height: 5' 6 1/2" - Born: Pennsylvania - Enlistment date: 10 May 1813 - Period: 18 Mos - Discharged at Black Rock, NY, on 10 Nov 1814.
- Clark, James - Private - 36th Infantry - Age: 28 - Height: 5' 6 1/2" - Born: Ireland - Enlistment date: 5 May 1813 - Place: Baltimore.
- Clark, James - Private - 5th Infantry - Age: 21 - Height: 5' 8" - Born: Cecil Cty, MD - Trade: Blacksmith - Enlistment date: 18 Mar 1814 - Place: Lancaster, PA - Period: 5 Yrs - Discharged for being a minor.
- Clark, James - Private - 14th Infantry - Company: Reuben Gilder - Age: 42 - Height: 5' 5 1/2" - Born: England - Trade: Glass blower - Enlistment date: 18 Apr 1813 - Period: War - Prisoner of War (Halifax), captured at Beaver Dams on 24 Jun 1813, exchanged on 15 Apr 1814; discharged on 30 Oct 1814 on Surgeon's Certificate.
- Clark, James - Private - 28th Infantry - Company: George Stockton - Other regiment: 3rd Infantry (New) - Age: 21 - Height: 5' 7" - Born: Maryland - Trade: Farmer - Enlistment date: 7 May 1814 - Place: Kentucky - Period: War - BLW 13773-160-12 - Discharged at Detroit on 30 Jun 1815.
- Clark, James M. - Seaman - US Sea Fencibles - Company: William Addison.
- Clark, John - Private - 38th Infantry - Company: John Rothrock - Enlistment date: 13 Jul 1813 - Period: 1 Yr - Deserted on 17 Jul 1813.
- Clark, John - Private - 35th Infantry - Company: Benjamin Hardaway - Age: 39 - Height: 6' 1/2" - Born: Maryland - Trade: Farmer - Enlistment date: 3 Jun 1814 - Place: Petersburg, VA - Period: War - Discharged at Norfolk, VA, on 15 Mar 1815.
- Clark, John - Second Lieutenant - 36th Infantry - Company: Joseph Hook - Born: Maryland - Commissioned as an ensign on 30 Apr 1813; promoted to 3rd lieutenant on 15 Aug 1813; promoted to 2nd lieutenant on 1 May 1814; discharged on 15 Jun 1815; died on 16 May 1820.
- Clark, John M. - Private - 2nd Light Dragoons - Company: William Littlejohn - Age: 25 - Height: 5' 8 3/4" - Born: Talbot Cty, MD - Trade: Farmer -

Enlistment date: 27 Nov 1813 - Place: Alexandria, DC - Period: War - BLW 105-160-12 - Discharged at Carlisle Barracks, PA, on 4 May 1815.

Clark, Joshua - Private - 36th Infantry - Company: Thomas Carbery - Age: 22 - Height: 5' 8 1/2" - Born: St. Mary's Cty, MD - Trade: Farmer - Enlistment date: 20 Feb 1814 - Place: Leonardstown, MD - Period: War - BLW 3520-160-12 - Discharged at Baltimore on 31 Mar 1815.

Clark, Joshua - Private - 2nd Artillery - Company: Nathan Towson - Pension: Land bounty to Elizabeth Sprouce, sister and only heirs at law of Joshua Clark - BLW 23860-160-12.

Clark, Joshua - Artificer - 2nd Artillery - Company: Frederick Evans - Other regiment: Corps of Artillery - Age: 35 - Height: 5' 7" - Born: Ireland - Trade: Blacksmith - Enlistment date: 15 Aug 1812 - Place: Annapolis - Period: 5 Yrs - BLW 10256-160-12 - Discharged at Greenleaf Point, DC, on 15 Aug 1817.

Clark, Mathias - Private - 2nd Artillery - Company: Thomas Biddle - Age: 22 - Height: 5' 6" - Born: Baltimore - Trade: Farmer - Enlistment date: 8 Jan 1813 - Period: 5 Yrs - Pension: Land bounty to Catherine Clark, daughter & other heirs at law of Mathias Clark - BLW 24089-160-12 - Transferred to Captain Nathan Towson's Company; died on 12 Dec 1815 by drowning.

Clark, Peter - Seaman - US Sea Fencibles - Company: John Gill - Age: 22 - Height: 5' 7" - Born: Baltimore City - Enlistment date: 10 Jan 1814.

Clark, Richard - Private - 5th Infantry - Company: Richard Dale - Other regiment: 3rd Infantry (New) - Height: 5' 4 1/2" - Born: Frederick Cty, MD - Trade: Farmer - Enlistment date: 1808 - BLW 10721-160-12 - Re-enlisted at Frederick, MD, on 6 Apr 1813 ; re-enlisted again after the war.

Clark, Samuel - Private - 36th Infantry - Age: 21 - Height: 5' 7 3/4" - Born: Centreville, MD - Enlistment date: 11 May 1813 - Place: Centreville, MD.

Clark, Sheldon - Private - 14th Infantry - Prisoner of War (Halifax), discharged from Halifax on 3 Feb 1814.

Clark, Stephen - Private - 14th Infantry - Company: David Cummings - Enlistment date: 25 Jan 1813 - Period: 5 Yrs - Prisoner of War (Halifax), sent to England, captured at Beaver Dams on 24 Jun 1813.

Clark, Thomas S. - Corporal - 22nd Infantry - Company: Joseph Henderson - Age: 25 - Height: 5' 8" - Born: Frederick, MD or Franklin, PA - Trade: Laborer - Enlistment date: 28 Mar 1814 - Place: Chambersburg, PA - Period: 5 Yrs.

Clark, William - Private - 14th Infantry - Company: David Cummings - Age: 24 - Height: 5' 7 1/2" - Born: Pennsylvania - Trade: Tailor - Enlistment date: 19 Mar 1814 - Place: Baltimore - Period: War - Discharged at Baltimore on 14 Mar 1815.

Clark, Zachariah - Private - 3rd Artillery - Company: James Boyle - Age: 16 - Height: 6' - Born: Baltimore or Harford County, MD - Trade: Blacksmith - Enlistment date: 5 May 1812 - Place: Richmond, VA - Period: 5 Yrs - Re-enlisted on 8 Jul 1817.

Clarke, Delia - Private - 14th Infantry - Discharged on 30 Sep 1814.

Clarke, John - Private - 14th Infantry - Prisoner of War (Halifax), discharged from Halifax on 3 Feb 1814.

Clarke, Joseph - Private - 14th Infantry - Company: Thomas Montgomery - Enlistment date: 22 May 1812 - Period: 5 Yrs - BLW 18143-160-12.

Clarke, Lemuel B. - Post Surgeon - Medical Department - Born: New York - Commissioned as a surgeon's mate on 4 Jan 1808; promoted to post surgeon on 24 Apr 1816; resigned on 18 Oct 1817.

Clarke, Thomas - Private - 38th Infantry - Company: John Mowton - BLW 4750-160-12.

Clarke, Thomas - Private - 12th Infantry - Company: Thomas Sangster - Age: 24 - Height: 5' 9 1/2" - Born: New Jersey - Trade: Laborer - Enlistment date: 20 Apr 1812 - Place: Havre de Grace, MD - Period: 5 Yrs - Deserted at Buffalo on 10 Mar 1815.

Clarke, William - Private - 36th Infantry - Company: Samuel Raisin - Age: 27 - Height: 5' 10" - Born: Carolina Cty, VA - Trade: Carpenter - Enlistment date: 18 Mar 1814 - Place: Stevensburg or Richmond, VA - Period: War - BLW 415-160-12 - Discharged at Washington, DC, on 20 Mar 1815.

Clary, Charles - Private - 8th Infantry - Other regiment: 2nd Infantry (New) - Age: 25 - Height: 5' 8 1/2" - Born: Baltimore Cty - Trade: Sailor - Enlistment date: 12 Dec 1814 - Place: Washington, DC - Period: 5 Yrs - Pension: Heirs received half pay for five years in lieu of land bounty - Discharged at Ripley Barracks, Bay of St. Louis, MS, on 31 May 1819.

Clary, Nathaniel - Private - 38th Infantry - Company: James Hook - Enlistment date: 7 Feb 1814 - Period: 1 Yr - Pension: Land bounty to John Clary, brother and other heirs at law of Nathaniel Clary - BLW 21508-160-12 - Re-enlisted on 14 Jun 1814 for the war; died on 3 Sep 1814.

Clary, Zachariah - Private - 19th Infantry - Company: George Kesling - Other regiment: 3rd Infantry (New) - Age: 34 - Height: 5' 11" - Born: Frederick Cty, MD - Trade: Tailor - Enlistment date: 29 Sep 1814 - Place: Franklinton, Franklin Cty, OH - Period: 5 Yrs - Deserted at Springfield, OH, on 16 Sep 1815.

Clay, Moses M. - Private - 14th Infantry - Company: Joseph Marshall.

Clay, Walter - Private - 14th Infantry - Prisoner of War (Halifax), discharged from Halifax on 3 Feb 1814.

Clayton, Andrew J. - Private - 12th Infantry - Company: Thomas Sangster - Other regiment: 8th Infantry (New) - Age: 22 - Height: 5' 4" - Born: Cecil Cty, MD - Trade: Laborer - Enlistment date: 17 Jun 1812 - Place: Elkton, MD - Period: 5 Yrs - Discharged on 17 Jul 1817.

Clayton, Henry - Private - 14th Infantry - Age: 23 - Height: 5' 10" - Born: Baltimore - Enlistment date: 13 Mar 1814 - Place: Towsontown, MD.

Clayton, Thomas - Private - 14th Infantry - Prisoner of War (Halifax), sent to England.

Cleaver, Seth - Corporal - 14th Infantry - Prisoner of War (Halifax), captured at Beaver Dams on 24 Jun 1813; died at Halifax on 16 Oct 1813.

Clellan, Samuel - Private - 14th Infantry - Company: Reuben Gilder - Age: 19 - Height: 5' 8" - Born: Virginia - Enlistment date: 7 Apr 1813 - Period: 18 Mos - Discharged at Black Rock, NY, on 7 Oct 1814.

Maryland Regulars in War of 1812

- Clemans, Michael - Private - 36th Infantry - Company: Joseph Hook - Age: 22 - Height: 5' 8" - Born: Charles Cty, MD - Trade: Blacksmith - Enlistment date: 29 Jul 1814 - Place: Georgetown, DC - Period: War - BLW 111-160-12 - Discharged at Fort Covington, MD, on 30 Mar 1815.
- Clemens, Thomas - Private - 14th Infantry - Company: Reuben Gilder - Other regiment: 4th Infantry (New) - Age: 18 - Height: 5' 1" - Born: Maryland - Trade: Laborer - Enlistment date: 6 May 1813 - Place: Maryland - Period: 5 Yrs.
- Clements, James - Private - 32nd Infantry - Company: Robert Patterson - Age: 29 - Height: 5' 8 1/2" - Born: Maryland - Trade: Wagoner - Enlistment date: 8 Feb 1814 - Place: Shippensburg, PA - Period: 1 Yr - Re-enlisted on 9 Feb 1815; discharged on 20 Jun 1815 at Philadelphia after furnishing a substitute.
- Clements, John - Corporal - 14th Infantry - Company: Thomas Montgomery - Other regiment: 4th Infantry (New) - Age: 26 - Height: 5' 9 1/2" - Born: Wilmington, DE - Trade: Seaman - Enlistment date: 2 Jul 1812 - Place: Fort Cumberland, MD - Period: 5 Yrs - Discharged on 2 Jul 1817.
- Clements, Samuel - Private - 38th Infantry - Company: John Rothrock - Age: 39 - Height: 5' 2" - Born: Maryland - Enlistment date: 6 Apr 1814 - Place: Richmond, VA - Period: War - Pension: Land bounty to James A. Clements, son and other heirs at law of Samuel Clements - BLW 26853-160-12 - Died on 8 Feb 1815.
- Clements, William - Private - 1st Infantry - Company: John Whistler - Age: 48 - Height: 5' 8" - Born: Charles Cty, MD - Trade: Laborer - Enlistment date: 28 Nov 1805 - Place: Fort Dearborn, IL - Re-enlisted at Detroit on 22 Aug 1810; died on 9 Jan 1814 at Fort Wayne, IN, of typhus fever.
- Clements, William G. - Private - 14th Infantry - Company: Samuel Lane - Discharged on 12 Apr 1814.
- Clements, Zachariah - Private - 14th Infantry - Company: Clement Sullivan - Pension: Land bounty to Thomas Clements, son and other heirs at law of Zachariah Clements - BLW 15054-160-12.
- Cleming, Francis - Private - 19th Infantry - Company: David Holt - Age: 22 - Height: 6' - Born: Charles Cty, MD - Trade: Laborer - Enlistment date: 10 Jun 1814 - Place: Beardstown, Nelson County, KY - Period: War - BLW 14244-160-12 - Discharged at Chillicothe, OH, on 14 Jun 1815.
- Cleming, James - Private - 17th Infantry - Age: 22 - Height: 6' - Born: Charles Cty, MD - Trade: Farmer - Enlistment date: 16 Jun 1814 - Period: War.
- Clendennin, Peter - Private - 38th Infantry - Age: 35 - Height: 5' 9" - Born: County Donegal, Ireland - Trade: Laborer - Enlistment date: 11 Dec 1814 - Place: Fredericktown, MD - Period: War.
- Clever, Seth - Private - 14th Infantry - Company: Samuel Lane - Age: 32 - Height: 5' 5 1/2" - Born: Germantown, PA - Enlistment date: 26 Aug 1812 - Period: 18 Mos.
- Clifford, James - Private - 38th Infantry - Company: Isaac Aldridge - Enlistment date: 5 Aug 1813 - Period: 1 Yr - Discharged on 5 Aug 1814.
- Cline, John - Private - 38th Infantry - Company: John Rothrock - Pension: SO-

By War of 1812 Soc. in MD

5297, SC-10249 - BLW 8619-160-12.

Clink, Andrew - Private - 22nd Infantry - Company: Jacob Carmack - Age: 39 - Height: 5' 7" - Born: Frederick Cty, MD - Trade: Weaver - Enlistment date: 3 Oct 1814 - Period: 5 Yrs - Deserted on 12 Nov 1814.

Clinton, James - Private - 2nd Artillery - Company: Nathan Towson - Age: 44 - Height: 5' 7 3/4" - Born: Ireland - Trade: Laborer - Enlistment date: 2 Jul 1812 - Place: Baltimore - Period: 5 Yrs.

Cloherty, John - Private - 5th Infantry - Company: Richard Bell - Age: 27 - Height: 5' 8 1/2" - Born: Baltimore or Gloucester - Trade: Shoemaker - Enlistment date: 26 Feb 1814 - Place: Plattsburgh, NY - Period: War - BLW 16198-160-12 - Discharged at Buffalo on 31 Mar 1815.

Cloud, Cornelius - Private - 14th Infantry - Company: Reuben Gilder - Age: 15 - Height: 4' 1 1/4" - Born: Maryland or Virginia - Trade: Blacksmith - Enlistment date: 5 Feb 1813 - Place: Maryland - Period: War - BLW 8766-160-12 - Discharged at Greenbush, NY, on 4 May 1815.

Clubb, Richard - Private - 36th Infantry - Company: Thomas Carbery - Deserted on 8 Sep 1814.

Clubb, Thomas - Private - 1st Artillery - Company: George Armistead - Other regiment: Corps of Artillery - Height: 5' 6" - Born: Prince George's Cty, MD - Trade: Brick layer - Enlistment date: 6 Feb 1809 - Period: 5 Yrs - BLW 10044-160-12 - Discharged on 1 Aug 1816.

Coal, Insor - Private - 36th Infantry - Age: 18 - Height: 5' 3 3/4" - Born: Baltimore - Enlistment date: 3 May 1813 - Place: Westminster, MD - Period: 1 Yr.

Coates, Daniel - Private - US Volunteers - Company: Stephen Moore - Enlistment date: 8 Sep 1812 - Period: 1 Yr - Discharged on 10 Dec 1812, unfit for service.

Coates, William - Private - 36th Infantry - Age: 22 - Height: 5' 10" - Born: Charles Cty, MD - Enlistment date: 25 Jun 1813 - Place: Waynesburg, PA - Period: 1 Yr.

Coats, George - Private - 38th Infantry - Company: John Rothrock - Age: 33 - Height: 5' 6 1/2" - Born: Richmond, VA - Trade: Tanner - Enlistment date: 14 Mar 1814 - Place: Norfolk, VA - Period: War - BLW 2866-160-12 - Discharged at Craney Island, VA, on 15 Mar 1815.

Coats, Henry - Private - 1st Light Dragoons - Company: Thomas Harrison - Age: 28 - Height: 5' 6" - Born: Charles Cty, MD or VA - Trade: Farmer - Enlistment date: 28 Mar 1814 - Place: Alexandria, DC - Period: War - Discharged at Fort Washington, MD, on 7 Mar 1815.

Coats, John - Private - 2nd Infantry - Company: Perrin Willis - Age: 21 - Height: 5' 10 1/12' - Born: Charles Cty, MD - Enlistment date: 5 Dec 1814 - Period: 5 Yrs.

Coats, William - Corporal - 20th Infantry - Company: William Jett - Age: 24 - Height: 5' 9 3/4" - Born: Charles Cty, MD - Trade: Ship's carpenter - Enlistment date: 21 Aug 1814 - Place: Alexandria, DC - Period: War - BLW 173-160-12 - Discharged at Norfolk, VA, on 31 Mar 1815.

Cobb, Henry - Private - 14th Infantry - Company: James Haslett - Enlistment date: 5 Jun 1813 - Period: 1 Yr.

Cobb, William - Private - 1st Artillery - Company: Heman Fay - Other regiment: Corps of Artillery - Age: 25 - Height: 5' 6" - Born: Drummington, MD - Trade: Laborer - Enlistment date: 1 Aug 1812 - Period: 5 Yrs - BLW 10680-160-12 - Discharged at Fort McHenry on 15 Aug 1817.

Cochran, David - Private - 14th Infantry - Company: Samuel Lane - Age: 36 - Height: 5' 5" - Born: Ireland - Enlistment date: 22 May 1812 - Place: W. M., MD - Period: 18 Mos - Pension: Heirs received half pay for five years in lieu of land bounty - Wounded at Beaver Dams, UC, on 24 Jun 1813 in his left arm.

Cochran, George - Sergeant - 38th Infantry - Company: Anthony Miltenberger - Height: 5' 6" - Born: Philadelphia - Trade: Printer - Pension: No pension claim - BLW 48-160-50 - Discharged at Fort McHenry on 28 Mar 1815.

Cochran, George - Private - 38th Infantry - Company: John Mowton - Enlistment date: 16 Jul 1813 - Period: 1 Yr - Discharged at Craney Island, VA, on 16 Jul 1814.

Cochran, George - Sergeant - 38th Infantry - Company: Shepperd Leakin - BLW 9768-160-12.

Cochran, Isaac - Private - 4th Rifles - Company: Matthew Magee - Age: 38 - Height: 5' 7" - Born: Harford Cty, MD - Trade: Miller - Enlistment date: 1 Jun 1814 - Period: War - BLW 12641-160-12 - Discharged at Carlisle Barracks, PA, on 25 Mar 1815.

Cochran, James - Third Lieutenant - 38th Infantry - Company: Shepperd Leakin - Commissioned as an ensign on 2 Mar 1814; promoted to 3rd lieutenant on 20 May 1814; discharged on 15 Jun 1815.

Cochran, James - Sergeant - 14th Infantry - Company: Edward Wilson - Age: 21 - Height: 5' 8 1/4" - Born: Pennsylvania - Trade: Printer - Enlistment date: 26 Feb 1814 - Period: War - BLW 4948-160-12 - Discharged at Washington, DC, on 17 Mar 1815.

Cochran, James J. L. M. - Private - US Volunteers - Company: Stephen Moore - Enlistment date: 8 Sep 1812 - Period: 1 Yr.

Cochran, John - Private - 14th Infantry - Company: Reuben Gilder - Age: 32 - Height: 5' 8" - Born: Dover, Kent Cty, DE - Trade: Clock maker - Enlistment date: 7 Mar 1814 - Place: Baltimore - Period: War - BLW 10088-160-12 - Discharged at Greenbush, NY, on 4 May 1815.

Cochran, Thomas - Private - 12th Infantry - Company: James Paxton - Age: 30 or 70 - Height: 5' 11" - Born: York, MD - Trade: Blacksmith - Enlistment date: 26 Sep 1812 - Place: Wheeling, VA - Period: 5 Yrs - Discharged at Buffalo or Fort Erie, UC, on 15 Jun 1815 for old age.

Cockley, John - Private - 36th Infantry - Age: 43 - Height: 5' 10" - Born: Pennsylvania - Enlistment date: 14 Jul 1813 - Place: Havre de Grace, MD.

Cocklin, James - Private - 32nd Infantry - Company: James Clark - Age: 29 - Height: 5' 8" - Born: Caroline Cty, MD - Trade: Laborer - Enlistment date: 31 May 1813 - Place: Lewistown - Period: 1 Yr - Discharged on 4 Jul 1814.

Coffee, Richard - Private - 24th Infantry - Other regiment: 2nd Rifles - Age: 25 - Height: 5' 5" - Born: Baltimore - Trade: Miller - Enlistment date: 9 Jun

1814 - Place: Michigan - Period: War - Discharged at Detroit on 30 Jun 1815.

Coffman, Christian - Private - 38th Infantry - Company: John Rothrock - Height: 5' 10" - Born: Fredericktown, MD - Trade: Butcher - Enlistment date: 16 Jun 1813 - Period: 1 Yr - BLW 377-160-12 - Re-enlisted on 20 Feb 1814 for the war; discharged on 15 Mar 1815 at Craney Island, VA.

Cogland, John - Private - 27th Infantry - Company: Aaron Craine - Age: 37 - Height: 5' 8" - Born: Maryland - Trade: Bookbinder - Enlistment date: 20 Aug 1814 - Place: New York - Period: 5 Yrs - BLW 347-160-12 - Discharged at New York City on 29 Jun 1815.

Cohen, William H. - Private - 14th Infantry - Company: Thomas Montgomery - Age: 23 - Height: 5' 6" - Born: Westchester, PA - Trade: Carpenter - Enlistment date: 12 Apr 1813 - Place: Salisbury, MD - Period: War - Discharged at Greenbush, NY, on 4 May 1815.

Coin, William - Private - 38th Infantry - Company: Shepperd Leakin - Pension: Land bounty to James Coin, brother and only heir at law of William Coin - BLW 24471-160-12.

Colbert, Henry - Private - 38th Infantry - Company: Isaac Aldridge - Age: 18 - Height: 5' 5" - Born: Somerset Cty, MD - Trade: Farmer - Enlistment date: 25 Oct 1814 - Place: Newtown, MD - Period: War - BLW 502-160-12 - Discharged at Baltimore on 6 Apr 1815.

Colder, William - Private - 38th Infantry - Company: James Haslett - Enlistment date: 20 Apr 1813 - Period: 1 Yr.

Cole, Andrew - Private - 14th Infantry - Prisoner of War (Halifax), sent to England.

Cole, George L. - Private - 26th Infantry - Company: William Bezeau - Age: 20 - Height: 5' 6" - Born: Baltimore - Trade: Cooper - Enlistment date: 24 Jan 1815 - Place: Baltimore - Period: War - BLW 16-320-12.

Cole, John - Private - 36th Infantry - Age: 35 - Height: 5' 6" - Born: St. Mary's Cty, MD - Enlistment date: 2 Jun 1813 - Place: Baltimore.

Cole, Lambert - Musician - 36th Infantry - Company: Thomas Carbery - Age: 42 - Height: 5' 7 1/2" - Born: Lenate, RI - Trade: Seaman - Enlistment date: 1 May 1813 - Place: Baltimore - BLW 24857-160-12 - Discharged at Baltimore on 31 Mar 1815.

Cole, Morris - Private - 36th Infantry - Age: 25 - Height: 5' 5" - Born: Ireland - Trade: Laborer - Enlistment date: 21 Dec 1814 - Period: War - BLW 82-320-12 - Discharged at Georgetown, DC, on 13 Mar 1815.

Cole, Thomas D. - Private - 22nd Infantry - Company: John Pentland - Age: 26 - Height: 5' 11" - Born: Maryland - Trade: Blacksmith - Enlistment date: 5 Feb 1813 - Place: Harrisburg, PA - Period: 5 Yrs - Died on 7 Jan 1814 near Ogdensburg, NY.

Coleman, John - Private - 36th Infantry - Company: Thomas Carbery - Age: 33 - Height: 5' 6" - Born: Harford Cty, MD - Trade: Farmer - Enlistment date: 3 May 1813 - Place: Havre de Grace, MD - Period: War - BLW 22270-160-12 - Discharged at Baltimore on 31 Mar 1815.

Coleman, Joseph - Private - 5th Infantry - Company: Colin Buckner - Age: 37

- Height: 5' 11" - Born: Cumberland Cty, PA - Trade: Miller - Enlistment date: 22 Feb 1813 - Place: Baltimore - Period: 5 Yrs - Deserted on 29 Jan 1814.
- Coleman, Philip - Private - 5th Infantry - Company: Colin Buckner - Other regiment: 3rd Infantry (New) - Age: 19 - Height: 5' 4 1/2" - Born: Baltimore - Trade: Laborer - Enlistment date: 1 Apr 1813 - Place: Baltimore or Alexandria, DC - Period: 5 Yrs - Discharged on 1 Apr 1818.
- Coleman, Richard - Private - 36th Infantry - Age: 24 - Height: 5' 8 1/2" - Born: Newburgh, Orange County, NY - Enlistment date: 27 May 1813 - Place: Baltimore - Period: 1 Yr - Deserted on 9 Jun 1813.
- Coleman, Thomas C. - Private - 5th Infantry - Company: William Bird - Age: 31 - Born: Baltimore - Trade: Farmer - Enlistment date: 7 Apr 1813 - Place: Baltimore - Period: 5 Yrs.
- Collins, Daniel - Private - 38th Infantry - Company: James Smith - BLW 14309-160-12.
- Collins, Francis - Private - 7th Infantry - Company: Carey Nicholas - Age: 21 - Height: 5' 10 1/4" - Born: Maryland - Trade: Farmer - Enlistment date: 21 Jan 1809 - Period: 5 Yrs - BLW 13796-160-12 - Re-enlisted at Fort Claiborne, LA, on 20 Apr 1813 for the war; discharged on 8 Apr 1815 at New Orleans, LA.
- Collins, Francis - Private - 38th Infantry - Company: John Rothrock - Age: 35 - Height: 5' 10" - Born: Anne Arundel Cty, MD - Trade: Farmer - Enlistment date: 14 Jul 1814 - Place: Craney Island, VA - Period: War - BLW 9896-160-12 - Discharged at Craney Island, VA, on 15 Mar 1815.
- Collins, George - Private - 36th Infantry - Company: Joseph Hook - Age: 38 - Height: 5' 8" - Born: Prince George's Cty, MD - Trade: Laborer - Enlistment date: 15 Jul 1814 - Place: Georgetown, DC - Period: War - BLW 95-160-12 - Discharged at Fort Covington, MD, on 30 Mar 1815.
- Collins, James - Private - 12th Infantry - Company: Thomas Post - Age: 18 - Height: 5' 6" - Born: Frederick Cty, MD - Trade: Tanner - Enlistment date: 4 Feb 1814 - Place: Washington Cty, MD - Period: 5 Yrs - Discharged at Buffalo on 20 Jun 1815 after furnishing a substitute.
- Collins, John - Private - 17th Infantry - Age: 32 - Height: 5' 6" - Born: Maryland - Enlistment date: 17 May 1814 - Place: Zanesville, OH.
- Collins, Robert - Private - 14th Infantry - Company: David Cummings - Other regiment: 4th Infantry (New) - Age: 27 - Height: 5' 10" - Born: Georgetown, DC - Trade: Screw cutler - Enlistment date: 17 Feb 1813 - Place: Georgetown, DC - Period: 5 Yrs - BLW 20410-160-12 - Prisoner of War (Halifax), discharged from Halifax on 3 Feb 1814; discharged on 16 Feb 1818 at Fort Scott.
- Collins, Thomas B. - Sergeant - 14th Infantry - Company: David Cummings - Age: 21 - Height: 5' 7" - Born: Baltimore - Trade: Seaman - Enlistment date: 22 Oct 1814 - Place: Baltimore - Period: War - BLW 8836-160-12 - Discharged at Baltimore on 16 Mar 1815.
- Colman, John - Corporal - 36th Infantry - Company: Thomas Carbery - BLW 22270-160-12.

By War of 1812 Soc. in MD

Colony, James - Private - 14th Infantry - Company: Samuel Lane - Age: 33 - Height: 5' 5" - Born: Baltimore - Enlistment date: 5 Aug 1812 - Place: Baltimore - Period: 5 Yrs.

Columbus, Nicholas - Private - 36th Infantry - Company: Samuel Raisin - Age: 25 - Height: 5' 8" - Born: New Orleans - Trade: Seaman - Enlistment date: 21 Sep 1814 - Place: Norfolk, VA - Period: War - Discharged at Washington, DC, on 20 Mar 1815.

Colven, John - Private - 14th Infantry - Company: Joseph Marshall - Age: 26 - Height: 6' - Born: Dauphin or Lancaster County, PA - Trade: Cooper - Enlistment date: 22 Sep 1814 - Place: Baltimore - Period: War - Discharged at Greenbush, NY, on 3 May 1815.

Combs, Lawrence - Private - 36th Infantry - Age: 42 - Height: 5' 8" - Born: Jersey - Enlistment date: 2 Jun 1813 - Place: Baltimore - Period: 1 Yr.

Comegys Jr., Cornelius - Second Lieutenant - 14th Infantry - Born: Maryland - Commissioned as a 3rd lieutenant on 10 May 1813; promoted to 2nd lieutenant on 14 Nov 1813; died on 31 Mar 1815.

Comegys, William - Private - 36th Infantry - Age: 21 - Height: 5' 3 3/8" - Born: Kent Cty, MD - Enlistment date: 11 Jun 1813 - Place: Chester - Period: 1 Yr.

Conahy, Benjamin M. - Private - 14th Infantry - Prisoner of War (Halifax), sent to England.

Conaway, John - Private - 38th Infantry - Company: Shepperd Leakin - Age: 38 - Height: 5' 8 1/2" - Born: Queen Anne's Cty, MD - Trade: Laborer - Enlistment date: 6 Aug 1814 - Place: Baltimore - Period: War - Discharged at Fort McHenry on 28 Mar 1815.

Concklin, John - Seaman - US Sea Fencibles - Company: Simmones Bunbury - Enlistment date: 27 Sep 1814 - Period: 3 Mos.

Condon, James - Private - 17th Infantry - Company: Harris Hickman - Age: 32 - Height: 5' 6 1/4" - Born: Baltimore Cty - Trade: Weaver - Enlistment date: 1 Mar 1814 - Place: New Lisbon, OH - Period: War - BLW 6254-160-12 - Discharged at Chillicothe, OH, on 9 Jun 1815.

Conkey, Lunar - Private - 14th Infantry - Prisoner of War (Halifax), discharged from Halifax on 3 Feb 1814.

Conkey, Pliney - Private - 14th Infantry - Prisoner of War (Halifax), discharged from Halifax on 3 Feb 1814.

Conklin, William M. - Sergeant Major - 14th Infantry - Age: 28 - Height: 5' 11 1/2" - Born: Baltimore City - Trade: Carver - Enlistment date: 7 Apr 1814 - Place: Baltimore - Period: War - BLW 16541-160-12 - Discharged at Greenbush, NY, on 3 May 1815.

Conley, John - Private - 36th Infantry - Age: 25 - Height: 5' 4" - Born: Ireland - Trade: Weaver - Enlistment date: 1 Feb 1815 - Place: Georgetown, DC - Period: War - BLW 6-320-12 - Discharged at Georgetown, DC, on 13 Mar 1815.

Conley, John - Private - 14th Infantry - Company: Reuben Gilder - Other regiment: 4th Infantry (New) - Age: 22 - Height: 5' 5" - Born: Maryland - Trade: Laborer - Enlistment date: 20 Sep 1813 - Place: Baltimore - Period:

5 Yrs - Deserted on 19 Jul 1815.

Conley, Patrick - Private - 14th Infantry - Company: James McDonald - Age: 45 - Height: 5' 8 1/2" - Trade: Ditcher - Enlistment date: 11 Jun 1813 - Period: 18 Mos.

Conley, Patrick - Private - 38th Infantry - Company: Charles Stansbury - Age: 28 - Height: 5' 7" - Born: Ireland - Trade: Collier - Enlistment date: 10 Oct 1814 - Place: Baltimore - Period: War - Deserted on 4 Jan 1815.

Conley, Samuel - Private - 2nd Light Dragoons - Company: Samuel Hopkins - Other regiment: Corps of Artillery - Age: 17 - Height: 5' 3" - Born: Maryland - Trade: Farmer - Enlistment date: 27 Mar 1812 - Place: Cincinnati - Period: 5 Yrs - BLW 15287-160-12 - Discharged on 27 Mar 1817.

Conley, William - Private - 38th Infantry - Company: John Buck - Age: 19 - Height: 5' 6" - Born: Maryland - Trade: Laborer - Enlistment date: 25 May 1814 - Place: Baltimore - Period: War.

Conly, Thomas - Private - 38th Infantry - Company: John Brookes.

Connally, James - Seaman - US Sea Fencibles - Company: Simmones Bunbury - Age: 23 - Height: 5' 8" - Born: Ireland - Enlistment date: 6 Jul 1814 - Period: 1 Yr.

Connally, John - Seaman - US Sea Fencibles - Company: Simmones Bunbury - Enlistment date: 6 Jul 1814 - Period: 1 Yr - Discharged at Fort McHenry on 24 Mar 1815.

Connell, James - Private - 16th Infantry - Company: Robert Gray - Age: 25 - Height: 5' 10 1/2" - Born: Maryland - Trade: Joiner - Enlistment date: 12 Sep 1812 - Place: Columbia, MD - Period: 18 Mos - Discharged at Burlington, VT, on 12 Apr 1814.

Connell, Rezin - Private - 38th Infantry - Company: James Hook - Enlistment date: 21 Dec 1813 - Period: 1 Yr - Discharged on 21 Dec 1814.

Conner, Dennis - Private - 38th Infantry - Company: Isaac Aldridge - Enlistment date: 12 Jul 1813 - Period: 1 Yr - Discharged on 13 Jul 1814.

Conner, James - Private - Light Artillery - Company: Benjamin Branch - Other regiment: Corps of Artillery - Age: 24 - Height: 5' 10" - Born: Prince George's Cty, MD - Trade: Baker - Enlistment date: 16 Mar 1812 - Place: Washington, DC - Period: 5 Yrs - BLW 10800-160-12 - Discharged at Fort Trumbull, CT, on 15 Mar 1817.

Conner, James - Private - 14th Infantry - Company: Reuben Gilder - Age: 22 - Height: 5' 7 3/4" - Born: York Cty, PA - Trade: Cordwainer - Enlistment date: 27 Apr 1812 - Place: Hagerstown, MD - Period: 5 Yrs - Pension: Old War IF-3211 - BLW 5192-160-12 - Wounded at Fort Erie, UC; died on 30 May 1812 at 4 Mile Creek near Fort Niagara, NY.

Conner, Killum - Corporal - 17th Infantry - Company: John Chunn - Age: 27 - Height: 5' 7" - Born: Somerset Cty, MD - Trade: Mariner - Enlistment date: 16 Aug 1813 - Place: Maysville, KY - Period: War.

Conner, Levin - Private - 2nd Light Dragoons - Company: Samuel Hopkins - Other regiment: Light Dragoons - Age: 18 - Height: 5' - Born: Snow Hill, MD - Trade: Sail maker - Enlistment date: 6 Jul 1814 - Place: Carlisle, PA - Period: 5 Yrs - BLW 21730-160-12 - Discharged at Avon, NY, on 8 Jun

By War of 1812 Soc. in MD

1815.

Conner, Mark L. - Clerk - 38th Infantry - Company: James Smith - Age: 23 - Height: 5' 9" - Born: Reading, Berks County, PA - Trade: Printer - Enlistment date: 20 Aug 1813 - Period: 1 Yr - Pension: Land bounty to Mary Shadell, sister and other heirs at law of Mark L. Conner - BLW 26264-160-12 - Re-enlisted on 20 Aug 1814 for the war; discharged on 15 Mar 1815 at Craney Island, VA.

Conner, Michael - Private - 38th Infantry - Age: 27 - Height: 5' 7" - Born: Shenandoah Cty, VA - Trade: Blacksmith - Enlistment date: 21 Feb 1814 - Place: Craney Island, VA - Period: War - Discharged at Craney Island, VA, on 15 Mar 1815.

Conner, Patrick - Corporal - 22nd Infantry - Company: George Barker - Age: 47 - Height: 5' 9 1/2" - Born: York, York Cty, PA - Trade: Blacksmith - Enlistment date: 15 Jan 1813 - Place: Cumberland, MD - Period: 5 Yrs - Pension: Land bounty to Thomas Conner, son & other heirs at law of Patrick Conner - BLW 26219-160-12 - Wounded at Chippewa, UC, on 25 Jul 1814; died at Buffalo on 30 Jul 1814.

Conner, Thomas - Private - 1st Light Dragoons - Company: Alexander Cummings - Other regiment: Corps of Artillery - Age: 23 - Height: 5' 7 1/2" - Born: Prince George's Cty, MD - Trade: Shoemaker - Enlistment date: 19 Jun 1812 - Place: Baltimore - Period: 5 Yrs - Discharged on 19 Jun 1817.

Conner, William - Corporal - 17th Infantry - Company: John Chunn - Age: 27 - Height: 5' 7" - Born: Somerset Cty, MD - Trade: Mariner - Enlistment date: 16 Aug 1813 - Place: Maysville, KY - Period: War - BLW 11900-160-12 - Discharged at Chillicothe, OH, on 7 Jun 1815.

Conner, William - Private - 4th Rifles - Company: John Lytle - Age: 35 - Height: 5' 8" - Born: Maryland or Virginia - Trade: Hair dresser - Enlistment date: 21 May 1814 - Period: War - Prisoner of War (Halifax); discharged on 14 Jun 1815 at Conjocquita Creek, NY.

Connerly, Thomas - Private - 38th Infantry - Company: Isaac Aldridge - Pension: Land bounty to William Connerly, brother and other heirs at law of Thomas Connerly - BLW 26436-160-12.

Connor, George - Private - 14th Infantry - Company: David Cummings - Age: 25 - Height: 5' 7" - Born: Ireland - Trade: Tailor - Enlistment date: 24 Oct 1814 - Place: Hagerstown, MD - Period: 5 Yrs.

Conrad, Aquila - Private - 2nd Artillery - Company: Nathan Towson - Other regiment: Corps of Artillery - Age: 20 - Height: 5' 7 1/2" - Born: Baltimore - Trade: Cooper - Enlistment date: 20 Apr 1812 - Place: Baltimore - Period: 5 Yrs - BLW 11512-160-12 - Discharged at Fort Niagara, NY, on 20 Apr 1817.

Conrad, John - Seaman - US Sea Fencibles - Company: John Gill - Age: 21 - Height: 5' 8" - Born: New York City - Enlistment date: 4 Jan 1814 - Period: 1 Yr - Discharged on 3 Jan 1815.

Conrad, John - Private - 38th Infantry - Company: Isaac Aldridge - Age: 32 - Height: 5' 7" - Born: Baltimore - Trade: Chair maker - Enlistment date: 9 Aug 1814 - Period: War - Pension: Land bounty to Elizabeth and George

Conrad, children and only heirs at law of John Conrad - BLW 25715-160-12 - Discharged at Baltimore on 3 Apr 1815.

Conrad, John - Sergeant - 38th Infantry - Company: Isaac Aldridge - Enlistment date: 2 Jan 1813 - Period: 1 Yr - Pension: Land bounty to George Conrad, heir at law of William Conrad and Elizabeth Conrad, who were the only heirs at law of John Conrad - BLW 25715-160-12 - Discharged at Craney Island, VA, on 3 Jun 1814.

Conroy, James - Private - 7th Infantry - Company: Uriah Blue - Age: 39 - Height: 5' 8 1/2" - Born: Maryland - Trade: Bookbinder - Enlistment date: 7 Jun 1811 - Place: Baton Rouge, LA - Period: 5 Yrs.

Constantine, H. C. - Sergeant - 38th Infantry - Company: Shepperd Leakin - Enlistment date: 17 Oct 1813 - Period: 1 Yr - Discharged on 28 Oct 1814.

Conway, Michael - Private - 14th Infantry - Company: Clement Sullivan - Age: 56 - Height: 5' 6" - Born: County Galway, Ireland - Trade: Shoemaker - Enlistment date: 2 Jun 1812 - Period: 5 Yrs - Pension: Old War IF-10127 - BLW 1517-160-12 - Prisoner of War, exchanged on 15 Apr 1814; discharged on 30 Apr 1815 at Greenbush, NY, on Surgeon's Certificate.

Conwell, John - Private - 14th Infantry - Prisoner of War (Halifax), discharged from Halifax on 3 Feb 1814.

Cook, Haz. - Private - 14th Infantry - Prisoner of War (Halifax), sent to England.

Cook, James - Private - 2nd Artillery - Company: James Barker - Age: 42 - Height: 5' 6" - Born: Maryland - Trade: Farmer - Enlistment date: 12 Jul 1812 - Place: New Haven, CT - Period: 5 Yrs - Deserted at Fort Mifflin, PA, on 12 Jul 1814.

Cook, John - Corporal - 38th Infantry - Company: James Hook - BLW 24411-160-12.

Cook, John H. - Sergeant - 14th Infantry - Company: Thomas Montgomery - Pension: Old War IF-12710 - Prisoner of War (Halifax), discharged from Halifax on 9 Nov 1813.

Cook, Laurence C. - Private - 38th Infantry - Company: John Rothrock - Age: 23 - Height: 5' 8 1/2" - Born: Westmoreland Cty, VA - Trade: Miller - Enlistment date: 4 Apr 1814 - Place: Richmond, VA - Period: War - BLW 6242-160-12 - Discharged at Craney Island, VA, on 15 Mar 1815.

Cook, Samuel - Seaman - US Sea Fencibles - Company: John Gill - Enlistment date: 23 Dec 1813.

Cook, Thomas - Private - 38th Infantry - Age: 40 - Height: 5' 6" - Born: Prince George's Cty, MD - Enlistment date: 14 Jan 1814 - Place: Alexandria, DC - Period: 1 Yr.

Cook, William - Private - 16th Infantry - Company: Miles Greenwood - Other regiment: 2nd Infantry (New) - Age: 19 - Height: 5' 9" - Born: Maryland - Trade: Butcher - Enlistment date: 10 Aug 1812 - Place: Marietta, PA - Period: 5 Yrs - BLW 7510-160-12 - Discharged at Sackets Harbor, NY, on 13 Aug 1817.

Cook, William - Private - 14th Infantry - Company: Henry Grindage - Enlistment date: 10 Sep 1812 - Period: 5 Yrs - Pension: Heirs received half pay for five years in lieu of land bounty - Prisoner of War (Halifax), sent to England.

Cooke, Levin - Private - US Volunteers - Company: Stephen Moore - Enlistment date: 8 Sep 1812 - Period: 1 Yr.

Cooke, Levin - Private - 26th Infantry - Company: William Bezeau - Age: 28 - Height: 5' 6" - Born: Dorchester Cty, MD - Trade: Carpenter - Enlistment date: 31 Jan 1815 - Place: Baltimore - Period: War - BLW 410-320-12 - Discharged at Baltimore.

Cookes, Henry - Private - 38th Infantry - Company: John Rothrock - Enlistment date: 9 Aug 1813 - Period: 1 Yr.

Cookes, Henry - Sergeant - 38th Infantry - Company: John Rothrock - Age: 23 - Height: 5' 10" - Born: Hagerstown, MD or Washington Cty, PA - Trade: hatter - Enlistment date: 16 Jan 1814 - Place: Baltimore - Period: War - BLW 549-160-12 - Discharged at Craney Island, VA, on 15 Mar 1815.

Coomes, John - Gunner - US Sea Fencibles - Company: Simmones Bunbury - Age: 30 - Height: 5' 6 3/4" - Born: Trieste, Austria - Enlistment date: 16 Nov 1814 - Discharged on 28 Jul 1815.

Coon, John - Corporal - 2nd Light Dragoons - Company: John Burd - Age: 20 - Height: 5' 5" - Born: Washington Cty, MD - Trade: Shoemaker - Enlistment date: 2 Jun 1812 - Period: 5 Yrs - Pension: Land bounty to Margaret Marshall, mother & only heir at law of John Coon - BLW 27376-260-42 - Died at Utica, NY, on 6 Jun 1813.

Cooney, Isaac - Private - 12th Infantry - Company: Charles Page - Age: 38 - Height: 5' 5" - Born: Maryland - Trade: Forgeman - Enlistment date: 13 Sep 1814 - Place: Lexington - Period: 5 Yrs.

Cooper, Ambrose - Private - 38th Infantry - Company: Isaac Aldridge - Enlistment date: 5 Aug 1813 - Period: 1 Yr - Discharged at Craney Island, VA, on 5 Aug 1814.

Cooper, Ambrose - Private - 38th Infantry - Company: James Haslett - Age: 36 - Height: 5' 10" - Born: Baltimore - Trade: Cooper - Enlistment date: 16 Jan 1815 - Period: War - BLW 125-320-12 - Discharged at Baltimore on 6 Apr 1815.

Cooper, Archibald - Private - 28th Infantry - Other regiment: 3rd Infantry (New) - Age: 24 - Height: 5' 8 1/2" - Born: Maryland - Trade: Farmer - Enlistment date: 9 May 1814 - Place: Kentucky - Period: 5 Yrs - Pension: Land bounty to James Cooper and other heirs at law of Archibald Cooper - BLW 20173-160-12 - Died on 30 Dec 1816.

Cooper, Hezekiah - Seaman - US Sea Fencibles - Company: William Addison.

Cooper, James - Private - 14th Infantry - Company: Thomas Montgomery - Other regiment: 4th Infantry (New) - Age: 14 - Height: 4' 9" - Born: Harford Cty, MD - Trade: Cooper - Enlistment date: 21 Dec 1812 - Place: Baltimore - Period: 5 Yrs - Died in Nov 1817 on board a schooner.

Cooper, John - Seaman - US Sea Fencibles - Company: Simmones Bunbury - Age: 23 - Height: 5' 7 1/4" - Born: Baltimore - Enlistment date: 9 Apr 1814 - Period: 1 Yr - Discharged at Fort McHenry on 24 Mar 1815.

Cooper, John - Private - 36th Infantry - Company: Thomas Carbery - Age: 20

- Height: 5' 4" - Born: Dumfries, Prince Williams Cty, VA - Trade: Sailor - Enlistment date: 29 Mar 1814 - Place: Georgetown, DC - Period: War - BLW 6624-160-12 - Discharged at Baltimore on 31 Mar 1815.

Cooper, Thornton Q. - Private - 38th Infantry - Company: John Mowton - Age: 21 - Height: 6' - Born: Frederick Cty, VA - Trade: Miller - Enlistment date: 26 Mar 1814 - Place: Craney Island, VA - Period: War - Pension: Land bounty to George Cooper, father and heir at law of Thornton Q. Cooper - BLW 25536-160-12 - Discharged at Craney Island, VA, on 15 Mar 1815.

Cooper, William - Private - 1st Infantry - Company: Horatio Stark - Other regiment: 3rd Infantry (New) - Age: 18 - Height: 5' 6 1/4" - Born: Montgomery Cty, MD - Trade: Farmer - Enlistment date: 2 Jul 1812 - Period: 5 Yrs - Pension: Land bounty to Micajah Cooper, father and only heir at law of William Cooper - BLW 18440-160-12 - Died at Sackets Harbor, NY, on 2 Feb 1815, natural death.

Cope, Henry - Private - 14th Infantry - Company: Thomas Montgomery - Age: 20 - Height: 5' 5 1/2" - Born: Germany - Trade: Sailor - Enlistment date: 12 Jul 1812 - Period: 5 Yrs - Pension: Old War IF-10469 - BLW 17396-160-12 - Discharged at Greenbush, NY, on 6 May 1815 on Surgeon's Certificate; wounded on Beaver's Dam on 24 Jun 1813.

Copeland, Samuel - Private - 14th Infantry - Company: Samuel Lane - Age: 32 - Height: 5' 9 3/4" - Born: Ireland - Enlistment date: 24 Jun 1812 - Place: Williamsport, MD - Period: 5 Yrs - Discharged at Plattsburgh, NY, on 11 Jul 1814 for infirmities.

Copeland, William - Private - 3rd Rifles - Age: 20 - Height: 5' 8" - Born: Harford Cty, MD - Enlistment date: 25 Jul 1814 - Place: Guilford Court House, NC.

Copenhaver, John - Private - 2nd Light Dragoons - Company: John Burd - Age: 37 - Height: 5' 9" - Born: Pennsylvania - Trade: Chair maker - Enlistment date: 25 Apr 1814 - Place: Baltimore - Period: War - Discharged at Carlisle Barracks, PA, on 4 May 1815.

Copsey, Cornelius - Private - 2nd Infantry - Company: Edmund Gaines - Other regiment: 44th Infantry - Height: 6' - Born: Leonardstown, MD - Trade: Farmer - Enlistment date: 1805 - Re-enlisted on 23 Jul 1808; discharged on 23 Jul 1813 at Mobile Point, AL; re-enlisted on 1 Dec 1813 in the 44th Infantry, Captain Isaac Baker's Company.

Corbin, Joshua - Private - 14th Infantry - Company: Reuben Gilder - Age: 24 - Height: 5' 11" - Born: Maryland - Trade: Miller - Enlistment date: 8 Oct 1813 - Period: 18 Mos.

Corbitt, John - Private - 2nd Artillery - Company: Nathan Towson - Other regiment: Corps of Artillery - Age: 34 - Height: 5' 10" - Born: Baltimore - Trade: Stone mason - Enlistment date: 12 Jul 1812 - Place: Baltimore - Period: 5 Yrs - BLW 12787-160-12 - Discharged at Fort Niagara, NY, on 11 Jul 1817.

Corboley, John R. - Captain - 5th Infantry - Born: Maryland - Commissioned as a 2nd lieutenant on 3 Jan 1812; promoted to 1st lieutenant on 6 Jul 1812; promoted to captain on 28 Jun 1814; discharged on 15 Jun 1815.

By War of 1812 Soc. in MD

Corcoran, Thomas - Captain - 36th Infantry - Company: Thomas Corcoran - Pension: Land bounty to Emily Corcoran, widow of Thomas Corcoran - BLW 25406-160-50 - Commissioned as a captain, 36th Infantry, on 30 Apr 1813; discharged on 15 Jun 1815; died on 26 Oct 1846.

Cord, Stephen - Private - 5th Infantry - Company: James Dorman - Other regiment: 3rd Infantry (New) - Age: 27 - Height: 5' 8" - Born: Frederick, MD - Trade: Shoemaker - Enlistment date: 29 Jan 1812 - Place: Hagerstown, MD - Period: 5 Yrs - BLW 13629-160-12 - Discharged at Fort Mackinaw, MI, on 28 Jan 1817.

Cordery, Isaac - Seaman - US Sea Fencibles - Company: Simmones Bunbury - Age: 22 - Height: 5' 6 1/2" - Born: Salisbury, CT - Enlistment date: 25 Feb 1814 - Period: 1 Yr - Discharged on 24 Feb 1815.

Cordery, William - Private - 38th Infantry - Age: 33 - Height: 5' - Born: Worcester Cty, MD - Trade: Blacksmith - Enlistment date: 17 Apr 1814 - Period: War - Pension: Land bounty to Mary Maddox, daughter & other heirs at law of William Cordery - BLW 25041-160-12 - Died on 20 Apr 1814.

Cords, Jacob - Private - 14th Infantry - Company: David Cummings - Enlistment date: 24 Oct 1812 - Period: 5 Yrs - Prisoner of War, sent to England; deserted on 10 Apr 1815 at Baltimore.

Corkery, Arthur - Private - 22nd Infantry - Company: John Pentland - Age: 22 - Height: 5' 7 1/2" - Born: Maryland - Trade: Hatter - Enlistment date: 20 May 1812 - Place: Gettysburg, PA.

Cornelius, John - Private - 38th Infantry - Age: 31 - Height: 5' 8 1/2" - Born: Franklin Cty, PA - Trade: Carpenter - Enlistment date: 15 Jan 1814 - Place: Baltimore - Period: War - Pension: Old War WF-13400 Rejected; heirs received half pay for five years in lieu of land bounty - Died on 9 Jul 1814.

Corrons, George - Private - 2nd Light Dragoons - Company: William Littlejohn - Age: 25 - Height: 5' 9" - Enlistment date: 1 Jan 1814 - Place: Baltimore.

Cost, Philip - Private - 38th Infantry - Company: John Brookes - BLW 8778-160-12.

Costigan, Michael - Private - 14th Infantry - Pension: Wife Mary Conroy, Old War WF-12752.

Costin, Dennis - Private - 2nd Light Dragoons - Company: William Littlejohn - Other regiment: Corps of Artillery - Age: 25 - Height: 5' 3 1/2" - Born: Amsterdam, Holland - Trade: Seaman - Enlistment date: 24 Dec 1813 - Place: Baltimore - Discharged on 1 May 1818, unfit for service.

Cotteral, Richard - Private - 12th Infantry - Age: 48 - Height: 5' 4" - Born: Maryland - Trade: Shoemaker - Enlistment date: 5 Jul 1814 - Period: War - BLW 4103-160-12 - Discharged at Fort Covington, MD, on 10 Feb 1815 for inability.

Cotts, Daniel - Private - 32nd Infantry - Company: Robert Patterson - Age: 25 - Height: 6' 3/4" - Born: Hagerstown, MD - Trade: Chair maker - Enlistment date: 27 Aug 1814 - Place: Trenton, NJ - Period: 5 Yrs.

Coulter, Henry - Second Lieutenant - 38th Infantry - Company: Isaac Aldridge

Maryland Regulars in War of 1812

- Commissioned as a 2nd lieutenant on 20 May 1813; resigned on 5 Oct 1813.

Coulter, Nathaniel - Private - 5th Infantry - Company: John Corboley - Age: 50 - Height: 5' 8" - Born: Lancaster Cty, PA - Trade: Farmer - Enlistment date: 18 Apr 1813 - Place: Hagerstown, MD or Leesburg, VA - Period: 5 Yrs - Discharged at Buffalo on 21 Jun 1815 due to old age.

Counselman, John - Private - 17th Infantry - Company: John Chunn - Age: 31 - Height: 5' 5 1/2" - Born: Baltimore Cty - Trade: Millwright - Enlistment date: 13 Jan 1814 - Place: Nashville, TN - Period: War - BLW 7018-160-12 - Discharged at Chillicothe, OH, on 7 Jun 1815.

Courtney, George - Private - 14th Infantry - Company: Thomas Montgomery - Age: 41 - Height: 5' 4" - Born: Ireland - Trade: Blacksmith - Enlistment date: 2 Apr 1812 - Place: Baltimore - Period: 5 Yrs - BLW 13962-160-12 - Prisoner of War (Halifax), sent to England, discharged on 20 Apr 1814 on Surgeon's Certificate.

Courtney, John - Private - 36th Infantry - Company: Samuel Raisin - BLW 2675-160-12.

Covington, Leonard - Brigadier General - 1st Light Dragoons - Born: Maryland - Commissioned as a cornet, Light Dragoons, on 14 Mar 1792; promoted to lieutenant on 25 Oct 1792; promoted to captain on 11 Jul 1794; resigned on 22 Sep 1795; commissioned as a lieutenant colonel, light dragoons, on 9 Jan 1809; promoted to colonel on 15 Feb 1809; promoted to brigadier general on 1 Aug 1813; died on 14 Nov 1813 from wounds received during the Battle of Crystler's Field on 11 Nov 1813.

Cowan, John - Private - 38th Infantry - Company: John Rothrock - Age: 24 - Height: 5' 9 1/2" - Born: Salem, Salem Cty, MA - Trade: Seaman - Enlistment date: 5 Apr 1814 - Place: Craney Island, VA - Period: War - BLW 4043-160-12 - Discharged at Craney Island, VA, on 15 Mar 1815.

Coward, William - Private - 2nd Light Dragoons - Company: Samuel Harris - Other regiment: Corps of Artillery - Age: 25 - Height: 5' 9" - Born: Easton, MD - Trade: Mariner - Enlistment date: 11 Aug 1813 - Place: Easton, MD - Period: 5 Yrs - Deserted at Fort Columbus, NY, on 21 Feb 1816.

Cowley, Charles - Private - 44th Infantry - Company: Ferdinand Amelung - Age: 40 - Height: 5' 8 1/2" - Born: Maryland - Trade: Blacksmith - Enlistment date: 12 Mar 1814 - Period: War - BLW 21882-160-12 - Discharged at New Orleans on 11 Apr 1815.

Cox, John - Private - 36th Infantry - Age: 24 - Height: 5' 10" - Born: Chester Cty, PA - Enlistment date: 1 May 1813 - Place: Baltimore.

Cox, John - Private - 35th Infantry - Company: James Belches - Age: 22 - Height: 5' 10" - Born: Caroline Cty, MD - Trade: Cooper - Enlistment date: 18 Oct 1814 - Place: Norfolk, VA - Period: War - BLW 8779-160-12 - Discharged at Norfolk, VA, on 15 Mar 1815.

Cox, Levi - Private - 38th Infantry - Age: 21 - Height: 5' 10" - Born: Baltimore - Trade: Farmer - Enlistment date: 19 Aug 1814 - Place: Baltimore - Period: War - Deserted.

Cox, William L. - Private - 17th Infantry - Company: Henry Crittenden - Age:

37 - Height: 5' 9" - Born: Charles Cty, MD - Trade: Laborer - Enlistment date: 8 Jun 1814 - Period: War.

Coxen, Washington - Private - 36th Infantry - Company: Thomas Carbery - Age: 26 - Height: 5' 9 1/2" - Born: Prince George's Cty, MD - Trade: Farmer - Enlistment date: 5 Apr 1814 - Place: Georgetown, DC - Period: War.

Coxton, George - Private - 14th Infantry - Company: David Cummings - Age: 23 - Height: 5' 10" - Born: Kentucky - Trade: Ship's carpenter - Enlistment date: 6 Feb 1815 - Place: Baltimore - Period: War - Deserted on 6 Feb 1815.

Cozart, Joseph - Private - 7th Infantry - Company: Jacob Miller - Age: 36 - Height: 5' 7" - Born: Maryland - Enlistment date: 4 May 1814 - Place: Hopkinsville, KY - Period: War - BLW 20300-160-12 - Discharged at Hopkinsville, KY, on 14 Apr 1815.

Cozens, George - Private - 14th Infantry - Company: Reuben Gilder - Enlistment date: 24 Mar 1813 - Period: 5 Yrs - Pension: Heirs received half pay for five years in lieu of land bounty - Discharged on 12 Jul 1814 on Surgeon's Certificate because of wounds.

Crabb, Richard J. - Captain - 12th Infantry - BLW 47638-160-50 - Commissioned as a captain on 20 Mar 1812; resigned on 23 Aug 1812.

Crabbin, Henry - Private - 1st Artillery - Company: Ethan Allen - Other regiment: Corps of Artillery - Age: 27 - Height: 5' 8 1/2" - Born: Kent Cty, MD - Trade: Sailor - Enlistment date: 27 Feb 1813 - Place: Fort Nelson, VA - Period: 5 Yrs - BLW 15739-160-12 - Discharged at Fort Nelson, VA, on 27 Feb 1818.

Crabbin, Hynson - Third Lieutenant - 38th Infantry - Commissioned as a 3rd lieutenant on 20 May 1813; resigned on 7 Jan 1814.

Crabtree, John - Private - 14th Infantry - Prisoner of War (Halifax), discharged from Halifax on 3 Feb 1814.

Craft, George B. - Private - 12th Infantry - Company: Andrew Madison - Other regiment: 8th Infantry (New) - Age: 21 - Height: 5' 8 1/2" - Born: Baltimore - Trade: Blacksmith - Enlistment date: 16 Aug 1812 - Place: Winchester - Period: 5 Yrs - BLW 15040-160-12 - Discharged at Fort Osage, MO, on 16 Aug 1817.

Craig, Benjamin - Private - 2nd Artillery - Company: Frederick Evans - Age: 25 - Height: 5' 7" - Born: Maryland - Trade: Weaver - Discharged on 24 Mar 1815 and re-enlisted on 12 Nov 1814 at Fort McHenry for the war.

Craig, George - Corporal - US Volunteers - Company: Stephen Moore - Enlistment date: 8 Sep 1812 - Period: 1 Yr.

Craig, James - Private - 14th Infantry - Company: James Britton - Age: 39 - Height: 5' 11" - Born: Philadelphia - Enlistment date: 11 May 1813 - Period: 5 Yrs - BLW 8213-160-12 - Discharged at Burlington, VT, on 14 Apr 1814 due to lameness of his left knee.

Craig, John - Seaman - US Sea Fencibles - Company: John Gill - Age: 37 - Height: 5' 1/2" - Born: Dorchester Cty, MD - Enlistment date: 5 Jan 1814 - Period: 1 Yr - Deserted on 6 Oct 1815.

Craig, Samuel - Private - 19th Infantry - Company: Joel Collins - Age: 44 - Height: 5' 6" - Born: Maryland - Trade: Laborer - Enlistment date: 19 Mar

1814 - Place: Detroit - Period: War - BLW 4390-160-12 - Discharged at Detroit on 20 Jul 1815.

Cramer, Baltzer - Sergeant - 4th Rifles - Company: Robert Scott - Other regiment: Rifles - Age: 36 - Height: 5' 6" - Born: Maryland - Trade: Saddler - Enlistment date: 13 Apr 1814 - Period: 5 Yrs - Discharged on 12 Apr 1819.

Cramer, Richard - Private - 19th Infantry - Age: 36 - Height: 5' 10 1/2" - Born: Baltimore Cty - Enlistment date: 10 Jul 1812 - Place: St. Clairsville, OH - Period: 5 Yrs.

Crandell, Joseph - Private - 1st Infantry - Company: Moses Hook - Age: 37 - Height: 5' 6" - Born: Maryland - Trade: Shoemaker - Enlistment date: 25 May 1803 - Period: 5 Yrs - Re-enlisted on 27 Feb 1808; prisoner of war at Fort Independence, MA, 31 Mar 1813.

Crane, Joseph L. - Corporal - US Volunteers - Company: Stephen Moore - Enlistment date: 8 Sep 1812 - Period: 1 Yr.

Cranny, Edward - Private - 14th Infantry - Prisoner of War (Halifax), sent to England.

Crasser, Jacob - Private - 19th Infantry - Company: Asabeal Nearing - Other regiment: 17th Infantry - Age: 21 - Height: 5' 8 1/2" - Born: Maryland - Enlistment date: 7 Jul 1812 - Place: Warren, Trumbull Cty, OH - Period: 5 Yrs - Pension: Land bounty to Henry Crasser, father & only heir at law of Jacob Crasser - BLW 25408-160-12 - Died in Oct 1813.

Craventon, George - Private - 14th Infantry - Prisoner of War (Halifax), discharged from Halifax on 3 Feb 1814.

Crawford, James - Private - 36th Infantry - Age: 26 - Height: 5' 6" - Born: Ireland - Enlistment date: 21 May 1813 - Place: Baltimore.

Crawford, John - Private - 12th Infantry - Company: Thomas Sangster - Other regiment: 8th Infantry (New) - Age: 40 - Height: 5' 6" - Born: Scotland - Trade: Barber - Enlistment date: 14 Jun 1812 - Place: Baltimore or Alexandria, DC - Period: 5 Yrs - Discharged on 17 Jun 1817.

Crawford, Thomas - Private - 2nd Artillery - Company: Nathan Towson - Pension: Old War IF-28503.

Cray, James - Private - 14th Infantry - Company: David Cummings - Age: 28 - Height: 5' 7 1/2" - Born: Ireland - Trade: Shoemaker - Enlistment date: 24 Sep 1814 - Place: Baltimore - Period: War - Deserted on 1 Oct 1814.

Craycraft, Balden - Sergeant - 2nd Infantry - Company: Robert Turner - Other regiment: 1st Infantry (New) - Height: 5' 8 1/2" - Born: Charles Cty, MD - Trade: Farmer - Enlistment date: 1803 - Re-enlisted on 1 Apr 1808; discharged on 1 Feb 1818.

Crayton, Hugh - Private - 12th Infantry - Company: Andrew Madison - Other regiment: 8th Infantry (New) - Age: 25 - Height: 5' 8" - Born: Frederick Cty, MD - Trade: Plasteror - Enlistment date: 1 Aug 1812 - Place: Charles Town, VA - Period: 5 Yrs - Discharged on 18 Aug 1817.

Crea, Hugh - Seaman - US Sea Fencibles - Company: Simmones Bunbury - Age: 29 - Height: 5' 11" - Born: Baltimore City - Enlistment date: 27 Nov 1813 - Period: 1 Yr - Discharged on 26 Nov 1814.

Creager, Daniel - Musician - 38th Infantry - Company: John Rothrock -

By War of 1812 Soc. in MD

Enlistment date: 2 Aug 1813 - Period: 1 Yr - Discharged at Craney Island, VA, on 2 Aug 1814.

Creager, George - Drummer - 38th Infantry - Company: John Rothrock - Enlistment date: 2 Aug 1813 - Period: 1 Yr.

Creagh, John - Sergeant - 38th Infantry - Company: George Keyser - Enlistment date: 7 Aug 1813 - Period: 1 Yr.

Creamer, David - Private - 1st Infantry - Company: Eli Clemson - Other regiment: 3rd Infantry (New) - Age: 22 - Height: 5' 11" - Born: Maryland - Trade: Carpenter - Enlistment date: 12 Jan 1807 - Place: Little York, PA - Period: 5 Yrs.

Creaven, Nathaniel - Private - 36th Infantry - Age: 35 - Height: 5' 4 1/2" - Born: Chester, Chester Cty, PA - Enlistment date: 3 May 1813 - Place: Baltimore - Period: 1 Yr.

Creback, Joseph - Private - 38th Infantry - Company: James Haslett - Enlistment date: 28 May 1813 - Period: 1 Yr - Died on 9 Sep 1813.

Creighton, William - Private - 38th Infantry - Company: John Buck - Enlistment date: 19 Aug 1813 - Period: 1 Yr - Discharged at Camp Winder, VA, on 19 Aug 1814.

Cremer, David - Private - 14th Infantry - Company: Samuel Lane - Age: 30 - Height: 5' 9" - Born: Baltimore - Enlistment date: 27 May 1812 - Period: 5 Yrs - Pension: Heirs received half pay for five years in lieu of land bounty - Killed on 1 Dec 1812 at Conjocta Creek, NY, accident.

Cremonthy, John - Private - 36th Infantry - Company: Samuel Raisin - Age: 18 - Height: 5' 5" - Born: France - Trade: Mariner - Enlistment date: 21 Oct 1814 - Place: Georgetown, DC - Period: War - Pension: Land bounty to Susan Cremonthy, widow & other heirs at law of John Cremonthy - BLW 1064-160-12 - Discharged at Washington, DC, on 20 Mar 1815.

Cressell, George - Corporal - 38th Infantry - Company: James Haslett - Enlistment date: 14 Jun 1813 - Period: 1 Yr.

Creswell, Hartford - Private - 38th Infantry - Company: John Rothrock - Enlistment date: 28 May 1813 - Period: 1 Yr - Died on 22 Jan 1814.

Creswell, John - Sergeant - 14th Infantry - Company: David Cummings - Enlistment date: 15 Oct 1812 - Period: 5 Yrs - Discharged on 17 Jan 1815 for lameness.

Crisher, Conrad - Private - 36th Infantry - Age: 21 - Height: 5' 6" - Born: Frederick Cty, MD - Enlistment date: 4 May 1813 - Place: Westminster, MD.

Crisman, John M. - 14th Infantry - Company: Kenneth McKenzie - Discharged on 6 Jul 1814.

Crissel, James - Private - 38th Infantry - Company: Charles Stansbury - Age: 26 - Height: 5' 9" - Born: Baltimore - Trade: Cooper - Enlistment date: 13 Oct 1814 - Place: Baltimore - Period: War - Discharged at Baltimore on 6 Apr 1815.

Crissel, Samuel - Drummer - 17th Infantry - Company: John Chunn - Age: 18 - Height: 5' 4" - Born: Harford Cty, MD - Trade: Plasterer - Enlistment date: 15 Jul 1814 - Place: Lancaster, Fairfield County, OH - Period: War - BLW

Maryland Regulars in War of 1812

7098-160-12 - Discharged at Chillicothe, OH, on 7 Jun 1815.

Crissell, George W. - Private - 38th Infantry - Company: James Smith - Age: 28 - Height: 5' 11" - Born: Bellaire, MD - Trade: Cooper - Enlistment date: 8 Mar 1814 - Place: Craney Island, VA - Period: War - BLW 9914-160-12 - Discharged at Craney Island, VA, on 15 Mar 1815.

Crissell, James - Private - 38th Infantry - Company: Charles Stansbury - Enlistment date: 13 Oct 1814 - Pension: Wife Temperance Gordon, WO-2060, WC-3175; married on 27 Oct 1812, Harford Cty, MD; soldier died on 2 May 1861, Baltimore Cty, MD - BLW 15363-160-12 - Discharged on 31 Dec 1814.

Crittendon, Lewis L. - Private - 38th Infantry - Company: James Smith - Age: 25 - Height: 5' 7" - Born: Essex, VA - Trade: Carpenter - Enlistment date: 8 Mar 1814 - Place: Norfolk, VA - Period: War - BLW 4347-160-12 - Discharged at Craney Island, VA, on 15 Mar 1815.

Crock, David - Private - 36th Infantry - Company: Thomas Carbery - Enlistment date: 17 Jan 1814 - Period: 1 Yr - Discharged on 16 Jan 1815.

Crocken, James - Seaman - US Sea Fencibles - Company: William Addison.

Croft, John - Private - 5th Infantry - Company: William Henshaw - Other regiment: 3rd Infantry (New) - Age: 22 - Height: 5' 6" - Born: Chambersburg, Franklin County, PA - Trade: Saddler - Enlistment date: 24 Jul 1812 - Place: Hagerstown, MD - Period: 5 Yrs - Discharged on 24 Jul 1817.

Cromley, George - Private - 14th Infantry - Company: Samuel Lane - Age: 43 - Height: 5' 8" - Born: Virginia - Enlistment date: 13 May 1812 - Place: Virginia - Period: 5 Yrs - Died at Sackets Harbor, NY, on 21 Dec 1813.

Crosby, John - Private - 36th Infantry - Company: Thomas Carbery - Enlistment date: 4 Oct 1814 - Place: Georgetown, DC - Period: War.

Crosby, Rufus - Sergeant - 26th Infantry - Company: William Bezeau - Age: 29 - Height: 5' 10 1/2" - Born: Bennington, VT - Trade: Shoemaker - Enlistment date: 14 Oct 1814 - Place: Baltimore - Period: War - BLW 19567-160-12 - Discharged at Philadelphia on 21 Mar 1815.

Croskey, Arthur - Private - 14th Infantry - Company: David Cummings - Age: 23 - Height: 5' 7" - Born: Maryland - Trade: Hatter - Enlistment date: 6 May 1814 - Place: Baltimore - Period: War - Discharged on 16 Mar 1815.

Crosley, Abraham - Sergeant - 38th Infantry - Company: Thomas Sangster - Other regiment: 4th Infantry (New) - Age: 33 - Height: 6' - Born: Delaware Cty, PA - Trade: Chair maker - Enlistment date: 9 Dec 1814 - Place: Baltimore - Period: 5 Yrs - BLW 24961-160-12 - Discharged on 8 Nov 1819.

Crosley, James - Musician - 38th Infantry - Company: Thomas Sangster - Other regiment: 4th Infantry (New) - Age: 12 - Height: 4' - Born: Philadelphia - Trade: Drummer - Enlistment date: 9 Nov 1814 - Place: Baltimore - Period: 5 Yrs - BLW 24962-160-12 - Discharged at Montpelier, AL, on 8 Nov 1819.

Cross, Archibald - Corporal - 2nd Infantry - Company: Edmund Gaines - Age: 21 - Height: 6' - Born: Prince George's Cty, MD - Trade: Blacksmith - Enlistment date: 13 Jul 1807 - Place: Washington, DC - Period: 5 Yrs.

By War of 1812 Soc. in MD

Cross, George - Private - 38th Infantry - Age: 34 - Height: 5' 7 1/2" - Born: England - Trade: Baker - Enlistment date: 11 Mar 1814 - Period: War - Deserted on 13 Mar 1184.

Cross, George - Private - 36th Infantry - Age: 20 - Height: 5' 6" - Born: Manchester, Essex Cty, MA - Enlistment date: 28 Apr 1813 - Place: Baltimore - Period: 1 Yr - Discharged on 5 Nov 1813 and transferred to the U.S. Navy.

Cross, Horatio - Private - 2nd Artillery - Company: Frederick Evans - Other regiment: Corps of Artillery - Age: 38 - Height: 5' 9" - Born: Maryland - Trade: Brick layer - Enlistment date: 8 Apr 1812 - Place: Greenleaf Point, DC - Period: 5 Yrs - Discharged at Washington, DC, on 9 Apr 1817.

Cross, Israel - Private - 27th Infantry (OH) - Company: Alexander Hill - Other regiment: 19th Infantry - Age: 21 - Height: 5' 10 1/2" - Born: Maryland or Virginia - Trade: Farmer - Enlistment date: 21 May 1814 - Place: Athens, Athens Cty, OH - Period: War - BLW 16414-160-12 - Discharged at Chillicothe, OH, on 6 Jun 1815.

Cross, Jesse - Private - 38th Infantry - Company: James Hook - Age: 18 - Height: 5' 2" - Born: Washington, DC - Trade: Painter - Enlistment date: 29 Apr 1814 - Place: Baltimore - Period: War - BLW 5496-160-12 - Discharged at Fort Covington, MD, on 31 Mar 1815.

Cross, Jonathan - Private - 2nd Rifles - Company: Benjamin Johnson - Age: 22 - Height: 5' 6" - Born: Montgomery Cty, MD - Trade: Farmer - Enlistment date: 9 May 1814 - Place: Michigan - Period: War - BLW 25664-160-12 - Discharged at Detroit on 30 Jun 1815.

Cross, Middleton - Sergeant - 36th Infantry - Age: 32 - Height: 6' 2 1/4" - Born: Maryland - Trade: Sawyer - Enlistment date: 26 Jan 1815 - Period: War - Pension: Land bounty to Mary Barkley, daughter & other heirs at law of William Cross, who was the father & only heir at law of Middleton Cross - BLW 1068-320-14 - Died on 28 Feb 1815.

Cross, Robert - Private - 22nd Infantry - Company: William Morrow - Other regiment: 2nd Infantry (New) - Age: 47 - Height: 6' - Born: Prince George's Cty, MD - Trade: Laborer - Enlistment date: 18 Nov 1813 - Place: Waynesburgh, PA - Period: 5 Yrs - Discharged on 17 Sep 1816 after furnishing a substitute.

Cross, Thomas - Private - 14th Infantry - Company: Samuel Lane - Age: 25 - Height: 5' 8" - Born: F. County, PA - Enlistment date: 30 Jun 1812 - Place: F. County, PA - Period: 18 Mos - Discharged on 30 Dec 1813.

Cross, Thomas - Private - 12th Infantry - Company: Thomas Sangster - Born: Ireland - Enlistment date: 17 Jul 1812 - Place: Baltimore - Period: 18 Mos - Discharged on 30 Dec 1813.

Crossley, William - Sergeant - 12th Infantry - Company: Thomas Sangster - Other regiment: 4th Infantry (New) - Age: 25 - Height: 5' 8 1/2" - Born: Washington Cty, MD - Trade: Carpenter - Enlistment date: 5 Mar 1814 - Place: Lexington, VA - Period: 5 Yrs.

Crossman, Joseph - Matross - 2nd Artillery - Company: Nathan Towson - BLW 15482-160-12.

Maryland Regulars in War of 1812

Crow, Edward - Corporal - 17th Infantry - Company: David Holt - Age: 20 - Height: 5' 6 1/2" - Born: Baltimore City - Trade: Tanner - Enlistment date: 2 Mar 1814 - Place: Franklinton, Franklin Cty, OH - Period: War - BLW 10778-160-12 - Discharged at Chillicothe, OH, on 4 Jun 1814.

Crow, James - Sergeant - 38th Infantry - Company: John Mowton - Enlistment date: 8 Jun 1813 - Period: 1 Yr - Pension: Land bounty to Owen Crow, brother and other heirs at law of James Crow - BLW 25521-160-12 - Re-enlisted on 8 Mar 1814; died on 5 Feb 1815 at Craney Island, VA.

Crowell, Peter - Private - 14th Infantry - Company: David Cummings - Age: 20 - Height: 5' 5" - Born: Martinsburg, VA - Trade: Farmer - Enlistment date: 10 Nov 1814 - Place: Snowden's Camp, VA - Period: 5 Yrs.

Crowley, Edward - Private - 14th Infantry - Company: Thomas Montgomery - Other regiment: 4th Infantry (New) - Age: 22 - Height: 5' 9" - Born: Ireland - Trade: Wheelwright - Enlistment date: 29 May 1812 - Place: Bladensburg, MD - Period: 5 Yrs - BLW 12779-160-12 - Deserted at Fort Severn, MD, on 18 Jul 1815.

Crozier, John - Private - 1st Infantry - Company: John Whistler - Other regiment: 3rd Infantry (New) - Height: 5' 7" - Born: Hagerstown, MD - Trade: Clerk - Enlistment date: 1803 - Re-enlisted on 2 Jul 1808, 18 Sep 1813 and 17 Jun 1818.

Crummel, Joseph - Private - 14th Infantry - Company: Thomas Montgomery - Enlistment date: 8 Jun 1812 - Period: 5 Yrs - Pension: Land bounty to John E. Crummel nephew and other heirs at law of Joseph Crummel - BLW 27562-160-42.

Cudmore, Paul - Private - 36th Infantry - Company: Thomas Carbery - Enlistment date: 22 Nov 1813 - Period: 1 Yr - Discharged on 21 Nov 1814.

Culbert, James - Private - 2nd Artillery - Company: Nathan Towson - Pension: Placed on the pension rolls on 3 Jul 1823.

Culbertson, Joseph - Private - 14th Infantry - Company: William McIlvane - Age: 32 - Height: 5' 10" - Born: Chester Cty, PA - Trade: Farmer - Enlistment date: 18 Mar 1813 - Period: War - Pension: Old War IF-2499 - BLW 553-160-12 - Discharged at Greenbush, NY, on 3 May 1815.

Cullany, James - Private - 14th Infantry - Company: Thomas Montgomery - Pension: Land bounty to Ann Cullary, daughter and only heir at law of James Cullany - BLW 25385-160-12.

Cullins, James M. - Sergeant - 14th Infantry - Company: Joseph Marshall - Age: 24 - Height: 5' 6 1/4" - Born: Carolina Cty, VA - Trade: Farmer - Enlistment date: 20 Jun 1814 - Place: Baltimore - Period: War - Discharged at Greenbush, NY, on 3 May 1815.

Cullins, Thomas - Private - 14th Infantry - Age: 27 - Height: 5' 8" - Born: Baltimore - Trade: Paper maker - Enlistment date: 25 Oct 1814 - Place: Baltimore - Period: War - BLW 494-160-12 - Discharged at Baltimore on 16 Mar 1815.

Cully, George - Private - 36th Infantry - Age: 19 - Height: 6' 6 3/4" - Born: Mathews Cty, VA - Enlistment date: 26 Apr 1813 - Place: Baltimore - Period: 1 Yr - Deserted on 9 May 1813.

By War of 1812 Soc. in MD

- Cummings, David - Captain - 14th Infantry - Age: 36 - Pension: Old War IF-3242 - Commissioned as a 1st lieutenant on 12 Mar 1812; promoted to captain on 13 Mar 1813; Prisoner of War (Dartmouth), sent to Prison Ship Success; discharged on 15 Sep 1815.
- Cummings, Mark - Private - 36th Infantry - Other regiment: 38th Infantry - Age: 38 - Height: 5' 5" - Born: Anne Arundel Cty, MD - Trade: Weaver - Enlistment date: 30 Jul 1813 - Place: Havre de Grace, MD - Period: 1 Yr - Re-enlisted at Baltimore on 26 Sep 1814 for the war in Captain Stansbury's Company, 38th Infantry.
- Cummings, William - Private - 17th Infantry - Company: William Bradford - Age: 33 - Height: 5' 1/4" - Born: Maryland - Trade: Farmer - Enlistment date: 1 Mar 1813 - Period: 18 Mos - Dishonorable discharged on 11 Sep 1814 at Fort Malden, UC, for cowardly conduct at Fort Meigs, OH.
- Cunard, Jarrett - Corporal - 38th Infantry - Company: John Mowton - Age: 24 - Height: 5' 7" - Born: Loudoun Cty, VA - Trade: Tanner - Enlistment date: 20 Jun 1814 - Place: Norfolk, VA - Period: War - Pension: Wife Leare Pyle, SO-1189, SC-713, WO-13294, WC-7590; married on 16 Mar 1829 or 13 Mar 1835, Jackson Cty, OH; soldier died on 24 Oct 1871, Jackson Cty, OH - BLW 1612-160-12 - Discharged at Craney Island, VA, on 2 Jun 1815.
- Cunning, John - Private - 14th Infantry - Company: Thomas Montgomery - Enlistment date: 25 May 1812 - Period: 5 Yrs - Pension: Land bounty to Barney Cunning, brother and other heirs at law of John Cunning - BLW 25482-160-12 - Died at Black Rock, NY, on 28 Dec 1812.
- Cunningham, John - Private - 38th Infantry - Age: 20 - Height: 5' 10" - Born: Pendleton, VA - Enlistment date: 18 Sep 1814 - Place: Craney Island, VA - Period: War - BLW 25640-160-12 - Discharged at Craney Island, VA, on 15 Mar 1815.
- Cunningham, John - Private - 12th Infantry - Company: Thomas Post - Other regiment: 8th Infantry (New) - Age: 40 - Height: 5' 10" - Born: Baltimore - Trade: Shoemaker - Enlistment date: 9 Mar 1814 - Place: Wheeling, VA - Period: 5 Yrs - Discharged at Fort Charlotte, AL, on 9 Mar 1819.
- Cunningham, John - Musician - 38th Infantry - Company: John Buck - BLW 25640-160-12.
- Cunningham, Kelly - Private - 14th Infantry - Prisoner of War (Halifax), sent to England.
- Cunningham, William - Private - 38th Infantry - Company: John Rothrock - Age: 25 - Height: 5' 9" - Born: Elk Ridge, MD - Trade: Tobacconist - Enlistment date: 3 Jun 1813 - Period: 1 Yr - BLW 15172-160-12 - Re-enlisted on 2 Jun 1814 for the war; discharged on 15 Mar 1815 at Craney Island, VA.
- Curlington, James - Private - 26th Infantry - Company: William Bezeau - Age: 23 - Height: 5' 6" - Born: Philadelphia - Trade: Tailor - Enlistment date: 13 Jan 1815 - Place: Baltimore - Period: War - Deserted on 31 Jan 1815.
- Curll, William - Private - 42nd Infantry - Company: Jonathan Robinson - Other regiment: Corps of Artillery - Age: 27 - Height: 5' 10" - Born: Nottingham, MD - Trade: Laborer - Enlistment date: 24 Nov 1814 - Place:

Wilmington, DE - Period: 5 Yrs - Discharged at Fort Severn, MD, on 23 Nov 1819 and then re-enlisted.

Curran, Joshua - Private - 38th Infantry - Company: Shepperd Leakin - Age: 20 - Height: 5' 9" - Born: Jefferson Cty, VA - Trade: Gunsmith - Enlistment date: 11 Mar 1814 - Period: War - Discharged at Baltimore on 30 Sep 1814 for being a minor.

Curry, Joseph - Private - 36th Infantry - Company: Thomas Carbery - Age: 17 - Height: 5' 2 1/4" - Born: St. Mary's Cty, MD - Trade: Farmer - Enlistment date: 29 Mar 1814 - Place: Leonardstown, MD - Period: War - BLW 369-160-12 - Discharged at Baltimore on 31 Mar 1815.

Curtin, William - Private - 36th Infantry - Company: Samuel Raisin - Age: 22 - Height: 5' 6 1/2" - Born: Prince George's Cty, MD - Trade: Laborer - Enlistment date: 11 Nov 1814 - Period: War - Pension: SC-22960, SO-10812 - BLW 2017-160-12 - Discharged at Washington, DC, on 20 Mar 1815.

Curtis, George - Sergeant - 14th Infantry - Company: Samuel Lane - Enlistment date: 11 Jan 1812 - Period: 18 Mos - Died at Lewiston, NY, from neck wounds received on 24 Jun 1813 at Beaver's Dam.

Curtis, Hezekiah - Private - 36th Infantry - Company: Samuel Raisin - Age: 25 - Height: 5' 7 1/2" - Born: Prince George's Cty, MD - Trade: Laborer - Enlistment date: 21 Aug 1814 - Period: War - Discharged at Washington, DC, on 20 Mar 1815.

Curtis, John - Seaman - US Sea Fencibles - Company: William Addison.

Curtis, Uria - Private - 14th Infantry - Prisoner of War (Halifax), discharged from Halifax on 3 Feb 1814.

Curtis, William - Seaman - US Sea Fencibles - Company: Simmones Bunbury - Age: 29 - Height: 5' 9 1/4" - Born: Somerset Cty, MD - Enlistment date: 30 Dec 1813 - Deserted on 2 Jan 1814.

Cutting, Lewis - Recruit - 26th Infantry - Company: William Bezeau - Race: 21 - Age: 21 - Height: 5' 3" - Born: Maryland - Trade: Laborer - Enlistment date: 4 Jan 1815 - Place: Philadelphia - Period: War - Listed on the Descriptive Roll of Colored Men, 1 Apr 1815.

D

Dailey, Christopher - Private - 38th Infantry - Company: John Rothrock - Age: 35 - Height: 5' 8 1/2" - Born: Baltimore - Trade: Farmer - Enlistment date: 29 Aug 1814 - Place: Craney Island, VA - Period: War - BLW 22524-160-12 - Discharged at Craney Island, VA, on 15 Mar 1815.

Dailey, Jehu (or John) - Corporal - 19th Infantry - Company: Angus Langham - Age: 25 - Height: 5' 8" - Born: Baltimore Cty - Trade: Laborer - Enlistment date: 23 Feb 1814 - Place: New Lisbon, OH - Period: War - BLW 1088-160-12 - Prisoner of War, exchanged 28 Apr 1814; discharged on 6 Jun 1815 at Chillicothe, OH.

Dailey, William - Private - 14th Infantry - Company: James McDonald - Age: 38 - Height: 6' - Born: Cecil Cty, MD - Trade: Carpenter - Enlistment date: 24 Feb 1814 - Period: 5 Yrs.

Daily, James - Private - 38th Infantry - Company: John Brookes - Enlistment

By War of 1812 Soc. in MD

date: 17 Jul 1813 - Period: 1 Yr.

Daily, John - Private - 38th Infantry - Company: James Smith - Age: 49 - Height: 5' 7 1/2" - Born: County Monahan, Ireland - Trade: Weaver - Enlistment date: 9 Jul 1814 - Place: Norfolk, VA - Period: War.

Daily, William - Private - 22nd Infantry - Company: Thomas Lawrence - Age: 36 - Height: 5' 9" - Born: Maryland - Trade: Hammerman - Enlistment date: 18 Feb 1813 - Place: Uniontown, PA - Period: 5 Yrs - Discharged at Pittsburgh on 10 Oct 1813 for a rupture.

Daley, John - Private - US Volunteers - Company: Stephen Moore - Enlistment date: 8 Sep 1812 - Period: 1 Yr.

Dalton, Edward - Seaman - US Sea Fencibles - Company: John Gill - Age: 25 - Height: 5' 8" - Born: Baltimore City - Enlistment date: 11 Jan 1814 - Period: 1 Yr.

Danford, James - Private - 36th Infantry - Pension: Placed on the pension roll on 28 Nov 1817.

Danford, James - Private - 36th Infantry - Company: Joseph Hook - Pension: Old War IF-20716 - BLW 13343-160-12.

Daniels, Benjamin - Private - 38th Infantry - Company: John Buck - Enlistment date: 27 Jul 1813 - Period: 1 Yr - Discharged at Camp Marlborough on 27 Jul 1814.

Daniels, John - Artificer - 2nd Artillery - Company: Frederick Evans - Age: 34 - Height: 5' 7 1/2" - Born: Virginia - Trade: Carpenter - Enlistment date: 25 Nov 1813 - Place: Fort McHenry - Period: 5 Yrs - Discharged on 27 Feb 1817.

Daniels, William - Private - 38th Infantry - Company: James Smith - BLW 4272-160-12.

Darley, Thomas - Private - 38th Infantry - Company: John Mowton - Age: 30 - Height: 5' 4" - Born: Maryland - Trade: Cooper - Enlistment date: 25 Apr 1814 - Place: Craney Island, VA - Period: War - BLW 2710-160-12 - Discharged at Craney Island, VA, on 15 Mar 1815.

Darling, John - Private - 15th Infantry - Company: Henry Van Dalsem - Age: 32 - Height: 5' 10 1/2" - Born: Baltimore - Trade: Coppersmith - Enlistment date: 13 Feb 1810 - Place: Philadelphia - Period: 5 Yrs - Discharged at Batavia, NY, on 13 Feb 1815.

Darling, Joseph - Private - 36th Infantry - Company: Joseph Merrick - Age: 20 - Height: 5' 5" - Born: Boston - Enlistment date: 3 May 1813 - Place: Baltimore - Transferred to the U.S. Navy on 26 Oct 1813.

Darling, Robert - Private - 5th Infantry - Company: Richard Whartenby - Enlistment date: 2 Apr 1812 - Place: Baltimore - Period: 5 Yrs.

Darnell Jr., John - Second Lieutenant - 5th Infantry - Other regiment: 2nd Infantry - Commissioned as an ensign, 5th Infantry, on 13 Nov 1812; promoted to 3rd lieutenant on 12 Mar 1813; resigned on 7 Dec 1813; commissioned as 3rd lieutenant, 2nd Infantry, on 31 Mar 1814; promoted to 2nd lieutenant on 1 May 1814; discharged on 15 Jun 1815.

Darrette, Simon - Recruit - 26th Infantry - Company: William Bezeau - Age: 31 - Height: 5' 5" - Born: Maryland - Trade: Mason - Enlistment date: 2 Feb

1815 - Place: Philadelphia - Period: War.

Darringer, James - Private - 5th Infantry - Company: Leroy Opie - Other regiment: 3rd Infantry (New) - Age: 30 - Height: 5' 7 1/2" - Born: Pennsylvania - Trade: Gunsmith - Enlistment date: 30 Jun 1813 - Place: Baltimore - Period: 5 Yrs - Died at Fort Howard, WI, on 16 or 26 Nov 1817.

Dasey, James - Private - 36th Infantry - Place: Cumberland, MD.

Dashell, Levi - Private - 38th Infantry - Company: Charles Stansbury - Age: 26 - Height: 5' 8 1/2" - Born: Somerset Cty, MD - Trade: Blacksmith - Enlistment date: 24 May 1814 - Period: War - Pension: Old War MC- 980; heirs received half pay for five years in lieu of land bounty - Died on 12 Feb 1815.

Dashield, Peter - Private - 14th Infantry - Company: Kenneth McKenzie - Enlistment date: 25 Oct 1812 - Period: 5 Yrs - Prisoner of War (Quebec); died at Quebec on 13 Oct 1813.

Datis, Lewis - Private - 14th Infantry - Company: James McDonald - Age: 36 - Height: 5' 4 1/2" - Born: Isle of France - Enlistment date: 14 Feb 1814 - Period: 5 Yrs - Deserted at Batavia, NY, on 30 Sep 1814.

Daugherty, John - Private - 38th Infantry - Company: Charles Stansbury - BLW 18215-160-12.

Daughtry, John - Private - 35th Infantry - Company: Benjamin Hardaway - Age: 22 - Height: 5' 11" - Born: Maryland or Northampton, NC - Trade: Silversmith - Enlistment date: 23 Feb 1814 - Place: Norfolk, VA - Period: War - Deserted on 22 Oct 1814.

Davage, Thomas - Ensign - 14th Infantry - Company: Isaac Barnard - Age: 28 - Height: 5' 11" - Born: England - Trade: Cordwainer - Enlistment date: 22 Apr 1812 - Place: Chester Cty, PA - Period: 5 Yrs - BLW 25660-160-12 - Promoted to ensign from sergeant on 12 Mar 1813; resigned at Chateauguay, NY, on 1 Feb 1814.

Davenport, Francis - Private - 36th Infantry - Company: Joseph Hook - Age: 18 - Height: 5' 9" - Born: Lancaster, PA - Trade: Farmer - Enlistment date: 1 Mar 1814 - Place: Annapolis - Period: War - Discharged at Fort Covington, MD, on 30 Mar 1815.

Davenport, Joseph - Private - 38th Infantry - BLW 16865-160-12.

Davenport, Robert - Private - 36th Infantry - Age: 42 - Height: 5' 8" - Born: Ireland - Enlistment date: 27 Jul 1813 - Place: Baltimore.

David, Elijah T. - Private - 20th Infantry - Company: John Stanard - Age: 39 - Height: 5' 7" - Born: Prince George's Cty, MD - Enlistment date: 21 Jul 1812 - Period: 5 Yrs - Died at Buffalo on 3 Nov 1812.

David, John - Private - 14th Infantry - Prisoner of War (Halifax), sent to England.

Davidge, Thomas - Third Lieutenant - 14th Infantry - Commissioned as an ensign on 12 May 1813; promoted to 3rd lieutenant on 11 Nov 1813; resigned on 1 Feb 1814.

Davidson, Daniel - Private - 38th Infantry - Company: Shepperd Leakin - Age: 38 - Height: 5' 9" - Born: Belfast, Ireland - Trade: Carpenter - Enlistment date: 17 Jul 1814 - Place: Baltimore - Period: War - BLW 231-160-12 -

By War of 1812 Soc. in MD

Discharged at Fort McHenry on 28 Mar 1815.

Davidson, John - Private - 38th Infantry - Age: 39 - Height: 5' 9" - Born: Pennsylvania - Trade: Cordwainer - Enlistment date: 13 Apr 1814 - Period: War - Deserted.

Davidson, John - Private - 14th Infantry - Company: William McIlvane - Enlistment date: 19 Nov 1812 - Period: 5 Yrs - Died on 20 Jun 1814 by drowning.

Davidson, John - Private - 14th Infantry - Company: Thomas Montgomery - Other regiment: 4th Infantry (New) - Age: 29 - Height: 5' 4" - Born: Baltimore - Trade: Carpenter - Enlistment date: 24 Jul 1812 - Place: Baltimore - Period: 5 Yrs - BLW 10329-160-12 - Prisoner of War (Halifax), captured at Beaver Dams on 24 Jun 1813, exchanged on 15 Apr 1814; discharged on Fort Hawkins on 24 Jul 1817.

Davidson, John - Private - 36th Infantry - Company: Samuel Raisin - Age: 22 - Height: 6' - Born: Botetourt Cty, VA - Trade: Shoemaker - Enlistment date: 1 Oct 1814 - Place: Richmond or Manchester, VA - Period: War - BLW 20271-160-12 - Discharged at Washington, DC, on 20 Mar 1815.

Davidson, William - Private - 19th Infantry - Company: William Gill - Other regiment: 3rd Infantry (New) - Age: 23 - Height: 5' 7" - Born: Baltimore Cty - Trade: Farmer - Enlistment date: 22 Jul 1814 - Place: New Lisbon, OH - Period: 5 Yrs - Discharged at Fort Howard, WI, on 22 Jul 1819.

Davis, Amos - Private - Light Artillery - Company: Benjamin Branch - Age: 34 - Height: 5' 6" - Born: Montgomery Cty, MD - Trade: Tailor - Enlistment date: 25 Mar 1812 - Place: Washington, DC - Period: 5 Yrs - BLW 8787-160-12 - Discharged at Fort Trumbull, CT, on 24 Mar 1817.

Davis, Amos - Private - 5th Infantry - Company: William Henshaw - Other regiment: 3rd Infantry (New) - Age: 21 - Height: 5' 6" - Born: Baltimore - Trade: Laborer - Enlistment date: 17 Feb 1813 - Place: Baltimore - Period: 5 Yrs - Discharged on 17 Feb 1818.

Davis, Archibald - Private - 36th Infantry - Company: Joseph Hook - Age: 23 - Height: 5' 7 3/4" - Born: Charles Cty, MD - Trade: Farmer - Enlistment date: 16 May 1814 - Place: Georgetown, DC - Period: War - BLW 107-160-12 - Discharged at Fort Covington, MD, on 30 Mar 1815.

Davis, Benjamin - Private - 14th Infantry - Prisoner of War (Halifax), discharged from Halifax on 3 Feb 1814.

Davis, Benjamin - Private - 17th Infantry - Company: Benjamin Sanders - Other regiment: 3rd Infantry (New) - Age: 37 - Height: 5' 10" - Born: Montgomery Cty, MD - Trade: Laborer - Enlistment date: 1 Feb 1814 - Place: Kentucky - Period: 18 Mos - Discharged on 1 Aug 1815.

Davis, Caleb - Sergeant - 36th Infantry - Company: Joseph Hook - Age: 23 - Height: 5' 7" - Born: Baltimore - Trade: Blacksmith - Enlistment date: 31 May 1813 - Place: Frederick Cty, MD - Period: 1 Yr - Re-enlisted at Annapolis on 1 Mar 1814 for the war.

Davis, Conner - Private - 14th Infantry - Company: David Cummings - Age: 26 - Height: 5' 9" - Born: Ireland - Trade: Laborer - Enlistment date: 30 Jun 1814 - Place: Baltimore - Period: War - Discharged at Baltimore on 14 Mar

Maryland Regulars in War of 1812

1815.

Davis, Cornelius - Private - 1st Artillery - Company: Heman Fay - Other regiment: Corps of Artillery - Age: 21 - Height: 5' 8" - Born: Fairfax Cty, VA - Trade: Hatter - Enlistment date: 28 Apr 1814 - Place: Annapolis - Period: War - BLW 22034-160-12 - Discharged at Washington, DC, on 9 May 1815.

Davis, David - Private - 20th Infantry - Company: John Duval - Age: 39 - Height: 5' 3" - Born: Maryland - Trade: Farmer - Enlistment date: 27 Mar 1814 - Period: War - BLW 22879-160-12 - Discharged at Buffalo on 20 Mar 1815.

Davis, David - Private - 38th Infantry - Company: John Mowton - Enlistment date: 19 Aug 1813 - Period: 1 Yr - Discharged at Craney Island, VA, on 19 Aug 1814.

Davis, Ezra - Private - 14th Infantry - Prisoner of War (Halifax), discharged from Halifax on 3 Feb 1814.

Davis, Henry - Corporal - 38th Infantry - Company: John Brookes - BLW 8419-160-12.

Davis, Hezekiah - Private - 43rd Infantry - Company: John Goodwyn - Age: 21 - Height: 5' 9" - Born: Charles Cty, MD - Trade: Farmer - Enlistment date: 2 Nov 1813 - Place: Laurensville, SC - Period: War - BLW 5917-160-12 - Discharged at Camp Greenfield, NC, on 17 May 1815.

Davis, Hugh - Private - 14th Infantry - Company: Henry Grindage - Age: 55 - Height: 5' 5 1/4" - Born: Ireland or Wales - Trade: Silk weaver - Enlistment date: 25 Aug 1812 - Place: Nottingham, MD - Period: 5 Yrs - BLW 23567-160-12 - Prisoner of War (Halifax), discharged from Halifax on 3 Feb 1814; discharged on 10 Apr 1815 at Greenbush, NY, for age and disability.

Davis, Jacob - Private - 14th Infantry - Company: Samuel Lane - Age: 23 - Height: 5' 10 1/2" - Born: Jersey - Enlistment date: 13 Jun 1812 - Place: Cumberland, MD - Period: 5 Yrs - Pension: Land bounty to John Davis and other heirs at law of Jacob Davis - BLW 23542-160-12 - Died at Black Rock, NY, on 5 Dec 1812.

Davis, John - Private - 14th Infantry - Pension: No pension claim - BLW 15222-160-12.

Davis, John - Corporal - 1st Artillery - Company: William Wilson - Age: 28 - Height: 5' 8 3/4" - Born: Sandersville, MD - Trade: Sailor - Enlistment date: 14 Jan 1812 - Place: Fort Johnston, NC - Period: 5 Yrs - Re-enlisted.

Davis, John - Private - 14th Infantry - Company: Joseph Marshall - Age: 28 - Height: 5' 5" - Born: Harford Cty, MD - Trade: Miller - Enlistment date: 14 Oct 1814 - Place: Baltimore - Period: War - BLW 438-160-12 - Discharged at Baltimore on 16 Mar 1815.

Davis, John - Private - 14th Infantry - Age: 33 - Height: 5' 8" - Born: Chester, MD - Trade: Laborer - Enlistment date: 2 Aug 1814 - Place: Baltimore - Period: War - BLW 15222-160-12 - Discharged at Greenbush, NY, on 3 May 1815.

Davis, John - Private - 14th Infantry - Enlistment date: 25 Dec 1812 - Period: 18 Mos - Prisoner of War (Halifax), sent to England; discharged on 31 Mar

By War of 1812 Soc. in MD

1815.

Davis, John - Private - 14th Infantry - Company: Joseph Marshall - Pension: No pension claim - BLW 438-160-12 - Prisoner of War (Halifax); died at Halifax on 25 Dec 1813.

Davis, John - Private - 2nd Artillery - Company: Frederick Evans - Age: 32 - Height: 5' 10 1/2" - Born: Maryland - Trade: Distiller - Enlistment date: 12 Feb 1814 - Place: Columbia, PA - Period: 5 Yrs - Discharged on 22 Apr 1815 because of pulmonary consumption.

Davis, John - Private - 36th Infantry - Company: Thomas Carbery - Pension: No pension claim - BLW 23127-160-12.

Davis, Josiah - Private - 38th Infantry - Company: John Brookes - Age: 21 - Height: 5' 8" - Born: Prince George's Cty, MD - Trade: Farmer - Enlistment date: 25 Apr 1814 - Place: Craney Island, VA - Period: War - Pension: Wife Mary Hyde, SO-3712, SC-1211; married on 28 Feb 1816, Prince George's Cty, MD; soldier died on 17 Jun 1865 - BLW 11177-160-12 - Discharged at Craney Island, VA, on 15 Mar 1815; also served in Captain Ford's Company, MD Militia.

Davis, Lawrence - Private - 38th Infantry - Company: Shepperd Leakin - Age: 34 - Height: 5' 9 1/2" - Born: Frederick, MD - Trade: Stone mason - Enlistment date: 11 Aug 1814 - Place: Liberty - Period: War - BLW 14730-160-12 - Discharged at Fort Covington, MD, on 30 Mar 1815.

Davis, Nathaniel - Private - 20th Infantry - Company: John Stanard - Other regiment: US Artificers - Age: 42 - Height: 5' 8" - Born: Prince George's Cty, MD - Trade: Joiner - Enlistment date: 5 Sep 1812 - Place: Washington, DC - Period: 5 Yrs - BLW 10877-160-12 - Transferred to the U.S. Artificers on 30 Aug 1812; discharged 14 Mar 1814 because of a rupture.

Davis, Peter L. - Private - 36th Infantry - Company: Joseph Hook - Age: 31 - Height: 5' 3" - Born: France - Trade: Farmer - Enlistment date: 12 May 1814 - Place: Georgetown, DC - Period: War - BLW 1038-160-12 - Discharged at Fort Covington, MD, on 12 Mar 1815.

Davis, Reuben - Private - 24th Infantry - Company: Robert Butler - Other regiment: 17th Infantry then 3rd Infantry (New) - Age: 37 - Height: 5' 8" - Born: Worcester Cty, MD - Trade: Farmer - Enlistment date: 4 Aug 1812 - Period: 5 Yrs - Transferred to 17th Infantry, Captain John Chunn's Company; discharged at Fort Harrison, IN, on 4 Aug 1817.

Davis, Richard - Fifer - 38th Infantry - Company: Shepperd Leakin - Age: 18 - Height: 5' 2" - Born: Maryland - Trade: Seaman - Fifer - Enlistment date: 31 May 1813 - Period: 1 Yr - BLW 14700-160-12 - Discharged at Fort McHenry on 28 Mar 1815.

Davis, Robert - Private - 14th Infantry - Company: Thomas Montgomery - Enlistment date: 17 Oct 1812 - Period: 18 Mos - Discharged at Plattsburgh, NY, on 17 Apr 1814.

Davis, Robert - Private - 14th Infantry - Company: Kenneth McKenzie - Age: 37 - Height: 5' 4" - Born: Baltimore - Trade: Farmer - Enlistment date: 17 Oct 1812 - Period: 18 Mos - Pension: Old War IF-25010 - BLW 6849-160-12 - Discharged at Greenbush, NY, on 24 May 1815 on Surgeon's

Maryland Regulars in War of 1812

Certificate, wounded in arm at Beaver Dams on 24 June 1813.

Davis, Shaddock - Private - 14th Infantry - Prisoner of War (Halifax), died at Halifax on 15 Sep 1813.

Davis, Thomas - Private - 14th Infantry - Company: Thomas Montgomery - Other regiment: 4th Infantry (New) - Age: 28 - Height: 5' 8 1/2" - Born: Baltimore - Trade: Laborer - Enlistment date: 25 Jul 1812 - Place: Baltimore - Period: 5 Yrs - Deserted at Fort Severn, MD, on 14 Jul 1815.

Davis, Thomas - Private - 38th Infantry - Company: Isaac Aldridge - Pension: Land bounty to Washington Davis, brother and other heirs at law of Thomas Davis - BLW 24637-160-12.

Davis, Thomas - Private - 18th Infantry - Company: Elias Dick - Age: 19 - Height: 5' 7" - Born: Prince George's Cty, MD - Trade: Carpenter - Enlistment date: 10 Aug 1813 - Place: Columbia, SC - Period: War - BLW 327-160-12 - Discharged at Fort Johnson, SC, on 28 Mar 1815.

Davis, Thomas - Private - 14th Infantry - Enlistment date: 13 Jul 1812 - Period: 18 Mos - Prisoner of War (Halifax), sent to England, captured at Beaver Dams on 24 Jun 1813; discharged on 31 Mar 1815.

Davis, Thomas - Private - 14th Infantry - Company: Thomas Montgomery - Age: 25 - Height: 5' 6" - Born: Bedford Cty, MA - Trade: Sailor - BLW 3236-160-12 - Discharged at Baltimore on 31 Mar 1815.

Davis, William - Private - 1st Artillery - Company: George Armistead - Age: 25 - Height: 5' 9" - Born: Virginia - Trade: Hatter - Enlistment date: 15 Jan 1810 - Place: Fort McHenry - Period: 5 Yrs - Discharged at Fort Columbus, NY, on 14 Jan 1815.

Davis, William - Private - 14th Infantry - Company: David Cummings - Age: 27 - Height: 5' 8" - Born: England - Trade: Sailor - Enlistment date: 16 Aug 1814 - Place: Baltimore - Period: War - Deserted.

Davis, William - Private - 6th Infantry - Company: John McChesney - Age: 22 - Height: 5' 7 1/2" - Born: Snow Hill, MD - Trade: Hatter - Enlistment date: 18 Dec 1808 - Period: 5 Yrs - Prisoner of War, captured at Queenstown, UC; paroled on 25 Mar 1813; discharged at Greenbush, NY, on 19 Jul 1814.

Davison, Elijah B. - Private - 38th Infantry - Company: John Brookes - Enlistment date: 22 Feb 1814 - Pension: Wife Elizabeth Bretty, SO-7702, SC-11422, WO-32352, WC-26354; married on 4 Nov 1847, Franklin Square, OH; soldier died on 20 Apr 1875, Leetonia, OH - BLW 27307-160-12 - Discharged on 15 Mar 1815.

Davison, John - Sergeant - 32nd Infantry - Company: Jonathan Robinson - Age: 23 - Height: 5' 6" - Born: Cecil Cty, MD - Trade: Cordwainer - Enlistment date: 15 Jun 1813 - Place: Wilmington, DE - Period: 1 Yr - Discharged on 15 Jun 1814.

Davy, David - Private - 5th Infantry - Company: James Dorman - Other regiment: 3rd Infantry (New) - Age: 20 - Height: 5' 4 1/4" - Born: Kent Cty, DE - Trade: Shoemaker - Enlistment date: 20 Apr 1813 - Place: Hagerstown, MD - Period: 5 Yrs - Re-enlisted at Fort Mackinaw, MI, on 28 Feb 1818.

Dawes, James - Private - 14th Infantry - Age: 34 - Height: 5' 8" - Born: Baltimore - Trade: Shoemaker - Enlistment date: 13 Jul 1814 - Place:

By War of 1812 Soc. in MD

Baltimore - Period: War - Discharged at Baltimore on 16 Mar 1815.

Dawes, James - Private - US Volunteers - Company: Stephen Moore - Enlistment date: 8 Sep 1812 - Period: 1 Yr.

Dawes, Samuel - Private - 14th Infantry - Company: Reuben Gilder - Age: 32 - Height: 5' 9" - Born: Baltimore - Trade: Butcher - Enlistment date: 9 Mar 1814 - Place: Baltimore - Period: War - BLW 2386-160-12 - Discharged at Greenbush, NY, on 4 May 1815.

Daws, William H. - Private - 14th Infantry - Company: Robert Kent - Enlistment date: 2 Jul 1812 - Period: 5 Yrs - Pension: Land bounty to Ann Daws and other heirs at law of William H. Daws - BLW 16685-160-12 - Died on 25 Nov 1812.

Dawson, Edward - Private - 14th Infantry - Company: David Cummings - Age: 18 - Height: 5' 5" - Born: Kentucky - Trade: Laborer - Enlistment date: 3 Feb 1815 - Place: Baltimore - Period: War - BLW 19845-160-12 Cancelled - Discharged at Baltimore on 18 Mar 1815.

Dawson, James - Seaman - US Sea Fencibles - Company: John Gill - Age: 24 - Height: 5' 6" - Born: Baltimore or Talbot Cty, MD - Enlistment date: 2 Jan 1814 - Period: 1 Yr.

Dawson, Robert - Private - 40th Infantry - Company: John Bailey - Age: 26 - Height: 5' 5" - Born: Cambridge, MD - Trade: Farmer - Enlistment date: 2 Mar 1814 - Place: Salisbury, MD - Period: War - Discharged on 30 Apr 1815.

Dawson, William - Corporal - 14th Infantry - Company: Thomas Montgomery - Other regiment: 4th Infantry (New) - Age: 19 - Height: 6' - Born: Dorchester Cty, MD - Trade: Farmer - Enlistment date: 9 May 1812 - Place: Hicksburgh - Period: 5 Yrs - BLW 9827-160-12 - Discharged at Raleigh, NC, on 12 May 1817.

Day, Cornelius - Seaman - US Sea Fencibles - Company: John Gill - Age: 28 - Height: 5' 7 1/2" - Born: Tappan, NY - Enlistment date: 29 Dec 1813.

Day, Cornelius - Private - US Volunteers - Company: Stephen Moore - Enlistment date: 8 Sep 1812 - Period: 1 Yr.

Day, Eder - Sergeant - 38th Infantry - Company: James Hook - Age: 23 - Height: 5' 7 1/4" - Born: New York City - Trade: Cabinet maker - Enlistment date: 18 Feb 1814 - Place: Baltimore - Period: War - BLW 11237-160-12 - Discharged at Fort Covington, MD, on 3 Mar 1815.

Day, Everett - Sergeant - 38th Infantry - Company: James Hook - Age: 24 - Height: 5' 5 1/2" - Born: Dedham, MA or CT - Trade: Butcher - Enlistment date: 21 Jun 1814 - Place: Baltimore - Period: War.

Day, James - Private - 14th Infantry - Company: David Cummings - Age: 25 - Height: 5' 6" - Born: Washington, MA - Trade: Farmer - Enlistment date: 4 Nov 1814 - Place: Baltimore - Period: War - BLW 14248-160-12 - Discharged on 16 Mar 1815.

Day, James - Private - 26th Infantry - Company: William Bezeau - Age: 25 - Height: 5' 5 1/2" - Born: Washington, MA - Enlistment date: 18 Sep 1814 - Place: Baltimore - Period: War - Deserted on 28 Nov 1814.

Day, William - Private - 36th Infantry - Company: Thomas Carbery - Age: 25

Maryland Regulars in War of 1812

- Height: 5' 11" - Born: Stafford Cty, VA - Trade: Farmer - Enlistment date: 25 Mar 1814 - Place: Leonardstown, MD - Period: War - BLW 116-160-12 - Discharged at Baltimore on 31 Mar 1815.

Dayhuff, Jacob - Private - 38th Infantry - BLW 23406-160-12.

Dayley, John - Private - 36th Infantry - Age: 37 - Height: 5' 7" - Born: Ireland - Enlistment date: 7 Jun 1813 - Place: Havre de Grace, MD.

de Noon, Charles - Private - 14th Infantry - Prisoner of War (Halifax), sent to England.

Deal, George - Private - 38th Infantry - Company: John Buck - Age: 19 - Height: 5' 5" - Born: Montgomery Cty, MD - Trade: Miller - Enlistment date: 17 Mar 1814 - Place: Baltimore - Period: War - BLW 20525-160-12 - Discharged at Baltimore on 30 Mar 1815.

Deal, Peter - Corporal - 15th Infantry - Company: Zachariah Rossell - Other regiment: Artillery - Age: 23 - Height: 6' 9 3/4" - Born: Baltimore - Trade: Shoemaker - Enlistment date: 8 Jul 1812 - Place: Philadelphia - Period: 5 Yrs - Re-enlisted on 24 May 1817.

Deamer, John - Private - 14th Infantry - Company: Reuben Gilder - Age: 22 - Height: 5' 10" - Born: Pennsylvania - Enlistment date: 10 Mar 1814 - Period: War - Died at Black Rock, NY, on 13 Dec 1814 from sickness.

Dean, James - Private - 1st Artillery - Company: Hopley Yeaton - Age: 22 - Height: 5' 6 3/4" - Born: Desert, MD - Trade: Laborer - Enlistment date: 19 Jan 1810 - Place: Fort Nelson, VA - Period: 5 Yrs - Discharged on 19 Jan 1815.

Dear, Isaac - Seaman - US Sea Fencibles - Company: Simmones Bunbury - Age: 27 - Height: 5' 10" - Born: Somerset Cty, MD - Enlistment date: 23 Dec 1813 - Period: 1 Yr - Discharged on 22 Dec 1814.

Deaver, John C. - Private - 28th Infantry - Company: Thomas Edmondson - Age: 29 - Height: 6' - Born: Maryland - Trade: Farmer - Enlistment date: 10 Mar 1814 - Place: Flemingsburg, KY - Period: War - BLW 7685-160-12 - Discharged at Detroit on 30 Jun 1815.

Deaver, Phillip - Private - 14th Infantry - Company: Reuben Gilder - Age: 39 - Height: 5' 11" - Born: Baltimore - Trade: Fisherman - Enlistment date: 14 Mar 1813 - Place: Baltimore - Period: War - BLW 15294-160-12 - On board the fleet on Lake Ontario; discharged at Greenbush, NY, on 4 May 1815.

Deaver, Stephen - Private - 24th Infantry - Company: Frank Hampton - Age: 26 - Height: 5' 9" - Born: Baltimore - Trade: Blacksmith - Enlistment date: 11 May 1813 - Period: War - Prisoner of War (Montreal), captured at Fort Niagara; discharged at Greenbush, NY, on 8 May 1815.

Declary, Noel - Private - 14th Infantry - Company: Reuben Gilder - Age: 50 - Height: 5' 5" - Born: France - Enlistment date: 7 Jan 1813 - Period: 18 Mos - Discharged at Plattsburgh, NY, on 7 Jul 1814.

Dedmont, Edward - Seaman - US Sea Fencibles - Company: John Gill - Age: 21 - Height: 5' 8" - Born: Charles Cty, MA - Enlistment date: 13 Jan 1814 - Period: 1 Yr.

Dee, Charles - Private - 36th Infantry - Company: Joseph Merrick - Transferred to the U.S. Navy on 5 Nov 1813.

By War of 1812 Soc. in MD

Deeds, Michael - Second Lieutenant - 24th Infantry - Company: James Campbell - Age: 40 - Height: 5' 7 1/2" - Born: Baltimore - Trade: Millwright - Enlistment date: 23 Jun 1814 - Place: North Carolina - Period: 5 Yrs - Promoted to 2nd Lieutenant on 11 Feb 1815.

Deems, George - Private - 38th Infantry - Enlistment date: 11 Jul 1814 - Pension: Wife Mary Uler, WO-35377, WC-22876; married on 20 Mar 1851, Bellville, Richland Cty, OH; soldier died on 27 Jun 1870 near Mansfield, OH - BLW 11590-160-12 - Discharged on 30 Apr 1815.

Deffendeffer, Frederick - Private - 2nd Artillery - Company: Matthew Massy - Height: 5' 9" - Born: Baltimore - Trade: Coppersmith - Enlistment date: 1806 - Place: Philadelphia - Period: 5 Yrs - Re-enlisted at Philadelphia for five years on 8 Mar 1811; prisoner of war, exchanged on 11 May 1814.

Deffendeffer, John - Private - 36th Infantry - Age: 35 - Height: 5' 8 3/4" - Born: Annapolis - Enlistment date: 25 Apr 1813 - Place: Westminster, MD.

Deford, George W. - Private - 36th Infantry - Age: 22 - Height: 5' 8 1/4" - Born: Chester, Kent Cty, MD - Enlistment date: 24 Aug 1813 - Place: Chestertown, MD - Period: 1 Yr.

Delahay, Henry - Private - US Volunteers - Company: Stephen Moore - Enlistment date: 8 Sep 1812 - Period: 1 Yr.

Deliney, George - Private - 14th Infantry - Company: Samuel Lane - Other regiment: 4th Infantry (New) - Age: 24 - Height: 6' - Born: Cecil Cty, MD - Trade: Farmer - Enlistment date: 29 Sep 1812 - Place: Rising Sun, MD - Period: 5 Yrs - Prisoner of War (Halifax), exchanged on 15 Apr 1814; deserted on 12 Jul 1815 from Fort Severn.

Delliard, Louis - Sergeant - 38th Infantry - Company: John Buck - BLW 10178-160-12.

Dellsher, George - Seaman - US Sea Fencibles - Company: John Gill - Age: 18 - Height: 5' 6" - Born: Baltimore City - Enlistment date: 9 Jan 1814.

Delozier, George - Private - 1st Rifles - Company: William Smyth - Age: 24 - Height: 6' 1" - Born: Charles Cty, MD - Trade: Mill digger - Enlistment date: 22 Feb 1812 - Place: Port Tobacco - Period: 5 Yrs - Discharged on 22 Feb 1817.

Delsaver, Michael - Private - 14th Infantry - Company: William McIlvane - Age: 37 - Height: 5' 5" - Born: Bridgetown, NJ - Trade: Laborer - Enlistment date: 23 Jan 1813 - Period: 18 Mos - Prisoner of War (Halifax), discharged from Halifax on 3 Feb 1814; discharged on 5 Jan 1815 for loss of right eye.

Dement, Edward - Private - 2nd Artillery - Company: Frederick Evans - Age: 21 - Height: 5' 7 1/4" - Born: Maryland - Trade: Wheelwright - Enlistment date: 4 May 1814 - Place: Fort McHenry - Period: 5 Yrs - Deserted on 30 Mar 1815.

Dempsey, George - Private - 14th Infantry - Company: Samuel Lane - Age: 25 - Height: 5' 10 1/2" - Born: Pennsylvania - Enlistment date: 8 May 1812 - Place: Williamsport, MD - Period: 5 Yrs - Prisoner of War (Halifax), sent to England.

Dempsey, John - Private - 36th Infantry - Age: 30 - Height: 5' 6" - Born: Baltimore - Enlistment date: 20 May 1813 - Place: Baltimore - Period: 1 Yr.

Maryland Regulars in War of 1812

Demry, John - Private - 38th Infantry - Age: 22 - Height: 5' 7" - Born: Frederick, MD - Trade: Farmer - Enlistment date: 25 Dec 1814 - Place: Baltimore - Period: War.

Demsey, John - Recruit - 26th Infantry - Company: William Bezeau - Age: 32 - Height: 5' 7" - Born: Baltimore - Trade: Chair maker - Enlistment date: 4 Nov 1814 - Place: Baltimore - Period: War - BLW 15227-160-12 - Discharged at Philadelphia on 23 Mar 1815.

Demster, Noah - Private - 27th Infantry (OH) - Company: Alexander Hill - Age: 27 - Height: 5' 9 1/2" - Born: Maryland - Enlistment date: 23 Apr 1814 - Place: Steubenville, OH.

Deneale, Hugh W. - Captain - 36th Infantry - Commissioned as a captain on 30 Apr 1813; discharged on 15 Jun 1815.

Denmade, Edward - Private - 14th Infantry - Prisoner of War (Halifax), sent to England.

Denney, William H. - Sergeant - 2nd Artillery - Company: Nathan Towson - BLW 7844-160-12.

Denning, James - Private - 17th Infantry - Company: Harris Hickman - Age: 40 - Height: 5' 8" - Born: Maryland - Trade: Farmer - Enlistment date: 23 Feb 1814 - Place: Hamilton, Butler Cty, OH - Period: War - BLW 9288-160-12 - Discharged at Chillicothe, OH, on 9 Jun 1815.

Denningsburgh, William - Private - 17th Infantry - Company: Angus Langham - Age: 23 - Height: 5' 4 1/2" - Born: Baltimore Cty - Trade: Wheelwright - Enlistment date: 1 Apr 1814 - Place: Detroit - Period: War - BLW 12224-160-12 - Prisoner of War (Halifax), captured on Lake Huron on 3 Sep 1814; released on 30 Mar 1815; discharged at Washington, DC, on 27 Jun 1815.

Dennis, John - Private - 5th Infantry - Company: James Dorman - Other regiment: 3rd Infantry (New) - Age: 21 - Height: 5' 6" - Trade: Tailor - Enlistment date: 1 Jan 1812 - Place: Hagerstown, MD - Period: 5 Yrs - Deserted at Erie, NY, on 1 Sep 1815.

Dennis, Thomas - Private - 38th Infantry - Company: Shepperd Leakin - Age: 21 - Height: 5' 5" - Born: Kent Cty, MD - Trade: Laborer - Enlistment date: 17 Aug 1814 - Place: Baltimore - Period: War - Deserted at Camp Baltimore on 22 Sep 1814.

Dennis, William - Private - 14th Infantry - Company: Richard Bennett - Enlistment date: 5 Apr 1813 - Period: 5 Yrs - Pension: Land bounty to Thomas Dennis, brother and other heirs at law of William Dennis - BLW 18939-160-12 - On board the fleet on Lake Ontario; died on 22 Aug 1813.

Denny, Samuel - Private - 1st Light Dragoons - Company: Arthur Hayne - Age: 31 - Height: 5' 7 1/2" - Born: Caroline Cty, MD - Trade: Carpenter - Enlistment date: 1 May 1813 - Place: Centreville, MD - Period: 5 Yrs.

Denson, Charles - Corporal - 14th Infantry - Company: Isaac Barnard - Enlistment date: 8 Feb 1813 - Period: 5 Yrs - Prisoner of War, sent to England; discharged at Baltimore on 31 Mar 1815 abscess of lungs.

Denson, Samuel - Private - 17th Infantry - Age: 21 - Height: 5' 8 1/2" - Born: Maryland - Trade: Laborer - Enlistment date: 16 Jan 1815 - Period: War - BLW 9734-160-12 Cancelled (entitled to 320 acres); BLW 403-320-14 -

By War of 1812 Soc. in MD

Discharged at Chillicothe, OH, on 11 Jul 1815.

Depass, George - Private - 14th Infantry - Enlistment date: 4 Mar 1813 - Period: 5 Yrs - Pension: Heirs received half pay for five years in lieu of land bounty - Died on 15 Jun 1813.

Depone, Charles - 5th Infantry - Age: 37 - Height: 5' 6" - Born: France - Trade: Laborer - Enlistment date: 9 Mar 1813 - Place: Baltimore - Period: 5 Yrs.

Deppisen, John C. - Private - US Volunteers - Company: Stephen Moore - Enlistment date: 8 Sep 1812 - Period: 1 Yr.

Deshong, James - Private - 38th Infantry - Company: Isaac Aldridge - Age: 20 - Height: 5' 3" - Born: Baltimore - Trade: Cooper - Enlistment date: 15 Jan 1815 - Place: Baltimore - Period: War - Discharged at Baltimore on 6 Apr 1815.

Desidnor, George - Private - 24th Infantry - Age: 30 - Height: 5' 8" - Born: Maryland - Trade: Shoemaker - Enlistment date: 11 Jul 1814 - Place: Knoxville, TN - Period: 5 Yrs.

Devanghan, Jonathan - Private - 12th Infantry - Company: Thomas Sangster - Age: 40 - Height: 5' 9 1/4" - Born: Maryland - Trade: Farmer - Enlistment date: 25 May 1812 - Place: Alexandria, DC - Period: 5 Yrs - Discharged at Pittsburgh on 18 Aug 1815, Surgeon's Certificate for old age.

Devault, Edward - Private - 19th Infantry - Age: 18 - Height: 5' 7" - Born: Prince George's Cty, MD - Enlistment date: 22 Jun 1812 - Period: 5 Yrs.

Devenport, Francis - Private - 36th Infantry - Age: 18 - Height: 5' 6" - Born: Berwick, England - Enlistment date: 6 May 1813 - Place: Havre de Grace, MD.

Dever, James - Quartermaster Sergeant - 1st Artillery - Other regiment: Corps of Artillery - Age: 29 - Height: 5' 9" - Born: Harford Cty, MD - Trade: Brick layer - Enlistment date: 28 Jan 1814 - Period: 5 Yrs - Discharged at New Orleans on 16 Apr 1816 after furnishing a substitute.

Dever, Michael - Private - 36th Infantry - Age: 27 - Height: 5' 4 1/2" - Born: Ireland - Trade: Stone mason - Enlistment date: 13 Jul 1814 - Period: War - Discharged at Washington, DC, on 23 Mar 1815.

Devers, John - Private - 26th Infantry (OH) - Company: Joel Collins - Other regiment: 19th Infantry - Age: 29 - Height: 5' 9" - Born: Maryland - Trade: Cooper - Enlistment date: 23 Sep 1813 - Period: 1 Yr - BLW 7100-160-12 - Re-enlisted on 21 Apr 1814 in the 19th Infantry, Lt Delorac, during the war enlistment, Sandwich, UC, discharged 20 Jul 1815 at Detroit, MI.

Devers, William - Private - 16th Infantry - Company: James McElroy - Other regiment: 2nd Infantry (New) - Age: 24 - Height: 5' 9" - Born: Maryland - Trade: Laborer - Enlistment date: 26 May 1812 - Place: Pennsylvania - Period: 5 Yrs - BLW 13041-160-12 - Discharged at Sackets Harbor, NY, on 21 Oct 1817.

Devin, Edward - Private - 14th Infantry - Company: Samuel Lane - Age: 42 - Height: 5' 10" - Born: Ireland - Trade: Farmer - Enlistment date: 29 Jun 1812 - Place: Delaware - Period: 18 Mos - BLW 8265-160-12 - Prisoner of War, sent to England; discharged on 31 Mar 1815 as a supernumerary.

Devinport, Francis - Private - 36th Infantry - Company: Joseph Hook - BLW

Maryland Regulars in War of 1812

14560-160-12.

Devlin, John - Corporal - 38th Infantry - Company: James Smith - Age: 21 - Height: 5' 10 1/2" - Born: Harford Cty, MD - Trade: Brick layer - Enlistment date: 2 Mar 1814 - Place: Craney Island, VA - Period: War - BLW 11114-160-12 - Discharged at Craney Island, VA, on 15 Mar 1815.

Devoe, William - Private - 38th Infantry - Company: Isaac Aldridge - Age: 21 - Height: 5' 5" - Born: Harford Cty, MD - Trade: Carpenter - Enlistment date: 14 Dec 1814 - Place: Baltimore - Period: War - Pension: Wife Elizabeth F. Rust, SO-25532, SC-19172, WO-16774, WC-13141; married on 4 Dec 1834, Winchester, Frederick Cty, VA; soldier died on 26 Dec 1875, Burlington, IA; first wife Sara Bowman - BLW 195-32014 - Discharged on 6 Apr 1815.

Devon, William J. - Seaman - US Sea Fencibles - Company: Simmones Bunbury - Age: 24 - Height: 5' 7 3/4" - Born: New Castle, DE - Enlistment date: 2 Apr 1814 - Period: 1 Yr.

Devore, David - Private - 24th Infantry - Company: Walter Wilkinson - Age: 23 - Born: Cumberland Cty, MD - Trade: Shoemaker - Enlistment date: 17 Jun 1814 - Period: War - BLW 19317-160-12 - Discharged at Belle Fontaine, MO, on 25 Jul 1815.

Devore, Henry - Private - 2nd Light Dragoons - Company: John Burd - Age: 21 - Height: 5' 9" - Born: Allegany Cty, MD - Trade: Blacksmith - Enlistment date: 20 Aug 1812 - Period: 5 Yrs - Deserted at Trenton, NJ, on 4 Oct 1812.

Dewer, Peter - Sergeant - 2nd Light Dragoons - Company: George Haig - Other regiment: Artillery - Age: 30 - Height: 5' 11" - Born: Snow Hill, MD - Trade: Hatter - Enlistment date: 27 Apr 1812 - Place: Richmond, VA - Period: 5 Yrs - Transferred to Capt Robert Hites Company, Artillery, on 31 Dec 1814.

Dickerson, James - Seaman - US Sea Fencibles - Company: Simmones Bunbury - Age: 50 - Height: 5' 5 1/4" - Born: Philadelphia - Enlistment date: 21 Feb 1814 - Discharged on 7 Mar 1814, unfit for service.

Dickerson, Martin - Servant - 36th Infantry - Company: Joseph Hook.

Dickey, Leonard W. - Private - 38th Infantry - Company: John Mowton - Age: 26 - Height: 5' 9 1/4" - Born: Hillsborough, NH - Enlistment date: 20 Apr 1813 - Place: Baltimore - Period: 1 Yr - BLW 22098-160-12.

Dickinson, John - Private - 14th Infantry - Tried by court-martial on 26 Jul 1813 for desertion; death by hanging approved on 10 Aug 1813 (alias John Thompson).

Dickinson, Terry - Private - 36th Infantry - Company: Samuel Raisin - Age: 33 - Height: 6' 2" - Born: Caswell, NC - Trade: Laborer - Enlistment date: 18 Jun 1814 - Place: Lynchburg - Period: War.

Dickison, John - Corporal - 5th Infantry - Company: John Corboley - Age: 24 - Height: 5' 5 3/4" - Born: Wilmington, DE - Trade: Carver - Enlistment date: 20 Oct 1813 - Place: Baltimore - Period: 5 Yrs.

Dicus, Absalom - Private - 26th Infantry - Company: William Bezeau - Age: 25 - Height: 6' - Born: Baltimore - Trade: Farmer - Enlistment date: 14 Oct

By War of 1812 Soc. in MD

1814 - Place: Baltimore - Period: War - Discharged at Philadelphia on 25 May 1815.

Dilbranche, John Patis - Private - 5th Infantry - Age: 39 - Height: 5' 3" - Born: France - Trade: Gardener - Enlistment date: 23 Feb 1813 - Place: Baltimore - Period: 5 Yrs.

Dill, Peter - Private - 14th Infantry - Company: Reuben Gilder - Age: 33 - Height: 6' - Born: Lancaster - Trade: Tailor - Enlistment date: 30 Aug 1812 - Place: Maryland - Period: 5 Yrs - Prisoner of War (Halifax).

Dilley, Jonathan - Private - 38th Infantry - Company: James Smith - BLW 12396-160-12.

Dilliard, Lewis - Sergeant - 38th Infantry - Company: Shepperd Leakin - Age: 21 - Height: 5' 8 1/2" - Born: Carolina Cty, VA - Trade: Inn keeper - Enlistment date: 28 Feb 1814 - Place: Baltimore - Period: War - Discharged at Baltimore on 30 Mar 1815.

Dinkins, James Edward - Major - 36th Infantry - Born: South Carolina - Commissioned as a 1st Lieutenant, 3rd Infantry, on 1 Jul 1808; promoted Captain on 6 Feb 1811; promoted to Major, 44th Infantry, on 15 May 1814; transferred to 36th Infantry on 18 Nov 1814; promoted to brevet Major, 4th Infantry, on 17 May 1815; promoted to major, 8th Infantry, 8 May 1818; transferred to 4th Infantry on 27 Jan 1819; transferred to 5th Infantry on 1 Jun 1821; transferred to 4th Infantry on 24 Oct 1821; died on 6 Oct 1822.

Dioxin, Charles - Musician - 38th Infantry - Company: Isaac Aldridge - BLW 12504-160-12.

Disbrey, Joseph - Corporal - 41st Infantry - Company: Mangle Quackenbos - Age: 24 - Height: 5' 8 3/4" - Born: Baltimore - Trade: Cabinet maker - Enlistment date: 21 Dec 1814 - Place: New York - Period: War - Discharged on 17 Jun 1815.

Disheron, Stephen - Private - 38th Infantry - Company: John Brookes - Age: 21 - Height: 5' 10 1/2" - Born: Somerset Cty, MD - Trade: Waterman - Enlistment date: 5 Apr 1814 - Place: Craney Island, VA - Period: War.

Disney, Joseph - Private - 38th Infantry - Company: Shepperd Leakin - Enlistment date: 12 Aug 1813 - Period: 1 Yr - Died at Fort McHenry on 15 or 18 Apr 1814.

Disney, Samuel - Recruit - 26th Infantry - Company: William Bezeau - Age: 37 - Height: 6' 1" - Born: Baltimore - Trade: Carpenter - Enlistment date: 18 Oct 1814 - Place: Baltimore - Period: War - BLW 6736-160-12 - Discharged at Philadelphia on 23 Mar 1815.

Ditman, John - Private - 35th Infantry - Company: Walter Cocke - Age: 16 - Height: 5' 1" - Born: Stafford Cty, VA - Trade: Planter - Enlistment date: 2 Jun 1814 - Place: Norfolk, VA - Period: War - Discharged at Norfolk, VA, on 20 Mar 1815.

Ditter, Abraham - Private - 26th Infantry - Company: William Bezeau - Age: 21 - Height: 5' 8" - Born: Harford Cty, MD - Enlistment date: 8 Nov 1814 - Place: Baltimore.

Dix Jr., Timothy - Lieutenant Colonel - 14th Infantry - Born: New Hampshire - Place: New Hampshire - Pension: Land bounty to Lucy Hartwell Dix, widow of Timothy Dix - BLW 10501-160-50 - Commissioned as a major on 12 Mar 1812; promoted to lieutenant colonel on 20 Jun 1813; died on 14 Nov 1813.

Dix, John Adams - Captain - 14th Infantry - Other regiment: Corps of Artillery - Born: New Hampshire - BLW 4830-160-50 - Commissioned as an ensign, 14th Infantry, on 10 May 1813; promoted to 3rd lieutenant on 7 Mar 1814; promoted to 2nd lieutenant on 8 Mar 1814; transferred to Corps of Artillery on 9 Aug 1814; promoted to 1st lieutenant on 23 Mar 1818; transferred to 1st Artillery on 1 Jun 1821; transferred to 3rd Artillery on 16 Aug 1821; promoted to captain on 30 Aug 1825; resigned on 31 Dec 1828.

Dixon, James - Private - 14th Infantry - Company: Samuel Lane - Age: 19 - Height: 5' 7" - Born: Frederick, MD - Enlistment date: 27 May 1812 - Place: Cumberland, MD - Period: 5 Yrs - Prisoner of War (Halifax), captured at Stoney Creek, UC, on 12 Jun 1813; deserted at Schenectady, NY, on 19 Jan 1815.

Dixon, James - Private - 21st Infantry - Company: Benjamin Ropes - Age: 25 - Height: 5' 4 1/2" - Born: Baltimore - Trade: Mariner - Enlistment date: 26 Mar 1814 - Place: Portsmouth, NH - Period: War - Discharged on 29 May 1815.

Dixon, James - Private - 2nd Artillery - Company: John Goodall - Other regiment: Corps of Artillery - Age: 23 - Height: 5' 6" - Born: Queen Anne's Cty, MD - Trade: Tailor - Enlistment date: 21 Aug 1813 - Place: Norfolk, VA - Period: 5 Yrs - BLW 19038-160-12 - Discharged at Fort Nelson, VA, on 21 Aug 1818.

Dixon, John - Private - US Volunteers - Company: Stephen Moore - Enlistment date: 8 Sep 1812 - Period: 1 Yr.

Dixon, William - Sergeant - 36th Infantry - Company: Samuel Raisin - Age: 31 - Height: 5' 10" - Born: England - Enlistment date: 3 Sep 1813 - Place: Baltimore - BLW 19889-160-12.

Dize, Levin - Private - 14th Infantry - BLW 18596-160-12.

Dize, Levin - Private - 2nd Infantry - Company: William Boots - Other regiment: 1st Infantry (New) - Age: 24 - Height: 5' 6" - Born: Somerset Cty, MD - Trade: Sailor - Enlistment date: 11 Dec 1802 - Place: Fort Wilkinson, GA - Period: 5 Yrs - Re-enlisted in 1807 and again on 11 Jun 1812; discharged at Baton Rouge, LA, on 1 Aug 1817.

Dobbins, James - Private - 36th Infantry - Company: Thomas Carbery - Age: 20 - Height: 5' 8" - Born: District of Columbia - Trade: Shoemaker - Enlistment date: 5 Apr 1814 - Place: Georgetown, DC - Period: War - Pension: Land bounty to Eleanor Marsteller, mother and heir at law of James Dobbins - BLW 20580-160-12 - Discharged at Baltimore on 31 Mar 1815.

Dobson, James - Private - 5th Infantry - Company: Richard Dale - Age: 32 - Height: 5' 5 1/2" - Born: Talbot Cty, MD - Trade: Laborer - Enlistment date: 28 Sep 1808 - Place: Easton, MD - Period: 5 Yrs - Discharged on 18 Jan 1809 for inability.

Dodd, Aaron - Private - 36th Infantry - Company: Samuel Raisin - Age: 23 - Height: 6' - Born: Westmoreland Cty, VA - Trade: Brick layer - Enlistment date: 29 Sep 1814 - Place: Manchester or Richmond, VA - Period: War.

Dodd, James - Private - 28th Infantry - Other regiment: 3rd Infantry (New) - Age: 22 - Height: 5' 5" - Born: Maryland - Trade: Farmer - Enlistment date: 18 Oct 1814 - Place: Augusta, MD - Period: 5 Yrs.

Dodd, Stephen - Private - 36th Infantry - Age: 21 - Height: 5' 5 1/2" - Born: Greensborough, MD - Enlistment date: 4 Aug 1813 - Place: Greensboro, MD - Period: 1 Yr.

Dogherty, Hamilton - Private - 14th Infantry - Pension: Land bounty to John Dogherty and other heirs at law of Hamilton Dogherty - BLW 4052-160-12 Cancelled.

Doherty, Daniel - Private - 23rd Infantry - Company: Richard Goodell - Age: 39 - Height: 5' 8" - Born: Frederick, MD - Trade: Farmer - Enlistment date: 26 Jan 1814 - Place: Utica, NY - Period: War - Discharged at Sackets Harbor, NY, on 5 May 1815.

Doize, Charles - Musician - 38th Infantry - Company: James Hook - Enlistment date: 5 Jan 1813 - Period: 1 Yr.

Dolby, Stephen - Private - 38th Infantry - Company: John Mowton - BLW 4314-160-12.

Doleman, George - Private - 14th Infantry - Other regiment: 4th Infantry (New) - Enlistment date: 2 May 1812 - Period: 5 Yrs - Deserted at Annapolis on 9 Oct 1815.

Doling, Thomas - Private - 36th Infantry - Age: 25 - Height: 5' 6" - Born: Ireland - Trade: Tailor - Enlistment date: 14 Nov 1814 - Period: War - Deserted.

Donaldson, George - Private - 14th Infantry - Company: Reuben Gilder - Age: 28 - Height: 5' 6" - Born: Baltimore City - Trade: Shoemaker - Enlistment date: 11 Jan 1813 - Place: Baltimore - Period: 5 Yrs - BLW 452-160-12 - Discharged at Greenbush, NY, on 25 Apr 1815 on Surgeon's Certificate.

Donaldson, Jesse - Private - 38th Infantry - Company: Charles Stansbury - Age: 25 - Height: 5' 7" - Born: Calvert Cty, MD - Trade: Carpenter - Enlistment date: 30 Nov 1814 - Place: Baltimore - Period: War - BLW 3218-160-12 - Discharged at Baltimore on 6 Apr 1815.

Donaldson, Stephen F. - First Lieutenant - 14th Infantry - Born: Maryland - Commissioned as an ensign on 12 Mar 1812; promoted to 2nd lieutenant on 13 Mar 1813; promoted to 1st lieutenant on 14 Nov 1814; discharged on 15 Jun 1815.

Donally, John - Private - 20th Infantry - Company: John Duval - Age: 21 - Height: 5' 6 1/2" - Born: Washington, MD - Trade: Joiner - Enlistment date: 22 May 1814 - Period: War.

Donaville, James - Private - 5th Infantry - Company: Richard Whartenby - Other regiment: Artillery - Age: 22 - Height: 5' 6" - Born: Ireland - Trade: Cabinet maker - Enlistment date: 7 Jun 1814 - Place: Baltimore - Period: 5 Yrs - Transferred to Corps of Artillery, Captain Isaac Roach's company; discharged on 16 Jul 1816.

Donelly, John - Private - 2nd Infantry - Age: 22 - Height: 5' 6" - Born: Washington, MD - Enlistment date: 13 Feb 1814 - Place: Hagerstown, MD.

Maryland Regulars in War of 1812

Donnell, Joshua P. - 36th Infantry - Pension: Wife Agnes, Old War WI-13837 Rejected.

Donovan, Carval - Private - 23rd Infantry - Company: Henry Whiting - Other regiment: 2nd Infantry (New) - Age: 22 - Height: 5' 10 1/2" - Born: Havre de Grace, MD - Trade: Stone cutter - Enlistment date: 28 Feb 1814 - Place: Albany, NY - Period: War - BLW 176-160-12 - Discharged at Sackets Harbor, NY, on 5 Jun 1815.

Dores, Hugh - Private - 14th Infantry - Company: Isaac Barnard - Enlistment date: 18 Jan 1813 - Period: 5 Yrs - Discharged at Greenbush, NY, on 28 Jan 1815.

Dorman, William - Private - 38th Infantry - Company: James Hook - Age: 24 - Height: 5' 8" - Born: Somerset Cty, MD - Enlistment date: 4 Jan 1814 - Place: Princess Anne, MD - Period: 1 Yr - Discharged on 4 Jan 1815.

Dornbrooke, Henry - Musician - 38th Infantry - Company: James Smith - Age: 21 - Height: 5' 1" - Born: Amsterdam, Holland - Trade: Tailor - Enlistment date: 5 Mar 1814 - Place: Norfolk, VA - Period: War.

Dorney, John - Private - 1st Artillery - Company: Louis Walbach - Age: 40 - Height: 5' 8" - Born: Old Hundred, Harford Cty, MD - Trade: Cordwainer - Enlistment date: 1 Aug 1811 - Place: Fort Constitution, MA - Period: 5 Yrs - Discharged at Fort Constitution, MA, on 1 Apr 1816.

Dorrim, John - Private - 36th Infantry - Age: 21 - Height: 5' 7" - Born: Charlotte, VA - Enlistment date: 30 Sep 1814 - Place: Norfolk, VA.

Dorsey, Daniel D. - Private - 14th Infantry - Company: James McDonald - Age: 26 - Height: 5' 10" - Born: Maryland - Enlistment date: 17 Nov 1813 - Period: 5 Yrs - Pension: Land bounty to William Dorsey, brother & other heirs at law of Daniel D. Dorsey - BLW 24256-160-12 - Died at Plattsburgh, NY, on 28 Jan or Jun 1814.

Dorsey, Henry K. - Private - 38th Infantry - Company: James Hook - Age: 32 - Height: 5' 8 1/2" - Born: Baltimore - Trade: Cabinet maker - Enlistment date: 11 Jun 1814 - Place: Baltimore - Period: War - BLW 11353-160-12 - Previously served in the US Sea Fencibles; discharged on 31 Mar 1815 at Fort Covington.

Dorsey, Henry K. - Seaman - US Sea Fencibles - Company: John Gill - Other regiment: 38th Infantry - Age: 32 - Height: 5' 8 1/2" - Born: Baltimore - Trade: Cabinet maker - Enlistment date: 27 Dec 1813 - Re-enlisted in 38th Infantry.

Dorsey, James - Private - 14th Infantry - Company: Samuel Lane - Age: 34 - Height: 5' 6" - Born: Ireland - Enlistment date: 9 Jun 1812 - Place: Cumberland, MD - Period: 5 Yrs.

Dorsey, Nicholas - Private - 38th Infantry - Company: John Rothrock - Age: 25 - Height: 5' 9" - Born: Elk Ridge, MD - Trade: Shoemaker - Enlistment date: 10 Aug 1814 - Place: Norfolk, VA - Period: War - BLW 225-160-12 - Discharged at Craney Island, VA, on 15 Mar 1815.

Dorsey, Samuel - Private - 36th Infantry - Age: 43 - Height: 5' 10" - Born: Lancaster, PA - Enlistment date: 26 Aug 1813 - Place: Waynesburg, PA - Period: 1 Yr.

By War of 1812 Soc. in MD

Dorsey, Samuel H. - Private - 38th Infantry - Company: Shepperd Leakin - Age: 19 - Height: 5' 8" - Born: Anne Arundel Cty, MD - Trade: Cooper - Enlistment date: 2 Mar 1814 - Place: Baltimore - Period: War - BLW 2005-160-12 - Discharged at Baltimore on 30 Mar 1815.

Doubts, George - Private - 38th Infantry - Age: 38 - Height: 5' 7 1/2" - Born: Lancaster, PA - Trade: Farmer - Enlistment date: 12 Mar 1814 - Period: War - Deserted.

Doud, John - Private - 14th Infantry - Company: David Cummings - Age: 52 - Height: 5' 4" - Trade: Farmer - Enlistment date: 6 Aug 1812 - Period: 5 Yrs - BLW 17756-160-12 - Prisoner of War, sent to England; discharged at Baltimore on 20 Apr 1815 due to old age (alias John David).

Dougherty, Charles - Private - 12th Infantry - Company: Andrew Madison - Other regiment: 8th Infantry (New) - Age: 36 - Height: 5' 9" - Born: Frederick Cty, MD - Trade: Farmer - Enlistment date: 13 Feb 1813 - Place: Strasburg - Period: 5 Yrs - BLW 16900-160-12 - Discharged at Camp Jackson, AL, on 13 Feb 1818.

Dougherty, David - Private - 14th Infantry - Company: Thomas Montgomery - Other regiment: 4th Infantry (New) - Age: 34 - Height: 5' 11" - Born: Derry, Ireland - Trade: Laborer - Enlistment date: 6 Aug 1812 - Place: Wilmington, DE - Period: 5 Yrs - Deserted at Annapolis on 3 Oct 1815.

Dougherty, Hamilton - Private - 14th Infantry - Company: Samuel Lane - Age: 25 - Height: 5' 6" - Born: Jefferson Cty, VA - Enlistment date: 21 May 1812 - Place: Virginia - Period: 5 Yrs - Pension: Land bounty to Patrick Dougherty, father and heir at law of Hamilton Doughtery - BLW 21686-160-12 - Prisoner of War (Chatham), died at Chatham on 22 May 1814.

Dougherty, James - Private - 7th Infantry - Company: Thornton Posey - Born: Cumberland Cty, MD - Trade: Farmer - Enlistment date: 16 Feb 1814 - Period: War - BLW 15531-160-12 - Discharged at Belle Fontaine, MO, on 25 Jul 1815.

Dougherty, William - Corporal - 14th Infantry - Company: Samuel Lane - Other regiment: 4th Infantry (New) - Age: 28 - Height: 5' 8 1/2" - Born: West Town, Chester Cty, PA - Trade: Carpenter - Enlistment date: 13 Jul 1812 - Place: Maryland - Period: 5 Yrs - Prisoner of War (Halifax); discharged on 13 Jul 1817.

Doughtery, John - Private - 38th Infantry - Company: Charles Stansbury - Age: 23 - Height: 5' 10 1/4" - Born: Ireland - Trade: Stone cutter - Enlistment date: 28 Sep 1814 - Place: Baltimore - Period: War - BLW 18215-160-12 - Discharged at Baltimore on 6 Apr 1815.

Doughty, Joseph - Private - US Volunteers - Company: Stephen Moore - Enlistment date: 8 Sep 1812 - Period: 1 Yr.

Douglas, George - Corporal - 5th Infantry - Company: James Dorman - Other regiment: 3rd Infantry (New) - Age: 23 - Height: 5' 7 1/3" - Born: Little York, PA - Trade: Shoemaker - Enlistment date: 29 Feb 1812 - Place: Baltimore - Period: 5 Yrs - Deserted at Erie, NY, on 29 Jun 1815.

Dove, John - Private - 1st Artillery - Company: George Armistead - Age: 24 - Height: 5' 6 1/2" - Born: Annapolis - Trade: Clerk - Enlistment date: 6 Feb

Maryland Regulars in War of 1812

1810 - Place: Fort McHenry - Period: 5 Yrs - Re-enlisted for the war; on 6 Feb 1815; discharged on 23 May 1815.

Dowlan, John - Private - 26th Infantry - Company: William Bezeau - Age: 29 - Height: 5' 8 3/4" - Born: Ireland - Trade: Shoemaker - Enlistment date: 2 Jan 1815 - Place: Baltimore - Period: War.

Dowler, Francis - Private - 1st Rifles - Company: Joshua Hamilton - Other regiment: Rifles - Age: 28 - Height: 5' 9" - Born: Washington Cty, MD - Trade: Gunsmith - Enlistment date: 26 May 1812 - Place: Hiwassee Garrison, TN - Period: 5 Yrs - Discharged on 26 May 1817.

Dowler, Thomas - Private - 1st Rifles - Company: Daniel Appling - Other regiment: Rifles - Age: 25 - Height: 5' 9 3/4" - Born: Washington Cty, MD - Trade: Gunsmith - Enlistment date: 25 May 1812 - Place: Hiwassee Garrison, TN - Period: 5 Yrs - Wounded at Lyon's Creek, UC, on 18 Oct 1814; discharged at Buffalo on 1 Aug 1815.

Downes, Nicey - Private - 1st Artillery - Company: Francis Newman - Age: 25 - Height: 5' 9" - Born: Maryland - Trade: Sailor - Enlistment date: 6 Sep 1814 - Place: New Orleans - Period: War - Discharged at New Orleans on 20 Apr 1815.

Downes, Richard C. - Surgeon's Mate - 14th Infantry - Commissioned as a surgeon's mate, 14th Infantry, on 12 May 1813; resigned on 7 Feb 1814.

Downs, Bernard - Private - 36th Infantry - Company: Thomas Carbery - Age: 36 - Height: 5' 9" - Born: St. Mary's Cty, MD - Trade: Blacksmith - Enlistment date: 24 Mar 1814 - Place: Leonardstown, MD or Washington, DC - Period: War - BLW 121-160-12 - Discharged at Baltimore on 31 Mar 1815.

Downs, Noah - Fifer - 19th Infantry - Company: Carey Trimble - Age: 23 - Height: 6' 1" - Born: Maryland - Trade: Farmer - Enlistment date: 22 Mar 1814 - Place: Chillicothe, OH - Period: War - BLW 9471-160-12 - Discharged at Chillicothe, OH, on 6 Jun 1815.

Downs, William - Private - 5th Infantry - Company: Leroy Opie - Age: 35 - Height: 5' 8" - Born: Baltimore - Trade: Farmer - Enlistment date: 5 Apr 1812 - Place: Baltimore - Period: 5 Yrs - Discharged at Greenbush, NY, on 1 Jun 1815, because of wounds.

Downs, William D. - Private - 5th Infantry - Company: Leroy Opie - Age: 38 - Height: 5' 8" - Born: Ireland - Trade: Turner - Enlistment date: 24 Feb 1813 - Place: Baltimore - Period: 5 Yrs - BLW 970-160-12.

Doyle, Hazard - Private - 26th Infantry - Company: William Bezeau - Age: 34 - Height: 6' - Born: Maryland - Trade: Blacksmith - Enlistment date: 11 Jan 1815 - Place: Philadelphia - Period: War.

Doyle, Michael - Private - 38th Infantry - Company: James Hook - Age: 30 - Height: 5' 8 3/4" - Born: Ballamoma, Whitlow, Ireland - Trade: Weaver - Enlistment date: 20 Oct 1814 - Place: Fredericktown, MD - Period: War - Pension: Wife Ann G. Sproul, WO-22860, WC-20328; married in Oct 1827, Bath Cty, VA; soldier died on 20 Apr 1860, Highland Cty, VA - BLW 2668-160-55 - Discharged at Fort Covington, MD, on 30 Mar 1815.

Doyle, Peter - Private - 2nd Artillery - Company: Nathan Towson - Height: 5'

- Trade: Laborer - Enlistment date: 23 Apr 1812 - Place: Reisterstown, MD - Period: 5 Yrs - BLW 25156-160-12 - Discharged at Greenbush, NY, on 17 Sep 1812 for old age.
- Doyle, William - Private - 22nd Infantry - Company: Willis Foulk - Age: 39 - Height: 5' 9" - Born: Baltimore - Trade: Distiller - Enlistment date: 5 Apr 1814 - Place: Washington, PA - Period: War - Discharged at Sackets Harbor, NY, on 24 May 1815.
- Doyle, William - Private - 36th Infantry - Company: Joseph Hook.
- Dozier, Charles - Musician - 38th Infantry - Company: Isaac Aldridge - Age: 19 - Height: 5' 4" - Born: West Indies - Trade: Drummer - Enlistment date: 18 Nov 1814 - Place: Baltimore - Period: War - Discharged at Baltimore on 6 Apr 1815.
- Drain, William - Private - 32nd Infantry - Company: Jonathan Robinson - Age: 24 - Height: 5' 8 1/2" - Born: Maryland - Trade: Miller - Enlistment date: 3 Nov 1813 - Place: Philadelphia - Period: 1 Yr - Discharged on 4 Nov 1814.
- Drake, Henry - Private - 36th Infantry - Company: Thomas Carbery - BLW 5511-160-12.
- Draper, Samuel - Recruit - 26th Infantry - Company: William Bezeau - Race: 19 - Age: 19 - Height: 5' 6" - Born: Maryland - Trade: Laborer - Enlistment date: 11 Oct 1814 - Place: Philadelphia - Period: 5 Yrs - Listed as "Col'd" in his service record; discharged for being a minor.
- Draper, Seyers - Private - Artillery - Other regiment: Corps of Artillery - Age: 21 - Height: 5' 5 1/2" - Born: Maryland - Trade: Laborer - Enlistment date: 1 Feb 1815 - Place: Philadelphia - Period: War - BLW 216-320-12 - Discharged at Philadelphia on 13 Mar 1815.
- Drean, Charles - Private - 36th Infantry - Age: 21 - Height: 5' 8" - Born: Leesburg, VA - Enlistment date: 26 Jun 1813 - Place: Cumberland, MD.
- Drear, Joseph - Seaman - US Sea Fencibles - Company: Simmones Bunbury - Age: 24 - Height: 5' 10" - Born: Germany - Enlistment date: 7 Jan 1814 - Period: 1 Yr - Discharged on 6 Jan 1815.
- Drummond, John - Private - 14th Infantry - Prisoner of War (Halifax), sent to England.
- Dubois, David - Private - 14th Infantry - Prisoner of War (Halifax), discharged from Halifax on 3 Feb 1814.
- Duckett, Richard - Private - 38th Infantry - Company: John Mowton - Age: 39 - Height: 6' - Born: Baltimore - Enlistment date: 14 Apr 1814 - Place: Craney Island, VA - Period: War - Pension: Heirs received half pay for five years in lieu of land bounty - Died on 8 Sep 1814.
- Duckett, Thomas W. - Private - 28th Infantry - Company: Henry Gist - Age: 25 - Height: 5' 11" - Born: Maryland - Trade: Carpenter - Enlistment date: 3 Jun 1814 - Place: Kentucky - Period: War - BLW 24439-160-12 - Discharged at Lower Sandusky, OH (Fort Stephenson), on 25 Jun 1815.
- Duff, David - Private - 14th Infantry - Company: Samuel Lane - Age: 35 - Height: 5' 9" - Born: Ireland - Enlistment date: 26 May 1812 - Place: Cumberland - Period: 5 Yrs - Pension: Land bounty to Henry Duff, heir at law to David Duff - BLW 23614-160-12 - Died at Black Rock, NY, on 16

Maryland Regulars in War of 1812

Dec 1812.

Duffield, John - Private - 25th Infantry - Company: George Howard - Age: 32 - Height: 5' 9" - Born: Frederick, MD - Trade: Cordwainer - Enlistment date: 9 Jul 1813 - Period: 1 Yr - Discharged at Sackets Harbor, NY, on 9 Jul 1814.

Duffy, Andrew - Private - 36th Infantry - BLW 2669-160-12.

Duffy, Owen - Private - 5th Infantry - Age: 34 - Height: 5' 7" - Born: Ireland - Trade: Cooper - Enlistment date: 22 Feb 1813 - Place: Baltimore - Period: 5 Yrs.

Dugan, James - Private - 5th Infantry - Company: Leroy Opie - Other regiment: 3rd Infantry (New) - Age: 25 - Height: 5' 10" - Born: Emmetsburg, MD - Trade: Chair maker - Enlistment date: 27 May 1812 - Place: Frederick, MD - Period: 5 Yrs - Prisoner of War (Halifax), captured at Stoney Creek, UC, on 8 Jun 1813; deserted at Detroit on 26 Jun 1816.

Duganor, Charles - Private - 14th Infantry - Prisoner of War (Halifax), captured at Beaver Dams on 24 Jun 1813, discharged from Halifax on 9 Nov 1813.

Dukes, Robert - Private - 14th Infantry - Company: David Cummings - Enlistment date: 1812 - Period: 5 Yrs - Deserted at Salisbury, MD, on 18 Nov 1814.

Dulany, Henry R. - Captain - Light Artillery - Born: Virginia - Graduated from the US Military Academy on 2 Mar 1815; commissioned as a 3rd lieutenant, Light Artillery, on 2 Mar 1815; transferred to Corps of Artillery on 17 May 1815; promoted to 2nd lieutenant, 4th Infantry, on 5 Mar 1817; promoted to 1st lieutenant on 10 Feb 1818; promoted to captain on 3 Feb 1822; resigned on 31 May 1825; died on 27 Nov 1838.

Dulin, Lewis E. - Corporal - 38th Infantry - Company: John Mowton - Pension: Land bounty to Susanna Dulin and other heirs at law of Lewis E. Dulin - BLW 22039-160-12.

Dulman, George - Private - 14th Infantry - Prisoner of War (Halifax), discharged from Halifax on 3 Feb 1814.

Dumeste, Jacob Adrian - First Lieutenant - Corps of Artillery - Born: Maryland - Graduated from the US Military Academy on 6 Jun 1814; commissioned as a 2nd lieutenant, Corps of Artillery, on 1 Jul 1819; transferred to 2nd Artillery on 1 Jun 1821; promoted to 1st lieutenant on 13 Jan 1831; died on 10 Oct 1831.

Dumour, Gideon - Private - 36th Infantry - Company: Samuel Raisin - Age: 24 - Height: 5' 7" - Born: Halifax, NC - Trade: Farmer - Enlistment date: 14 Jun 1814 - Period: War - Died on 28 Jan 1815.

Duncan, David W. - First Lieutenant - 38th Infantry - Commissioned as a 2nd lieutenant on 20 May 1813; promoted to 1st lieutenant on 15 Aug 1813; discharged on 15 Jun 1815.

Duncan, James - Private - 37th Infantry - Company: Samuel Northrop - Age: 33 - Height: 5' 8" - Born: Centreville, MD - Trade: Mariner - Enlistment date: 4 Oct 1814 - Place: New Haven, CT - Period: War - BLW 1127-160-12 - Discharged at New London, CT, on 12 May 1815.

Duncan, James - Private - 2nd Artillery - Company: Nathan Towson -

Enlistment date: 24 Apr 1812 - Period: 5 Yrs - Killed in action on 25 Jul 1814.

Dunfre, George - Private - 38th Infantry - Company: John Mowton - BLW 22955-160-12.

Dunlap, Henry - Third Lieutenant - 36th Infantry - Commissioned as an ensign on 1 Sep 1813; promoted to 3rd lieutenant on 17 Mar 1814; dropped on 27 Dec 1814.

Dunn, John - Private - 2nd Infantry - Company: Peter Schuyler - Age: 25 - Height: 5' 9" - Born: Morris, NJ - Trade: Cooper - Enlistment date: 7 Aug 1807 - Place: Fredericktown, MD - Period: 5 Yrs - Died at Fort Charlotte, AL, on 24 Dec 1814 of fever.

Dunn, Thomas L. - Private - 38th Infantry - Company: James Smith - BLW 20710-160-12.

Dunnavon, George - Private - 24th Infantry - Company: James Campbell - Other regiment: 7th Infantry (New) - Age: 30 - Height: 5' 9" - Born: Maryland - Trade: Farmer - Enlistment date: 11 Jul 1814 - Place: McMinnville, TN - Period: 5 Yrs - Deserted at Fort Montgomery, AL, on 21 Mar 1817.

Dunneal, Edward - Private - 14th Infantry - Company: David Cummings - Age: 43 - Height: 5' 3 1/2" - Born: Ireland - Trade: Carpenter - Enlistment date: 3 Jul 1812 - Place: Baltimore - Period: 5 Yrs - Prisoner of War, sent to England; discharged at Washington, DC, on 20 Apr 1815 on Surgeon's Certificate, lost trigger finger of right hand.

Dunnevin, Walter - Private - 14th Infantry - Company: Isaac Barnard - Enlistment date: 7 Jan 1813 - Pension: Land bounty to Mary Buck, niece and other heirs at law of Walter Dunnevin - BLW 25573-160-12 - Died on 12 May 1813.

Duquesney, Charles - Private - 14th Infantry - Company: David Cummings - Enlistment date: 19 Mar 1812 - Period: 5 Yrs - Prisoner of War, sent to England; deserted on 30 Apr 1815 from Fort McHenry.

Durcas, James - Private - 2nd Artillery - Company: Nathan Towson - Pension: Land bounty to Susanna Kerns, sister and other heirs at law of James Durcas - BLW 19405-160-12.

Duskey, Elijah - Private - 14th Infantry - Company: Joseph Marshall - Age: 19 - Height: 5' 5" - Born: Salisbury, MD - Trade: Farmer - Enlistment date: 30 May 1814 - Place: Salisbury, MD - Period: War - Pension: Wife Polly H. Shockley, SO-18883, SC-21199, WO-33195, WC-21946; married on 24 Dec 1817, Saulsbury, MD; died on 14 Jun 1871, Manchester, IN - BLW 16817-160-12 - Discharged at Greenbush, NY, on 3 May 1815.

Dutton, Stephen - Private - 2nd Light Dragoons - Company: Beverly Turpin - Other regiment: Light Dragoons - Age: 21 - Height: 5' 8" - Born: Maryland - Trade: Farmer - Enlistment date: 18 Jun 1812 - Place: Granville, NC - Period: 5 Yrs - Discharged on 17 Jun 1817.

Duval, Alvin - Corporal - 14th Infantry - Enlistment date: 21 Jul 1812 - Period: 5 Yrs - Prisoner of War, captured at Beaver's Dam on 24 Jun 1813; paroled at Halifax.

Duvall, James - Private - 36th Infantry - Company: Thomas Carbery - Age: 21 - Height: 5' 7 1/2" - Born: District of Columbia - Enlistment date: 7 Apr 1814 - Place: Georgetown, DC - Period: War - BLW 24167-160-12 - Discharged at Baltimore on 3 Mar 1815.

Duvall, John - Sergeant - 1st Infantry - Company: Hugh Moore - Other regiment: 3rd Infantry (New) - Age: 21 - Height: 5' 9" - Born: Fredericktown, MD - Trade: Blacksmith - Enlistment date: 16 Jan 1812 - Place: Pittsburgh - Period: 5 Yrs - Discharged at Detroit on 18 Jan 1817.

Dyer, Henry - Private - 36th Infantry - Other regiment: Ordnance Department - Age: 32 - Height: 5' 5 1/2" - Born: Prince George's Cty, MD - Trade: Carpenter - Enlistment date: 2 Jan 1815 - Place: Georgetown, DC - Period: 5 Yrs - BLW 849-320-12 - Transferred to the U.S. Ordnance Department on 21 Jun 1815; discharged on 1 Jan 1820 at Richmond, VA.

Dyer, John B. - Private - 14th Infantry - Company: Clement Sullivan - Enlistment date: 29 May 1812 - Period: 5 Yrs - Pension: Heirs received half pay for five years in lieu of land bounty - Died at Fort Niagara, NY, on 1 Oct 1812.

Dyson Francis - Private - 7th Infantry - Company: Elijah Montgomery - Age: 22 - Height: 5' 6" - Born: Maryland - Trade: Farmer - Enlistment date: 25 Jul 1814 - Place: Kentucky - Period: War - Discharged at New Orleans on 1 Apr 1815.

Dyson, Jonathan - Private - 38th Infantry - Company: John Mowton - BLW 3612-160-12.

Dyson, Samuel T. - Captain - 1st Artillery - Other regiment: Corps of Artillery - Commissioned as a lieutenant, 1st Artillery and Engineers, on 19 Dec 1796; transferred to Artillery on 1 Apr 1802; promoted to captain on 15 Sep 1803; transferred to 1st Artillery on 12 Mar 1812; transferred to Corps of Artillery on 12 May 1814; dismissed on 17 Nov 1814.

E

Eades, John - Private - 38th Infantry - Company: John Brookes - Enlistment date: 27 Mar 1814 - Period: 1 Yr - Pension: SO-14828, SC-5299 - BLW 3164-160-12 - Discharged on 15 Mar 1815.

Eades, John - Private - 20th Infantry - Company: John Duval - Age: 24 - Height: 5' 7 3/4" - Born: Maryland - Trade: Laborer - Enlistment date: 7 Sep 1813 - Period: 5 Yrs - Died on 30 Jan 1815.

Eadlen, James - Private - 36th Infantry - Company: Thomas Carbery - Age: 19 - Height: 5' 5" - Born: Anne Arundel Cty, MD - Trade: Farmer - Enlistment date: 2 Mar 1814 - Place: Leonardstown, MD - Period: War - Discharged at Baltimore on 31 Mar 1815.

Eagle, James - Private - 5th Infantry - Age: 22 - Height: 5' 4" - Born: Maryland - Trade: Cabinet maker - Enlistment date: 7 Apr 1813 - Place: Baltimore - Period: 5 Yrs.

Eaglehart, Nathan - Private - 5th Infantry - Company: Colin Buckner - Age: 34 - Height: 5' 7" - Born: Accomack Cty, VA - Trade: Nailor - Enlistment date: 4 Apr 1813 - Place: Baltimore - Period: 5 Yrs - Died on 28 Nov 1813.

By War of 1812 Soc. in MD

Eagleston, Benjamin - Private - 2nd Artillery - Company: Frederick Evans - Other regiment: Corps of Artillery - Age: 38 - Height: 5' 6" - Born: Baltimore - Trade: Shoemaker - Enlistment date: 7 May 1812 - Place: Baltimore - Period: 5 Yrs - Discharged at Washington, DC, on 13 May 1817.

Eagleston, William C. - Private - 5th Infantry - Company: Richard Whartenby - Age: 32 - Height: 5' 9" - Born: York, PA - Trade: Printer - Enlistment date: 20 Mar 1812 - Place: Baltimore - Period: 5 Yrs - Deserted.

Eakin, William - Private - 38th Infantry - Age: 44 - Height: 5' 11 3/4" - Born: Bedford, VA - Enlistment date: 10 Dec 1814 - Place: Baltimore.

Earle, Caleb - Captain - Quartermasters Department - Born: Maryland - Commissioned as a captain, assistant deputy quartermaster general, on 2 Aug 1814; resigned on 18 Aug 1814.

Earle, Daniel - Private - 1st Artillery - Company: George Armistead - Other regiment: Corps of Artillery - Age: 28 - Height: 5' 11 1/2" - Born: New York - Trade: Painter - Enlistment date: 23 Mar 1808 - Place: Fort McHenry - Period: 5 Yrs - BLW 4794-160-12 - Re-enlisted for the war on 27 Sep 1814; discharged at New York, NY, on 23 May 1815.

Earle, William N. - First Lieutenant - 36th Infantry - Born: Maryland - Commissioned as a 2nd lieutenant on 30 Apr 1813; promoted to 1st lieutenant on 15 Aug 1813; discharged on 15 Jun 1815.

Early, Fielding - Private - 10th Infantry - Company: Philip Brittain - Other regiment: 8th Infantry (New) - Age: 24 - Height: 5' - Born: Prince George's Cty, MD - Trade: Farmer - Enlistment date: 13 Jul 1812 - Place: Wilkesboro, NC - Period: 5 Yrs - Pension: Land bounty to Samuel Early, brother & other heirs at law of Fielding Early - BLW 2500-160-12 - Died on 31 Aug 1816.

Early, Samuel - Private - 38th Infantry - Company: John Brookes - Height: 5' 6" - Born: Prince George's Cty, MD - Enlistment date: 23 Jul 1813 - Period: 1 Yr - Pension: Land bounty to Benoni Early, brother and other heirs at law of Samuel Early - BLW 17678-160-12 - Re-enlisted for the war on 12 Aug 1814; died on 27 Oct 1814.

East, Ezekiel - Private - 16th Infantry - Company: Robert Gray - Other regiment: 2nd Regiment (New) - Age: 21 - Height: 5' 9" - Born: Maryland - Trade: Laborer - Enlistment date: 9 Feb 1813 - Place: New Brunswick, NJ - Period: 5 Yrs - BLW 16038-160-12 - Discharged at Greenbush, NY, on 8 Feb 1818.

Eastham, Robert - Private - 5th Infantry - Company: Richard Whartenby - Enlistment date: 28 May 1814 - Place: Baltimore - Period: 5 Yrs.

Easton, Hezekiah - Private - 36th Infantry - Company: Thomas Carbery - Age: 22 - Height: 5' 8" - Born: Montgomery Cty, MD - Trade: Wagoner - Enlistment date: 3 Apr 1814 - Place: Georgetown, DC - Period: War - BLW 884-160-12 - Discharged at Baltimore on 31 Mar 1815.

Easton, John - Private - 36th Infantry - Company: Samuel Raisin - Age: 40 - Height: 5' 9" - Born: District of Columbia - Trade: Laborer - Enlistment date: 10 Jul 1813 - Place: Waynesburg, PA - Period: War - Pension: Land bounty to Thomas Easton, son and other heirs at law of John Easton - BLW 11635-160-12 - Re-enlisted on 14 Dec 1814 for the war; died at Georgetown, DC, on 19 Jan 1815.

Eaton, John - Private - 14th Infantry - Age: 26 - Height: 5' 10" - Born: York, PA - Trade: Sugar baker - Enlistment date: 14 Dec 1814 - Period: War - Pension: Land bounty to Justina Eaton, widow and other heirs at law of John Eaton - BLW 1050-320-12 - Died at Baltimore on 22 Feb 1815.

Eberline, John - Private - 12th Infantry - Company: Thomas Post - Age: 22 - Height: 5' 10" - Born: Lancaster, PA - Trade: Butcher - Enlistment date: 9 Jan 1814 - Place: Hagerstown, MD - Period: War - Discharged at Buffalo on 31 May 1815.

Echard, Frederick - Private - 9th Infantry - Company: Ebenezer Childs - Age: 25 - Height: 5' 6" - Born: Baltimore - Trade: Laborer - Enlistment date: 23 Dec 1813 - Place: Boston - Period: 5 Yrs - Wounded at Fort Erie, UC, on 17 Sep 1814 in the thigh; discharged at Pittsfield, on 13 Jun 1815, on Surgeon's Certificate.

Eckelleyer, Jonathan - Private - 2nd Light Dragoons - Company: Samuel Hopkins - Age: 23 - Height: 5' 11" - Born: Hagerstown, MD - Trade: Hatter - Enlistment date: 2 Jun 1814 - Place: Carlisle, PA - Period: 5 Yrs.

Eckford, James - Private - 38th Infantry - Company: John Rothrock - BLW 21529-160-12.

Eckman, Jacob - Private - 38th Infantry - Company: Charles Stansbury - Enlistment date: 25 Aug 1813 - Period: 1 Yr - Pension: Wife Mary Raney, SO-3340, SC-5670; married in 1844, Cleves, OH; soldier died about 1877 - BLW 44498-80-50 & 28135-80-55 - Discharged on 30 Apr 1814.

Edeer, William - Private - 38th Infantry - Company: John Brookes - BLW 12631-160-12.

Edelin, Richard - Private - 1st Artillery - Company: James Reed - Age: 21 - Height: 5' 5" - Born: Charles Cty, MD - Trade: Farmer - Enlistment date: 2 May 1814 - Place: Alexandria, DC - Period: War - Discharged at Fort Washington, MD, on 7 Mar 1815.

Edelin, Stanislane - Private - 43rd Infantry - Company: John Goodwyn - Age: 24 - Height: 5' 8" - Born: Charles Cty, MD - Trade: Carpenter - Enlistment date: 28 Dec 1813 - Place: Camden, SC - Period: War - Discharged at Camp Greenfield, NC, on 17 May 1815.

Edgan, John - Private - 14th Infantry - Prisoner of War (Halifax), sent to England.

Edgell, Zedehiah - Private - 14th Infantry - Company: Reuben Gilder - Age: 20 - Height: 5' 10 1/2" - Born: Maryland - Enlistment date: 12 Apr 1813 - Period: 18 Mos - Discharged at Black Rock, NY, on 12 Oct 1814.

Edging, Martin - Private - 19th Infantry - Company: John Chunn - Other regiment: 17th Infantry - Age: 32 - Height: 5' 11" - Born: Easton, MD - Enlistment date: 18 Jul 1812 - Place: St. Clairsville, OH - Period: 5 Yrs - Prisoner of War (Halifax), captured at Fort Erie, UC, on 15 Aug 1815; released on 8 Apr 1815.

Edmonds, Peter - Private - 2nd Light Dragoons - Company: William Littlejohn - Age: 22 - Height: 5' 4" - Born: St. Mary's Cty, MD - Trade: Seaman - Enlistment date: 7 Feb 1814 - Place: Alexandria, DC - Period: War -

By War of 1812 Soc. in MD

Deserted on 7 Dec 1814.

Edmunds, Abijah - Seaman - US Sea Fencibles - Company: Simmones Bunbury - Age: 22 - Height: 5' 8" - Born: Vermont - Enlistment date: 1 Jul 1814 - Period: 1 Yr - Deserted on 21 Oct 1814.

Edwards, Aquila - Corporal - US Volunteers - Company: Stephen Moore - Enlistment date: 8 Sep 1812 - Period: 1 Yr.

Edwards, Edward - Private - US Volunteers - Company: Stephen Moore - Enlistment date: 8 Sep 1812 - Period: 1 Yr - Pension: Land bounty to Aquila Edwards & other heirs at law of Edward Edwards - BLW 26987-160-12 - Killed at York, UC, on 27 Apr 1813.

Edwards, Henry - Artificer - Light Artillery - Company: Arthur Thornton - Other regiment: Light Artillery - Age: 34 - Height: 5' 11" - Born: St. Mary's Cty, MD - Trade: Carpenter - Enlistment date: 10 Mar 1812 - Place: District of Columbia - Period: 5 Yrs - BLW 9481-160-12 - Discharged on 18 Mar 1817.

Edwards, James B. - Private - 36th Infantry - Age: 25 - Height: 5' 10" - Born: Amherst Cty, VA - Enlistment date: 2 Apr 1814 - Place: Fort Washington, MD.

Edwards, John - Private - 1st Infantry - Company: Hezekiah Johnson - Age: 30 - Height: 5' 6" - Born: Maryland - Trade: Farmer - Enlistment date: 10 Jun 1814 - Place: Frenchtown, NJ - Period: War - Died at Brownsville, NY, on 8 or 9 Sep 1815 of fever.

Edwards, Thomas - Private - 14th Infantry - Company: Reuben Gilder - Age: 38 - Height: 5' 8" - Born: Baltimore Cty - Trade: Laborer - Enlistment date: 30 Mar 1814 - Place: Towsontown, MD - Period: War - Pension: No pension claim - BLW 11785-160-12 - Discharged at Greenbush, NY, on 22 Apr 1815 for inability.

Eggleston, Benjamin - Private - 2nd Light Dragoons - Company: William Littlejohn - Age: 18 - Height: 5' 4 1/2" - Born: Baltimore - Enlistment date: 20 Nov 1813 - Place: Baltimore - Period: 5 Yrs.

Eichelberger, George S. - First Lieutenant - 38th Infantry - Pension: Land bounty to Eliza Ann Eichelberger, widow of George S. Eichelberger - BLW 22528-160-50 - Commissioned as a 1st lieutenant on 20 May 1813; resigned on 22 Apr 1814.

Eichelberger, Peter - Private - US Volunteers - Company: Stephen Moore - Enlistment date: 8 Sep 1812 - Period: 1 Yr.

Eicholtz, John - Private - 5th Infantry - Enlistment date: 29 Apr 1812 - Place: Baltimore - Period: 5 Yrs - Prisoner of War.

Elbert, John L. - Second Lieutenant - 2nd Light Dragoons - Other regiment: Light Dragoons - Born: Maryland - Pension: Placed on the pension rolls on 15 Jan 1816 - Commissioned as a cornet, 2nd Light Dragoons, on 28 Apr 1813; promoted to 3rd lieutenant on 20 Jun 1813; transferred to Light Dragoons on 12 May 1814; promoted to 2nd lieutenant on 18 Jul 1814; discharged on 15 Jun 1815; died on 13 Feb 1838.

Eldridge, Simon - Private - 26th Infantry - Company: William Bezeau - Age: 41 - Height: 6' - Born: Maryland - Trade: Laborer - Enlistment date: 8 Feb

Maryland Regulars in War of 1812

1814 - Place: Philadelphia - Period: War.

Elliot, James W. - Sergeant - 14th Infantry - Company: David Cummings - Age: 22 - Height: 6' - Born: Baltimore - Trade: Hatter - Enlistment date: 24 Feb 1814 - Place: Baltimore - Period: War - BLW 1615-160-12 - Discharged at Baltimore on 18 Mar 1815.

Elliott, Benjamin - Seaman - US Sea Fencibles - Company: John Gill - Enlistment date: 23 Dec 1813.

Elliott, Benjamin - Private - US Volunteers - Company: Stephen Moore - Enlistment date: 8 Sep 1812 - Period: 1 Yr.

Elliott, Thomas - Private - 38th Infantry - Company: John Brookes - BLW 5428-160-12.

Elliott, William - Private - 1st Artillery - Company: Hopley Yeaton - Other regiment: Corps of Artillery - Age: 29 - Height: 5' 11" - Born: Queenstown, MD - Trade: Bookkeeper - Enlistment date: 16 Nov 1812 - Place: Fort Nelson, VA - Period: 5 Yrs - BLW 14080-160-12 - Discharged at Fort Nelson, VA, on 16 Nov 1817.

Ellis, Charles - Private - 20th Infantry - Age: 30 - Height: 5' 8" - Born: Charles Cty, MD - Trade: Sailor - Enlistment date: 10 Dec 1814 - Place: Alexandria, DC - Period: War - BLW 15967-160-12 - Discharged on 15 Mar 1815.

Ellis, John - Private - 20th Infantry - Age: 38 - Height: 6' 3" - Born: St. Mary's, MD - Trade: Farmer - Enlistment date: 20 Jun 1814 - Place: Haymaker - Period: 5 Yrs - Discharged on 6 Apr 1815.

Elmore, Philip - Private - 14th Infantry - Prisoner of War (Halifax), discharged from Halifax on 3 Feb 1814.

Elms, Benjamin - Private - 38th Infantry - Company: Isaac Aldridge - Age: 25 - Height: 5' 5 3/4" - Born: Scituate, MA - Trade: Competioner - Enlistment date: 11 Oct 1814 - Place: Baltimore - Period: War - BLW 2011-160-12 - Discharged at Baltimore on 6 Apr 1815.

Elsey, Thomas - Private - 38th Infantry - Company: John Rothrock - Enlistment date: 6 Apr 1814 - Period: War - Pension: Wife Nancy Ellis, SO-2637, SC-14761; married on 20 Sep 1824, Martin Cty, IN - BLW 6893-160-12 - Discharged on 15 Mar 1815.

Elson, Joseph P. - Private - 36th Infantry - Age: 43 - Height: 5' 11" - Born: North Marlboro, MD - Trade: Farmer - Enlistment date: 3 Feb 1815 - Place: Georgetown, DC - Period: War - Pension: No pension claim - BLW 4-320-12 - Discharged at Georgetown, DC, on 13 Mar 1815.

Emerson, Richard - Private - 36th Infantry - Age: 25 - Height: 5' 7 3/4" - Born: Hunterdon Cty, NJ - Enlistment date: 7 May 1813 - Place: Baltimore - Period: 1 Yr.

Emmons, Henry - Private - 14th Infantry - Company: Samuel Lane - Age: 26 - Height: 5' 11 1/2" - Born: Jersey - Enlistment date: 28 Apr 1812 - Place: Hagerstown, MD - Period: 5 Yrs - Prisoner of War (Halifax), discharged from Halifax on 3 Feb 1814; deserted at Plattsburgh, NY, on 20 Aug 1814.

Enalds, Thomas - Private - 39th Infantry - Company: Alfred Douglass - Age: 28 - Height: 5' 10" - Born: Dorchester Cty, MD - Enlistment date: 20 Sep 1813 - Place: Gallatin, TN - Period: 1 Yr.

By War of 1812 Soc. in MD

Enfield, George - Private - 36th Infantry - Age: 25 - Height: 5' 7" - Born: Lancaster, PA - Enlistment date: 1 Sep 1813 - Place: Havre de Grace, MD.

Engles, Friday - Private - 14th Infantry - Company: Thomas Montgomery - Height: 5' 8" - Born: Maryland - Enlistment date: 17 Nov 1812 - Period: 18 Mos - Re-enlisted on 21 Jan 1814 for the war; deserted at Sackets Harbor, NY, on 14 Sep 1814.

English, John - Private - 14th Infantry - Enlistment date: 24 Jan 1813 - Period: 18 Mos - Prisoner of War (Halifax).

English, Patrick - Private - 14th Infantry - Company: Kenneth McKenzie - Age: 36 - Height: 5' 8" - Trade: Farmer - Enlistment date: 27 Jul 1813 - Period: 1 Yr - Prisoner of War, captured during the Battle of Stoney Creek; discharged on 7 Aug 1814.

Ennels, Henry - Private - 2nd Artillery - Company: James Barker - Age: 37 - Height: 5' 9 3/4" - Born: Maryland - Trade: Cabinet maker - Enlistment date: 27 Apr 1813 - Place: Philadelphia - Period: 5 Yrs - Discharged on 21 Jul 1815 on Surgeon's Certificate.

Ennis, William - Private - 36th Infantry - Company: Joseph Hook - Age: 21 - Height: 5' 11" - Born: Worcester Cty, MD - Trade: Farmer - Enlistment date: 3 Aug 1813 - Place: Snow Hill, MD - Period: 1 Yr - BLW 6189-160-12 - Discharged at Fort Covington, MD, on 30 Mar 1815.

Enos, Augustus - Quartermaster Sergeant - 36th Infantry - Age: 22 - Height: 5' 7" - Born: Hartford, CT - Trade: Hatter - Enlistment date: 20 Jul 1813 - Place: Waynesburg, PA - Period: War - BLW 16941-160-12 - Discharged at Washington, DC, on 23 Mar 1815.

Ent, John - Corporal - 14th Infantry - Company: Reuben Gilder - Pension: Placed on the pension roll on 23 Sep 1817 - BLW 1649-160-12 - Died on 2 Jul 1823.

Ent, John - Corporal - 14th Infantry - Company: Reuben Gilder - Pension: Old War IF-21179.

Ernest, John - Corporal - 38th Infantry - Company: Thomas Sangster - Age: 21 - Height: 5' 3" - Born: Somerset Cty, MD - Trade: Planter - Enlistment date: 18 Mar 1814 - Period: War.

Ervin, Robert - Private - 1st Artillery - Company: Henry Craig - Age: 27 - Height: 5' 6" - Born: Baltimore - Trade: Laborer - Enlistment date: 6 Apr 1814 - Place: Pittsburgh - Period: War - Discharged at Buffalo on 3 Jun 1815.

Etlopp, Charles - Private - 26th Infantry - Company: William Bezeau - Age: 18 - Height: 5' 4" - Born: Maryland - Trade: Laborer - Enlistment date: 24 Jan 1815 - Place: Philadelphia - Period: War.

Eubanks, George - Private - 3rd Infantry - Company: James Dinkins - Age: 30 - Height: 5' 10" - Born: Maryland - Trade: Carpenter - Enlistment date: 7 Apr 1809 - Place: York - Period: 5 Yrs - Re-enlisted for the war on 23 May 1813.

Evans, Edward - Private - 14th Infantry - Prisoner of War (Halifax), discharged from Halifax on 27 Aug 1813.

Evans, Elijah - Private - 14th Infantry - Company: David Cummings - Age:

26 - Height: 5' 4 1/2" - Born: Montgomery Cty, MD - Trade: Farmer - Enlistment date: 9 Nov 1814 - Place: Snowden's Camp, VA - Period: 5 Yrs - Pension: Wife Elizabeth, WO-629.

Evans, George - Sergeant - US Volunteers - Company: Stephen Moore - Enlistment date: 8 Sep 1812 - Period: 1 Yr.

Evans, Jacob - Private - 38th Infantry - Company: John Mowton - Age: 25 - Height: 5' 6" - Born: Philadelphia - Enlistment date: 4 May 1813 - Place: Baltimore - BLW 13760-160-12.

Evans, James - Private - 14th Infantry - Prisoner of War (Halifax), discharged from Halifax on 27 Aug 1813.

Evans, John - Private - 36th Infantry - Company: Thomas Carbery - Age: 40 - Height: 5' 8" - Born: Maryland - Trade: Planter - Enlistment date: 24 Feb 1814 - Place: Leonardstown, MD - Period: War - Pension: Land bounty to Susanna Drake, sister and heir at law of John Evans - BLW 11701-160-12 - Died in regimental hospital on 30 Jan 1815.

Evans, John - Private - 4th Rifles - Company: John Lytle - Age: 43 - Height: 5' 7" - Born: Baltimore - Trade: Sailor - Enlistment date: 12 Jul 1814 - Place: Canandaigua, NY - Period: War - Discharged at Buffalo on 13 Jun 1815.

Evans, Patrick - Seaman - US Sea Fencibles - Company: Simmones Bunbury - Age: 26 - Height: 5' 6" - Born: Dublin, Ireland - Enlistment date: 2 Jun 1814 - Period: 1 Yr - Deserted on 5 Nov 1814.

Evans, Patrick - Private - 38th Infantry - Age: 26 - Height: 5' 7" - Born: Ireland - Trade: Currier - Enlistment date: 2 Nov 1814 - Place: Baltimore - Period: 5 Yrs.

Evans, Patrick C. - Private - 36th Infantry - Age: 26 - Height: 5' 6" - Born: Ireland - Enlistment date: 27 May 1813 - Place: Baltimore - Period: 1 Yr.

Evans, Philip - Private - 14th Infantry - Company: Thomas Montgomery - Discharged at Burlington, VT, on 21 Apr 1814 for inability.

Evans, Samuel - Private - 3rd Infantry - Company: William Butler - Age: 30 - Height: 5' 7" - Born: Baltimore - Trade: Laborer - Enlistment date: 11 Jul 1808 - Period: 5 Yrs - Re-enlisted on 19 Apr 1813.

Evans, Thomas - Private - Light Dragoons - Age: 33 - Height: 5' 9" - Born: Baltimore - Trade: Craftsman - Enlistment date: 5 Apr 1814 - Period: War - Deserted at Baltimore.

Evans, William - Private - 2nd Artillery - Company: Alexander Williams - Age: 39 - Height: 5' 11" - Born: Maryland - Trade: Carpenter - Enlistment date: 7 Sep 1813 - Place: Philadelphia - Period: 5 Yrs - Missing after the attack on Fort Erie, UC, on 15 or 16 Aug 1814.

Evans, William - Private - 36th Infantry - Age: 24 - Height: 5' 6" - Born: Wales - Trade: Seaman - Enlistment date: 26 Dec 1814 - Period: War - Discharged on 13 Mar 1815.

Everett, John - Private - 38th Infantry - Company: James Haslett - Age: 21 - Height: 5' 6" - Born: Baltimore - Trade: Cooper - Enlistment date: 6 Jul 1813 - Period: 1 Yr - Re-enlisted at Craney Island, VA, on 8 Mar 1814; discharged at Craney Island, VA, on 15 Mar 1815.

Everett, John - Private - 8th Infantry - Company: Thomas Farrar - Age: 30 -

By War of 1812 Soc. in MD

Height: 5' 6" - Born: Maryland - Trade: Farmer - Enlistment date: 31 May 1813 - Place: Georgia - Period: War - Discharged at Camp Flournoy, GA, on 4 Mar 1815.

Everhard, Jacob - Corporal - 38th Infantry - Company: Shepperd Leakin - Age: 23 - Height: 6' - Born: Baltimore - Trade: Silversmith - Enlistment date: 26 Jul 1814 - Place: Baltimore - Period: War - BLW 11156-160-12 - Discharged at Fort McHenry on 28 Mar 1815.

Everhart, John - Private - 38th Infantry - Company: Shepperd Leakin - Age: 23 - Height: 5' 8" - Born: Baltimore - Trade: Tobacconist - Enlistment date: 6 Apr 1814 - Place: Baltimore - Period: War - BLW 25113-160-12 - Discharged at Washington, DC, on 19 Mar 1815.

Everton, Thomas - Private - 8th Infantry - Company: Felix Warley - Other regiment: 7th Infantry (New) - Age: 17 - Height: 4' 10" - Born: Baltimore - Trade: Farmer - Enlistment date: 20 Jan 1813 - Period: 5 Yrs - Discharged at Camp Montgomery, AL, on 5 Oct 1817 for inability.

Evett, William - Private - 36th Infantry - Company: Joseph Hook - Age: 30 - Height: 5' 10" - Born: Harford Cty, MD - Trade: Farmer - Enlistment date: 10 Aug 1813 - Place: Havre de Grace, MD - Period: 1 Yr - BLW 19230-160-12 - Re-enlisted at Annapolis on 16 Apr 1814; discharged at Fort Covington, MD, on 30 Mar 1815.

Evins, David - Seaman - US Sea Fencibles - Company: William Addison.

Ewings, John - Private - 44th Infantry - Company: Joseph Miles - Age: 28 - Height: 6' - Born: Maryland - Trade: Shoemaker - Enlistment date: 2 Oct 1814 - Period: War - Discharged at New Orleans on 8 Apr 1815.

F

Fagan, James - Private - 14th Infantry - Company: James McDonald - Other regiment: 4th Infantry (New) - Age: 35 - Height: 5' 7" - Born: Ireland - Trade: Gardener - Enlistment date: 15 Feb 1814 - Period: 5 Yrs - BLW 26303-160-12 - Discharged at Fort Moultrie, SC, on 3 May 1816, wounded at Fort George, UC.

Fagan, James - Private - 3rd Artillery - Company: Alexander Brookes - Age: 32 - Height: 5' 7" - Born: Ireland - Trade: Gardener - Enlistment date: 23 Jun 1813 - Place: Havre de Grace, MD - Period: 18 Mos - Discharged on 23 Dec 1813.

Fagunday, Benjamin - Private - 38th Infantry - Company: Shepperd Leakin - Age: 25 - Height: 5' 10 1/4" - Born: Bucks Cty, PA - Trade: Brick layer - Enlistment date: 27 Jul 1814 - Place: Baltimore - Period: War - BLW 3161-160-12 - Discharged at Fort McHenry on 28 Mar 1815.

Fair, George - Recruit - 26th Infantry - Company: William Bezeau - Age: 29 - Height: 5' 4 1/2" - Born: Berks Cty, PA - Trade: Shoemaker - Enlistment date: 16 Oct 1814 - Place: Baltimore - Period: War - BLW 15137-160-12 - Discharged at Philadelphia on 23 Mar 1815.

Fairbanks, John - Private - 5th Infantry - Company: Colin Buckner - Age: 37 - Height: 5' 8" - Born: Talbot Cty, MD - Trade: Carpenter - Enlistment date: 9 Apr 1813 - Place: Baltimore - Period: 5 Yrs.

Fairfield, Solomon - Private - 14th Infantry - Prisoner of War (Halifax), died at Halifax on 24 Jun 1813 from small pox.

Fait, Peter - Private - 7th Infantry - Company: Zachery Taylor - Age: 30 - Born: Baltimore - Trade: Shoemaker - Enlistment date: 1 May 1814 - Period: War - BLW 15533-160-12 - Discharged at Belle Fontaine, MO, on 25 Jul 1815.

Falconer, Jonathan H. - Second Lieutenant - 14th Infantry - Commissioned as a 3rd lieutenant on 4 May 1813; promoted to 2nd lieutenant on 14 Nov 1813; resigned on 4 Mar 1814.

Falconer, Joseph - Private - 36th Infantry - Company: Samuel Raisin - Age: 22 - Height: 5' 6" - Born: England - Trade: Carpenter - Enlistment date: 31 Oct 1814 - Place: Fredericktown, MD - Period: 5 Yrs - Deserted on 17 Jan 1815.

Falconer, Richard - Private - 36th Infantry - Company: Samuel Raisin - Age: 21 - Height: 5' 6" - Born: Loudoun Cty, VA - Trade: Rope maker - Enlistment date: 6 Aug 1814 - Period: War - BLW 6669-160-12 - Discharged at Washington, DC, on 20 Mar 1815.

Falians, Francis - Private - 14th Infantry - Company: Samuel Lane - Age: 36 - Height: 5' 7 1/2" - Born: Maryland - Enlistment date: 6 May 1812 - Place: Williamsport, MD - Period: 5 Yrs - Deserted at Williamsport, MD, on 23 Jun 1812.

Famity, Patrick - Private - 14th Infantry - Age: 26 - Height: 5' 2 1/2" - Born: Ireland - Trade: Weaver - Enlistment date: 22 Oct 1814 - Period: 5 Yrs.

Farman, Joshua - Private - 14th Infantry - Prisoner of War (Halifax), sent to England.

Farr, John B. - Corporal - 5th Infantry - Company: Richard Whartenby - Other regiment: 3rd Infantry (New) - Age: 22 - Height: 5' 9 1/2" - Born: Prince George's Cty, MD - Trade: Potter - Enlistment date: 11 Feb 1811 - Place: Baltimore - Period: 5 Yrs - Discharged on 11 Feb 1816.

Farr, Stephen - Private - 2nd Light Dragoons - Company: William Littlejohn - Age: 25 - Height: 5' 10" - Born: Maryland - Trade: Farmer - Enlistment date: 3 Sep 1813 - Place: Virginia - Period: 5 Yrs - Discharged at Baltimore on 3 Sep 1818.

Farrell, Michael - Private - 14th Infantry - Company: Thomas Montgomery - Age: 48 - Height: 5' 6" - Born: Galway, Ireland - Trade: Gardener - Enlistment date: 12 May 1812 - Place: Baltimore - Period: 5 Yrs - BLW 10165-160-12 - Prisoner of War (Halifax); discharged at Greenbush, NY, on 10 Apr 1815 for inability and old age.

Farrell, Samuel - Private - 38th Infantry - Company: Charles Stansbury - Age: 26 - Height: 5' 7 1/2" - Born: Ireland - Trade: Laborer - Enlistment date: 9 Nov 1814 - Period: War - BLW 5490-160-12 - Discharged at Baltimore on 6 Apr 1815.

Farris, William - Drummer - 36th Infantry - Company: Samuel Raisin - Age: 26 - Height: 5' 9" - Born: Ireland - Trade: Laborer - Enlistment date: 14 Nov 1814 - Place: Fredericktown, MD - Period: War - BLW 139-160-12 - Discharged at Washington, DC, on 20 Mar 1815.

By War of 1812 Soc. in MD

- Farry, Locklin - Private - 14th Infantry - Company: Reuben Gilder - Age: 30 - Height: 5' 2" - Born: Ireland - Enlistment date: 5 Apr 1813 - Period: War - Pension: Old War IF-25077 - BLW 10041-160-12 - Discharged at Burlington, VT, on 9 Apr 1814 for inability, incurable bruise received at Chateauguay.
- Fatherbrige, John Colear - Private - 38th Infantry - Company: John Brookes - BLW 17490-160-12.
- Faulk, William - Private - 4th Rifles - Company: Robert Scott - Age: 35 - Height: 5' 10" - Born: Harford Cty, MD - Trade: Laborer - Enlistment date: 24 Nov 1814 - Period: War - Discharged at Carlisle Barracks, PA, on 26 Mar 1815.
- Faulkner, Asa - Private - 14th Infantry - Company: Reuben Gilder - Pension: Land bounty to William Faulkner, brother and only heir at law of Asa Faulkner - BLW 17248-160-12.
- Fawcett, Rowland - Private - 38th Infantry - Company: James Hook - Age: 29 - Height: 5' 8 1/2" - Born: County Donegal, Ireland - Trade: Shoemaker - Enlistment date: 22 Mar 1814 - Place: Baltimore - Period: War - BLW 12685-160-12 Cancelled - Discharged at Fort Covington, MD, on 31 Mar 1815.
- Fazier, Foley - Private - 36th Infantry - Company: Joseph Hook - BLW 6250-160-12.
- Fearon, Daniel - Private - 14th Infantry - Company: David Cummings - Age: 28 - Height: 5' 5" - Born: Ireland - Trade: Weaver - Enlistment date: 29 Oct 1814 - Place: Williamsport, MD - Period: 5 Yrs.
- Feasker, Jacob - Private - 14th Infantry - Pension: Old War WF-14119 Rejected.
- Fee, Frederick - Private - 36th Infantry - Company: Samuel Raisin - Age: 21 - Height: 5' 4 1/2" - Born: Maryland - Trade: Shoemaker - Enlistment date: 8 Aug 1814 - Period: War - BLW 4379-160-12 - Discharged at Washington, DC, on 30 Mar 1815.
- Feeling, Jeremiah - Private - 14th Infantry - Company: David Cummings - Age: 29 - Height: 5' 9" - Born: Pennsylvania - Trade: Baker - Enlistment date: 9 Aug 1814 - Place: Baltimore - Period: War - Deserted.
- Felby, John B. - Private - 36th Infantry - Company: Joseph Nelson - Age: 23 - Height: 5' 10 1/2" - Born: France - Enlistment date: 29 Jun 1813 - Place: Cumberland, MD - Period: 1 Yr.
- Felix, John - Private - 1st Artillery - Company: George Armistead.
- Fellers, William - Private - 36th Infantry - Age: 21 - Height: 5' 6" - Born: Baltimore - Enlistment date: 28 Apr 1813 - Place: Baltimore - Period: 1 Yr.
- Felty, Jacob - Private - 14th Infantry - Company: Isaac Barnard - Pension: Land bounty to Leonard Felty, brother and other heirs at law of Jacob Felty - BLW 21106-160-12.
- Felvey, John - Corporal - 36th Infantry - Company: Joseph Hook - Age: 24 - Height: 5' 11 1/2" - Born: Paris, France - Trade: Tailor - Enlistment date: 6 Mar 1814 - Place: Sandy Point, VA - Period: War - BLW 15733-160-12 - Discharged at Fort Covington, MD, on 30 Mar 1815.

Fendall, John - First Lieutenant - 5th Infantry - Born: Maryland - Pension: Land bounty to Penelope C. D. Fendall, widow of John Fendall - BLW 12300-160-50 - Commissioned as an ensign on 27 May 1812; promoted to 2nd lieutenant on 30 Apr 1813; promoted to 1st lieutenant on 28 Jun 1814; discharged on 15 Jun 1815.

Fenning, Thomas - Private - 36th Infantry - Company: Joseph Hook - Age: 40 - Height: 5' 6" - Born: Ireland - Enlistment date: 16 Jul 1813 - Place: Baltimore - Period: 1 Yr.

Fenton, Joseph - Private - 14th Infantry - Company: Reuben Gilder - Other regiment: 4th Infantry (New) - Age: 38 - Height: 5' 3" - Born: Pennsylvania - Trade: Laborer - Enlistment date: 20 Apr 1813 - Place: Frederick, MD - Period: 5 Yrs - Pension: Old War IF-25085 - Discharged on 5 Aug 1815.

Feres, John - Private - 14th Infantry - Prisoner of War (Halifax), sent to England, captured at Beaver Dams on 24 Jun 1813.

Ferguson, Benjamin - Private - 22nd Infantry - Age: 37 - Height: 5' 7" - Born: Harford Cty, MD - Enlistment date: 27 Apr 1814 - Place: Pittsburgh.

Ferguson, John - Private - 1st Artillery - Company: George Armistead - Age: 33 - Height: 5' 9" - Born: Pennsylvania - Trade: Tinner - Enlistment date: 17 Apr 1812 - Place: Fort McHenry - Period: 5 Yrs - Discharged at Fort Columbus, NY, on 20 Sep 1813 for disability.

Ferguson, William - Private - 36th Infantry - Company: Samuel Raisin - Age: 21 - Height: 5' 4" - Born: Botetourt Cty, VA - Trade: Shoemaker - Enlistment date: 18 Sep 1814 - Place: Richmond, VA - Period: War - BLW 278-160-12 - Discharged at Washington, DC, on 20 Mar 1815.

Fernders, Adam - Recruit - 26th Infantry - Company: William Bezeau - Age: 23 - Height: 5' 6" - Born: Holland - Enlistment date: 19 Nov 1814 - Place: Baltimore - Period: War.

Ferrell, George L. - Private - 36th Infantry - Company: Joseph Hook - BLW 13543-160-12.

Ferrell, George L. - Private - 36th Infantry - Company: Joseph Hook - Age: 22 - Height: 5' 7 1/2" - Born: Kent Cty, MD - Trade: Shoemaker - Enlistment date: 25 Aug 1813 - Place: Chestertown, MD - Period: 1 Yr - BLW 13543-160-12 - Re-enlisted at Sandy Point, VA, 21 Mar 1814 for the war; discharged at Fort Covington, MD, on 30 Mar 1815.

Fetty, Jacob F. - Private - 14th Infantry - Company: Isaac Barnard - Enlistment date: 12 Nov 1812 - Period: 5 Yrs - Killed at Beaver Dams, UC, on 24 Jun 1813.

Fibbins, Francis - Private - 14th Infantry - Company: Reuben Gilder - Age: 32 - Height: 5' 5" - Born: Colbert, MD - Trade: Distiller - Enlistment date: 18 Mar 1814 - Place: Virginia - Period: War - BLW 5123-160-12 - Discharged at Greenbush, NY, on 4 May 1815.

Fickle, Isaac - Sergeant - 39th Infantry - Company: John Long - Age: 29 - Height: 5' 6" - Born: Maryland - Enlistment date: 11 Oct 1813 - Place: McMinnville, TN - Period: 1 Yr.

Fiddy, William - Musician - 12th Infantry - Company: Thomas Post - Age: 17 - Height: 5' 1 1/4" - Born: Allegany Cty, MD - Trade: Laborer - Enlistment date: 22 Aug 1813 - Place: Virginia - Period: 5 Yrs - Pension: Land bounty to Thompson Fiddy, father and heir at law of William Fiddy - BLW 25469-160-12 - Died at Sackets Harbor, NY, on 29 Oct or Nov 1814.

Fielder, Robert - Private - 1st Artillery - Company: Hopley Yeaton - Age: 31 - Height: 5' 6 1/2" - Born: St. Mary's, MD - Trade: Carpenter - Enlistment date: 19 Feb 1812 - Place: Fort Nelson, VA - Period: 5 Yrs - Died on 18 May 1815.

Fields, Edward C. - Musician - 5th Infantry - Company: William Henshaw - Age: 14 - Height: 5' 7" - Born: Baltimore - Trade: Laborer - Enlistment date: 8 May 1813 - Place: Maryland - Period: 5 Yrs.

Fields, George - Corporal - 1st Rifles - Company: Daniel Appling - Age: 20 - Height: 5' 10 3/4" - Born: Maryland - Trade: Mason - Enlistment date: 5 Apr 1813 - Place: Fort Hampton, AL - Period: 5 Yrs - Pension: Land bounty to John Fields, brother and other heirs at law of George Fields - BLW 21539-160-12 - Killed at Fort Erie, UC, on 6 Aug 1814.

Fields, James - Corporal - 7th Infantry - Company: Edward Hord - Age: 25 - Height: 5' 10 1/2" - Born: Maryland - Trade: Shoemaker - Enlistment date: 8 Feb 1809 - Period: 5 Yrs - Discharged on 7 Feb 1814.

Fields, John - Private - 1st Rifles - Company: Daniel Appling - Other regiment: Rifles - Age: 27 - Height: 5' 8" - Born: Maryland - Trade: Stone mason - Enlistment date: 21 Aug 1813 - Place: Knoxville, TN - Period: War - BLW 22718-160-12 - Discharged on 11 Jun 1815.

Fields, Thomas - Private - 14th Infantry - Company: William McIlvane - Enlistment date: 11 May 1812 - Period: 5 Yrs - Prisoner of War (Halifax), exchanged on 15 Apr 1814; died on 30 Jun 1814, shot by a sentinel.

Fields, William - Private - 36th Infantry - Company: Charles Randolph - Age: 21 - Height: 6' 1" - Born: Washington, MD - Enlistment date: 19 May 1813 - Place: Frederick, MD - Period: 1 Yr - Wagoner for the regimental quartermaster.

Fife, Andrew H. - Private - US Volunteers - Company: Stephen Moore - Enlistment date: 8 Sep 1812 - Period: 1 Yr.

Fife, Andrew H. - Gunner - US Sea Fencibles - Company: John Gill - Enlistment date: 20 Dec 1813 - Period: 1 Yr.

Fife, John - Private - 38th Infantry - Company: John Buck - BLW 20524-160-12.

Fife, William - Private - 36th Infantry - Company: Samuel Raisin - Age: 18 - Height: 5' 6" - Born: Scotland - Trade: Laborer - Enlistment date: 10 Oct 1814 - Place: Georgetown, DC - Period: War - BLW 106-160-12 - Discharged at Washington, DC, on 20 Mar 1815.

Fifer, Frederick - Private - 14th Infantry - Company: David Cummings - Age: 35 - Height: 5' 7" - Born: Germany - Trade: Laborer - Enlistment date: 24 Feb 1813 - Place: Baltimore - Period: 18 Mos - Discharged at Chazy, NY, on 24 Aug 1814; born as a Peifer.

Finity, Joseph - Sergeant - 26th Infantry - Company: William Bezeau - Age: 22 - Height: 5' 6 1/2" - Born: Fannard, Franklin Cty, PA - Trade: Saddler - Enlistment date: 8 Oct 1814 - Place: Baltimore - Period: War - Discharged at Philadelphia on 23 Mar 1815.

Finn, James - Private - 38th Infantry - Company: Charles Stansbury - Age: 28 - Height: 5' 5 1/2" - Born: Isle of Wexford, Ireland - Trade: Laborer - Enlistment date: 9 Nov 1814 - Place: Cumberland - Period: War - BLW 5492-160-12 - Discharged at Baltimore on 6 Apr 1815.

Fintin, William - Private - 26th Infantry - Company: William Bezeau - Age: 21 - Height: 5' 4" - Born: Baltimore - Enlistment date: 18 Nov 1814 - Place: Baltimore.

Fish, Levin - Private - 12th Infantry - Pension: Heirs received half pay for five years in lieu of land bounty - Died on 11 Feb 1814.

Fish, Peter - Private - 19th Infantry - Company: William Gill - Age: 18 - Height: 5' 5 1/2" - Born: Maryland - Enlistment date: 17 Aug 1814 - Place: New Lisbon, OH - Period: War.

Fish, Richard - Corporal - 26th Infantry (OH) - Company: Joel Collins - Other regiment: 19th Infantry - Age: 41 - Height: 5' 10" - Born: Maryland - Trade: Farmer - Enlistment date: 9 Jun 1814 - Place: Detroit - Period: War - BLW 15847-160-12 - Discharged at Detroit on 20 Jul 1815; re-enlisted.

Fisher, Anthony - Private - 2nd Artillery - Company: Nathan Towson - Age: 23 - Prisoner of War, captured during the Battle of Stoney Creek.

Fisher, Anthony - Private - 2nd Artillery - Company: James Barker - Other regiment: Corps of Artillery - Age: 22 - Height: 5' 2" - Born: Baltimore - Trade: Butcher - Enlistment date: 4 Aug 1812 - Place: Philadelphia - Period: 5 Yrs - BLW 15503-160-12 - Prisoner of War (Halifax), captured at Stoney Creek, UC, on 6 Jun 1813; discharged at Fort Niagara, NY, on 3 Aug 1817.

Fisher, Brice - Private - 26th Infantry (OH) - Company: Joel Collins - Other regiment: 19th Infantry - Age: 35 - Height: 5' 10" - Born: Maryland - Trade: Farmer - Enlistment date: 21 May 1814 - Place: Detroit - Period: War - BLW 10748-160-12 - Discharged at Detroit on 19 Jul 1815.

Fisher, Jacob - Private - 5th Infantry - Company: James Dorman - Age: 35 - Height: 5' 6" - Born: Pennsylvania - Trade: Comb maker - Enlistment date: 1 May 1812 - Place: Fredericktown, MD - Period: 5 Yrs - Discharged at Plattsburgh, NY, on 4 Jul 1814 for old age and debility.

Fisher, John - Private - 38th Infantry - Company: John Mowton - BLW 3017-160-12.

Fisher, Levin - Private - 19th Infantry - Company: Joel Collins - Age: 21 - Height: 5' 6" - Born: Maryland - Trade: Laborer - Enlistment date: 16 Mar 1814 - Place: Detroit - Period: War - BLW 2955-160-12 - Discharged at Detroit on 19 Jul 1815.

Fisher, Michael - Private - 5th Infantry - Company: Richard Bell - Age: 38 - Height: 5' 6 1/2" - Born: Germany - Trade: Stone cutter - Enlistment date: 11 Feb 1813 - Place: Baltimore - Period: 5 Yrs - Discharged at Washington, DC, on 19 Apr 1817.

Fisher, Peter - Private - 26th Infantry - Company: William Bezeau - Age: 21 - Height: 5' 6" - Born: Lancaster Cty, PA - Enlistment date: 27 Oct 1814 - Place: Baltimore - Period: War.

Fisher, Philip - Second Lieutenant - 36th Infantry - Pension: Wife Sarah

By War of 1812 Soc. in MD

Houston, SO-18753, SC-4592; married on 22 Jun 1817, Wheeling, VA - BLW 11685-160-50 - Commissioned as a 3rd lieutenant on 30 Apr 1813; promoted to 2nd lieutenant on 1 May 1814; discharged on 15 Jun 1815.

Fisher, Robert - Private - 14th Infantry - Company: Reuben Gilder - Other regiment: 4th Infantry (New) - Age: 35 - Height: 5' 9" - Born: Harford Cty, MD - Trade: Currier - Enlistment date: 10 May 1813 - Place: Maryland - Period: 5 Yrs - BLW 18146-160-12 - Discharged at Fort Gadsden, FL, on 10 May 1818.

Fisher, Samuel - Private - 27th Infantry (OH) - Company: Alexander Hill - Other regiment: 19th Infantry - Age: 25 - Height: 5' 10 1/2" - Born: Maryland - Trade: Blacksmith - Enlistment date: 22 Apr 1814 - Place: Steubenville, OH - Period: War - Re-enlisted at Erie, PA, on 1 Apr 1815 for 5 years.

Fisher, William - Private - 2nd Artillery - Company: Nathan Towson - Age: 35 - Height: 5' 6" - Born: Baltimore - Trade: Laborer - Enlistment date: 30 Apr 1812 - Place: Baltimore - Period: 5 Yrs - Discharged on 30 Jun 1817.

Fister, Jacob - Private - 14th Infantry - Company: Richard Arell - Other regiment: 4th Infantry (New) - Age: 23 - Height: 5' 7 1/2" - Born: Frederick, MD - Trade: Farmer - Enlistment date: 22 Jan 1813 - Place: Fredericktown, MD - Period: 5 Yrs - Prisoner of War (Halifax), discharged from Halifax on 3 Feb 1814; discharged on 22 Jan 1818.

Fitch, Henry - Private - US Volunteers - Company: Stephen Moore - Enlistment date: 8 Sep 1812 - Period: 1 Yr.

Fitch, Thomas - Private - 5th Infantry - Company: Richard Bell - Age: 27 - Height: 5' 6" - Born: Baltimore - Trade: Laborer - Enlistment date: 20 Feb 1811 - Place: Baltimore - Period: 5 Yrs - Deserted at Detroit on 28 Jul 1816.

Fitsler, William - Private - 14th Infantry - Company: Reuben Gilder - Other regiment: 4th Infantry (New) - Age: 20 - Height: 5' 5" - Born: Womelsdorf, PA - Trade: Stone mason - Enlistment date: 14 Nov 1812 - Place: Fredericktown, MD - Period: 5 Yrs - Deserted on 17 Jul 1815.

Fitz, Charles - Private - 14th Infantry - Prisoner of War (Halifax), discharged from Halifax on 3 Feb 1814.

Fitzgerald, James - Private - 14th Infantry - Company: Samuel Lane - Age: 19 - Height: 5' 6" - Born: Ireland - Enlistment date: 7 Jul 1812 - Place: Carlisle, PA - Period: 5 Yrs - Prisoner of War (Halifax), captured at Stoney Creek, UC, on 12 Jun 1813.

Fitzgerald, Samuel - Private - 8th Infantry - Company: William Jones - Age: 26 - Height: 5' 9 1/2" - Born: Charles Cty, MD - Trade: Farmer - Enlistment date: 18 Jul 1812 - Place: Georgia - Period: 5 Yrs - Discharged on 21 Aug 1815 for debility.

Fitzhugh, Samuel - Sergeant - 38th Infantry - Company: James Smith - Age: 21 - Height: 5' 3 1/2" - Born: Cambridge, MD - Trade: Tailor - Enlistment date: 21 Jun 1813 - Period: 1 Yr - Re-enlisted at Craney Island, VA, on 8 Mar 1814 for the war at Craney Island, VA.

Fitzhugh, William - Private - 14th Infantry - Company: Reuben Gilder - Other regiment: 4th Infantry (New) - Age: 19 - Height: 5' 7" - Born: Court Guard, North Hampton Cty, VA - Trade: Waterman - Enlistment date: 27 Apr 1813

Maryland Regulars in War of 1812

- Place: Maryland - Period: 5 Yrs.

Fitzjeffrey, Richard - Private - 6th Infantry - Company: Thomas Davis - Age: 45 - Height: 5' 9" - Born: Maryland - Trade: Laborer - Re-enlisted at New Orleans on 31 Mar 1813 for the war; discharged at New Orleans on 8 Apr 1815.

Fitzpatrick, John - Private - 14th Infantry - Prisoner of War (Halifax), discharged from Halifax on 27 Aug 1813.

Fitzpatrick, Thomas - Private - 36th Infantry - Company: Mortimer Hall - Age: 35 - Height: 5' 6" - Born: Omah, Ireland - Enlistment date: 12 May 1813 - Place: Dublin or Havre de Grace, MD - Period: 1 Yr.

Flaherty, Mathias - Private - 14th Infantry - Company: David Cummings - Enlistment date: 28 Jul 1812 - Period: 5 Yrs - Prisoner of War (Halifax), sent to England.

Flake, Richard - Private - 36th Infantry - Company: Joseph Merrick - Enlistment date: 4 Mar 1813 - Period: 1 Yr - Transferred to the U.S. Navy on 26 Oct 1813.

Flanagan, James - Private - 5th Infantry - Age: 26 - Height: 5' 6 1/2" - Born: Pennsylvania - Trade: Laborer - Enlistment date: 25 Feb 1813 - Place: Baltimore - Period: 5 Yrs.

Flanagan, Thomas - Private - 1st Rifles - Company: William Smyth - Age: 21 - Height: 5' 9 1/4" - Born: Long Green, MD - Trade: Laborer - Enlistment date: 5 Oct 1812 - Place: Sheppardstown, VA - Period: 5 Yrs.

Flanary, Peter - Private - 5th Infantry - Age: 21 - Height: 5' 6" - Born: Ireland - Trade: Seaman - Enlistment date: 11 Feb 1813 - Place: Baltimore - Period: 5 Yrs.

Fleming, Henry - Captain - 14th Infantry - Enlistment date: 6 Jul 1812 - BLW 12318-160-50 - Commissioned as a captain 8 Apr 1812; discharged on 15 Jun 1815.

Fleming, James - Sergeant - 14th Infantry - Company: David Cummings - Age: 30 - Height: 5' 6" - Born: Worcester Cty, MD - Trade: Merchant - Enlistment date: 29 Jul 1814 - Place: Snow Hill, MD - Period: 5 Yrs.

Flenner, John - Private - 3rd Rifles - Company: Edward Carrington - Age: 30 - Height: 5' 7" - Born: Frederick, MD - Trade: Blacksmith - Enlistment date: 7 Jun 1814 - Place: Beverly, VA - Period: 5 Yrs - Died on 24 Oct 1814.

Fletcher, Andrew - Private - 36th Infantry - Age: 23 - Height: 5' 8 1/4" - Born: Baltimore - Enlistment date: 1 Jul 1814 - Place: Westminster, MD - Period: War.

Fletcher, Frederick - Sergeant - 38th Infantry - Company: John Rothrock - Age: 21 - Height: 5' 5" - Born: Baltimore - Trade: Carpenter - Enlistment date: 2 Jun 1813 - Period: 1 Yr - BLW 66-160-12 - Re-enlisted on 20 Feb 1814 for the war; discharged at Craney Island, VA, on 15 Mar 1815.

Fletcher, George - First Lieutenant - 38th Infantry - Company: James Hook - Enlistment date: 3 Jun 1813 - Commissioned as a 2nd lieutenant on 20 May 1813; promoted to 1st lieutenant on 20 May 1814; discharged on 15 Jun 1815.

Fletcher, James - Private - 14th Infantry - Company: Joseph Marshall - Age:

By War of 1812 Soc. in MD

16 - Height: 5' 8" - Born: Alexandria, DC - Enlistment date: 9 Jul 1812 - Place: Baltimore - Period: 5 Yrs - Discharged at Burlington, VT, on 17 Mar 1814 for contracted arm, disabled when enlisted.

Fletcher, James - Private - 14th Infantry - Company: Joseph Marshall - Age: 21 - Height: 5' 8" - Born: Fairfax, VA - Trade: Cooper - Enlistment date: 5 May 1814 - Place: Baltimore - Period: War - Deserted at Utica, NY, on 12 Jan 1815.

Fletcher, John - Seaman - US Sea Fencibles - Company: Simmones Bunbury - Enlistment date: 2 May 1814 - Period: 1 Yr - Discharged at Fort McHenry on 24 Mar 1815.

Fletcher, John - Private - 36th Infantry - Age: 19 - Height: 5' 6" - Born: Boston - Enlistment date: 28 Apr 1813 - Place: Baltimore - Period: 1 Yr.

Fletcher, John R. - Sergeant - 15th Infantry - Company: Jacob Howell - Age: 32 - Height: 5' 10 1/2" - Born: Baltimore - Trade: Printer - Enlistment date: 26 Sep 1814 - Place: New York - Period: War - BLW 9957-160-12 - Discharged at Trenton, NJ, on 16 Mar 1815.

Fletcher, Joseph - Fifer - 38th Infantry - Company: John Rothrock - Age: 23 - Height: 5' 5" - Born: Baltimore - Trade: Blacksmith - Enlistment date: 22 Jun 1813 - Period: 1 Yr - Pension: Land bounty to Eliza Bintzell, sister and only heir at law of Joseph Fletcher - BLW 26778-160-12 - Re-enlisted for the war; discharged on 15 Mar 1815.

Fletcher, Joseph - Sergeant - 38th Infantry - Company: James Smith - Other regiment: 4th Infantry (New) - Age: 35 - Height: 5' 6" - Born: Fredericktown, MD - Trade: Miller - Enlistment date: 22 Feb 1814 - Place: Petersburg, VA - Period: 5 Yrs - Discharged on 22 Jul 1816 after furnishing a substitute.

Fletcher, Levin - Private - 14th Infantry - Company: Thomas Kearney - Enlistment date: 29 May 1812 - Period: 5 Yrs - Pension: Land bounty to James Fletcher, brother and other heirs at law of Levin Fletcher - BLW 26377-160-12 - Died on 12 Jan 1813.

Fletcher, Lewis - Private - 1st Light Dragoons - Company: Alexander Cummings - Age: 22 - Height: 5' 7" - Born: New York - Trade: Shoemaker - Enlistment date: 19 Jun 1812 - Place: Baltimore - Period: 5 Yrs - Deserted on 5 Oct 1815.

Fletcher, Richard - First Lieutenant - 14th Infantry - Born: Maryland - Commissioned as an ensign on 12 Mar 1812; promoted to 2nd lieutenant on 12 May 1813; promoted to 1st lieutenant on 1 Dec 1814; discharged on 15 Jun 1815; died on 8 Oct 1843.

Fletcher, Thomas G. - Private - 14th Infantry - Company: David Cummings - Age: 20 - Height: 5' 6" - Born: Sussex Cty, DE - Trade: Tanner - Enlistment date: 3 Dec 1814 - Place: Salisbury, MD - Period: War - BLW 16993-160-12 - Discharged at Baltimore on 16 Mar 1815.

Flinn, James - Private - 38th Infantry - Company: John Rothrock - Enlistment date: 5 Jun 1813 - Period: 1 Yr.

Flinn, James - Private - 38th Infantry - Company: James Hook - Age: 22 - Height: 5' 5 1/2" - Born: Georgetown, DC - Trade: Baker - Enlistment date: 22 Mar 1814 - Period: War.

Maryland Regulars in War of 1812

Flinn, James - Corporal - 38th Infantry - Company: James Hook - Age: 25 - Height: 5' 2 3/4" - Born: Georgetown, DC - Trade: Baker - Enlistment date: 27 Feb 1814 - Place: Baltimore - Period: War - Discharged at Fort Covington, MD, on 31 Mar 1815.

Flinn, William - Private - 5th Infantry - Company: William Henshaw - Other regiment: 3rd Infantry (New) - Age: 15 - Height: 4' 10" - Born: Baltimore - Trade: Brick layer - Enlistment date: 7 Jul 1813 - Place: Baltimore - Period: 5 Yrs - BLW 26411-160-12 - Discharged at Fort Howard, WI, on 7 Jul 1818.

Flint, John Jr. - Private - 2nd Rifles - Company: Batteal Harrison - Age: 23 - Height: 5' 11" - Born: Worcester Cty, MD - Trade: Farmer - Enlistment date: 12 Jun 1814 - Place: Brookville, Franklin Cty, IN - Period: War - BLW 22452-160-12 - Discharged on 30 Jun 1815.

Flood, John - Private - 5th Infantry - Company: William Henshaw - Other regiment: 3rd Infantry (New) - Age: 41 - Height: 5' 10" - Born: Baltimore - Trade: Shoemaker - Enlistment date: 17 Feb 1813 - Place: Baltimore - Period: 5 Yrs - BLW 18169-160-12 - Discharged at Fort Howard, WI, on 19 Feb 1818.

Flowers, Andrew - Private - 14th Infantry - Company: Thomas Montgomery - Enlistment date: 8 Jun 1812 - Period: 5 Yrs - Pension: Land bounty to Henry Flower, brother and other heirs at law of Andrew Flower - BLW 25982-160-12 - Died at Burlington, VT, of 25 Feb or 25 Apr, 1814, from sickness.

Floyd, Jonathan - Private - 3rd Infantry - Company: James Dinkins - Age: 21 - Height: 5' 5 1/2" - Born: Maryland - Trade: Farmer - Enlistment date: 24 Dec 1808 - Place: Talbot Cty, MD - Period: 5 Yrs - BLW 16681-160-12 - Re-enlisted at Mount Vernon on 24 Dec 1813; discharged at New Orleans on 9 Apr 1815.

Floyd, Samuel - Private - 14th Infantry - Company: Henry Fleming - Pension: Old War IF-9884.

Fogarty, Joseph - Private Servant - 36th Infantry - Company: Thomas Carbery - Private servant to Lieutenants Rogers and Latimer.

Fogger, Archibald - Private - 14th Infantry - Prisoner of War (Halifax), captured at Beaver Dams on 24 Jun 1813, discharged from Halifax on 19 Nov 1813.

Foley, James - Private - 36th Infantry - Company: Joseph Hook - Age: 20 - Height: 5' 7 1/2" - Born: England - Trade: Barber - Enlistment date: 17 Mar 1814 - Place: Annapolis - Period: War.

Foley, Lloyd - Private - 5th Infantry - Age: 40 - Height: 5' 10" - Born: Ireland - Enlistment date: 18 Mar 1813 - Place: Baltimore - Period: 5 Yrs.

Fonder, Joseph - Recruit - 26th Infantry - Company: William Bezeau - Age: 23 - Height: 5' 7" - Born: Baltimore - Trade: Cooper - Enlistment date: 15 Oct 1814 - Place: Baltimore - Period: War - Discharged at Philadelphia on 23 Mar 1815.

Fonder, Joseph - Private - US Volunteers - Company: Stephen Moore - Enlistment date: 8 Sep 1812 - Period: 1 Yr.

Foot, Henry - Private - 44th Infantry - Age: 27 - Height: 5' 6" - Born:

By War of 1812 Soc. in MD

Washington, MD - Enlistment date: 3 May 1814 - Place: Powder Magazine Barracks, LA - Period: War - Killed on 23 Dec 1814.

Fop less, Michael - Sergeant - 26th Infantry - Company: William Bezeau - Age: 26 - Height: 5' 10 1/2' - Born: Newcastle, DE - Trade: Hatter - Enlistment date: 11 Nov 1814 - Place: Baltimore - Period: War - BLW 527-160-12 - Discharged at Philadelphia on 23 Mar 1815.

Forbes, William - Seaman - US Sea Fencibles - Company: John Gill - Age: 18 - Height: 5' 2 1/2" - Born: Scotland - Enlistment date: 3 Jan 1814 - Period: 1 Yr.

Ford, Edward - Private - 14th Infantry - Age: 37 - Height: 5' 6" - Born: Frederick, MD - Trade: Color maker - Enlistment date: 2 Aug 1814 - Period: War.

Ford, Elijah B. - Private - 16th Infantry - Company: Thomas Harrell - Age: 25 - Height: 5' 7" - Born: Baltimore - Enlistment date: 7 Apr 1814 - Place: Philadelphia - Period: 5 Yrs - Deserted at Province Island Barracks, PA, on 17 May 1814.

Ford, George - Private - 36th Infantry - Company: Joseph Hook - Race: - Enlistment date: 4 Oct 1813 - Period: 1 Yr - Re-enlisted on 17 Aug 1814 for the war; discharged at Camp Snowden, claimed by his master as a minor.

Ford, George S. - Drummer - 38th Infantry - Company: James Hook - Other regiment: 4th Infantry (New) - Age: 14 - Height: 4' 10" - Born: Philadelphia - Trade: Painter - Enlistment date: 10 Nov 1814 - Place: Baltimore - Period: 5 Yrs - BLW 23117-160-12 - Discharged at Montpelier, AL, on 9 Nov 1819.

Ford, Joseph - Private - 36th Infantry - Company: Joseph Hook - Age: 39 - Height: 5' 8" - Born: Baltimore - Enlistment date: 10 May 1813 - Place: Baltimore - Period: 1 Yr - Deserted on 14 May 1813.

Ford, Joshua - Private - 14th Infantry - Company: Thomas Montgomery - Pension: Land bounty to Rosanna Ford, daughters and only heir at law of Joshua Ford - BLW 25773-160-12.

Ford, Samuel J. - Private - 36th Infantry - Age: 37 - Height: 6' 2" - Born: St. Mary's, MD - Enlistment date: 9 Jun 1813 - Place: Baltimore - Period: 1 Yr - Discharged on 12 Oct 1813.

Ford, Timothy - Private - 2nd Artillery - Company: Thomas Biddle - Other regiment: Corps of Artillery - Age: 48 - Height: 5' 6" - Born: Baltimore Cty - Trade: Farmer - Enlistment date: 1 Aug 1812 - Period: 5 Yrs - BLW 5677-160-12 - Wounded at Fort George, UC, on 27 Mar 1813; discharged at Boston, MA, on 13 Jun 1815 on Surgeon's Certificate.

Ford, William - First Lieutenant - 38th Infantry - Commissioned as a 1st lieutenant on 20 May 1813; discharged on 15 Jun 1815; died on 11 Sep 1834.

Ford, William - Private - 38th Infantry - Company: John Buck - Height: 5' 9" - Born: Cecil Cty, MD - Trade: Carpenter - Enlistment date: 24 Nov 1813 - Period: 1 Yr - Pension: Old War MC-877, heirs received half pay for five years in lieu of land bounty - Re-enlisted on 7 Mar 1814 for the war; died at Cantonment Baltimore on 10 Dec 1814.

Ford, William - Private - 36th Infantry - Company: Joseph Hook - Enlistment date: 4 Oct 1813 - Period: 1 Yr - Discharged at Camp Snowden on 4 Oct

Maryland Regulars in War of 1812

1814.

Fording, Jacob - Private - 2nd Infantry - Company: John Campbell - Age: 21 - Height: 5' 9 1/2" - Born: Waynesborough, MD - Trade: Hatter - Enlistment date: 27 Jul 1808 - Place: Washington, MS - Period: 5 Yrs - Re-enlisted at Mobile, AL, on 7 Apr 1813 for the war; Discharged at New Orleans on 9 Apr 1815.

Foreman, Daniel - Private - 5th Infantry - Company: William Henshaw - Age: 21 - Height: 5' 7" - Born: Blacksmith, MD - Trade: Blacksmith - Enlistment date: 19 Oct 1813 - Place: Baltimore - Period: 5 Yrs - Deserted on 26 Mar 1816.

Foreman, John - Private - 5th Infantry - Age: 23 - Height: 5' 11" - Born: Anne Arundel Cty, MD - Trade: Sail maker - Enlistment date: 15 Apr 1815 - Place: Maryland - Period: 5 Yrs.

Forester, George - Private - 2nd Artillery - Company: Daniel Cushing - Age: 42 - Height: 5' 5" - Born: Baltimore - Trade: Hatter - Enlistment date: 1 Nov 1812 - Place: Lexington, KY - Period: 5 Yrs - Discharged at Detroit on 31 Oct 1817.

Forester, Robert - Sergeant - 17th Infantry - Company: Henry Crittenden - Age: 40 - Height: 5' 10" - Born: Queen Anne's Cty, MD - Trade: Soldier - Enlistment date: 30 Dec 1813 - Place: Lexington, KY - Period: War - BLW 15710-160-12 - Discharged at Chillicothe, OH, on 9 Jun 1815.

Forkom, Emery - Private - 36th Infantry - Age: 19 - Height: 5' 3 1/4" - Born: Queen Anne's Cty, MD - Enlistment date: 29 May 1813 - Place: Centreville, MD.

Forman, Aaron - Private - 2nd Infantry - Company: Mathew Arbuckle - Age: 20 - Height: 5' 7" - Born: Berkley, VA - Trade: Farmer - Enlistment date: 13 Jun 1808 - Place: Fredericktown, MD - Period: 5 Yrs - Died at Fort Charlotte, AL, on 11 Dec 1813.

Forman, Daniel - Private - US Volunteers - Company: Stephen Moore - Enlistment date: 8 Sep 1812 - Period: 1 Yr.

Fornett, Charles - Private - 14th Infantry - Company: William McIlvane - Enlistment date: 11 Jul 1812 - Period: 5 Yrs.

Forney, Thomas - Private - 26th Infantry - Company: William Bezeau - Age: 21 - Height: 5' 4" - Born: Baltimore - Trade: Farmer - Enlistment date: 16 Nov 1814 - Place: Baltimore - Period: War - Discharged at Philadelphia on 23 Mar 1815.

Forrester, Charles - Private - 2nd Infantry - Company: Peter Schuyler - Age: 39 - Height: 5' 6" - Born: Queen Anne's Cty, MD - Trade: Sawyer - Enlistment date: 1806 - Re-enlisted at Fort Stoddert on 20 Jul 1811; discharged at Pass Christian on 31 Aug 1815 on Surgeon's Certificate.

Forrester, George - Private - 14th Infantry - Company: David Cummings - Age: 25 - Height: 5' 7 1/2" - Born: Anne Arundel Cty, MD - Trade: Blacksmith - Enlistment date: 4 Nov 1814 - Place: Baltimore - Period: War - BLW 5865-160-12 - Discharged at Baltimore on 18 Mar 1815.

Forrester, Jacob - Private - 1st Artillery - Company: George Armistead - Other regiment: 38th Infantry - Enlistment date: 10 Feb 1810 - Pension: Wife

By War of 1812 Soc. in MD

Catharine Tippet, WO-27870, WC-25026; married on 18 Oct 1823, Baltimore, MD; soldier died on 11 Jan 1830, Baltimore, MD - BLW 17679-160-12 - Discharged on 20 May 1813 and re-enlisted in 38th Infantry, Captain John Buck's Company.

Forrester, Jacob - Private - 38th Infantry - Company: Shepperd Leakin - Height: 5' 8" - Born: Baltimore - Trade: Baker - Enlistment date: 18 Aug 1813 - Place: Baltimore - Period: 1 Yr - BLW 17679-160-12 - Discharged at Baltimore on 30 Mar 1815.

Forrester, John - Sergeant - 22nd Infantry - Company: Jacob Carmack - Age: 19 - Height: 5' 5" - Born: Maryland - Trade: Saddler - Enlistment date: 2 Sep 1814 - Period: 5 Yrs.

Forsey, Elias P. - Seaman - US Sea Fencibles - Company: Simmones Bunbury - Age: 49 - Height: 5' 2 1/2" - Born: Baltimore - Enlistment date: 2 Apr 1814 - Period: 1 Yr - Discharged at Fort McHenry on 24 Mar 1815.

Forsythe, William - Private - 38th Infantry - Company: John Brookes - BLW 12722-160-12.

Fortine, Zebedee - Private - 14th Infantry - Company: Joseph Marshall - Age: 19 - Height: 5' 5" - Born: Harford Cty or Baltimore, MD - Trade: Cooper - Enlistment date: 1 May 1814 - Place: Baltimore - Period: War - BLW 14356-160-12 - Discharged at Greenbush, NY, on 3 May 1815.

Fortner, Richard - Private - 36th Infantry - Age: 21 - Height: 5' 6" - Born: Loudoun Cty, VA - Enlistment date: 6 Oct 1814 - Place: Manchester.

Fortune, Nicholas - Sergeant - 38th Infantry - Company: Isaac Aldridge - Age: 32 - Height: 5' 8 1/2" - Born: Baltimore - Trade: Baker - Enlistment date: 10 Feb 1814 - Period: War - Pension: Wife Ann M., WO-316 - BLW 13720-160-12 - Discharged at Baltimore on 3 Apr 1815.

Foss, Joseph - Ensign - 38th Infantry - Commissioned as an ensign on 20 May 1813; dismissed on 3 Oct 1813.

Fossett, John - Private - 36th Infantry - Company: Joseph Hook - Age: 31 - Height: 5' 6" - Born: Alexandria, DC - Trade: Laborer - Enlistment date: 15 Aug 1814 - Place: Georgetown, DC - Period: War - BLW 157-160-12 - Discharged at Fort Covington, MD, on 30 Mar 1815.

Fossett, William - Private - 36th Infantry - Company: Thomas Carbery - Age: 20 - Height: 5' 3" - Born: Fairfax Cty, VA - Trade: Laborer - Enlistment date: 22 Feb 1814 - Place: Leonardstown, MD - Period: War - BLW 158-160-12 - Discharged at Baltimore on 31 Mar 1815.

Foster, Caleb - Private - 1st Infantry - Company: Simon Owens - Age: 18 - Height: 5' 6" - Born: Baltimore - Trade: Farmer - Enlistment date: 5 Apr 1810 - Place: Chambersburg, PA - Period: 5 Yrs - Deserted at Fort Erie, UC, on 24 Sep 1814.

Foster, George - Private - 2nd Artillery - Company: Nathan Towson - Age: 39 - Prisoner of War (Quebec), captured during the Battle of Stoney Creek and released on 31 Oct 1813.

Foster, George - Private - 1st Artillery - Company: George Armistead - Age: 40 - Height: 5' 10" - Born: England - Trade: Soldier - Enlistment date: 8 Feb 1810 - Place: Fort McHenry - Period: 5 Yrs - Discharged at New York on 7

Maryland Regulars in War of 1812

Feb 1815.

Foster, George - Private - 38th Infantry - Company: Shepperd Leakin - Age: 23 - Height: 5' 7" - Born: County Donegal, Ireland - Trade: Farmer - Enlistment date: 21 Jul 1814 - Place: Baltimore - Period: War - BLW 9796-160-12 - Discharged at Fort McHenry on 28 Mar 1815.

Foster, Jacob - Private - 14th Infantry - Prisoner of War, exchanged on 15 Apr 1814.

Foster, James B. - Private - 38th Infantry - Company: John Brookes - Enlistment date: 23 Jul 1813 - Period: 1 Yr - Died on 31 Jul 1813.

Foster, John - Private - 29th Infantry - Company: Peter Van Buren - Other regiment: 6th Infantry (New) - Age: 23 - Height: 5' 4" - Born: Baltimore or Havre de Grace, MD - Trade: Mariner - Enlistment date: 15 May 1814 - Place: Plattsburgh, NY - Period: 5 Yrs - Deserted at Rouse's Point on 26 Jun 1818.

Foster, John S. - Private - 14th Infantry - Company: James Britton - Age: 27 - Height: 5' 5" - Born: Marblehead, MA - Trade: Sailor - Enlistment date: 10 Apr 1813 - Period: 5 Yrs - Discharged at Washington, DC, on 25 Jul 1815 for inability.

Foster, William - Private - 36th Infantry - Age: 23 - Height: 5' 7 1/2" - Born: Talbot Cty, MD - Enlistment date: 4 Jun 1813 - Place: Cambridge, MD - Period: 1 Yr.

Foster, William - Private - 38th Infantry - Company: John Rothrock - BLW 26152-160-12.

Foulton, James - Private - 14th Infantry - Company: Thomas Montgomery - Enlistment date: 8 Dec 1812 - Period: 18 Mos - Discharged on 14 Apr 1814.

Fountain, Henry - Private - 38th Infantry - Company: John Brookes - BLW 4256-160-12.

Fowell, Daniel - Private - 38th Infantry - Company: Shepperd Leakin - Age: 28 - Height: 5' 8" - Born: Dorchester Cty, MD - Trade: Mariner - Enlistment date: 22 Feb 1814 - Place: Baltimore - Period: War.

Fowler, George - Corporal - 2nd Light Dragoons - Company: John Burd - Age: 33 - Height: 5' 7 1/4" - Born: Baltimore - Trade: Shoemaker - Enlistment date: 19 Jun 1812 - Place: Bedford, PA - Period: 5 Yrs - Discharged on 19 Jun 1817.

Fowler, James - Private - 36th Infantry - Age: 20 - Height: 5' 5" - Born: Ireland - Enlistment date: 29 Aug 1813 - Place: Baltimore.

Fowler, James - Private - 14th Infantry - Company: David Cummings - Age: 23 - Height: 5' 8" - Born: Calvert, MD - Trade: Farmer - Enlistment date: 1 Oct 1814 - Place: Baltimore - Period: 5 Yrs.

Fowler, Richard - Private - 38th Infantry - Company: Charles Stansbury - Age: 33 - Height: 5' - Born: Annapolis - Trade: Carpenter - Enlistment date: 24 Sep 1814 - Place: Annapolis - Period: War - BLW 5491-160-12 - Discharged at Baltimore on 6 Apr 1815.

Fowler, Thomas - Private - 14th Infantry - Company: Thomas Montgomery - Age: 22 - Height: 5' 6" - Born: Maryland - Enlistment date: 29 Apr 1812 - Period: 5 Yrs - Pension: Land bounty to Mary Martin, aunt and other heirs at law of Thomas Fowler - BLW 25705-160-12 - Killed during the Battle of Lyon's Creek on 19 Oct 1814.

Fowler, Thomas - Private - Light Artillery - Company: Benjamin Branch - Age: 42 - Height: 6' - Born: Georgetown, MD - Trade: Laborer - Enlistment date: 19 May 1812 - Place: Georgetown, DC - Period: 5 Yrs - Discharged on 19 Mar 1816.

Fowler, Zachariah - Private - 14th Infantry - Company: Reuben Gilder - Enlistment date: 21 Apr 1813 - Period: 5 Yrs - Pension: Land bounty to Mary Martin, sister and other heirs at law of Zachariah Fowler - BLW 25381-160-12.

Fox, Henry - Private - 36th Infantry - Company: Joseph Merrick - Age: 20 - Height: 5' 8" - Born: Wilmington, DE - Enlistment date: 1 May 1813 - Place: Baltimore - Period: 1 Yr - Discharged on 19 Dec 1813.

Fox, Joseph - Private - 19th Infantry - Company: Joel Collins - Age: 27 - Height: 5' 7" - Born: Maryland - Trade: Farmer - Enlistment date: 15 Apr 1814 - Period: War - Discharged at Detroit on 20 Jul 1815.

Fox, Samuel - Sergeant - 1st Rifles - Company: William Smyth - Age: 19 - Height: 5' 8" - Born: Frederick, MD - Trade: Weaver - Enlistment date: 5 Aug 1812 - Place: Fredericktown, MD - Period: 5 Yrs - Discharged at New Orleans on 4 Aug 1817.

Foy, Edward - Private - 5th Infantry - Age: 23 - Height: 5' 5 1/2" - Born: Baltimore Cty - Trade: Shoemaker - Enlistment date: 9 May 1813 - Place: Maryland - Period: 5 Yrs.

Foy, Gregory - First Lieutenant - US Sea Fencibles - Company: Simmones Bunbury - Commissioned as a 1st lieutenant on 1 Oct 1813; discharged at Fort McHenry on 15 Jun 1815.

Foy, Gregory - Sergeant - US Volunteers - Company: Stephen Moore - Enlistment date: 8 Sep 1812 - Period: 1 Yr.

Foy, Jacob - Private - 2nd Light Dragoons - Company: William Littlejohn - Age: 27 - Height: 5' 6" - Born: Baltimore - Trade: Baker - Enlistment date: 1 Jul 1814 - Period: 5 Yrs - Discharged on 1 Jan 1819.

Fracewand, John - Private - 36th Infantry - Age: 32 - Height: 5' 9" - Born: France - Enlistment date: 2 Sep 1813 - Place: Baltimore - Period: 1 Yr.

Fraher, Edward - Private - 14th Infantry - Company: Thomas Montgomery - Age: 55 - Height: 5' 8" - Born: Ireland - Enlistment date: 28 May 1812 - Period: 5 Yrs - Prisoner of War (Halifax).

Frailey, Leonard - Major - 38th Infantry - Born: Maryland - Place: Maryland - Pension: Old War IF-25094 - BLW 324-160-50 - Commissioned as a major on 19 May 1813; resigned on 1 May 1814; died on 23 Jul 1864.

Frame, Robinson - Private - 2nd Artillery - Company: Frederick Evans - Age: 40 - Height: 5' 11" - Born: Delaware - Trade: Tanner - Enlistment date: 20 Dec 1814 - Place: Fort McHenry - Period: War - Discharged on 24 Mar 1815.

Franc, Christian - Private - 14th Infantry - Prisoner of War (Halifax), sent to England.

Francis, William - Private - 14th Infantry - Company: Robert Kent - Other regiment: 4th Infantry (New) - Enlistment date: 1 May 1813 - Period: 5 Yrs - On board the fleet on Lake Ontario; deserted from Fort Severn on 18 Jul 1815.

Francisco, John - Private - 14th Infantry - Prisoner of War (Halifax), discharged from Halifax on 3 Feb 1814.

Francy, John - Private - 14th Infantry - Company: David Cummings - Age: 25 - Height: 5' 9" - Born: France - Trade: Gardener - Enlistment date: 12 Feb 1815 - Place: Salisbury, MD - Period: War - Discharged on 18 Mar 1815.

Franklin, James D. - Sergeant - 38th Infantry - Company: James Smith - BLW 16589-160-12.

Franklin, Jeremiah - Private - 3rd Infantry - Company: Mossman Houston - Age: 39 - Height: 5' 9 1/2" - Born: Baltimore - Trade: Shoemaker - Enlistment date: 28 Aug 1808 - Place: Durien - Period: 5 Yrs - Re-enlisted at English Turn, LA, on 20 Apr 1813 for the war.

Frazier, Daniel - Private - 5th Infantry - Company: Richard Whartenby - Age: 26 - Height: 5' 8 3/4" - Born: Anne Arundel Cty, MD - Trade: Mariner - Enlistment date: 22 Feb 1812 - Place: Baltimore - Period: 5 Yrs.

Frazier, Hubbard - Private - 14th Infantry - Other regiment: 4th Infantry (New) - Height: 5' 8" - Born: Dorchester Cty, MD - Trade: Farmer - Enlistment date: 22 May 1812 - Period: 5 Yrs - BLW 12462-160-12 - Prisoner of War (Halifax), captured at Beaver Dams on 24 Jun 1813, discharged from Halifax on 9 Nov 1813; discharged on 22 May 1817.

Frazier, John - Private - 42nd Infantry - Company: John Carty - Age: 21 - Height: 5' 3" - Born: Maryland - Trade: Seaman - Enlistment date: 9 Nov 1813 - Period: War.

Frazier, Thomas - Private - 40th Infantry - Age: 22 - Height: 5' 5" - Born: Cambridge, MD - Trade: Brush maker - Enlistment date: 3 Feb 1815 - Place: Boston - Period: War - Deserted on 11 Feb 1815.

Frazier, William - Private - 36th Infantry - Company: Joseph Hook - Age: 23 - Height: 6' 1/4" - Born: Philadelphia - Enlistment date: 3 Sep 1813 - Period: 1 Yr - BLW 15237-160-12 - Re-enlisted at Annapolis on 16 Apr 1814 for the war.

Freaner, William - Private - 6th Infantry - Age: 21 - Height: 5' 8" - Born: Hagerstown, MD - Enlistment date: 16 May 1814 - Place: Reading, PA.

Frederick McCombs - Private - US Volunteers - Company: Stephen Moore - Enlistment date: 8 Sep 1812 - Period: 1 Yr.

Frederick, Lewis - Private - Light Artillery - Company: John McIntosh - Race: 25 - Age: 25 - Height: 5' 7 1/2" - Born: Baltimore - Trade: Mariner - Enlistment date: 14 Jul 1814 - Place: Boston - Period: War - Possible Black (not verified); discharged at Plattsburgh, NY, on 31 May 1815.

Frederick, Paul - Private - 36th Infantry - Age: 35 - Height: 5' 5" - Born: Sweden - Enlistment date: 7 May 1813 - Place: Baltimore - Period: 1 Yr.

Frederick, Paul - Seaman - US Sea Fencibles - Company: Simmones Bunbury - Enlistment date: 23 May 1814 - Period: 1 Yr - Discharged at Fort McHenry on 24 Mar 1815.

Fredericks, Jacob - Private - 5th Infantry - Company: Colin Buckner - Other regiment: 3rd Infantry (New) - Age: 18 - Height: 5' 5" - Born: Baltimore - Trade: Blacksmith - Enlistment date: 11 Mar 1813 - Place: Baltimore - Period: 5 Yrs - BLW 18270-160-12 - Discharged on 1 Mar 1818.

Free, John R. - Private - 36th Infantry - Age: 25 - Height: 5' 9" - Born: Maryland - Trade: Miller - Enlistment date: 8 Feb 1815 - Place: Georgetown, DC - Period: War - Pension: Land bounty to John Free, uncle and heir at law of John R. Free - BLW 1020-160-12 - Discharged on 13 Mar 1815.

Freebrook, George - Private - 36th Infantry - Company: Joseph Merrick - Age: 25 - Height: 5' 4" - Born: Elfert, Germany - Enlistment date: 28 Apr 1813 - Place: Baltimore - Transferred to the U.S. Navy on 5 Nov 1813.

Freeman, John - Private - 36th Infantry - Company: Joseph Hook - Enlistment date: 18 Jun 1814 - Period: War - Pension: Land bounty to Elizabeth Campbell, sister and other heirs at law of John Freeman - BLW 11863-160-12 - Died at Baltimore on 26 Dec 1814.

Freeman, John - Private - 32nd Infantry - Company: James Clark - Age: 23 - Height: 5' 8" - Born: Maryland - Trade: Laborer - Enlistment date: 1 Jun 1813 - Place: Lewistown, DE - Period: 1 Yr - Discharged on 30 Jun 1814.

Freeman, John - Private - 1st Light Dragoons - Company: Alexander Cummings - Other regiment: Corps of Artillery - Age: 22 - Height: 5' 9 1/2" - Born: Maryland - Trade: Farmer - Enlistment date: 2 Jun 1813 - Place: Plattsburgh, NY - Period: 5 Yrs - Pension: Land bounty to James Freeman, Elizabeth Decamps, Mary Colwell and Hannah Fritzue, brothers and sisters and only heirs at law of John Freeman - BLW 27195-160-12 - Drowned on 1 Jan 1817.

Freeman, Thomas - Private - 26th Infantry - Company: William Bezeau - Age: 28 - Height: 5' 6 1/2" - Born: Churchill, Somerset, MD - Enlistment date: 16 Oct 1814 - Place: Baltimore.

Freeman, William - Seaman - US Sea Fencibles - Company: John Gill - Age: 38 - Height: 5' 3" - Born: Talbot Cty, MD - Enlistment date: 7 Jan 1814 - Period: 1 Yr.

Freeny, Peter - Private - 44th Infantry - Company: Anatole Peychand - Age: 30 - Height: 5' 10 1/2" - Born: Maryland - Trade: Farmer - Enlistment date: 27 Nov 1814 - Place: Wharton, St. Tammany Parish, LA - Period: War - Discharged at New Orleans on 8 Apr 1815.

French, James - Private - 14th Infantry - Company: David Cummings - Age: 18 - Height: 5' 4" - Born: Cecil Cty, MD - Trade: Shoemaker - Enlistment date: 31 Oct 1814 - Place: Baltimore - Period: War - BLW 9704-160-12 - Discharged at Baltimore on 16 Mar 1815.

French, Samuel - Private - 20th Infantry - Company: John Stanard - Other regiment: 8th Infantry (New) - Age: 18 - Height: 4' 8" - Born: Maryland - Trade: Sailor - Enlistment date: 21 Jul 1812 - Place: Alexandria, DC - Period: 5 Yrs - Discharged at Grand Ecor on 21 Jul 1817.

Frenchaw, Joseph H. - Private - 14th Infantry - Company: Joseph Marshall - Enlistment date: 8 Feb 1814 - Period: War - Drowned in Delaware River on 9 Oct 1814.

Frequy, John - Private - 36th Infantry - Age: 21 - Height: 5' 9" - Born: New

Maryland Regulars in War of 1812

York - Enlistment date: 4 Dec 1814 - Place: Fredericksburg, VA.

Freshman, Jacob - Private - 38th Infantry - Company: Charles Stansbury - Age: 26 - Height: 5' 11" - Born: Frederick Cty, MD - Trade: Nail smith - Enlistment date: 21 Nov 1814 - Period: War - BLW 12604-160-12 - Discharged at Baltimore on 6 Apr 1815.

Friel, Barney - Private - 38th Infantry - Company: Charles Stansbury - BLW 21507-160-12.

Frost, Campbell J. - Private - 38th Infantry - Company: James Hook - Age: 22 - Height: 5' 8 1/2" - Born: Alexandria, DC - Trade: Baker - Enlistment date: 12 Mar 1814 - Place: Baltimore - Period: War - BLW 11587-160-12 - Discharged at Fort Covington, MD, on 31 Mar 1815.

Fry, Benjamin - Private - 2nd Infantry - Company: Francis Johnston - Other regiment: 1st Infantry (New) - Age: 21 - Height: 5' 9" - Born: Maryland - Trade: Shoemaker - Enlistment date: 24 Apr 1808 - Place: Frederick, MD - Period: 5 Yrs - Re-enlisted at Fort Stoddard on 12 Jan 1813; died at New Orleans on 20 Mar 1816.

Fry, Jacob - Private - 36th Infantry - Company: Joseph Hook - BLW 7049-160-12.

Fry, Patrick - Private - 38th Infantry - Company: John Mowton - Age: 26 - Height: 5' 4" - Born: Baltimore - Trade: Carpenter - Enlistment date: 12 Jul 1814 - Place: Norfolk, VA - Period: War - Pension: Land bounty to Mary Cochran and other heirs at law of Patrick Fry - BLW 10067-160-12 - Died at Craney Island, VA, on 9 Mar 1815.

Fry, Robert - Musician - 36th Infantry - Company: Samuel Raisin - Age: 16 - Height: 4' 9" - Born: Washington, DC - Trade: Ship's carpenter - Enlistment date: 21 Oct 1814 - Period: War - BLW 5384-160-12 - Discharged at Washington, DC, on 20 Mar 1815.

Fugua, William - Private - 38th Infantry - Company: John Mowton - BLW 13992-160-12.

Fuller, Daniel - Private - 5th Infantry - Age: 41 - Height: 5' 5" - Born: Philadelphia - Trade: Sailor - Enlistment date: 5 Feb 1813 - Place: Baltimore - Period: 5 Yrs.

Fuller, Jacob - Recruit - 26th Infantry - Company: William Bezeau - Race: 26 - Age: 26 - Height: 5' 5" - Born: Maryland - Trade: Farmer - Enlistment date: 18 Jan 1815 - Place: Philadelphia - Period: War - Listed on the Descriptive Roll of Colored Men, 1 Apr 1815; deserted on 22 Jan 1815.

Fullerton, John - Private - 14th Infantry - Prisoner of War (Halifax), discharged from Halifax on 9 Nov 1813.

Fulting, Jeremiah - Recruit - 26th Infantry - Company: William Bezeau - Race: 34 - Age: 34 - Height: 5' 11" - Born: Maryland - Trade: Laborer - Enlistment date: 24 Jan 1815 - Place: Philadelphia - Period: War - BLW 832-320-14 Cancelled; BLW 1039-320-14 - Listed on the Descriptive Roll of Colored Men, 1 Apr 1815; land bounty to Jeremiah Hubbard, cousin & only heir at law of Jeremiah Fulting, deceased; first bounty was cancelled by an order from the War Department, 19 May 1820.

Fulton, James - Private - 14th Infantry - Company: Reuben Gilder - Enlistment date: 28 Dec 1812 - Period: 18 Mos - Discharged at Burlington, VT, on 14 Apr 1814 for old age (over 50).

Fulton, John - Private - 36th Infantry - Company: Samuel Raisin - Died in hospital on 17 Feb 1815.

Fulton, Josiah - Private - 38th Infantry - Company: Shepperd Leakin - Age: 25 - Height: 5' 6" - Born: Boston - Trade: Distiller - Enlistment date: 8 Sep 1814 - Place: Baltimore - Period: War - Deserted at Camp Baltimore on 13 Sep 1814.

Funderville, John - Private - Light Dragoons - Age: 26 - Height: 5' 7" - Born: Baltimore - Trade: Blacksmith - Enlistment date: 26 Feb 1813 - Place: Baltimore - Period: 18 Mos - Prisoner of War, sent to England; discharged on 31 Mar 1815.

Furnace, Littleton - Private - 38th Infantry - Company: Charles Stansbury - Age: 21 - Height: 5' 10" - Born: Somerset Cty, MD - Trade: Plasterer - Enlistment date: 26 Oct 1814 - Place: Newtown - Period: War - BLW 5493-160-12 - Discharged at Baltimore on 6 Apr 1815.

G

Gaines, James - Private - 36th Infantry - Company: Joseph Hook - Age: 38 - Height: 5' 10 1/4" - Born: Culpeper Cty, VA - Trade: Laborer - Enlistment date: 2 Apr 1814 - Place: Fort Washington, MD - Period: War - BLW 3221-160-12 - Discharged at Fort Covington, MD, on 30 Mar 1815.

Gaines, John - Private - 24th Infantry - Company: James Stuart - Other regiment: 7th Infantry (New) - Age: 44 - Height: 5' 3" - Born: Baltimore - Trade: Farmer - Enlistment date: 13 Jun 1813 - Period: 5 Yrs - Pension: Wife Dorcas, Old War IF-14381 Rejected - Prisoner of War (Quebec), captured at Fort Niagara, NY, on 19 Dec 1813 and exchanged on 11 May 1814; discharged at Fort Hawkins, GA, on 31 Dec 1815 on Surgeon's Certificate.

Gaines, John - Private - 38th Infantry - Age: 30 - Height: 5' 5" - Born: England - Trade: Tailor - Enlistment date: 30 May 1814 - Period: War.

Gaither, Lott - Private - 38th Infantry - Company: John Mowton - Age: 23 - Height: 5' 9" - Born: Maryland - Trade: Farmer - Enlistment date: 21 Feb 1814 - Place: Craney Island, VA - Period: War - BLW 2859-160-12 - Discharged at Craney Island, VA, on 15 Mar 1815.

Gale, James H. - Captain - 14th Infantry - Other regiment: 4th Infantry (New) - Born: Maryland - Pension: Placed on the pension rolls on 22 Aug 1831 - BLW 6986-160-50 - Commissioned as an ensign, 14th Infantry, on 12 Mar 1812; promoted to 2nd lieutenant on 13 Mar 1813; promoted to 1st lieutenant on 29 Jun 1814; transferred to 4th Infantry on 17 May 1815; promoted to captain on 31 Jul 1817; transferred to 1st Infantry on 1 Jun 1821; resigned on 28 Jul 1831.

Gallagher, John - Corporal - 14th Infantry - Company: David Cummings - Age: 23 - Height: 5' 8 1/4" - Born: Washington, MD - Enlistment date: 2 May 1812 - Place: Williamsport, MD - Period: 5 Yrs - Pension: Land bounty to Elizabeth Gallagher, sister and other heirs at law of John Gallagher - BLW 26227-160-12 - Died at Lewiston, NY, in Jul 1813.

Maryland Regulars in War of 1812

Gallagher, Patrick - Private - 3rd Infantry - Company: Robert Moore - Age: 33 - Height: 5' 9 1/4" - Born: Tyrone, Ireland - Trade: Weaver - Enlistment date: 9 Dec 1808 - Place: Baltimore - Period: 5 Yrs - Discharged at Fort Claiborne, AL, on 8 Dec 1813.

Gallasphy, James - Private - Light Dragoons - Age: 25 - Height: 5' 9 1/2" - Born: York Cty, PA - Trade: Blacksmith - Enlistment date: 21 Dec 1812 - Place: Fredericktown, MD - Period: 5 Yrs.

Gallop, William - Third Lieutenant - 36th Infantry - Other regiment: 4th Rifles - Commissioned as an ensign, 36th Infantry, on 30 Apr 1813; promoted to 3rd lieutenant on 21 Feb 1814; transferred to 4th Rifles on 31 May 1814; discharged on 15 Jun 1814.

Galloway, Joseph - Private - 6th Infantry - Company: Gad. Humphreys - Age: 22 - Height: 5' 7 1/2" - Born: Maryland - Trade: Tailor - Enlistment date: 7 Oct 1813 - Period: War - Deserted at Governor's Island, NY, on 10 Jun 1814.

Gamble, John - Private - 38th Infantry - Company: Charles Stansbury - Age: 35 - Height: 5' 5" - Born: Londonderry, Ireland - Trade: Farmer - Enlistment date: 11 Oct 1814 - Place: Baltimore - Period: War - BLW 23350-160-12 - Discharged at Baltimore on 6 Apr 1815.

Gamble, Nicholas - Private - US Volunteers - Company: Stephen Moore - Enlistment date: 8 Sep 1812 - Period: 1 Yr.

Gant, Kehelm - Private - 36th Infantry - Company: Thomas Carbery - Age: 21 - Height: 5' 5" - Born: Charles Cty, MD - Trade: Brick layer - Enlistment date: 11 Oct 1814 - Period: War - BLW 226-160-12 - Discharged at Georgetown, DC, on 15 Mar 1815.

Gardiner, John - Private - 38th Infantry - Company: Shepperd Leakin - Age: 28 - Height: 5' 7" - Born: Ireland - Trade: Soldier - Enlistment date: 31 Aug 1814 - Place: Baltimore - Period: 1 Yr.

Gardiner, Peter - Private - US Volunteers - Company: Stephen Moore - Enlistment date: 8 Sep 1812 - Period: 1 Yr.

Gardiner, Samuel - Seaman - US Sea Fencibles - Company: John Gill - Age: 35 - Height: 5' 7 1/2" - Born: Baltimore - Enlistment date: 1 Jan 1814 - Period: 1 Yr.

Gardiner, Samuel - Private - US Volunteers - Company: Stephen Moore - Enlistment date: 8 Sep 1812 - Period: 1 Yr.

Gardner, George - Private - 5th Infantry - Company: William Bird - Age: 31 - Height: 5' 8" - Born: Baltimore - Trade: Laborer - Enlistment date: 25 Apr 1813 - Place: Maryland - Period: 5 Yrs.

Gardner, George - Private - 38th Infantry - Company: Charles Stansbury - Age: 30 - Height: 5' 6 3/4" - Born: Anne Arundel Cty, MD - Trade: Farmer - Enlistment date: 23 Sep 1814 - Place: Annapolis - Period: War - Died on 2 Feb 1815.

Gardner, Jacob - Private - 2nd Artillery - Company: Stanton Sholes - Other regiment: Corps of Artillery - Age: 39 - Height: 5' 7" - Born: Maryland - Trade: Carpenter - Enlistment date: 16 Mar 1814 - Place: Greensburg, Beaver Cty, PA - Period: 5 Yrs - Discharged at Detroit on 26 Oct 1815, Surgeon's Certificate.

By War of 1812 Soc. in MD

Gardner, James - Private - 17th Infantry - Company: Caleb Holder - Age: 25 - Height: 5' 8" - Born: Baltimore Cty - Trade: Shoemaker - Enlistment date: 27 Sep 1814 - Period: War - BLW 25488-160-12 - Discharged at Detroit on 30 Jul 1815.

Gardner, John - Private - 14th Infantry - Company: Reuben Gilder - Age: 23 - Height: 5' 6" - Born: Dorchester Cty, MD - Trade: Shoemaker - Enlistment date: 27 Feb 1814 - Place: Plattsburgh, NY - Period: War.

Gardner, John - Private - Light Dragoons - Age: 22 - Height: 5' 6 1/2" - Born: Maryland - Trade: Shoemaker - Enlistment date: 22 Feb 1813 - Place: Baltimore - Period: 18 Mos.

Gardner, Mathew - Private - 36th Infantry - Company: Joseph Hook - Enlistment date: 7 Sep 1813 - Period: 1 Yr - Pension: Land bounty to Grace Gardner, widow of Mathew Gardner - BLW 16617-160-12 - Re-enlisted on 15 Aug 1814 for the war; died at Baltimore, MD, on 16 Nov 1814.

Gardner, Silas T. - Private - 14th Infantry - Company: Thomas Montgomery - Enlistment date: 26 Dec 1812 - Place: Pennsylvania - Period: 5 Yrs - Pension: Old War IF-27844 - BLW 24805-160-12 - Discharged at Burlington, VT, on 21 Apr 1814 on Surgeon's Certificate, inability and lame foot.

Garland, Levi - Private - 14th Infantry - Company: Richard Arell - Other regiment: 4th Infantry (New) - Age: 20 - Height: 5' 2 1/2" - Born: Mifflin, PA - Trade: Farmer - Enlistment date: 6 Jun 1812 - Place: Lewiston, NY - Period: 5 Yrs - Prisoner of War (Halifax), captured at Beaver Dams on 24 Jun 1813, exchanged on 15 Apr 1814; deserted at Fort Severn, MD, on 14 Jul 1815.

Garnder, Michael - Private - 1st Light Dragoons - Company: Thomas Harrison - Age: 20 - Height: 5' 7" - Born: Prince George's Cty, MD - Enlistment date: 26 Mar 1814.

Garner, Isidore - Private - 1st Artillery - Company: James Reed - Age: 21 - Height: 5' - Born: Charles Cty, MD - Trade: Farmer - Enlistment date: 14 Apr 1814 - Place: Alexandria, DC - Period: War - Discharged at Fort Washington, MD, on 7 Mar 1815.

Garner, Michael - Private - 1st Artillery - Company: James Reed - Age: 23 - Height: 5' 7" - Born: Prince George's Cty, MD - Trade: Farmer - Enlistment date: 20 Mar 1814 - Place: Alexandria, DC - Period: War - Discharged at Fort Washington, MD, on 7 Mar 1815.

Garraway, James - Private - US Volunteers - Company: Stephen Moore - Enlistment date: 8 Sep 1812 - Period: 1 Yr.

Garren, John - Private - 22nd Infantry - Company: Thomas Lawrence - Other regiment: 2nd Infantry (New) - Age: 21 - Height: 5' 9" - Born: Baltimore Cty - Trade: Laborer - Enlistment date: 1 Jul 1813 - Place: Uniontown, PA - Period: 5 Yrs - Discharged at Sackets Harbor, NY, on 11 Aug 1815 for inability and chronic rheumatism.

Garrett, Levi - Private - 38th Infantry - Company: Shepperd Leakin - Age: 20 - Height: 5' 11" - Born: Harford Cty, MD - Trade: Farmer - Enlistment date: 14 Apr 1814 - Place: Elkton, MD - Period: War - BLW 1621-160-12 -

Maryland Regulars in War of 1812

Discharged at Fort McHenry on 28 Mar 1815.

Garrion, Thomas - Private - 14th Infantry - Company: Reuben Gilder - Age: 40 - Height: 5' 2 1/2" - Born: Ireland - Trade: Baker - Enlistment date: 14 Mar 1813 - Period: 18 Mos - Discharged at White Hall, NY, on 28 Feb 1815.

Garrison, John - Private - 38th Infantry - Company: Charles Stansbury - Age: 37 - Height: 5' 7 1/2' - Born: Maryland - Trade: Cordwainer - Enlistment date: 22 Aug 1814 - Place: Baltimore - Period: War - BLW 23220-160-12 - Discharged at Baltimore on 6 Apr 1815.

Garver, Samuel - Private - 12th Infantry - Company: Lewis Willis - Age: 34 - Height: 5' 8" - Born: York, PA - Trade: Miller - Enlistment date: 3 Feb 1814 - Place: Hagerstown, MD - Period: 5 Yrs.

Garvin, Thomas - Private - 26th Infantry - Company: William Bezeau - Age: 21 - Height: 5' 6" - Born: Baltimore - Trade: Tailor - Enlistment date: 7 Oct 1814 - Place: Baltimore - Period: War - BLW 5315-60-12 - Discharged at Philadelphia on 23 Mar 1815.

Gary, Asa - Private - 2nd Artillery - Company: Frederick Evans - Other regiment: Corps of Artillery - Age: 21 - Height: 5' 10 1/2" - Born: South Carolina - Trade: Carpenter - Enlistment date: 26 Jan 1815 - Place: Fort McHenry - Period: 5 Yrs - BLW 859-320-12 - Discharged at Fort McHenry on 26 Jan 1820.

Gassaway, John - First Lieutenant - 5th Infantry - Company: James Dorman - Born: Maryland - Pension: Land bounty to John Allen Gassaway, minor child of John Gassaway - BLW 11476-160-50 - Commissioned as a 2nd lieutenant, U.S. Marine Corps, on 18 Jun 1810; resigned on 29 Dec 1810; commissioned as an 2nd lieutenant, 5th Infantry, on 3 Jan 1812; promoted to 1st lieutenant on 15 Aug 1813; discharged on 15 Jun 1815.

Gates, George - Private - 2nd Infantry - Company: Robert Purdy - Other regiment: 1st Infantry (New) - Age: 25 - Height: 5' 9" - Born: Prince George's Cty, MD - Trade: Farmer - Enlistment date: 23 Jun 1798 - Period: 5 Yrs - BLW 21711-160-12 - Re-enlisted in 1803, 1808 and 1813; discharged at Baton Rough, LA, on 30 Jun 1818.

Gates, Jacob - Private - 14th Infantry - Prisoner of War (Halifax), discharged from Halifax on 3 Feb 1814.

Gates, Richard - Private - 36th Infantry - Company: Samuel Raisin - Age: 19 - Height: 5' 11" - Born: Charles Cty, MD - Trade: Farmer - Enlistment date: 2 Mar 1814 - Place: Annapolis - Period: War - BLW 137-160-12 - Discharged at Washington, DC, on 20 Mar 1815.

Gattin, Lewis - Private - 22nd Infantry - Company: Jacob Carmack - Other regiment: 2nd Infantry (New) - Age: 21 - Height: 5' 7 1/2" - Born: Montgomery Cty, MD - Trade: Laborer - Enlistment date: 30 Sep 1814 - Period: 5 Yrs - Discharged on 30 Sep 1819.

Gatton, Bennett - Private - 36th Infantry - Company: Thomas Carbery - Age: 20 - Height: 5' 4" - Born: St. Mary's Cty, MD - Trade: Farmer - Enlistment date: 27 Feb 1814 - Place: Georgetown, DC - Period: 5 Yrs - BLW 10628-160-12 - Discharged at Baltimore on 31 May 1815.

Gatton, Edward - Private - 36th Infantry - Company: Thomas Carbery - Age:

By War of 1812 Soc. in MD

18 - Height: 5' 2" - Born: St. Mary's, MD - Trade: Farmer - Enlistment date: 25 Feb 1814 - Place: Georgetown, DC - Period: War - Discharged at Baltimore on 31 Mar 1815.

Gavin, James - Private - 14th Infantry - Prisoner of War (Halifax), discharged from Halifax on 27 Aug 1813.

Gavin, Mathias - Private - 5th Infantry - Company: James Dorman - Other regiment: 3rd Infantry (New) - Age: 29 - Height: 5' 6" - Born: Philadelphia - Trade: Silversmith - Enlistment date: 21 May 1812 - Place: Baltimore - Period: 5 Yrs - Discharged at Fort Mackinaw, MI, on 20 May 1817.

Gaynor, Patrick - Private - 36th Infantry - Company: Mortimer Hall - Age: 43 - Height: 5' 7" - Born: Ireland - Enlistment date: 26 Jul 1813 - Place: Havre de Grace, MD - Period: 1 Yr.

Gearhart, Jacob - Private - 14th Infantry - Company: James McDonald - Age: 17 - Height: 5' - Born: Pennsylvania - Enlistment date: 11 Oct 1813 - Period: 5 Yrs - Pension: Land bounty to John Gearhart, brother and other heirs at law of Jacob Gearhart - BLW 26303-160-12 - Died at Plattsburgh, NY, on 9 Aug 1814 from sickness.

Geers, John - Farrier - 1st Light Dragoons - Company: Alexander Cummings - Age: 36 - Height: 6' - Born: Kent Cty, MD - Trade: Farmer - Enlistment date: 4 Aug 1812 - Place: Pittsburgh - Period: 5 Yrs - Discharged at Greenbush, NY, on 10 Jun 1815 for rheumatism and debility.

Genoad, Peter (Jenoad) - Private - 36th Infantry - Company: Joseph Hook - Age: 28 - Height: 5' 6 1/2" - Born: France - Trade: Currier - Enlistment date: 14 Jun 1814 - Period: War - Discharged at Fort Covington, MD, on 30 Mar 1815.

Gent, Thomas - Private - 14th Infantry - Company: David Cummings - Age: 29 - Height: 5' 8 1/2" - Born: Ireland or Maryland - Enlistment date: 3 Mar 1814 - Place: Towsontown, MD - Period: War - Deserted on 26 Dec 1814.

Geoghegan, Anthony - Musician - 19th Infantry - Company: George Kesling - Age: 40 - Height: 5' 11 3/4" - Born: Maryland - Trade: Farmer - Enlistment date: 15 Jun 1814 - Place: Lebanon, Warren Cty, OH - Period: War - BLW 3485-160-12 - Discharged at Zanesville, OH, on 27 Mar 1815.

Geoghegan, Thomas - Private - 14th Infantry - Pension: Placed on the pension rolls on 26 Nov 1828.

George, Ezekiel - Private - US Volunteers - Company: Stephen Moore - Enlistment date: 8 Sep 1812 - Period: 1 Yr.

George, Ezekiel C. - Seaman - US Sea Fencibles - Company: John Gill - Age: 22 - Height: 5' 10" - Born: Baltimore - Enlistment date: 10 Jan 1814 - Period: 1 Yr.

George, James - Seaman - US Sea Fencibles - Company: John Gill - Age: 20 - Height: 5' 5" - Born: Baltimore - Enlistment date: 11 Jan 1814 - Period: 1 Yr.

Germaine, Julius - Ensign - 36th Infantry - Company: Thomas Corcoran - Commissioned as an ensign on 30 Apr 1813; resigned on 23 Nov 1813.

Gerring, George - Private - 38th Infantry - Company: Charles Stansbury - Age: 29 - Height: 5' 10 1/2" - Born: Baltimore - Trade: Carpenter - Enlistment date: 17 Oct 1814 - Place: Baltimore - Period: War - BLW 22909-160-12 - Discharged at Baltimore on 6 Apr 1815.

Getty, Frederick - Private - 32nd Infantry - Company: Robert Patterson - Other regiment: 2nd Infantry (New) - Age: 19 - Height: 5' 6" - Born: Baltimore Cty - Trade: Mason - Enlistment date: 11 Apr 1814 - Place: Shippensburg, PA - Period: 5 Yrs - Deserted on 15 Nov 1815.

Geyer, Henry S. - First Lieutenant - 38th Infantry - Commissioned as a 1st lieutenant on 20 May 1813; discharged on 15 Jun 1815.

Gibbins, Thomas - Private - 36th Infantry - Age: 36 - Height: 5' 7 1/2" - Born: Pennsylvania - Enlistment date: 24 Jul 1813 - Place: Havre de Grace, MD - Period: 1 Yr.

Gibbons, Richard - Private - 5th Infantry - Company: George Brooks - Other regiment: 3rd Infantry (New) - Age: 32 - Height: 5' 7" - Born: Baltimore - Trade: Laborer - Enlistment date: 10 May 1813 - Place: Baltimore - Period: 5 Yrs - Discharged at Fort Howard, WI, on 10 May 1818.

Gibbs, John - Private - 5th Infantry - Company: Leroy Opie - Other regiment: 3rd Infantry (New) - Age: 31 - Height: 5' 11" - Born: Worcester Cty, MD - Trade: Farmer - Enlistment date: 6 Apr 1812 - Period: 5 Yrs - BLW 10542-160-12 - Discharged on 6 Apr 1817.

Gibbs, Levin - Seaman - US Sea Fencibles - Company: John Gill - Age: 25 - Height: 5' 8" - Born: Worcester Cty, MD - Enlistment date: 8 Jan 1814 - Place: Baltimore.

Gibbs, Levin - Private - 2nd Light Dragoons - Company: William Littlejohn - Age: 25 - Height: 5' 10" - Born: Baltimore - Trade: Hatter - Enlistment date: 1 Feb 1814 - Place: Baltimore - Period: 5 Yrs.

Gibbs, William - Sergeant - 36th Infantry - Company: Joseph Hook - Age: 32 - Height: 5' 8" - Born: England - Trade: Seaman - Enlistment date: 1 Sep 1813 - Place: Baltimore - Period: 1 Yr - BLW 14830-160-12 - Re-enlisted at Annapolis on 1 Mar 1814 for the war; discharged at Fort Covington, MD, on 30 Mar 1815.

Gibbs, William - Private - 14th Infantry - Company: David Cummings - Age: 23 - Height: 5' 7 1/2" - Born: Loudoun Cty, VA - Trade: Stone mason - Enlistment date: 12 Nov 1814 - Place: Snowden's Camp, VA - Period: 5 Yrs.

Gibbs, William - Private - 2nd Rifles - Company: Batteal Harrison - Age: 21 - Height: 5' 5" - Born: Worcester Cty, MD - Trade: Sailor - Enlistment date: 12 Jun 1814 - Place: Brookville, Franklin Cty, IN - Period: War - Discharged at Detroit on 30 Jul 1815.

Gibson, Edward - Private - 14th Infantry - Company: Thomas Montgomery - Enlistment date: 23 May 1812 - Period: 5 Yrs - Died in Sep 1813.

Gibson, Garrard - Private - 38th Infantry - Company: Isaac Aldridge - Age: 30 - Height: 5' 8" - Born: Ireland - Trade: Farmer - Enlistment date: 13 Oct 1814 - Place: Baltimore - Period: War - BLW 13358-160-12 - Discharged at Baltimore on 6 Apr 1815.

Gibson, James - Colonel - 4th Rifles - Other regiment: ex Light Artillery - Born: Maryland - Pension: Land bounty to Matilda H. Evans, widow of James Gibson - BLW 5191-160-50 - Graduated from the US Military

By War of 1812 Soc. in MD

Academy on 12 Dec 1808; commissioned as a 1st lieutenant, Light Artillery, on 12 Dec 1808; promoted to captain on 2 May 1810; promoted to major on 2 Apr 1813; promoted to colonel on 13 Jul 1813; transferred to 4th Rifles on 21 Feb 1814; killed during the Battle of Fort Erie, UC, on 17 Sep 1814.

Gibson, John - Private - 38th Infantry - Company: James Smith - Age: 19 - Height: 5' 6" - Born: Charles Cty, MD - Trade: Farmer - Enlistment date: 1 Mar 1814 - Place: Drumfires, MD - Period: War - BLW 21307-160-12 - Discharged at Craney Island, VA, on 15 Mar 1815.

Gibson, Robert - Private - 36th Infantry - Company: Hugh Deneale - Age: 20 - Height: 6' - Born: Gloucester, VA - Trade: Farmer - Enlistment date: 15 May 1813 - Period: 1 Yr - Re-enlisted at Annapolis on 20 Mar 1814 for the war; discharged on 30 Mar 1815.

Gibson, Thomas - Private - 14th Infantry - Company: Richard Arell - Age: 28 - Height: 5' 1 1/2" - Born: London, England - Trade: Shoemaker - Enlistment date: 18 May 1812 - Place: Baltimore - Period: 5 Yrs - Prisoner of War (Halifax), exchanged on 5 Apr 1814.

Gibson, Thomas - Seaman - US Sea Fencibles - Company: Simmones Bunbury - Age: 27 - Height: 5' 9 1/4" - Born: Baltimore - Enlistment date: 17 Feb 1814 - Period: 1 Yr - Discharged at Fort McHenry on 16 Feb 1815.

Gibson, William - Second Lieutenant - 36th Infantry - Company: Hugh Deneale - Commissioned as a 2nd lieutenant on 30 Apr 1813; resigned on 22 Oct 1813.

Gidelman, John - Private - 5th Infantry - Company: Richard Whartenby - Age: 32 - Height: 5' 9" - Born: Maryland - Trade: Cordwainer - Enlistment date: 8 Jan 1812 - Period: 5 Yrs - Pension: Heirs received half pay for five years in lieu of land bounty - Discharged at Washington, DC, on 25 Jan 1816, lost his left eye while in the service.

Gilbert, Benjamin - Sergeant - 20th Infantry - Company: Matthew Payne - Other regiment: 4th Infantry (New) - Age: 20 - Height: 5' 7" - Born: Charles or King George Cty, MD - Trade: Cooper - Enlistment date: 10 Oct 1814 - Place: Fredericksburg, VA - Period: 5 Yrs - BLW 23036-160-12 - Discharged at Baltimore on 9 Oct 1819.

Gilbert, David - Private - 12th Infantry - Company: Andrew Madison - Other regiment: 8th Infantry (New) - Age: 21 - Height: 5' 2 1/2" - Born: Charles Cty, MD - Trade: Cooper - Enlistment date: 4 May 1812 - Place: Front Royal, VA - Period: 5 Yrs - Discharged at Fort Edwards, IL, on 4 May 1817.

Gilbert, Henry - Private - 36th Infantry - Company: Joseph Merrick - Age: 26 - Height: 5' 10" - Born: Kent Cty, MD - Enlistment date: 22 May 1813 - Place: Centreville, MD - Period: 1 Yr.

Gilbert, John - Private - 38th Infantry - Company: Charles Stansbury - Age: 23 - Height: 5' 7 1/2" - Born: Massachusetts - Trade: Farmer - Enlistment date: 17 Oct 1814 - Place: Baltimore - Period: War - Pension: Wife Mary Smice, WO-42552, WC-34399; married 18 Sep 1830, Greencastle, PA; soldier died on 25 Feb 1864, Grand Detour, IL - BLW 27894-160-42 - Discharged on 3 Apr 1815.

Gilberthorp, James - Private - 38th Infantry - Age: 21 - Height: 5' 6" - Born:

Baltimore - Trade: Shoemaker - Enlistment date: 4 Aug 1814 - Period: War.

Gilchrist, John - Private - 36th Infantry - Age: 24 - Height: 5' 4" - Born: Scotland - Trade: Mariner - Enlistment date: 20 Dec 1814 - Period: War - Died on 16 Feb 1815.

Gilder, Reuben - Captain - 14th Infantry - Other regiment: 4th Infantry (New) - Born: Maryland - Pension: Land bounty to Margaret Gilder, widow of Reuben Gilder - BLW 30339-80-50 & 11517-80-55 - Commissioned as a 1st lieutenant on 12 Mar 1812; promoted to captain on 26 Jun 1813; transferred to 4th Infantry on 17 May 1815; resigned on 30 Nov 1815.

Giles, Thomas - Sergeant - 26th Infantry - Company: William Bezeau - Age: 24 - Height: 5' 4" - Born: Elkton, MD - Trade: Merchant - Enlistment date: 12 Nov 1814 - Place: Baltimore - Period: War - BLW 5310-160-12 - Discharged at Philadelphia on 31 Mar 1815.

Gill, Cooper - Private - 5th Infantry - Company: William Bird - Age: 25 - Height: 5' 8 3/4" - Born: Baltimore - Enlistment date: 15 May 1813 - Period: 5 Yrs.

Gill, Cooper - Private - 14th Infantry - Died at Fort McHenry on 10 Sep 1814.

Gill, Henry - Drummer - 32nd Infantry - Company: William Smith - Height: 5' 7" - Born: Baltimore - Trade: Drummer - Enlistment date: 31 Jul 1813 - Period: 1 Yr - BLW 23605-160-12 - Re-enlisted at New Utrecht, NY, on 27 Feb 1814 for the war; discharged at New York on 12 May 1815.

Gill, John - Captain - US Sea Fencibles - Other regiment: ex US Volunteers - Commissioned as a 1st lieutenant, US Volunteers, Captain Stephen Moore's Company; commissioned as a captain on 25 Nov 1813; negative by the U.S. Senate 17 Mar 1814.

Gill, John P. - Private - US Volunteers - Company: Stephen Moore - Enlistment date: 8 Sep 1812 - Period: 1 Yr - Discharged as unfit for service.

Gillespie, James - Private - 14th Infantry - Company: Thomas Montgomery - Other regiment: 4th Infantry (New) - Enlistment date: 31 Dec 1812 - Place: Fredericktown, MD - Period: 5 Yrs - Discharged at Fort Hawkins, GA, on 31 Dec 1817.

Gillespie, James - Private - 22nd Infantry - Company: Joseph Henderson - Age: 27 - Height: 5' 9" - Born: Middletown, MD - Trade: Hatter - Enlistment date: 14 Apr 1814 - Place: Chambersburg, PA - Period: 5 Yrs - Discharged at Fort Niagara, NY, on 14 Apr 1819.

Gillen, James - Private - 14th Infantry - Died at Sackets Harbor, NY, on 3 Dec 1813.

Gillen, William - Private - 36th Infantry - Company: Samuel Raisin - Age: 28 - Height: 5' 7" - Born: Scotland - Trade: Shoemaker - Enlistment date: 26 Sep 1814 - Place: Manchester - Period: War - Discharged on 14 Jan 1815.

Gillis, William - Private - 14th Infantry - Company: Joseph Marshall - Age: 35 - Height: 5' 6 3/4" - Born: County Donegal, Ireland - Trade: Weaver - Enlistment date: 29 Jul 1814 - Place: Baltimore - Period: War - BLW 17386-160-12 - Discharged at Greenbush, NY, on 3 May 1815.

Gilmore, John - Private - 26th Infantry - Company: William Bezeau - Age: 27 - Height: 5' 6 1/2" - Born: Ireland - Trade: Seaman - Enlistment date: 16 Oct

By War of 1812 Soc. in MD

1814 - Place: Baltimore - Period: War - BLW 14824-160-12 - Discharged at Philadelphia on 23 Mar 1815.

Ginnerman, Edward - Private - 38th Infantry - Company: James Smith - Age: 35 - Height: 5' 8" - Born: Baltimore - Trade: Laborer - Enlistment date: 4 Mar 1814 - Place: Craney Island, VA - Period: War - BLW 25632-160-12 - Discharged at Craney Island, VA, on 15 Mar 1815.

Girdinston, William - Corporal - 36th Infantry - Company: Thomas Carbery - Age: 21 - Height: 5' 7" - Born: Germany - Trade: Laborer - Enlistment date: 29 May 1813 - Period: 1 Yr - Pension: Wife Lucresia Aldridge, WO-32680, WC-20832; married on 9 Dec 1819, Georgetown, DC; soldier died on 5 Jan 1844, Washington, DC - BLW 63-160-12 - Re-enlisted at Leonardstown, MD, on 25 Mar 1814 for the war; discharged at Baltimore on 31 Mar 1815.

Girvin, David - Private - Light Dragoons - Company: John Erving - Age: 20 - Height: 5' 9 1/2" - Born: Maryland - Enlistment date: 14 Aug 1812 - Place: North Carolina - Period: 5 Yrs - Deserted on 1 Feb 1817.

Gist, Charles - Sergeant - 38th Infantry - Company: James Hook - Age: 32 - Height: 6' - Born: Frederick, MD - Trade: Cabinet maker - Enlistment date: 24 Aug 1814 - Place: Baltimore - Period: War - BLW 12791-160-12 - Discharged at Fort Covington, MD, on 31 Mar 1815.

Gist, Jesse - Private - 24th Infantry - Company: William Allen - Age: 45 - Height: 5' 10 1/2" - Born: Baltimore Cty - Enlistment date: 9 Sep 1812 - Period: 5 Yrs - Died at Camp Russell, TN, on 3 Jun 1815.

Gist, John - Corporal - 38th Infantry - Company: James Hook - Age: 22 - Height: 5' 4 1/4" - Born: Frederick, MD - Trade: Tailor - Enlistment date: 23 Aug 1813 - Period: 1 Yr - Re-enlisted at Baltimore on 15 Jun 1814 for the war; discharged on 31 Mar 1815.

Gist, Mordecai - Sergeant - 28th Infantry - Company: Joseph Belt - Age: 33 - Height: 6' 1/2" - Born: Maryland - Trade: Farmer - Enlistment date: 20 Oct 1814 - Place: Winchester, KY - Period: War - BLW 1728-160-12 - Discharged at Olympian Springs, KY, on 31 Mar 1815.

Gist, Thomas - First Lieutenant - 12th Infantry - Commissioned as a 1st lieutenant on 12 Mar 1812; dismissed on 25 Nov 1812.

Gladden, Jenkins - Private - 38th Infantry - Company: Alexander Cummings - Other regiment: 4th Infantry (New) - Age: 22 - Height: 5' 7 1/2' - Born: Virginia - Trade: Sailor - Enlistment date: 6 Jul 1814 - Place: Princess Anne, MD - Period: 5 Yrs.

Gladden, William - Private - 22nd Infantry - Company: Willis Foulk - Age: 18 - Height: 5' 7" - Born: Maryland - Trade: Stone cutter - Enlistment date: 29 Apr 1814 - Place: Washington, PA - Period: War - Died at Williamsville, NY, on 20 Feb 1815.

Gladding, George - Private - 14th Infantry - Enlistment date: 19 Oct 1812 - Period: 5 Yrs - Died at Sackets Harbor, NY, on 27 Dec 1813.

Gladhill, Thomas - Private - 38th Infantry - Company: John Mowton - Age: 19 - Height: 5' 6" - Born: Maryland - Trade: Farmer - Enlistment date: 21 Feb 1814 - Place: Craney Island, VA - Period: War - BLW 21482-160-12 - Discharged at Craney Island, VA, on 15 Mar 1815.

Maryland Regulars in War of 1812

Gladman, Michael - Private - 19th Infantry - Company: William Gill - Age: 21 - Height: 5' 8" - Born: Maryland - Enlistment date: 13 Aug 1814 - Place: Steubenville, OH - Period: 5 Yrs - Discharged at St. Clairsville, OH, on 10 Dec 1814 by writ of habeas corpus.

Glanville, Stephen - Private - 19th Infantry - Age: 33 - Height: 5' 7" - Born: Kent Cty, MD - Enlistment date: 24 Mar 1814 - Place: Columbus, OH - Period: War.

Gleason, John - Seaman - US Sea Fencibles - Company: John Gill - Enlistment date: 21 Jan 1814 - Pension: Wife Catharine Firsch, WO-21885, WC-20993; married on 10 Sep 1820 in York, PA; seaman died on 30 Nov 1844 in York, PA - BLW 2697-160-50 - Discharged on 28 Feb 1814.

Glenn, James - Seaman - US Sea Fencibles - Company: Simmones Bunbury - Age: 42 - Height: 5' 7 1/2" - Born: Ireland - Enlistment date: 9 Jan 1815 - Period: 1 Yr - Discharged at Fort McHenry on 24 Mar 1815.

Glenn, John - Private - 14th Infantry - Company: Thomas Montgomery - Age: 24 - Height: 5' 8" - Born: Maryland - Trade: Blacksmith - Enlistment date: 7 Nov 1812 - Period: 18 Mos - Discharged at Plattsburgh, NY, on 10 Mar 1814.

Glenn, Peregrine - Private - 36th Infantry - Company: Joseph Merrick - Age: 24 - Height: 5' 6" - Born: Queen Anne's Cty, MD - Enlistment date: 7 May 1813 - Place: Centreville, MD - Period: 1 Yr.

Glenn, William - Private - 2nd Light Dragoons - Company: Samuel Hopkins - Age: 21 - Height: 5' 10" - Born: Baltimore - Trade: Shoemaker - Enlistment date: 17 Mar 1813 - Place: Philadelphia - Period: 5 Yrs - Discharged on 28 Mar 1821.

Glisson, Charles - Private - 12th Infantry - Company: Lewis Willis - Age: 42 - Height: 5' 8" - Born: Anne Arundel Cty, MD - Trade: Carpenter - Enlistment date: 12 Feb 1815 - Place: Washington, DC - Period: 5 Yrs - Discharged at Pittsburgh on 19 Aug 1815, Surgeon's Certificate, old age.

Glover, Archibald - Private - 20th Infantry - Company: William Jett - Age: 35 - Height: 5' 9" - Born: Montgomery Cty, MD - Trade: Laborer - Enlistment date: 8 May 1813 - Period: War - Deserted on 15 Mar 1814.

Godard, Elias - Private - 36th Infantry - Company: Joseph Hook - Age: 27 - Height: 5' 6" - Born: Berkley, VA - Trade: Cooper - Enlistment date: 6 Mar 1814 - Place: Fredericktown, MD - Period: War - BLW 1525-160-12 - Discharged at Fort Covington, MD, on 30 Mar 1815.

Goddard, John - Private - 14th Infantry - Company: Thomas Montgomery - Age: 22 - Height: 6' - Born: Leonardstown, MD - Trade: Farmer - Enlistment date: 3 Apr 1813 - Place: Frederick, MD - Period: War - BLW 1450-160-12 - Discharged at Greenbush, NY, on 4 May 1815.

Goddard, Robert - Sergeant - 36th Infantry - Company: Thomas Carbery - Age: 22 - Height: 5' 7" - Born: Berkshire, England - Trade: Laborer - Enlistment date: 3 May 1813 - Place: Dublin - Period: 1 Yr - BLW 12785-160-12 - Re-enlisted at Leonardstown, MD, on 24 Feb 1814 for the war; discharged at Baltimore on 31 Mar 1815.

Godshall, George - Private - 14th Infantry - Company: William McIlvane -

By War of 1812 Soc. in MD

Age: 26 - Height: 5' 7" - Born: Pennsylvania - Trade: Carpenter - Enlistment date: 19 Dec 1812 - Period: 5 Yrs - BLW 13948-160-12 - Discharged at Greenbush, NY, on 4 May 1815 for lameness.

Godwin, Kimmel - First Lieutenant - 14th Infantry - Age: 25 - Pension: Land bounty to Alanora Jones, widow of Kimmel Godwin - BLW 29572-160-55.

Goetz, John - Private - 2nd Light Dragoons - Company: William Littlejohn - Age: 35 - Height: 5' 7" - Born: Germany - Enlistment date: 13 Feb 1814 - Place: Baltimore.

Golden, Walter - Private - 2nd Infantry - Company: Perrin Willis - Age: 23 - Height: 5' 10" - Born: Charles Cty, MD - Trade: Laborer - Enlistment date: 11 May 1814 - Place: Washington, DC - Period: 5 Yrs - Deserted on 20 Mar 1815.

Golder, Edmund - Private - 38th Infantry - BLW 11280-160-12.

Golder, James - Private - 38th Infantry - Company: James Smith - Pension: Land bounty to Edmund Golder, only heir at law of James Golder - BLW 22456-160-12.

Goldsberg, Samuel - Private - 36th Infantry - Company: Joseph Hook - Age: 20 - Height: 5' 9" - Born: St. Mary's Cty, MD - Trade: Farmer - Enlistment date: 2 Mar 1814 - Place: Annapolis - Period: War - Discharged at Fort Covington, MD, on 30 Mar 1815.

Goldsmith, Joseph - Private - 36th Infantry - Company: Joseph Hook - Age: 35 - Height: 5' 11" - Born: New York - Enlistment date: 14 Jul 1813 - Place: Baltimore - Period: 1 Yr.

Gongers, Peter C. - Private - US Volunteers - Company: Stephen Moore - Enlistment date: 8 Sep 1812 - Period: 1 Yr.

Gooden, William - Private - 38th Infantry - Company: John Brookes - BLW 8559-160-12.

Goodger, Joseph - Private - 1st Infantry - Company: Eli Clemson - Other regiment: 3rd Infantry (New) - Age: 29 - Height: 5' 7" - Born: Baltimore - Trade: Butcher - Enlistment date: 10 Apr 1813 - Place: St. Louis, MO - Period: 5 Yrs - BLW 18463-160-12 - Present at the Battle of Chippewa, UC, on 25 Jul 1814; discharged on 10 Apr 1818.

Goodin, Joseph - Private - 14th Infantry - Company: David Cummings - Age: 28 - Height: 5' 4 3/4" - Born: Cecil Cty, MD - Trade: Hatter - BLW 9839-160-12 - Re-enlisted at Baltimore on 25 Oct 1814 for the war; discharged at Baltimore on 16 Mar 1815.

Goodin, Peter - Private - 14th Infantry - Company: David Cummings - Age: 23 - Height: 5' 7" - Born: Carlisle, Caroline Cty, MD - Trade: Carpenter - Enlistment date: 20 Oct 1814 - Place: Baltimore - Period: War - BLW 23951-160-12 - Discharged on 18 Mar 1815.

Goodmanson, Peter - Seaman - US Sea Fencibles - Company: Simmones Bunbury - Age: 37 - Height: 5' 7 1/2" - Born: Hamburg, Germany - Enlistment date: 23 Dec 1813 - Period: 1 Yr - Discharged at Fort McHenry on 24 Mar 1815.

Goodrich, A. C. - Private - 14th Infantry - Prisoner of War (Halifax), discharged from Halifax on 9 Nov 1813.

Maryland Regulars in War of 1812

Goodwin, George - Private - 36th Infantry - Company: Thomas Carbery - Age: 26 - Height: 5' 11" - Born: Philadelphia - Trade: Cabinet maker - Enlistment date: 18 Apr 1813 - Place: Baltimore - Period: 1 Yr - BLW 25363-160-12 - Re-enlisted at Leonardstown, MD, on 24 Feb 1814 for the war; Discharged at Baltimore on 31 Mar 1815.

Goodwin, Joseph - Private - 14th Infantry - Prisoner of War (Halifax), discharged from Halifax on 9 Nov 1813.

Goodwin, Moses - Private - 2nd Artillery - Company: Nathan Towson - Other regiment: Corps of Artillery - Age: 25 - Height: 5' 9 1/4" - Born: Baltimore - Trade: Laborer - Enlistment date: 11 Apr 1812 - Period: 5 Yrs - BLW 10357-160-12 - Discharged at Fort Niagara, NY, on 11 Apr 1817.

Gordon, Aaron - Private - 5th Infantry - Company: Richard Whartenby - Age: 24 - Height: 5' 10" - Born: Harford Cty, MD - Trade: Tobacconist - Enlistment date: 24 Apr 1812 - Place: Baltimore - Period: 5 Yrs - Discharged at Fort McHenry on 26 Apr 1817.

Gordon, John - Seaman - US Sea Fencibles - Company: William Addison.

Gordon, John - Private - 36th Infantry - Company: Joseph Merrick - Age: 24 - Height: 5' 6" - Born: Charlestown, SC - Enlistment date: 3 May 1813 - Place: Baltimore - Transferred to the U.S. Navy on 5 Nov 1813.

Gordon, William - Private - 14th Infantry - Company: David Cummings - Age: 37 - Height: 5' 10" - Born: Harford Cty, MD - Enlistment date: 15 Jan 1814 - Place: Baltimore - Period: War - Discharged on 14 Mar 1815.

Gordon, William - Private - 14th Infantry - Company: Robert Kent - Enlistment date: 3 Aug 1812 - Period: 5 Yrs - Pension: Heirs received half pay for five years in lieu of land bounty - Died on 20 Nov 1812.

Gordon, William - Private - Light Dragoons - Age: 30 - Height: 5' 11" - Born: Maryland - Trade: Farmer - Enlistment date: 26 Feb 1813 - Place: Baltimore - Period: 18 Mos.

Gordon, William - Private - 2nd Infantry - Company: John Pemberton - Age: 29 - Height: 5' 7" - Born: Frederick, MD - Trade: Carpenter - Enlistment date: 2 Sep 1810 - Period: 5 Yrs.

Gorley, James - Private - 38th Infantry - Company: James Hook - Age: 33 - Height: 5' 7" - Born: Carlisle, PA - Trade: Farmer - Enlistment date: 11 Mar 1814 - Place: Hagerstown, MD - Period: War - BLW 24628-160-12 - Discharged at Fort Covington, MD, on 31 Mar 1815.

Gorrell, Andrew - Private - 2nd Artillery - Company: Frederick Evans - Age: 25 - Height: 5' 7" - Born: Harford Cty, MD - Trade: Laborer - Enlistment date: 5 May 1813 - Place: Annapolis - Period: 5 Yrs - Pension: Placed on the pension rolls on 31 Aug 1824 - Discharged at Washington, DC, on 5 May 1818.

Gorsuch, Gerard - Third Lieutenant - US Sea Fencibles - Company: Simmones Bunbury - BLW 14039-160-50 - Commissioned as a 3rd lieutenant on 1 Oct 1813; discharged on 15 Jun 1815.

Gorsuch, Stephen - Artificer - 2nd Artillery - Company: Nathan Towson - Age: 30 - Height: 5' 5 3/4" - Born: Maryland - Trade: Blacksmith - Enlistment date: 3 Jun 1812 - Place: Baltimore - Period: 5 Yrs - Pension: No

By War of 1812 Soc. in MD

pension claim - BLW 44-160-50 - Prisoner of War, captured during the Battle of Stoney Creek; discharged at Fort Niagara, NY, on 27 Nov 1817.

Gortz, John - Private - 38th Infantry - Company: John Buck - BLW 9843-160-12.

Gosnell, Adam - Private - 2nd Rifles - Other regiment: 24th Infantry - Age: 32 - Height: 5' 2 1/2" - Born: Baltimore - Trade: Shoemaker - Enlistment date: 14 Jul 1814 - Period: War - Transferred to 24th Infantry; discharged at Mobile, AL, on 8 Apr 1815.

Goss, Ezekiel C. - Private - 36th Infantry - Company: Joseph Merrick - Age: 19 - Height: 5' 6" - Born: Marblehead, MA - Enlistment date: 13 May 1813 - Place: Baltimore - Transferred to the U.S. Navy on 5 Nov 1813.

Gossage, Garrard (Jared) - Private - 5th Infantry - Company: James Dorman - Other regiment: 3rd Infantry (New) - Age: 35 - Height: 5' 6" - Born: Baltimore - Trade: Cooper - Enlistment date: 18 Mar 1813 - Place: Baltimore - Period: 5 Yrs - Wounded at Lyon's Creek, UC, on 19 Oct 1814; discharged at Pittsburgh, PA, on 16 Aug 1815.

Gosswel, Anthony - Seaman - US Sea Fencibles - Company: John Gill - Age: 20 - Height: 6' - Born: Baltimore - Enlistment date: 12 Jan 1814 - Period: 1 Yr - Taken away by civil authority on 28 Feb 1814.

Gotear, Charles - Private - 36th Infantry - Company: Samuel Raisin - Age: 17 - Height: 5' 6" - Born: Fairfax Cty, VA - Trade: Engineer - Enlistment date: 29 Sep 1814 - Place: District of Columbia - Period: War - BLW 6846-160-12 - Discharged at Washington, DC, on 20 Mar 1815.

Gould, John - Private - 38th Infantry - Company: James Hook - Age: 22 - Height: 5' 6 1/4" - Born: New York - Trade: Laborer - Enlistment date: 8 Dec 1814 - Place: Baltimore - Period: War.

Govan, Patrick - Private - 36th Infantry - Company: Joseph Hook - Age: 25 - Height: 5' 11" - Born: Ireland - Enlistment date: 18 May 1813 - Place: Baltimore.

Gowers, John - Fife Major - 1st Artillery - Company: George Armistead - Age: 32 - Height: 5' 10" - Born: Philadelphia - Trade: Musician - Enlistment date: 2 Aug 1811 - Place: Baltimore - Period: 5 Yrs - Discharged at New York City on 2 Aug 1816.

Gozler, Henry - Private - 36th Infantry - Company: Thomas Carbery - Age: 34 - Height: 5' 6 1/2" - Born: Washington Cty, MD - Trade: Carpenter - Enlistment date: 20 Mar 1814 - Place: Georgetown, DC - Period: War - BLW 11873-160-12 - Discharged at Baltimore on 31 Mar 1815.

Grace, Bennet - Private - 14th Infantry - Company: Joseph Marshall - Age: 31 - Height: 5' 7" - Born: Harford Cty, MD - Trade: Laborer - Enlistment date: 20 Jan 1814 - Place: Baltimore - Period: War - BLW 2664-160-12 - Discharged at Greenbush, NY, on 3 May 1815.

Graham, George - Private - 36th Infantry - Company: Samuel Raisin - Age: 18 - Height: 5' 4 1/2" - Born: District of Columbia - Trade: Ship's carpenter - Enlistment date: 17 Oct 1814 - Place: District of Columbia - Period: War - BLW 162-160-12 - Discharged at Washington, DC, on 20 Mar 1815.

Graham, John - Private - 1st Artillery - Company: Heman Fay - Other regiment: Corps of Artillery - Age: 38 - Height: 5' 8" - Born: Ireland - Trade: Merchant - Enlistment date: 30 May 1814 - Place: Annapolis - Period: War - BLW 508-160-12 - Discharged at Washington, DC, on 9 Mar 1815.

Graham, John - Private - 1st Light Dragoons - Age: 30 - Height: 5' 9 1/2" - Born: Tyrone, Ireland - Trade: Farmer - Enlistment date: 18 Aug 1812 - Place: Baltimore - Period: 5 Yrs - Deserted on 5 Sep 1815.

Graine, Frederick - Private - 14th Infantry - Prisoner of War (Halifax), captured at Beaver Dams on 24 Jun 1813, discharged from Halifax on 3 Feb 1813.

Grainer, Jonas - Private - 5th Infantry - Company: Richard Bell - Age: 20 - Height: 5' 5" - Born: Baltimore - Trade: Laborer - Enlistment date: 20 Mar 1813 - Period: 5 Yrs.

Granger, William - Private - 14th Infantry - Company: Reuben Gilder - Age: 25 - Height: 5' 9" - Born: Baltimore - Trade: Shoemaker - Enlistment date: 22 Apr 1814 - Place: Baltimore - Period: War - BLW 7285-160-12 - Discharged at Philadelphia on 10 Apr 1815.

Grant, Charles - Private - 38th Infantry - Company: James Smith - Age: 24 - Height: 5' 7 1/2" - Born: Hookstown, MD - Trade: Butcher - Enlistment date: 8 Mar 1814 - Place: Craney Island, VA - Period: War - BLW 17207-160-12 - Discharged at Craney Island, VA, on 15 Mar 1815.

Grant, John - Private - 14th Infantry - Pension: Wife Susan, Old War WF-11078.

Granville, Patrick - Sergeant - 36th Infantry - Company: Joseph Hook - Age: 27 - Height: 5' 7 1/2" - Born: Ireland - Enlistment date: 18 May 1813 - Place: Baltimore - Period: 1 Yr - Pension: Wife Frances "Fanny" Hartnett, WO-4043, WC-5349; married on 25 Nov 1810, Aghada, County Cork, Ireland; soldier died on 12 Feb 1857, Franklin Cty, KS - BLW 25929-160-50 - Discharged on 18 May 1814.

Grapevine, Frederick - Private - 14th Infantry - Company: Joseph Marshall - Age: 34 - Height: 5' - Born: New Jersey - Trade: Laborer - Enlistment date: 10 Apr 1813 - Period: 5 Yrs - Pension: Heirs received half pay for five years in lieu of land bounty - Died at Buffalo on 26 Jan 1815.

Graves, Darens - Private - 14th Infantry - Prisoner of War (Halifax), discharged from Halifax on 3 Feb 1814.

Graves, George B. - Private - 14th Infantry - Company: Joseph Marshall - Age: 27 - Height: 5' 11 1/2" - Born: Portsmouth, VA - Trade: Tailor - Enlistment date: 2 Aug 1814 - Place: Alexandria, DC - Period: War.

Gray, David - Private - Light Dragoons - Company: James Reed - Age: 22 - Height: 5' 7 1/2" - Born: Prince George's Cty, MD - Trade: Farmer - Enlistment date: 1 Mar 1814 - Place: Alexandria, DC - Period: War - Discharged at Fort Washington, MD, on 7 Mar 1815.

Gray, Henry - Private - 36th Infantry - Age: 25 - Height: 5' 8 1/2" - Born: Savannah, GA - Enlistment date: 28 Apr 1813 - Place: Baltimore - Period: 1 Yr.

Gray, James - Private - 14th Infantry - Prisoner of War.

Gray, John - Private - 14th Infantry - Age: 28 - Height: 5' 9 1/2" - Born:

By War of 1812 Soc. in MD

Virginia - Enlistment date: 25 Jul 1814 - Place: Alexandria, DC.

Gray, John - Private - 36th Infantry - Age: 24 - Height: 5' 8 1/2" - Born: Montgomery Cty, MD - Enlistment date: 11 Apr 1814 - Place: Georgetown, DC.

Gray, John - Private - 14th Infantry - Company: Samuel Lane - Other regiment: 4th Infantry (New) - Age: 30 - Height: 5' 9 1/4" - Born: Rupert, Bennington Cty, VT - Trade: Farmer - Enlistment date: 31 Dec 1812 - Place: Buffalo - Period: 5 Yrs - BLW 4045-160-12 - Discharged at Fort Scott, GA, on 31 Dec 1817.

Gray, John D. - Sergeant - 14th Infantry - Enlistment date: 22 May 1812 - Period: 18 Mos - Prisoner of War (Halifax), sent to England; discharged on 31 Mar 1815.

Gray, Nathan - Private - 14th Infantry - Company: Samuel Lane - Age: 26 - Height: 5' 10 1/2" - Born: Jersey - Enlistment date: 6 Jun 1812 - Period: 18 Mos - Died at Buffalo in Nov 1812.

Gray, Thomas - Drummer - 36th Infantry - Company: Joseph Hook - Age: 21 - Height: 5' 2" - Born: Baltimore - Enlistment date: 25 May 1813 - Place: Annapolis - Period: 1 Yr - Pension: Wife Catharine Swaggard, SO-3822, SC-3337, WO-13876, WC-22663; married on 12 Jan 1828, Franklin Cty, PA; soldier died on 30 Aug 1874, Bloomington, IL - BLW 15-160-50 - Discharged on 20 May 1814.

Gray, William - Private - 14th Infantry - Company: Thomas Montgomery.

Gray, Zachariah - Private - 22nd Infantry - Company: Jacob Carmack - Age: 25 - Height: 5' 8" - Born: Anne Arundel Cty, MD - Trade: Tailor - Enlistment date: 27 Feb 1814 - Period: War.

Grayson, John - Ensign - US Volunteers - Company: Stephen Moore - Enlistment date: 8 Sep 1812 - Period: 1 Yr - BLW 5138-160-50 - Promoted from private to ensign on 1 Mar 1813.

Greaves, William - Private - 1st Light Dragoons - Company: Arthur Hayne - Other regiment: Light Dragoons - Age: 18 - Height: 5' 6" - Born: Centreville, MD - Trade: Tailor - Enlistment date: 28 Apr 1813 - Place: Centreville, MD - Period: War - BLW 25740-160-12 - Discharged at Greenbush, NY, on 4 May 1815.

Green, Andrew - Private - 14th Infantry - Company: Reuben Gilder - Height: 5' 8" - Born: Maryland - Trade: Joiner - Enlistment date: 10 Jul 1812 - Period: 5 Yrs - BLW 12792-160-12 - Discharged at New York on 23 May 1814 on Surgeon's Certificate.

Green, Anthony - Seaman - US Sea Fencibles - Company: Simmones Bunbury - Enlistment date: 17 Jan 1814 - Period: 1 Yr - Discharged at Baltimore on 16 Jan 1815.

Green, Eleazer E. - Private - 14th Infantry - Company: Joseph Marshall - Age: 33 - Height: 5' 6 1/2" - Born: Cecil Cty, MD - Trade: Veterinarian - Enlistment date: 10 Mar 1814 - Place: Chestertown, MD - Period: War.

Green, George - Private - 36th Infantry - Company: Thomas Carbery - Age: 33 - Height: 5' 10" - Born: Mathews Cty, VA - Trade: Farmer - Enlistment date: 19 May 1814 - Place: Cool Springs, St. Mary's Cty, MD - Period: War

Maryland Regulars in War of 1812

- Pension: Land bounty to Simon Green, uncle and heir at law of George Green - BLW 8986-160-12 - Died at regimental hospital on 18 Feb 1815.

Green, Hillary - Private - 26th Infantry - Company: William Bezeau - Age: 21 - Height: 5' 4" - Born: Alexandria, DC - Trade: Stone mason - Enlistment date: 11 Nov 1814 - Place: Baltimore - Period: War - BLW 2875-160-12 - Discharged at Philadelphia on 23 Mar 1815.

Green, Isidore - Artificer - 3rd Artillery - Company: James House - Age: 34 - Height: 5' 11" - Born: Charles Cty, MD - Trade: Carpenter - Enlistment date: 1806 - Period: 5 Yrs - Re-enlisted at Fort Wolcott, on 28 Jan 1811 for five years; discharged on 27 Jan 1816.

Green, James - Private - 36th Infantry - Company: Joseph Merrick - Age: 39 - Height: 5' 7 1/2" - Born: Pennsylvania - Enlistment date: 11 May 1813 - Place: Baltimore - Period: 1 Yr - Transferred to the U.S. Navy on 26 Oct 1813.

Green, Jeremiah - First Lieutenant - 38th Infantry - Company: Shepperd Leakin - Commissioned as a 2nd lieutenant on 20 May 1813; promoted to 1st lieutenant on 1 May 1814; resigned on 9 Jul 1814.

Green, John - Private - 14th Infantry - Company: Samuel Lane - Age: 26 - Height: 5' 6 1/2" - Born: Cecil Cty, MD - Trade: Farmer - Enlistment date: 24 Jun 1812 - Place: Lewiston, NY - Period: 5 Yrs - Prisoner of War, exchanged on 15 Apr 1814; deserted at Fort Severn on 16 Jul 1815.

Green, John - Seaman - US Sea Fencibles - Company: Simmones Bunbury - Age: 39 - Height: 5' 6" - Born: Ireland - Enlistment date: 16 Apr 1814 - Period: 1 Yr - Died at Fort McHenry on 23 Oct 1814.

Green, Joseph - Private - 38th Infantry - Company: Thomas Sangster - Age: 32 - Height: 5' 6" - Born: Massachusetts - Trade: Seaman - Enlistment date: 29 Oct 1814 - Place: Baltimore - Period: 5 Yrs.

Green, Lewis - Private - 38th Infantry - Company: John Brookes - BLW 3193-160-12.

Green, Robert - Seaman - US Sea Fencibles - Company: Simmones Bunbury - Age: 37 - Height: 5' 5" - Born: Charles Cty, MA - Enlistment date: 13 Jan 1814 - Period: 1 Yr - Discharged on 13 Jan 1815.

Green, Thomas - Sergeant - 38th Infantry - Company: John Brookes - BLW 16829-160-12.

Green, Thomas - Private - 14th Infantry - Company: Joseph Marshall - Age: 39 - Height: 5' 10 1/2" - Born: Cumberland, Luzerne Cty, PA - Trade: Shoemaker - Enlistment date: 26 Jul 1814 - Place: Baltimore - Period: War.

Green, Thomas - Sergeant - 14th Infantry - Company: Reuben Gilder - Other regiment: 4th Infantry (New) - Age: 27 - Height: 5' 9 1/4" - Born: Virginia - Trade: Carpenter - Enlistment date: 6 Aug 1812 - Place: Cumberland - Period: 5 Yrs - Prisoner of War (Halifax), captured at Stoney Creek, UC, on 12 Jun 1813; deserted on 18 Jul 1815.

Green, Thomas - Private - 1st Artillery - Company: George Armistead - Age: 30 - Height: 5' 11" - Born: Maryland - Trade: Tailor - Enlistment date: 8 Aug 1812 - Place: Fort McHenry - Period: 5 Yrs - Discharged on 9 Apr 1817.

Green, Walter A. - Private - 1st Artillery - Company: George Armistead - Age:

By War of 1812 Soc. in MD

27 - Height: 5' 6" - Born: Maryland - Trade: Farmer - Enlistment date: 4 Oct 1809 - Period: 5 Yrs - Re-enlisted at Fort Columbus on 9 Jul 1814 for five years; discharged at Fort Shelby, MI, on 9 Jul 1819 and re-enlisted.

Green, William - Private - 26th Infantry - Company: William Bezeau - Age: 24 - Height: 5' 1" - Born: Ireland - Trade: Weaver - Enlistment date: 30 Jan 1815 - Place: Baltimore - Period: War - BLW 6419-160-12 - Discharged at Baltimore.

Green, William - Private - 38th Infantry - Company: John Buck - Age: 19 - Height: 5' 4" - Born: Queen Anne's Cty, MD - Trade: Farmer - Enlistment date: 12 May 1814 - Place: Baltimore - Period: War - Discharged at Buffalo on 30 Mar 1815.

Green, William - Private - 20th Infantry - Company: John Thornton - Other regiment: 8th Infantry (New) - Age: 21 - Height: 5' 6" - Born: Washington, MD - Trade: Miller - Enlistment date: 20 Mar 1812 - Place: Leesburg, VA - Period: 5 Yrs - Discharged at New Orleans on 20 Mar 1817.

Greenfield, James - Private - 14th Infantry - Company: William McIlvane - Height: 5' 8" - Born: Maryland - Trade: Carpenter - Enlistment date: 10 Apr 1813 - Period: 5 Yrs - BLW 9105-160-12 - Discharged at Greenbush, NY, on 4 May 1814 for old age.

Greenfield, John B. - Private - 36th Infantry - Age: 23 - Height: 5' 2" - Born: Ireland - Enlistment date: 2 Dec 1813 - Place: Baltimore.

Greenwell, Michael - Private - 38th Infantry - Company: John Rothrock - Pension: Land bounty to Jacob Greenwell, brother and other heirs at law of Michael Greenwell - BLW 22617-160-12.

Greenwell, Philip P. - First Lieutenant - 5th Infantry - Commissioned as a 2nd lieutenant on 3 Jan 1812; promoted to 1st lieutenant on 30 Apr 1813; resigned on 13 Aug 1814.

Greenwood, Henry - Private - 1st Artillery - Company: Hopley Yeaton - Other regiment: Corps of Artillery - Age: 30 - Height: 5' 6 3/4" - Born: Baltimore - Trade: Smelter - Enlistment date: 11 Oct 1813 - Place: Petersburg, VA - Period: 5 Yrs - BLW 19964-160-12 - Discharged on 11 Oct 1818.

Greenwood, Henry - Private - 2nd Artillery - Company: Nathan Towson - Enlistment date: 19 Nov 1812 - Period: 5 Yrs.

Gregg, James - Private - 14th Infantry - Company: William McIlvane - Enlistment date: 24 Jan 1813 - Period: 5 Yrs - Prisoner of War (Halifax), sent to England.

Gregg, James - Sergeant - 14th Infantry - Company: David Cummings - Enlistment date: 30 Apr 1813 - Period: 5 Yrs.

Gregory, James - Private - 20th Infantry - Company: William Jett - Age: 43 - Height: 5' 9 1/2" - Born: Prince George's Cty, MD - Trade: Carpenter - Enlistment date: 9 Aug 1814 - Place: Alexandria, DC - Period: War - Pension: Land bounty to Matilda Donaldson, daughter and other heirs at law of James Gregory - BLW 23301-160-12 - Discharged at Norfolk, VA, on 15 Mar 1815.

Gregory, Jesse - Private - 36th Infantry - Company: Joseph Merrick - Age: 18 - Height: 5' 4 3/4" - Born: Queen Anne's Cty, MD - Enlistment date: 1 Jun

1813 - Place: Centreville, MD - Period: 1 Yr.

Gravel, Benjamin - Sergeant - 14th Infantry - Company: Reuben Gilder - Other regiment: 4th Infantry (New) - Age: 26 - Height: 5' 9" - Born: Baltimore Cty - Trade: Painter - Enlistment date: 13 Mar 1813 - Place: Baltimore - Period: 5 Yrs - Deserted on 12 Jul 1815.

Grey, Isaiah - Private - 38th Infantry - Company: Charles Stansbury - Age: 28 - Height: 5' 9" - Born: Chester Cty, PA - Trade: Farmer - Enlistment date: 30 Nov 1814 - Place: Baltimore - Period: War - BLW 2860-160-12 - Discharged at Baltimore on 6 Apr 1815.

Grey, Nathaniel - Private - 14th Infantry - Company: Thomas Montgomery.

Grey, Samuel - Private - 14th Infantry - Company: Samuel Lane - Enlistment date: 18 May 1812 - Period: 5 Yrs - Prisoner of War (Halifax), sent to England; discharged on 18 May 1817.

Griffin, James - Private - 14th Infantry - Company: Joseph Marshall - Age: 28 - Height: 5' 11" - Born: Baltimore - Trade: Sailor - Enlistment date: 1 Jul 1814 - Place: Elkton, MD - Period: War - BLW 19578-160-12 - Discharged at Greenbush, NY, on 3 May 1815.

Griffin, James - Private - 36th Infantry - Company: Joseph Hook - Age: 27 - Height: 5' 10" - Born: Baltimore - Enlistment date: 22 Apr 1813 - Place: Baltimore - Period: 1 Yr.

Griffin, James S. - First Lieutenant - 38th Infantry - Company: James Smith - Commissioned as a 2nd lieutenant on 20 May 1813; promoted to 1st lieutenant on 20 May 1814; discharged on 15 Jun 1815.

Griffin, John - Private - 14th Infantry - Prisoner of War (Halifax), sent to England.

Griffin, John - Private - 5th Infantry - Company: Leroy Opie - Age: 35 - Height: 5' 6 1/4" - Born: Stafford, Harford Cty, MD - Trade: Sawyer - Enlistment date: 15 May 1814 - Place: Lancaster, PA - Period: 5 Yrs.

Griffin, John - Private - 38th Infantry - Company: Isaac Aldridge - Age: 22 - Height: 5' 6" - Born: Westmoreland Cty, PA - Trade: Farmer - Enlistment date: 28 Nov 1814 - Place: Baltimore - Period: War - Pension: Land bounty to Ann Griffin, sister & only heir at law of John Griffin - BLW 363-160-12 - Discharged at Baltimore on 6 Apr 1815.

Griffin, Thomas - Private - 38th Infantry - Age: 26 - Height: 5' 9" - Born: Frederick, MD - Trade: Farmer - Enlistment date: 26 Oct 1814 - Place: Baltimore - Period: War.

Griffith, Benjamin - Private - 17th Infantry - Company: William Adair - Age: 24 - Height: 5' 10" - Born: Harford Cty, MD - Enlistment date: 21 May 1814.

Griffith, Edward - Private - 1st Artillery - Company: Heman Fay - Other regiment: Corps of Artillery - Age: 24 - Height: 6' 6 1/2" - Born: Queen Anne's Cty, MD - Trade: Cabinet maker - Enlistment date: 29 Aug 1814 - Place: Annapolis - Period: War - BLW 509-160-12 - Discharged at Washington, DC, on 9 May 1815.

Griffith, John - Private - 38th Infantry - Company: Shepperd Leakin - Age: 27 - Height: 5' 4" - Born: Ireland - Trade: Brick layer - Enlistment date: 18 Aug 1814 - Place: Baltimore - Period: 1 Yr - Deserted at Camp Baltimore on 2

Sep 1814.

Griffith, W. R. - Private - 28th Infantry - Company: Joseph Belt - Other regiment: 3rd Infantry (New) - Age: 35 - Height: 5' 11 1/2" - Born: Montgomery Cty, MD - Trade: Farmer - Enlistment date: 6 Sep 1814 - Place: Fleming Cty, KY - Period: 5 Yrs - Discharged on 24 Jul 1815 after furnishing a substitute (Abijah Davis).

Griffiths, Thomas B. - Seaman - US Sea Fencibles - Company: William Addison.

Grimes, Elijah - Sergeant - 2nd Light Dragoons - Company: Samuel Harris - Other regiment: Light Dragoons - Age: 22 - Height: 5' 10" - Born: Baltimore - Trade: Farmer - Enlistment date: 20 Mar 1813 - Period: 5 Yrs - BLW 15617-160-12 - Discharged at Avon, NY, on 8 Jun 1815.

Grimes, Ephraim - Private - 17th Infantry - Age: 46 - Height: 5' 9" - Born: Maryland - Enlistment date: 12 Jan 1815 - Place: Cincinnati - Period: War.

Grimes, Jacob - Private - 38th Infantry - Company: Thomas Sangster - Other regiment: 4th Infantry (New) - Age: 21 - Height: 5' 6" - Born: Frederick, MD - Trade: Laborer - Enlistment date: 14 Nov 1814 - Place: Fredericktown, MD - Period: 5 Yrs - Died at Fort Gadsden, FL, on 14 or 15 Jul 1819.

Grimes, John - Private - 36th Infantry - Company: Thomas Carbery - Age: 37 - Height: 5' 10" - Born: Fairfax, VA - Enlistment date: 20 Mar 1814 - Place: Georgetown, DC - Period: War.

Grimes, John - Private - 14th Infantry - Company: Isaac Barnard - Other regiment: 4th Infantry (New) - Age: 19 - Height: 5' 5 1/2" - Born: Baltimore - Trade: Cordwainer - Enlistment date: 28 Dec 1812 - Place: Baltimore - Period: 5 Yrs - Pension: Old War IF-25113 - Prisoner of War, exchanged on 15 Apr 1814; discharged at Raleigh, NC, on 8 Jan 1818.

Grimes, Joshua - Private - 38th Infantry - Company: Shepperd Leakin - Age: 21 - Height: 5' 10" - Born: Maryland - Trade: Farmer - Enlistment date: 21 Mar 1814 - Period: War.

Grimes, Nicholas - Private - 14th Infantry - Company: Samuel Lane - Other regiment: 4th Infantry (New) - Age: 21 - Height: 5' 11" - Born: Frederick, MD - Trade: Laborer - Enlistment date: 25 May 1812 - Place: Cumberland, MD - Period: 5 Yrs - Prisoner of War (Halifax), sent to England; discharged on 25 May 1817.

Grimes, Thomas - Private - 12th Infantry - Age: 44 - Height: 5' 9" - Born: Prince George's Cty, MD - Trade: Laborer - Enlistment date: 6 Jun 1812 - Period: 5 Yrs - Discharged on 23 Mar 1814 on Surgeon's Certificate, rheumatism and lame back.

Grimes, Thomas - Private - 16th Infantry - Company: Miles Greenwood - Age: 40 - Height: 5' 6 1/2" - Born: Maryland - Trade: Carpenter - Enlistment date: 22 Apr 1814 - Place: Burlington, VT - Period: War - Died at Plattsburgh, NY, on 12 Oct 1814.

Grimes, William - Private - 38th Infantry - Age: 31 - Height: 5' 5" - Born: Ireland - Trade: Shoemaker - Enlistment date: 2 Jan 1815 - Place: Baltimore - Period: War.

Grimes, William - Drummer - 14th Infantry - Company: Thomas Montgomery

- Age: 22 - Height: 5' 8 1/2" - Born: Anne Arundel Cty, MD - Trade: Shoemaker - Enlistment date: 22 May 1812 - Place: Baltimore - Period: 5 Yrs - Deserted at Annapolis on 19 Sep 1815.

Grindage, Henry - Major - 14th Infantry - Other regiment: 15th Infantry - Born: Maryland - Commissioned as a captain, 14th Infantry, on 12 Mar 1812; promoted to major, 15th Infantry, on 14 Nov 1813; discharged on 15 Jun 1815.

Grindle, Richard - Private - 14th Infantry - Company: Reuben Gilder - Enlistment date: 21 Feb 1813 - Period: 5 Yrs - Discharged on 17 Jan 1814 for inability; died at Greenbush, NY, on 15 Nov 1814.

Griner, George - Private - 14th Infantry - Prisoner of War (Halifax), discharged from Halifax on 3 Feb 1814.

Groce, Jacob - Private - 36th Infantry - Company: Joseph Hook - Enlistment date: 2 Jun 1814 - Period: War - Pension: Land bounty to Mary Hunter, sister & other heirs at law of Jacob Groce - BLW 25589-160-12 - Killed on 28 Sep 1814 while attempting to desert.

Grodon, Thomas - Private - 36th Infantry - Age: 21 - Height: 6' - Born: Baltimore - Enlistment date: 14 Jun 1813 - Place: Annapolis.

Gromly, Constantine - Private - 14th Infantry - Company: Joseph Marshall - Enlistment date: 16 Sep 1812 - BLW 12445-160-12 - Discharged at Greenbush, NY, on 7 Jun 1815 for lameness.

Grones, Andrew - Private - 36th Infantry - Age: 20 - Height: 5' 5 3/4" - Born: Northampton Cty, PA - Enlistment date: 10 May 1813 - Place: Westminster, MD - Period: 1 Yr.

Groom, John - Private - 4th Rifles - Company: Matthew Magee - Age: 22 - Height: 5' 7" - Born: Maryland - Trade: Brick layer - Enlistment date: 21 Jun 1814 - Period: War.

Gross, John - Private - 7th Infantry - Company: Edward Hord - Age: 22 - Height: 6' - Born: Maryland - Trade: Farmer - Enlistment date: 7 Jan 1809 - Period: 5 Yrs - Re-enlisted at Fort Claiborne on 5 Jul 1813 for the war.

Grouse, Frederick - Private - 14th Infantry - Company: Richard Arell - BLW 3243-160-12.

Grove, Jacob - Private - 38th Infantry - Company: Shepperd Leakin - Age: 21 - Height: 5' 7" - Born: Franklin Cty, PA - Trade: Cooper - Enlistment date: 5 Jul 1814 - Place: Hagerstown, MD - Period: War - BLW 27724-160-42 - Discharged on 28 Mar 1815.

Groves, Isaac - Private - 14th Infantry - Other regiment: 4th Infantry (New) - Age: 25 - Height: 5' 6 1/2" - Born: Providence, RI - Trade: Shoemaker - Enlistment date: 3 Jun 1812 - Place: Baltimore - Period: 5 Yrs - Prisoner of War (Halifax), exchanged on 15 Apr 1814; deserted at Fort Severn, MD, on 14 Jul 1815.

Grow, John - Private - 38th Infantry - Company: Shepperd Leakin - Enlistment date: 5 Jul 1814 - Period: War - Pension: Wife Mary Miller, WO-22314, WC-25087; married 23 Dec 1819, Huntingdon Cty, PA; soldier died on 18 Dec 1855, Strongstown, Indiana Cty, PA - BLW 27724-160-12 - Discharged on 31 Mar 1815.

By War of 1812 Soc. in MD

Grozier, William - Private - 36th Infantry - Company: Joseph Hook - Age: 25 - Height: 5' 5 1/2" - Born: Eastham, MA - Trade: Seaman - Enlistment date: 16 Apr 1814 - Place: Annapolis - Period: War - BLW 368-160-12 - Discharged at Fort Covington, MD, on 30 Mar 1815.

Grubb, Joseph - Sergeant - 14th Infantry - Company: Joseph Marshall - Age: 24 - Height: 5' 11" - Born: Chester, PA - Trade: Stone mason - Enlistment date: 18 Jun 1813 - Place: Baltimore - Period: War - BLW 3238-160-12 - Discharged at Greenbush, NY, on 3 May 1815.

Grubs, Jacob - Private - 38th Infantry - Age: 25 - Height: 6' 1" - Born: Pennsylvania - Enlistment date: 29 Dec 1813 - Place: Cumberland - Period: 1 Yr.

Gruet, James - Private - 14th Infantry - Prisoner of War (Halifax), discharged from Halifax on 3 Feb 1814.

Gruet, William - Private - 14th Infantry - Prisoner of War (Halifax), discharged from Halifax on 3 Feb 1814.

Guess, Thomas - Private - 38th Infantry - Company: Isaac Aldridge - BLW 5494-160-12.

Guess, Thomas - Seaman - US Sea Fencibles - Company: Simmones Bunbury - Other regiment: 38th Infantry - Age: 21 - Height: 5' 5" - Born: Calvert Cty, MD - Trade: Cooper - Enlistment date: 21 Oct 1814 - Period: 1 Yr - BLW 234-320-12 - Re-enlisted in 38th Infantry on 26 Jan 1815 for the war; discharged at Baltimore on 6 Apr 1815.

Guinn, John - Private - 38th Infantry - Company: John Buck - BLW 24318-160-12.

Gurney, Leonard - Private - 2nd Light Dragoons - Company: John Burd - Age: 20 - Height: 5' 7 3/4" - Born: Somerset Cty, MD - Trade: Shoemaker - Enlistment date: 16 Jun 1812 - Period: 5 Yrs - Deserted at Trenton, NJ, on 4 or 9 Oct 1812.

Gurts, John E. - Private - 14th Infantry - Company: James McDonald - Age: 36 - Height: 5' 11" - Born: Maryland - Enlistment date: 7 Jan 1813 - Period: 18 Mos - Discharged at Plattsburgh, NY, on 7 Jul 1814.

Guthrie, William - Artillery driver - 2nd Artillery - Company: Nathan Towson - Pension: Old War IF-25157 - BLW 6551-160-12.

Guthrow, Joseph - Clerk - 2nd Artillery - Company: Frederick Evans - Age: 39 - Height: 5' 8" - Born: Baltimore - Trade: Farmer - Enlistment date: 30 Mar 1810 - Place: Newport, RI - Period: 5 Yrs - Discharged on 30 Mar 1815.

Guy, Bailey (Grey) - Private - 36th Infantry - Age: 30 - Height: 5' 8" - Born: Norfolk, VA - Enlistment date: 22 Sep 1814 - Place: Norfolk, VA - Period: War.

Guy, Thomas - Private - 36th Infantry - Company: Thomas Carbery - Age: 24 - Height: 5' 10" - Born: Charles Cty, MD - Trade: Wheelwright - Enlistment date: 14 Aug 1814 - Place: St. Mary's, MD - Period: War - BLW 539-160-12 - Discharged at Baltimore on 31 Mar 1815.

Gwinn, James - Private - 14th Infantry - Company: David Cummings - BLW 25489-160-12.

Gwinn, John - Private - 38th Infantry - Company: Shepperd Leakin - Age: 28

- Height: 5' 6" - Born: Pennsylvania - Trade: Blacksmith - Enlistment date: 28 Jan 1814 - Place: Cumberland - Period: 1 Yr - Discharged at Baltimore on 30 Mar 1815.

Gwinn, William R. - First Lieutenant - 38th Infantry - Company: Anthony Miltenberger - BLW 29249-160-55 - Commissioned as a 1st lieutenant on 20 May 1813; resigned on 31 Dec 1813.

Gwynne, David - Major - 19th Infantry - Other regiment: 2nd Rifles - Born: Maryland - Commissioned as a 1st lieutenant, 19th Infantry, on 12 Mar 1812; promoted to captain on 30 Mar 1813; promoted to major, 2nd Rifles, on 21 Feb 1814; discharged on 17 May 1815; died in 1849.

H

Hacket, Thomas H. - Private - 38th Infantry - Company: John Brookes - BLW 15160-160-12.

Haddaway, John - Private - 14th Infantry - Company: Thomas Montgomery - Other regiment: 4th Infantry (New) - Age: 22 - Height: 5' 7" - Born: St. Michaels, MD - Trade: Hatter - Enlistment date: 14 May 1812 - Place: Surkeville, MD - Period: 5 Yrs - BLW 12286-160-12 - Discharged at Montpelier, VT, on 13 May 1817.

Hadley, Joseph - Seaman - US Sea Fencibles - Company: William Addison.

Hadricks, Michael - Private - 20th Infantry - Company: Walter Hayes - Age: 32 - Height: 5' 6" - Born: Maryland - Trade: Tailor - Enlistment date: 3 Oct 1814 - Period: War - Discharged at Norfolk, VA, on 15 Mar 1815.

Hagan, Bernard - Private - 38th Infantry - Company: Anthony Miltenberger - Pension: No pension claim - BLW 38-160-50.

Hagan, Charles - Corporal - 17th Infantry - Company: Caleb Holder - Age: 25 - Height: 5' 8" - Born: Maryland - Trade: Laborer - Enlistment date: 31 Aug 1814 - Place: Louisville, KY - Period: War - BLW 8941-160-12 - Discharged at Chillicothe, OH, on 30 Jun 1815.

Hagan, Isaac - Private - 14th Infantry - Died at Cheektowaga Hospital, NY, on 8 Feb 1815 from dysentery.

Hagan, John - Private - 14th Infantry - Company: Samuel Lane - Other regiment: 4th Infantry (New) - Age: 38 - Height: 5' 5" - Born: New Jersey - Enlistment date: 17 Jun 1812 - Place: Cumberland, MD - Period: 5 Yrs - Pension: Land bounty to James Agan, brother and other heirs at law of John Hagan - BLW 21174-160-12 - Prisoner of War (Halifax), exchanged on 15 Apr 1814.

Hagan, Joseph - Private - 14th Infantry - Deserted on 12 Jul 1813.

Hagan, Thomas H. - Private - 17th Infantry - Company: Caleb Holder - Other regiment: 3rd Infantry (New) - Age: 45 - Height: 5' 9" - Born: Charles Cty, MD - Trade: Carpenter - Enlistment date: 4 Jun 1812 - Place: Charlestown, Clark Cty, IN - Period: 5 Yrs - Discharged at Fort Wayne, IN, on 4 Jun 1817.

Hagerman, John - Private - 36th Infantry - Company: Thomas Carbery - Age: 22 - Height: 5' 8 1/2" - Born: Galesburgh, Adams Cty, PA - Trade: Weaver - Enlistment date: 28 Feb 1814 - Place: Baltimore - Period: War - Pension: Wife Elizabeth, WO-41767, WC-33652 - BLW 20081-160-12 - Discharged

By War of 1812 Soc. in MD

at Baltimore on 31 Mar 1815.

Hagerman, Stephen - Private - 36th Infantry - Company: Samuel Raisin - Age: 18 - Height: 5' 4" - Born: Baltimore - Enlistment date: 18 Aug 1813 - Place: Baltimore - Period: 1 Yr - BLW 22319-160-12 - Discharged on 20 Mar 1815.

Hagerman, Stephen - Private - 36th Infantry - Company: Samuel Raisin - Age: 17 - Height: 5' 5 1/2" - Born: Baltimore - Trade: Laborer - Enlistment date: 1 Mar 1814 - Place: Annapolis - Period: War.

Haggerty, Levi - Private - 14th Infantry - Company: Kenneth McKenzie - Enlistment date: 24 Nov 1812 - Period: 18 Mos - Pension: Wife Martha E. Herring (second wife), SO-5254, SC-2433, WO-14641, WC-8669; married on 25 Jul 1838, Baltimore, MD; died on 29 Aug 1871, Washington, DC - BLW 27369-160-42 - Prisoner of War (Halifax), discharged from Halifax on 3 Feb 1814, exchanged on 15 Apr 1814; discharged at Greenbush, NY, on 23 Jun 1814.

Haggerty, Mahlon - Private - 5th Infantry - Company: Colin Buckner - Age: 24 - Height: 5' 10 1/2" - Born: Maryland - Trade: Printer - Enlistment date: 24 Mar 1813 - Place: Baltimore - Period: 5 Yrs - Died on 9 Dec 1813.

Hagner, Frederick - Private - 44th Infantry - Company: Anatole Peychand - Age: 30 - Height: 5' 8" - Born: Maryland - Trade: Soldier - Enlistment date: 9 May 1814 - Place: New Orleans - Period: War.

Hailey, James - Private - 25th Infantry - Company: Edward White - Age: 27 - Height: 5' 7" - Born: Winchester, MD - Trade: Shoemaker - Enlistment date: 2 Jul 1814 - Place: Buffalo - Period: War - Discharged at Sackets Harbor, NY, on 17 Mar 1815.

Hailey, James - Corporal - 16th Infantry - Company: William Davenport - Age: 25 - Height: 5' 9 1/2" - Born: Maryland - Trade: Shoemaker - Enlistment date: 20 Nov 1813 - Place: Easton, MD - Period: 18 Mos.

Hailey, John - Private - 1st Infantry - Company: Simon Owens - Age: 23 - Height: 5' 10" - Born: Baltimore - Trade: Laborer - Enlistment date: 6 Nov 1809 - Place: Winchester, MD - Period: 5 Yrs - Discharged at Williamsville, NY, on 6 Nov 1814.

Haines, Benjamin - Private - 23rd Infantry - Company: Richard Goodell - Age: 38 - Height: 5' 5" - Born: Maryland - Enlistment date: 15 Feb 1814 - Period: 5 Yrs - Died.

Haines, Daniel - Corporal - 14th Infantry - Company: Reuben Gilder - Age: 21 - Height: 5' 9" - Born: Maryland - Enlistment date: 14 Apr 1814 - Period: 18 Mos - Discharged at Fort Erie, UC, on 14 Oct 1814.

Haines, James - Musician - 36th Infantry - Age: 38 - Height: 5' 5 1/2" - Born: Ireland - Trade: Skin dresser - Enlistment date: 7 Dec 1814 - Period: War - Discharged at Georgetown, DC, on 13 Mar 1815.

Hake, Richard - Private - 36th Infantry - Age: 24 - Height: 5' 5 1/2" - Born: Salem, Salem Cty, NJ - Enlistment date: 4 May 1813 - Place: Baltimore.

Hale, Samuel - Private - 38th Infantry - Company: John Buck - Age: 27 - Height: 5' 11 1/4" - Born: Baltimore City - Trade: Carpenter - Enlistment date: 22 Sep 1814 - Place: Baltimore - Period: War - BLW 3560-160-12 - Discharged at Baltimore on 30 Mar 1815.

Maryland Regulars in War of 1812

Halkett, George - Private - 26th Infantry - Company: William Bezeau - Age: 21 - Height: 5' 5" - Born: Lancaster, PA - Trade: Tailor - Enlistment date: 27 Oct 1814 - Place: Baltimore - Period: War - Discharged at Philadelphia on 23 Mar 1815.

Hall Jr., Caleb - Seaman - US Sea Fencibles - Company: John Gill - Enlistment date: 29 Feb 1814 - Period: 1 Yr - Discharged on 6 Jun 1814.

Hall, Andrew - Private - 3rd Rifles - Company: John Blount - Age: 36 - Height: 5' 7 1/2" - Born: Maryland - Trade: Farmer - Enlistment date: 16 Jul 1814 - Place: Guilford Court House, NC - Period: War - BLW 2652-160-12 - Discharged at Washington, DC, on 7 Apr 1815.

Hall, Benjamin - Private - 27th Infantry (OH) - Company: Isaac Van Horne - Other regiment: 19th Infantry - Age: 25 - Height: 5' 10" - Born: Maryland - Trade: Farmer - Place: Jefferson - Period: War - BLW 6036-160-12 - On board the US Schooner Somers, Lake Erie Squadron; re-enlisted on 11 Apr 1814; discharged at Chillicothe, OH, on 6 Jun 1815.

Hall, Caleb - Sergeant - 38th Infantry - Company: Shepperd Leakin - Age: 23 - Height: 6' 4 1/2" - Born: Baltimore - Trade: Plasterer - Enlistment date: 2 Jun 1813 - Period: 1 Yr - Re-enlisted on 13 Mar 1814 for the war; deserted on 1 Mar 1815.

Hall, Caleb - Seaman - US Sea Fencibles - Company: Simmones Bunbury - Age: 21 - Height: 5' 11" - Born: Baltimore - Enlistment date: 1 Mar 1814 - Period: 1 Yr.

Hall, Henry - Captain - 2nd Light Dragoons - Born: Maryland - Commissioned as a captain, 2nd Light Dragoons, on 12 Mar 1812; transferred to the Light Dragoons on 12 May 1814; discharged on 15 Jun 1815.

Hall, Henry D. - Private - 7th Infantry - Company: Samuel Vail - Other regiment: 8th Infantry (New) - Age: 27 - Height: 5' 10 1/2" - Born: Maryland - Trade: Shoemaker - Enlistment date: 15 Jan 1811 - Place: Pittsburgh - Period: 5 Yrs - Prisoner of War at Prairie de Chien on 20 Jul 1814; discharged on 7 Jul 1816.

Hall, Henry F. - Surgeon - 32nd Infantry - Other regiment: 42nd Infantry - Born: Maryland - Commissioned as a surgeon's mate, 32nd Infantry, on 17 May 1813; promoted to surgeon on 15 Apr 1814; transferred to 42nd Infantry on 26 Apr 1814; discharged on 15 Jun 1815.

Hall, James - Private - 38th Infantry - Company: George Keyser - Age: 20 - Height: 5' 2 1/2" - Born: Worcester Cty, MD - Enlistment date: 5 Jan 1814 - Place: Princess Anne, MD - Period: 1 Yr.

Hall, John - Private - 38th Infantry - Company: James Smith - BLW 14396-160-12.

Hall, John - Private - 14th Infantry - Company: Henry Grindage.

Hall, John S. - Private - 36th Infantry - Company: Joseph Hook - Age: 21 - Height: 5' 11 1/2" - Born: Calvert, MD - Enlistment date: 17 May 1813 - Place: Baltimore - Period: 1 Yr - Deserted on 10 Jul 1813.

Hall, Joseph - Private - 17th Infantry - Age: 37 - Height: 5' 11" - Born: Maryland - Trade: Farmer - Enlistment date: 4 Jan 1815 - Place: West Union, Adams County, OH - Period: War - BLW 484-320-14 - Discharged at

By War of 1812 Soc. in MD

Chillicothe, OH, on 18 Apr 1815.

Hall, Joseph - Seaman - US Sea Fencibles - Company: Simmones Bunbury - Age: 42 - Height: 5' 5 1/2" - Born: Baltimore - Enlistment date: 17 Feb 1814 - Period: 1 Yr - Discharged on 16 Feb 1815.

Hall, Levi - Sergeant - 1st Rifles - Company: Lodowick Morgan - Age: 24 - Height: 5' 11" - Born: Washington, MD - Trade: Blacksmith - Enlistment date: 16 Jun 1812 - Place: Shepherdstown, VA - Period: 5 Yrs.

Hall, Mortimer D. - Captain - 36th Infantry - Commissioned as a captain on 30 Apr 1813; dropped on 30 Sep 1814.

Hall, Robert - Private - 2nd Light Dragoons - Company: William Littlejohn - Age: 20 - Height: 5' 6" - Born: Baltimore - Trade: Blacksmith - Enlistment date: 19 Feb 1814 - Period: War - Died on 3 or 10 Jan 1815.

Hall, Thomas P. - Surgeon - 36th Infantry - Born: Maryland - Commissioned as a surgeon, 36th Infantry, on 10 Jul 1813; discharged on 15 Jun 1815; commissioned as a post surgeon on 12 Dec 1820, promoted to assistant surgeon on 1 Jun 1821; died on 21 Sep 1825.

Hall, William - First Lieutenant - 38th Infantry - Commissioned as a 1st lieutenant on 20 May 1813; discharged on 15 Jun 1815.

Hall, William - Private - 14th Infantry - Company: Joseph Marshall - Age: 22 - Height: 5' 5" - Born: Montreal, LC - Trade: Tailor - Enlistment date: 17 Jul 1814 - Place: Baltimore - Period: War - Deserted at Greenbush, NY, on 15 Feb 1815.

Hall, William - Private - 1st Rifles - Company: William Smyth - Age: 26 - Height: 5' 7 1/4" - Born: Cecil Cty, MD - Trade: Blacksmith - Enlistment date: 24 Aug 1813 - Place: Shepherdstown - Period: 5 Yrs.

Hall, William Wilmont - Surgeon - 1st Rifles - Born: Maryland - Commissioned as a surgeon, Rifles, on 24 May 1812; promoted to hospital surgeon's mate on 21 Apr 1814; discharged on 15 Jun 1815.

Hall, Wright - Second Lieutenant - 36th Infantry - Commissioned as a 3rd lieutenant on 30 Apr 1813; promoted to 2nd lieutenant on 21 Feb 1814; discharged on 15 Jun 1815.

Hallen, John - Private - 14th Infantry - Company: David Cummings - Enlistment date: 1 Sep 1812 - Period: 5 Yrs - Prisoner of War, sent to England; deserted at Fort McHenry on 20 Apr 1815.

Halll, Alexander F. - Clerk - 7th Infantry - Company: Narcissus Broutin - Age: 39 - Height: 5' 7" - Born: Baltimore - Enlistment date: 19 Mar 1813 - Period: War - Pension: Land bounty to William Biays Hall and George Ord Hall, heirs at law of Alexander F. Hall - BLW 130-160-12 - Discharged at Fort St. Philip, LA, on 30 Apr 1815.

Halson, James - Private - 14th Infantry - Prisoner of War (Halifax), discharged from Halifax on 3 Feb 1814.

Hamaker, Adam - Private - 36th Infantry - Age: 19 - Height: 5' 11" - Born: Pennsylvania - Enlistment date: 10 Jul 1814 - Place: West Minster.

Hambley, James - Seaman - US Sea Fencibles - Company: John Gill - Age: 24 - Height: 5' 4" - Born: Baltimore City - Enlistment date: 11 Jan 1814 - Period: 1 Yr.

Hamilton, David - Private - 36th Infantry - Age: 38 - Height: 5' 6" - Born: Lancaster Cty, PA - Trade: Farmer - Enlistment date: 2 Nov 1814 - Period: War - BLW 8837-160-12 - Discharged at Georgetown, DC, on 13 Mar 1815.

Hamilton, John - Seaman - US Sea Fencibles - Company: William Addison.

Hamilton, Robert - Sergeant - 38th Infantry - Company: John Mowton - Pension: Land bounty to Rachel B. Hilyeard, daughter and only heir at law of Robert Hamilton - BLW 26943-160-12.

Hamilton, William - Private - 14th Infantry - Company: Henry Fleming - Age: 29 - Height: 5' 5 1/2" - Born: Prince George's Cty, MD - Trade: Brick layer - Enlistment date: 22 Nov 1812 - Period: 5 Yrs - BLW 8718-160-12 - Discharged at Washington, DC, on 7 Feb 1817 because of wounds received at Beaver Dams.

Hamilton, William - Private - 38th Infantry - Company: Charles Stansbury - Age: 28 - Height: 5' 8 1/2" - Born: Ireland - Trade: Shoemaker - Enlistment date: 5 Oct 1814 - Place: Baltimore - Period: War - BLW 5488-160-12 - Discharged at Baltimore on 6 Apr 1815.

Hamilton, William - Private - 14th Infantry - Company: Thomas Montgomery - Enlistment date: 25 Nov 1812 - Period: 5 Yrs - Pension: Placed on the pension roll on 17 Feb 1814.

Hammel, George - Private - 38th Infantry - Company: John Brookes - BLW 16852-160-12.

Hammers, Francis - Private - 38th Infantry - Company: John Buck - Age: 23 - Height: 5' 9 1/2" - Born: Maryland - Enlistment date: 8 Jan 1814 - Place: Cumberland - Period: 1 Yr - Discharged on 7 Jan 1815.

Hammill, Daniel - Private - 38th Infantry - Company: James Hook - Age: 26 - Height: 5' 10" - Born: Belfast, Ireland - Trade: Baker - Enlistment date: 4 Apr 1814 - Place: Baltimore - Period: War - BLW 6540-160-12 - Discharged at Fort Covington, MD, on 31 Mar 1815.

Hammond, Christopher - Private - 28th Infantry - Company: Henry Gist - Age: 40 - Height: 5' 10" - Born: Maryland - Trade: Seaman - Enlistment date: 12 Aug 1814 - Place: Maysville, KY - Period: War - BLW 24569-160-12 - Discharged at Lower Sandusky, OH (Fort Stephenson), on 25 Jun 1815.

Hammond, Hezekiah - Private - 1st Artillery - Company: Hopley Yeaton - Age: 34 - Height: 5' 9" - Born: Frederick, MD - Trade: Laborer - Enlistment date: 31 Oct 1810 - Place: Washington, DC - Period: 5 Yrs - Died at Fort Nelson, VA, on 4 Mar 1815.

Hammond, Roger - Clerk - 44th Infantry - Company: Joseph Miles - Age: 29 - Height: 6' 2" - Born: Anne Arundel Cty, MD - Trade: Merchant - Enlistment date: 25 Apr 1814 - Place: New Orleans - Period: War - Discharged on 8 Apr 1815.

Hampson, James - Private - 38th Infantry - Company: Shepperd Leakin - Age: 24 - Height: 5' - Born: Huntington Cty, PA - Trade: Laborer - Enlistment date: 21 Mar 1814 - Place: Baltimore - Period: War - Discharged at Baltimore on 30 Mar 1815.

Hanalin, Patrick - Gunner - US Sea Fencibles - Company: William Addison - Enlistment date: 23 Dec 1813 - Period: 1 Yr - Pension: Old War IF-25141.

By War of 1812 Soc. in MD

Hancock, Daniel - Private - 19th Infantry - Company: William Gill - Age: 18 - Height: 5' 7" - Born: Maryland - Trade: Farmer - Enlistment date: 9 Jun 1814 - Period: War - BLW 17630-160-12 - Discharged at Chillicothe, OH, on 5 Jun 1815.

Hancock, Thomas - Private - 1st Rifles - Company: William Smyth - Age: 21 - Height: 5' 6" - Born: Charles Cty, MD - Trade: Farmer - Enlistment date: 27 Mar 1812 - Place: Maryland - Period: 5 Yrs - Discharged at Baton Rouge, LA, on 31 Mar 1817.

Handle, John - Private - 2nd Light Dragoons - Company: William Littlejohn - Other regiment: Corps of Artillery - Age: 21 - Height: 5' 5 3/4" - Born: Philadelphia - Trade: Paper hanger - Enlistment date: 26 Dec 1813 - Place: Baltimore - Period: 5 Yrs - Deserted on 10 Jun 1815.

Hands, Ephraim - Seaman - US Sea Fencibles - Company: William Addison.

Hands, Nicholas - Seaman - US Sea Fencibles - Company: William Addison.

Hane Jr., Jacob - Seaman - US Sea Fencibles - Company: William Addison.

Hanes, David - Private - 14th Infantry - Company: Joseph Marshall - Age: 24 - Height: 5' 7 1/2" - Born: Sussex Cty, DE - Trade: Farmer - Enlistment date: 12 Apr 1814 - Place: Baltimore - Period: War - BLW 5363-160-12 - Discharged at Greenbush, NY, on 3 May 1815.

Hanes, James - Seaman - US Sea Fencibles - Company: Simmones Bunbury - Enlistment date: 1 Mar 1814 - Period: 1 Yr - Discharged on 31 Jan 1815.

Hanham, James R. - Captain - 1st Artillery - Other regiment: Corps of Artillery - Born: England - BLW 23116-183-50 - Commissioned as a 2nd lieutenant, Artillery, on 17 Jan 1805; promoted to 1st lieutenant on 1 Nov 1806; promoted to captain on 10 Jul 1811; transferred to Corps of Artillery on 12 May 1814; discharged on 15 Jun 1815; died on 2 Nov 1865.

Hanks, Porter - First Lieutenant - 1st Artillery - Enlistment date: 17 Jan 1805 - Place: Maryland - Pension: Land bounty to Margaret Hanks, widow of Porter Hanks - BLW 4955-160-50 - Commissioned as a 2nd lieutenant on 17 Jan 1805; promoted to 1st lieutenant on 31 Dec 1806; killed at Detroit on 16 Aug 1812.

Hanley, William - Private - 6th Infantry - Company: Henry Shell - Other regiment: 2nd Infantry (New) - Age: 31 - Height: 5' 6 1/2" - Born: Baltimore - Trade: Printer - Enlistment date: 28 May 1814 - Place: Philadelphia - Period: 5 Yrs - Discharged at Plattsburgh, NY, on 28 May 1819.

Hanley, William - Private - 14th Infantry - Company: Reuben Gilder - Age: 50 - Height: 5' 7" - Born: Devonshire County, England - Trade: Sailor - Enlistment date: 31 Oct 1812 - Period: 5 Yrs - BLW 4174-160-12 - Discharged at Greenbush, NY, on 1 May 1815 for old age.

Hann, Frederick - Private - 38th Infantry - Company: Shepperd Leakin - Age: 24 - Height: 5' 3 1/2" - Born: Frederick Cty, MD - Trade: Farmer - Enlistment date: 18 Aug 1814 - Place: Baltimore - Period: War - Discharged at Fort McHenry on 28 Mar 1815.

Hann, Jacob - Drummer - 38th Infantry - Company: John Rothrock - Pension: SO-30596.

Hanna, John - Private - US Volunteers - Company: Stephen Moore -

Maryland Regulars in War of 1812

Enlistment date: 8 Sep 1812 - Period: 1 Yr - Died at Buffalo on 13 Jan 1813.

Hanna, William - Sergeant - 26th Infantry - Company: William Bezeau - Age: 27 - Height: 5' 6" - Born: Ireland - Trade: Clerk - Enlistment date: 14 Oct 1814 - Place: Baltimore - Period: War - Discharged at Baltimore.

Hanna, William - Private - 2nd Artillery - Company: Nathan Towson - Pension: Wife Eleanor, Old War MC-4; heirs received half pay for five years in lieu of land bounty - Died on 25 Mar 1813.

Hannah, James - Private - 5th Infantry - Company: Leroy Opie - Age: 21 - Height: 6' - Born: Talbot Cty, MD - Trade: Chair maker - Enlistment date: 27 Dec 1808 - Place: Baltimore - Period: 5 Yrs.

Hannah, Thomas - Private - 2nd Infantry - Company: John Brahan - Age: 25 - Height: 5' 8" - Born: Baltimore - Trade: Tailor - Enlistment date: 8 Mar 1808 - Place: Staunton, VA - Period: 5 Yrs - BLW 10526-160-12 - Re-enlisted at New Orleans on 1 Mar 1813 for the war; discharged at New Orleans on 9 Apr 1815.

Hannifan, John - Private - 5th Infantry - Company: Leroy Opie - Other regiment: 3rd Infantry (New) - Age: 22 - Height: 5' 9" - Born: Baltimore - Trade: Shoemaker - Enlistment date: 18 Mar 1814 - Place: Plattsburgh, NY - Period: War - Drowned at Erie, PA, on 5 Sep 1815.

Hannon, George C. - Private - 36th Infantry - Age: 27 - Height: 5' 9" - Born: Philadelphia - Enlistment date: 26 Apr 1813 - Place: Baltimore - Period: 1 Yr.

Hannum, Peter - Private - 14th Infantry - Company: David Cummings - Age: 44 - Height: 5' 7 1/2" - Born: Pennsylvania - Enlistment date: 13 Aug 1812 - Period: 5 Yrs - Pension: Land bounty to Ann H. Hannum, daughter and other heirs at law of Peter Hannum - BLW 17139-160-12 - Killed at Beaver Dams, UC, on 24 Jun 1813.

Hansell, Philip - Private - 14th Infantry - Company: Samuel Lane - Other regiment: 4th Infantry (New) - Age: 21 - Height: 5' 9 1/3" - Born: Shephendsown, VA - Trade: Miller - Enlistment date: 20 Jun 1812 - Place: Cumberland, MD - Period: 5 Yrs - Prisoner of War, exchanged on 15 Apr 1814; discharged at Montpelier, MS, on 20 Jun 1817.

Hanson, John - Private - 38th Infantry - Company: James Smith - BLW 11452-160-12.

Hanson, John - Private - 22nd Infantry - Company: Jacob Carmack - Age: 39 - Height: 5' 5" - Born: Frederick, MD - Trade: Saddler - Enlistment date: 4 Jul 1814 - Period: War - BLW 16680-160-12 - Discharged at Fort Fayette, PA, on 24 Mar 1815.

Hanson, William - Quarter Gunner - US Sea Fencibles - Company: John Gill - Age: 38 - Height: 5' 6 1/2" - Born: Baltimore - Enlistment date: 28 Dec 1813 - Period: 1 Yr.

Harbaugh, Charles - Private - 5th Infantry - Company: Colin Buckner - Age: 23 - Height: 5' 5" - Born: Baltimore - Trade: Shoemaker - Enlistment date: 18 Feb 1813 - Place: Baltimore - Period: 5 Yrs - Pension: Land bounty to Leonard Harbaugh, father and heir at law of Charles Harbaugh - BLW 9764-160-12 - Died on 1 Jul 1814.

By War of 1812 Soc. in MD

Harbaugh, Frederick - Sergeant - 38th Infantry - Company: Shepperd Leakin - Enlistment date: 12 Jun 1813 - Period: 1 Yr - Pension: Wife Mary Bolton, WO-1719, WC-2712; married on 24 Dec 1811, Hanover, York Cty, PA; soldier died on 17 Dec 1860, Eaton, OH - BLW 143-160-50 - Discharged on 12 Jun 1814.

Harbin, Miley - Private - 10th Infantry - Company: Emanuel Leigh - Other regiment: Corps of Artillery - Age: 32 - Height: 5' 10" - Born: Charles Cty, MD - Trade: Tailor - Enlistment date: 9 Mar 1814 - Place: Louisburg, Franklin - Period: 5 Yrs.

Harbough, Jacob - Private - 2nd Infantry - Company: Francis Johnston - Age: 25 - Height: 5' 8" - Born: Frederick, MD - Trade: Farmer - Enlistment date: 14 May 1808 - Place: Newtown - Period: 5 Yrs - Discharged at Camp Perdido, FL, on 14 Mar 1813.

Harden, Robert - Private - 36th Infantry - Company: Samuel Raisin - BLW 2609-160-12.

Harder, Henry - Private - 5th Infantry - Company: James Dorman - Other regiment: 3rd Infantry (New) - Age: 25 - Height: 5' 7 3/4" - Born: Frederick, MD - Trade: Laborer - Enlistment date: 3 Apr 1812 - Place: Fredericktown, MD - Period: 5 Yrs - Deserted on 28 Feb 1816.

Hardesty, George - Private - 12th Infantry - Age: 36 - Height: 6' 1" - Born: Anne Arundel Cty, MD - Trade: Nailor - Enlistment date: 15 Nov 1814 - Place: Winchester - Period: 5 Yrs - Pension: Land bounty to James Hardesty, son & other heirs at law of George Hardesty - BLW 26749-160-12 - Died on 28 Mar or 6 Apr 1815.

Harding, Charles - Private - 2nd Light Dragoons - Company: George Haig - Other regiment: 1st Rifles - Age: 25 - Height: 5' 6" - Trade: Saddler - Enlistment date: 16 Jul 1812 - Place: Baltimore - Period: 5 Yrs - Transferred to 1st Rifles on 4 Mar 1814; killed at Fort Erie, UC, on 17 Sep 1814.

Harding, Thomas - Private - 38th Infantry - Company: James Smith - BLW 11186-160-12.

Hardman, George - Musician - 5th Infantry - Company: William Henshaw - Other regiment: 3rd Infantry (New) - Age: 26 - Height: 5' 9" - Born: Frederick, MD - Trade: Shoemaker - Enlistment date: 5 Apr 1812 - Place: Hagerstown, MD - Period: 5 Yrs - Discharged on 4 Apr 1817.

Hardy, Amos - Private - 14th Infantry - Company: Thomas Montgomery - Other regiment: 4th Infantry (New) - Age: 24 - Height: 6' - Born: Loudoun Cty, VA - Trade: Blacksmith - Enlistment date: 15 Jan 1813 - Place: Alexandria, DC - Period: 5 Yrs - Prisoner of War (Halifax), discharged from Halifax on 3 Feb 1814, exchanged on 15 Apr 1814; deserted at Fort Severn on 12 Jul 1815.

Hardy, George - Private - 5th Infantry - Company: James Paxton - Age: 24 - Height: 5' 2 1/4" - Born: Frederick, MD - Trade: Laborer - Enlistment date: 31 Dec 1812 - Place: Lewisburg, VA - Period: 5 Yrs - Died at French Mills, NY, on 3 Feb 1814.

Hardy, John - Corporal - 12th Infantry - Company: Thomas Moore - Age: 22 - Height: 5' 3" - Born: Frederick Cty, MD - Enlistment date: 22 Mar 1813 -

Maryland Regulars in War of 1812

Place: Wheeling, VA - Period: 5 Yrs - Killed in action at Williamsburg, NY, on 11 Nov 1813.

Hardy, Samuel T. - Private - 14th Infantry - Company: Thomas Montgomery - Enlistment date: 16 Apr 1812 - Period: 5 Yrs.

Hargood, George - Private - 14th Infantry - Company: Isaac Barnard - Pension: Wife Elizabeth, Old War WF-14932 Rejected; Old War WF-13372 - Prisoner of War (Halifax), sent to England.

Harker, Mary - Woman - 14th Infantry - Entered hospital on 7 Sep 1814 and discharged same day; probably a washerwomen.

Harman, John - Private - 2nd Artillery - Company: Spotswood Henry - Other regiment: Corps of Artillery - Age: 29 - Height: 5' 10 1/2" - Born: Maryland - Trade: Laborer - Enlistment date: 21 Jul 1812 - Place: Sullivan, NC - Period: 5 Yrs - Pension: Land bounty to James Harman, son and other heirs at law of John Harman - BLW 15513-160-12 - Died on wounds received in action in the Mediterranean on 21 Jul 1815.

Harman, William - Private - 36th Infantry - Company: Joseph Hook - Age: 39 - Height: 5' 6" - Born: Norfolk, VA - Enlistment date: 6 May 1813 - Place: Baltimore - Period: 1 Yr - Died on 20 Jun 1813.

Harn, Jacob - Private - 1st Artillery - Company: George Armistead - Age: 21 - Height: 5' 11 1/2" - Born: Little York, PA - Trade: Farmer - Enlistment date: 19 Mar 1812 - Place: Fort McHenry - Period: 5 Yrs - Discharged on 18 Mar 1819.

Harn, Jacob - Corporal - 36th Infantry - Company: Joseph Hook - Age: 33 - Height: 5' 9" - Born: Northampton Cty, PA - Trade: Hatter - Enlistment date: 15 Jul 1814 - Place: Baltimore - Period: War - Discharged at Fort Covington, MD, on 30 Mar 1815.

Harney, Manlove - Private - 2nd Artillery - Company: John Ritchie - Age: 27 - Height: 5' 10 1/2" - Born: Maryland - Trade: Farmer - Enlistment date: 1 Apr 1813 - Place: Richmond, VA - Period: War - Discharged at Buffalo on 3 Jun 1815.

Harp, Leonard - Private - 36th Infantry - Company: Joseph Hook - Age: 42 - Height: 6' 1/2" - Born: Washington Cty, MD - Trade: Farmer - Enlistment date: 10 Oct 1810 - Place: Boston - Period: 5 Yrs.

Harper, Benjamin - Private - 14th Infantry - Company: Thomas Montgomery - Other regiment: 4th Infantry (New) - Age: 25 - Height: 5' 8" - Born: District of Columbia - Trade: Shoemaker - Enlistment date: 7 Nov 1812 - Place: Baltimore - Period: 5 Yrs - BLW 15658-160-12 - Discharged at Montpelier, VT, on 6 Nov 1817.

Harper, John - Private - 36th Infantry - Company: Thomas Carbery - Age: 19 - Height: 5' 6" - Born: Georgetown, MD - Trade: Butcher - Enlistment date: 27 Feb 1814 - Place: Leonardstown, MD - Period: War - Discharged at Baltimore on 31 Mar 1815.

Harrell, Dempsey - Private - 38th Infantry - Other regiment: 4th Infantry (New) - Pension: Wife Mary, SO-15625, SC-5784.

Harrigan, Elisha - Private - 14th Infantry - Prisoner of War (Halifax), discharged from Halifax on 3 Feb 1814.

By War of 1812 Soc. in MD

Harrigan, John - Private - 38th Infantry - Company: James Hook - Age: 18 - Height: 4' 11" - Born: Baltimore - Trade: Cooper - Enlistment date: 4 May 1814 - Place: Baltimore - Period: War - Pension: Wife Elizabeth, WO-12986, WC-21585 - BLW 7146-160-12 - Discharged at Fort Covington, MD, on 31 Mar 1815.

Harrington, Richard - Private - 36th Infantry - Company: Samuel Raisin - Age: 24 - Height: 5' 10" - Born: Talbot Cty, MD - Enlistment date: 14 Jun 1813 - Place: Chester - Period: 1 Yr.

Harrington, Robert - Gunner - US Sea Fencibles - Company: Simmones Bunbury - Enlistment date: 30 Jan 1815 - Period: 1 Yr - Discharged at Fort McHenry on 24 Mar 1815.

Harrington, Samuel - Private - 38th Infantry - Company: James Smith - Age: 25 - Height: 5' 7 1/2" - Born: Nanticoke, MD - Trade: Shoemaker - Enlistment date: 11 Mar 1814 - Place: Norfolk, VA - Period: War - BLW 6342-160-12 - Discharged at Craney Island, VA, On 15 Mar 1815.

Harrington, Thomas - Private - 14th Infantry - Company: Thomas Kearney - Enlistment date: 20 May 1812 - Period: 5 Yrs - Pension: Land bounty to Elizabeth Harrington, sister & other heirs at law of Thomas Harrington - BLW 22181-160-12 - Died on 8 Feb 1813.

Harris, Andrew - Private - 14th Infantry - Other regiment: 38th Infantry - Enlistment date: 18 May 1812 - Period: 5 Yrs - Deserted from the 14th Infantry, reassigned to Captain Leakin's Company, 38th Infantry; died in hospital on 10 Apr 1814.

Harris, Hiram - Private - 2nd Light Dragoons - Company: John Burd - Age: 20 - Height: 5' 6" - Born: Loudoun Cty, VA - Trade: Chandler - Enlistment date: 22 Dec 1814 - Place: Baltimore - Period: War - BLW 24928-160-12 - Discharged at Carlisle Barracks, PA, on 4 May 1815.

Harris, John - Seaman - US Sea Fencibles - Company: William Addison.

Harris, John - Private - 6th Infantry - Company: Gad. Humphreys - Age: 26 - Height: 5' 10 1/2" - Born: Prince George's Cty, MD - Trade: Farmer - Enlistment date: 5 Oct 1813 - Place: New York - Period: War - Discharged at Plattsburgh, NY, on 24 Jul 1815.

Harris, John - Private - 12th Infantry - Company: Thomas Sangster - Other regiment: 8th Infantry (New) - Age: 20 - Height: 5' 5" - Born: Sussex, DE - Trade: Blacksmith - Enlistment date: 17 Jun 1812 - Place: Elkton, MD - Period: 5 Yrs - Discharged on 17 Jun 1817.

Harris, Josiah - Private - 14th Infantry - Company: Thomas Montgomery - Enlistment date: 17 Feb 1813 - Period: 5 Yrs - Pension: Land bounty to William Harris and Zephaniah Harris, brothers and only heirs at law of Josiah Harris - BLW 27428-160-42 - Died at Burlington, VT, in Mar 1814 from sickness.

Harris, Littleton - Private - 36th Infantry - Company: Joseph Hook - Age: 22 - Height: 5' 10" - Born: Dorchester Cty, MD - Trade: Weaver - Enlistment date: 1 Mar 1814 - Place: Annapolis - Period: War - BLW 13239-160-12 - Discharged at Fort Covington, MD, on 30 Mar 1815.

Harris, Thomas - Private - 5th Infantry - Company: William Bird - Age: 24 -

Height: 5' 8" - Born: Ireland - Trade: Seaman - Enlistment date: 17 Apr 1813 - Place: Baltimore - Period: 5 Yrs.

Harris, William - Sergeant - 2nd Artillery - Company: Frederick Evans - Age: 26 - Height: 5' 8" - Born: New York - Trade: Shoemaker - Enlistment date: 4 Apr 1814 - Place: Baltimore - Period: 5 Yrs - BLW 16860-160-12 - Discharged at Fort McHenry on 6 Apr 1819.

Harris, William - Private - 2nd Artillery - Company: Nathan Towson - Age: 22 - Height: 6' 2" - Born: New York - Trade: Blacksmith - Enlistment date: 10 Mar 1813 - Place: Buffalo - Period: 5 Yrs - Discharged at Fort Niagara, NY, on 8 Mar 1818 and re-enlisted.

Harris, William - Private - 2nd Rifles - Company: William Adair - Other regiment: 17th Infantry - Age: 27 - Height: 5' 10 1/2" - Born: Baltimore City - Enlistment date: 22 May 1814 - Place: Cincinnati - Transferred to the 17th Infantry on 1 Jul 1814, Captain Harris Hickman's Company.

Harris, William - Private - 38th Infantry - Company: John Mowton - BLW 24254-160-12.

Harris, William - Private - 14th Infantry - Company: James McDonald - Age: 21 - Height: 5' 5 1/2" - Born: Harford Cty, MD - Trade: Sailor - Enlistment date: 5 Oct 1812 - Period: 18 Mos - Re-enlisted at Baltimore on 14 Mar 1814 for the war.

Harrison, Ephraim - Private - 22nd Infantry - Age: 21 - Height: 5' 10" - Born: Maryland - Enlistment date: 3 Jan 1814.

Harrison, George - Private - 26th Infantry - Company: William Bezeau - Age: 22 - Height: 5' 8 1/2" - Born: Anne Arundel Cty, MD - Trade: Mason - Enlistment date: 15 Nov 1814 - Place: Baltimore - Period: War - BLW 1811-160-12 - Discharged at Philadelphia on 23 Mar 1815.

Harrison, Josiah - Private - 14th Infantry - Company: Richard Bennett.

Harrison, Samuel - Private - 20th Infantry - Company: William Jett - Age: 22 - Height: 5' 10" - Born: Calvert Cty, MD - Trade: Hatter - Enlistment date: 26 Mar 1813 - Period: 5 Yrs - Pension: Land bounty to Isaac Harrison and other heirs at law of Samuel Harrison - BLW 12236-160-12 - Died in Jan 1814.

Harrison, William - Private - 15th Infantry - Company: Joseph Barton - Age: 25 - Height: 5' 8 1/2" - Born: Calvert, MD - Trade: Farmer - Enlistment date: 6 Sep 1814 - Place: Philadelphia - Period: War - BLW 5317-160-12 - Discharged at Trenton, NJ, on 6 Mar 1815.

Hart, Caleb - Private - 25th Infantry - Company: John Murdock - Age: 36 - Height: 5' 10" - Born: Maryland - Trade: Farmer - Enlistment date: 24 Oct 1812 - Period: 18 Mos - Discharged at Sackets Harbor, NY, on 9 Jun 1814.

Hart, Joshua - Private - 22nd Infantry - Company: John Pentland - Other regiment: 2nd Infantry (New) - Age: 20 - Height: 6' - Born: Baltimore - Trade: Cooper - Enlistment date: 31 Dec 1812 - Place: Harrisburg, PA - Period: 5 Yrs - Deserted at Sackets Harbor, NY, on 26 Jan 1816.

Harter, Lawrence - Private - 38th Infantry - Company: John Brookes - BLW 11451-160-12.

Hartley, David - Private - 14th Infantry - Company: Samuel Lane - Age: 39 -

By War of 1812 Soc. in MD

Height: 5' 6" - Born: Ireland - Enlistment date: 22 Jun 1812 - Place: O Bourg - Period: 5 Yrs - Died at Black Rock, NY, on 5 Dec 1812.

Hartley, Simeon - Private - 38th Infantry - Company: James Hook - Age: 19 - Height: 5' 10" - Born: Baltimore - Trade: Cooper - Enlistment date: 20 Dec 1814 - Place: Baltimore - Period: War - BLW 139-320-12 - Discharged at Fort Covington, MD, on 31 Mar 1815.

Harvey, Benjamin - Private - 38th Infantry - Company: Shepperd Leakin - Age: 38 - Height: 5' 8" - Born: Cecil Cty, MD - Trade: Shoemaker - Enlistment date: 6 Aug 1814 - Place: Baltimore - Period: War - BLW 5280-160-12 - Discharged at Fort McHenry on 28 Mar 1815.

Harvey, James - Private - 14th Infantry - Company: Reuben Gilder - Age: 42 - Height: 5' 5 3/4" - Born: Ireland - Enlistment date: 27 May 1813 - Period: 18 Mos - Discharged at Black Rock, NY, on 27 Sep 1814.

Harvey, John - Private - 14th Infantry - Company: William McIlvane - Other regiment: 4th Infantry (New) - Enlistment date: 14 Apr 1812 - Period: 5 Yrs - Prisoner of War, exchanged on 15 Apr 1814; discharged in Jul 1815.

Harvey, William - Sergeant - 2nd Light Dragoons - Company: John Burd - Age: 22 - Height: 5' 10 1/4" - Born: Washington Cty, MD - Trade: Miller - Enlistment date: 11 May 1812 - Period: 18 Mos.

Harvy, John - Private - 5th Infantry - Company: William Bird - Age: 25 - Height: 5' 8" - Born: Baltimore - Enlistment date: 17 May 1813 - Place: Baltimore - Period: 5 Yrs - Deserted on 24 Aug 1813.

Harwood, William - Second Lieutenant - 38th Infantry - Company: John Brookes - Born: Maryland - BLW 13807-160-50 - Commissioned as a 3rd lieutenant on 20 May 1813; promoted to 2nd lieutenant on 1 May 1814; discharged on 15 Jun 1815.

Hasaman, William - Private - 36th Infantry - Company: Samuel Raisin - Age: 25 - Height: 5' 3" - Born: Somerset Cty, MD - Enlistment date: 15 Jun 1813 - Place: Cambridge, MD - Period: 1 Yr.

Haselton, Joseph - Private - 19th Infantry - Company: William Gill - Age: 28 - Height: 5' 9" - Born: Maryland - Trade: Hatter - Enlistment date: 24 Apr 1814 - Place: Jefferson (Pickaway Plains), Pickaway Cty, OH - Period: War - Discharged at Chillicothe, OH, on 5 Jun 1815.

Hash, Peter - Seaman - US Sea Fencibles - Company: Simmones Bunbury.

Haskins, Robert - Private - 36th Infantry - Company: Joseph Hook - Age: 33 - Height: 6' - Born: St. Mary's Cty, MD - Trade: Farmer - Enlistment date: 1 Mar 1814 - Place: Annapolis - Period: War - BLW 8075-160-12 - Discharged at Fort Covington, MD, on 30 Mar 1815.

Haslett, James - Captain - 38th Infantry - Commissioned as a captain, 38th Infantry resigned on 20 May 1814.

Haslett, Joseph - Private - 1st Infantry - Company: John Whistler - Age: 34 - Height: 5' 10" - Born: Maryland - Trade: Laborer - Enlistment date: 23 Sep 1806 - Place: Lancaster - Period: 5 Yrs - Re-enlisted on 29 Sep 1811; Prisoner of War; died at Boston, MA, on 23 Dec 1812.

Haslett, William - Private - 2nd Infantry - Age: 39 - Height: 5' 6" - Born: Ireland - Trade: Laborer - Enlistment date: 30 Mar 1807 - Place: Fort

McHenry - Period: 5 Yrs.

Haslip, John - Private - 14th Infantry - Company: Clement Sullivan - Enlistment date: 2 May 1812 - Period: 5 Yrs - Pension: Land bounty to Nelly Haslip, sister and other heirs at law of John Haslip - BLW 16485-160-12 - Died on 4 Dec 1812.

Hassan, James - Private - 5th Infantry - Company: William Bird - Age: 31 - Height: 5' 11 1/2" - Born: Cecil Cty, MD - Trade: Farmer - Enlistment date: 3 Mar 1814 - Place: Elizabethtown - Period: 5 Yrs.

Hastar, John - Private - 36th Infantry - Company: Mortimer Hall - Age: 38 - Height: 5' 7 1/2" - Born: York, PA - Enlistment date: 10 May 1813 - Place: Havre de Grace, MD - Period: 1 Yr - Deserted on 16 Aug 1813.

Hatfield, Samuel - Private - 2nd Artillery - Company: Isaac Roach - Other regiment: Corps of Artillery - Age: 22 - Height: 5' 6 1/2" - Born: Anne Arundel Cty, MD - Trade: Shoemaker - Enlistment date: 14 Apr 1814 - Place: Annapolis - Period: 5 Yrs.

Haukpal, Conrad - Private - 19th Infantry - Age: 18 - Height: 5' 5" - Born: Allegany Cty, MD - Enlistment date: 13 Jul 1812 - Period: 5 Yrs.

Haupt, George - Private - 1st Artillery - Company: Moses Swett - Age: 23 - Height: 5' 7" - Born: Baltimore - Trade: Tobacconist - Enlistment date: 13 Aug 1814 - Place: New York - Period: War - Discharged on 12 May 1815.

Householder, William - Private - 22nd Infantry - Company: Joseph Henderson - Age: 18 - Height: 5' 6" - Born: Huntington, MD - Trade: Laborer - Enlistment date: 3 Mar 1814 - Place: Huntington, MD - Period: 5 Yrs.

Hawkenbrok, Henry - Private - 14th Infantry - Prisoner of War (Halifax), captured at Beaver Dams on 24 Jun 1813, discharged from Halifax on 3 Feb 1813.

Hawkersmith, William - Private - 28th Infantry - Company: Thomas Edmondson - Age: 20 - Height: 5' 10" - Born: Frederick, MD - Trade: Carpenter - Enlistment date: 30 Mar 1814 - Place: Detroit - Period: War - BLW 23680-160-12 - Discharged on 30 Jun 1815.

Hawkins, Daniel E. - Private - 26th Infantry - Company: William Bezeau - Age: 21 - Height: 5' 7" - Born: New Haven, CT - Trade: Shoemaker - Enlistment date: 3 Oct 1814 - Place: Baltimore - Period: War - BLW 7194-160-12 - Discharged at Philadelphia on 23 Mar 1815.

Hawkins, John - Private - 28th Infantry - Age: 23 - Height: 6' 1" - Born: Maryland - Enlistment date: 3 Aug 1814 - Place: Maysville, KY.

Hawkins, Thomas - Private - 26th Infantry - Company: Isaac Finch - Age: 28 - Height: 5' 4" - Born: Maryland - Trade: Cabinet maker - Enlistment date: 16 Nov 1814 - Place: Peru, NY - Period: 5 Yrs - Discharged at Burlington, VT, on 11 Jun 1815 after furnishing a substitute (Joseph Young).

Hawkins, Thomas - Private - 12th Infantry - Company: James Paxton - Age: 18 - Height: 5' 4 1/2" - Born: Baltimore - Trade: Laborer - Enlistment date: 4 Jan 1813 - Place: Lewisburg, VA - Period: 18 Mos - Discharged at Champlain, NY, on 5 Jul 1814.

Hawkins, William - Private - 36th Infantry - Company: Samuel Raisin - Other regiment: Ordnance Department - Age: 21 - Height: 5' 5" - Born: North

Carolina - Trade: Farmer - Enlistment date: 23 Dec 1814 - Place: Norfolk, VA - Period: 5 Yrs - Transferred to Ordnance Department on 30 Mar 1815; deserted from Greenleaf Point, DC, on 25 Mar 1815.

Hay, Judson - Private - 14th Infantry - Company: Joseph Marshall - Age: 18 - Height: 5' 3" - Born: Prince George's Cty, MD - Trade: Shoemaker - Enlistment date: 30 Mar 1814 - Place: Upper Marlboro, MD - Period: War - Discharged at Greenbush, NY, on 3 May 1815.

Hayden, Thomas - Private - 36th Infantry - Company: Samuel Raisin - Age: 19 - Height: 5' 10" - Born: St. Mary's Cty, MD - Trade: Hatter - Enlistment date: 18 Sep 1814 - Place: District of Columbia - Period: War - Discharged at Washington, DC, on 20 Mar 1815.

Hayes, Adam - Seaman - US Sea Fencibles - Company: Simmones Bunbury - Age: 40 - Height: 5' 10" - Born: Hartford, CT - Enlistment date: 5 Mar 1814 - Period: 1 Yr - Discharged at Fort McHenry on 24 Mar 1815.

Hayes, George - Private - US Volunteers - Company: Stephen Moore - Enlistment date: 8 Sep 1812 - Period: 1 Yr.

Hayes, Nicholas - Private - US Volunteers - Company: Stephen Moore - Enlistment date: 8 Sep 1812 - Period: 1 Yr.

Hayes, William - Private - 14th Infantry - Company: Reuben Gilder - Age: 29 - Height: 6' - Born: Baltimore - Trade: Collier - Enlistment date: 27 Mar 1814 - Place: Frederick, MD - Period: War - Pension: Wife Sarah, WO-39072 - BLW 13800-160-12 - Discharged at Greenbush, NY, on 4 May 1815.

Hayley, Edward - Private - US Volunteers - Company: Stephen Moore - Enlistment date: 8 Sep 1812 - Period: 1 Yr.

Haynes, Jacob - Private - 4th Rifles - Company: Matthew Magee - Age: 32 - Height: 5' 6" - Born: Baltimore - Trade: Shoemaker - Enlistment date: 6 Dec 1814 - Place: Shepherdstown, VA - Period: 5 Yrs.

Haynes, John - Private - 1st Rifles - Company: Daniel Appling - Other regiment: Rifles - Age: 24 - Height: 5' 6" - Born: Maryland - Trade: Farmer - Enlistment date: 1 Apr 1813 - Place: Fort Hampton, AL - Period: 5 Yrs - Discharged on 1 Apr 1818.

Haynes, John - Private - 1st Rifles - Company: Daniel Appling - Other regiment: Rifles - Age: 24 - Height: 5' 6" - Born: Maryland - Trade: Farmer - Enlistment date: 1 Apr 1813 - Place: Fort Hampton, AL - Period: 5 Yrs - BLW 19313-160-12 - Discharged on 1 Apr 1818.

Hays, Bernard - Private - 14th Infantry - Company: Reuben Gilder - Age: 22 - Height: 5' 9" - Born: Hagerstown, MD - Trade: Laborer - Enlistment date: 11 Mar 1814 - Place: Frederick, MD - Period: War - Pension: Old War IF-10228 - BLW 2712-160-12 - Wounded at Lyon's Creek, UC; discharged at Greenbush, NY, on 11Apr 1815 because of wounds.

Hays, George - Corporal - 38th Infantry - Company: Shepperd Leakin - Age: 28 - Height: 6' - Born: Harford Cty, MD - Trade: Farmer - Enlistment date: 28 Feb 1814 - Period: War - Pension: Land bounty to Abraham Hays, brother and other heirs at law of George Hays - BLW 12803-160-12 - Discharged at Fort McHenry on 28 Mar 1815.

Maryland Regulars in War of 1812

Hays, George - Private - 38th Infantry - Age: 24 - Height: 5' 9" - Born: Cork, Ireland - Trade: Brush maker - Enlistment date: 31 Aug 1814 - Place: Baltimore - Period: War - Deserted at Camp Douglass, Ellicott's Mills, MD, on 12 Nov 1814.

Hays, Michael - Private - 38th Infantry - Company: Charles Stansbury - Age: 36 - Height: 5' 9" - Born: Dublin, Ireland - Trade: Spinster - Enlistment date: 22 Nov 1814 - Place: Fredericktown, MD - Period: War - Discharged at Baltimore on 6 Apr 1815.

Haythorn, Samuel - Private - 12th Infantry - Company: Robert Houston - Age: 39 - Height: 5' 9" - Born: Maryland - Trade: Shoemaker - Enlistment date: 1 Sep 1814 - Place: Clarksburg, VA - Period: War - Discharged at Winchester, VA, on 20 Mar 1815.

Hayward, George - Private - 14th Infantry - Company: Thomas Montgomery - Enlistment date: 19 May 1812 - Period: 5 Yrs - Pension: Heirs received half pay for five years in lieu of land bounty - Died in 1814.

Hayworth, John - Private - 5th Infantry - Company: Richard Bell - Other regiment: 3rd Infantry (New) - Age: 20 - Height: 5' 5" - Born: Middletown, VA - Trade: Shoemaker - Enlistment date: 4 May 1813 - Place: Baltimore - Period: 5 Yrs - Discharged on 4 May 1818.

Hazle, Jeremiah - Musician - 14th Infantry - Company: David Cummings - Age: 16 - Height: 4' 8" - Born: St. Mary's, MD - Trade: Laborer - Enlistment date: 9 Nov 1814 - Place: Camp Snowden - Period: 5 Yrs.

Hazledine, John - Corporal - 38th Infantry - Company: John Hook - Age: 21 - Height: 5' 7" - Born: Talbot Cty, MD - Trade: Cabinet maker - Enlistment date: 5 Dec 1814 - Place: Baltimore - Period: War.

Hazletine, Thomas - Private - US Volunteers - Company: Stephen Moore - Enlistment date: 8 Sep 1812 - Period: 1 Yr - Wounded at York, UC, on 27 Apr 1813.

Hazlett, Jacob - Private - 44th Infantry - Company: Joseph Miles - Age: 24 - Height: 5' 11" - Born: Frederick Cty, MD - Trade: Farmer - Enlistment date: 21 Nov 1814 - Place: New Orleans - Period: 5 Yrs - Discharged at Baton Rouge, LA, on 20 Nov 1819.

Hazlewood, Thomas - Private - 2nd Infantry - Company: Thomas Swaine - Age: 24 - Height: 5' 6" - Born: Berkley, VA - Trade: Farmer - Enlistment date: 10 Apr 1808 - Place: Fredericktown, MD - Period: 5 Yrs.

Head, Benjamin P. - Ensign - 38th Infantry - Company: John Rothrock - Pension: Wife Margaret Stembler, WO-762, WC-1316; married on 2 Dec 1810 in Baltimore, MD; soldier died on 22 Apr 1814 in Baltimore. MD - BLW 6740-160-50 - Commissioned as an ensign on 20 May 1813; dismissed on 19 Apr 1814; previously a private in Captain Keller's Company, MD Militia.

Headman, Simon M. - Private - 14th Infantry - Company: Joseph Marshall - Age: 36 - Height: 5' 5 1/2" - Born: Sweden - Trade: Manufacturer - Enlistment date: 16 May 1814 - Place: Baltimore - Period: War - BLW 3244-160-12 - Discharged at Greenbush, NY, on 8 May 1815.

Heafer, Frederick - Private - 14th Infantry - Company: David Cummings -

Age: 16 - Height: 4' 10" - Born: Saxony, Germany - Trade: Laborer - Enlistment date: 19 Jan 1815 - Place: Baltimore - Period: 5 Yrs - Deserted at Baltimore on 25 or 28 Mar 1815.

Heard, Benedict J. - Third Lieutenant - 14th Infantry - Commissioned as an ensign on 16 Sep 1814; promoted to 3rd lieutenant on 1 Oct 1814; discharged on 15 Jun 1815.

Hearn, William - Private - 8th Infantry - Company: Matthew Keith - Other regiment: 7th Infantry (New) - Age: 22 - Height: 5' 8 1/4" - Born: Leniar, MD - Trade: Farmer - Enlistment date: 15 May 1813 - Place: Eatontown, GA - Period: 5 Yrs - Discharged at Fort Scott, GA, on 16 May 1818.

Heater, William - Private - 38th Infantry - Company: John Mowton - Age: 27 - Height: 5' 8" - Born: Baltimore - Trade: Hatter - Enlistment date: 24 Jul 1814 - Place: Norfolk, VA - Period: War - BLW 20701-160-12 - Discharged at Craney Island, VA, on 15 Mar 1815.

Heaton, Joseph - Private - 27th Infantry (OH) - Company: Alexander Hill - Other regiment: 19th Infantry - Age: 44 - Height: 5' 10" - Born: Maryland - Trade: Blacksmith - Enlistment date: 24 Feb 1814 - Place: Zanesville, OH - Period: War - BLW 11015-160-12 - Discharged at Chillicothe, OH, on 6 Jun 1815.

Hebner, Frederick - Private - 36th Infantry - Age: 23 - Height: 5' 4" - Born: Prussia - Trade: Weaver - Enlistment date: 13 Dec 1814 - Period: War - BLW 165-160-12 - Discharged at Georgetown, DC, on 13 Mar 1815.

Hedden, Amos - Private - 14th Infantry - Prisoner of War (Halifax), discharged from Halifax on 9 Nov 1813.

Hedgebeth, William - Private - 38th Infantry - Company: John Mowton - BLW 12139-160-12.

Hedrick, George - Private - 2nd Artillery - Company: Nathan Towson - Other regiment: Corps of Artillery - Age: 18 - Height: 5' 5 1/4" - Born: Baltimore - Trade: Laborer - Enlistment date: 12 Apr 1812 - Place: Baltimore - Period: 5 Yrs - BLW 23139-160-12 - Prisoner of War (Halifax), captured at Stoney Creek, UC, on 12 Jun 1813; discharged from Fort Niagara, NY, on 30 Apr 1817.

Hedricks, George - Private - 2nd Artillery - Company: Nathan Towson - Age: 18 - Prisoner of War (Quebec), captured during the Battle of Stoney Creek and released on 31 Oct 1813.

Heffley, Jacob - Private - 2nd Artillery - Company: James Barker - Age: 26 - Height: 5' 5" - Born: Harford Cty, MD - Trade: Butcher - Enlistment date: 20 Mar 1814 - Place: Philadelphia - Period: 5 Yrs - Discharged at Fort Mifflin, PA, on 19 Aug 1815 by civil authority.

Heffner, Adam - Sergeant - 1st Artillery - Company: Francis Newman - Age: 32 - Height: 5' 4 3/4" - Born: Maryland - Trade: Tailor - Enlistment date: 15 Oct 1813 - Period: War - Discharged at New Orleans on 20 Apr 1815.

Heifer, John - Private - 14th Infantry - Company: David Cummings - Age: 31 - Height: 5' 10" - Born: Adams Cty, PA - Enlistment date: 20 Jun 1814 - Place: Towsontown, MD - Period: War - Deserted on 3 Jul 1814.

Heininger, Caleb - Private - 14th Infantry - Company: Henry Grindage - Age:

29 - Height: 5' 1" - Born: Germany - Trade: Baker - Enlistment date: 12 Sep 1812 - Period: 5 Yrs - Discharged at Sackets Harbor, NY, on 11 Oct 1813 on Surgeon's Certificate.

Heinnan, Dennis - Corporal - 38th Infantry - Company: George Keyser - Pension: Land bounty to William Heinnan, father and heir at law of Dennis Heinnan - BLW 22952-160-12.

Heinnan, Dennis - Corporal - 38th Infantry - Company: George Keyser - Age: 27 - Height: 5' 4 1/2" - Born: Ireland - Trade: Laborer - Enlistment date: 24 Sep 1814 - Place: Baltimore - Period: War - Pension: Land bounty to William Heinnan, father and heir at law of Dennis Heinnan - BLW 22952-160-12 - Drowned on 2 Dec 1814 in pursuit of a deserter.

Heiser, Jacob - Private - 38th Infantry - Company: George Keyser - Other regiment: Ordnance Department - Age: 30 - Height: 5' 11" - Born: Carlisle, PA - Trade: Blacksmith - Enlistment date: 21 Nov 1814 - Place: Baltimore - Period: 5 Yrs - BLW 23175-160-12 - Transferred to Ordnance Department on 21 Jun 1815; discharged from Bellona Arsenal on 21 Nov 1819.

Helms, Abraham - Private - 12th Infantry - Company: James Paxton - Age: 30 - Height: 5' 7 1/2" - Born: Maryland - Trade: Tailor - Enlistment date: 8 Jul 1812 - Place: Botetourt Cty, VA - Period: 5 Yrs - Pension: Land bounty to John D. Helms, father and heir at law of Abraham Helms - BLW 16498-160-12 - Severely wounded during the Battle of La Cole, UC, on 30 Mar 1814; died from wounds at Burlington, VT, on 14 or 18 May 1814.

Helms, John - Private - 14th Infantry - Company: James Britton.

Helvenstine, George W. - Sergeant - 1st Rifles - Company: George Seviers - Age: 30 - Height: 5' 10" - Born: Washington, MD - Trade: Farmer - Enlistment date: 1 Jan 1808 - Period: 5 Yrs - Died at Sacket's Harbor, NY, on 14 Dec 1813.

Hemming, Samuel - Private - 19th Infantry - Company: Carey Trimble - Age: 19 - Height: 5' 7 1/2" - Born: Maryland - Trade: Farmer - Enlistment date: 9 May 1814 - Place: Salem, Columbiana County, OH - Period: War - BLW 18101-160-12 - Discharged at Chillicothe, OH, on 6 Jun 1815.

Hemmingway, John - Private - 2nd Artillery - Company: Frederick Evans - Other regiment: Corps of Artillery - Age: 39 - Height: 5' 6" - Born: Boston - Trade: Tailor - Enlistment date: 18 Jul 1814 - Place: Fort McHenry - Period: 5 Yrs - BLW 440-160-12 - Discharged at Elkton, MD, on 10 Aug 1815 for inability.

Henderson, David - Private - 36th Infantry - Company: Joseph Hook - Age: 37 - Height: 5' 9 3/4" - Born: Scotland - Enlistment date: 20 May 1813 - Place: Annapolis - Period: 1 Yr - Died on 25 or 26 1813.

Henderson, George - Private - 36th Infantry - Age: 24 - Height: 5' 9" - Born: Richmond, VA - Enlistment date: 4 Jun 1813 - Place: Baltimore - Period: 1 Yr - Deserted on 9 Jun 1813.

Henderson, James - Private - 38th Infantry - Company: John Buck - Age: 31 - Height: 5' 10" - Born: Newtown, MD - Enlistment date: 1 Jan 1814 - Place: Princess Anne, MD - Period: 1 Yr - Discharged on 31 Dec 1814.

Henderson, John - Corporal - 38th Infantry - Company: Charles Stansbury -

By War of 1812 Soc. in MD

Age: 21 - Height: 5' 6 1/2" - Born: Dorchester Cty, MD - Trade: Cooper - Enlistment date: 6 May 1814 - Period: War - BLW 10058-160-12 - Discharged at Baltimore on 6 Apr 1815.

Henderson, Levin - Private - 14th Infantry - Company: Joseph Marshall - Age: 24 - Height: 5' 6" - Born: Worcester Cty, MD - Trade: Farmer - Enlistment date: 27 Jul 1814 - Place: Snow Hill, MD - Period: War - BLW 4400-160-12 - Discharged at Greenbush, NY, on 3 May 1815.

Henderson, Samuel - Private - 14th Infantry - Company: David Cummings - Age: 31 - Height: 5' 10" - Born: Chester, PA - Trade: Laborer - Enlistment date: 5 Dec 1814 - Place: Baltimore - Period: War - Deserted on 9 Feb 1815.

Henderson, Stephen - Private - 14th Infantry.

Hendrick, William - Private - 26th Infantry - Company: William Bezeau - Age: 21 - Height: 5' 11" - Born: Baltimore - Trade: Laborer - Enlistment date: 16 Nov 1814 - Place: Baltimore - Period: War - Discharged at Philadelphia on 28 Mar 1815.

Hendricks, William - Private - 38th Infantry - Company: Charles Stansbury - Age: 36 - Height: 5' 7" - Born: County Waterford, Ireland - Trade: Laborer - Enlistment date: 25 Nov 1814 - Place: Cumberland - Period: War - BLW 6807-160-12 - Discharged at Baltimore on 6 Apr 1815.

Hendrickson, Daniel - Private - 5th Infantry - Company: William Bird - Age: 21 - Height: 5' 9" - Born: Holland - Trade: Hatter - Enlistment date: 12 May 1813 - Place: Baltimore - Period: 5 Yrs.

Hennenger, Caleb - Private - 14th Infantry - Pension: Placed on the pension roll on 4 Apr 1821.

Henry, George - Private - 14th Infantry - Prisoner of War (Halifax), discharged from Halifax on 9 Nov 1813.

Henry, James - Private - 14th Infantry - Company: David Cummings - Enlistment date: 1 Jun 1812 - Period: 5 Yrs - Prisoner of War (Halifax), sent to England; deserted at Fort McHenry on 20 Apr 1815.

Henry, John - Seaman - US Sea Fencibles - Company: John Gill - Age: 19 - Height: 5' 6" - Born: New York City - Enlistment date: 15 Jan 1814 - Place: Baltimore - Period: 1 Yr - BLW 29292-160-55 - Discharged on 14 Jan 1815.

Henry, John - Private - 5th Infantry - Company: James Dorman - Other regiment: 3rd Infantry (New) - Age: 21 - Height: 5' 6 1/2" - Born: Baltimore - Trade: Tallow chandler - Enlistment date: 14 Jun 1813 - Place: Baltimore - Period: 5 Yrs - Discharged on 13 Jul 1818.

Henry, Samuel - Private - 5th Infantry - Company: James Dorman - Other regiment: 3rd Infantry (New) - Age: 21 - Height: 5' 9" - Born: Adams Cty, PA - Trade: Farmer - Enlistment date: 27 Feb 1812 - Place: Baltimore - Period: 5 Yrs - Prisoner of War (Halifax), captured at Stoney Creek, UC, on 10 Jun 1813; exchanged on 15 Apr 1815; deserted at Detroit on 15 Feb 1816.

Henshaw, Lewis - Private - 5th Infantry - Company: William Bird - Age: 25 - Height: 5' 11 1/4" - Born: Rhode Island - Trade: Shoemaker - Enlistment date: 22 Feb 1813 - Place: Baltimore - Period: 5 Yrs.

Herd, Samuel - Servant - US Sea Fencibles - Company: Simmones Bunbury.

Herrick, Thomas - Private - 2nd Artillery - Company: Nathan Towson -

Maryland Regulars in War of 1812

Enlistment date: 13 Apr 1812 - Period: 5 Yrs - Died at Black Rock, NY, on 11 Jan 1815.

Herring, Gardner - Private - 14th Infantry - Company: Clement Sullivan - Age: 28 - Height: 5' 6 1/4" - Born: Kent Cty, MD - Trade: Laborer - Enlistment date: 3 Jun 1812 - Place: Cambridge, MD - Period: 5 Yrs - Pension: Wife Maria, Old War WF-14999 Rejected, Old War IF-15027 - BLW 9275-160-12 - Discharged at Greenbush, NY, on 2 Dec 1813 on Surgeon's Certificate.

Herring, John - Private - 14th Infantry - Company: Clement Sullivan - Pension: Land bounty to Gardner Herring and other heirs at law of John Herring - BLW 26805-160-12 - Prisoner of War (Halifax), captured at Beaver Dams on 24 Jun 1813; died at Halifax on 11 Jan 1814.

Herrings, Joshua - Private - 22nd Infantry - Company: David Millikin - Age: 25 - Height: 5' 11" - Born: Baltimore - Trade: Laborer - Enlistment date: 27 Jul 1813 - Place: Waynesburg, PA - Period: 5 Yrs - Discharged at Sackets Harbor, NY, on 11 Aug 1815 for partial blindness.

Herron, James - Private - 14th Infantry - Company: Henry Grindage - Age: 32 - Height: 5' 7" - Trade: Blacksmith - Enlistment date: 18 Sep 1812 - Period: 5 Yrs - Pension: Old War IF-25151 - BLW 6691-160-12 - Prisoner of War (Halifax), captured at Beaver Dams on 24 Jun 1813; discharged at Boston, MA, on 7 Aug 1814, because of wounds.

Herrwood, William - Private - 2nd Light Dragoons - Company: George Haig - Age: 21 - Height: 5' 5" - Born: Annapolis - Trade: Barber - Enlistment date: 5 Jun 1812 - Place: Richmond, VA - Period: 5 Yrs - Discharged on 5 Jun 1817.

Herry, Tobias - Private - 22nd Infantry - Company: Daniel McFarland - Age: 38 - Height: 5' 7" - Born: Maryland - Trade: Farmer - Enlistment date: 21 May 1812 - Place: Washington, PA - Period: 5 Yrs - Wounded at Chippewa, UC, on 25 Jul 1814; discharged at Buffalo, NY, on 1 Aug 1815.

Hersey, John H. - Private - 38th Infantry - Company: Isaac Aldridge - BLW 3162-160-12.

Hersey, William - Private - 14th Infantry - Company: Thomas Montgomery - Age: 30 - Height: 5' 7 1/2" - Born: Maryland - Enlistment date: 15 Jan 1813 - Period: 18 Mos - Discharged at Plattsburgh, NY, on 15 Jul 1814.

Hersey, William - Private - 29th Infantry - Company: Matthew Danvers - Age: 31 - Height: 5' 8" - Born: Salisbury, MD - Trade: Farmer - Enlistment date: 23 Dec 1814 - Place: Albany, NY - Period: War - Discharged at Plattsburgh, NY, on 29 Jun 1815.

Hess, George - Private - 36th Infantry - Age: 23 - Height: 5' 11 1/2" - Born: Hagerstown, MD - Enlistment date: 2 Mar 1814 - Place: Georgetown, DC - Period: War - Deserted on 14 Jan 1815.

Hess, Jacob - Private - 38th Infantry - Company: John Mowton - BLW 20012-160-12.

Hess, Jeremiah - Private - 2nd Light Dragoons - Company: John Burd - Other regiment: Corps of Artillery - Age: 30 - Height: 6' - Born: Bedford, Washington Cty, MD - Trade: Farmer - Enlistment date: 12 Jun 1812 - Place: Bedford Cty, PA - Period: 5 Yrs - BLW 9916-160-12 - Discharged at Fort

By War of 1812 Soc. in MD

McHenry on 30 Jun 1817.

Hess, John - Private - 38th Infantry - Company: John Rothrock - Age: 30 - Height: 5' 4" - Born: Maryland - Trade: Farmer - Enlistment date: 24 Feb 1814 - Place: Craney Island, VA - Period: War - Pension: Wife Catharine, WO-40727, WC-32123 - BLW 14428-160-12 - Discharged at Craney Island, VA, on 15 Mar 1815.

Hetzer, George - Sergeant - 14th Infantry - Company: Samuel Lane - Age: 21 - Height: 5' 6" - Born: Washington, MD - Enlistment date: 2 May 1812 - Place: Maryland - Period: 5 Yrs - Pension: Wife Matilda, WO-28071, WC-28376 - Slightly wounded at Black Rock, NY, on 28 or 29 Nov 1812; Prisoner of War (Halifax), captured at Beaver Dams on 24 Jun 1813.

Hewey, John - Private - 14th Infantry - Prisoner of War (Halifax), discharged from Halifax on 3 Feb 1814.

Heyden, Daniel - Private - 14th Infantry - Prisoner of War (Halifax), discharged from Halifax on 3 Feb 1814.

Hickey, Edward - Private - 38th Infantry - Company: James Smith - Age: 24 - Height: 5' 9" - Born: Anne Arundel Cty, MD - Trade: Blacksmith - Enlistment date: 5 Apr 1814 - Place: Richmond, VA - Period: War - Discharged at Craney Island, VA, on 15 Mar 1815.

Hickey, Michael - Private - 38th Infantry - Company: Charles Stansbury - Age: 27 - Height: 5' 6" - Born: Ireland - Trade: Laborer - Enlistment date: 26 Nov 1814 - Place: Fredericktown - Period: War - BLW 5489-160-12 - Discharged at Baltimore on 6 Apr 1815.

Hickman, Isaac - Private - 2nd Artillery - Company: Alexander Williams - Age: 20 - Height: 5' 6" - Born: Frederick, MD - Trade: Sailor - Enlistment date: 29 May 1813 - Place: Lewisbury - Period: War - Discharged at Buffalo on 3 Jun 1815.

Hicks, Isaac - Private - 14th Infantry - Company: William McIlvane - Other regiment: 4th Infantry (New) - Age: 32 - Height: 5' 10 1/2" - Born: Frederick Cty, VA - Trade: Clock maker - Enlistment date: 13 Jun 1812 - Period: 5 Yrs - BLW 14644-160-12 - Prisoner of War, exchanged on 15 Apr 1814; discharged on 18 Jun 1819.

Hicks, Michael - Musician - 36th Infantry - Age: 17 - Height: 4' 9" - Born: Columbia, VA - Trade: Shoemaker - Enlistment date: 6 Feb 1815 - Period: War - Discharged on 13 Mar 1815.

Hickson, Thomas - Private - 14th Infantry - Company: David Cummings - Age: 21 - Height: 5' 8" - Born: Frederick Cty, MD - Trade: Merchant - Enlistment date: 28 Nov 1814 - Place: Baltimore - Period: War - BLW 9869-160-12 - Discharged at Baltimore on 16 Mar 1815.

Higby, Noah - Gunner - US Sea Fencibles - Company: Simmones Bunbury - Age: 42 - Height: 5' 7" - Born: Middletown, MD - Enlistment date: 8 Dec 1813 - Period: 1 Yr - Discharged on 13 Oct 1814.

Higdon, Bernard - Private - 36th Infantry - Age: 21 - Height: 5' 1/2" - Born: Prince George's Cty, MD - Enlistment date: 14 May 1814 - Place: Georgetown, DC.

Higdon, John - Private - 14th Infantry - Company: Thomas Montgomery -

Maryland Regulars in War of 1812

Other regiment: 4th Infantry (New) - Age: 29 - Height: 5' 5" - Born: Charles Cty, MD - Trade: Hatter - Enlistment date: 28 Apr 1813 - Place: Baltimore - Period: 5 Yrs - Discharged on 27 Apr 1818.

Higgins, Daniel - Private - 2nd Artillery - Company: James Barker - Age: 28 - Height: 5' 8" - Born: Maryland - Trade: Tailor - Enlistment date: 7 May 1814 - Place: Philadelphia - Period: War - Deserted at Fort Mifflin, PA, on 5 Jun 1814.

Higgs, James - Private - 36th Infantry - Company: Joseph Hook - Age: 38 - Height: 5' 6" - Born: Charles Cty, MD - Trade: Laborer - Enlistment date: 21 Jun 1814 - Place: Georgetown, DC - Period: War - BLW 955-160-12 - Discharged at Fort Covington, MD, on 30 Mar 1815.

Higgs, James - Private - 36th Infantry - Company: Thomas Carbery - Age: 24 - Height: 5' 6" - Born: Washington, DC - Trade: Sailor - Enlistment date: 25 Feb 1814 - Place: Georgetown, DC - Period: War - BLW 956-160-12 - Discharged at Baltimore on 31 Mar 1815.

Higsby, Andrew W. - Private - 2nd Rifles - Age: 20 - Height: 5' 3 1/2" - Born: St. Mary's, MD - Trade: Laborer - Enlistment date: 18 May 1814 - Place: Georgetown, KY - Period: War - Discharged on 30 Jun 1815.

Hill, Allen - Private - 14th Infantry - Company: Isaac Barnard - BLW 2815-160-12.

Hill, George - Private - 3rd Rifles - Company: Edward Carrington - Age: 39 - Height: 5' 9" - Born: Harford Cty, MD - Trade: Blacksmith - Enlistment date: 28 Sep 1814 - Period: 5 Yrs.

Hill, Henry Oswell - First Lieutenant - 5th Infantry - Commissioned as a 2nd lieutenant on 3 Jan 1812; promoted to 1st lieutenant on 21 Jan 1814; discharged on 15 Jun 1815.

Hill, John - Private - 36th Infantry - Company: Thomas Carbery - Age: 38 - Height: 5' 11 1/2" - Born: Montgomery Cty, MD - Trade: Joiner - Enlistment date: 12 Mar 1814 - Place: Georgetown, DC - Period: War - BLW 2463-160-12 - Discharged at Baltimore on 31 Mar 1815.

Hill, John - Private - 2nd Infantry - Company: Reuben Chamberlain - Other regiment: 1st Infantry (New) - Age: 34 - Height: 5' 10" - Born: Cecil Cty, MD - Trade: Farmer - Enlistment date: 19 Jul 1808 - Place: Columbia Springs - Period: 5 Yrs - Re-enlisted at Fort Bowyer on 24 Jul 1813 for five years; died at Mobile, AL.

Hill, Joshua - Private - 14th Infantry - Company: Thomas Montgomery - Age: 20 - Born: Worcester Cty, MD - Trade: Farmer - Enlistment date: 21 May 1812 - Place: Salisbury, MD - Period: 5 Yrs - Discharged on 21 Jul 1815, invalid.

Hill, Levi - Private - 5th Infantry - Company: John Corboley - Other regiment: 3rd Infantry (New) - Age: 21 - Height: 5' 4" - Born: Cecil Cty, MD - Trade: Wagon maker - Enlistment date: 23 Sep 1814 - Place: Lancaster, PA - Period: 5 Yrs - Discharged at Fort Howard, WI, on 28 Sep 1819.

Hill, Stephen - Corporal - 14th Infantry - Enlistment date: 12 Nov 1812 - Period: 18 Mos - Prisoner of War (Halifax), sent to England; discharged on 31 Mar 1815.

By War of 1812 Soc. in MD

Hill, Thomas - Clerk - 38th Infantry - Company: John Brookes - BLW 17467-160-12.

Hill, Thomas - Private - 2nd Artillery - Company: Alexander Williams - Age: 29 - Height: 5' 7 1/2" - Born: Baltimore - Trade: Laborer - Enlistment date: 2 May 1813 - Period: 5 Yrs - Pension: Heirs received half pay for five years in lieu of land bounty - Wounded at Chippewa, UC, on 5 Jul 1814 in left leg; discharged at Greenbush, NY, on 24 Apr 1815 because of wounds; died in 1822.

Hill, William - Private - 16th Infantry - Company: George Steele - Age: 23 - Height: 6' - Born: Dorchester Cty, MD - Trade: Shoemaker - Enlistment date: 3 Mar 1814 - Place: Philadelphia - Period: War - Discharged at Philadelphia on 19 May 1815.

Hill, William - Private - 38th Infantry - Company: Shepperd Leakin - Age: 22 - Height: 5' 8" - Born: Maryland - Trade: Shoemaker - Enlistment date: 22 Feb 1814 - Period: War - BLW 3227-160-12 - Discharged at Fort McHenry on 28 Mar 1815.

Hilliard, Benjamin R. - Private - 3rd Artillery - Company: Horace Watson - Enlistment date: 12 Dec 1813 - Pension: Wife Harriet McNeil, SO-9989, SC-15899, WO-12007, WC-8445; married on 25 Feb 1816, Baltimore, MD; soldier died on 17 May 1877, Baltimore, MD - BLW 11916-160-12 - Discharged on 25 May 1816.

Hilton, James - Private - US Volunteers - Company: Stephen Moore - Enlistment date: 8 Sep 1812 - Period: 1 Yr.

Hilton, John - Private - 17th Infantry - Age: 23 - Height: 5' 10" - Born: St. Mary's, MD - Enlistment date: 17 Mar 1814 - Period: 5 Yrs.

Hindman, Jacob - Captain - 2nd Artillery - Other regiment: Corps of Artillery - Born: Maryland - Pension: Land bounty to Susan Harwood, widow of Jacob Hindman - BLW 2886-156-50 - Commissioned as a 2nd Lieutenant, 5th Infantry, on 3 May 1808; promoted to 1st lieutenant on 1 May 1810; promoted to captain, 2nd Artillery, on 2 Jul 1812; promoted to major on 26 Jun 1813l transferred to Corps of Artillery on 12 May 1814; transferred to 2nd Artillery on 1 Jun 1821; died on 17 Feb 1827.

Hindman, Jr., William - First Lieutenant - 36th Infantry - Company: Samuel Raisin - Other regiment: 3rd Rifles - Born: Maryland - BLW 14385-160-50 - Commissioned as a 3rd lieutenant, 36th Infantry, on 30 Apr 1813; promoted to 2nd lieutenant on 15 Aug 1815; promoted to 1st lieutenant, 3rd Rifles, on 17 Mar 1814; discharged on 15 Jun 1815.

Hines, Herbert S. - Private - 36th Infantry - Company: Joseph Hook - Age: 23 - Height: 5' 9 1/4" - Born: Huntington, NY - Enlistment date: 10 May 1813 - Place: Westminster, MD - Period: 1 Yr.

Hines, John - Private - 36th Infantry - Company: Joseph Hook - Age: 55 - Height: 5' 8" - Born: Frederick, MD - Enlistment date: 20 May 1813 - Place: Baltimore - Period: 1 Yr - Pension: No pension claim - BLW 89-160-50.

Hines, William - Private - 38th Infantry - Company: James Smith - Age: 21 - Height: 5' 7" - Born: Harford Cty, MD - Trade: Seaman - Enlistment date: 20 Feb 1814 - Place: Craney Island, VA - Period: War - BLW 6565-160-12

Maryland Regulars in War of 1812

- Discharged at Craney Island, VA, on 15 Mar 1815.

Hinkle, John - Corporal - 36th Infantry - Company: Thomas Carbery - Age: 36 - Height: 5' 6 1/4" - Born: Pennsylvania - Enlistment date: 27 Apr 1814 - Place: Georgetown, DC - Period: War - BLW 9912-160-12 - Discharged on 31 Mar 1815.

Hinkle, John - Sergeant - 38th Infantry - Company: John Buck - Age: 32 - Height: 5' 6" - Born: Pennsylvania - Trade: Carpenter - Enlistment date: 2 Apr 1814 - Place: Annapolis - Period: War - BLW 11008-160-12 - Discharged at Baltimore on 30 Mar 1815.

Hinton, Benjamin - Private - 28th Infantry - Company: George Stockton - Age: 33 - Height: 6' 2" - Born: Maryland - Trade: Farmer - Enlistment date: 10 Jun 1814 - Place: Flemingsburg, KY - Period: War - BLW 9460-160-12 - Discharged at Detroit on 30 Jun 1815.

Hinton, Reason - Private - 38th Infantry - Company: Shepperd Leakin - Age: 32 - Height: 5' 10" - Born: Upper Marlboro, MD - Trade: Farmer - Enlistment date: 25 Jan 1814 - Place: Baltimore - Period: War - Pension: Heirs received half pay for five years in lieu of land bounty - Died at Baltimore on 22 Aug 1814.

Hintz, John - Private - 38th Infantry - Company: John Rothrock - Enlistment date: 19 Jun 1813 - Period: 1 Yr - Discharged on 19 Jun 1814.

Hitch, William - Private - 14th Infantry - Company: Joseph Marshall - Age: 21 - Height: 5' - Born: Worcester Cty, MD - Enlistment date: 15 Jul 1814 - Place: Salisbury, MD - Period: War - Deserted at New Castle, DE, on 9 Oct 1814.

Hixon, Benjamin - Private - 38th Infantry - Company: John Buck - Age: 39 - Height: 5' 9" - Born: Loudoun Cty, VA - Trade: Post railer - Enlistment date: 24 May 1814 - Place: Baltimore - Period: War.

Hobbs, Brice - Private - 1st Infantry - Company: Simon Owens - Age: 35 - Height: 5' 10 1/2" - Born: Frederick, MD - Trade: Shoemaker - Enlistment date: 5 Apr 1810 - Place: Chambersburg, PA - Period: 5 Yrs - Died at Belle Fontaine, MO, on 1 Nov 1812.

Hobbs, James - Private - 36th Infantry - Company: Joseph Hook - Age: 25 - Height: 5' 8 1/2" - Born: Somerset Cty, MD - Trade: Carpenter - Enlistment date: 22 Sep 1813 - Re-enlisted at Westmoreland on 19 Mar 1814 for the war; discharged on 30 Mar 1815.

Hobbs, Shelby - Private - 14th Infantry - Company: Joseph Marshall - Other regiment: 4th Infantry (New) - Age: 21 - Height: 5' 6" - Born: Worcester Cty, MD - Trade: Shoemaker - Enlistment date: 1 May 1812 - Place: Salisbury, MD - Period: 5 Yrs - Discharged on 1 May 1817.

Hobbs, William - Private - 1st Artillery - Company: William Wilson - Age: 34 - Height: 6' - Born: Sandersville, MD - Trade: Carpenter - Enlistment date: 14 Nov 1811 - Place: Fort Johnson, SC - Period: 5 Yrs - Discharged on 14 Nov 1816.

Hobbs, William C. - Captain - 36th Infantry - Born: Maryland - Pension: Land bounty to Christina Hobbs, widow of William C. Hobbs - BLW 11342-160-12 - Commissioned as a 1st lieutenant on 30 Apr 1813; promoted to captain on 30 Sep 1814; died on 8 Feb 1815.

Hockenbrough, Henry - Private - 14th Infantry - Company: Richard Arell - Other regiment: 4th Infantry (New) - Age: 38 - Height: 5' 5 3/4" - Born: Germany - Trade: Tanner - Enlistment date: 15 Aug 1812 - Place: Baltimore - Period: 5 Yrs - Prisoner of War, exchanged on 15 Apr 1814; discharged at Annapolis on 18 Oct 1815, unfit for service.

Hodges, John F. - Corporal - 36th Infantry - Company: Samuel Raisin - Age: 34 - Height: 5' 10" - Born: Norfolk, VA - Trade: Carpenter - Enlistment date: 15 Jul 1814 - Place: Norfolk, VA - Period: War - BLW 14576-160-12 - Discharged at Washington, DC, on 20 Mar 1815.

Hoff, Benjamin - Private - 19th Infantry - Company: Carey Trimble - Age: 25 - Height: 5' 5" - Born: Maryland - Trade: Printer - Enlistment date: 5 Apr 1814 - Place: Chillicothe, OH - Period: War.

Hoff, Michael - Corporal - 14th Infantry - Enlistment date: 24 Nov 1812 - Period: 18 Mos - Prisoner of War, sent to England; discharged on 31 Mar 1815.

Hoffman, Frederick W. - Second Lieutenant - 38th Infantry - Company: James Hook - Pension: Wife Mary Desire Lieutaud, WO-14949, WC-9096; married on 28 Nov 1816, Baltimore, MD; soldier died on 16 May 1840 in Frederick, MD - BLW 12401-160-55 - Commissioned as a 3rd lieutenant on 20 May 1813; promoted to 2nd lieutenant on 1 May 1814; discharged on 15 Jun 1815.

Hoffman, Henry - Private - 14th Infantry - Company: James McDonald - Age: 41 - Height: 5' 6" - Born: Virginia - Enlistment date: 7 Feb 1814 - Period: War - Pension: Land bounty to Jacob Hoffman, son and other heirs at law of Henry Hoffman - BLW 25344-160-12 - Killed during the Battle of Lyon's Creek on 19 Oct 1814.

Hoffman, John - Private - 38th Infantry - Company: John Rothrock - Age: 21 - Height: 5' 4" - Born: Baltimore - Trade: Cooper - Enlistment date: 4 Apr 1814 - Place: Craney Island, VA - Period: War - BLW 26399-160-12 - Discharged at Craney Island, VA, on 15 Mar 1815.

Hoffman, John - Musician - 38th Infantry - Company: Shepperd Leakin - Age: 27 - Height: 5' 2" - Born: Prussia - Trade: Musician - Enlistment date: 3 Aug 1814 - Place: Annapolis - Period: War - BLW 13636-160-12 - Discharged at Fort McHenry on 28 Mar 1815.

Hoffman, John - Private - 36th Infantry - Age: 25 - Height: 5' 8" - Born: Pennsylvania - Enlistment date: 19 May 1813 - Place: Frederick, MD - Period: 1 Yr.

Hoffman, Philip - Private - 14th Infantry - Prisoner of War (Halifax), discharged from Halifax on 3 Feb 1814.

Hogan, William R. (Hogue) - Private - 1st Rifles - Company: Thomas Ramsey - Age: 19 - Height: 5' 7 1/2" - Born: Harford Cty, MD - Trade: Farmer - Enlistment date: 15 Dec 1813 - Place: Cincinnati - Period: War - Pension: SC-34971 - Deserted at Cincinnati on 27 May 1814; also served as a fifer in Captain Thomas Ramsey's Company, OH Militia.

Hogart, William - Private - Light Artillery - Company: James Gibson - Age:

33 - Height: 5' 10" - Born: Baltimore - Trade: Brick layer - Enlistment date: 26 Jan 1812 - Period: 5 Yrs - BLW 10006-160-12 - Prisoner of War, captured at Queenston Heights, UC, on 3 Oct 1812; discharged at Fort Independence, MA, on 26 Jan 1817.

Hogg, George - Private - 14th Infantry - Company: Richard Arell - Other regiment: 4th Infantry (New) - Age: 36 - Height: 5' 8" - Born: Lancaster, PA - Trade: Farmer - Enlistment date: 6 Sep 1812 - Place: Oxford, PA - Period: 5 Yrs - Prisoner of War (Halifax), captured at Beaver Dams on 24 Jun 1813, exchanged on 15 Apr 1814; deserted on 12 Jul 1815.

Holden, James - Private - 37th Infantry - Company: Elizur Warner - Age: 24 - Height: 5' 2 1/4" - Born: Baltimore City - Trade: Seaman - Enlistment date: 31 Oct 1814 - Place: Fort Griswold, CT - Period: War - BLW 25253-160-12 - Discharged at New London, CT, on 10 May 1815.

Holder, Daniel - Private - 38th Infantry - Company: John Mowton - BLW 6489-160-12.

Holder, Robert - Private - 36th Infantry - Company: Samuel Raisin - Age: 26 - Height: 5' 4" - Born: England - Trade: Shoemaker - Enlistment date: 19 Jan 1814 - Period: War - BLW 2609-160-12 - Discharged at Washington, DC, on 20 Mar 1815.

Holland, Daniel - Private - 2nd Artillery - Company: Nathan Towson - Pension: Land bounty to Mary Holland, sister and other heirs at law of Daniel Holland - BLW 14233-160-12.

Holland, James - Private - 2nd Infantry - Company: Reuben Chamberlain - Other regiment: 1st Infantry (New) - Age: 36 - Height: 5' 9" - Born: England - Trade: Laborer - Enlistment date: 1 Oct 1812 - Place: Washington Cantonment, MD - Period: 5 Yrs - Deserted at New Orleans on 15 Jun 1816.

Holland, John - Private - 38th Infantry - Company: John Rothrock - Enlistment date: 20 Jun 1813 - Period: 1 Yr - Discharged on 20 Jun 1814.

Holland, John - Private - 14th Infantry - Age: 42 - Prisoner of War (Quebec), died at Quebec on 17 Jul 1813.

Holland, Joseph - Fifer - 1st Artillery - Company: Michael Walsh - Other regiment: Corps of Artillery - Age: 25 - Height: 5' 11" - Born: Maryland - Trade: Musician - Enlistment date: 4 Jan 1813 - Period: 5 Yrs - BLW 19378-160-12 - Discharged at New Orleans on 14 Jun 1817 and re-enlisted.

Hollings, John - Seaman - US Sea Fencibles - Company: William Addison.

Hollings, Thomas - Private - 14th Infantry - Company: Isaac Barnard - Enlistment date: 15 Jan 1813 - Period: 5 Yrs - Pension: Land bounty to Jesse Hollings, father and only heir at law of Thomas Hollings - BLW 22920-160-12 - Died on 7 Aug 1813.

Hollingsworth, Jacob - Private - 38th Infantry - Enlistment date: 26 Oct 1813 - Period: 1 Yr - Discharged or died at Baltimore on 31 Oct 1813.

Hollis, Henry - Private - 36th Infantry - Company: Thomas Carbery - Age: 28 - Height: 5' 5" - Born: New England - Trade: Trader - Enlistment date: 19 Feb 1814 - Place: Georgetown, DC - Period: 5 Yrs.

Hollis, William - Private - 36th Infantry - Company: Samuel Raisin - Age: 22 - Height: 5' 4 1/2" - Born: England - Trade: Mariner - Enlistment date: 4 Oct

By War of 1812 Soc. in MD

1814 - Place: Georgetown, DC - Period: War - BLW 10301-160-12.

Hollowell, James - Drum Major - 2nd Infantry - Company: Edmund Gaines - Other regiment: 1st Infantry (New) - Age: 14 - Height: 5' 3" - Born: Annapolis - Trade: Soldier - Enlistment date: 1 Oct 1807 - Place: Fort Adams, MS - Period: 5 Yrs - Re-enlisted at Fort Stoddert on 2 Oct 1812 for five years; discharged at Pass Christian, MS, on 31 Aug 1815.

Holmes, Ezekiel - Private - 2nd Light Dragoons - Company: William Littlejohn - Other regiment: Artillery - Age: 22 - Height: 5' 7" - Born: Baltimore - Trade: Plasterer - Enlistment date: 22 Jan 1814 - Place: Baltimore - Period: 5 Yrs - Deserted on 28 Jul 1185.

Holmes, James - First Lieutenant - 38th Infantry - Commissioned as a 2nd lieutenant on 20 May 1813; promoted to 1st lieutenant on 31 Dec 1813; discharged on 15 Jun 1815.

Holmes, John - Private - 2nd Artillery - Company: Frederick Evans - Age: 19 - Height: 5' 7" - Born: Maryland - Trade: Shoemaker - Enlistment date: 20 Aug 1812 - Place: Annapolis - Period: 5 Yrs - Deserted at Fort McHenry on 28 Jul 1815.

Holmes, Thomas - Private - 5th Infantry - Company: William Henshaw - Age: 30 - Height: 5' 9" - Born: Kent Cty, MD - Trade: Laborer - Enlistment date: 5 Jun 1813 - Place: Baltimore - Period: 5 Yrs.

Holmes, Walter - Private - 38th Infantry - Company: Shepperd Leakin - Age: 29 - Height: 5' 9 1/2" - Born: Philadelphia - Trade: Blacksmith - Enlistment date: 30 Aug 1814 - Place: Baltimore - Period: 5 Yrs - Died at Camp Snowden on 19 Oct or 19 Nov 1814.

Holmes, William - Sergeant - 38th Infantry - Company: John Rothrock - Age: 21 - Height: 5' 6" - Born: Boston - Trade: Seaman - Enlistment date: 23 Feb 1814 - Place: Craney Island, VA - Period: War - BLW 26237-160-12 - Discharged at Craney Island, VA, on 15 Mar 1815.

Holmes, William - Private - 38th Infantry - Company: John Brookes - Enlistment date: 9 Oct 1812 - Period: 1 Yr.

Holston, James - Private - 14th Infantry - Company: Kenneth McKenzie - Enlistment date: 28 Dec 1812 - Period: 18 Mos - Prisoner of War, exchanged on 15 Apr 1814; discharged at Greenbush, NY, on 23 Jun 1814.

Holstone, James - Private - 14th Infantry - Company: Joseph Marshall - Age: 19 - Height: 5' 5" - Born: Kent Cty, MD - Trade: Baker - Enlistment date: 9 Aug 1814 - Place: Baltimore - Period: War - Discharged at Greenbush, NY, on 3 May 1815.

Holt, John - Private - 38th Infantry - Company: John Buck - Age: 27 - Height: 5' 9 1/2" - Born: Washington Cty, MD - Trade: Gunsmith - Enlistment date: 11 Mar 1814 - Place: Charles Town, VA - Period: War - BLW 1678-160-12 - Discharged at Baltimore on 30 Mar 1815.

Holton, John - Private - 36th Infantry - Company: Thomas Carbery - BLW 23706-160-12.

Hood, Benjamin - Private - 36th Infantry - Company: Joseph Hook - Age: 19 - Height: 5' 6" - Born: Baltimore - Trade: Laborer - Enlistment date: 14 Jul 1814 - Place: Baltimore - Period: War - Pension: SO-17212, SC-17718 -

Maryland Regulars in War of 1812

BLW 14124-160-12 - Discharged at Fort Covington, MD, on 30 Mar 1815.

Hood, Frederick - Private - 14th Infantry - Company: George Steele - Other regiment: 4th Infantry (New) - Age: 30 - Height: 5' 8 1/2" - Born: Philadelphia - Trade: Hatter - Enlistment date: 7 Sep 1812 - Place: Baltimore - Period: 5 Yrs - Prisoner of War (Halifax), captured at Stoney Creek, UC, on 12 Jun 1813; deserted on 16 Jul 1815.

Hook, Greenbury - Sergeant - 22nd Infantry - Company: Willis Foulk - Age: 25 - Height: 5' 7" - Born: Maryland - Trade: Chair maker - Enlistment date: 26 May 1812 - Place: Washington, DC - Period: 5 Yrs - Pension: Land bounty to James Hook, brother & other heirs at law of Greenbury Hook - BLW 22096-160-12 - Died at Williamsville, NY, on 14 Oct 1814 from fever.

Hook, James Harvey - Captain - 5th Infantry - Other regiment: 38th Infantry - Born: Maryland - Pension: Land bounty to Mary B. Hook, widow of James H. Hook - BLW 15791-160-50 - Commissioned as an ensign, 5th Infantry, on 30 Apr 1812; promoted to 2nd lieutenant on 1 Sep 1812; promoted to captain, 38th Infantry, on 20 May 1813; discharged on 15 Jun 1815; died on 30 Nov 1841.

Hook, Joseph - Captain - 36th Infantry - Other regiment: ex 2nd Artillery - Born: Maryland - Pension: Land bounty to Mary Ann Hook, widow of Joseph Hook - BLW 1722-160-50 - Commissioned as a 2nd lieutenant, 2nd Artillery, on 12 Mar 1812; resigned on 1 Jan 1813; commissioned as a captain, 36th Infantry, on 30 Apr 1813; discharged on 15 Jun 1815; died on 18 Sep 1837.

Hooker, Moses - Private - 2nd Light Dragoons - Company: John Burd - Age: 21 - Height: 5' 10" - Born: Baltimore Cty - Trade: Laborer - Enlistment date: 29 Jun 1812 - Period: 18 Mos - Deserted on 14 Aug 1812.

Hoop, Henry - Sergeant - 36th Infantry - Company: Henry Neale - Age: 38 - Height: 5' 10" - Born: Winchester, VA - Trade: Hatter - Enlistment date: 7 Jun 1814 - Period: War - BLW 11888-160-12 - Discharged at Washington, DC, on 21 Mar 1815.

Hooper, Barton - Private - 12th Infantry - Company: Andrew Madison - Age: 25 - Height: 5' 11" - Born: Frederick Cty, MD - Trade: Distiller - Enlistment date: 11 Aug 1812 - Place: Ch's Town - Period: 5 Yrs - BLW 8711-160-12 - Discharged at Pittsburgh on 17 Aug 1815 for ulcerated leg.

Hooper, Erestus - Seaman - US Sea Fencibles - Company: Simmones Bunbury - Age: 24 - Height: 5' 8 1/2" - Born: Massachusetts - Enlistment date: 8 Mar 1814 - Period: 1 Yr - Deserted on 18 Apr 1814.

Hoover, George - Private - 14th Infantry - Age: 23 - Prisoner of War (Quebec), died at Quebec on 9 Aug 1813.

Hope, Henry - Private - 36th Infantry - Company: Thomas Carbery - Age: 22 - Height: 5' 8" - Born: Chester Cty, PA - Trade: Blacksmith - Enlistment date: 19 Feb 1814 - Place: Georgetown, DC - Period: War - BLW 15374-160-12 - Discharged at Baltimore on 31 Mar 1815.

Hopkins, ----- - Private - 14th Infantry - Died at Sackets Harbor, NY, on 4 Nov 1813.

Hopkins, Edward - Ensign - 36th Infantry - Company: Joseph Merrick - Born:

By War of 1812 Soc. in MD

Maryland - Commissioned as an ensign on 30 Apr 1813; killed in a duel on 26 May 1814.

Hopkins, John - Private - 14th Infantry - Company: Thomas Montgomery - Enlistment date: 5 Aug 1812 - Period: 5 Yrs - Died at Lewiston, NY, on 15 Aug 1813.

Hopkins, Leroy - Private - 36th Infantry - Age: 21 - Height: 5' 11" - Born: Richmond, VA - Enlistment date: 6 Jul 1814 - Place: Norfolk, VA.

Hopkins, Lewis E. - Private - 22nd Infantry - Company: Jacob Carmack - Age: 28 - Height: 6' 1/4" - Born: Baltimore - Trade: Hatter - Enlistment date: 8 Oct 1814 - Period: War - Discharged at Fort Fayette, PA, on 8 Mar 1815.

Hopkins, Philip - Private - 14th Infantry - Company: Thomas Montgomery - Enlistment date: 11 Jan 1813 - Period: 5 Yrs - Pension: Heirs received half pay for five years in lieu of land bounty - Died in Aug 1813.

Hopkins, Thomas - Private - 14th Infantry - Company: David Cummings - Age: 26 - Height: 5' 11" - Born: Massachusetts - Trade: Farmer - Enlistment date: 17 Aug 1814 - Place: Baltimore - Period: War - Deserted on 5 or 30 Aug 1814.

Hopkins, William - Private - 32nd Infantry - Age: 21 - Height: 5' 7 1/2" - Born: Easton, MD - Trade: Farmer - Enlistment date: 11 Jun 1814 - Place: Long Island, NY - Period: 5 Yrs - Discharged at Philadelphia on 19 May 1815.

Hoppell, George - Private - 36th Infantry - Age: 20 - Height: 5' 6" - Born: Philadelphia - Enlistment date: 3 May 1813 - Place: Baltimore - Period: 1 Yr.

Hopwood, David - Private - 38th Infantry - Company: John Rothrock - Pension: Land bounty to Nancy Hopwood and other heirs at law of David Hopwood - BLW 24054-160-12.

Horne, Jacob - Corporal - 36th Infantry - Company: Joseph Hook - BLW 5683-160-12.

Horner, Nathan - Private - 38th Infantry - Company: Isaac Aldridge - BLW 5806-160-12.

Horner, Nathan - Private - 38th Infantry - Company: Isaac Aldridge - Age: 38 - Height: 5' 6 1/2" - Born: Anne Arundel Cty, MD - Trade: Ship's carpenter - Enlistment date: 8 Nov 1814 - Place: Baltimore - Period: War - BLW 5866-160-12 - Discharged at Baltimore on 6 Apr 1815.

Horney, David - Private - 5th Infantry - Company: Richard Whartenby - Other regiment: Corps of Artillery - Age: 30 - Height: 5' 8 1/2" - Born: Kent Cty, MD - Trade: Butcher - Enlistment date: 6 Jan 1812 - Place: Baltimore - Period: 5 Yrs - Prisoner of War (Halifax), captured at Stoney Creek, UC, on 12 Jun 1813; discharged at Fort McHenry on 30 Jun 1817.

Horush, George - Private - 38th Infantry - Company: John Buck - BLW 16381-160-12.

Hosanna, William - Private - Artillery - Other regiment: Corps of Artillery - Age: 24 - Height: 5' 3 1/4" - Born: Maryland - Trade: Laborer - Enlistment date: 18 Jan 1815 - Place: Annapolis - Period: War - BLW 271-320-12 - Discharged at Washington, DC, on 9 May 1815.

Hose, Frederick - Private - 14th Infantry - Company: Thomas Montgomery -

Age: 45 - Height: 5' 7" - Born: Hagerstown, MD - Trade: Farmer - Enlistment date: 4 Oct 1812 - Place: Hagerstown, MD - Period: 5 Yrs - BLW 883-160-12 - Discharged at Washington, DC, on 8 Nov 1815 for disability.

Hoseleborough, Philip - Private - 14th Infantry - Company: Henry Fleming - Enlistment date: 1 Jun 1812 - Period: 18 Mos - Discharged at Greenbush, NY, on 28 Jun 1814.

Hoskins, John - Private - 14th Infantry - Company: Thomas Montgomery - Age: 24 - Height: 6' - Born: Sharpsburg, MD - Enlistment date: 19 May 1813 - Place: Sharpsburg, MD - Period: 5 Yrs - BLW 8844-160-12 - Discharged at Fort Niagara, NY, on 31 Aug 1813 for head wound.

Hoss, John - Private - 16th Infantry - Company: James McElroy - Other regiment: 2nd Infantry (New) - Age: 43 - Height: 5' 11" - Born: Maryland - Trade: Hatter - Enlistment date: 21 Sep 1812 - Place: Harrisburg, PA - Period: 5 Yrs - Discharged at Sackets Harbor, NY, on 15 Aug 1815 for old age and rheumatism.

Hotchkiss, Caleb - Sergeant - 14th Infantry - Company: Thomas Montgomery - Age: 21 - Height: 5' 8 1/2" - Born: North Haven, CT - Trade: Farmer - Enlistment date: 12 Jun 1813 - Place: Herkimer Cty, NY - Period: War - BLW 8115-160-12 - Discharged at Greenbush, NY, on 4 May 1815.

Hough, Henry - Private - 36th Infantry - Company: Samuel Raisin - Enlistment date: 7 Jun 1814 - Place: Georgetown, DC - Period: War - Pension: Land bounty to Margaret Murrey, Elizabeth Simmes and Susannah Bayless, children and only heirs at law of Henry Hough - BLW 27747-160-42 - Discharged at Washington, DC, on 30 Mar 1815.

Hough, John - Private - 36th Infantry - Company: Samuel Raisin - Age: 27 - Height: 5' 11" - Born: Loudoun Cty, VA - Trade: Carpenter - Enlistment date: 30 Jul 1814 - Place: Norfolk, VA - Period: War - Discharged at Washington, DC, on 20 Mar 1815.

Hough, Thomas - Private - 36th Infantry - Company: Joseph Nelson - Age: 39 - Height: 6' 1" - Enlistment date: 28 Jun 1814 - Period: War - Pension: Land bounty to Joseph Hough, son and other heirs at law of Thomas Hough - BLW 26182-160-12 - Died on or about 31 Jan 1815.

House, John - Private - 5th Infantry - Company: Richard Whartenby - Age: 22 - Height: 5' 6" - Born: Norfolk, VA - Trade: Farmer - Enlistment date: 13 Mar 1812 - Place: Baltimore - Period: 5 Yrs.

Housholder, Nicholas - Private - 14th Infantry - Company: Reuben Gilder - Age: 36 - Height: 5' 6" - Born: Washington, PA - Trade: Carpenter - Enlistment date: 7 Sep 1813 - Place: Hagerstown, MD - Period: 18 Mos - Discharged at Greenbush, NY, on 7 Mar 1815.

Houston, Philip - Musician - 22nd Infantry - Company: John Pentland - Age: 49 - Height: 5' 7" - Born: Kent Cty, MD - Trade: Tailor - Enlistment date: 2 Jun 1814 - Place: Washington Cty, PA - Period: 5 Yrs - Discharged at Sackets Harbor, NY, on 1 Feb 1816 for old age and rheumatism.

Hovey, Samuel - Private - 14th Infantry - Company: Henry Fleming - Prisoner of War (Halifax), discharged from Halifax on 3 Feb 1814; died in Feb 1814.

Howard, Benjamin - Private - 5th Infantry - Company: Richard Whartenby -

By War of 1812 Soc. in MD

Age: 41 - Height: 5' 7" - Born: Maryland - Trade: Tailor - Enlistment date: 6 Mar 1812 - Period: 5 Yrs - Discharged at Plattsburgh, NY, on 24 Mar 1814 for wounds received in service.

Howard, Benjamin L. - Private - 3rd Artillery - Company: John Gookin - Age: 37 - Height: 5' 7 1/2" - Born: Charles Cty, MD - Trade: Tailor - Enlistment date: 5 May 1814 - Place: Pokeepsic, NY - Period: 5 Yrs - BLW 16690-160-12 - Discharged at Sackets Harbor, NY, on 20 Jun 1815.

Howard, Charles - Private - 14th Infantry - Company: David Cummings - Other regiment: 4th Infantry (New) - Age: 18 - Height: 5' 6" - Born: Washington, MD - Trade: Weaver - Enlistment date: 2 Feb 1815 - Place: Hagerstown, MD - Period: 5 Yrs - Discharged on 1 Feb 1820.

Howard, Jesse - Private - 36th Infantry - Company: Thomas Carbery - Age: 41 - Height: 6' 2 1/2" - Born: St. Mary's Cty, MD - Trade: Blacksmith - Enlistment date: 4 Aug 1814 - Place: Washington, DC - Period: War - BLW 171-160-12 - Discharged at Baltimore on 28 Feb 1815.

Howard, John - Private - 5th Infantry - Company: Richard Bell - Other regiment: 3rd Infantry (New) - Age: 35 - Height: 5' 10" - Born: Baltimore - Trade: White smith - Enlistment date: 22 Jun 1813 - Place: Baltimore - Period: 5 Yrs - Discharged at Fort Howard, WI, on 22 Jun 1818.

Howard, John - Corporal - US Volunteers - Company: Stephen Moore - Enlistment date: 8 Sep 1812 - Period: 1 Yr.

Howard, John G. - Sergeant - 14th Infantry - Company: Reuben Gilder - Other regiment: 4th Infantry (New) - Age: 19 - Height: 5' 6 1/2" - Born: Connecticut - Trade: Laborer - Enlistment date: 7 Apr 1813 - Period: 5 Yrs - BLW 24760-160-12 - On board the fleet on Lake Champlain; discharged at Fort St. Marks, FL, on 7 Apr 1818.

Howard, Lawrence - Private - 38th Infantry - Company: John Rothrock - Age: 19 - Height: 5' 9" - Born: Hagerstown, MD - Trade: Tinner - Enlistment date: 20 Aug 1814 - Place: Craney Island, VA - Period: War.

Howard, Perry - Private - 26th Infantry - Company: William Bezeau - Age: 21 - Height: 5' 6" - Trade: Tailor - Enlistment date: 24 Sep 1814 - Place: Baltimore - Period: War - Discharged at Philadelphia on 23 Mar 1815.

Howard, Philip - Private - 36th Infantry - Company: Thomas Carbery - Age: 30 - Height: 6' - Born: St. Mary's Cty, MD - Trade: Blacksmith - Enlistment date: 18 Aug 1814 - Place: Piscataway - Period: War - BLW 93-160-12 - Discharged at Baltimore on 31 Mar 1815.

Howard, Thomas - Private - 36th Infantry - Company: Joseph Hook - Age: 21 - Height: 5' 5" - Born: Baltimore - Trade: Shoemaker - Enlistment date: 8 May 1813 - Place: Baltimore - Period: 1 Yr - BLW 25436-160-12 - Re-enlisted at Annapolis on 1 Mar 1814 for the war; discharged at Fort Covington, MD, on 30 Mar 1815.

Howard, William - Private - 3rd Rifles - Company: Alexander Brandon - Age: 28 - Height: 5' 9" - Born: Baltimore - Trade: Sailor - Enlistment date: 19 Jul 1814 - Place: Ashe Courthouse, NC - Period: 5 Yrs - Discharged at Alexandria, DC.

Howe, Ira - Private - 36th Infantry - Company: Ezra Turner - Pension: SO-591

Maryland Regulars in War of 1812

Rejected.

Howe, James W. - Private - 38th Infantry - Company: Shepperd Leakin - Age: 21 - Height: 5' 4" - Born: Connecticut - Enlistment date: 2 Sep 1814 - Place: Baltimore - Period: War - BLW 24798-160-12 - Discharged on 27 Mar 1815.

Howell, John - Private - 5th Infantry - Company: William Henshaw - Age: 25 - Height: 5' 11" - Born: Newtown, MD - Trade: Tailor - Enlistment date: 3 Jun 1813 - Place: Baltimore - Period: 5 Yrs.

Howell, Thomas - Private - 5th Infantry - Company: George Brooks - Age: 50 - Height: 5' 6" - Born: Dorchester Cty, MD - Trade: Seaman - Enlistment date: 18 Jun 1813 - Period: 5 Yrs - Discharged on 22 Apr 1814 for old age.

Howington, William - Private - 1st Rifles - Company: William Smyth - Age: 26 - Height: 4' 4" - Born: Nottingham, MD - Trade: Carpenter - Enlistment date: 20 Dec 1812 - Place: Shepherdstown - Period: 5 Yrs - Discharged at Fort McHenry on 8 Apr 1817 after furnishing a substitute (Charles W. Rogers).

Howland, George - Private - 36th Infantry - Company: Joseph Merrick - Age: 18 - Height: 5' 4" - Born: Providence, MD - Enlistment date: 14 May 1813 - Place: Baltimore - Period: 1 Yr - Pension: SO-40, SC-2251 - Discharged on 19 Dec 1813.

Hoy, Barney - Private - 14th Infantry - Prisoner of War (Halifax), sent to England.

Hoy, Judson - Private - 14th Infantry - Company: Joseph Marshall.

Hoye, Enoch - Private - 2nd Infantry - Company: Perrin Willis - Age: 17 - Height: 5' 2 1/2" - Born: Prince George's Cty, MD - Trade: Laborer - Enlistment date: 19 Oct 1814 - Place: Washington, DC - Period: 5 Yrs - Discharged on 19 Feb 1819 after furnishing a substitute.

Hubbard, Daniel - Private - 10th Infantry - Company: George Vashon - Other regiment: 8th Infantry (New) - Age: 19 - Height: 5' 7" - Born: Caroline Cty, MD - Trade: Farmer - Enlistment date: 23 Apr 1813 - Place: Guilford Cty, NC - Period: 5 Yrs - Discharged at Fort Crawford. IL, on 23 Apr 1818.

Hubbard, Hugh - Private - 14th Infantry - Company: Thomas Montgomery - Age: 18 - Height: 5' 6 1/2" - Born: Newmarket, MD - Trade: Farmer - Enlistment date: 11 Jul 1812 - Place: Cambridge, MD - Period: 5 Yrs - BLW 3849-160-12 - Discharged at Greenbush, NY, on 20 Mar 1815 for rupture on right side.

Huber, Henry - Second Lieutenant - 38th Infantry - Company: John Brookes - Commissioned as a 3rd lieutenant on 20 May 1813; promoted to 2nd lieutenant on 1 May 1814; discharged on 15 Jun 1815.

Huber, Jacob - Corporal - 14th Infantry - Company: James Green - Enlistment date: 15 Jun 1812 - Period: 5 Yrs - Prisoner of War (Halifax), captured at Beaver Dams on 24 Jun 1813.

Huddleston, Solomon - Private - 2nd Infantry - Company: Mathew Arbuckle - Age: 34 - Height: 6' 1" - Born: Baltimore - Trade: Laborer - Enlistment date: 23 Feb 1810 - Place: Columbia Springs - Period: 5 Yrs - Re-enlisted at Fort Bowyer on 24 Jan 1815 for the war; discharged on 9 Apr 1815.

Hudson Jr.,, James - Private - 12th Infantry - Company: Thomas Sangster -

By War of 1812 Soc. in MD

Other regiment: 8th Infantry (New) - Age: 21 - Height: 5' 10" - Born: Maryland - Trade: Blacksmith - Enlistment date: 20 May 1812 - Place: Elkton, MD - Period: 5 Yrs - Discharged on 20 May 1817.

Hudson Sr.,, James - Private - 12th Infantry - Company: Thomas Sangster - Age: 32 - Height: 5' 5 1/2" - Born: Ireland - Trade: Laborer - Enlistment date: 9 May 1812 - Place: Maryland - Period: 5 Yrs.

Hudson, Littleton - Private - 14th Infantry - Company: David Cummings - Age: 21 - Height: 5' 7 1/2" - Born: Worcester Cty, MD - Trade: Shoemaker - Enlistment date: 6 Aug 1814 - Place: Salisbury, MD - Period: War - BLW 3787-160-12 - Discharged at Baltimore on 16 Mar 1815.

Hudson, William - Private - 36th Infantry - Age: 24 - Height: 5' 5 1/4" - Born: Carolina Cty, VA - Enlistment date: 8 May 1813 - Place: Baltimore - Period: 1 Yr - Deserted on 1 Aug 1813.

Huff, Michael - Private - 14th Infantry - Prisoner of War (Halifax), sent to England.

Huffman, John - Fifer - 38th Infantry - Company: Shepperd Leakin - BLW 13636-160-12.

Hughes Jr., John - Musician - 1st Artillery - Company: George Armistead - Age: 10 - Height: 4' 8" - Born: Baltimore Cty - Trade: Drummer - Enlistment date: 1 Jul 1812 - Place: Bedloe's Island, NY - Period: 5 Yrs - BLW 11188-160-12 - Discharged on 1 Jul 1817.

Hughes, Archibald - Private - Light Dragoons - Company: James Reed - Age: 26 - Height: 5' 9" - Born: Charles Cty, MD - Trade: Seaman - Enlistment date: 11 May 1814 - Place: Alexandria, DC - Period: War - Discharged at Fort Washington, MD, on 7 May 1815.

Hughes, Boyle - Private - 38th Infantry - Company: John Mowton - BLW 23639-160-12.

Hughes, Daniel - Major - 1st Infantry - Born: Maryland - Commissioned as an ensign, 9th Infantry, on 8 Jan 1799; promoted to 2nd Lieutenant on 3 Mar 1799; discharged on 15 Jun 1800; commissioned as a 2nd lieutenant, 2nd Infantry, on 16 Feb 1801; transferred to 1st Infantry on 1 Apr 1802; promoted to 1st lieutenant on 23 Mar 1805; promoted to captain on 15 Dec 1808; promoted to major on 21 Feb 1814; discharged on 15 Jun 1815.

Hughes, Daniel Boyle - Private - 38th Infantry - Company: John Mowton - Age: 23 - Height: 5' 2" - Born: Frederick, MD - Trade: Hatter - Enlistment date: 21 Apr 1814 - Place: Craney Island, VA - Period: War - BLW 23639-160-12 - Discharged at Craney Island, VA, on 15 Mar 1815.

Hughes, Isaac - Private - 36th Infantry - Company: Samuel Raisin - Age: 22 - Height: 5' 9" - Born: Virginia - Trade: Blacksmith - Enlistment date: 20 Jan 1814 - Place: Norfolk, VA - Period: War - BLW 5169-160-12 - Discharged at Washington, DC, on 20 Mar 1815.

Hughes, James - Private - 29th Infantry - Company: John Rochester - Age: 24 - Height: 5' 4" - Born: Emmetsburg, MD - Trade: Laborer - Enlistment date: 21 Apr 1814 - Place: Sackets Harbor, NY - Period: War - Discharged at Fort McHenry on 21 Apr 1819.

Hughes, James - Private - 36th Infantry - Company: Joseph Hook - Age: 22 -

Height: 5' 10" - Born: New York City - Trade: Mariner - Enlistment date: 5 Jul 1814 - Place: Baltimore - Period: War - Deserted at Fort Covington, MD, on 17 Jan 1815.

Hughes, James - Private - 12th Infantry - Company: Andrew Madison - Age: 23 - Height: 5' 6" - Born: Emmetsburg, MD - Trade: Farmer - Enlistment date: 4 Mar 1813 - Place: Charles Town - Period: 18 Mos - Discharged at Buffalo.

Hughes, John - Private - 3rd Infantry - Company: James Denking - Age: 40 - Height: 6' - Born: Ireland - Trade: Soldier - Enlistment date: 26 Aug 1808 - Place: Baltimore - Period: 5 Yrs - Discharged on 25 Aug 1818.

Hughes, John F. - Private - 36th Infantry - Age: 34 - Height: 6' - Born: Virginia - Enlistment date: 15 Jul 1814 - Place: Norfolk, VA.

Hughes, Samuel - Private - 1st Artillery - Company: Hopley Yeaton - Other regiment: Corps of Artillery - Age: 26 - Height: 5' 6 1/2" - Born: Frederick, MD - Trade: Saddle tree maker - Enlistment date: 18 Sep 1810 - Period: 5 Yrs - Discharged on 18 Sep 1815.

Hughes, William - Private - 2nd Infantry - Company: Peter Schuyler - Age: 36 - Height: 5' 7" - Born: Maryland - Trade: Brick layer - Enlistment date: 18 Oct 1807 - Place: Columbia Springs - Period: 5 Yrs - Died at Mount Vernon on 3 Jul 1812.

Hughey, William - Private - 19th Infantry - Company: George Kesling - Age: 23 - Height: 6' 1" - Born: Maryland - Trade: Shoemaker - Enlistment date: 28 Jan 1815 - Place: Zanesville, OH - Period: War - BLW 466-320-14 - Discharged at Zanesville, OH, on 27 Mar 1815.

Hugle, John - Private - 38th Infantry - Company: Shepperd Leakin - BLW 23517-160-12.

Hugo, Samuel B. - Surgeon's Mate - 14th Infantry - Other regiment: Rifles - Pension: Wife Mary Ann Spangler, WO-33908, WC-19505; married on 10 May 1825, York, York Cty, PA; soldier died at York, PA, on 1 Mar 1861 - BLW 16140-160-50 - Commissioned as a surgeon's mate, 14th Infantry, on 12 Mar 1812; transferred to Rifle on 26 Jan 1814; promoted to post surgeon on 27 Aug 1818; dismissed on 7 Sep 1818.

Hukill, Levi - Captain - 1st Light Dragoons - Born: Maryland - Commissioned as a cornet, Light Dragoons, on 3 May 1808; promoted to 1st lieutenant on 18 Jan 1810; served as a major, assistant inspector general, from 6 Aug to 5 Dec 1813; promoted to captain on 7 Jun 1813; died on 5 Dec 1813.

Hukins, Thomas - Private - 28th Infantry - Company: Henry Gist - Age: 22 - Height: 5' 11" - Born: Maryland - Trade: Farmer - Enlistment date: 23 Sep 1814 - Place: Flemingsburg, KY - Period: War - BLW 9312-160-12 - Discharged at Lower Sandusky, OH (Fort Stephenson), on 25 Jun 1815.

Hukins, Thomas - Private - 28th Infantry - Company: Henry Gist - Age: 28 - Height: 6' 1" - Born: Maryland - Trade: Farmer - Enlistment date: 3 Aug 1814 - Place: Kentucky - Period: War.

Hull, James - Musician - 38th Infantry - Company: James Hook - Pension: Wife Nancy, WO-13988, WC-28896.

Hult, Richard - Private - 14th Infantry - Company: Reuben Gilder - Age: 18 -

By War of 1812 Soc. in MD

Height: 5' 7 1/2" - Born: Baltimore City - Trade: Laborer - Enlistment date: 17 Feb 1813 - Place: Baltimore - Period: 5 Yrs - Deserted on 16 Jul 1815.

Humphreys, Mathias - Private - 1st Rifles - Company: William Smyth - Other regiment: 4th Infantry then 5th Infantry (New) - Age: 20 - Height: 5' 5" - Born: Baltimore - Trade: Cordwainer - Enlistment date: 26 Feb 1814 - Place: Baltimore - Period: 5 Yrs - Discharged at Fort Wayne, IN, on 26 Feb 1819.

Hunt, Bazil - Private - 2nd Light Dragoons - Company: George Haig - Other regiment: Light Dragoons - Age: 27 - Height: 6' - Born: Bryan, Charles Cty, MD - Trade: Carpenter - Enlistment date: 13 Sep 1812 - Place: Wheeling, VA - Period: War - BLW 20734-160-12 - Discharged at Greenbush, NY, on 3 May 1815.

Hunt, Elisha - Private - 14th Infantry - Enlistment date: 20 Aug 1812 - Pension: Wife Sally Butler, WO-25131, WC-14719; married on 4 Jan 1817, Fluvanna Cty, VA; soldier died on May 1842, Rockbridge Cty, VA - BLW 7627-160-50 - Discharged on 20 Feb 1814.

Hunter, James - Private - 14th Infantry - Enlistment date: 30 Nov 1812 - Period: 18 Mos - Prisoner of War (Halifax), sent to England; discharged on 31 Mar 1815.

Hunter, Joseph - Private - 38th Infantry - Company: James Hook - Age: 22 - Height: 5' 9 1/2" - Born: Harford Cty, MD - Trade: Cooper - Enlistment date: 28 Jan 1814 - Place: Baltimore - Period: War - BLW 3222-160-12 - Discharged at Fort Covington, MD, on 31 Mar 1815.

Hunter, Joseph - Private - 14th Infantry - Company: David Cummings - Enlistment date: 25 Jul 1812 - Period: 5 Yrs - Pension: Land bounty to Robert Hunter, brother and only heir at law of Joseph Hunter - BLW 25878-160-12 - Prisoner of War (Halifax), captured at Beaver Dams; died at Halifax on 30 Aug 1813.

Hunter, Samuel - Private - 36th Infantry - Company: Joseph Hook - Age: 26 - Height: 5' 11 1/4" - Born: Dauphin Cty, PA - Trade: Carpenter - Enlistment date: 20 May 1813 - Place: Annapolis - Period: 1 Yr - BLW 14501-160-12 - Re-enlisted at Annapolis on 16 Mar 1814 for the war; discharged at Fort Covington, MD, on 30 Mar 1815.

Hunter, William - Corporal - 4th Infantry - Company: Oliver Burton - Age: 29 - Height: 5' 6 1/2" - Born: Bellaire, MD - Trade: Cooper - Enlistment date: 19 Dec 1808 - Period: 5 Yrs - Prisoner of War; discharged on 18 Dec 1813.

Huran, John - Private - 14th Infantry - Age: 29 - Height: 5' 6" - Born: Maryland - Trade: Tailor - Enlistment date: 22 Mar 1810 - Place: Washington, DC - Period: 5 Yrs - Discharged on 22 Mar 1815.

Hurd, John - Private - 14th Infantry - Prisoner of War (Halifax), discharged from Halifax on 9 Nov 1813.

Hurdle, John - Musician - 36th Infantry - Age: 14 - Height: 4' 10" - Born: Georgetown, DC - Trade: Musician - Enlistment date: 14 Feb 1815 - Place: Georgetown, DC - Period: War - BLW 398-320-12 - Discharged at Georgetown, DC, on 13 Mar 1815.

Hurdle, Robert - Sergeant - 36th Infantry - Company: Samuel Raisin - Age: 20 - Height: 5' 9" - Born: Montgomery Cty, MD - Trade: Shoemaker -

Maryland Regulars in War of 1812

Enlistment date: 17 Jul 1814 - Place: Annapolis - Period: War - BLW 1513-160-12 - Discharged at Washington, DC, on 20 Mar 1815.

Hurley, Thomas - Private - 6th Infantry - Company: Alexander Thompson - Other regiment: 2nd Infantry (New) - Age: 22 - Height: 5' 8" - Born: Leesburg, VA - Trade: Hatter - Enlistment date: 22 Feb 1812 - Place: Hagerstown, MD - Period: 5 Yrs - Discharged at Fort Niagara, NY, on 22 Feb 1817.

Hurry, Leonard - Private - 1st Rifles - Company: William Smyth - Other regiment: Rifles - Age: 23 - Height: 5' 7" - Born: Charles Cty, MD - Trade: Farmer - Enlistment date: 5 Mar 1812 - Place: Maryland - Period: 5 Yrs - BLW 20250-120-12 - Discharged at Baton Rouge, LA, on 27 May 1817.

Hursan, John - Private - 14th Infantry - Company: Reuben Gilder - Age: 32 - Height: 5' 6" - Born: Ireland - Trade: Laborer - Enlistment date: 17 Sep 1813 - Period: 18 Mos.

Hush, John - Private - 28th Infantry - Company: Henry Gist - Age: 28 - Height: 5' 4 1/3" - Born: Maryland - Trade: Hatter - Enlistment date: 24 Jul 1814 - Place: Greenupsburg, KY - Period: War - Discharged at Lower Sandusky, OH (Fort Stephenson), on 25 Jun 1815.

Hush, Peter - Seaman - US Sea Fencibles - Company: Simmones Bunbury - Enlistment date: 31 Jan 1814 - Period: 1 Yr - Discharged at Fort McHenry on 24 Mar 1815.

Hutchcraft, John - Private - 36th Infantry - Height: 5' 9" - Born: Washington, MD - Enlistment date: 26 Jul 1812 - Place: Wayesburg, PA - Period: 5 Yrs.

Hutchings, John A. - Musician - 38th Infantry - Company: Thomas Sangster - Other regiment: 4th Infantry (New) - Age: 16 - Height: 5' 2" - Born: Eastern Shore, MD - Trade: Painter - Enlistment date: 5 Oct 1814 - Place: Baltimore - Period: 5 Yrs - Deserted at Augusta, GA, on 20 Jul 1816.

Hutchinson, Francis - Private - 14th Infantry - Company: Reuben Gilder - Enlistment date: 8 May 1813 - Period: 5 Yrs - Pension: Old War IF-25173.

Hutchinson, Michael - Private - 36th Infantry - Company: Mortimer Hall - Age: 41 - Height: 5' 6" - Born: Ireland - Enlistment date: 10 Sep 1813 - Place: Baltimore - Period: 1 Yr.

Hutchinson, William - Private - 5th Infantry - Age: 37 - Height: 5' 10" - Born: Wilmington, NC - Trade: Carpenter - Enlistment date: 4 Feb 1812 - Place: Baltimore - Period: 5 Yrs.

Hutton, George - Private - 36th Infantry - Company: Samuel Raisin - Born: Caroline Cty, MD - Enlistment date: 10 Jun 1813 - Place: Cambridge, MD - Period: 1 Yr - Discharged, being under age.

Hutton, Samuel - Seaman - US Sea Fencibles - Company: John Gill - Enlistment date: 21 Dec 1813.

Hutton, William - Private - 1st Rifles - Company: George Seviers - Age: 23 - Height: 5' 6" - Born: Maryland - Trade: Laborer - Enlistment date: 29 Sep 1808 - Place: Greensburg - Period: 5 Yrs - Discharged on 29 Sep 1813.

Hutton, William - Seaman - US Sea Fencibles - Company: John Gill - Age: 26 - Height: 5' 10 1/2" - Born: Queen Anne, MD - Enlistment date: 27 Dec 1813 - Period: 1 Yr.

By War of 1812 Soc. in MD

Huver, Peter - Private - 36th Infantry - Company: Charles Randolph - Age: 43 - Height: 5' 5" - Born: Maryland - Enlistment date: 15 Aug 1813 - Place: Waynesburg, PA - Period: 1 Yr - Deserted on 20 Aug 1813.

Hyatt, Jose - Private - Light Dragoons - Age: 28 - Height: 5' 7" - Born: Frederick Cty, MD - Trade: Farmer - Enlistment date: 19 Oct 1812 - Period: 5 Yrs.

Hyatt, Joseph - Private - 14th Infantry - Company: Reuben Gilder - Age: 42 - Height: 5' 7 1/2" - Born: Frederick, MD - Trade: Farmer - Enlistment date: 1 Oct 1812 - Place: Maryland - Period: 5 Yrs - Pension: Old War IF-25174 - Prisoner of War (Halifax); discharged at Greenbush, NY, on 28 Mar 1815, wounded in right breast at Cook's Mills on 19 Oct 1814.

Hyde, Constant - Private - 5th Infantry - Company: Leroy Opie - Other regiment: 3rd Infantry (New) - Age: 26 - Height: 6' 1 1/4" - Born: Rhode Island - Trade: Shoemaker - Enlistment date: 9 Feb 1813 - Place: Baltimore - Period: 5 Yrs - BLW 19290-160-12 - Discharged at Fort Howard, WI, on 9 Feb 1818.

Hyde, John - Private - 14th Infantry - Company: Thomas Montgomery - Age: 22 - Height: 5' 6" - Born: Rhode Island - Enlistment date: 31 Jan 1813 - Period: 18 Mos - Discharged at Plattsburgh, NY, on 31 Jul 1814.

Hysinger, John - Private - 16th Infantry - Company: James McElroy - Age: 28 - Height: 5' 11" - Born: Maryland - Trade: Distiller - Enlistment date: 26 Feb 1813 - Place: Columbia, MD - Period: 18 Mos - Discharged at Chazy, NY, on 26 Aug 1814.

Hysler, Anthony - Private - 44th Infantry - Company: Anatole Peychand - Age: 32 - Height: 5' 4" - Born: Baltimore - Enlistment date: 12 Jul 1814 - Place: Powder Magazine Barracks, LA - Period: 5 Yrs.

I

Icenogal, David - Private - 26th Infantry (OH) - Company: Benjamin Watson - Other regiment: 25th Infantry - Age: 22 - Height: 5' 7" - Born: Frederick Cty, MD - Trade: Farmer - Enlistment date: 3 Apr 1813 - Place: Ohio - Period: War - Discharged at Sackets Harbor, NY, on 2 Apr 1815.

Igo, Jacob - Private - 17th Infantry - Company: Benjamin Sanders - Age: 31 - Height: 6' 1 1/2" - Born: Maryland - Trade: Farmer - Enlistment date: 19 Apr 1814 - Place: Kentucky - Period: War - BLW 24864-160-12 - Discharged at Chillicothe, OH, on 7 Jun 1815.

Ijams, Isaac - Sergeant - 1st Infantry - Company: Hugh Moore - Age: 28 - Height: 5' 7" - Born: Washington, MD - Trade: Farmer - Enlistment date: 3 Mar 1812 - Place: Pittsburgh - Period: 5 Yrs - Discharged at Fort Wayne, IN, on 30 Apr 1817.

Ijams, William - Sergeant - 1st Infantry - Company: Hugh Moore - Other regiment: 3rd Infantry (New) - Age: 32 - Height: 5' 7 1/2" - Born: Washington, MD - Trade: Laborer - Enlistment date: 2 Mar 1812 - Place: Pittsburgh - Period: 5 Yrs - BLW 20343-160-12 - Discharged at Fort Wayne, IN, on 2 Mar 1817.

Imes, Peter - Private - 36th Infantry - Company: Samuel Raisin - BLW 12239-

Maryland Regulars in War of 1812

160-12.

Ing, John - Seaman - US Sea Fencibles - Company: William Addison.

Ingram, John H. - Private - 43rd Infantry - Company: George Dabney - Age: 35 - Height: 5' 9 1/2" - Born: Baltimore - Trade: Wheelwright - Enlistment date: 4 Apr 1814 - Place: Roxboro - Period: War - BLW 2623-160-12 - Discharged at Fort Hampton on 1 Aug 1815.

Ireland, Archibald - Private - 5th Infantry - Company: Colin Buckner - Other regiment: 23rd Infantry then 4th Infantry (New) - Age: 24 - Height: 5' 6" - Born: Baltimore - Trade: Blacksmith - Enlistment date: 21 Feb 1812 - Place: Virginia - Period: 5 Yrs - Transferred to the 23rd Infantry on 3 Jun 1814; discharged on 28 Feb 1818.

Ireland, James - Private - 14th Infantry - Company: Thomas Montgomery - Other regiment: 4th Infantry (New) - Age: 21 - Height: 5' 8" - Born: Egg Harbor, NJ - Trade: Laborer - Enlistment date: 22 Feb 1813 - Place: Philadelphia - Period: 5 Yrs - Deserted on 25 Jul 1815.

Ireland, Jonathan - Private - 14th Infantry - Prisoner of War (Halifax), sent to England.

Irvin, Henry - Corporal - 42nd Infantry - Company: Edmund Duval - Age: 39 - Height: 5' 4 1/2" - Born: Prince George's Cty, MD - Trade: Millwright - Enlistment date: 13 Dec 1813 - Place: Bladensburg, MD - Period: War - Discharged at Fort Lewis, NY.

Irvin, Thomas - Fifer - 14th Infantry - Company: Reuben Gilder - Other regiment: 4th Infantry (New) - Enlistment date: 12 Aug 1812 - Period: 5 Yrs - Discharged on 11 Aug 1817.

Irvine, Baptist - Second Lieutenant - US Volunteers - Company: Stephen Moore - Enlistment date: 8 Sep 1812 - Period: 1 Yr - Wounded at York, UC, on 27 Apr 1813.

Iser, Richard - Private - 2nd Light Dragoons - Company: William Littlejohn - Age: 35 - Height: 5' 9" - Born: Baltimore City - Trade: Blacksmith - Enlistment date: 14 Sep 1814 - Place: Baltimore - Period: 5 Yrs - Deserted.

Iserminger, Henry - Private - 12th Infantry - Company: Andrew Madison - Age: 25 - Height: 5' 10" - Born: Middletown, MD - Trade: Joiner - Enlistment date: 19 Jan 1814 - Place: Hagerstown, MD - Period: War - Discharged at Buffalo on 31 May 1815.

Isgrig, William - Private - 36th Infantry - Company: Joseph Hook - Age: 33 - Height: 5' 9" - Born: Maryland - Enlistment date: 5 Apr 1814 - Place: Baltimore - Discharged at Fort Covington, MD, on 17 Dec 1814.

Isler, Anthony - Private - 44th Infantry - Company: William Butler - Other regiment: 4th Infantry (New) - Age: 32 - Height: 5' 4" - Born: Baltimore - Trade: Tailor - Enlistment date: 12 Jul 1814 - Place: New Orleans - Period: 5 Yrs - Discharged on 11 Jul 1819.

Isles, Thomas - Private - 26th Infantry - Company: William Bezeau - Age: 36 - Height: 5' 6" - Born: Baltimore - Trade: Laborer - Enlistment date: 11 Oct 1814 - Place: Baltimore - Period: War - BLW 9868-160-12 - Discharged at Baltimore.

Israel, Tolbert - Sergeant - 38th Infantry - Company: John Buck - Height: 6'

By War of 1812 Soc. in MD

3" - Born: Baltimore - Trade: Cooper - Enlistment date: 27 Oct 1813 - Period: 1 Yr - BLW 17209-160-12 - Re-enlisted at Baltimore on 28 Feb 1814 for the war; discharged at Baltimore on 31 Mar 1815.

Izer, Joshua - Seaman - US Sea Fencibles - Company: John Gill - Age: 20 - Height: 5' 6" - Born: Baltimore City - Enlistment date: 12 Jan 1814.

J

Jacks, Richard - Private - 10th Infantry - Company: Hugh Carson - Age: 40 - Height: 5' 7" - Born: Baltimore - Trade: Farmer - Enlistment date: 7 Oct 1814 - Place: Wilkesboro, NC - Period: War - BLW 4056-160-12 - Discharged at Raleigh, NC, on 24 Apr 1815.

Jackson, Elisha - Private - 36th Infantry - Company: Joseph Merrick - Age: 22 - Height: 5' 7 1/2' - Born: Albany, NY - Enlistment date: 1 Aug 1813 - Place: Havre de Grace, MD - Period: 1 Yr - Discharged on 19 Dec 1813.

Jackson, Francis - Private - 14th Infantry - Company: Samuel Lane - Age: 23 - Height: 5' 7 1/2" - Enlistment date: 3 Aug 1812 - Period: 18 Mos - Deserted at Black Rock, NY, 15 Dec 1813.

Jackson, George - Recruit - 26th Infantry - Company: William Bezeau - Race: 30 - Age: 30 - Height: 5' 5" - Born: Maryland - Trade: Farmer - Enlistment date: 25 Dec 1814 - Place: Philadelphia - Period: 5 Yrs - Listed as "Col'd" in his service record.

Jackson, John - Private - 36th Infantry - Age: 36 - Height: 5' 8" - Born: Ireland - Enlistment date: 9 Jul 1814 - Place: Westminster, MD - Period: War.

Jackson, John - Private - 2nd Light Dragoons - Company: William Littlejohn - Other regiment: Corps of Artillery - Age: 20 - Height: 5' 9 1/2" - Born: Philadelphia - Trade: Cabinet maker - Enlistment date: 17 Feb 1814 - Place: Baltimore - Period: 5 Yrs - Transferred to Corps of Artillery on 31 Oct 1815; discharged on 17 Feb 1819.

Jackson, John - Seaman - US Sea Fencibles - Company: Simmones Bunbury - Enlistment date: 7 Jun 1814 - Period: 1 Yr - Enlisted as a substitute for Caleb Hall; discharged at Fort McHenry on 29 Jan 1815.

Jackson, Robert - Private - 2nd Artillery - Company: Frederick Evans - Age: 44 - Height: 5' 11 1/4" - Born: Maryland - Trade: Laborer - Enlistment date: 2 Jun 1812 - Place: Annapolis - Period: 5 Yrs - Deserted on 1 Apr 1816.

Jackson, Robert - Private - 5th Infantry - Company: Richard Whartenby - Age: 22 - Height: 5' 6" - Born: Boston - Trade: Shoemaker - Enlistment date: 16 Jul 1812 - Place: Baltimore - Period: 5 Yrs.

Jackson, Robin - Private - 36th Infantry - Age: 21 - Height: 5' 8" - Born: Russell, VA - Enlistment date: 5 Oct 1814 - Place: Norfolk, VA.

Jackson, Samuel - Private - 5th Infantry - Company: William Bird - Age: 30 - Height: 5' 9" - Born: Carlisle, PA - Trade: Carpenter - Enlistment date: 9 Feb 1813 - Place: Maryland - Period: 5 Yrs.

Jackson, Samuel - Private - 26th Infantry - Company: William Bezeau - Age: 28 - Height: 5' 5" - Born: Baltimore - Enlistment date: 24 Sep 1814 - Place: Baltimore - Period: War - Deserted on 14 Dec 1814.

Jackson, Samuel - Private - Artillery - Other regiment: Corps of Artillery -

Maryland Regulars in War of 1812

Age: 34 - Height: 5' 8" - Born: Queen Anne's Cty, MD - Trade: Laborer - Enlistment date: 20 Dec 1814 - Place: Annapolis - Period: War - BLW 625-320-12 - Discharged at Washington, DC, on 9 May 1815.

Jackson, Sheldon - Sergeant - 38th Infantry - Company: John Buck - Age: 23 - Height: 5' 9 1/2" - Born: Litchfield Cty, CT - Trade: Trader - Enlistment date: 16 Apr 1814 - Place: Baltimore - Period: War - BLW 14701-160-12 - Discharged at Baltimore on 30 Mar 1815.

Jacobs, Benjamin - Sergeant - 8th Infantry - Company: Charles Crawford - Age: 28 - Height: 5' 8" - Born: Maryland - Enlistment date: 7 Dec 1812 - Period: 18 Mos - Discharged on 6 Jun 1814.

Jacobs, Henry - Private - 14th Infantry - Company: David Cummings - Age: 21 - Height: 5' 7 1/2" - Born: Cecil Cty, MD - Trade: Laborer - Enlistment date: 5 Nov 1814 - Place: Baltimore - Period: War - BLW 1645-160-12 - Discharged at Baltimore on 16 Mar 1815.

Jacobs, Henry - Private - 7th Infantry - Company: Alexander White - Age: 18 - Height: 5' 2" - Born: Baltimore - Trade: Seaman - Enlistment date: 11 Mar 1813 - Place: Balize - Period: War - BLW 1342-160-12 - Discharged at New Orleans on 8 Apr 1815.

Jacobs, John W. - Private - 14th Infantry - Company: Thomas Kearney - Enlistment date: 11 Jul 1812 - Period: 18 Mos - Prisoner of War (Halifax), sent to England; discharged on 31 Mar 1815.

James, Allen - Private - 14th Infantry - Company: Reuben Gilder - BLW 2883-160-12.

James, Charles - Private - 14th Infantry - Company: Robert Kent - Enlistment date: 1 May 1813 - Period: 5 Yrs - On board the fleet on Lake Ontario.

James, Edward T. - Private - 38th Infantry - Company: Thomas Sangster - Age: 33 - Height: 5' 3" - Born: England - Trade: Chandler - Enlistment date: 8 Nov 1814 - Place: Baltimore - Period: 5 Yrs.

James, Jesse - Private - 12th Infantry - Company: James Paxton - Age: 21 - Height: 5' 9" - Born: Baltimore - Trade: Miller - Enlistment date: 31 Jul 1812 - Place: Morgantown, VA - Period: 18 Mos - Discharged at French Mills on 31 Jan 1814.

James, John - Private - 17th Infantry - Age: 50 - Height: 6' - Born: Maryland - Enlistment date: 4 Aug 1813 - Period: War - BLW 18356-160-12 - Discharged at Newport, KY, on 24 May 1815.

James, Peter - Private - 36th Infantry - Age: 20 - Height: 5' 8" - Born: Georgetown, DC - Enlistment date: 21 Apr 1814 - Place: Georgetown, DC.

James, Seth - Corporal - 2nd Light Dragoons - Age: 22 - Born: Maryland.

James, Thomas - Private - 12th Infantry - Company: James Paxton - Other regiment: 8th Infantry (New) - Age: 20 - Height: 6' 1" - Born: Maryland - Trade: Farmer - Enlistment date: 9 Oct 1812 - Place: Wythe Cty, VA - Period: 5 Yrs - Discharged at Fort Claiborne, AL, on 15 Oct 1817.

James, William - Seaman - US Sea Fencibles - Company: John Gill - Age: 29 - Height: 5' 6 1/2" - Born: Baltimore - Enlistment date: 1 Jan 1814 - Period: 1 Yr.

Jameson, Nicholas - Private - 2nd Rifles - Company: Benjamin Desha - Age:

39 - Height: 5' 10 1/2" - Born: Charles Cty, MD - Trade: Laborer - Enlistment date: 15 Nov 1814 - Period: 5 Yrs - BLW 24410-160-12 - Discharged at Detroit on 30 Jul 1815.

Jamison, William - Private - 14th Infantry - Company: Reuben Gilder - Age: 28 - Height: 5' 10" - Born: Delaware - Trade: Wagoner - Enlistment date: 1 Jun 1813 - Place: Baltimore - Period: 5 Yrs - BLW 8098-160-12 - Discharged at Greenbush, NY, on 4 May 1815.

Jarivs, John - Private - 28th Infantry - Company: George Stockton - Age: 21 - Height: 5' 7" - Born: Maryland - Trade: Farmer - Enlistment date: 11 Feb 1814 - Place: Flemingsburg, KY - Period: War - BLW 23005-160-12 - Discharged at Detroit on 30 Jun 1815.

Jarrett, William - Private - 14th Infantry - Company: Reuben Gilder - Age: 35 - Height: 5' 11" - Born: Baltimore - Trade: Laborer - Enlistment date: 3 Mar 1814 - Place: Plattsburgh, NY - Period: War - BLW 10894-160-12 - Discharged at Greenbush, NY, on 4 May 1815.

Jarvis, William H. - Corporal - US Volunteers - Company: Stephen Moore - Enlistment date: 8 Sep 1812 - Period: 1 Yr.

Jasper, Peter - Recruit - 26th Infantry - Company: William Bezeau - Race: 18 - Age: 18 - Height: 5' 5" - Born: Maryland - Trade: Laborer - Enlistment date: 21 Jan 1815 - Place: Philadelphia - Period: War - Listed on the Descriptive Roll of Colored Men, 1 Apr 1815; deserted on 27 Jan 1815.

Jay, David - Private - 38th Infantry - Company: John Rothrock - BLW 8325-160-12.

Jefferson, Hamilton - Private - 36th Infantry - Company: Henry Redman - Pension: Wife Nancy, SO-7875, WO-4345, WC-34338.

Jefferson, Jeremiah - Private - 38th Infantry - Company: John Mowton - Age: 40 - Height: 5' 9" - Born: Calvert, MD - Trade: Laborer - Enlistment date: 7 Apr 1814 - Place: Alexandria, DC - Period: War - BLW 486-160-12 - Discharged at Craney Island, VA, on 15 Mar 1815.

Jeffrey, Thomas - Private - 36th Infantry - Company: Thomas Carbery - Age: 36 - Height: 6' - Born: St. Mary's, MD - Trade: Brick maker - Enlistment date: 23 Apr 1814 - Place: Leonardstown, MD - Period: War - BLW 7353-160-12 - Discharged at Baltimore on 31 Mar 1815.

Jeffries, James - Private - 12th Infantry - Company: James Charlton - Other regiment: 8th Infantry (New) - Age: 34 - Height: 5' 6" - Born: Maryland - Trade: Brick maker - Enlistment date: 5 May 1812 - Place: Alexandria, DC - Period: 5 Yrs - Discharged at Fort Clark, IL, on 6 May 1817.

Jeffries, Robert - Private - 38th Infantry - Company: Isaac Aldridge - Age: 25 - Height: 5' 8" - Born: Westmoreland, VA - Trade: Farmer - Enlistment date: 28 Dec 1814 - Place: Baltimore - Period: War - BLW 19-320-12 - Discharged at Baltimore on 6 Apr 1815.

Jeggey, James - Servant - 36th Infantry - Company: Joseph Hook.

Jellers, John G. - Private - 13th Infantry - Company: William Adams - Age: 27 - Height: 5' 2" - Born: Baltimore - Trade: Baker - Enlistment date: 8 Mar 1814 - Period: 5 Yrs - Died at Sackets Harbor, NY, on 19 Sep 1814 from fever.

Maryland Regulars in War of 1812

Jenkins, David - Private - 38th Infantry - Company: Shepperd Leakin - Age: 37 - Height: 5' 7" - Born: Chester, MD - Trade: Mason - Enlistment date: 26 Mar 1814 - Period: War - Deserted at Ellicott's Mills, MD, on 29 Apr 1814.

Jenkins, Elias - Private - 7th Infantry - Company: Walter Overton - Age: 27 - Height: 6' - Born: Maryland - Trade: Farmer - Enlistment date: 16 Nov 1808 - Period: 5 Yrs - Discharged on 15 Nov 1813.

Jenkins, James - Corporal - 2nd Infantry - Company: Mathew Arbuckle - Age: 29 - Height: 5' 8" - Born: England - Trade: Nailor - Enlistment date: 5 Jul 1807 - Place: Fredericktown, MD - Period: 5 Yrs.

Jenkins, Job - Private - 2nd Infantry - Company: Mathew Arbuckle - Other regiment: 1st Infantry (New) - Age: 24 - Height: 5' 6" - Born: Frederick Cty, MD - Trade: Farmer - Enlistment date: 14 Aug 1807 - Place: Fredericktown, MD - Period: 5 Yrs - Re-enlisted at Fort Stoddard on 14 May 1812 for five years; deserted from New Orleans on 18 Mar 1816.

Jenkins, John - Private - 36th Infantry - Company: Joseph Hook - Age: 25 - Height: 5' 7" - Born: Culpeper, VA - Enlistment date: 11 Apr 1814 - Place: Fort Washington, MD - Period: War - Deserted on 30 Apr 1814.

Jenkins, Richard - Private - 20th Infantry - Company: Henry Lewis - Age: 21 - Height: 5' 6 1/2" - Born: Charles Cty, MD - Trade: Farmer - Enlistment date: 13 Dec 1814 - Place: Alexandria, DC - Period: War - Pension: Land bounty to Charlotte Jenkins, daughter and sole heir at law of Richard Jenkins - BLW 1083-320-12 - Discharged at Fredericksburg, VA, on 15 Mar 1815.

Jenkins, Thomas - Private - 1st Artillery - Company: George Armistead - Age: 35 - Height: 5' 10" - Born: Maryland - Trade: Laborer - Enlistment date: 15 Feb 1810 - Place: Fort McHenry - Period: 5 Yrs - Discharged on 14 Feb 1815.

Jenkinson, George - Private - 4th Rifles - Company: Matthew Magee - Other regiment: Rifles - Age: 20 - Height: 5' 7 1/4" - Born: Washington Cty, PA - Trade: Farmer - Enlistment date: 25 Sep 1814 - Place: Bedford, PA - Period: 5 Yrs - BLW 23184-160-12 - Discharged at Martin's Cantonment on 25 Sep 1819.

Jenkinson, William - Private - 32nd Infantry - Company: Robert Patterson - Age: 19 - Height: 5' 6" - Born: Frederick Cty, MD - Trade: Laborer - Enlistment date: 4 Apr 1814 - Place: Bedford, PA - Period: War - BLW 1530-160-12 - Discharged at Philadelphia on 19 May 1815.

Jennett, Francis - Private - 38th Infantry - Company: Isaac Aldridge - Age: 16 - Height: 5' 2 1/2" - Born: Aux Cayes, St. Domingo (Haiti) - Trade: Boat maker - Enlistment date: 6 Aug 1814 - Place: Baltimore - Period: War - Pension: Wife Elizabeth, WO-32079, WC-22279 - BLW 20435-160-12 - Discharged at Baltimore on 5 Apr 1815.

Jenny, John - Private - 14th Infantry - Company: David Cummings - Enlistment date: 20 May 1812 - Period: 5 Yrs - Prisoner of War (Halifax), sent to England; deserted at Baltimore on 20 Apr 1815.

Jerome, James - Private - 3rd Infantry - Company: Matthew Arbuckle - Age: 27 - Height: 5' 7 1/2" - Born: Maryland - Trade: Shoemaker - Enlistment date: 7 Jan 1815 - Period: War - Discharged at New Orleans on 9 Apr 1815.

By War of 1812 Soc. in MD

Jerrard, Peter (Gerroad) - Private - 36th Infantry - Company: Joseph Hook - Age: 28 - Height: 5' 6 1/2" - Born: Paris, France - Trade: Currier - Enlistment date: 14 Jun 1814 - Place: Georgetown, DC - Period: War - Discharged on 30 Mar 1815.

Jobson, William - Private - 2nd Artillery - Company: Nathan Towson - Enlistment date: 1 Jul 1812 - Period: 5 Yrs - Deserted on 4 Aug 1812.

Johns, Jonathan - Private - 38th Infantry - Company: Isaac Aldridge - BLW 14806-160-12.

Johnson, Abel - Private - 17th Infantry - Company: Harris Hickman - Other regiment: 3rd Infantry (New) - Age: 29 - Height: 5' 6" - Born: Maryland - Trade: Farmer - Enlistment date: 1 May 1814 - Period: 5 Yrs - Discharged at Fort Dearborn, IL, on 1 May 1819.

Johnson, Abraham - Servant - 36th Infantry - Company: Joseph Hook.

Johnson, Andrew - Private - 3rd Artillery - Company: Horace Watson - Age: 30 - Height: 5' 10" - Born: Baltimore - Trade: Seaman - Enlistment date: 1 Feb 1814 - Place: New York - Period: War - Deserted on 29 Nov 1814.

Johnson, Burton - Private - 14th Infantry - Company: Thomas Montgomery - Enlistment date: 30 Dec 1812 - Period: 18 Mos - Discharged at Plattsburgh, NY, on 30 Jun 1814.

Johnson, Eli - Private - 26th Infantry - Company: William Bezeau - Age: 21 - Height: 5' 4" - Born: Montgomery Cty, MD - Trade: Tailor - Enlistment date: 8 Sep 1814 - Place: Baltimore - Period: War - BLW 19027-160-12 - Discharged at Philadelphia on 28 Mar 1815.

Johnson, Elisha - Private - 14th Infantry - Company: Reuben Gilder - Enlistment date: 4 Jan 1813 - Period: 18 Mos - Discharged at Plattsburgh, NY, on 4 Jul 1814.

Johnson, George - Third Lieutenant - 36th Infantry - Other regiment: 4th Rifles - Commissioned as an ensign, 36th Infantry, on 30 Apr 1813; promoted to 3rd lieutenant on 21 Feb 1814; transferred to 4th Rifles on 31 May 1814; discharged on 15 Jun 1815.

Johnson, George - Private - 5th Infantry - Company: Richard Whartenby - Other regiment: 3rd Infantry (New) - Age: 20 - Height: 5' 9 1/2" - Born: Baltimore - Trade: Cabinet maker - Enlistment date: 17 Mar 1811 - Place: Baltimore - Period: 5 Yrs - Deserted on 20 Aug 1815.

Johnson, Henry - Private - 36th Infantry - Company: Samuel Raisin - Age: 30 - Height: 5' 9" - Born: England - Trade: Farmer - Enlistment date: 3 Oct 1814 - Place: Fredericktown, MD - Period: War - BLW 7734-160-12 - Discharged at Washington, DC, on 20 Mar 1815.

Johnson, Henry - Seaman - US Sea Fencibles - Company: Simmones Bunbury - Age: 35 - Height: 5' 8" - Born: Little York, PA - Enlistment date: 6 Jan 1814 - Period: 1 Yr.

Johnson, Henry - Private - 36th Infantry - Company: Joseph Merrick - Age: 19 - Height: 5' 7" - Born: Baltimore - Enlistment date: 2 Aug 1813 - Place: Havre de Grace, MD - Period: 1 Yr - Discharged on 19 Dec 1813.

Johnson, Henry - Private - 38th Infantry - Company: James Hook - Age: 19 - Height: 5' 8" - Born: Baltimore - Trade: Cooper - Enlistment date: 20 Dec

Maryland Regulars in War of 1812

1814 - Place: Baltimore - Period: 5 Yrs.

Johnson, Hezekiah - Private - 17th Infantry - Company: John Chunn - Other regiment: 3rd Infantry (New) - Age: 33 - Height: 5' 6" - Born: Cecil Cty, MD - Trade: Laborer - Enlistment date: 27 Feb 1813 - Place: Newark, OH - Period: 5 Yrs - Discharged on 2 Feb 1816 on Surgeon's Certificate.

Johnson, Hezekiah - Captain - 1st Infantry - Born: Maryland - Pension: Land bounty to Catharine H. T. Johnson, widow of Hezekiah Johnson - BLW 20332-160-50 - Commissioned as an ensign, 1st Infantry, on 26 Mar 1804; promoted to 2nd lieutenant on 20 Jun 1806; promoted to 1st lieutenant on 18 Aug 1808; promoted to captain on 20 Jan 1813; discharged on 16 Jun 1818; died on 8 Sep 1837.

Johnson, Isaac - Private - 14th Infantry - Prisoner of War (Halifax), discharged from Halifax on 3 Feb 1814.

Johnson, James - Private - 14th Infantry - Company: Thomas Montgomery - Other regiment: 4th Infantry (New) - Age: 28 - Height: 5' 9" - Born: New Jersey - Trade: Laborer - Enlistment date: 1 Dec 1812 - Place: Baltimore - Period: 5 Yrs - Prisoner of War, exchanged on 11 May 1814; deserted on 13 Jul 1815.

Johnson, James - Private - 14th Infantry - Company: James McDonald - Other regiment: 4th Infantry (New) - Age: 26 - Height: 5' 8" - Born: St. Mary's, MD - Trade: Cooper - Enlistment date: 25 Jan 1813 - Period: 5 Yrs - Discharged on 24 Jan 1818.

Johnson, James - Private - 38th Infantry - Company: James Hook - Age: 36 - Height: 5' 10" - Born: Philadelphia - Trade: Carpenter - Enlistment date: 21 Mar 1814 - Period: War - BLW 517-160-12 - Discharged at Baltimore on 6 Apr 1815.

Johnson, James - Recruit - 26th Infantry - Company: William Bezeau - Race: 35 - Age: 35 - Height: 5' 9" - Born: Maryland - Trade: Laborer - Enlistment date: 24 Nov 1814 - Place: Philadelphia - Period: War - BLW 10293-160-12 - Listed as "Col'd" in his service record; discharged on 20 Mar 1815 at Philadelphia.

Johnson, James - Private - 14th Infantry - Company: David Cummings - Age: 19 - Height: 5' 5" - Born: Baltimore - Trade: Shoemaker - Enlistment date: 6 Nov 1814 - Place: Snowden's Camp, VA - Period: 5 Yrs.

Johnson, Jason - Private - 14th Infantry - Company: Isaac Barnard - Age: 30 - Height: 5' 8 3/4" - Enlistment date: 5 Jan 1813 - Period: 18 Mos - Prisoner of War (Halifax), captured at Stoney Creek, UC, on 12 Jun 1813; discharged on 7 Aug 1814.

Johnson, Jeremiah (Johnston) - Private - 19th Infantry - Company: George Kesling - Other regiment: 17th Infantry - Age: 23 - Height: 5' 9" - Born: Allegany Cty, MD - Trade: Farmer - Enlistment date: 10 Apr 1814 - Place: Chillicothe, OH - Period: War - BLW 8813-160-12 - Accidentally wounded after enlistment and never joined his regiment; discharged on 16 Feb 1815.

Johnson, John - Private - 19th Infantry - Company: Carey Trimble - Other regiment: 17th Infantry - Age: 23 - Height: 6' - Born: Maryland - Trade: Shoemaker - Enlistment date: 14 Mar 1814 - Place: Newark, OH - Period:

By War of 1812 Soc. in MD

War - BLW 7416-160-12 - Discharged at Chillicothe, OH, on 6 Jun 1815.

Johnson, John - Private - 5th Infantry - Company: John Clarke - Age: 37 - Height: 5' 7" - Born: Georgetown, MD - Trade: Shoemaker - Enlistment date: 10 Jun 1814 - Place: Millersburgh, PA - Period: 5 Yrs.

Johnson, John - Private - 14th Infantry - Company: James McDonald - Age: 45 - Height: 5' 11" - Born: Cecil Cty, MD - Trade: Laborer - Enlistment date: 28 Oct 1813 - Period: 5 Yrs - Discharged on 28 Oct 1818.

Johnson, John - Sergeant - 14th Infantry - Company: Kenneth McKenzie - Age: 27 - Height: 5' 7" - Born: Germany - Enlistment date: 23 Nov 1812 - Period: 18 Mos - Discharged at Black Rock, NY, on 1 Nov 1814.

Johnson, John - Drummer - 36th Infantry - Company: Thomas Carbery - Age: 17 - Height: 5' 2" - Born: St. Mary's, MD - Trade: Turner - Enlistment date: 23 Mar 1814 - Place: Leonardstown, MD - Period: War - BLW 1789-160-12 - Discharged at Baltimore on 31 Mar 1815.

Johnson, John - Private - 36th Infantry - Company: Samuel Raisin - Enlistment date: 1 Nov 1814 - Period: 5 Yrs - Died in barracks on 22 Jan 1815.

Johnson, John - Private - 14th Infantry - Company: Richard Arell - Age: 19 - Height: 5' 9" - Born: Somerset Cty, MD - Trade: Farmer - Enlistment date: 15 Sep 1812 - Place: Snow Hill, MD - Period: 5 Yrs - Deserted at Fort Severn, MD, on 12 or 14 Jul 1815.

Johnson, Joseph - Private - 2nd Infantry - Company: James Wilkinson - Age: 30 - Height: 5' 11" - Born: Frederick Cty, MD - Trade: Laborer - Enlistment date: 9 Apr 1819 - Place: Fort Stoddert, AL - Period: 5 Yrs - Discharged at New Orleans on 10 Apr 1815.

Johnson, Noble - Private - 14th Infantry - Company: Clement Sullivan - Other regiment: 4th Infantry (New) - Age: 20 - Height: 5' 8 1/2" - Born: Dorchester Cty, MD - Trade: Farmer - Enlistment date: 9 May 1812 - Place: Cambridge, MD - Period: 5 Yrs - Pension: Wife Matilda, WO-42378 - Prisoner of War (Halifax), captured at Beaver Dams on 24 Jun 1813; discharged on 9 May 1817.

Johnson, Peter W. - Private - 38th Infantry - Company: Shepperd Leakin - Age: 28 - Height: 5' 10" - Born: Baltimore City - Trade: Carpenter - Enlistment date: 25 Mar 1814 - Period: War - Pension: Old War IF-25187 - Discharged on 23 Nov 1814 on Surgeon's Certificate for a rupture.

Johnson, Samuel - Private - 17th Infantry - Company: Harris Hickman - Other regiment: 3rd Infantry (New) - Age: 40 - Height: 6' 1" - Born: Maryland - Trade: Blacksmith - Enlistment date: 16 May 1814 - Period: 5 Yrs - BLW 22289-160-12 - Discharged at Fort Dearborn, IL, on 16 May 1819.

Johnson, Thomas - Private - 38th Infantry - Company: Charles Stansbury - Age: 22 - Height: 5' 3" - Born: District of Columbia - Trade: Seaman - Enlistment date: 12 Oct 1814 - Place: Baltimore - Period: War - BLW 5495-160-12 - Discharged at Baltimore on 6 Apr 1815.

Johnson, Truman - Private - 2nd Artillery - Company: Nathan Towson - Age: 29 - Prisoner of War (Quebec), captured during the Battle of Stoney Creek and released on 31 Oct 1813.

Johnson, William - Corporal - 5th Infantry - Company: William McDonald -

Other regiment: 3rd Infantry (New) - Enlistment date: 8 Mar 1811 - Place: Baltimore - Period: 5 Yrs - Prisoner of War on parole; discharged on 3 Mar 1816.

Johnson, William - Private - 26th Infantry - Company: John Levakes - Other regiment: Light Dragoons - Age: 24 - Height: 5' 9 1/2" - Born: Montgomery Cty, MD - Trade: Farmer - Enlistment date: 1 Feb 1815 - Place: Peru - Period: 5 Yrs - Transferred to Light Dragoons; discharged on 1 Nov 1819 and re-enlisted.

Johnson, William - Private - 16th Infantry - Company: Miles Greenwood - Age: 18 - Height: 5' 9" - Born: Philadelphia - Enlistment date: 7 Jul 1812 - Place: Baltimore - Period: 5 Yrs - Discharged on 1 Apr 1815, a minor who enlisted without the consent of parents.

Johnson, William - Private - 14th Infantry - Company: Thomas Montgomery - Other regiment: 4th Infantry (New) - Age: 31 - Height: 5' 9" - Born: Fairfax, VA - Trade: Farmer - Enlistment date: 15 Dec 1812 - Place: Alexandria, DC - Period: 5 Yrs - Deserted on 6 Aug 1815.

Johnson, William - Corporal - 1st Artillery - Company: James Reed - Age: 26 - Height: 5' 6" - Born: Annapolis - Trade: Tailor - Enlistment date: 28 Aug 1814 - Place: Annapolis - Period: 5 Yrs - Discharged on 28 Aug 1819.

Johnson, William - Private - 14th Infantry - Prisoner of War (Halifax), captured at Beaver Dams on 24 Jun 1813, discharged from Halifax on 3 Feb 1813.

Johnson, William - Sergeant - 14th Infantry - Company: Thomas Montgomery - Age: 25 - Height: 5' 2 1/2" - Born: Belfast, Ireland - Trade: Shoemaker - Enlistment date: 29 May 1813 - Place: Baltimore - Period: War - BLW 4451-160-12 - Discharged at Greenbush, NY, on 4 May 1815.

Johnston, Albert - Private - 10th Infantry - Company: George Strother - Age: 28 - Height: 5' 10" - Born: Montgomery Cty, MD - Trade: Farmer - Enlistment date: 19 Jul 1814 - Place: Wilkesboro, NC - Period: War.

Johnston, John W. - Private - 2nd Light Dragoons - Company: William Littlejohn - Other regiment: Corps of Artillery - Age: 48 - Height: 5' 9 1/2" - Born: Queen Anne's Cty, MD - Trade: Farmer - Enlistment date: 7 Sep 1813 - Period: 5 Yrs - Discharged on 24 Jun 1815 as an invalid.

Johnston, Joshua - Private - 14th Infantry - Company: James McDonald - Age: 25 - Height: 5' 10" - Born: Maryland - Enlistment date: Aug 1813 - Period: War - Drowned in May 1814.

Johnston, William - Sergeant - 14th Infantry - Company: Thomas Montgomery - Pension: Wife Anna, SO-21527, SC-10910, WO-20692, WC-14833 - BLW 4451-160-12.

Johnston, William - Private - 17th Infantry - Company: John Chunn - Age: 26 - Height: 5' 8 1/2" - Born: Maryland - Trade: Farmer - Enlistment date: 13 Nov 1813 - Place: Chillicothe, OH - Period: War - BLW 19600-160-12 - Discharged at Chillicothe, OH, on 7 Jun 1815.

Johnston, William - Private - 3rd Artillery - Company: James House - Height: 5' 9" - Born: St. Mary's, MD - Trade: Wheelwright - Enlistment date: 1806 - Re-enlisted on 19 Aug 1811 for five years; deserted at North Battery or

By War of 1812 Soc. in MD

Eastern Point on 15 Oct 1812.

Joice, Asia - Private - 38th Infantry - Company: John Rothrock - BLW 12519-160-12.

Joice, Richard - Private - 2nd Infantry - Company: Perrin Willis - Age: 32 - Height: 6' - Born: Prince George's Cty, MD - Trade: Carpenter - Enlistment date: 6 Jan 1815 - Place: Washington, DC - Period: 5 Yrs.

Jones, Andrew - Private - 35th Infantry - Company: Benjamin Hardaway - Age: 21 - Height: 5' 9" - Born: Caroline Cty, MD - Trade: Farmer - Enlistment date: 25 Apr 1814 - Place: Norfolk, VA - Period: War - Discharged at Norfolk, VA, on 15 Mar 1815.

Jones, Bartholomew - Private - Light Dragoons - Other regiment: Corps of Artillery - Age: 42 - Height: 5' 2 1/2" - Born: Centreville, MD - Trade: Shoemaker - Enlistment date: 24 Apr 1813 - Place: Centreville, MD - Period: 5 Yrs - Transferred to Corps of Artillery on 26 Jul 1814; discharged on 23 Apr 1816 after furnishing a substitute.

Jones, Benjamin - Private - 36th Infantry - Company: Joseph Hook - Age: 24 - Height: 5' 3" - Born: Baltimore - Enlistment date: 7 May 1813 - Place: Baltimore - Period: 1 Yr.

Jones, Benjamin - Private - 38th Infantry - Company: John Rothrock - Enlistment date: 16 Aug 1813 - Period: 1 Yr - Discharged at Craney Island, VA, on 10 Oct 1814.

Jones, Benjamin - Private - 14th Infantry - Company: Reuben Gilder - Other regiment: 4th Infantry (New) - Age: 19 - Height: 5' 2" - Born: Maryland - Trade: Farmer - Enlistment date: 31 Aug 1812 - Place: Salisbury, MD - Period: 5 Yrs - Deserted on 23 Aug 1815.

Jones, Benjamin - Private - 7th Infantry - Company: Walter Overton - Age: 43 - Height: 5' 7" - Born: Maryland - Trade: Farmer - Enlistment date: 3 Mar 1813 - Place: Fort Claiborne, AL - Period: War - Discharged at Fort Claiborne, AL, on 30 Apr 1815.

Jones, Caleb - Private - 5th Infantry - Company: James Dorman - Other regiment: 3rd Infantry (New) - Age: 22 - Height: 5' 7" - Born: Maryland - Trade: Cooper - Enlistment date: 4 Aug 1812 - Place: Baltimore - Period: 5 Yrs - BLW 16692-160-12 - Discharged at Fort Mackinaw, MI, on 3 Aug 1817.

Jones, Elisha - Private - 36th Infantry - BLW 21329-160-12.

Jones, Enoch - Private - 2nd Artillery - Company: Nathan Towson - Other regiment: Corps of Artillery - Age: 29 - Height: 5' 6 1/2" - Born: Baltimore - Trade: Shoemaker - Enlistment date: 4 May 1812 - Place: Baltimore - Period: 5 Yrs - Prisoner of War (Quebec), captured during the Battle of Stoney Creek and released on 31 Oct 1813; discharged on 30 Jun 1817.

Jones, Henry - Private - 38th Infantry - Company: John Brookes - Age: 18 - Height: 5' 2 1/2" - Born: Somerset Cty, MD - Trade: Laborer - Enlistment date: 15 Apr 1814 - Place: Norfolk, VA - Period: War.

Jones, James - Fifer - 32nd Infantry - Company: William Gates - Age: 40 - Height: 5' 4" - Born: Prince George's Cty, MD - Trade: Farmer - Enlistment date: 21 Nov 1813 - Place: Wilkesboro, PA - Period: 5 Yrs - Discharged on

29 Aug 1817 for inability.

Jones, James - Private - 14th Infantry - Company: James Britton - On board the fleet on Lake Ontario.

Jones, James - Private - 1st Artillery - Company: Heman Fay - Age: 26 - Height: 5' 6" - Born: New Jersey - Trade: Laborer - Enlistment date: 21 Apr 1814 - Place: Annapolis - Period: 5 Yrs - Pension: Land bounty to William Jones, father and heir at law of James Jones - BLW 14338-160-12 - Died on 8 Mar 1815.

Jones, John - Private - 14th Infantry - Company: Reuben Gilder - Other regiment: 4th Infantry (New) - Enlistment date: 31 Dec 1812 - Period: 5 Yrs - On board the fleet on Lake Champlain; deserted at Fort Severn on 14 Jul 1815.

Jones, John - Private - 36th Infantry - Age: 23 - Height: 5' 7" - Born: Norwalk, CT - Enlistment date: 20 May 1813 - Place: Baltimore.

Jones, John - Private - 38th Infantry - Age: 21 - Height: 5' 2" - Born: Baltimore - Trade: Blacksmith - Enlistment date: 4 Dec 1814 - Period: War - Deserted on 6 Dec 1814.

Jones, John - Private - 36th Infantry - Company: Joseph Hook - Age: 24 - Height: 5' 6" - Born: Frederick, MD - Enlistment date: 10 May 1813 - Place: Westminster, MD - Period: 1 Yr.

Jones, Jonathan - Private - 38th Infantry - Age: 28 - Height: 5' 9" - Born: Bucks Cty, PA - Trade: Blacksmith - Enlistment date: 2 Nov 1814 - Place: Baltimore - Period: War - Discharged at Baltimore on 6 Apr 1815.

Jones, Joshua - Private - 12th Infantry - Company: Andrew Madison - Age: 39 - Height: 5' 8" - Born: Baltimore Cty - Trade: Laborer - Enlistment date: 16 Nov 1812 - Place: Morgantown, VA - Period: 5 Yrs - Died at Sackets Harbor, NY, on 17 Nov 1813.

Jones, Levin - Second Lieutenant - 21st Infantry - Company: Sullivan Burbank - Age: 17 - Height: 5' 6 1/2" - Born: Salisbury, MD - Trade: Sailor - Enlistment date: 16 May 1812 - Place: Boston - Period: 5 Yrs - BLW 1066-160-50 - Promoted to sergeant on 1 Nov 1814, to sergeant major on 23 Dec 1813, commissioned at an Ensign on 24 Jun 1814; promoted to 2nd Lieutenant on 1 Oct 1814; discharged at Sackets Harbor, NY, on 16 May 1815.

Jones, Levin - Private - 35th Infantry - Company: Meriwether Taliaferros - Age: 37 - Height: 5' 8 1/2" - Born: Prince George's Cty, MD - Trade: Farmer - Enlistment date: 21 Jun 1814 - Place: Norfolk, VA - Period: War - Discharged at Norfolk, VA, on 15 Mar 1815.

Jones, Lewis - Private - 38th Infantry - Company: James Smith - Age: 21 - Height: 5' 7" - Born: Anne Arundel Cty, MD - Trade: Shoemaker - Enlistment date: 25 Apr 1814 - Place: Norfolk, VA - Period: War - BLW 14729-160-12 - Discharged on 15 May 1815.

Jones, Richard - Private - 38th Infantry - Company: John Brookes - Pension: Wife Elizabeth, WO-31839, WC-18998.

Jones, Samuel - Private - 8th Infantry - Company: Roswell Johnson - Age: 35 - Height: 5' 7" - Born: Charles Cty, MD - Trade: Blacksmith - Enlistment date: 9 Jun 1814 - Period: 5 Yrs - Discharged at Fort Scott, GA, on 29 May 1819.

Jones, Theophilus - Private - 14th Infantry - Company: David Cummings - Age: 36 - Height: 5' 8" - Born: Pennsylvania - Trade: Farmer - Enlistment date: 15 Oct 1814 - Place: Baltimore - Period: 5 Yrs.

Jones, Thomas - Private - 4th Rifles - Company: Joseph Kean - Age: 39 - Height: 6' - Born: Maryland - Trade: Laborer - Enlistment date: 8 Aug 1814 - Place: McConnelstown - Period: 5 Yrs - Died on 8 Aug 1815.

Jones, Thomas - Private - US Volunteers - Company: Stephen Moore - Enlistment date: 8 Sep 1812 - Period: 1 Yr.

Jones, Thomas - Private - 19th Infantry - Company: William Gill - Other regiment: 17th Infantry - Age: 30 - Height: 5' 9" - Born: Maryland - Trade: Farmer - Enlistment date: 10 May 1814 - Place: Chillicothe, OH - Period: War - Pension: SO-19177, SC-19177 - BLW 1969-160-12 - Discharged at Chillicothe, OH, on 5 Jun 1815.

Jones, Thomas R. - Private - 36th Infantry - Company: Thomas Carbery - Age: 22 - Height: 5' 10" - Born: Prince George's Cty, MD - Trade: Farmer - Enlistment date: 3 May 1814 - Place: Leonardstown, MD - Period: War - BLW 551-160-12 - Discharged at Baltimore on 31 Mar 1815.

Jones, William - Private - 38th Infantry - Company: Isaac Aldridge - BLW 4658-160-12.

Jones, William - First Lieutenant - 38th Infantry - Commissioned as a 2nd lieutenant on 20 May 1813; promoted to 1st lieutenant on 1 May 1814; discharged on 15 Jun 1815.

Jones, William - Private - Unknown - Age: 20 - Born: Baltimore - Prisoner of War (Quebec), captured during the Battle of Stoney Creek and released on 31 Oct 1813.

Jones, William - Seaman - US Sea Fencibles - Company: Simmones Bunbury - Age: 23 - Height: 5' 7" - Born: Norfolk, VA - Enlistment date: 2 Nov 1813 - Period: 1 Yr - Discharged on 2 Nov 1814.

Jones, William - Private - 32nd Infantry - Company: John Steele - Age: 24 - Height: 5' 5" - Born: Montgomery Cty, MD - Trade: Farmer - Enlistment date: 5 Dec 1813 - Period: 1 Yr - Re-enlisted on 29 Mar 1814 for the war; discharged on 11 May 1815.

Jones, William - Sergeant - 14th Infantry - Company: David Cummings - Age: 25 - Height: 5' 11" - Born: Caswell, NC - Trade: Pilot - Enlistment date: 24 Jun 1812 - Place: Fort Cumberland, MD - Period: 5 Yrs - Prisoner of War (Halifax), escaped from Halifax on 27 Sep 1813; discharged in Mar 1815.

Jones, William - Private - 14th Infantry - Company: Samuel Lane - Age: 31 - Height: 5' 10" - Enlistment date: 15 May 1812 - Place: Shephardstown - Period: 5 Yrs - Pension: Land bounty to Joanna J. Mills, Martha H. Richards and Letitia Harvey, the only heirs at law of William L. Jones - BLW 27268-160-12 - Killed in camp near Buffalo, NY, on 11 Oct 1812.

Jones, William - Private - 12th Infantry - Company: Thomas Sangster - Age: 24 - Height: 5' 5" - Born: Maryland - Trade: Farmer - Enlistment date: 15 May 1812 - Place: Alexandria, DC - Period: 18 Mos - Discharged at Albany,

Maryland Regulars in War of 1812

NY, on 15 Nov 1813.

Jones, William - Private - 8th Infantry - Age: 19 - Height: 5' 4" - Born: Maryland - Trade: Farmer - Enlistment date: 1 Dec 1812 - Period: 5 Yrs - Discharged on 21 Feb 1817 for wounds received in action with Indians near Rock Island, IL.

Jones, William - Artificer - 2nd Artillery - Company: Nathan Towson - Age: 27 - Height: 5' 8 1/2" - Born: Baltimore - Trade: Wheelwright - Enlistment date: 25 Apr 1812 - Period: 5 Yrs - Discharged at Fort McHenry on 25 Apr 1817.

Jones, William - Hospital Surgeon's Mate - Medical Department - Commissioned as a hospital surgeon's mate on 2 Jul 1813; discharged on 15 Jun 1815.

Jones, William - Private - 5th Infantry - Company: James Dorman - Other regiment: 8th Infantry (New) - Age: 21 - Height: 5' 6 1/2" - Born: Maryland - Trade: Cotton spinner - Enlistment date: 4 Aug 1812 - Place: Baltimore - Period: 5 Yrs - Prisoner of War, captured on Lake Ontario; deserted at Pittsburgh, PA, on 18 Aug 1815.

Jones, William L. - Sergeant - 14th Infantry - Enlistment date: 5 Jan 1813 - Period: War - Pension: Land bounty to Joanna J. Mills, Martha H. Richards and Leitita Harvey, the only heirs at law of William L. Jones - BLW 27268-160-12 - Prisoner of War, sent to England; discharged on 31 Mar 1815.

Jones, Zachariah - Private - 20th Infantry - Company: Bernard Peyton - Other regiment: 35th Infantry - Age: 27 - Height: 5' 7" - Born: Frederick, MD - Trade: Shoemaker - Enlistment date: 4 Jul 1813 - Period: 5 Yrs - Pension: Land bounty to Thomas Jones, brother and other heirs at law of Zachariah Jones - BLW 21729-160-12 - Transferred to 35th Infantry on 20 Mar 1815; died on 30 Jun 1815.

Jonquin, Anthony - Private - 14th Infantry - Company: David Cummings - Age: 22 - Height: 5' 1" - Born: South America - Trade: Seaman - Enlistment date: 5 Dec 1814 - Place: Salisbury, MD - Period: War - BLW 4623-160-12 - Discharged at Baltimore on 16 Mar 1815.

Jonquin, Antony - Private - 14th Infantry - BLW 4623-160-12.

Jordan, Abraham - Private - 14th Infantry - Company: David Cummings - Age: 23 - Height: 5' 5 1/2" - Born: Berkley, VA - Trade: Laborer - Enlistment date: 30 Nov 1814 - Place: Williamsport, MD - Period: War - BLW 18834-160-12 - Discharged at Baltimore on 16 Mar 1815.

Jordan, James - Private - 7th Infantry - Company: Samuel Vail - Other regiment: 1st Infantry (New) - Age: 28 - Height: 5' 8" - Born: Baltimore - Trade: Baker - Enlistment date: 28 Dec 1811 - Place: Pittsburgh - Period: 5 Yrs - Discharged at Pass Christian on 23 Feb 1816.

Jordan, Jeremiah - Private - 36th Infantry - Company: Samuel Raisin - Age: 22 - Height: 5' 6" - Born: Stafford, VA - Trade: Farmer - Enlistment date: 9 Jun 1814 - Place: Richmond, VA - Period: War - BLW 9370-160-12 - Discharged at Washington, DC, on 20 Mar 1815.

Jordan, Joel - Private - 14th Infantry - Company: Joseph Marshall - Age: 28 - Height: 6' - Born: Chester, MD - Trade: Stone mason - Enlistment date: 19

By War of 1812 Soc. in MD

Jun 1813 - Place: Baltimore - Period: War - BLW 545-160-12 - Discharged at Greenbush, NY, on 3 May 1815.

Jordan, John - Private - 14th Infantry - Company: James McDonald - Age: 25 - Height: 5' 6" - Born: Ireland - Enlistment date: 21 Feb 1814 - Period: War - Killed during the Battle of Lyon's Creek on 19 Oct 1814.

Jordan, Samuel - Quarter Gunner - US Sea Fencibles - Company: William Addison - Pension: Old War IF-25186.

Jordan, William - Private - 36th Infantry - Company: Samuel Raisin - Height: 5' 9" - Born: Henrico Cty, VA - Enlistment date: 30 May 1814 - Place: Richmond, VA - Period: War - Pension: Heirs received half pay for five years in lieu of land bounty - Died in barracks on 14 Jan 1815.

Jordan, William - Sergeant - 38th Infantry - Company: James Smith - Age: 31 - Height: 5' 8 3/4" - Born: Harford Cty, MD - Trade: Tailor - Enlistment date: 8 Mar 1814 - Place: Craney Island, VA - Period: War - BLW 12559-160-12 - Discharged at Craney Island, VA, on 15 Mar 1815.

Jordon, George - Private - Light Artillery - Company: Luther Leonard - Other regiment: Corps of Artillery - Age: 17 - Height: 5' 9" - Born: St. Mary's, MD - Trade: Plasterer - Enlistment date: 31 May 1812 - Place: Washington, DC - Period: 5 Yrs - BLW 13785-160-12 - Discharged at Fort George, Castino, on 31 May 1817.

Jordon, Robert - Private - 38th Infantry - Pension: Heirs received half pay for five years in lieu of land bounty - Died on 22 Feb 1815.

Joseph, John - Private - 12th Infantry - Company: James Charlton - Age: 35 - Height: 5' 8" - Born: Bordeaux, France - Trade: Seaman - Enlistment date: 30 Jun 1812 - Place: Baltimore - Period: 5 Yrs - Prisoner of War (Halifax), captured at Stoney Creek, UC, on 12 Jun 1813; discharged at Buffalo, NY, on 10 Jun 1815, lost of right eye.

Joshihorn, Joseph - Private - 5th Infantry - Age: 33 - Height: 5' 9" - Born: Virginia - Trade: Shoemaker - Enlistment date: 6 Feb 1813 - Place: Baltimore - Period: 5 Yrs - Discharged at Greenbush, NY, on 9 Dec 1813 on Surgeon's Certificate.

Joshinhom, Aquila - Private - 5th Infantry - Company: William Bird - Age: 14 - Height: 4' 4" - Born: Baltimore - Trade: Boy - Enlistment date: 9 Feb 1813 - Place: Baltimore - Period: 5 Yrs.

Journey, John - Private - 6th Infantry - Company: John Walworth - Other regiment: 2nd Infantry (New) - Age: 21 - Height: 5' 8" - Born: Baltimore - Trade: Seaman - Enlistment date: 22 Jan 1814 - Place: New York - Period: 5 Yrs.

Joyce, Abel - Private - 36th Infantry - Company: Thomas Carbery - Age: 32 - Height: 5' 9" - Born: Maryland - Trade: Farmer - Enlistment date: 2 Mar 1814 - Place: Leonardstown, MD - Period: War - Died in regimental hospital on 5 Feb 1815.

Judifine, John - Private - 38th Infantry - Company: James Hook - BLW 11904-160-12.

Jump, Abel - Corporal - 36th Infantry - Company: Joseph Merrick - Age: 22 - Height: 5' 6 3/4" - Born: Milford, Kent Cty, DE - Enlistment date: 12 May

Maryland Regulars in War of 1812

1813 - Place: Centreville, MD - Period: 1 Yr.

K

Kady, Samuel - Sergeant - 14th Infantry - Company: Joseph Marshall - Other regiment: 4th Infantry (New) - Age: 27 - Height: 5' 8" - Born: County Donegal, Ireland - Trade: Cooper - Enlistment date: 5 Sep 1813 - Place: Baltimore - Period: 5 Yrs - Deserted at Fort Severn, MD, on 14 Jul 1815.

Kahill, Patrick - Private - 38th Infantry - Company: Charles Stansbury - Age: 25 - Height: 5' 9 1/2" - Born: County Wicklow, Ireland - Trade: Laborer - Enlistment date: 10 Nov 1814 - Place: Cumberland - Period: War - BLW 141-160-12 - Discharged at Baltimore on 6 Apr 1815.

Kain, James - Private - 12th Infantry - Company: Thomas Sangster - Age: 43 - Height: 5' 7" - Born: Ireland - Trade: Laborer - Enlistment date: 20 Jun 1812 - Place: Elkton, MD - Period: 5 Yrs.

Kain, Mathew - Private - 36th Infantry - Company: Thomas Carbery - Other regiment: 4th Infantry (New) - Enlistment date: 3 Aug 1814 - Place: Georgetown, DC - Period: 5 Yrs - Deserted on 24 Jul 1815.

Kain, William - Private - 22nd Infantry - Age: 30 - Height: 5' 9" - Born: Hagerstown, MD - Trade: Laborer - Enlistment date: 30 May 1814 - Place: Pittsburgh - Period: War - BLW 8358-160-12 - Discharged at Fort Fayette, PA, on 24 Mar 1815.

Kane Jr., Jacob - Seaman - US Sea Fencibles - Company: John Gill - Age: 29 - Height: 6' - Born: Fredericktown, MD - Enlistment date: 8 Feb 1814 - Period: 1 Yr.

Kane, Dennis - Private - 2nd Light Dragoons - Company: William Littlejohn - Age: 23 - Height: 5' 11 1/2" - Born: County Donegal, Ireland - Trade: Laborer - Enlistment date: 13 Jan 1814 - Place: Baltimore - Period: 5 Yrs - Discharged on 24 Jun 1815 as an invalid.

Kane, Jacob - Private - 38th Infantry - Company: James Smith - Age: 20 - Height: 5' 6" - Born: Fredericktown, MD - Trade: Wagoner - Enlistment date: 18 Feb 1814 - Place: Craney Island, VA - Period: War - BLW 18104-160-12 - Discharged at Craney Island, VA, on 15 Mar 1815.

Karmon, Peter - Recruit - 26th Infantry - Company: William Bezeau - Race: 17 - Age: 17 - Height: 5' 4" - Born: Maryland - Trade: Farmer - Enlistment date: 16 Jan 1815 - Place: Philadelphia - Period: War - Listed on the Descriptive Roll of Colored Men, 1 Apr 1815.

Kernigan, James - Private - 36th Infantry - Company: Joseph Hook - Other regiment: 14th Infantry - Age: 34 - Height: 5' 11" - Born: Ireland - Enlistment date: 20 May 1813 - Place: Baltimore - Period: 1 Yr - Enlisted in the 14th Infantry on 23 May 1814 at Baltimore for the war; discharged at Greenbush, NY, on 31 May 1815.

Karney, Thomas - Captain - 14th Infantry - Born: Ireland - Commissioned as a 1st lieutenant on 12 Mar 1812; promoted to captain on 13 May 1813; discharged on 15 Jun 1815; died on 18 Jul 1834.

Karsner, David - Private - 14th Infantry - Company: James McDonald - Other regiment: 4th Infantry (New) - Age: 35 - Height: 6' - Born: Washington, MD

By War of 1812 Soc. in MD

- Trade: Carpenter - Enlistment date: 21 Mar 1814 - Place: Williamsport, MD - Period: 5 Yrs - Deserted at Fort Severn, MD, on 14 Jul 1815.
- Karsner, John - Private - 14th Infantry - Company: Reuben Gilder - Other regiment: 4th Infantry (New) - Age: 24 - Height: 5' 1 1/2" - Born: Bedford, Bedford Cty, PA - Trade: Tailor - Enlistment date: 12 Nov 1813 - Place: Maryland - Period: 5 Yrs - Deserted on 16 Jul 1815.
- Karsner, John - Private - 2nd Artillery - Company: Alexander Williams - Age: 25 - Height: 5' 6" - Born: Maryland - Trade: Shoemaker - Enlistment date: 29 Oct 1813 - Place: Wilmington, DE - Period: War - Deserted on 24 Jan 1814.
- Kavenaugh, John - Sergeant - 14th Infantry - Company: Reuben Gilder - Other regiment: 4th Infantry (New) - Age: 21 - Height: 5' 7 1/2" - Born: Ireland - Trade: Stone cutter - Enlistment date: 13 Mar 1813 - Place: Baltimore - Period: 5 Yrs - Discharged on 12 Mar 1818.
- Kealer, Frederick - Musician - 14th Infantry - Company: Samuel Lane - Age: 22 - Height: 5' 8 1/2" - Born: Maryland - Enlistment date: 9 Jul 1812 - Place: Chambersburg, PA - Period: 18 Mos - Discharged on 9 Jan 1814.
- Kean, Samuel - Private - 12th Infantry - Age: 19 - Height: 5' 3" - Born: Baltimore City - Trade: Laborer - Enlistment date: 15 Nov 1814 - Place: Winchester, VA - Period: War - Discharged at Winchester, VA, on 18 Mar 1815.
- Kearls, Jacob - Private - 38th Infantry - Company: John Rothrock - BLW 7371-160-12.
- Kearnes, John - Private - 2nd Artillery - Company: Nathan Towson - Pension: Placed on the pension rolls on 5 Jul 1816 - Died on 11 Mar 1821.
- Kearnes, John - Private - 2nd Artillery - Company: Nathan Towson - Pension: Old War IF-15794.
- Kearney, John - Private - 38th Infantry - Company: John Rothrock - BLW 12992-160-12.
- Kearns, Jacob - Private - 38th Infantry - Company: John Rothrock - Age: 20 - Height: 5' 8" - Born: Lancaster, MD - Trade: Farmer - Enlistment date: 29 Feb 1814 - Place: Norfolk, VA - Period: War - Discharged at Craney Island, VA, on 15 Mar 1815.
- Kearns, John - Private - 2nd Artillery - Company: Nathan Towson - Age: 37 - Height: 5' 11" - Born: Ireland - Trade: Accountant - Enlistment date: 22 May 1812 - Place: Baltimore - Period: 5 Yrs - BLW 13753-160-12 - Wounded during the Battle of Chippewa, UC, on 5 Jul 1814; discharged at Baltimore on 1 May 1815.
- Kearson, John - Private - 14th Infantry - Age: 33 - Height: 5' 6" - Born: Ireland - Enlistment date: 31 Mar 1809 ?? - Re-enlisted.
- Keaton, John - Private - 14th Infantry - Company: Reuben Gilder - Other regiment: 4th Infantry (New) - Age: 33 - Height: 5' 7 1/2" - Born: Pasquotank, NC - Trade: Farmer - Enlistment date: 28 Aug 1812 - Place: Maryland - Period: 5 Yrs.
- Keefo, Patrick - Private - 36th Infantry - Company: Samuel Raisin - Age: 29 - Height: 5' 7" - Born: Ireland - Trade: Laborer - Enlistment date: 12 Nov

Maryland Regulars in War of 1812

1814 - Place: Frederick, MD - Period: War - Discharged at Washington, DC, on 20 Mar 1815.

Keen, Jesse - Third Lieutenant - 14th Infantry - Commissioned as an ensign on 19 Jul 1813; promoted to 3rd lieutenant on 14 Nov 1813; dismissed on 31 Jan 1814.

Keen, Samuel - Private - 2nd Light Dragoons - Company: Samuel Hopkins - Age: 18 - Height: 5' 4" - Born: Maryland - Enlistment date: 19 Jan 1814 - Place: Philadelphia - Period: 5 Yrs - Discharged at Philadelphia on 30 Apr 1814.

Keener, Melchor - Sergeant - 38th Infantry - Age: 30 - Height: 5' 10" - Born: Baltimore City - Trade: Clerk - Enlistment date: 3 Oct 1814 - Period: War - BLW 13692-160-12 - Discharged at Baltimore on 7 Apr 1815.

Keese, Benjamin - Private - 36th Infantry - Company: Mortimer Hall - Age: 25 - Height: 5' 8 1/2" - Born: Maryland - Enlistment date: 22 May 1813 - Place: Havre de Grace, MD - Period: 1 Yr.

Keeves, William - Private - US Volunteers - Company: Stephen Moore - Enlistment date: 8 Sep 1812 - Period: 1 Yr.

Keho, Patrick - Private - 14th Infantry - Company: David Cummings - Age: 34 - Height: 5' 4" - Born: Wexford, Ireland - Trade: Laborer - Enlistment date: 9 May 1812 - Place: Baltimore - Period: 5 Yrs - BLW 6624-160-12 - Prisoner of War, sent to England; discharged on 20 Apr 1815 on Surgeon's Certificate for ulcerated leg.

Keifleight, Jacob - Private - 5th Infantry - Age: 26 - Height: 5' 5 1/2" - Born: Baltimore - Trade: Butcher - Enlistment date: 23 Mar 1814 - Place: Philadelphia - Period: 5 Yrs.

Keiler, Daniel - Private - 14th Infantry - Company: David Cummings - Age: 28 - Height: 5' 7" - Born: Lancaster, PA - Trade: Farmer - Enlistment date: 24 Nov 1814 - Place: Baltimore - Period: War - BLW 24037-160-12 - Discharged at Baltimore on 16 Mar 1815.

Keiler, Frederick - Musician - 2nd Artillery - Company: Richard Zantzinger - Other regiment: Corps of Artillery - Age: 24 - Height: 5' 9" - Born: Maryland - Trade: Musician - Enlistment date: 5 Feb 1814 - Place: Reading, PA - Period: 5 Yrs - Pension: Land bounty to Jacob Kayler, brother & other heirs at law of Frederick Kaylor, alias Keiler - BLW 21252-160-12 - Died at Fort Mifflin, PA, on 14 Apr 1817.

Keith, Zachariah - Private - 5th Infantry - Company: Colin Buckner - Age: 37 - Height: 5' 10" - Born: Baltimore - Enlistment date: 28 Apr 1813 - Period: 5 Yrs.

Keizer, John - Private - 2nd Light Dragoons - Company: William Littlejohn - Age: 31 - Height: 5' 8" - Born: Baltimore - Trade: Coppersmith - Enlistment date: 13 Feb 1814 - Place: Baltimore - Period: 5 Yrs - Deserted on 11 Jun 1815.

Kell, Allen - Private - 14th Infantry - Company: Isaac Barnard - Age: 27 - Height: 5' 11" - Born: Westchester, PA - Trade: Farmer - Enlistment date: 29 Jun 1813 - Period: 5 Yrs - Pension: Placed on the pension rolls on 1 Mar 1816 - Discharged at Greenbush, NY, on 24 Jun 1815.

By War of 1812 Soc. in MD

Kellals, Henry - Private - 2nd Light Dragoons - Company: George Haig - Age: 26 - Height: 5' 5" - Born: Fredericktown, MD - Trade: Carpenter - Enlistment date: 4 Mar 1812 - Place: Winchester - Period: 5 Yrs - Discharged on 4 Mar 1817.

Keller, Conrad - Private - 7th Infantry - Company: Samuel Vail - Age: 26 - Height: 5' 11" - Born: Frederick, MD - Trade: Tinner - Enlistment date: 5 Aug 1811 - Place: Washington, MS - Period: 5 Yrs - Discharged on 5 Aug 1816.

Keller, Conrad - Private - US Volunteers - Company: Stephen Moore - Enlistment date: 8 Sep 1812 - Period: 1 Yr.

Kelley, Michael - Private - 2nd Artillery - Company: Nathan Towson - Pension: Heirs received half pay for five years in lieu of land bounty - Killed during the Battle of Stoney Creek on 6 Jun 1813.

Kelley, Patrick - Private - US Volunteers - Company: Stephen Moore - Enlistment date: 8 Sep 1812 - Period: 1 Yr.

Kellog, Isaac - Corporal - 14th Infantry - Enlistment date: 22 Dec 1813 - Period: 18 Mos - Prisoner of War, sent to England; discharged on 30 Mar 1815.

Kelly, Charles - Private - 14th Infantry - Company: Robert Kent - Age: 28 - Height: 5' 7" - Born: Ireland - Trade: Laborer - Enlistment date: 22 Jul 1812 - Period: 5 Yrs - Pension: Old War IF-1970 - BLW 7039-160-12 - Prisoner of War (Halifax), sent to England; discharged at Washington, DC, on 26 Jun 1815 because of wounds received in action.

Kelly, Edward - Sergeant - 36th Infantry - Company: Samuel Raisin - Age: 29 - Height: 5' 10" - Born: Ireland - Trade: Farmer - Enlistment date: 6 Nov 1814 - Place: Fredericktown, MD - Period: War - Discharged at Washington, DC, on 20 Mar 1815.

Kelly, Jeremiah - Corporal - 38th Infantry - Company: John Mowton - BLW 17123-160-12.

Kelly, John - Private - 1st Rifles - Company: Lewis Armistead - Age: 24 - Height: 5' 7" - Born: Fredericktown, MD - Trade: Coach maker - Enlistment date: 19 Dec 1814 - Place: Shepherdstown, VA - Period: 5 Yrs - Discharged at Carlisle Barracks, PA, on 13 Jul 1815 for disability.

Kelly, Michael - Private - Corps of Artillery - Pension: Heirs received half pay for five years in lieu of land bounty - Died on 6 Jun 1813.

Kelly, Michael - Private - 14th Infantry - Company: Reuben Gilder - Age: 49 - Height: 5' 10" - Born: Ireland - Trade: Weaver - Enlistment date: 24 Mar 1814 - Place: Baltimore - Period: War - Pension: Land bounty to Edward Kelly, son and other heirs at law of Michael Kelly - BLW 11933-160-12 - Died at Greenbush, NY, on 24 Mar 1815 from sickness.

Kelly, Patrick - Private - Light Dragoons - Age: 24 - Height: 5' 6" - Born: Baltimore City - Trade: Shoemaker - Enlistment date: 17 Feb 1814 - Period: 5 Yrs.

Kelly, Solomon - Private - 17th Infantry - Company: William Adair - Age: 25 - Height: 5' 7" - Born: Frederick Cty, MD - Enlistment date: 27 May 1814 - Period: War - Died at Fort Erie, UC, on 24 Oct 1814.

Maryland Regulars in War of 1812

Kelly, Thomas - Private - 14th Infantry - Age: 24 - Height: 5' 6 1/2" - Born: Somerset Cty, MD - Enlistment date: 4 Jul 1814 - Place: Salisbury, MD.

Kelly, William - Corporal - 14th Infantry - Company: David Cummings - Age: 40 - Height: 5' 7" - Trade: School master - Enlistment date: 2 Aug 1812 - Period: 5 Yrs - BLW 3228-160-12 - Prisoner of War (Halifax), sent to England; wounded in the mouth during the Battle of Beaver Dams, UC, on 24 Jul 1813; discharged at Baltimore on 30 Apr 1815 on Surgeon's Certificate.

Kelnor, George - Private - US Volunteers - Company: Stephen Moore - Enlistment date: 8 Sep 1812 - Period: 1 Yr.

Kendal, Samuel - Private - 38th Infantry - Company: James Smith - BLW 3128-160-12.

Kendrick, Noah - Private - 12th Infantry - Company: Andrew Madison - Age: 26 - Height: 5' 8 1/2" - Born: Talbot Cty, MD - Trade: Miller - Enlistment date: 27 Sep 1813 - Period: War - Died at Buffalo on 7 Dec 1814 from sickness.

Kendrick, William - Private - 36th Infantry - Company: Joseph Merrick - Age: 27 - Height: 5' 6" - Born: St. Mary's, MD - Enlistment date: 23 Apr 1813 - Place: Baltimore - Discharged on 19 Dec 1813.

Kennedy, Darby - Private - 36th Infantry - Company: Mortimer Hall - Age: 41 - Height: 5' 8" - Born: Ireland - Enlistment date: 3 Sep 1813 - Place: Baltimore - Period: 1 Yr - Prisoner of War (Halifax), sent to England.

Kennedy, David - Private - 2nd Artillery - Company: Nathan Towson - Age: 32 - Height: 5' 6 3/4" - Born: Ireland - Trade: Weaver - Enlistment date: 10 May 1812 - Place: Baltimore - Period: 5 Yrs - Discharged on 10 May 1817.

Kennedy, James M. - Private - 38th Infantry - Company: Shepperd Leakin - Age: 20 - Height: 5' 6 1/2" - Born: Baltimore - Trade: Blacksmith - Enlistment date: 17 Jul 1814 - Place: Baltimore - Period: War - BLW 24198-160-12 - Discharged at Fort McHenry on 28 Mar 1815.

Kennedy, John - Private - 5th Infantry - Company: James Dorman - Age: 25 - Height: 5' 6 3/4" - Born: Trenton, NJ - Trade: Shoemaker - Enlistment date: 19 Jun 1811 - Place: Baltimore - Period: 5 Yrs - Discharged at Pittsburgh on 17 Aug 1815 for incapacity.

Kennedy, Lawrence - Private - 38th Infantry - Company: Charles Stansbury - Age: 28 - Height: 6' 4 1/2' - Born: Ireland - Trade: Laborer - Enlistment date: 9 Nov 1814 - Place: Cumberland - Period: War - Died on 10 Feb 1815.

Kennedy, Thomas - Private - 14th Infantry - Company: David Cummings - Enlistment date: 9 May 1812 - Period: 5 Yrs - Pension: Land bounty to Richard Kennedy and other heirs at law of Thomas Kennedy - BLW 3229-160-12 - Died on 5 Apr 1813.

Kennedy, William - Private - 14th Infantry - Company: Thomas Montgomery - Age: 38 - Height: 5' 10" - Born: York, PA - Trade: Laborer - Enlistment date: 14 Jul 1812 - Place: Baltimore - Period: 5 Yrs - BLW 10180-160-12 - Discharged at Greenbush, NY, on 16 Apr 1815 for consumption and inability.

Kenoth, Thomas - Private - 17th Infantry - Age: 38 - Height: 5' 1" - Born:

By War of 1812 Soc. in MD

Montgomery Cty, MD - Enlistment date: 17 May 1814.

Kent, Robert - Sergeant - US Volunteers - Company: Stephen Moore - Enlistment date: 8 Sep 1812 - Period: 1 Yr - Wounded at Newark on 27 May 1813.

Kent, Robert Wheeler - Captain - 14th Infantry - Commissioned as a captain on 12 Mar 1812; resigned on 15 Aug 1813.

Kepler, Abijah - Private - 14th Infantry - Company: Samuel Lane - Age: 24 - Height: 6' - Born: York, PA - Enlistment date: 25 May 1812 - Place: Cumberland - Period: 5 Yrs - Killed at Carlisle Barracks, PA, on 20 Jul 1812.

Keplinger, George - Seaman - US Sea Fencibles - Company: John Gill - Age: 19 - Height: 5' 11" - Born: Baltimore City - Enlistment date: 4 Jan 1814 - Period: 1 Yr.

Keplinger, John - Private - 14th Infantry - Company: Thomas Montgomery - Enlistment date: 24 Dec 1812 - Period: 5 Yrs - Pension: Land bounty to Catherine Hauer, Margaret Currall and Elizabeth Mount, sisters and only heirs at law of John Keplinger - BLW 27181-160-42 - Died on 27 Dec 1813.

Keppold, John - Private - 14th Infantry - Company: Joseph Marshall - Age: 25 - Height: 5' 10 1/2" - Born: Baltimore City - Trade: Wheelwright - Enlistment date: 29 Jun 1814 - Place: Baltimore - Period: War - BLW 433-160-12 - Discharged at Greenbush, NY, on 3 May 1815.

Kerchner, Henry - Private - 26th Infantry - Company: William Bezeau - Age: 21 - Height: 5' 6" - Born: Lancaster, PA - Trade: Weaver - Enlistment date: 22 Oct 1814 - Place: Baltimore - Period: War - Discharged at Philadelphia on 23 Mar 1815.

Kerns, William - Private - 38th Infantry - Company: Charles Stansbury - Age: 30 - Height: 5' 7 1/2" - Born: Baltimore City - Trade: Farmer - Enlistment date: 6 Oct 1814 - Place: Baltimore - Period: War - BLW 14100-160-12 - Discharged at Baltimore on 6 Apr 1815.

Kerr, Thomas - Private - 26th Infantry - Company: William Bezeau - Age: 19 - Height: 5' 9" - Born: Maryland - Trade: Nailor - Enlistment date: 21 Nov 1814 - Place: Philadelphia - Period: War - BLW 7287-160-12 - Discharged at Philadelphia on 20 Mar 1815.

Kersey, John H. - Private - 38th Infantry - Company: Isaac Aldridge - Age: 25 - Height: 5' 7" - Born: Talbot Cty, MD - Trade: Carpenter - Enlistment date: 16 Oct 1814 - Place: Baltimore - Period: War - Discharged at Craney Island, VA, on 31 May 1814.

Kesner, Josiah - Private - 26th Infantry - Company: William Bezeau - Age: 21 - Height: 5' 7" - Born: Montgomery Cty, MD - Trade: Shoemaker - Enlistment date: 8 Oct 1814 - Place: Baltimore - Period: War - Discharged at Philadelphia on 23 Mar 1815.

Kessner, John - Private - US Volunteers - Company: Stephen Moore - Enlistment date: 8 Sep 1812 - Period: 1 Yr.

Keys, Benjamin - Private - 36th Infantry - Age: 32 - Height: 5' 8 1/4" - Born: Harford Cty, MD - Enlistment date: 1 Mar 1814 - Place: Annapolis.

Keys, Benjamin - Private - 36th Infantry - Company: Joseph Hook - Age: 27 - Height: 5' 7" - Born: Maryland - Enlistment date: 10 Jul 1813 - Place:

Maryland Regulars in War of 1812

Baltimore - Period: 1 Yr.

Keys, Benjamin - Private - 14th Infantry - Company: Joseph Marshall - Age: 29 - Height: 5' 8" - Born: Kent Cty, MD - Trade: Cooper - Enlistment date: 25 Jul 1814 - Place: Baltimore - Period: War.

Keys, Joseph - Private - 14th Infantry - Company: Reuben Gilder - Other regiment: 4th Infantry (New) - Enlistment date: 27 Feb 1813 - Period: 5 Yrs - Deserted at Fort Severn, MD, on 15 Jul 1815.

Keyser, George - Major - 38th Infantry - Born: Maryland - Commissioned as a major on 19 May 1813; discharged on 15 Jun 1815.

Keyser, Peter - Second Lieutenant - 38th Infantry - Commissioned as a 3rd lieutenant on 20 May 1813; promoted to 2nd lieutenant on 22 Apr 1814; died on 1 Oct 1814.

Keysler, George - Private - 20th Infantry - Company: John Stanard - Other regiment: 12th Infantry - Age: 14 - Height: 5' - Born: Baltimore - Trade: Planter - Enlistment date: 23 Jan 1813 - Place: Drumfires, MD - Period: 5 Yrs - Transferred to 12th Infantry; discharged at Pittsburgh on 17 Aug 1815 on Surgeon's Certificate for epilepsy.

Kidwell, Parker - Private - 36th Infantry - Company: Thomas Carbery - Enlistment date: 1 Aug 1814 - Place: Georgetown, DC - Period: War.

Kidwell, Theodore - Private - 20th Infantry - Company: William Jett - Age: 23 - Height: 5' 6" - Born: Prince George's Cty, MD - Trade: Farmer - Enlistment date: 21 Aug 1814 - Period: 5 Yrs.

Kidwell, Theodore - Private - 36th Infantry - Company: Joseph Hook - Age: 41 - Height: 5' 6" - Born: Charles Cty, MD - Trade: Laborer - Enlistment date: 6 Aug 1814 - Place: Georgetown, DC - Period: War - BLW 211-160-12 - Discharged at Fort Covington, MD, on 30 Mar 1815.

Kildue, George M. - Seaman - US Sea Fencibles - Company: Simmones Bunbury - Enlistment date: 6 Feb 1815 - Period: 1 Yr - Missing at Fort McHenry on 28 Feb 1815.

Kiley, John - Private - 38th Infantry - Company: Charles Stansbury - Age: 34 - Height: 5' 5" - Born: County Cork, Ireland - Trade: Farmer - Enlistment date: 9 Nov 1814 - Place: Baltimore - Period: War - BLW 22419-160-12 - Discharged at Baltimore on 6 Apr 1815.

Killmam, Henry C. - Private - 40th Infantry - Company: Robert Neal - Age: 27 - Height: 5' 9" - Born: Baltimore - Trade: Mariner - Enlistment date: 11 Sep 1812 - Place: Portsmouth, NH - Period: War - Discharged on 20 Apr 1815.

Kilson, John - Private - 5th Infantry - Age: 34 - Height: 6' - Born: Baltimore - Trade: Cooper - Enlistment date: 17 Apr 1814.

Kim, John - Seaman - US Sea Fencibles - Company: John Gill - Age: 22 - Height: 5' 6" - Born: Baltimore - Enlistment date: 1 Jan 1814 - Period: 1 Yr.

Kincaid, George William - Sergeant - 14th Infantry - Company: Isaac Barnard - Age: 28 - Height: 5' 5 1/2" - Born: Philadelphia - Trade: Tailor - Enlistment date: 28 Dec 1812 - Place: Little York, PA - Period: 5 Yrs - Pension: Old War WF-15681 Rejected - BLW 26713-160-12 - Discharged on 14 Mar 1815 for diseased liver.

By War of 1812 Soc. in MD

Kincaid, Myers - Seaman - US Sea Fencibles - Company: Simmones Bunbury - Age: 26 - Height: 5' 5 1/4" - Born: Nottingham, MD - Enlistment date: 12 Feb 1814 - Period: 1 Yr - Discharged on 11 Feb 1815.

Kindale, Samuel - Corporal - 14th Infantry - Company: William McIlvane - Age: 24 - Height: 5' 8 3/4" - Born: New Jersey - Enlistment date: 29 Jan 1813 - Period: 18 Mos - Prisoner of War, exchanged on 15 Apr 1814; discharged at Plattsburg, NY, on 28 Jul 1814.

Kindle, Samuel - Private - 14th Infantry - Prisoner of War (Halifax), discharged from Halifax on 3 Feb 1814.

Kindley, Samuel - Private - 2nd Artillery - Company: Sander Donoho - Age: 40 - Height: 5' 7 1/2" - Born: Baltimore City - Trade: Gunsmith - Enlistment date: 5 Nov 1812 - Place: Rockingham Cty, NC - Period: 5 Yrs - Discharged on 1 Sep 1815.

King, Abraham - Private - 2nd Light Dragoons - Company: William Littlejohn - Age: 21 - Height: 5' 8" - Born: Baltimore - Trade: Stage driver - Enlistment date: 22 Dec 1813 - Place: Alexandria, DC - Period: War - BLW 23763-160-12 - Discharged at Carlisle Barracks, PA, on 4 May 1815.

King, Benjamin - Private - 1st Artillery - Company: George Armistead - Age: 25 - Height: 5' 10" - Born: Maryland - Trade: Carpenter - Enlistment date: 4 Jan 1810 - Place: Washington, DC - Period: 5 Yrs - Discharged at New York on 3 Jan 1815.

King, Edward - Private - 2nd Infantry - Company: Francis Johnston - Age: 20 - Height: 5' 8" - Born: Prince George's Cty, MD - Trade: Blacksmith - Enlistment date: 22 Oct 1807 - Place: Fredericktown, MD - Period: 5 Yrs - Re-enlisted at Fort Stoddert on 12 Oct 1812; died at Dauphin Island on 18 Jul 1813.

King, Francis - Private - 2nd Light Dragoons - Company: William Littlejohn - Age: 22 - Height: 6' - Born: Maryland - Enlistment date: 19 Oct 1813 - Place: Leesburg - Period: 5 Yrs - Deserted at Baltimore in 1814.

King, John - Private - 38th Infantry - Company: John Rothrock - Pension: Land bounty to Lydia M. Perkins and other heirs at law of John King - BLW 21513-160-12.

King, John - Corporal - 14th Infantry - Company: William McIlvane - Age: 22 - Height: 5' 8" - Born: Delaware - Enlistment date: 19 Jan 1814 - Place: Baltimore - Period: 5 Yrs.

King, Joseph - Corporal - 38th Infantry - Company: John Mowton - Age: 28 - Height: 5' 10" - Born: Baltimore City - Trade: Cordwainer - Enlistment date: 27 Feb 1814 - Place: Alexandria, DC - Period: War - BLW 247-160-12 - Discharged at Craney Island, VA, on 15 Mar 1815.

King, Martin - Private - 36th Infantry - Company: Thomas Carbery - Age: 19 - Height: 5' 10" - Born: Montgomery Cty, MD - Trade: Farmer - Enlistment date: 27 Feb 1814 - Place: Leonardstown, MD - Period: War - BLW 142-160-12 - Discharged at Baltimore on 31 Mar 1815.

King, Michael - Private - 22nd Infantry - Company: John Foster - Age: 37 - Height: 5' 9" - Born: Maryland - Trade: Nailor - Enlistment date: 19 Apr 1813 - Period: 5 Yrs - Deserted on 1 Jun 1814.

Maryland Regulars in War of 1812

King, Thomas B. - Private - 36th Infantry - Age: 20 - Height: 5' - Born: Dinwiddie, VA - Enlistment date: 16 Sep 1814 - Place: Petersburg, VA.

King, William - Private - 2nd Infantry - Company: John Brahan - Age: 21 - Height: 5' 7" - Born: Maryland - Trade: Laborer - Enlistment date: 14 Oct 1807 - Place: Fredericksburg, VA - Period: 5 Yrs - Re-enlisted at Fort Stoddert on 14 Jul 1812 for five years; discharged on 17 Feb 1817 and re-enlisted.

King, William - Colonel - 5th Infantry - Other regiment: 15th Infantry then 3rd Rifles then 4th Infantry (New) - Born: Delaware - Commissioned as a 2nd lieutenant, 5th Infantry, on 3 May 1808; promoted to 1st lieutenant on 30 Sep 1810; promoted to captain, 15th Infantry, on 2 May 1812; promoted to major on 3 May 1813; promoted to colonel, 3rd Rifles, on 21 Feb 1814; transferred to 4th Infantry on 17 May 1815; discharged on 1 Jun 18121; died on 1 Jan 1826.

King, William - Private - 14th Infantry - Company: David Cummings - Enlistment date: 1 Jun 1812 - Period: 5 Yrs - Prisoner of War (Halifax), sent to England, captured at Beaver Dams on 24 Jun 1813.

Kingsbury, Thomas - Private - 12th Infantry - Company: Thomas Sangster - Age: 22 - Height: 5' 10" - Born: Hagerstown, MD - Trade: Laborer - Enlistment date: 3 May 1814 - Period: 5 Yrs - Pension: Land bounty to Domilian Kingsbury, father and heir at law of Thomas Kingsbury - BLW 3952-160-12 - Died at Washington, DC, on 9 Jan 1815.

Kingsley, Samuel - Recruit - 26th Infantry - Company: William Bezeau - Race: 21 - Age: 21 - Height: 5' 6" - Born: Havre de Grace, MD - Enlistment date: 8 Sep 1814 - Place: Baltimore - Period: War - Listed as "Col'd" in his service record.

Kinnear, William - Private - 14th Infantry - Company: Joseph Marshall - Age: 38 - Height: 5' 10" - Born: Philadelphia - Trade: Miller - Enlistment date: 18 Jul 1814 - Place: Baltimore - Period: War - Pension: Wife Elizabeth, Navy IF-893 - BLW 18531-160-50 - Served in the U.S. Marine Corps, Captain Williams' Company between 1811 and 1814; discharged on 3 Dec 1814 for a rupture.

Kinny, John - Private - 38th Infantry - Company: John Brookes - BLW 19881-160-12.

Kirby, Ezekiel - Private - 6th Infantry - Company: Robert Patterson - Age: 15 - Height: 5' 2" - Born: Maryland - Trade: Currier - Enlistment date: 12 Aug 1814 - Place: Philadelphia - Period: 5 Yrs.

Kirk, John - Private - 2nd Infantry - Company: Francis Johnston - Age: 23 - Height: 5' 6 1/4" - Born: Baltimore - Trade: Tailor - Enlistment date: 28 Jan 1808 - Place: Columbia Springs - Period: 5 Yrs - Died at Fort Charlotte, AL, on 1 Feb 1815.

Kirk, Joseph - Private - 43rd Infantry - Company: George Dabney - Age: 39 - Height: 5' 5" - Born: Cecil Cty, MD - Trade: Farmer - Enlistment date: 19 Feb 1814 - Place: Lancaster - Period: War - BLW 14292-160-12 - Discharged at Fort Hampton on 1 Aug 1815.

Kiser, John - Private - 14th Infantry - Company: Samuel Lane - Age: 25 -

By War of 1812 Soc. in MD

Height: 5' 8" - Born: Cumberland, MD - Enlistment date: 11 Jun 1812 - Place: Shippensburg - Period: 18 Mos - Killed at Beaver Dams, UC, on 24 Jun 1813.

Kisling, James - Recruit - 26th Infantry - Company: William Bezeau - Race: 27 - Age: 27 - Height: 5' 6" - Born: Maryland - Trade: Farmer - Enlistment date: 19 Jan 1815 - Place: Philadelphia - Period: War - Listed on the Descriptive Roll of Colored Men, 1 Apr 1815.

Kisner, Jacob - Private - 36th Infantry - Company: Samuel Raisin - Age: 23 - Height: 5' 6" - Born: Montgomery Cty, MD - Trade: Farmer - Enlistment date: 17 Aug 1814 - Period: War - BLW 872-160-12 - Discharged at Washington, DC, on 20 Mar 1815.

Kisner, John - Private - 36th Infantry - Company: Samuel Raisin - Age: 19 - Height: 5' 8 1/2" - Born: Montgomery Cty, MD - Trade: Farmer - Enlistment date: 9 Sep 1814 - Period: War - Discharged at Washington, DC, on 20 Mar 1815.

Kleinhans, George H. - Private - US Volunteers - Company: Stephen Moore - Enlistment date: 8 Sep 1812 - Period: 1 Yr.

Kline, Peter - Private - 14th Infantry - Company: Richard Arell - Age: 38 - Height: 5' 6" - Born: Frederick, MD - Trade: Cooper - Enlistment date: 14 May 1812 - Place: Shepherdstown, VA - Period: 5 Yrs - BLW 20509-160-12 - Discharged on 25 Apr 1815.

Knight Jr., Absalom - Private - 14th Infantry - Company: Reuben Gilder - Age: 24 - Height: 5' 5" - Born: Maryland - Enlistment date: 7 May 1813 - Period: 5 Yrs - Pension: Land bounty to Absalom Knight, father and heir at law of Absalom Knight, Jr. - BLW 15385-160-12 - Died at Plattsburgh, NY, on 18 Nov 1814.

Knight, Elitus - Private - 2nd Artillery - Company: Frederick Evans - Age: 21 - Height: 5' 9" - Born: Maryland - Enlistment date: 5 May 1813 - Place: Annapolis - Period: 5 Yrs - Deserted on 8 Mar 1816.

Knight, John - Sergeant - 4th Infantry - Company: John Smith - Other regiment: 5th Infantry (New) - Age: 18 - Height: 5' 7" - Born: Worcester Cty, MD - Trade: Hatter - Enlistment date: 9 Apr 1812 - Period: 5 Yrs - Discharged at Washington, DC, on 11 Apr 1817.

Knight, Joshua - Private - 14th Infantry - Company: Henry Fleming - Enlistment date: 11 Feb 1813 - Period: 5 Yrs - Pension: Wife Mary, Old War Widow File 10818; heirs received half pay for five years in lieu of land bounty - Died on 12 Jun 1813.

Knight, Joshua - Private - Light Dragoons - Age: 44 - Height: 5' 7" - Born: Maryland - Trade: Shoemaker - Enlistment date: 11 Feb 1813 - Place: Baltimore - Period: 5 Yrs.

Knight, William - Corporal - 36th Infantry - Company: Joseph Hook - Age: 23 - Height: 5' 11 1/2" - Born: Stafford, VA - Trade: Farmer - Enlistment date: 21 Jul 1814 - Place: Georgetown, DC - Period: War - BLW 23284-160-12 - Discharged on 30 Mar 1815.

Knight, William - Private - 5th Infantry - Company: Richard Whartenby - Other regiment: 3rd Infantry (New) - Age: 12 - Height: 4' 5" - Born:

Annapolis - Enlistment date: 18 May 1813 - Place: Georgetown, DC - Period: 5 Yrs - Prisoner of War (Halifax), captured at Stoney Creek, UC, on 12 Jun 1813; discharged on 18 May 1817.

Knott, George - Private - 14th Infantry - Company: Joseph Marshall - Other regiment: 4th Infantry (New) - Age: 28 - Height: 5' 7" - Born: Baltimore City - Trade: Butcher - Enlistment date: 2 Jun 1812 - Place: Baltimore - Period: 5 Yrs - Deserted at Fort Severn, MD, on 18 Jul 19815.

Knott, George - Private - US Volunteers - Company: Stephen Moore - Enlistment date: 8 Sep 1812 - Period: 1 Yr.

Knower, John - Seaman - US Sea Fencibles - Company: Simmones Bunbury - Age: 40 - Height: 5' 8 1/4" - Born: Middletown, MD - Enlistment date: 9 Dec 1813 - Period: 1 Yr - Died at Baltimore on 26 Dec 1813.

Knowlton, David - Private - 36th Infantry - Transferred to the U.S. Navy on 26 Oct 1812.

Knox, John - Private - 38th Infantry - Company: Shepperd Leakin - Age: 38 - Height: 5' 7 1/2" - Born: Baltimore City - Trade: Chair maker - Enlistment date: 11 Apr 1814 - Period: War - BLW 12875-160-12 - Discharged at Fort McHenry on 28 Mar 1815.

Knuckles, Charles - Private - 36th Infantry - Company: Samuel Raisin - Other regiment: Ordnance Department - Age: 21 - Height: 5' 10" - Born: Rockbridge, VA - Trade: Farmer - Enlistment date: 7 Oct 1814 - Place: Richmond, VA - Period: 5 Yrs - Transferred to Ordnance Department on 23 Mar 1815; discharged in disgrace on 17 May 1820.

Kohlstradst, Benjamin - Private - US Volunteers - Company: Stephen Moore - Enlistment date: 8 Sep 1812 - Period: 1 Yr.

Kolb, John - Private - 14th Infantry - Company: Reuben Gilder - Age: 22 - Height: 5' 8" - Born: Baltimore - Trade: Butcher - Enlistment date: 3 Mar 1814 - Place: Fredericktown, MD - Period: War - Pension: Land bounty to Daniel Kolb, brother and other heirs at law of John Kolb - BLW 25287-160-12 - Died at Greenbush, NY, on 9 Apr 1815.

Kolp, Jacob - Private - 2nd Infantry - Company: John Brahan - Age: 24 - Height: 5' 6" - Born: Frederick Cty, MD - Trade: Shoemaker - Enlistment date: 8 Apr 1808 - Place: Staunton, VA - Period: 5 Yrs - Died at Mount Vernon on 29 Aug 1812.

Koog, Martin - Seaman - US Sea Fencibles - Company: Simmones Bunbury - Enlistment date: 27 Jan 1814 - Period: 1 Yr - Discharged on 26 Jan 1815.

Koontz, Jacob - Private - 12th Infantry - Company: Andrew Madison - Age: 34 - Height: 5' 7 1/2" - Born: Frederick Cty, MD - Trade: Brick layer - Enlistment date: 8 Jun 1813 - Place: New Market, VA - Period: 18 Mos - Prisoner of War at Chazy; discharged at Buffalo on 8 Dec 1814.

Kountz, David - Private - 14th Infantry - Prisoner of War (Halifax), sent to England.

Kow, Frederick - Private - 14th Infantry - Company: Samuel Lane - Age: 35 - Height: 5' 8" - Born: Hagerstown, MD - Trade: Farmer - Enlistment date: 25 Jun 1812 - Place: Hagerstown, MD - Period: 5 Yrs - Discharged at Greenbush, NY, on 4 May 1815 for inability.

By War of 1812 Soc. in MD

Krause, Frederick Charles - Private - 36th Infantry - Company: Thomas Carbery - Age: 40 - Height: 5' 4" - Born: Germany - Enlistment date: 4 Oct 1813 - Place: Baltimore - Period: 1 Yr - Discharged on 4 Oct 1814.

Kreps, George - Private - 38th Infantry - Company: John Rothrock - Age: 26 - Height: 5' 9" - Born: Hagerstown, MD - Trade: Gunsmith - Enlistment date: 28 Jul 1813 - Place: Craney Island, VA - Period: War - BLW 2861-160-12 - Discharged at Craney Island, VA, on 15 Mar 1815.

Kuntz, David - Private - 14th Infantry - Company: Samuel Lane - Age: 24 - Height: 5' 8 1/2" - Born: Lancaster, PA - Enlistment date: 9 Jun 1812 - Place: Shephardstown - Period: 5 Yrs - Pension: Wife Elizabeth, WO-17639, WC-16817 - Prisoner of War, sent to England.

Kurty, Henry - Private - 2nd Artillery - Company: Nathan Towson - BLW 19343-160-12.

Kymes, Ferdinand - Private - 38th Infantry - Company: John Brookes - BLW 21858-160-12.

L

La Pierre, George - Private - 36th Infantry - Company: Samuel Raisin - Age: 29 - Height: 5' 10" - Born: Italy - Trade: Soldier - Enlistment date: 14 Nov 1814 - Place: Georgetown, DC - Period: War.

La Pierre, Peter - Private - 36th Infantry - Company: Samuel Raisin - Deserted on 19 Feb 1815.

Lacey, William - Seaman - US Sea Fencibles - Company: William Addison.

Lacey, William - Private - 17th Infantry - Company: Caleb Holder - Age: 35 - Height: 5' 9 1/2" - Born: Harford Cty, MD - Enlistment date: 13 Aug 1814 - Place: Steubenville, OH - Period: War.

Lackey, Pierre - Private - 38th Infantry - Company: John Brookes - Age: 21 - Height: 5' 2" - Born: Worcester Cty, MD - Trade: Farmer - Enlistment date: 23 Dec 1814 - Place: Craney Island, VA - Period: War - Discharged at Craney Island, VA, on 15 Mar 1815.

Lacompt, William - Corporal - 26th Infantry - Company: William Bezeau - Age: 23 - Height: 5' 10 1/2" - Born: Dorchester Cty, MD - Trade: Trader - Enlistment date: 12 Oct 1814 - Place: Philadelphia - Period: War - BLW 2931-160-12 - Discharged at Philadelphia on 22 Mar 1815.

Lacoumpt, Nathan - Sergeant - 5th Infantry - Company: Benjamin Wallace - Age: 20 - Height: 5' 7" - Born: Talbot Cty, MD - Trade: Farmer - Enlistment date: 8 Sep 1808 - Period: 5 Yrs - Discharged at Fort George, UC, on 8 Sep 1813.

Ladyman, Benjamin - Private - 43rd Infantry - Company: Henry Garrett - Other regiment: Corps of Artillery - Age: 21 - Height: 5' 6" - Born: Charles Cty, MD - Trade: Distiller - Enlistment date: 16 Aug 1814 - Place: Bethel, MD - Period: 5 Yrs - Deserted at Fort Powhattan, VA, on 19 Jan 1817.

Ladyman, Samuel - Private - 43rd Infantry - Company: Henry Garrett - Other regiment: Corps of Artillery - Age: 25 - Height: 5' 10" - Born: Charles Cty, MD - Trade: Blacksmith - Enlistment date: 12 Aug 1814 - Place: Bethel, MD - Period: 5 Yrs - Deserted on 20 Oct 1817.

Maryland Regulars in War of 1812

Laffell, Charles - Private - 14th Infantry - Company: Joseph Marshall - BLW 3224-160-12.

Laind, William - Private - 26th Infantry - Company: William Bezeau - Age: 34 - Height: 5' 7 1/2" - Born: Washington, MD - Enlistment date: 3 Oct 1814 - Place: Baltimore - Period: War - Deserted on 9 Oct 1814.

Lake, Benjamin - Private - 14th Infantry - Age: 25 - Born: Plattsburgh, NY - Prisoner of War (Quebec), died at Quebec on 22 Jul 1813.

Lake, James - Private - 36th Infantry - Company: Henry Neale - Enlistment date: 23 Jul 1813 - Period: 1 Yr - Died on 26 Jul 1813.

Lamb, Joseph - Private - 5th Infantry - Company: Richard Bell - Other regiment: 8th Infantry (New) - Age: 31 - Height: 5' 8 1/2" - Born: Burlington, NJ - Trade: Accountant - Enlistment date: 11 Mar 1811 - Place: Baltimore - Period: 5 Yrs - Discharged on 11 Nov 1816.

Lamb, William - Private - 14th Infantry - Company: Richard Arell - Other regiment: 4th Infantry (New) - Age: 31 - Height: 5' 7" - Born: Dinwiddie, VA - Trade: Watch maker - Enlistment date: 8 Sep 1812 - Place: Baltimore - Period: 5 Yrs - Prisoner of War, exchanged on 15 Apr 1814; discharged on 8 Sep 1817.

Lambuson, Samuel - Private - 38th Infantry - Company: Shepperd Leakin - Age: 24 - Height: 5' 8 1/2" - Born: Worcester Cty, MD - Enlistment date: 7 Feb 1814 - Place: Princess Anne, MD - Period: 1 Yr.

Lamott, John - Private - 7th Infantry - Company: Carey Nicholas - Age: 30 - Height: 5' 4" - Born: Baltimore - Trade: Farmer - Enlistment date: 29 Jul 1814 - Place: New Orleans - Period: War.

Lanberg, John - Private - 14th Infantry - Company: David Cummings - Age: 20 - Height: 5' 6" - Born: Philadelphia - Trade: Seaman - Enlistment date: 5 Dec 1814 - Place: Salisbury, MD - Period: War - Discharged at Baltimore on 16 Mar 1815.

Landon, Zorababale - Private - 38th Infantry - Company: James Hook - Age: 26 - Height: 5' 9" - Born: Somerset Cty, MD - Trade: Carpenter - Enlistment date: 18 Mar 1814 - Place: Princess Anne, MD - Period: 5 Yrs - BLW 3890-160-12 - Discharged at Baltimore on 17 May 1815.

Lane, Benjamin - Private - 14th Infantry - Company: Samuel Lane - Age: 28 - Height: 5' 8" - Enlistment date: 6 Jan 1813 - Pension: Land bounty to Daniel Lane and other heirs at law of Benjamin Lane - BLW 15331-160-12 - Prisoner of War (Quebec), died at Quebec on 23 Jul 1813.

Lane, Daniel - Private - 38th Infantry - Company: John Rothrock - BLW 7888-160-12.

Lane, John - Musician - 36th Infantry - Age: 15 - Height: 4' 11" - Born: St. Mary's Cty, MD - Trade: Drummer - Enlistment date: 28 Dec 1814 - Period: War - BLW 595-160-12 - Discharged at Georgetown, DC, on 13 May 1815.

Lane, Samuel - Lieutenant Colonel - 14th Infantry - Other regiment: 32nd Infantry - Born: Maryland - Place: Maryland - Pension: Old War IF-20840 - Served as an officer in the U.S. Army between 3 Mar 1799 and 12 Aug 1802; commissioned as a captain, 14th Infantry, on 12 Mar 1812; promoted to major on 3 Mar 1813; promoted to lieutenant colonel, 32nd Infantry, on 15

By War of 1812 Soc. in MD

Jun 1814; discharged on 15 Jun 1815; died on 10 Sep 1822.

Lane, Samuel M. - Private - 3rd Artillery - Company: Alexander Fanning - Age: 29 - Height: 5' 6" - Born: Baltimore - Trade: Coach maker - Enlistment date: 12 Feb 1812 - Period: 5 Yrs - Discharged on 12 Feb 1817.

Lang, Benjamin - Private - 38th Infantry - Company: Shepperd Leakin - Age: 35 - Height: 5' 6" - Born: Baltimore - Trade: Hatter - Enlistment date: 19 Feb 1814 - Period: War - BLW 530-150-12 - Discharged at Fort McHenry on 28 Mar 1815.

Lang, John L. B. - Corporal - 38th Infantry - Company: Isaac Aldridge - Age: 36 - Height: 5' 9" - Born: Prince George's Cty, MD - Trade: Coppersmith - Enlistment date: 14 Feb 1814 - Period: War - BLW 6134-160-12 - Discharged at Baltimore on 6 Apr 1815.

Langalier, Lewis - Private - 38th Infantry - Company: Isaac Aldridge - Age: 31 - Height: 5' 7" - Born: France - Trade: Gardener - Enlistment date: 25 Dec 1814 - Place: Baltimore - Period: 5 Yrs - Discharged at Baltimore on 6 Apr 1815.

Langburgh, John - Private - 14th Infantry - BLW 4622-160-12.

Langford, Stephen - Private - 38th Infantry - Company: John Buck - Age: 25 - Height: 5' 7" - Born: Somerset Cty, MD - Trade: Planter - Enlistment date: 11 Mar 1814 - Place: Princess Anne, MD - Period: 5 Yrs - BLW 13459-160-12 - Discharged at Fort Covington, MD, on 31 Mar 1815.

Langley, George - Private - 1st Light Dragoons - Company: Alexander Cummings - Other regiment: Corps of Artillery - Age: 25 - Height: 5' 6" - Born: Charles Cty, MD - Trade: Shoemaker - Enlistment date: 11 Nov 1812 - Place: Frederick, MD - Period: 5 yrs - Discharged on 11 Nov 1817.

Languell, Robert - Private - 38th Infantry - Company: Shepperd Leakin - Age: 25 - Height: 5' 4" - Born: Baltimore - Trade: Stone mason - Enlistment date: 23 Mar 1814 - Place: Baltimore - Period: War - Discharged at Baltimore on 30 Mar 1815.

Langwell, Thomas - Private - 10th Infantry - Company: Emanuel Leigh - Age: 38 - Height: 5' 5 1/2" - Born: Dorchester Cty, MD - Trade: Farmer - Enlistment date: 14 May 1814 - Place: Caswell Cty, NC - Period: War - Discharged at Fredericksburg, VA, on 7 Apr 1815.

Lanham, John - Sergeant - 38th Infantry - Company: Thomas Sangster - Other regiment: 4th Infantry (New) - Age: 19 - Height: 5' 6" - Born: Baltimore - Trade: Ship joiner - Enlistment date: 13 Dec 1814 - Place: Baltimore - Period: 5 Yrs - BLW 843-320-12 - Discharged on 13 Dec 1819.

Lanham, Norman - Private - 36th Infantry - Company: Thomas Carbery - Age: 27 - Height: 5' 10 1/4" - Born: Washington, MD - Trade: Laborer - Enlistment date: 26 Feb 1814 - Place: Georgetown, DC - Period: War - BLW 1986-160-12 - Discharged at Baltimore on 31 Mar 1815.

Lanman, Elisha - Private - 36th Infantry - Company: Samuel Raisin - Age: 22 - Height: 5' 6 1/4" - Born: Prince George's Cty, MD - Trade: Blacksmith - Enlistment date: 7 Oct 1814 - Place: Georgetown, DC - Period: War - Discharged at Washington, DC, on 20 Mar 1815.

Lapayre, William - Sergeant - 2nd Artillery - Company: Nathan Towson -

Maryland Regulars in War of 1812

Age: 21 - Height: 5' 8" - Born: Baltimore - Trade: Cabinet maker - Enlistment date: 8 Feb 1813 - Place: Philadelphia - Period: 5 Yrs.

Lapier, George L. - Private - 36th Infantry - Company: Samuel Raisin - BLW 18757-160-12.

Lappin, Robert - Private - 22nd Infantry - Age: 39 - Height: 5' 8" - Born: Hagerstown, MD - Trade: Carpenter - Enlistment date: 2 Jul 1814 - Period: 5 Yrs - Deserted on 21 Nov 1814.

Larner, Matthew - Private - 36th Infantry - Company: Thomas Carbery - Age: 28 - Height: 5' 8" - Born: Ireland - Trade: Laborer - Enlistment date: 25 Jul 1814 - Place: Washington, DC - Period: War - BLW 1823-160-12 - Discharged at Baltimore on 30 Mar 1815.

Larovett, Philip - Private - 36th Infantry - Company: Joseph Hook - Age: 34 - Height: 5' 6" - Born: Germany - Enlistment date: 9 Jun 1813 - Place: Baltimore - Period: 1 Yr.

Larrere, Joseph - Private - 38th Infantry - Company: John Mowton - BLW 6985-160-12.

Larson, William - Private - 36th Infantry - Age: 27 - Height: 5' 8" - Born: Virginia - Enlistment date: 18 Jul 1813 - Place: Baltimore - Transferred to the U.S. Navy on 26 Oct 1813.

Larwood, William - Private - 14th Infantry - Company: William McIlvane - Enlistment date: 20 May 1813 - Period: 5 Yrs - Prisoner of War, exchanged on 15 Apr 1814; discharged at Greenbush, NY, on 1 Jun 1815, disability.

Lashley, Samuel - Private - 14th Infantry - Company: Thomas Montgomery - Other regiment: 4th Infantry (New) - Age: 24 - Height: 5' 10" - Born: Georgetown, MD - Trade: Stone mason - Enlistment date: 5 May 1813 - Place: Frederick, MD - Period: 5 Yrs - Died at Fort Hawkins, GA, in Nov or Dec 1819.

Latham, Edward - Private - 14th Infantry - Prisoner of War (Halifax), sent to England.

Latharn, John - Private - 10th Infantry - Company: William Bailey - Other regiment: 8th Infantry (New) - Age: 23 - Height: 5' 10 1/2" - Born: Cecil Cty, MD - Trade: Painter - Enlistment date: 14 Jul 1813 - Place: Rowan Cty, NC - Period: 5 Yrs - Pension: Land bounty to Robert Latham, brother & other heirs at law of John Latham - BLW 25717-160-12 - Died at Fort Crawford, IL, on 4 Jun 1818.

Latimer, Marcus - Second Lieutenant - 36th Infantry - Company: Thomas Carbery - Pension: Land bounty to Clementine Latimer, widow and heir at law of Marcus Latimer - BLW 813-80-50 - Commissioned as an ensign on 25 Oct 1813; promoted to 3rd lieutenant on 17 Mar 1814; promoted to 2nd lieutenant on 30 Sep 1814; discharged on 15 Jun 1815.

Lattimore, Jesse - Private - 2nd Light Dragoons - Company: Samuel Hopkins - Other regiment: Corps of Artillery - Age: 30 - Height: 5' 9" - Born: Baltimore - Trade: Farmer - Enlistment date: 12 Mar 1813 - Period: 5 Yrs - Discharged on 7 Oct 1816 after furnishing a substitute.

Laughlin, James - Private - 2nd Artillery - Company: Frederick Evans - Other regiment: Corps of Artillery - Age: 39 - Height: 5' 9 1/4" - Born: Virginia -

Trade: Laborer - Enlistment date: 9 Nov 1813 - Place: Annapolis - Period: 5 Yrs - Discharged on 9 Nov 1818.

Laughlin, John - Private - 36th Infantry - Age: 28 - Height: 5' 9 1/2" - Born: Scotland - Trade: Soldier - Enlistment date: 25 Jan 1815 - Period: War - BLW 658-320-12 - Discharged at Georgetown, DC, on 13 Mar 1815.

Laurence, Joseph - Private - US Volunteers - Company: Stephen Moore - Enlistment date: 8 Sep 1812 - Period: 1 Yr.

Laurence, Paul - Private - 36th Infantry - Company: Samuel Raisin - Other regiment: 4th Infantry (New) - Age: 30 - Height: 5' 5" - Born: England - Trade: Tailor - Enlistment date: 26 Oct 1814 - Place: Frederick, MD - Period: 5 Yrs.

Law, Thomas - Private - 38th Infantry - Company: Shepperd Leakin - Age: 20 - Height: 5' 6 1/2" - Born: Prince George's Cty, MD - Trade: Cooper - Enlistment date: 17 Feb 1814 - Period: War - Deserted at Fort Covington, MD, on 20 Nov 1814.

Lawless, John - Private - 14th Infantry - Company: Joseph Marshall - Age: 34 - Height: 5' 6 1/2" - Born: Baltimore - Trade: Chandler - Enlistment date: 4 Oct 1812 - Period: 5 Yrs - Pension: Old War IF-20027 - Wounded at Beaver Dams, UC, on 24 Jun 1813; discharged at Greenbush, NY< on 24 May 1815, on Surgeon's Certificate.

Lawless, Philip - Private - 14th Infantry - Company: Reuben Gilder - Age: 37 - Height: 5' 10 1/2" - Born: Frederick, MD - Trade: Carpenter - Enlistment date: 7 Mar 1814 - Place: Maryland - Period: 5 Yrs - Pension: Old War IF-26214 - BLW 6389-160-12 - Wounded in left thigh by accident; discharged at Greenbush, NY, on 8 Jun 1815, wounded.

Lawrence, Jacob - Private - 38th Infantry - Company: James Smith - Age: 23 - Height: 5' 4 1/2" - Born: Baltimore City - Trade: Blacksmith - Enlistment date: 18 Feb 1814 - Place: Craney Island, VA - Period: War - BLW 1515-160-12 - Discharged at Craney Island, VA, on 15 Mar 1815.

Lawrence, James - Sergeant - 38th Infantry - Company: Anthony Miltenberger - Pension: No pension claim - BLW 29-160-50.

Lawrence, James - Boatswain - US Sea Fencibles - Company: Simmones Bunbury - Age: 24 - Height: 5' 4 1/2" - Born: New Haven, CT - Enlistment date: 29 Oct 1813 - Period: 1 Yr - Discharged at Fort McHenry on 24 Mar 1815.

Lawrence, Joseph - Private - Light Artillery - Company: Luther Leonard - Age: 25 - Height: 5' 7" - Born: Baltimore - Trade: Baker - Enlistment date: 28 Apr 1812 - Place: Washington, DC - Period: 5 Yrs.

Lawrence, Richard - Private - 12th Infantry - Company: Thomas Sangster - Other regiment: 8th Infantry (New) - Age: 27 - Height: 5' 9 1/2" - Born: St. Mary's, MD - Trade: Tailor - Enlistment date: 16 May 1812 - Place: Fort Royal, VA - Period: 5 Yrs - Discharged on 16 May 1817.

Lawrence, William - Colonel - 2nd Infantry - Other regiment: 8th Infantry (New) - Born: Maryland - Commissioned as a 2nd lieutenant, 4th Infantry, on 8 Jun 1801; transferred to 2nd Infantry on 1 Apr 1802; promoted to 1st lieutenant on 12 Oct 1804; promoted to captain on 1 Jan 1810; promoted to major on 19 Apr 1814; transferred to 8th Infantry on 17 May 1815; promoted to lieutenant colonel on 8 May 1818; promoted to colonel on 20 Aug 1828; resigned on 15 Jul 1831; died on 7 Jan 1841.

Lawrence, William - Private - 5th Infantry - Company: John Corboley - Other regiment: 3rd Infantry (New) - Age: 21 - Height: 5' 4" - Born: Baltimore - Trade: Brick layer - Enlistment date: 6 Sep 1814 - Place: York, PA - Period: 5 Yrs.

Laws, Charles - Private - 38th Infantry - Company: John Rothrock - Age: 26 - Height: 5' 8" - Born: Worcester Cty, MD - Trade: Ship joiner - Enlistment date: 15 Jun 1814 - Place: Craney Island, VA - Period: War - BLW 10102-160-12 - Discharged at Craney Island, VA, on 15 Mar 1815.

Laxius, Michael - Private - 38th Infantry - Company: Isaac Aldridge - BLW 12974-160-12.

Layland, John S. - Private - 36th Infantry - Company: Joseph Hook - Age: 29 - Height: 6' - Born: Trenton, NJ - Trade: Cooper - Enlistment date: 5 Jul 1813 - Place: Georgetown, DC - Period: War - BLW 86-160-12 - Discharged at Fort Covington, MD, on 30 Mar 1815.

Lazare, Jacob - Private - 14th Infantry - Company: James McDonald - Age: 19 - Height: 5' 11" - Born: Maryland - Enlistment date: 2 Dec 1813 - Place: Clarksburg, VA - Period: 18 Mos - Deserted at Batavia, NY, on 1 Oct 1814.

Lazelear, Benjamin - Private - 14th Infantry - Prisoner of War (Halifax), died at Halifax on 22 Sep 1813.

Lazure, Jacob W. - Private - 14th Infantry - Company: Reuben Gilder - Age: 21 - Height: 6' 4" - Born: Maryland - Trade: Laborer - Enlistment date: 9 Jun 1813 - Period: 5 Yrs - Died at Brownsville, NY, on 29 Dec 1814.

Lazure, John - Private - 14th Infantry - Company: Reuben Gilder - Age: 20 - Height: 5' 10" - Born: Montgomery Cty, MD - Trade: Farmer - Enlistment date: 2 Dec 1813 - Place: Maryland - Period: 18 Mos - Discharged at Greenbush, NY, on 3 Jun 1815.

Lea, George W. - Third Lieutenant - 38th Infantry - Company: James Haslett - Born: Maryland - Pension: SO-16984, SC-10955 - BLW 3734-160-50 - Commissioned as a 3rd lieutenant on 15 Aug 1813; resigned on 19 May 1814.

Leach, Benjamin - Private - 1st Infantry - Company: Simon Owens - Age: 31 - Height: 5' 8" - Born: Frederick, MD - Trade: Farmer - Enlistment date: 14 Apr 1809 - Place: Winchester - Period: 5 Yrs - Re-enlisted at Winchester of 8 Jun 1814 for the war; died on 27 Jan 1815.

Leach, Jesse - Private - 2nd Infantry - Company: Bartholomew Armistead - Age: 25 - Height: 5' 11" - Born: Baltimore - Trade: Farm laborer - Enlistment date: 20 Feb 1810 - Place: Columbia Springs - Period: 5 Yrs - Discharged on 9 Apr 1815.

Leach, Samuel - Private - 38th Infantry - Company: John Rothrock - Pension: Wife Mary, Old War IF-12097, Old War Minor 106, received half pay for five years in lieu of land bounty - BLW 855-160-50.

League, Daniel - Private - 38th Infantry - Company: John Mowton - Age: 34 - Height: 5' 8 3/4" - Born: Baltimore - Trade: Cooper - Enlistment date: 13

By War of 1812 Soc. in MD

Mar 1814 - Place: Craney Island, VA - Period: War - BLW 3575-160-12 - Discharged at Craney Island, VA, on 15 Mar 1815.

Leahy, John - Sergeant - 15th Infantry - Company: Joseph Barton - Age: 29 - Height: 6' - Born: Hunterdon Cty, NJ - Trade: Blacksmith - Enlistment date: 19 Dec 1812 - Place: Baltimore - Period: 5 Yrs - Discharged at Fort Erie, UC, on 16 Jun 1815 on Surgeon's Certificate.

Leakin, Sheppard Church - Captain - 38th Infantry - BLW 16729-160-50 - Commissioned as a captain on 20 May 1813; discharged on 15 Jun 1815.

Lear, William - Sergeant - 2nd Light Dragoons - Company: Samuel Hopkins - Age: 28 - Height: 5' 10" - Born: Havre de Grace, MD - Trade: Carpenter - Enlistment date: 18 May 1812 - Place: Havre de Grace, MD - Period: 5 Yrs - Discharged on 17 May 1817.

Learey, James - Musician - 36th Infantry - Age: 14 - Height: 4' 7" - Born: Maryland - Enlistment date: 7 Feb 1815 - Place: Georgetown, DC - Period: War - Discharged at Georgetown, DC, on 13 Mar 1815.

Leary, William C. - Private - 43rd Infantry - Company: Robert Love - Age: 22 - Height: 5' 8 1/2" - Born: Baltimore - Trade: Shoemaker - Enlistment date: 8 Jun 1814 - Place: Fayetteville, NC - Period: War - Discharged at Camp Greenfield, NC, on 17 May 1815.

Leazer, Elisha - Private - 14th Infantry - Prisoner of War (Halifax), captured at Beaver Dams on 24 Jun 1813, discharged from Halifax on 3 Feb 1813.

Leblance, Allen - Private - 26th Infantry - Company: William Bezeau - Age: 18 - Height: 5' 3" - Born: Philadelphia - Trade: Tobacconist - Enlistment date: 9 Sep 1814 - Place: Baltimore - Period: War - BLW 3926-160-12 - Discharged at Philadelphia on 23 Mar 1815.

Lecompte, William - Private - 14th Infantry - Company: Richard Arell - Enlistment date: 19 May 1812 - Period: 5 Yrs - Pension: Land bounty to Joseph Lecompte, uncle and other heirs at law of William Lecompte - BLW 26733-160-12 - Prisoner of War, exchanged on 15 Apr 1814; died at Williamsville, NY, on 10 Jan 1815.

Lecumpte, Levin - Private - 14th Infantry - Company: Richard Arell - Other regiment: 4th Infantry (New) - Age: 26 - Height: 6' 2" - Born: Dorchester Cty, MD - Trade: Farmer - Enlistment date: 19 May 1812 - Place: Vienna, MD - Period: 5 Yrs - Deserted on 14 Jul 1815.

Ledman, Francis - Corporal - 11th Infantry - Age: 29 - Height: 6' - Born: Maryland - Trade: Blacksmith - Enlistment date: 6 Jul 1814 - Place: Burlington, VT - Period: 5 Yrs - Died on 25 Dec 1814 from sickness.

Lee Jr., Henry - Major - 36th Infantry - Place: Maryland - Commissioned as a major on 8 Apr 1813; discharged on 15 Jun 1815; died on 30 Jan 1837.

Lee, Charles - Private - 36th Infantry - Age: 28 - Height: 5' 6" - Born: Manchester, Essex Cty, MA - Enlistment date: 28 Apr 1813 - Place: Baltimore - Period: 1 Yr - Transferred to the U.S. Navy on 30 Oct 1813.

Lee, David - Private - 1st Artillery - Company: Henry Craig - Age: 20 - Height: 5' 8 3/4" - Born: Cecil Cty, MD - Trade: Shoemaker - Enlistment date: 2 Jun 1814 - Place: Youngstown, PA - Period: 5 Yrs - Wounded at Fort Erie, UC, on 18 Oct 1814; discharged at Buffalo, NY, on 2 Aug 1815.

Maryland Regulars in War of 1812

Lee, James - Private - 20th Infantry - Company: William Jett - Age: 35 - Height: 5' 7 1/2" - Born: St. Mary's Cty, MD - Trade: Farmer - Enlistment date: 15 Jul 1813 - Place: Virginia - Period: War - Discharged at Buffalo on 31 May 1815.

Lee, John - Private - 5th Infantry - Company: Colin Buckner - Age: 38 - Height: 5' 10" - Born: Bucks Cty, PA - Trade: Farmer - Enlistment date: 17 Feb 1813 - Place: Baltimore - Period: 5 Yrs - Discharged at Buffalo on 21 Jun 1815 on Surgeon's Certificate.

Lee, Philip - Corporal - 14th Infantry - Company: Joseph Marshall - Age: 39 - Height: 6' 3" - Born: Harford Cty, MD - Trade: Farmer - Enlistment date: 23 Jul 1814 - Place: Baltimore - Period: War - BLW 9639-160-12 - Discharged at Greenbush, NY, on 3 May 1815.

Lee, Richard H. - First Lieutenant - 36th Infantry - Other regiment: 4th Rifles - Born: Maryland - Commissioned as an ensign, 36th Infantry, on 30 Apr 1813; promoted to 3rd lieutenant, 4th Rifles, on 17 Mar 1814; transferred to Corps of Artillery, on 17 May 1815; promoted to 2nd lieutenant on 17 May 1816; promoted to 1st lieutenant on 20 Apr 1818; resigned on 1 Jan 1820.

Leech, Joseph - Private - 2nd Artillery - Company: Nathan Towson - Pension: Old War IF-9750 - BLW 2497-160-12.

Leeman, Isaac - Private - 12th Infantry - Company: William Howell - Age: 19 - Height: 5' 7" - Born: Frederick, MD - Trade: Farmer - Enlistment date: 5 Feb 1814 - Period: 5 Yrs.

Leggett, Benjamin - Private - 19th Infantry - Company: George Kesling - Age: 40 - Height: 5' 9 1/2" - Born: Baltimore - Trade: Farmer - Enlistment date: 31 Jan 1815 - Place: Lebanon, Warren Cty, OH - Period: War - BLW 382-320-14 - Discharged at Zanesville, OH, on 27 Mar 1815.

Leidenstricker, Daniel - Private - 38th Infantry - Company: John Mowton - BLW 3239-160-12.

Leisure, Hezekiah - Private - 17th Infantry - Company: William Adair - Age: 31 - Height: 5' 11 3/4" - Born: Maryland - Trade: Carpenter - Enlistment date: 3 Jun 1814 - Place: Louisville, KY - Period: War - BLW 18887-160-12 - Discharged at Chillicothe, OH, on 4 Jun 1814.

Leizer, Peter - Private - 22nd Infantry - Company: Joseph Henderson - Age: 23 - Height: 6' 1" - Born: Frederick, MD - Trade: Laborer - Enlistment date: 23 Mar 1814 - Period: 5 Yrs - Killed on 25 Jul 1814.

Leland, Johns - Private - Light Artillery - Age: 38 - Height: 5' 8" - Born: Baltimore - Enlistment date: 14 Nov 1814 - Place: Providence, RI.

Lemand, James - Private - 14th Infantry - Company: Thomas Montgomery - Age: 40 - Height: 5' 5" - Born: Baltimore - Trade: Shoemaker - Enlistment date: 21 Apr 1812 - Period: 5 Yrs - Pension: Land bounty to Mary Lemand, guardian of the children of James Lemand - BLW 2865-160-12 - Killed at Chippewa, UC, on 15 Oct 1814.

Lemmy, John - Private - 14th Infantry - BLW 2672-160-12.

Lepechey, Michael - Corporal - 38th Infantry - Company: John Buck - BLW 6023-160-12.

Lescure, John - Sergeant - 14th Infantry - Company: David Cummings - Age:

22 - Height: 5' 6" - Born: Philadelphia - Trade: Soldier - Enlistment date: 10 May 1814 - Place: Baltimore - Period: War - Discharged on 16 Mar 1815.

Letts, Thomas - Seaman - US Sea Fencibles - Company: John Gill - Age: 22 - Height: 5' 10" - Born: Perth Amboy, NJ - Enlistment date: 13 Jan 1814 - Place: Maryland - Period: 1 Yr - Deserted on 9 Jun 1814.

Letzinger, George - Private - 12th Infantry - Company: Thomas Sangster - Other regiment: 8th Infantry (New) - Age: 21 - Height: 5' 7" - Born: Oldtown, Baltimore Cty, MD - Trade: Brick layer - Enlistment date: 12 May 1812 - Place: Alexandria, DC - Period: 5 Yrs - Discharged on 12 May 1817.

Leverton, John - Private - 14th Infantry - Company: Clement Sullivan - Other regiment: 4th Infantry (New) - Enlistment date: 20 May 1812 - Period: 5 Yrs - Pension: Old War WF-15857 Rejected - Prisoner of War (Halifax), captured at Beaver Dams on 24 Jun 1813, discharged from Halifax on 9 Nov 1813; discharged on 19 May 1817.

Lewis, Barall - Seaman - US Sea Fencibles - Company: William Addison.

Lewis, Bazel - Private - 2nd Artillery - Company: Joseph Philips - Age: 44 - Height: 5' 6" - Born: Maryland - Enlistment date: 11 Feb 1814 - Place: Fort Massac, IL - Period: War - BLW 25073-160-12 - Discharged at Fort Clark, IL, on 21 Aug 1815.

Lewis, David R. - Private - 1st Artillery - Company: James Many - Age: 30 - Height: 5' 11" - Born: Maryland - Trade: Miller - Enlistment date: 10 Aug 1806 - Place: Baltimore - Period: 5 Yrs - Re-enlisted on 31 Dec 1812; discharged at New York on 17 Nov 1814 on Surgeon's Certificate, deafness.

Lewis, Dudley - Private - 14th Infantry - Prisoner of War (Halifax), discharged from Halifax on 3 Feb 1814.

Lewis, Edward - Private - 38th Infantry - Company: Isaac Aldridge - BLW 16454-160-12.

Lewis, Isaac - Private - 1st Artillery - Company: Hopley Yeaton - Age: 27 - Height: 5' 7" - Born: Prince George's Cty, MD - Trade: Laborer - Enlistment date: 21 Feb 1814 - Place: Fort Nelson, VA - Period: 5 Yrs - Pension: Land bounty to Samuel W. Hughes and Carl Conrad Luning, heirs at law of Isaac Lewis - BLW 27427-160-12 - Died at Fort Nelson, VA, on 25 Feb 1815.

Lewis, John - Private - 36th Infantry - Company: Joseph Merrick - Enlistment date: 12 May 1813 - Period: 1 Yr - Deserted on 21 Aug 1813.

Lewis, John - Corporal - 36th Infantry - Company: Charles Randolph - Age: 30 - Height: 5' 7" - Born: Hagerstown, MD - Enlistment date: 15 May 1813 - Place: Frederick, MD - Period: 1 Yr.

Lewis, John W. - Private - 2nd Infantry - Company: John Pemberton - Other regiment: 1st Infantry (New) - Age: 21 - Height: 5' 9" - Born: Harford Cty, MD - Trade: Tailor - Enlistment date: 28 Feb 1808 - Place: Frederick, MD - Period: 5 Yrs - Re-enlisted on 12 Dec 1812 for five years; discharged on 10 Sep 1817 and re-enlisted.

Lewis, Samuel - Private - 36th Infantry - Company: Joseph Hook - Age: 22 - Height: 5' 8 1/4" - Born: Frederick, MD - Trade: Laborer - Enlistment date: 5 Jul 1813 - Place: New Town, VA - Period: 5 Yrs - BLW 18705-160-12 - Discharged at Fort Covington, MD, on 30 Mar 1815.

Maryland Regulars in War of 1812

Lewis, Samuel - Private - 38th Infantry - Company: John Rothrock - Age: 21 - Height: 5' 8 1/2" - Born: Northumberland Cty, VA - Trade: Joiner - Enlistment date: 18 Aug 1814 - Place: Craney Island, VA - Period: War - BLW 26610-160-12 - Discharged at Craney Island, VA, on 15 Mar 1815.

Lewis, Samuel - Private - 36th Infantry - Company: Joseph Hook - Height: 5' 8 1/4" - Born: Frederick, MD - Enlistment date: 7 Apr 1814 - Place: Fort Washington, MD - Period: War.

Lewis, Thomas - Private - 38th Infantry - Company: Isaac Aldridge - Age: 21 - Height: 5' 5 1/2" - Born: England - Trade: Mariner - Enlistment date: 7 Nov 1814 - Place: Baltimore - Period: War - Pension: Land bounty to Martha Lewis, daughter and only heir at law of Thomas Lewis - BLW 25289-160-12 - Discharged on 3 Apr 1815.

Lewis, William - Private - 38th Infantry - Company: John Brookes - BLW 11352-160-12.

Lewis, Zachariah - Private - 2nd Artillery - Company: Nathan Towson - BLW 13641-160-12.

Light, Leonard - Private - 38th Infantry - Company: Shepperd Leakin - Age: 25 - Height: 5' 11" - Born: Baltimore City - Trade: Wheelwright - Enlistment date: 15 Feb 1814 - Period: War - Deserted on 15 Feb 1814.

Lightbody, John - Private - 38th Infantry - Company: John Buck - Age: 37 - Height: 5' 9" - Born: Ireland - Trade: Millwright - Enlistment date: 7 Jun 1814 - Place: Baltimore - Period: War - BLW 16814-160-12 - Discharged at Baltimore on 30 Mar 1815.

Lightell, John - Private - 36th Infantry - Company: Thomas Carbery.

Lighter, Jacob - Private - 14th Infantry - Company: Samuel Lane - Other regiment: 4th Infantry (New) - Age: 23 - Height: 5' 11 1/2" - Born: Hagerstown, MD - Trade: Butcher - Enlistment date: 11 May 1812 - Place: Hagerstown, MD - Period: 5 Yrs - BLW 10368-160-12 - Prisoner of War (Halifax), captured at Stoney Creek, UC, on 12 Jun 1813; discharged at Montpelier, VT, on 11 May 1817.

Lightfoot, Thomas - Private - 36th Infantry - Company: Thomas Carbery - Age: 19 - Height: 5' 2 1/2" - Born: Alexandria, DC - Trade: Trader - Enlistment date: 19 Feb 1814 - Place: Georgetown, DC - Period: War - Pension: Land bounty to William Lightfoot, father and heir at law of Thomas Lightfoot - BLW 14354-160-12 - Discharged at Baltimore on 31 Mar 1815.

Lightfoot, Timothy - Private - 38th Infantry - Company: John Brookes - BLW 11064-160-12.

Lighton, Jacob - Private - 14th Infantry - Company: Samuel Lane - Other regiment: 4th Infantry (New) - Age: 30 - Height: 5' 6" - Born: Maryland - Trade: Laborer - Enlistment date: 5 Jun 1812 - Place: Cumberland - Period: 5 Yrs - Discharged on 5 Jun 1817.

Lilly, Ephraim - Private - 14th Infantry - Company: David Cummings - Age: 38 - Height: 5' 10" - Born: Maryland - Trade: Laborer - Enlistment date: 1 May 1814 - Place: Baltimore - Period: War - Deserted on 5 May 1814.

Limmer, Terrance - Seaman - US Sea Fencibles - Company: John Gill - Age: 28 - Height: 5' 7" - Born: Nore, County Nore, Ireland - Enlistment date: 4

By War of 1812 Soc. in MD

Jan 1814 - Period: 1 Yr.

Linard, Alfred - Private - 14th Infantry - Prisoner of War (Halifax), sent to England.

Linch, Elijah - Private - 14th Infantry - Pension: Old War IF-20440.

Lindemoore, Henry - Private - 14th Infantry - Company: Reuben Gilder - Age: 30 - Height: 5' 8" - Born: Baltimore - Trade: Shoemaker - Enlistment date: 3 Mar 1814 - Place: Towsontown, MD - Period: War - BLW 9961-160-12 - Discharged at Greenbush, NY, on 4 May 1815.

Lindsay, John - Private - 26th Infantry (OH) - Company: Joel Collins - Other regiment: 19th Infantry - Age: 45 - Height: 5' 4 3/4" - Born: Maryland - Trade: Weaver - Enlistment date: 18 Sep 1813 - Place: Sandwich, UC - Period: War - BLW 8319-160-12 - Discharged at Detroit on 20 Jul 1815.

Lingard, Noah - Private - 1st Rifles - Company: Joshua Hamilton - Age: 23 - Height: 5' 9 1/2" - Born: Dorchester Cty, MD - Trade: Farmer - Enlistment date: 1 Apr 1813 - Place: Fort Hampton, AL - Period: 5 Yrs - Deserted on 22 Sep 1813.

Lingenfetter, Peter - Private - 36th Infantry - Age: 35 - Height: 5' 8" - Born: Germany - Enlistment date: 21 Nov 1814 - Place: Norfolk, VA - Deserted.

Lingerall, Thomas (Lingrell) - Private - 19th Infantry - Company: Carey Trimble - Other regiment: 17th Infantry - Age: 18 - Height: 5' 9" - Born: Maryland - Trade: Laborer - Enlistment date: 13 Apr 1814 - Place: Adelphi, Ross Cty, OH - Period: War - Pension: Old War IF-26705, SO-17267, SC-18720; lived in Ohio. - BLW 17728-160-12 - Discharged at Chillicothe, OH, on 7 Jun 1815.

Lingo, Elijah - Private - 1st Rifles - Company: Thomas Ramsey - Other regiment: Rifles - Age: 18 - Height: 5' 10" - Born: Maryland - Trade: Farmer - Enlistment date: 13 May 1813 - Place: Elizabethtown - Period: 5 Yrs - BLW 18862-160-12 - Discharged at Belle Fontaine, MO, on 18 Mar 1818.

Lingo, William - Private - 1st Rifles - Company: Thomas Ramsey - Other regiment: Rifles - Age: 18 - Height: 5' 6" - Born: Maryland - Trade: Farmer - Enlistment date: 19 May 1813 - Place: Tennessee - Period: 5 Yrs - Discharged at Buffalo on 1 Aug 1815, wounded on 12 Aug 1814 at Fort Erie, UC.

Lingo, William - Private - 1st Rifles - Company: George Ramsey - Age: 18 - Height: 5' 6" - Born: Maryland - Trade: Farmer - Enlistment date: 19 May 1813 - Period: 5 Yrs - Wounded at Fort Erie, UC, on 12 Aug 1814; discharged at Buffalo, NY, on 1 Aug 1815.

Lingold, William - Private - 14th Infantry - Company: Samuel Lane - Other regiment: 4th Infantry (New) - Age: 21 - Height: 5' 6" - Born: Kent Cty, DE - Trade: Saddler - Enlistment date: 29 May 1812 - Place: Cumberland, MD - Period: 5 Yrs - Discharged on 28 May 1817.

Lingrell, Thomas - Private - 19th Infantry - Company: Carey Trimble - Age: 18 - Height: 5' 9" - Born: Maryland - Trade: Laborer - Enlistment date: 13 Apr 1814 - Place: Adelphi, Ross Cty, OH - Period: War - Discharged at Chillicothe, OH, on 7 Jun 1815.

Linnenburger, John - Seaman - US Sea Fencibles - Company: Simmones

Maryland Regulars in War of 1812

Bunbury - Age: 21 - Height: 5' 6" - Born: Maryland - Enlistment date: 26 Oct 1814 - Period: 1 Yr.

Linsey, Joseph - Seaman - US Sea Fencibles - Company: Simmones Bunbury - Age: 34 - Height: 5' 6 1/2" - Born: Ireland - Enlistment date: 31 Aug 1814 - Period: 1 Yr - Discharged at Fort McHenry on 24 Mar 1815.

Lintain, Robert - Private - 36th Infantry - Company: Thomas Carbery - BLW 4130-160-12.

Linthicomb, William - Private - 38th Infantry - Company: James Hook - Age: 33 - Height: 5' 9" - Born: Anne Arundel Cty, MD - Trade: Shoemaker - Enlistment date: 31 Mar 1814 - Place: Baltimore - Period: War - BLW 10905-160-12 - Discharged at Fort Covington, MD, on 31 May 1815.

Linton, George - Private - 16th Infantry - Company: Thomas Horrell - Other regiment: 2nd Infantry (New) - Age: 26 - Height: 5' 5" - Born: Maryland - Trade: Shoemaker - Enlistment date: 9 Dec 1813 - Place: Carlisle, PA - Period: 5 Yrs - BLW 25753-160-12 - Discharged on 10 Jun 1816 after furnishing a substitute (Samuel W. Wright).

Linton, William - Private - 1st Artillery - Company: James Reed - Age: 39 - Height: 5' 6" - Born: Charles Cty, MD - Trade: Farmer - Enlistment date: 1 Aug 1814 - Place: Fort Washington, MD - Period: 5 Yrs - Discharged on 29 Mar 1817, inability.

List, John - Private - 14th Infantry - Company: Joseph Marshall - Age: 25 - Height: 5' 11 1/4" - Born: Reading, PA - Trade: Paper hanger - Enlistment date: 23 Mar 1814 - Place: Baltimore - Period: War - BLW 1729-160-12 - Discharged at Greenbush, NY, on 3 May 1815.

Litchlighten, Conrad - Private - 12th Infantry - Company: Thomas Moore - Other regiment: 8th Infantry (New) - Age: 35 - Height: 5' 10" - Born: Frederick, MD - Trade: Saddler - Enlistment date: 28 Sep 1813 - Place: Staunton, VA - Period: 5 Yrs - Discharged at Pensacola, FL, on 23 Sep 1818.

Little, Ezekiel - Private - 5th Infantry - Company: Colin Buckner - Age: 35 - Height: 6' - Born: Anne Arundel Cty, MD - Trade: Laborer - Enlistment date: 18 Feb 1813 - Period: 5 Yrs - Died on 18 Aug 1813.

Little, Peter - Colonel - 38th Infantry - Born: Pennsylvania - Place: Maryland - Pension: Land bounty to Catherine Little, widow of Peter Little - BLW 8564-160-50 - Commissioned as a colonel, 38th Infantry, on 19 May 1813; discharged on 15 Jun 1815; died on 5 Feb 1830.

Little, William - Private - 7th Infantry - Company: George Allen - Other regiment: 1st Infantry (New) - Age: 36 - Height: 5' 6" - Born: Maryland - Enlistment date: 19 Aug 1814 - Place: Livingston, KY - Period: 5 Yrs - Discharged on 18 Aug 1819.

Littleton, John - Private - 14th Infantry - Company: David Cummings - Age: 24 - Height: 5' 5" - Born: Delaware - Trade: Laborer - Enlistment date: 24 Dec 1814 - Period: War - Pension: Wife Elizabeth, WO-23773, WC-31556 - Discharged on 16 Mar 1815.

Livers, John - Private - 14th Infantry - Company: Joseph Marshall - Age: 38 - Height: 5' 6" - Born: Maryland - Enlistment date: 30 Oct 1813 - Period: 5 Yrs - Pension: Heirs received half pay for five years in lieu of land bounty -

By War of 1812 Soc. in MD

Died at Greenbush, NY, on 9 Feb 1815.

Lives, William G. - Seaman - US Sea Fencibles - Company: Simmones Bunbury - Age: 20 - Height: 5' 9" - Born: New York - Enlistment date: 3 Jan 1814 - Period: 1 Yr - Died on 16 Jul 1814.

Livingston, John - Private - 2nd Rifles - Company: John O'Fallon - Other regiment: Rifles - Age: 27 - Height: 5' 9" - Born: Maryland - Trade: Laborer - Enlistment date: 28 Jun 1814 - Place: Kentucky - Period: War - BLW 4123-160-12 - Discharged at Detroit on 30 Jun 1815.

Livingston, Thomas - Private - 20th Infantry - Company: William Jett - Other regiment: 8th Infantry (New) - Age: 28 - Height: 6' 1" - Born: Baltimore - Trade: Sailor - Enlistment date: 1 May 1813 - Period: 5 Yrs - BLW 4917-160-12 - Discharged at Greenbush, NY, on 4 May 1815 on Surgeon's Certificate.

Lloyd, Stephen L. - Private - 19th Infantry - Company: Harris Hickman - Other regiment: 17th Infantry - Age: 30 - Height: 5' 6" - Born: Baltimore Cty - Trade: Brewer - Enlistment date: 1 Apr 1814 - Place: Limestone, KY - Period: War - BLW 4501-160-12 - Discharged at Chillicothe, OH, on 9 Jun 1815.

Loachelle, Joseph - Private - Light Dragoons - Age: 21 - Height: 5' 7" - Born: Bordeaux, France - Trade: Carpenter - Enlistment date: 10 May 1814 - Place: Baltimore - Period: 5 Yrs - Deserted at Baltimore.

Loar, John - Fifer - 17th Infantry - Company: Asabael Nearing - Age: 26 - Height: 5' 10" - Born: Allegany Cty, MD - Trade: Stone mason - Enlistment date: 7 Mar 1813 - Period: War - Pension: Wife Elizabeth, WO-25425, WC-26364 - BLW 6162-160-12 - Discharged at Chillicothe, OH, on 9 Jun 1815.

Lobenstine, Fredrick - Private - 14th Infantry - Company: Reuben Gilder - Age: 28 - Height: 5' 6" - Born: Montgomery Cty, MD - Trade: Shoemaker - Enlistment date: 3 Mar 1814 - Place: Plattsburgh, NY - Period: War - Discharged at Greenbush, NY, on 4 May 1815.

Lobstein, Jacob - Private - 14th Infantry - Company: Reuben Gilder - Age: 35 - Height: 5' 9" - Born: Philadelphia - Trade: Baker - Enlistment date: 24 Mar 1814 - Place: Baltimore - Period: War - BLW 5431-160-12 - Discharged at Greenbush, NY, on 4 May 1815.

Locey, William - Seaman - US Sea Fencibles - Company: John Gill - Age: 27 - Height: 5' 11 1/2" - Born: Hartford, CT - Enlistment date: 11 Feb 1814 - Place: Baltimore - Period: 1 Yr.

Logan, Hugh - Corporal - 14th Infantry - Company: David Cummings - Age: 20 - Height: 5' 8" - Born: Alexandria, DC - Trade: Cabinet maker - Enlistment date: 8 Nov 1814 - Place: Alexandria, DC - Period: 5 Yrs.

Logan, Otho - Private - 22nd Infantry - Company: John Pentland - Age: 23 - Height: 5' 7" - Born: Maryland - Trade: Laborer - Enlistment date: 17 Jan 1813 - Place: Mercersburg, PA - Period: War - Discharged at Sackets Harbor, NY, on 24 May 1815.

Logsdan, Archibald - Private - 36th Infantry - Company: Joseph Nelson - Age: 25 - Height: 5' 8 1/2" - Born: Cumberland, MD - Enlistment date: 25 Jun 1813 - Place: Cumberland, MD - Period: 1 Yr.

Maryland Regulars in War of 1812

Logsden, James - Private - 17th Infantry - Company: John Chunn - Other regiment: 3rd Infantry (New) - Age: 49 - Height: 5' 6" - Born: Baltimore City - Trade: Silversmith - Enlistment date: 13 Mar 1813 - Place: Elizabethtown, Hardin County, KY - Period: 5 Yrs.

Loller, David C. - Sergeant - 1st Infantry - Company: William Oliver - Age: 25 - Height: 5' 3 1/2" - Born: Baltimore - Trade: Plasterer - Enlistment date: 13 May 1813 - Place: Harrisburg, PA - Period: 5 Yrs - Discharged at Philadelphia on 9 Sep 1815.

Lomax, Lawson - Private - 12th Infantry - Company: Thomas Sangster - Age: 23 - Height: 5' 7" - Born: Frederick Cty, MD - Trade: Farmer - Enlistment date: 15 May 1812 - Place: Alexandria, DC - Period: 18 Mos - BLW 529-320-12 - Re-enlisted at Alexandria, DC, on 28 Feb 1814 for the war; discharged at Fredericksburg on 15 Mar 1815.

London, William - Private - 38th Infantry - Company: John Rothrock - BLW 3982-160-12.

Long, Charles - Private - 36th Infantry - Company: Thomas Carbery - Age: 27 - Height: 6' 3/4" - Born: St. Mary's Cty, MD - Trade: Blacksmith - Enlistment date: 7 Mar 1814 - Place: Leonardstown, MD - Period: War - BLW 232-160-12 - Discharged at Baltimore on 31 Mar 1815.

Long, Daniel - Private - 14th Infantry - Company: Samuel Lane - Age: 27 - Height: 5' 8" - Enlistment date: 15 May 1812 - Place: Shephardstown, MD - Period: 5 Yrs - Pension: Land bounty to John Long and other heirs at law of Daniel Long - BLW 7241-160-12 - Died at Black Rock, NY, on 16 Dec 1812 from wounds received in action on 28 Nov 1812.

Long, Isaac - Private - 14th Infantry - Company: David Cummings - Age: 23 - Height: 5' 5 1/2" - Born: Harford Cty, MD - Trade: Shoemaker - Enlistment date: 7 Oct 1814 - Place: Baltimore - Period: War - BLW 21410-160-12 - Discharged at Baltimore on 16 Mar 1815.

Long, John - Private - 2nd Infantry - Company: Perrin Willis - Age: 22 - Height: 5' 5" - Born: Frederick Cty, MD - Trade: Laborer - Enlistment date: 24 Oct 1814 - Place: Washington, DC - Period: 5 Yrs - Deserted on 24 Oct 1814.

Long, Joseph - Private - 2nd Infantry - Company: John Campbell - Age: 22 - Height: 5' 7 1/4" - Born: Newcastle, DE - Trade: Farmer - Enlistment date: 30 Mar 1808 - Place: Fredericktown, MD - Period: 5 Yrs - Discharged on 30 Mar 1813.

Long, Major - Private - 36th Infantry - Company: Joseph Bryan - Discharged on Surgeon's Certificate on 29 Nov 1814.

Long, Mark - Private - 36th Infantry - Company: Joseph Hook - Age: 17 - Height: 5' 1" - Born: Montgomery Cty, MD - Trade: Laborer - Enlistment date: 21 Jul 1814 - Place: Georgetown, DC - Period: War - BLW 16352-160-12 - Discharged at Fort Covington, MD, on 30 Mar 1815.

Long, Mark - Private - 36th Infantry - Company: Joseph Hook - BLW 2604-160-12.

Long, Peter - Private - 1st Rifles - Company: Thomas Ramsey - Other regiment: Rifles - Age: 19 - Height: 5' 5" - Born: Maryland - Trade: Farmer

By War of 1812 Soc. in MD

- Enlistment date: 4 Apr 1814 - Place: Cincinnati - Period: 5 Yrs - Discharged at Martin's Cantonment on 24 Mar 1819 and re-enlisted.

Long, Robert W. - Ensign - 14th Infantry - Commissioned as an ensign on 27 May 1812; resigned on 15 Feb 1813.

Longley, Peter - Private - 12th Infantry - Company: Thomas Sangster - Other regiment: 8th Infantry (New) - Age: 35 - Height: 5' 4" - Born: Baltimore - Trade: Blacksmith - Enlistment date: 22 Jun 1814 - Place: Abingdon, VA - Period: 5 Yrs.

Longly, Edmond - Corporal - 14th Infantry - Company: Reuben Gilder - Age: 41 - Height: 5' 9" - Born: Groton, MA - Trade: Rope maker - Enlistment date: 12 May 1813 - Place: Fort Independence, MA - Period: 18 Mos - Dropped on rolls on 25 Feb 1815 for not having been legally enlisted.

Longwell, Robert - Private - 38th Infantry - Company: John Buck - BLW 9599-160-12.

Loog, Michael (Lough) - Seaman - US Sea Fencibles - Company: Simmones Bunbury - Age: 22 - Height: 5' 11" - Born: Ireland - Enlistment date: 25 Oct 1814 - Period: 1 Yr - Discharged at Fort McHenry on 24 Mar 1815.

Lord, Samuel - Private - 2nd Artillery - Company: Nathan Towson - Pension: SO-20700 Rejected - Also served in Captain Thomas Densmore's Company, Lieutenant Stephen Woodman's Company and Captain James Banks' Company, plus naval service.

Loudenslager, John - Private - 14th Infantry - Company: Reuben Gilder - Pension: Land bounty to Jacob Loudenslager, brother and other heirs at law of John Loudenslager - BLW 21905-160-12.

Louis, John - Private - 36th Infantry - Age: 40 - Height: 5' 6" - Born: Leghorn, Italy - Enlistment date: 14 May 1813 - Place: Baltimore.

Louis, Louis - Private - 2nd Artillery - Company: Nathan Towson - BLW 13359-160-12.

Love, David - Private - 12th Infantry - Company: Thomas Moore - Age: 32 - Height: 5' 10" - Born: Frederick, MD - Enlistment date: 22 May 1813 - Place: Wheeling, VA - Period: 1 Yr - Died at French Mills, NY, on 2 Jan 1814, from sickness.

Love, James - Private - 5th Infantry - Company: James Dorman - Other regiment: 3rd Infantry (New) - Age: 21 - Height: 5' 8" - Born: Maryland - Trade: Carpenter - Enlistment date: 12 May 1812 - Place: Hagerstown, MD - Period: 5 Yrs - Deserted at Pittsburgh on 18 Aug 1815.

Loveday, John - Private - 14th Infantry - Company: David Cummings - Age: 23 - Height: 5' 7 1/'2" - Born: Talbot Cty, MD - Trade: Seaman - Enlistment date: 25 Jun 1814 - Place: Baltimore - Period: War - Discharged on 14 Mar 1815.

Lovegrove, George - Corporal - 12th Infantry - Company: Thomas Sangster - Age: 24 - Height: 5' 4 1/4" - Born: Delaware - Trade: Tailor - Enlistment date: 30 Apr 1812 - Place: Havre de Grace, MD - Period: 5 Yrs - Discharged on 30 Apr 1817.

Loveless, Elijah - Private - 36th Infantry - Company: Joseph Hook - Age: 27 - Height: 5' 7" - Born: Prince George's Cty, MD - Trade: Farmer - Enlistment date: 19 Jul 1814 - Place: Georgetown, DC - Period: War - BLW 183-160-12 - Discharged at Fort Covington, MD, on 30 Mar 1815.

Loveless, Hezekiah - Private - 10th Infantry - Company: Philip Britain - Other regiment: 8th Infantry (New) - Age: 21 - Height: 5' 6" - Born: Charles Cty, MD - Trade: Blacksmith - Enlistment date: 5 Aug 1812 - Place: Caswell Cty, NC - Period: 5 Yrs - Discharged on 10 May 1817 and re-enlisted.

Lovett, Joseph - Sergeant - 21st Infantry - Company: Sullivan Burbank - Age: 36 - Height: 6' - Born: Baltimore City - Trade: Trader - Enlistment date: 12 May 1814 - Place: Batavia, NY - Period: 5 Yrs - Furlough at Williamsville, NY, an account of a severe wound.

Lovett, Thomas - Private - 36th Infantry - Company: Joseph Hook - Age: 23 - Height: 5' 7" - Born: Harford Cty, MD - Trade: Shoemaker - Enlistment date: 15 Apr 1814 - Place: Annapolis - Period: War - BLW 18562-160-12 - Discharged at Fort Covington, MD, on 30 Mar 1815.

Low, Nathaniel - Private - 38th Infantry - Company: John Brookes - BLW 18915-160-12.

Low, Peter - Private - 38th Infantry - Company: John Brookes - Age: 22 - Height: 5' 8 1/2" - Born: Somerset Cty, MD - Trade: Tailor - Enlistment date: 25 Mar 1814 - Place: Norfolk, VA - Period: War - BLW 506-160-12 - Discharged at Craney Island, VA, on 15 Mar 1815.

Lowe, Bradley S. A. - First Lieutenant - Light Artillery - Other regiment: Corps of Artillery - Born: Maryland - Graduated from the US Military Academy on 21 Jul 1814; commissioned as a 3rd lieutenant, Light Artillery, on 21 Jul 1814; transferred to the Corps of Artillery on 17 May 1815; promoted to 2nd lieutenant on 31 Oct 1816; promoted to 1st lieutenant on 17 Sep 1818; resigned on 30 Sep 1819; died in Aug 1857.

Lowe, George - Private - 14th Infantry - Company: Reuben Gilder - Age: 26 - Height: 5' 5 1/2" - Born: Brunswick, NJ - Trade: Coach maker - Enlistment date: 14 Mar 1814 - Place: Plattsburgh, NY - Period: War - BLW 3693-160-12 - Discharged at Greenbush, NY, on 4 May 1815.

Lowe, Henry T. - Private - 14th Infantry - Company: Reuben Gilder - Age: 25 - Height: 5' 7" - Born: Easton, MD - Trade: Tailor - Enlistment date: 3 Sep 1813 - Place: Maryland - Period: War - BLW 14282-160-12 - Discharged at Greenbush, NY, on 4 May 1815.

Lowe, Samuel - Private - 14th Infantry - Company: Isaac Barnard - Age: 22 - Height: 5' 5 1/4" - Trade: Cooper - Enlistment date: 1 Feb 1813 - Period: 18 Mos - Prisoner of War, captured at Beaver Dams on 24 Jun 1813; discharged on 7 Aug 1814.

Lowman, Dennis - Private - 38th Infantry - Company: James Smith - Age: 32 - Height: 6' 1 1/4" - Born: Elk Ridge, MD - Trade: Farmer - Enlistment date: 19 Feb 1814 - Place: Craney Island, VA - Period: War - BLW 19432-160-12 - Discharged at Craney Island, VA, on 15 Mar 1815.

Lowman, James - Private - 2nd Infantry - Company: John Miller - Age: 20 - Height: 5' 6" - Born: Washington, MD - Enlistment date: 18 Jan 1813 - Period: 5 Yrs.

Lowman, James - Corporal - 5th Infantry - Company: Richard Bell - Other

By War of 1812 Soc. in MD

- regiment: 3rd Infantry (New) - Age: 20 - Height: 5' 6 1/2" - Born: Washington, MD - Trade: Shoemaker - Enlistment date: 1 May 1813 - Place: Williamsport, MD - Period: 5 Yrs - Discharged on 18 May 1818.
- Lowrey, James - Private - 14th Infantry - Prisoner of War (Halifax), sent to England.
- Lowry, Obadiah - Private - 36th Infantry - Company: Samuel Raisin - Age: 35 - Height: 5' 7" - Born: Halifax, VA - Trade: Farmer - Enlistment date: 10 Jul 1814 - Place: Norfolk, VA - Period: War - Died in hospital on 6 Mar 1815.
- Loyal, Seniel - Private - 38th Infantry - Company: Thomas Sangster - Age: 19 - Height: 5' - Born: Lisbon, Portugal - Trade: Seaman - Enlistment date: 17 Dec 1814 - Place: Baltimore - Period: 5 Yrs.
- Loyd, John - Private - 14th Infantry - Company: David Cummings - Age: 22 - Height: 5' 8" - Born: England - Enlistment date: 18 Jan 1815 - Place: Salisbury, MD - Period: War - Discharged on 18 Mar 1815.
- Loyd, Samuel - Private - 42nd Infantry - Company: Jonathan Robinson - Age: 30 - Height: 6' - Born: Maryland - Trade: Laborer - Enlistment date: 2 Nov 1814 - Place: Wilmington, DE - Period: 5 Yrs - Discharged on 10 Aug 1815, lameness prior to enlistment.
- Lozier, Elisha - Private - 14th Infantry - Company: Samuel Lane - Other regiment: 4th Infantry (New) - Age: 27 - Height: 5' 7 3/4" - Born: Bedford, PA - Trade: Brick maker - Enlistment date: 6 Jun 1812 - Place: Allegany Cty, MD - Period: 5 Yrs - Prisoner of War, exchanged on 15 Apr 1814; deserted at Fort Severn on 15 Jul 1815.
- Lubstins, Jacob - Private - 36th Infantry - Age: 31 - Height: 5' 7" - Born: Pennsylvania - Enlistment date: 13 Dec 1813 - Place: Baltimore.
- Lucas, Benjamin - Private - 14th Infantry - Company: Joseph Marshall - Age: 32 - Height: 5' 10 1/2" - Born: Cecil Cty, MD - Trade: Laborer - Enlistment date: 2 Sep 1814 - Period: 5 Yrs - BLW 16058-160-12 - Discharged at Greenbush, NY, on 25 May 1815, a cripple.
- Lucas, Bowen - Private - 36th Infantry - Company: Thomas Carbery - Age: 30 - Height: 5' 3" - Born: Maryland - Enlistment date: 25 Nov 1814 - Place: Leonardstown, MD - Period: War - BLW 462-160-12 - Discharged at Baltimore on 31 Mar 1815.
- Lucas, David - Private - 1st Light Dragoons - Company: George Birch - Other regiment: Corps of Artillery - Age: 26 - Height: 5' 4" - Born: Centreville, MD - Trade: Shoemaker - Enlistment date: 7 Jun 1813 - Place: Centreville, MD - Period: 5 Yrs - Discharged on 4 Sep 1815 on Surgeon's Certificate.
- Lucas, Ephraim - Private - 27th Infantry (OH) - Company: Alexander Hill - Other regiment: 19th Infantry then 17th Infantry - Age: 49 - Height: 6' 3/4" - Born: Maryland - Trade: Farmer - Enlistment date: 5 May 1814 - Place: Zanesville, OH - Period: War - BLW 7232-160-12 - Discharged at Chillicothe, OH, on 6 Jun 1815.
- Lucas, Ichabod - Private - 27th Infantry (OH) - Company: George Kesling - Other regiment: 19th Infantry - Age: 18 - Height: 5' 5" - Born: Maryland - Trade: Farmer - Enlistment date: 25 Apr 1814 - Place: Zanesville, OH - BLW 331-320-14 - Discharged at Zanesville, OH, on 27 Mar 1815; re-

enlisted.

Lucas, Richard - Private - 10th Infantry - Company: Jesse Copeland - Age: 17 - Height: 5' 2" - Born: Sussex Cty, MA - Trade: Farmer - Enlistment date: 10 Aug 1813 - Place: Annapolis - Period: 5 Yrs.

Luck, James - Private - 36th Infantry - Company: Joseph Hook - Age: 36 - Height: 5' 9" - Born: Hanover, VA - Trade: Laborer - Enlistment date: 3 Oct 1812 - Place: Richmond, VA - Period: 5 Yrs.

Lucket, William - Corporal - 12th Infantry - Company: James Charlton - Other regiment: 8th Infantry (New) - Age: 26 - Height: 5' 9" - Born: Charles Cty, MD - Trade: Farmer - Enlistment date: 9 May 1812 - Place: Alexandria, DC - Period: 5 Yrs - Discharged on 9 May 1817.

Luckett, Alexander - Private - 1st Infantry - Company: James Rhea - Age: 31 - Height: 5' 6 1/2" - Born: Maryland - Trade: Laborer - Enlistment date: Mar 1809 - Place: Detroit - Period: 5 Yrs - Prisoner of War on parole.

Luckett, Alexander - Corporal - 13th Infantry - Company: Israel Turner - Age: 29 - Height: 5' 7" - Born: Port Tobacco, MD - Trade: Farmer - Enlistment date: 26 Jan 1814 - Place: Greenbush, NY - Period: War - Prisoner of War on parole; discharged at Plattsburg, NY, on 20 Aug 1815.

Luckett, John - Sergeant - 2nd Artillery - Company: Frederick Evans - Other regiment: Corps of Artillery - Age: 32 - Height: 6' 1" - Born: Charles Cty, MD - Trade: Soldier - Enlistment date: 14 Dec 1812 - Place: Greenleaf Point, DC - Period: 5 Yrs - BLW 13808-160-12 - Discharged at Fort McHenry on 14 Dec 1817.

Luckett, John Roger Nelson - Captain - 2nd Infantry - Born: Maryland - Commissioned as an ensign on 26 Mar 1804; promoted to 2nd lieutenant on 30 Nov 1805; promoted to 1st lieutenant on 17 Aug 1807; promoted to captain on 6 Jul 1812; died on 5 May 1813.

Luckett, Nelson - Lieutenant Colonel - 1st Light Dragoons - Born: Maryland - BLW 787-160-50 - Commissioned as a 2nd lieutenant, U.S. Marine Corps, on 5 Jun 1807; promoted to 1st lieutenant on 26 Jan 1809; promoted to captain, light dragoons, on 23 Feb 1809; transferred to 1st Light Dragoons on 12 Mar 1812; promoted to major on 20 Jan 1813; promoted to lieutenant colonel on 1 Aug 1813; discharged on 1 Jun 1814.

Luckett, Valentine P. - Second Lieutenant - 14th Infantry - Other regiment: 1st Light Dragoons - Born: Maryland - Commissioned as an ensign, 14th Infantry, on 12 Mar 1812; promoted to 2nd lieutenant, 1st Light Dragoons, on 9 Oct 1812; resigned on 14 Jan 1814.

Luckey, Samuel - Private - 14th Infantry - Prisoner of War (Halifax), died at Halifax on 13 Sep 1813.

Luke, Abraham - Matross - 2nd Artillery - Company: Nathan Towson - BLW 15366-160-12.

Luley, Charles - Seaman - US Sea Fencibles - Company: Simmones Bunbury - Enlistment date: 18 Jul 1814 - Period: 1 Yr.

Lundgrade, Andrew - Artificer - 2nd Artillery - Company: Nathan Towson - Enlistment date: 1 Jun 1812 - Period: 18 Mos.

Lusk, John B. - Private - 14th Infantry - Company: Samuel Lane - Age: 29 -

Height: 6' 1" - Born: Pennsylvania - Enlistment date: 14 Jun 1812 - Place: Hagerstown, MD - Period: 5 Yrs - Killed in camp on 27 Nov 1812.

Lutz, Daniel - Private - 38th Infantry - Company: Charles Stansbury - Age: 26 - Height: 5' 7 3/4" - Born: Cumberland, MD - Trade: Farmer - Enlistment date: 28 Dec 1814 - Place: Baltimore - Period: War - Discharged on 3 Apr 1815.

Lutz, John - Private - 14th Infantry - Company: Joseph Marshall - Age: 24 - Height: 5' 8" - Born: Baltimore - Trade: Farmer - Enlistment date: 6 Sep 1814 - Period: War - BLW 13527-160-12 - Discharged at Greenbush, NY, on 3 May 1815.

Lutz, Joseph - Private - 14th Infantry - Company: Joseph Marshall.

Lutz, Joseph - Corporal - 2nd Artillery - Company: Nathan Towson - Age: 20 - Height: 5' 7 1/4" - Born: Baltimore - Trade: Brick maker - Enlistment date: 31 May 1812 - Place: Baltimore - Period: 5 Yrs - Pension: Old War IF-25234 - Wounded at Chippewa, UC, on 5 Jul 1814; discharged at Baltimore on 1 May 1815.

Luxon, John - Private - 36th Infantry - Company: Samuel Raisin - Age: 21 - Height: 5' 7" - Born: Prince George's Cty, MD - Trade: Laborer - Enlistment date: 29 Oct 1814 - Place: Georgetown, DC - Period: War - Pension: Land bounty to William Luxon, brother and other heirs at law of John Luxon - BLW 24854-160-12.

Lyda, Samuel - Drummer - 14th Infantry - Company: Reuben Gilder - Other regiment: 4th Infantry (New) - Age: 19 - Height: 5' 4" - Born: Maryland - Trade: Chair maker - Enlistment date: 27 Apr 1813 - Place: Liberty - Period: 5 Yrs.

Lydic, John - Private - 38th Infantry - Company: James Hook - BLW 4225-160-12.

Lydie, John - Private - 38th Infantry - Company: Shepperd Leakin - Age: 40 - Height: 5' 8" - Born: Fredericktown, MD - Trade: Blacksmith - Enlistment date: 14 Feb 1814 - Place: Hagerstown, MD - Period: War - Discharged at Fort Covington, MD, on 31 Mar 1815.

Lyle, John - Private - 5th Infantry - Age: 32 - Height: 5' 9" - Born: Ireland - Trade: Nailor - Enlistment date: 19 May 1813 - Place: Baltimore - Period: 5 Yrs.

Lynch, Daniel - Private - 5th Infantry - Company: Richard Whartenby - Age: 19 - Height: 5' 11" - Born: Baltimore - Trade: Cordwainer - Enlistment date: 7 Mar 1812 - Place: Baltimore - Period: 5 Yrs.

Lynch, Elijah - Private - 14th Infantry - Company: Robert Kent - Enlistment date: 12 Aug 1812 - Period: 5 Yrs - Pension: Placed on the pension rolls on 30 Oct 1818 - Discharged at Burlington, VT, on 6 Apr 1814, wounded in hand in action on 11 Nov 1813.

Lynch, John - Second Lieutenant - 14th Infantry - Born: Maryland - Commissioned as an ensign on 9 Oct 1812; promoted to 3rd lieutenant on 13 Mar 1813; promoted to 2nd lieutenant on 15 Aug 1813; killed during the Battle of Crystler's Field, UC, on 11 Nov 1813.

Lynch, John - Private - 14th Infantry - Prisoner of War (Halifax), sent to

Maryland Regulars in War of 1812

England.

Lynch, John - Sergeant - 12th Infantry - Company: Andrew Madison - Other regiment: 8th Infantry (New) - Age: 25 - Height: 5' 10" - Born: Frederick Cty, MD - Trade: Hatter - Enlistment date: 5 Aug 1812 - Place: Millwood - Period: 5 Yrs - BLW 16807-160-12 - Discharged at Fort Osage, MO, on 5 Aug 1817.

Lynch, Samuel - Clerk - 19th Infantry - Company: Carey Trimble - Other regiment: 17th Infantry - Age: 23 - Height: 5' 11 1/2" - Born: Baltimore - Trade: Laborer - Enlistment date: 23 Feb 1814 - Place: Franklinton, Franklin Cty, OH - Period: War - BLW 11111-160-12 - Discharged at Chillicothe, OH, on 6 Jun 1815.

Lynch, William - Private - 5th Infantry - Company: Richard Bell - Age: 40 - Height: 5' 10" - Born: Ireland - Trade: Farmer - Enlistment date: 26 Feb 1813 - Place: Baltimore - Period: 5 Yrs - Pension: Placed on the pension rolls on 4 Jun 1816 - Discharged at Plattsburgh, NY, on 5 Jul 1815 on Surgeon's Certificate.

Lynn, Philip - Private - 36th Infantry - Company: Samuel Raisin - Age: 39 - Height: 5' 6" - Born: Smyrna, DE - Enlistment date: 29 Sep 1813 - Place: Chestertown, MD - Period: 1 Yr - BLW 2946-160-12 - Re-enlisted on 15 Oct 1814 for the war; discharged on Washington, DC, on 20 Mar 1815.

Lyon, Samuel - Fifer - 2nd Artillery - Company: Frederick Evans - Other regiment: Corps of Artillery - Age: 15 - Height: 5' - Born: Ireland - Trade: Baker - Enlistment date: 8 Jul 1811 - Place: Fort Washington, MD - Period: 5 Yrs - Discharged on 8 Jul 1816.

Lyons, Jonathan - Private - 39th Infantry - Age: 44 - Height: 5' 7" - Born: Queen Anne's Cty, MD - Trade: Farmer - Enlistment date: 15 Feb 1815 - Period: 5 Yrs - Pension: Land bounty to Richard Lyons, son and other heirs at law on Jonathan Lyons - BLW 850-320-14 - Died on 17 Apr 1815.

Lyons, Robert - Private - 38th Infantry - Company: Isaac Aldridge - Age: 18 - Height: 5' 4" - Born: Harford Cty, MD - Trade: Cooper - Enlistment date: 20 Jan 1815 - Place: Baltimore - Period: War - BLW 719-320-12 - Discharged at Baltimore on 6 Apr 1815.

Lysaght, Richard - Private - 2nd Artillery - Company: Nathan Towson - Enlistment date: 22 Apr 1812 - Period: 5 Yrs.

M

Mack, David - Private - 14th Infantry - Company: Joseph Marshall - Age: 27 - Height: 5' 2" - Born: Pennsylvania - Trade: Farmer - Enlistment date: 21 Sep 1814 - Period: War - Died at Batavia, NY, on 29 Nov 1814.

Mack, George - Private - 14th Infantry - Company: Joseph Marshall - Age: 31 - Height: 5' 4" - Born: Berks Cty, PA - Trade: Farmer - Enlistment date: 27 Sep 1814 - Place: Baltimore - Period: War.

Mackelfresh, John H. - Private - Light Dragoons - Other regiment: 5th Infantry then 4th Infantry (New) - Age: 29 - Height: 5' 7" - Born: Baltimore - Trade: Harness maker - Enlistment date: 6 Dec 1808 - Period: 5 Yrs - Transferred to 5th Infantry on 4 Mar 1809, Captain Whaternby's Company as a musician;

By War of 1812 Soc. in MD

Prisoner of War, exchanged on 15 Apr 1814; re-enlisted at Pittsfield, MA, on 19 Apr 1814 for five years; died on 14 Aug 1819.

Mackey, Benjamin - Private - 22nd Infantry - Company: John Foster - Other regiment: 2nd Infantry (New) - Age: 28 - Height: 5' 8" - Born: Baltimore - Trade: Millwright - Enlistment date: 8 Jul 1813 - Place: Uniontown, PA - Period: 5 Yrs - Discharged at Sackets Harbor, NY, on 8 Jul 1815, wounded.

Mackey, John - Private - 14th Infantry - Enlistment date: 24 Dec 1812 - Period: 18 Mos - Prisoner of War (Halifax), sent to England; discharged on 31 Mar 1815.

Mackey, Robert - Seaman - US Sea Fencibles - Company: William Addison.

Madden, Stacy - Private - 5th Infantry - Age: 31 - Height: 5' 8 1/2" - Born: New Jersey - Trade: Shoemaker - Enlistment date: 30 Mar 1813 - Place: Baltimore - Period: 5 Yrs.

Madden, Thomas - Sergeant - 2nd Infantry - Company: William Platt - Age: 29 - Height: 5' 8" - Born: Montgomery Cty, MD - Trade: Blacksmith - Enlistment date: 25 Feb 1810 - Place: Columbia Springs - Period: 5 Yrs - Discharged at New Orleans on 31 Mar 1815.

Madding, Daniel - Private - 38th Infantry - Company: Shepperd Leakin - Age: 21 - Height: 5' 5" - Born: Derry, Ireland - Trade: Shoemaker - Enlistment date: 1 Sep 1814 - Place: Baltimore - Period: War.

Maddochs, Daniel - Private - 5th Infantry - Age: 43 - Height: 5' 8" - Born: England - Trade: Brick layer - Enlistment date: 29 Mar 1813 - Place: Baltimore - Period: 5 Yrs.

Maddocks, Samuel - Private - 5th Infantry - Company: Richard Bell - Other regiment: 3rd Infantry (New) - Age: 43 - Height: 5' 8" - Born: England - Trade: Brick layer - Enlistment date: 29 May 1813 - Place: Baltimore - Period: 5 Yrs - Discharged on 29 May 1818.

Maddon, John - Sergeant - 5th Infantry - Company: James Dorman - Other regiment: 3rd Infantry (New) - Age: 21 - Height: 5' 6 3/4" - Born: Baltimore - Trade: Tailor - Enlistment date: 28 Apr 1812 - Place: Alexandria, DC - Period: 5 Yrs - Discharged on 27 Apr 1817 and re-enlisted.

Maddon, Michael - Private - 36th Infantry - Age: 36 - Height: 5' 7" - Born: Baltimore - Enlistment date: 6 May 1813 - Place: Baltimore - Period: 1 Yr.

Maddox, George - Private - 2nd Infantry - Company: Francis Johnston - Age: 30 - Height: 5' 10" - Born: Charles Cty, MD - Trade: Farmer - Enlistment date: 22 Aug 1807 - Place: Fredericktown, MD - Period: 5 Yrs - Discharged at Fort Stoddard, AL, on 22 Aug 1812.

Maddy, Levin - Recruit - 26th Infantry - Company: William Bezeau - Race: 30 - Age: 30 - Height: 5' 6" - Born: Maryland - Trade: Laborer - Enlistment date: 9 Oct 1814 - Place: Philadelphia - Period: War - BLW 10295-160-12 - Listed as "Col'd" in his service record; discharged on 20 Mar 1815 at Philadelphia.

Madison, George - 14th Infantry - Company: Joseph Marshall - Waiter to Lieutenant William Thompson.

Madison, Jacob - Private - 14th Infantry - Company: Thomas Montgomery - Age: 21 - Height: 5' 8" - Born: Harford, Washington Cty, NY - Trade:

Laborer - Enlistment date: 16 Sep 1812 - Place: Tioga Cty, PA - Period: 5 Yrs.

Madison, John - Private - 5th Infantry - Company: Leroy Opie - Age: 22 - Height: 5' 6 1/2" - Born: Maryland - Trade: Hatter - Enlistment date: 14 Jun 1812 - Place: Harrisburg, PA - Period: 5 Yrs.

Maeyer, Henry - Private - 38th Infantry - Company: John Buck - Age: 45 - Height: 5' 11" - Born: Pennsylvania - Trade: Laborer - Enlistment date: 3 Jun 1813 - Place: Baltimore - Period: 1 Yr.

Magee, John - Private - 1st Artillery - Company: William Wilson - Age: 26 - Height: 5' 10" - Born: Baltimore - Trade: Shoemaker - Enlistment date: 26 Jun 1812 - Place: Charleston, SC - Period: 5 Yrs - Discharged on 26 Jun 1817.

Magee, Robert - Private - 14th Infantry - Company: David Cummings - Enlistment date: 3 Jun 1812 - Period: 5 yrs - Prisoner of War (Halifax), sent to England, captured at Beaver Dams on 24 Jun 1813; deserted at Fort McHenry on 20 Apr 1815.

Magee, Roger - Private - 38th Infantry - Company: John Brookes - BLW 7067-160-12.

Magness, William - Private - 14th Infantry - Company: Henry Grindage - Enlistment date: 1 Jul 1813 - Period: 5 Yrs - Pension: Land bounty to Mary Ann Clark, sister and other heirs at law of William Magness - BLW 17407-160-12 - Died on 23 Feb 1813.

Magrew, John - Private - 36th Infantry - Age: 21 - Height: 5' 7" - Born: Conawaga, PA - Enlistment date: 26 Jul 1813 - Place: Waynesburg, PA.

Magruder, Henry B. - First Lieutenant - 36th Infantry - Company: Joseph Hook - Born: Maryland - Commissioned as a 1st lieutenant on 7 Aug 1813; resigned on 28 Oct 1813.

Magruder, Peter - Second Lieutenant - 12th Infantry - Company: James Paxton - Born: Maryland - BLW 12005-160-50 - Commissioned as a 2nd lieutenant on 12 Mar 1812; resigned on 20 Nov 1813.

Magruder, Samuel W. - Surgeon's Mate - 14th Infantry - Born: Maryland - Commissioned as a surgeon's mate, 14th Infantry, on 28 Mar 1813; resigned on 27 Jun 1814.

Maguire, Hugh - Private - 14th Infantry - Company: Kenneth McKenzie - Age: 22 - Height: 5' 6" - Born: Ireland - Enlistment date: 25 Oct 1812 - Period: 18 Mos - Discharged at Black Rock, NY, on 1 Nov 1814.

Maher, Samuel - Private - 14th Infantry - Company: William McIlvane - Enlistment date: 23 Nov 1813 - Period: 5 Yrs - Prisoner of War (Halifax), discharged from Halifax on 3 Feb 1814, exchanged on 15 Apr 1814;.

Maison, Lewis - Private - 2nd Rifles - Age: 27 - Height: 5' 10" - Born: Maryland - Enlistment date: 4 Jun 1814 - Place: Louisville, KY.

Majore, John - Private - 25th Infantry - Company: Peter Bradley - Age: 25 - Height: 6' - Born: Baltimore - Trade: Tobacconist - Enlistment date: 23 Apr 1813 - Place: New York - Period: War - Discharged at Sackets Harbor, NY, on 17 May 1815.

Majors, William - Corporal - 38th Infantry - Company: John Buck - Age: 23

- Height: 5' 8 1/2" - Born: Baltimore - Trade: Tobacconist - Enlistment date: 14 Feb 1814 - Place: Baltimore - Period: War - BLW 24592-160-12 - Discharged at Baltimore on 30 Mar 1815.
Malog, Thomas (Maloy) - Private - 38th Infantry - Company: John Mowton - Age: 14 - Height: 4' 7" - Born: Baltimore - Trade: Tailor - Enlistment date: 13 Jul 1814 - Place: Craney Island, VA - Period: War - BLW 12798-160-12 - Discharged at Craney Island, VA, on 15 Mar 1815.
Malone, Caleb - Private - 14th Infantry - Enlistment date: 17 Aug 1813 - Period: 5 Yrs - Pension: Land bounty to Susannah Malone, widow of Caleb Malone - BLW 17845-160-50 - Died in Brownsville, NY, on 2 Feb 1815.
Malone, John - Private - 24th Infantry - Company: Frank Hampton - Other regiment: 7th Infantry (New) - Age: 35 - Height: 5' 10" - Born: Baltimore - Trade: Laborer - Enlistment date: 29 Jun 1813 - Period: 5 Yrs - Prisoner of War (Quebec), captured at Fort Niagara, NY, on 19 Dec 1813 and exchanged on 11 May 1814; discharged on 28 Jun 1818.
Maloney, Edward - Private - 2nd Artillery - Company: Nathan Towson - Age: 21 - Height: 5' 1 1/2" - Born: Baltimore - Trade: Shoemaker - Enlistment date: 3 May 1812 - Place: Baltimore - Period: 5 Yrs.
Maloney, James - Private - 5th Infantry - Company: George Brooks - Other regiment: 3rd Infantry (New) - Age: 25 - Height: 5' 9" - Born: Maryland - Trade: Cooper - Enlistment date: 16 Jun 1813 - Place: Baltimore - Period: 5 Yrs - Discharged on 16 Jun 1818.
Malure, William - Private - 17th Infantry - Company: Henry Crittenden - Other regiment: 3rd Infantry (New) - Age: 23 - Height: 5' 6 1/2" - Born: Baltimore - Trade: Shoemaker - Enlistment date: 16 Jul 1814 - Period: 5 Yrs - Discharged on 16 Jul 1819.
Manahan, David - Private - 14th Infantry - Company: Samuel Lane - Age: 25 - Height: 5' 4 1/2" - Born: Maryland - Enlistment date: 21 May 1812 - Place: Westminster, MD - Period: War - Prisoner of War (Halifax), sent to England; discharged on 30 Apr 1815.
Manley, William - Private - 26th Infantry - Company: William Bezeau - Age: 28 - Height: 5' 6" - Born: Ireland - Trade: Teller - Enlistment date: 31 Jan 1815 - Place: Baltimore - Period: War.
Manly, Joseph - Private - 36th Infantry - Company: Thomas Carbery - Age: 24 - Height: 5' 11" - Born: St. Mary's Cty, MD - Trade: Shoemaker - Enlistment date: 22 Mar 1814 - Place: Georgetown, DC - Period: War - BLW 4764-160-12 - Discharged at Baltimore on 31 Mar 1815.
Mann, John - Corporal - 26th Infantry - Company: William Bezeau - Age: 32 - Height: 5' 4" - Born: Lancaster Cty, PA - Trade: Tin plater - Enlistment date: 27 Sep 1814 - Place: Baltimore - Period: War - Discharged at Baltimore.
Mannah, Francis - Private - 14th Infantry - Company: Reuben Gilder - Other regiment: 4th Infantry (New) - Age: 25 - Height: 5' 9" - Born: Germany - Trade: Laborer - Enlistment date: 5 Mar 1813 - Place: Baltimore - Period: 5 Yrs - Deserted on 20 Jul 1815.
Manning, Samuel - Private - 14th Infantry - Company: Reuben Gilder - Age:

25 - Height: 5' 7" - Born: Baltimore - Enlistment date: 9 Aug 1813 - Period: 18 Mos - Pension: Land bounty to Elizabeth Smith, sister and heir at law of Samuel Maning - BLW 15618-160-12 - Wounded at Lyon's Creek, UC, on 19 Oct 1814; died at Buffalo on 9 Nov 1814.

Manning, William - Private - 38th Infantry - Company: Charles Stansbury - Age: 27 - Height: 5' 6 1/2" - Born: England - Trade: Mariner - Enlistment date: 23 Dec 1814 - Place: Fredericktown, MD - Period: War - BLW 1015-320-12 - Discharged at Baltimore on 6 Apr 1815.

Mannon, Hugh - Private - 36th Infantry - Company: Joseph Hook - Age: 27 - Height: 5' 3 1/2" - Born: Ireland - Enlistment date: 9 May 1813 - Place: Baltimore - Period: 1 Yr - Deserted on 10 Jun 1813.

Manor, Leonard - Private - 26th Infantry - Company: William Bezeau - Age: 27 - Height: 5' 5" - Born: Baltimore - Trade: Saddler - Enlistment date: 9 Nov 1814 - Place: Baltimore - Period: War - BLW 17995-160-12 - Discharged at Philadelphia on 23 Mar 1815.

Mansfield, Elijah - Private - 44th Infantry - Age: 23 - Height: 5' 11" - Born: Maryland - Enlistment date: 25 Feb 1815.

Mansfield, John - Second Lieutenant - 36th Infantry - Company: Joseph Hook - Born: Maryland - BLW 7984-160-60 - Commissioned as an ensign on 30 Apr 1813; promoted to 3rd lieutenant on 17 Mar 1814; promoted to 2nd lieutenant on 30 Sep 1814; discharged on 15 Jun 1815.

Mansfield, John - Private - 20th Infantry - Company: John Stanard - Age: 19 - Height: 5' 4 1/2" - Born: Frederick, MD - Enlistment date: 12 Aug 1812 - Period: 5 Yrs - BLW 20871-160-12 - Discharged at Greenbush, NY, on 14 Oct 1813 on Surgeon's Certificate.

Mansfield, Joseph - Corporal - 38th Infantry - Company: John Rothrock - BLW 7540-160-12.

Mansfield, Thomas - Private - 38th Infantry - Company: Charles Stansbury - Age: 35 - Height: 5' 7" - Born: Delaware - Trade: Farmer - Enlistment date: 8 Nov 1814 - Place: Baltimore - Period: 5 Yrs.

Mansfield, Vachael - Private - 38th Infantry - Company: Isaac Aldridge - Age: 19 - Height: 5' 1 1/2" - Born: Dorchester Cty, MD - Trade: Farmer - Enlistment date: 5 Dec 1814 - Place: Baltimore - Period: War - BLW 5504-160-12 - Discharged at Baltimore on 6 Apr 1815.

Manson, Henry - Seaman - US Sea Fencibles - Company: Simmones Bunbury - Age: 27 - Height: 5' 6 1/2" - Born: Denmark - Enlistment date: 28 Dec 1813 - Pension: Placed on the pension rolls on 1 Dec 1820 - Discharged on 22 Dec 1814; died on 27 Nov 1832.

Manville, Joshua - Private - 14th Infantry - Company: Reuben Gilder - Age: 26 - Height: 5' 6" - Born: South America - Trade: Tailor - Enlistment date: 21 Dec 1812 - Period: 18 Mos - Re-enlisted on 27 Feb 1814 for the war; died at Plattsburgh, NY, on 28 May 1814.

Marcelin, Joseph - Private - 5th Infantry - Age: 21 - Height: 5' 5" - Born: France - Enlistment date: 22 Dec 1813 - Place: Baltimore - Period: 5 Yrs - Deserted at York, PA, in Oct 1814.

March, William - Private - 14th Infantry - Prisoner of War (Halifax), died at

By War of 1812 Soc. in MD

Halifax on 7 Jan 1814.

Marks, John - Private - 38th Infantry - BLW 8719-160-12.

Marks, Nathaniel - Private - 17th Infantry - Company: Benjamin Sanders - Age: 22 - Height: 6' 1" - Born: Maryland - Trade: Laborer - Enlistment date: 16 Apr 1814 - Place: Kentucky - Period: War - BLW 20899-160-12 - Discharged at Chillicothe, OH, on 7 Jun 1815.

Marlow, Thomas - Sergeant - 14th Infantry - Company: Joseph Marshall - Age: 22 - Height: 5' 9" - Born: Prince George's Cty, MD - Trade: Farmer - Enlistment date: 23 Mar 1814 - Place: Upper Marlboro, MD - Period: War - Pension: Land bounty to Martha Ann Early, sister and other heirs at law of Thomas Marlow - BLW 24113-160-12 - Died at Canandaigua, NY, on 15 Jan 1815 from diarrhea.

Marr, Dennis - Private - 5th Infantry - Company: Leroy Opie - Other regiment: 3rd Infantry (New) - Age: 24 - Height: 5' 8" - Born: Baltimore - Trade: Tailor - Enlistment date: 26 Apr 1813 - Place: Baltimore - Period: 5 Yrs - Discharged on 26 Apr 1818.

Marsh, Henry - Private - 5th Infantry - Company: James Dorman - Age: 21 - Height: 5' 10 3/4" - Born: Maryland - Trade: Farmer - Enlistment date: 18 May 1812 - Place: Alexandria, DC - Period: 5 Yrs - BLW 9558-160-12 - Discharged at Greenbush, NY, on 30 May 1815 on Surgeon's Certificate, frost bite on right leg.

Marshall, Elias - Seaman - US Sea Fencibles - Company: Simmones Bunbury - Enlistment date: 31 Jan 1814 - Period: 1 Yr - Discharged on 30 Jan 1815.

Marshall, John - Corporal - 12th Infantry - Company: James Charlton - Other regiment: 8th Infantry (New) - Age: 22 - Height: 5' 8" - Born: Ireland - Trade: Weaver - Enlistment date: 20 Jun 1812 - Place: Baltimore - Period: 5 Yrs - Discharged on 20 Jun 1817.

Marshall, Joseph - Captain - 14th Infantry - Age: 38 - Pension: Placed on the pension roll on 11 May 1830 - Commissioned as a 1st lieutenant on 12 Mar 1812; promoted to captain on 15 Aug 1813; Prisoner of War (Dartmouth), released on 31 May 1814; discharged on 15 Jun 1815.

Marshall, Josias - Private - 36th Infantry - Company: Thomas Carbery - Enlistment date: 7 Jan 1814 - Period: 1 Yr - Discharged on 6 Jan 1815.

Marshall, Richard - Private - 38th Infantry - Company: Charles Stansbury - Age: 29 - Height: 5' 4 3/4" - Born: Sussex, England - Trade: Brick layer - Enlistment date: 4 Nov 1814 - Place: Baltimore - Period: War - BLW 2741-160-12 - Discharged at Baltimore on 6 Apr 1815.

Marshall, Thomas - Private - 38th Infantry - Company: Isaac Aldridge - Age: 37 - Height: 5' 9" - Born: Somerset Cty, MD - Trade: Clerk - Enlistment date: 5 Oct 1814 - Place: Baltimore - Period: War - BLW 6923-160-12 - Discharged at Baltimore on 6 Apr 1815.

Marshall, Thomas - Corporal - 14th Infantry - Company: Joseph Marshall - Other regiment: 4th Infantry (New) - Height: 5' 9 1/2" - Born: Northumberland Cty, VA - Trade: Ship joiner - Enlistment date: 24 Jul 1812 - Place: Cambridge, MD - Period: 5 Yrs - Deserted at Fort Severn, MD, on 14 Jul 1815.

Maryland Regulars in War of 1812

Marshall, Thomas - Private - 14th Infantry - Company: Thomas Montgomery - Enlistment date: 27 Jul 1812 - Period: 5 Yrs.

Martell, William N. - Private - 36th Infantry - Company: Samuel Raisin - Age: 30 - Height: 5' 9" - Born: L'Orient, France - Trade: Powdermaker - Enlistment date: 17 Jun 1814 - Place: Georgetown, DC - Period: War - Pension: Land bounty to Ann Montell, widow and other heirs at law of William Martell - BLW 418-160-12 - Discharged at Washington, DC, on 20 Mar 1815.

Martin, Andrew - Private - 17th Infantry - Company: Caleb Holder - Age: 25 - Height: 5' 8" - Born: Maryland - Trade: Brick layer - Enlistment date: 23 Aug 1814 - Place: Louisville, KY - Period: War - BLW 15885-160-12 - Discharged at Detroit on 30 Jul 1815.

Martin, Bain - Private - 14th Infantry - Company: Joseph Marshall - Pension: Old War IF-10074.

Martin, Caleb - Private - 14th Infantry - Company: David Cummings - Age: 20 - Height: 5' 10" - Born: Chester, PA - Trade: Shoemaker - Enlistment date: 24 Jul 1814 - Place: Baltimore - Period: War - Deserted on 3 Aug 1814.

Martin, David W. - Private - 14th Infantry - Company: David Cummings - Age: 33 - Height: 5' 10" - Born: Lancaster, PA - Trade: Laborer - Enlistment date: 21 Nov 1814 - Place: Baltimore - Period: War - Discharged on 18 Mar 1815.

Martin, Dory - Private - 2nd Infantry - Company: Francis Johnston - Other regiment: 1st Infantry (New) - Age: 23 - Height: 5' 11 1/2" - Born: Baltimore - Trade: Farmer - Enlistment date: 26 Jun 1803 - Period: 5 Yrs - BLW 17253-160-12 - Re-enlisted at Columbian Springs on 26 Mar 1808 for five years; re-enlisted at Mount Vernon on 28 Jan 1813 for five years; discharged at Camp Ann, LA, on 27 Jan 1818.

Martin, John - Private - 14th Infantry - Company: David Cummings - Race: 16 - Age: 16 - Height: 4' 4" - Born: Port Tobacco, MD - Trade: Soldier - Enlistment date: 22 Oct 1814 - Place: Alexandria, DC - Period: War - BLW 21317-160-12 - Discharged at Baltimore on 16 Mar 1815.

Martin, John - Private - 26th Infantry - Company: William Bezeau - Age: 21 - Height: 5' 3 1/2" - Born: Baltimore - Trade: Weaver - Enlistment date: 9 Oct 1814 - Place: Baltimore - Period: War - Pension: Wife Catharine, SO-7212, SC-8603, WO-37320, WC-29619 - Discharged at Philadelphia on 23 Mar 1815; also served in Captain William McLaughlin's Company, MD Militia.

Martin, John - 38th Infantry - Company: James Hook - Pension: SO-2480, SC-4933.

Martin, John - Sergeant - 16th Infantry - Company: William Davenport - Age: 25 - Height: 5' 8" - Born: Baltimore - Trade: Printer - Enlistment date: 3 Aug 1813 - Place: Troy - Period: 5 Yrs.

Martin, John - Private - 5th Infantry - Company: William Henshaw - Other regiment: 3rd Infantry (New) - Age: 17 - Height: 4' 11" - Born: Pittsburgh - Trade: Farmer - Enlistment date: 31 Jul 1813 - Place: Baltimore - Period: 5 Yrs - Discharged at Fort Howard, WI, on 11 Aug 1818.

Martin, John - Private - 2nd Rifles - Company: Benjamin Desha - Other regiment: Rifles - Age: 22 - Height: 5' 7 1/2" - Born: St. Mary's Cty, MD - Trade: Laborer - Enlistment date: 17 Nov 1814 - Period: War - Discharged at Detroit on 30 Jun 1815.

Martin, John B. - Second Lieutenant - 38th Infantry - Born: Maryland - BLW 4950-160-50 - Commissioned as an ensign on 20 May 1813; promoted to 3rd lieutenant on 1 May 1814; promoted to 2nd lieutenant on 9 Sep 1814; discharged on 15 Jun 1815.

Martin, Robert - Private - 2nd Light Dragoons - Company: William Littlejohn - Age: 27 - Height: 5' 5" - Born: Baltimore - Trade: Shoemaker - Enlistment date: 22 Feb 1814 - Place: Baltimore - Period: 5 Yrs - Deserted on 1 Oct 1815.

Martin, William - Corporal - 2nd Artillery - Company: Daniel Cushing - Other regiment: Corps of Artillery - Age: 26 - Height: 5' 7 3/4" - Born: Maryland - Trade: Laborer - Enlistment date: 7 Nov 1814 - Period: War - BLW 7512-160-12 - Discharged at Fort Meigs, OH, on 25 Jul 1815.

Martin, William - Private - 36th Infantry - Company: Samuel Raisin - Age: 36 - Height: 5' 6 1/2" - Born: Dorchester Cty, MD - Enlistment date: 17 May 1813 - Place: Cambridge, MD - Period: 1 Yr - Died at Camp Cold Spring on 15 Oct 1813.

Martin, William - Private - 14th Infantry - Company: David Cummings - Other regiment: 4th Infantry (New) - Age: 25 - Height: 5' 8" - Born: Baltimore - Trade: Tobacconist - Period: 5 Yrs - BLW 12890-160-12 - Prisoner of War (Halifax), sent to England, captured at Beaver Dams on 24 Jun 1813; discharged at Fort Hawkins on 9 Apr 1817.

Mason, Henry - Seaman - US Sea Fencibles - Company: Simmones Bunbury - Pension: Old War IF-25245 - Died on 27 Nov 1832.

Mason, Henry - Private - 15th Infantry - Company: White Young - Age: 40 - Height: 5' 5" - Born: Baltimore - Trade: Comb maker - Enlistment date: 12 May 1812 - Place: Bristol - Period: 5 Yrs - Discharged on 12 May 1817.

Mason, Ira - Private - 14th Infantry - Prisoner of War (Halifax), died at Halifax on 28 Oct 1813.

Mason, John T. - Third Lieutenant - 36th Infantry - Born: Virginia - Pension: Land bounty to Nancy T. Mason, widow of John T. Mason - BLW 10723-160-50 - Commissioned as an ensign on 10 Jun 1814; promoted to 3rd lieutenant on 30 Sep 1814; discharged on 15 Jun 1814.

Mason, Richard - Private - 2nd Light Dragoons - Company: William Littlejohn - Age: 36 - Height: 5' 6 1/2" - Born: Maryland - Trade: Farmer - Enlistment date: 9 Sep 1813 - Place: Head of Elk, MD - Period: 5 Yrs - Died on 3 Apr 1815.

Mason, Roderick R. - Private - 36th Infantry - Age: 32 - Height: 5' 7 1/2" - Born: Stansbury, RI - Enlistment date: 19 Apr 1813 - Place: Baltimore - Period: War.

Mason, Thomas - Private - 38th Infantry - Company: Shepperd Leakin - Age: 25 - Height: 5' 9 1/2" - Born: Caroline Cty, MD - Trade: Hatter - Enlistment date: 6 Aug 1814 - Place: Baltimore - Period: War - Discharged on 27 Mar 1815.

Maryland Regulars in War of 1812

Massey, Alexander - Private - 14th Infantry - Company: Joseph Marshall - Age: 28 - Height: 5' 5" - Born: New York City - Trade: Laborer - Enlistment date: 12 Sep 1814 - Place: Baltimore - Period: War - BLW 3398-160-12 - Discharged at Greenbush, NY, on 3 May 1815.

Massey, Benjamin - Corporal - 29th Infantry - Company: James Spencer - Age: 21 - Height: 5' 10" - Born: Queen Anne, MD - Trade: Blacksmith - Enlistment date: 10 Feb 1814 - Place: Plattsburgh, NY - Period: War - Discharged at Plattsburgh, NY, on 29 Jun 1815.

Matchett, James - Private - 38th Infantry - Company: Shepperd Leakin - Age: 25 - Height: 5' 3" - Born: Ireland - Trade: Weaver - Enlistment date: 29 Jul 1814 - Place: Baltimore - Period: 5 Yrs - Deserted on 6 Jan 1815.

Mathews, Samuel - Private - 1st Rifles - Age: 30 - Height: 5' 5" - Born: Baltimore Cty - Trade: Cooper - Enlistment date: 28 Aug 1814 - Period: 5 Yrs - Deserted on 10 Sep 1814.

Mathias, Charles - Recruit - 26th Infantry - Company: William Bezeau - Race: 26 - Age: 26 - Height: 5' - Born: Maryland - Trade: Shoemaker - Enlistment date: 8 Oct 1814 - Place: Philadelphia - Period: 5 Yrs - Listed as "Colored" in his service record; discharged at Philadelphia on 16 May 1815.

Mathing, Samuel - Private - 36th Infantry - BLW 22055-160-12.

Matney, George - Private - 12th Infantry - Company: Andrew Madison - Age: 20 - Height: 5' 5 1/2" - Born: Allegany Cty, MD - Trade: Farmer - Enlistment date: 11 Sep 1812 - Period: 5 Yrs - Discharged at Fort Erie, UC, on 17 Jun 1815 on Surgeon's Certificate, epilepsy.

Matthews, Jared - Private - 14th Infantry - Company: Thomas Montgomery - Enlistment date: 27 Jul 1812 - Period: 5 Yrs.

Matthews, John - Private - 14th Infantry - Company: David Cummings - Enlistment date: 30 Dec 1812 - Period: 5 Yrs - Prisoner of War, sent to England; deserted at Fort McHenry on 18 Apr 1815.

Matthews, Thomas - Private - 2nd Infantry - Company: John Bowyer - Age: 31 - Height: 5' 6 1/2" - Born: Baltimore - Trade: Farmer - Enlistment date: 2 Jan 1807 - Place: Fort Adams, MS - Period: 5 Yrs.

Mattingly, Lewis - Private - 36th Infantry - Company: Thomas Carbery - Age: 21 - Height: 5' 10" - Born: St. Mary's Cty, MD - Trade: Trader - Enlistment date: 27 Feb 1814 - Place: Leonardstown, MD - Period: War - Discharged at Baltimore on 3 Mar 1815.

Mattison, James - Private - 38th Infantry - Company: Isaac Aldridge - Age: 19 - Height: 5' 6" - Born: Cumberland Cty, NJ - Trade: Laborer - Enlistment date: 13 Nov 1814 - Place: Baltimore - Period: War - BLW 6135-160-12 - Discharged at Baltimore on 6 Apr 1815.

Mattock, Joseph - Private - 14th Infantry - Company: David Cummings - Enlistment date: 12 Jul 1812 - Period: 5 Yrs - Prisoner of War (Halifax), captured at Beaver Dams on 24 Jun 1813, discharged from Halifax on 9 Nov 1813; discharged 21 Apr 1815; still a minor, enlisted without the consent of his parents or guardians.

Mattox, Joshua (Maddox) - Private - 17th Infantry - Company: John Chunn - Age: 26 - Height: 5' 10" - Born: Somerset Cty, MD - Trade: Laborer -

By War of 1812 Soc. in MD

Enlistment date: 2 Apr 1814 - Place: Detroit - Period: War - BLW 17972-160-12 - Discharged at Chillicothe, OH, on 9 Jun 1815.

Mawe, Michael - Private - 36th Infantry - Company: Joseph Hook - Age: 43 - Height: 5' 8" - Born: Annapolis - Enlistment date: 24 May 1813 - Place: Baltimore - Period: 1 Yr - Deserted on 4 Sep 1813.

Maxfield, James - Private - 1st Light Dragoons - Company: Alexander Cummings - Age: 26 - Height: 5' 7" - Born: Baltimore - Trade: Joiner - Enlistment date: 22 Apr 1812 - Place: Pittsburgh - Period: 5 Yrs - Deserted on 15 Mar 1816.

Maxwell, John G. - Private - 4th Rifles - Company: Matthew Magee - Age: 26 - Height: 5' 10" - Born: Baltimore - Trade: Farmer - Enlistment date: 5 May 1814 - Period: 5 Yrs - Pension: Land bounty to Eliza Hustead, sister & other heirs at law of John G. Maxwell - BLW 24858-160-12 - Died on 29 Mar 1815.

Maxwell, John Y. - Private - US Volunteers - Company: Stephen Moore - Enlistment date: 8 Sep 1812 - Period: 1 Yr.

Maxwell, Robert - Private - 14th Infantry - Company: David Cummings - Other regiment: 4th Infantry (New) - Age: 29 - Height: 5' - Born: Ireland - Trade: Tanner - Enlistment date: 11 Feb 1813 - Period: 5 Yrs - BLW 24105-160-12 - Prisoner of War (Halifax), sent to England; discharged at Montgomery, AL, on 13 Feb 1818.

Maxwell, Samuel - Private - 38th Infantry - Company: Shepperd Leakin - Age: 20 - Height: 5' 7 1/2' - Born: Baltimore - Trade: Shoemaker - Enlistment date: 12 Aug 1814 - Place: Baltimore - Period: War - BLW 3591-160-12 - Discharged at Fort McHenry on 28 Mar 1815.

Maxwell, William - Quartermaster Sergeant - 14th Infantry - Other regiment: 4th Infantry (New) - Age: 21 - Height: 5' 8 1/2" - Born: Kent Cty, MD - Trade: Hatter - Enlistment date: 23 Mar 1814 - Place: Rising Sun, MD - Period: 5 Yrs - Died on 10 Feb 1819.

Maxwell, William - Private - US Volunteers - Company: Stephen Moore - Enlistment date: 8 Sep 1812 - Period: 1 Yr.

Maxwell, William F. - Private - 1st Artillery - Company: George Armistead - Age: 28 - Height: 5' 10" - Born: Ireland - Trade: Shoemaker - Enlistment date: 27 Apr 1812 - Place: Fort McHenry - Period: 5 Yrs - Died at Fort Columbus, NY, on 1 Apr 1813.

Maybe, Jacob - Private - 14th Infantry - Prisoner of War (Halifax), discharged from Halifax on 3 Feb 1814.

Mayers, John - Private - 12th Infantry - Age: 23 - Height: 5' 9" - Born: Washington Cty, MD - Enlistment date: 1 Jan 1814 - Place: Charles Town, VA - Period: War.

Mayhew, Nicholas - Private - 36th Infantry - Company: Thomas Carbery - Enlistment date: 23 Aug 1814 - Period: War - Deserted on 15 Sep 1814.

McAllister, Robert - Corporal - US Volunteers - Company: Stephen Moore - Enlistment date: 8 Sep 1812 - Period: 1 Yr.

McAnnall, John - Private - 36th Infantry - Company: Joseph Hook - Age: 32 - Height: 5' 6 1/2" - Born: Ireland - Trade: Carpenter - Enlistment date: 28

Jun 1812 - Place: Georgetown, DC - Period: War - BLW 14836-160-12 - Discharged at Fort Covington, MD, on 30 Mar 1815.

McArra, George - Private - 2nd Artillery - Company: Frederick Evans - Other regiment: Corps of Artillery - Age: 34 - Height: 5' 8" - Born: Scotland - Trade: Baker - Enlistment date: 2 Jun 1812 - Place: Annapolis - Period: 5 Yrs - BLW 9973-160-12 - Discharged at Washington, DC, on 19 Jun 1817.

McAter, Leonard - Private - 2nd Light Dragoons - Company: John Stokes - Age: 24 - Height: 5' 11 3/4" - Born: Frederick, MD - Enlistment date: 27 Dec 1812 - Place: Salisbury, NC - Period: 18 Mos.

McBier, Charles - Private - 36th Infantry - Company: Joseph Hook - Age: 29 - Height: 5' 3" - Born: Caresol, Ireland - Trade: Mariner - Enlistment date: 17 Jul 1812 - Place: Fredericktown, MD - Period: War - BLW 6848-160-12 - Discharged at Fort Covington, MD, on 30 Mar 1815.

McBlair, Hamilton - Private - 14th Infantry - Company: William McIlvane - Age: 38 - Height: 5' 10 1/2" - Born: Ireland - Trade: Laborer - Enlistment date: 30 Mar 1814 - Place: Towsontown, MD - Period: War - BLW 16220-160-12 - Discharged at Greenbush, NY, on 5 May 1815.

McBride, Hugh - Private - 14th Infantry - Company: Joseph Marshall - Age: 23 - Height: 5' 11 3/4" - Born: Ireland - Trade: Brick maker - Enlistment date: 22 Jun 1813 - Period: 5 Yrs - Deserted at Fort Severn, MD, on 14 Jul 1815.

McBrierty, James - Private - 14th Infantry - Company: David Cummings - Age: 24 - Height: 5' 8" - Born: Ireland - Trade: Cooper - Enlistment date: 12 Jan 1814 - Place: Towsontown, MD - Period: War - Deserted on 24 Aug 1814.

McBuckner, Patrick (Buckner) - 38th Infantry - Company: James Smith - Pension: SO-15737, SC-16944 - Also served in Captain John Hammon's Company, VA Militia.

McCahan, Samuel - Private - 36th Infantry - Company: Joseph Hook - Age: 37 - Height: 5' 10" - Born: Chester, PA - Trade: Carpenter - Enlistment date: 26 May 1814 - Place: Fredericktown, MD - Period: War - BLW 217-160-12 - Discharged at Fort Covington, MD, on 30 Mar 1815.

McCandless, John - Private - 28th Infantry - Company: Henry Gist - Age: 29 - Height: 5' 6" - Born: Maryland - Trade: Farmer - Enlistment date: 1 Dec 1814 - Place: Lexington, KY - Period: War - Discharged at Lower Sandusky, OH (Fort Stephenson), on 25 Jun 1815.

McCandra, John - Private - 13th Infantry - Age: 26 - Height: 5' 3 1/2" - Born: Maryland - Enlistment date: 4 May 1812 - Place: Auburn, MD - Period: 5 Yrs.

McCandy, William - Private - 26th Infantry - Company: William Bezeau - Age: 22 - Height: 5' 9" - Born: Baltimore - Enlistment date: 6 Oct 1814 - Place: Baltimore.

McCann, William - Private - 14th Infantry - Company: Reuben Gilder - Age: 33 - Height: 5' 8" - Born: Ireland - Trade: Blacksmith - Enlistment date: 2 Mar 1814 - Period: War.

McCanniker, Joseph - Private - 5th Infantry - Age: 22 - Height: 6' 2 1/2" -

By War of 1812 Soc. in MD

Born: Maryland - Trade: Carpenter - Enlistment date: 20 Oct 1814 - Place: York, PA - Period: 5 Yrs.

McCarrier, Stephen - Private - 14th Infantry - Company: Richard Arell - Age: 28 - Height: 5' 10" - Born: Vincent, Chester Cty, PA - Trade: Stone mason - Enlistment date: 13 Sep 1812 - Place: Baltimore - Period: 5 Yrs - Prisoner of War, exchanged on 15 Apr 1814; discharged at Greenbush, NY, on 20 Mar 1815 with loss of two finders on right hand.

McCarty, Anthony - Private - 3rd Infantry - Company: Robert Moore - Age: 28 - Height: 5' 5" - Born: Frederick Cty, MD - Trade: Boatman - Enlistment date: 8 Aug 1808 - Place: Fredericktown, MD - Period: 5 Yrs - Re-enlisted at English Turn, LA, on 30 Aug 1813 for the war.

McCarty, Charles - Private - 14th Infantry - Company: William McIlvane - Other regiment: 4th Infantry (New) - Age: 24 - Height: 5' 10 1/2" - Born: Caroline Cty, MD - Trade: Farmer - Enlistment date: 22 May 1812 - Place: Denton, MD - Period: 5 Yrs - Prisoner of War, exchanged on 15 Apr 1814; discharged on 22 May 1817.

McCarty, John - Private - 1st Artillery - Company: James Reed - Age: 39 - Height: 5' 7 1/2" - Born: Berks Cty, PA - Trade: Farmer - Enlistment date: 19 Apr 1814 - Place: Annapolis - Period: 5 Yrs - Discharged at Philadelphia on 26 Jan 1816 for infirmity.

McCarty, Joseph - Private - 2nd Artillery - Company: Sander Donoho - Age: 33 - Height: 5' 8" - Born: Fredericktown, MD - Trade: Laborer - Enlistment date: 23 Oct 1813 - Place: Savannah, GA - Period: 5 Yrs - Discharged on 22 Oct 1818.

McCarty, Michael - Private - 36th Infantry - Company: Samuel Raisin - Age: 20 - Height: 5' 6" - Born: Ireland - Trade: Weaver - Enlistment date: 13 Oct 1814 - Place: Georgetown, DC - Period: War - BLW 862-160-12 - Discharged at Washington, DC, on 20 Mar 1815.

McCarty, Patrick - Private - 2nd Light Dragoons - Company: William Littlejohn - Other regiment: Light Dragoons - Age: 19 - Height: 6' - Born: New London, CT - Trade: Farmer - Enlistment date: 22 Apr 1814 - Place: Baltimore - Period: War - BLW 19166-160-12 - Discharged at Carlisle Barracks, PA, on 4 May 1815.

McCarty, William - Private - 5th Infantry - Company: Colin Buckner - Age: 24 - Height: 5' 10" - Born: Harford Cty, MD - Trade: Cooper - Enlistment date: 12 May 1813 - Place: Baltimore - Period: 5 Yrs - Pension: Land bounty to James Everit, father and other heirs at law of William McCarty, alias John Everit - BLW 12825-160-12 - Died on 13 Jul 1813.

McCaulley, William C. - Private - 38th Infantry - Company: Shepperd Leakin - Age: 26 - Height: 5' 7" - Born: Lancaster Cty, PA - Trade: Tanner - Enlistment date: 6 Aug 1814 - Place: Baltimore - Period: War - Discharged at Fort McHenry on 28 Mar 1815.

McCausland, Hugh - Private - 5th Infantry - Company: James Dorman - Age: 34 - Height: 5' 6" - Born: Ireland - Trade: Weaver - Enlistment date: 5 May 1812 - Place: Baltimore - Period: 5 Yrs - Prisoner of War, captured at Fort Erie in Nov 1812, paroled; discharged on 4 May 1817.

Maryland Regulars in War of 1812

McCausland, William - Private - 1st Artillery - Company: George Armistead - Age: 34 - Height: 5' 10" - Born: Ireland - Trade: Farmer - Enlistment date: 10 Jan 1812 - Place: Baltimore - Period: 5 Yrs - Discharged on 10 Jan 1817.

McCaw, Robert - Private - 38th Infantry - Company: Charles Stansbury - Age: 40 - Height: 5' 8" - Born: County Antrim, Ireland - Trade: Weaver - Enlistment date: 2 Jan 1815 - Place: Fredericktown, MD - Period: War - BLW 261-320-12 - Discharged at Baltimore on 6 Apr 1815.

McClain, Andrew - Private - 36th Infantry - Age: 18 - Height: 5' 5" - Born: Ireland - Enlistment date: 24 Jun 1814 - Place: Westminster, MD.

McClary, David - Private - 14th Infantry - Company: Samuel Lane.

McClay, Moses - Private - 14th Infantry - Company: Joseph Marshall - Age: 21 - Height: 5' 3" - Born: Alexandria, DC - Trade: Gunsmith - Enlistment date: 9 Jun 1814 - Place: Alexandria, DC - Period: War - BLW 4542-160-12 - Discharged at Greenbush, NY, on 3 May 1815.

McCleary, Henry - Artificer - 1st Artillery - Company: George Armistead - Other regiment: Corps of Artillery - Age: 26 - Height: 5' 9 1/2" - Born: Maryland - Trade: Carpenter - Enlistment date: 16 Dec 1811 - Place: Fort McHenry - Period: 5 Yrs - Discharged on 22 Aug 1816.

McClellan, Robert - Sergeant - 36th Infantry - Company: Thomas Carbery - Age: 31 - Height: 5' 8" - Born: County Antrim, Ireland - Trade: Ship joiner - Enlistment date: 8 Apr 1814 - Place: Georgetown, DC - Period: War - BLW 11824-160-12 - Discharged on 7 Apr 1815.

McClement, Jonathan - Corporal - 36th Infantry - Company: Joseph Hook - Age: 39 - Height: 5' 7" - Born: Pennsylvania - Enlistment date: 30 Apr 1812 - Place: Baltimore - Period: 1 Yr.

McClintock, John - Private - 2nd Infantry - Company: William Boots - Age: 22 - Height: 5' 7 1/2" - Born: Cumberland Cty, PA - Trade: Cooper - Enlistment date: 22 Apr 1807 - Place: Baltimore - Period: 5 Yrs.

McCluer, John T. - Private - 3rd Artillery - Company: Horace Watson - Other regiment: Corps of Artillery - Age: 21 - Height: 5' 7" - Born: Kent Cty, MD - Trade: Merchant - Enlistment date: 29 Mar 1814 - Place: New York - Period: War - BLW 647-160-12 - Discharged at New York on 25 May 1815.

McClung, John - Private - 27th Infantry (OH) - Company: George Sanderson - Other regiment: 19th Infantry - Age: 21 - Height: 6' - Born: Maryland - Trade: Farmer - Enlistment date: 28 Apr 1813 - Period: 1 Yr - BLW 14145-160-12 - Re-enlisted at Detroit on 17 Mar 1814 for the war; discharged at Detroit on 20 Jul 1815.

McClure, John - Private - 22nd Infantry - Company: Thomas Lawrence - Age: 20 - Height: 6' 1" - Born: Maryland - Trade: Laborer - Enlistment date: 3 Apr 1813.

McClure, John - Private - 14th Infantry - Company: Samuel Lane - Enlistment date: 26 Jun 1812 - Period: 5 Yrs - Prisoner of War (Halifax), discharged from Halifax on 3 Feb 1814, exchanged on 15 Apr 1814; deserted at Easton on 19 Jan 1814.

McClure, John - Private - 2nd Light Dragoons - Company: William Littlejohn - Age: 19 - Height: 5' 6" - Born: Baltimore - Trade: Manufacturer -

By War of 1812 Soc. in MD

Enlistment date: 19 Mar 1814 - Place: Baltimore - Period: War - Discharged on 1 May 1815.

McClure, William - Private - 14th Infantry - Company: Thomas Montgomery - Age: 54 - Height: 5' 10" - Born: Ireland - Trade: Soap boiler - Enlistment date: 5 Jun 1812 - Period: 5 Yrs - BLW 6335-160-12 - Discharged at Plattsburgh, NY, on 18 Mar 1814, for old age inability.

McCollam, James - Artificer - 1st Artillery - Company: George Armistead - Other regiment: Corps of Artillery - Age: 43 - Height: 5' 8" - Born: Ireland - Trade: Carpenter - Enlistment date: 30 Mar 1811 - Place: Fort McHenry - Period: 5 Yrs - Discharged on 11 Sep 1815.

McComas, Charles - Seaman - US Sea Fencibles - Company: William Addison.

McConaghey, Benjamin - Private - 14th Infantry - Age: 25 - Height: 5' 10 1/2" - Trade: Plasterer - Enlistment date: 26 Jan 1813 - Period: 18 Mos - Discharged at Philadelphia on 2 Jan 1815.

McConity, Hugh - Private - 14th Infantry - Company: Reuben Gilder - Age: 33 - Height: 5' 6" - Born: Ireland - Enlistment date: 26 Jun 1813 - Period: War - Died at Buffalo on 4 Nov 1814 from wounds received during the Battle of Lyons Creek on 19 Oct 1814.

McConklin, William - Sergeant - 14th Infantry - BLW 16541-160-12.

McConky, David - Private - 17th Infantry - Company: Harris Hickman - Other regiment: 3rd Infantry (New) - Age: 28 - Height: 5' 7" - Born: Maryland - Trade: Tailor - Enlistment date: 9 Feb 1814 - Place: Steubenville, OH - Period: 5 Yrs - Deserted on 1 Sep 1815.

McConn, William - Private - 14th Infantry - Company: William McIlvane - Pension: Wife Rosanna, Old War WF-21127 Rejected, Old War WF-16181 Rejected.

McConville, Samuel - Private - 5th Infantry - Company: James Dorman - Other regiment: 3rd Infantry (New) - Age: 20 - Height: 5' 9" - Born: St. Mary's, MD - Trade: Tailor - Enlistment date: 5 Apr 1812 - Place: Hagerstown, MD - Period: 5 Yrs - Discharged on 4 Apr 1817.

McCool, James - Private - 5th Infantry - Company: Leroy Opie - Other regiment: 3rd Infantry (New) - Age: 40 - Height: 5' 4" - Born: Ireland - Trade: Laborer - Enlistment date: 20 Feb 1812 - Place: Baltimore - Period: 5 Yrs - Discharged on 20 Feb 1818.

McCord, Joseph - Private - 17th Infantry - Company: Martin Hawkins - Age: 37 - Height: 6' - Born: Maryland - Trade: Farmer - Enlistment date: 28 Apr 1814 - Period: War - BLW 8321-160-12 - Discharged at Chillicothe, OH, on 4 Jun 1815.

McCord, Thomas - Private - 38th Infantry - Company: John Buck - Age: 33 - Height: 5' 6" - Born: Newark, DE - Trade: Farmer - Enlistment date: 5 Mar 1814 - Place: Baltimore - Period: War - Discharged on 30 Mar 1815.

McCorkill, James - Private - 14th Infantry - Company: Richard Arell - Other regiment: 4th Infantry (New) - Age: 23 - Height: 5' 9" - Born: Chester, PA - Trade: Shoemaker - Enlistment date: 19 Jul 1812 - Place: Westchester, PA - Period: 5 Yrs - Prisoner of War (Halifax), captured at Beaver Dams on 24

Jun 1813, discharged from Halifax on 3 Feb 1813, exchanged on 15 Apr 1814; deserted from Fort Seven on 13 Jul 1815.

McCormack, John - Private - 36th Infantry - Company: Thomas Carbery - Age: 41 - Height: 5' 5 1/2" - Born: Baltimore - Trade: Laborer - Enlistment date: 24 Feb 1814 - Place: Leonardstown, MD - Period: War - BLW 14699-160-12 - Discharged at Baltimore on 31 Mar 1815.

McCormick, Barney - Private - 14th Infantry - Company: William McIlvane - Age: 40 - Height: 5' 5" - Born: Ireland - Trade: Farmer - Enlistment date: 11 Jun 1813 - Period: 5 Yrs - BLW 1113-160-12 - Discharged at Greenbush, NY, on 5 May 1815.

McCormick, Edward - Private - 12th Infantry - Company: Andrew Madison - Age: 27 - Height: 5' 7" - Born: Montgomery Cty, MD - Trade: Shoemaker - Enlistment date: 20 Mar 1814 - Place: New Market, VA - Period: War - Discharged at Buffalo on 31 May 1815.

McCormick, Patrick - Private - 14th Infantry - Company: Reuben Gilder - Age: 30 - Height: 5' 8" - Born: Ireland - Enlistment date: 15 Dec 1812 - Period: 18 Mos - Discharged at Plattsburgh, NY, on 15 Jun 1814.

McCotter, Willis - Private - 16th Infantry - Company: William Smith - Age: 51 - Height: 5' 10" - Born: Easton, MD - Trade: Joiner - Enlistment date: 21 May 1813 - Period: 5 Yrs.

McCowan, Michael - Private - 14th Infantry - Company: Thomas Montgomery - Enlistment date: 13 Apr 1813 - Period: 5 Yrs - Pension: Land bounty to John McCowan, brother and other heirs at law of Michael McCowan - BLW 13120-160-12 - Died in Feb 1814.

McCoy, Alexander - Seaman - US Sea Fencibles - Company: William Addison.

McCoy, Archibald - Private - 22nd Infantry - Company: John Pentland - Age: 35 - Height: 6' 1" - Born: Maryland - Trade: Farmer - Enlistment date: 25 Sep 1812 - Place: Sherlysburg - Period: 18 Mos - Discharged at Buffalo on 12 Oct 1814.

McCoy, Henry - Private - 36th Infantry - Company: Thomas Carbery - Age: 22 - Height: 5' 10" - Born: Maryland - Enlistment date: 27 Apr 1814 - Place: Washington, DC - Period: War - BLW 25586-160-12 - Discharged on 31 Mar 1815.

McCoy, James - Private - 5th Infantry - Company: Richard Whartenby - Age: 25 - Height: 5' 6 1/2" - Born: Ireland - Trade: Printer - Enlistment date: 29 May 1811 - Place: Baltimore - Period: 5 Yrs - Discharged at Greenbush, NY, on 3 May 1815 for general debility.

McCoy, John - Private - 7th Infantry - Company: Samuel Vail - Other regiment: 1st Infantry (New) - Age: 32 - Height: 5' 9" - Born: Maryland - Trade: Farmer - Enlistment date: 13 Jul 1814 - Period: 5 Yrs - Discharged at Fort Selden on 12 Jul 1819.

McCoy, John - Private - 1st Light Dragoons - Company: Alexander Cummings - Age: 24 - Height: 5' 8 1/2" - Born: Cecil Cty, MD - Trade: Cooper - Enlistment date: 22 Jun 1812 - Place: Baltimore - Period: 5 Yrs - Deserted at Fort Columbus, NY, on 26 Aug 1815.

By War of 1812 Soc. in MD

McCracken, John - Quarter Gunner - US Sea Fencibles - Company: William Addison.

McCracken, John - Private - US Volunteers - Company: Stephen Moore - Enlistment date: 8 Sep 1812 - Period: 1 Yr.

McCrackin, James - Seaman - US Sea Fencibles - Company: John Gill - Age: 25 - Height: 5' 8" - Born: Baltimore - Enlistment date: 10 Jan 1814 - Place: Baltimore.

McCray, Philip - Private - 14th Infantry - Company: Reuben Gilder - Age: 33 - Height: 5' 7 1/2" - Born: Ireland - Enlistment date: 5 Apr 1813 - Period: War - Pension: Land bounty to Joseph Sterret, nephew and other heirs at law of Philip McCray - BLW 25907-160-12 - Died on 20 Nov 1814 of sickness.

McCrea, Archibald F. - Private - 2nd Artillery - Company: Frederick Evans - Age: 30 - Height: 5' 5 1/2" - Born: Scotland - Enlistment date: 14 Jun 1812 - Place: Fort McHenry - Period: 5 Yrs - Discharged at Fort McHenry on 23 Jul 1817.

McCrimmin, Daniel - Second Lieutenant - 14th Infantry - Company: Thomas Montgomery - Born: Maryland - Enlistment date: 25 Jun 1812 - Period: 5 Yrs - Served as a sergeant and then sergeant major, 14th Infantry; commissioned as an ensign on 30 Apr 1813; promoted to 3rd lieutenant on 15 May 1813; promoted to 2nd lieutenant on 1 Oct 1814; discharged on 15 Jun 1815.

McCrory, James - Private - 14th Infantry - Prisoner of War (Halifax), captured at Beaver Dams on 24 Jun 1813, discharged from Halifax on 9 Nov 1813.

McCue, John - Private - 36th Infantry - Company: Samuel Raisin - Age: 26 - Height: 5' 3 1/2" - Born: Dublin, Ireland - Trade: Stone mason - Enlistment date: 17 Sep 1814 - Place: Stevensburg, VA - Period: 5 Yrs - Discharged at Greenleaf Point, DC, on 2 Apr 1815.

McCullin, James - Private - 14th Infantry - Company: Joseph Marshall - BLW 3220-160-12.

McCullock, Samuel - Private - 2nd Rifles - Company: John O'Fallon - Other regiment: Rifles - Age: 28 - Height: 5' 10" - Born: Maryland - Trade: Millwright - Enlistment date: 3 May 1814 - Place: Lexington, KY - Period: War - BLW 18361-160-12 - Discharged at Detroit on 30 Jun 1815.

McCullough, John - Private - 36th Infantry - Company: Joseph Hook - Age: 28 - Height: 5' 8" - Born: Cecil Cty, MD - Trade: Plasteror - Enlistment date: 7 Jul 1814 - Place: Baltimore - Period: War - BLW 5503-160-12 - Discharged at Fort Covington, MD, on 30 Mar 1815.

McCurd, James - Private - 13th Infantry - Company: William Adams - Other regiment: 5th Infantry (New) - Age: 27 - Height: 5' 8 1/2" - Born: Baltimore - Trade: Cordwainer - Enlistment date: 2 Feb 1814 - Place: New York - Period: 5 Yrs - Discharged on 9 Sep 1815 for debility.

McCutchen, Robert - Private - 38th Infantry - Company: John Mowton - BLW 13463-160-12.

McDade, William - Private - 36th Infantry - Company: Thomas Carbery - Age: 27 - Height: 5' 8" - Born: Montgomery Cty, MD - Trade: Blacksmith - Enlistment date: 25 Feb 1814 - Place: Georgetown, DC - Period: 5 Yrs.

Maryland Regulars in War of 1812

McDaniel, Thomas - Private - 36th Infantry - Company: Joseph Hook - Age: 18 - Height: 5' 4 1/2" - Born: Delaware - Trade: Blacksmith - Enlistment date: 16 Jun 1814 - Place: Georgetown, DC - Period: War - BLW 1644-160-12 - Discharged at Fort Covington, MD, on 30 Mar 1815.

McDaniels, Horatio - Private - 36th Infantry - Age: 22 - Height: 5' 8" - Born: Charles Cty, MD - Trade: Farmer - Enlistment date: 11 Nov 1814 - Period: War - BLW 2911-160-12 - Discharged at Georgetown, DC, on 13 Mar 1815.

McDaniels, John - Private - 14th Infantry - Company: Reuben Gilder - Age: 28 - Height: 5' 5" - Born: Denton, MD - Trade: Butcher - Enlistment date: 2 Mar 1814 - Place: Maryland - Period: War - BLW 11593-160-12 - Discharged at Greenbush, NY, on 4 May 1815.

McDaniels, John - Private - 38th Infantry - Company: Shepperd Leakin - Age: 25 - Height: 5' 5" - Born: Anne Arundel, MD - Enlistment date: 27 Aug 1813 - Place: Baltimore - Period: 1 Yr - Re-enlisted on 2 Apr 1814 for the war; discharged at Fort McHenry on 28 Mar 1815.

McDeamoth, Charles - Private - 36th Infantry - Age: 43 - Height: 5' 6" - Born: Kent Cty, MD - Enlistment date: 28 Aug 1813 - Place: Havre de Grace, MD.

McDennis, Patrick - Private - 14th Infantry - Enlistment date: 25 Jan 1813 - Period: 18 Mos - Prisoner of War (Halifax), sent to England; discharged on 31 Mar 1815.

McDevil, Cornelius - Private - 14th Infantry - Company: Samuel Lane - Age: 33 - Height: 5' 3" - Born: Ireland - Enlistment date: 1 Aug 1812 - Period: 5 Yrs.

McDevot, John - Private - 36th Infantry - Company: Samuel Raisin - Age: 25 - Height: 5' 7" - Born: Harford Cty, MD - Trade: Shoemaker - Enlistment date: 1 Mar 1814 - Place: Annapolis - Period: War.

McDonald, Barney - Private - 38th Infantry - Age: 25 - Height: 5' 6" - Born: Ireland - Trade: Barber - Enlistment date: 20 Nov 1814 - Place: Fredericktown, MD - Period: War - Deserted.

McDonald, Daniel - Seaman - US Sea Fencibles - Company: Simmones Bunbury - Age: 23 - Height: 5' 2" - Born: Scotland - Trade: Farmer - Enlistment date: 16 Feb 1814.

McDonald, James - Captain - 14th Infantry - Born: Maryland - Commissioned as a 1st lieutenant on 12 Mar 1812; promoted to captain on 1 Oct 1813; died at Buffalo on 11 Nov 1814.

McDonald, John - Sergeant - 38th Infantry - Company: Shepperd Leakin - Age: 28 - Height: 5' 4 1/2" - Born: Anne Arundel, MD - Trade: Painter - Enlistment date: 2 Apr 1814 - Period: War.

McDonald, Patrick - Private - 38th Infantry - Company: Charles Stansbury - Age: 27 - Height: 5' 8" - Born: Wexford, Ireland - Trade: Laborer - Enlistment date: 17 Nov 1814 - Place: Fredericktown, MD - Period: War - Discharged at Baltimore on 6 Apr 1815.

McDonald, Samuel - Gunner - US Sea Fencibles - Company: John Gill - Enlistment date: 21 Dec 1813 - Period: 1 Yr.

McDonald, Theopilus - Private - 14th Infantry - Prisoner of War (Halifax), discharged from Halifax on 3 Feb 1814.

By War of 1812 Soc. in MD

McDongle, Robert - Private - 38th Infantry - Company: John Rothrock - BLW 8931-160-12.

McDonough, Patrick - Private - 38th Infantry - Company: James Hook - Age: 35 - Height: 6' - Born: Ireland - Enlistment date: 26 Jan 1814 - Place: Baltimore - Period: 1 Yr - Discharged on 26 Jan 1815.

McDonough, Peter - Private - 14th Infantry - Company: Kenneth McKenzie - Age: 38 - Height: 5' 2" - Trade: Farmer - Enlistment date: 22 Dec 1812 - Period: 5 Yrs - Prisoner of War, captured at Beaver Dams on 24 Jun 1813; discharged on 7 Aug 1814.

McDowell, Thomas - Seaman - US Sea Fencibles - Company: John Gill - Age: 21 - Height: 5' 8 3/4" - Born: Baltimore City - Enlistment date: 28 Dec 1813 - Period: 1 Yr.

McElhatton, John - Private - 7th Infantry - Company: Walter Overton - Age: 32 - Height: 5' 10" - Born: Maryland - Trade: Farmer - Enlistment date: 11 Jun 1812 - Place: Fort Claiborne, AL - Period: 5 Yrs - Deserted at Fort Claiborne, AL, on 14 Mar 1814.

McElory, William - Private - 2nd Artillery - Company: Nathan Towson - Age: 30 - Prisoner of War, captured during the Battle of Stoney Creek.

McFarland, Joseph - Private - 5th Infantry - Company: Richard Whartenby - Age: 25 - Height: 5' 6 1/2" - Trade: Printer - Enlistment date: 14 Mar 1811 - Place: Baltimore - Period: 5 Yrs.

McFarlin, John - Private - 36th Infantry - Company: Thomas Carbery - Enlistment date: 14 Oct 1813 - Period: 1 Yr - Discharged on 12 Oct 1814.

McFarlin, Peter - Private - 36th Infantry - Company: Charles Randolph - Age: 27 - Height: 5' 5" - Born: Doron, Ireland - Enlistment date: 16 Jul 1813 - Place: Waynesburg, PA - Period: 1 Yr.

McGahy, Edward - Private - 5th Infantry - Company: James Dorman - Other regiment: 3rd Infantry (New) - Age: 32 - Height: 5' 5 1/2" - Born: Scotland - Trade: Carpenter - Enlistment date: 8 Jul 1812 - Place: Baltimore - Period: 5 Yrs - Discharged at Pittsburgh on 16 Aug 1815 on Surgeon's Certificate.

McGarvey, Patrick - Private - 38th Infantry - Company: Shepperd Leakin - Age: 24 - Height: 5' 6 1/2" - Born: County Donegal, Ireland - Trade: Laborer - Enlistment date: 14 Aug 1814 - Place: Baltimore - Period: War - BLW 24420-160-12 - Discharged at Fort McHenry on 28 Mar 1815.

McGee, James - Musician - 14th Infantry - Company: Reuben Gilder - Other regiment: 4th Infantry (New) - Age: 16 - Height: 4' 9" - Born: Annapolis - Trade: Laborer - Enlistment date: 28 Aug 1812 - Place: Alexandria, DC - Period: 5 Yrs - BLW 14820-160-12 - Discharged at Raleigh, NC, on 12 Sep 1817.

McGee, James - Private - 14th Infantry - Company: David Cummings - Other regiment: 4th Infantry (New) - Age: 28 - Height: 5' 6" - Born: Ireland - Trade: Laborer - Enlistment date: 29 Oct 1814 - Place: Williamsport, MD - Period: 5 Yrs - Died on 14 Aug 1815.

McGee, James - Private - 14th Infantry - Company: Reuben Gilder - Other regiment: 4th Infantry (New) - Age: 38 - Height: 5' 6" - Born: Strabane, County Tyrone, Ireland - Trade: Shoemaker - Enlistment date: 10 May 1813

- Place: Annapolis - Period: 5 Yrs - BLW 18407-160-12 - Discharged at Fort Gadsden, FL, on 13 May 1818.

McGee, James - Private - 26th Infantry - Company: William Bezeau - Age: 36 - Height: 6' 1" - Born: New Castle, DE - Trade: Shoemaker - Enlistment date: 30 Jun 1814 - Place: Baltimore - Period: War - BLW 493-320-12 - Discharged at Baltimore.

McGerry, John - Private - 5th Infantry - Company: James Dorman - Other regiment: 3rd Infantry (New) - Age: 34 - Height: 5' 9 1/2" - Born: Maryland - Trade: Farmer - Enlistment date: 5 Jun 1812 - Place: Creagerstown, MD - Period: 5 Yrs - Discharged on 14 Jun 1817.

McGill, Daniel - Private - 2nd Artillery - Company: Nathan Towson - Pension: Old War IF-9950.

McGill, James - Private - 2nd Light Dragoons - Company: Samuel Hopkins - Age: 25 - Height: 5' 5" - Born: Maryland - Enlistment date: 17 Jul 1813 - Place: Philadelphia - Period: 5 Yrs - Deserted on 15 Sep 1813.

McGilton, John - Private - 5th Infantry - Age: 34 - Height: 5' 10" - Born: Baltimore - Trade: Shoemaker - Enlistment date: 22 Feb 1813 - Place: Baltimore - Period: 5 Yrs.

McGinity, Nicholas - Private - 38th Infantry - Company: John Rothrock - BLW 22182-160-12.

McGinley, Andrew - Private - 14th Infantry - Company: Kenneth McKenzie - Wounded at Fort George, UC, on 27 May 1813.

McGinnis, Henry - Private - 36th Infantry - Age: 21 - Height: 5' 8" - Born: Lancaster, PA - Enlistment date: 13 Jul 1814 - Place: Baltimore - Period: 1 Yr.

McGlindey, Michael - Private - 36th Infantry - Company: Joseph Hook - Enlistment date: 14 Jun 1814 - Period: War - Pension: Land bounty to George McGlindey and other heirs at law of Michael McGlindey - BLW 27096-160-12 - Died on 20 Aug 1814.

McGlinn, Shederick - Private - 38th Infantry - Company: John Buck - Age: 34 - Height: 5' 6 1/2" - Born: Maryland - Trade: Laborer - Enlistment date: 19 Mar 1814 - Place: Westmoreland, VA ?? - Period: War - BLW 16085-160-12 - Discharged at Fort Covington, MD, on 30 mar 1815.

McGonegall, Daniel - Private - 36th Infantry - Age: 40 - Height: 5' 5 1/2" - Born: Ireland - Enlistment date: 26 Jul 1812 - Place: Havre de Grace, MD.

McGonigall, Daniel - Private - 1st Artillery - Company: Heman Fay - Other regiment: Corps of Artillery - Age: 36 - Height: 5' 7 1/4" - Born: Ireland - Trade: Farmer - Enlistment date: 18 Apr 1814 - Place: Annapolis - Period: War - Pension: Land bounty to Daniel McGonigall and Bridget Fulton, children and only heirs at law of Daniel McGonigall - BLW 27240-160-12 - Discharged at Washington, DC, on 9 May 1815.

McGowan, George - Private - 14th Infantry - Company: David Cummings - Age: 27 - Height: 5' 4 1/2" - Born: Ireland - Trade: Weaver - Enlistment date: 24 Oct 1814 - Place: Hagerstown, MD - Period: 5 Yrs.

McGowan, John - Private - 14th Infantry - Company: David Cummings - Other regiment: 4th Infantry (New) - Age: 39 - Height: 5' 5 1/2" - Born:

By War of 1812 Soc. in MD

Ireland - Trade: Tailor - Enlistment date: 1 Sep 1812 - Period: 5 Yrs - Prisoner of War (Halifax), sent to England; discharged at Montpelier, MS, on 1 Sep 1817.

McGowan, Levi - Private - 19th Infantry - Company: Carey Trimble - Other regiment: 17th Infantry - Age: 22 - Height: 5' 7" - Born: Maryland - Trade: Breeches maker - Enlistment date: 23 Feb 1814 - Place: Franklinton, Franklin Cty, OH - Period: 1 Yr - Discharged on 21 Apr 1815.

McGrath, Patrick - Private - 38th Infantry - Company: Shepperd Leakin - Age: 18 - Height: 5' 4 1/2' - Born: New York City - Trade: Bar keeper - Enlistment date: 30 Apr 1814 - Place: Baltimore - Period: 5 Yrs.

McGraw, John - Private - 5th Infantry - Age: 27 - Height: 5' 7" - Born: Frederick, MD - Trade: Laborer - Enlistment date: 18 Oct 1814 - Place: Lancaster, PA - Period: 5 Yrs - Deserted at Lancaster, PA, on 21 Oct 1814.

McGraw, John - Private - 36th Infantry - Company: Joseph Hook - Age: 21 - Height: 5' 6" - Born: Frederick, MD - Trade: Laborer - Enlistment date: 2 Mar 1814 - Place: Annapolis - Period: War - Discharged at Fort Covington, MD, on 30 Mar 1815.

McGraw, John - Private - 36th Infantry - Age: 20 - Height: 5' 6" - Born: Pennsylvania - Enlistment date: 30 Aug 1814 - Place: Norfolk, VA.

McGraw, Michael - Private - 12th Infantry - Company: Thomas Sangster - Other regiment: 8th Infantry (New) - Age: 30 - Height: 5' 3 1/2" - Born: Ireland - Trade: Brick maker - Enlistment date: 23 Apr 1812 - Place: Havre de Grace, MD - Period: 5 Yrs - Discharged on 23 Apr 1817.

McGrew, John - Private - 22nd Infantry - Age: 21 - Height: 5' 8" - Born: Maryland - Trade: Carpenter - Enlistment date: 10 Jun 1813 - Place: Hanover, York Cty, PA - Period: 5 Yrs - Deserted.

McGriffin, John - Private - 3rd Artillery - Company: Alexander Brookes - Other regiment: Corps of Artillery - Age: 32 - Height: 5' 7" - Born: Cecil Cty, MD - Trade: Farmer - Enlistment date: 29 Jan 1813 - Place: Caningsburgh - Period: 5 Yrs - Prisoner of War, exchanged on 28 Apr 1814; discharged on 29 Jan 1818.

McGuire, Barkley - Private - 1st Artillery - Company: George Armistead - Age: 30 - Height: 6' 3" - Born: Ireland - Trade: Laborer - Enlistment date: 9 Nov 1811 - Place: Fort McHenry - Period: 5 Yrs - Discharged on 15 May 1815 for disability.

McGuire, Hugh - Private - 14th Infantry - Prisoner of War (Halifax), sent to England.

McGuire, William - Private - 36th Infantry - Company: Charles Randolph - Age: 21 - Height: 5' 9" - Born: Monongahela, VA - Enlistment date: 21 Jun 1813 - Place: Frederick, MD - Period: 1 Yr.

McHenry, Patrick - Private - 36th Infantry - Company: Thomas Carbery - Age: 29 - Height: 5' 6" - Born: Ireland - Trade: Shoemaker - Enlistment date: 24 Jun 1814 - Place: Georgetown, DC - Period: War - BLW 5592-160-12 - Discharged at Baltimore on 31 Mar 1815.

McHugh, Patrick - Private - 36th Infantry - Company: Mortimer Hall - Age: 35 - Height: 5' 6" - Born: Ireland - Enlistment date: 25 Sep 1813 - Place:

Maryland Regulars in War of 1812

Baltimore - Period: 1 Yr.

McIlvaine, William - Captain - 14th Infantry - Born: Pennsylvania - Pension: Wife Wilhelmina S., WO-37384, WC-28662 - Commissioned as a captain on 12 Mar 1812; discharged on 15 Jun 1815.

McIver, William - Private - 14th Infantry - Prisoner of War (Halifax), sent to England.

McKartey, John - Private - 36th Infantry - Age: 44 - Height: 5' 8" - Born: Berks Cty, PA - Enlistment date: 15 Aug 1813 - Place: Waynesburg, PA.

McKee, Edward - Private - 14th Infantry - Company: Richard Arell - Other regiment: 4th Infantry (New) - Age: 25 - Height: 5' 6" - Born: New Castle, DE - Trade: Carpenter - Enlistment date: 26 May 1812 - Place: West Chester, PA - Period: 5 Yrs - Prisoner of War, exchanged on 15 Apr 1814; discharged on 26 May 1817.

McKee, William - Private - 36th Infantry - Company: Thomas Carbery - Enlistment date: 29 Oct 1813 - Period: 1 Yr.

McKee, William - Private - 36th Infantry - Age: 23 - Height: 5' 6" - Born: Georgetown, DC - Enlistment date: 3 Mar 1814 - Place: Georgetown, DC - Period: War - Deserted.

McKeever, Charles - Private - 5th Infantry - Age: 34 - Height: 5' 7" - Born: Ireland - Enlistment date: 1 Mar 1812 - Place: Baltimore - Period: 5 Yrs.

McKeever, Charles - Private - 14th Infantry - Prisoner of War (Halifax), sent to England.

McKelden, George - Private - 38th Infantry - Company: John Buck - Age: 22 - Height: 5' 6" - Born: Ireland - Trade: Laborer - Enlistment date: 13 Jun 1814 - Place: Baltimore - Period: War.

McKelip, Joseph - Private - 14th Infantry - Company: Samuel Lane - Enlistment date: 15 Jun 1812 - Period: 1 Yr - Died at Black Rock, NY, on 16 Dec 1812.

McKeller, Benjamin - Private - 14th Infantry - Company: William McIlvane - Enlistment date: 30 Nov 1812 - Period: 18 Mos - Prisoner of War, exchanged on 15 Apr 1815; discharged at Plattsburgh, NY, on 5 Jul 1814.

McKenna, Solomon - Private - 12th Infantry - Company: Thomas Sangster - Other regiment: 8th Infantry (New) - Age: 19 - Height: 5' 8" - Born: Alexandria, DC - Trade: Cooper - Enlistment date: 6 Jul 1812 - Place: Baltimore - Period: 5 Yrs - Discharged on 6 Jul 1817.

McKenney, John - Private - 36th Infantry - Age: 24 - Height: 5' 9 1/4" - Born: Ireland - Trade: Shoemaker - Enlistment date: 2 Jan 1815 - Period: War - BLW 11910-160-12 Cancelled - Discharged at Georgetown, DC, on 11 Mar 1815.

McKenney, William - Sergeant Major - 20th Infantry - Age: 20 - Height: 5' 8 1/2" - Born: Baltimore - Trade: Cooper - Enlistment date: 4 Jan 1814 - Place: Charlottesville, VA - Period: War - Discharged at Norfolk, VA, on 15 Mar 1815.

McKenny, John - Private - 12th Infantry - Company: Thomas Sangster - Age: 20 - Born: Ireland - Trade: Shoemaker - Enlistment date: 8 Jul 1812 - Place: Baltimore - Period: 18 Mos - Prisoner of War since 1 Jan 1814.

By War of 1812 Soc. in MD

McKensie, David - Private - 1st Infantry - Age: 21 - Height: 5' - Born: Baltimore - Enlistment date: 7 Oct 1814 - Place: New York - Recruit rejected.

McKenzie, David - Private - 14th Infantry - Enlistment date: 26 Jan 1813 - Period: 18 Mos - Prisoner of War (Halifax), discharged from Halifax on 3 Feb 1814, exchanged on 15 Apr 1814.

McKenzie, John - Private - 13th Infantry - Age: 21 - Height: 5' 10" - Born: Maryland - Enlistment date: 1 Nov 1814 - Place: Auburn, MD - Period: 5 Yrs.

McKenzie, Kenneth - Captain - 14th Infantry - Age: 36 - Born: Maryland - Commissioned as a captain on 6 Jul 1812; Prisoner of War (Dartmouth), sent to Prison Ship *Success*; sent to Ashburton, England, on 23 Jun 1814; discharged on 15 Jun 1815; died on 29 Sep 1817.

McKerver, Charles - Private - 14th Infantry - Company: David Cummings - Other regiment: 4th Infantry (New) - Enlistment date: 1 Jan 1813 - Period: 5 Yrs - Prisoner of War, sent to England; discharged on 31 Dec 1817.

McKim, William - Drummer - 16th Infantry - Company: George Steele - Age: 15 - Height: 4' 6" - Born: Centreville, MD - Trade: Musician - Enlistment date: 3 Jan 1814 - Place: Philadelphia - Period: 5 Yrs - Discharged on 6 Jan 1817 after furnishing a substitute.

McKinsey, James - Private - 42nd Infantry - Company: George Barker - Age: 39 - Height: 4' 1 1/2" - Born: Baltimore - Trade: Shoemaker - Enlistment date: 25 May 1814 - Place: New Castle, DE - Period: War - BLW 18147-160 - Discharged at Philadelphia on 19 May 1815.

McKinsey, John - Private - 19th Infantry - Age: 30 - Height: 5' 6 1/2" - Born: Allegany Cty, MD - Enlistment date: 2 Jul 1812 - Period: 5 Yrs.

McKinsey, John - Private - 14th Infantry - Company: David Cummings - Age: 19 - Height: 5' 5" - Born: Scotland - Trade: Sailor - Enlistment date: 28 Dec 1814 - Place: Salisbury, MD - Period: War - Discharged at Baltimore on 18 Mar 1815.

McKnight, John - Musician - 19th Infantry - Company: George Kesling - Age: 40 - Height: 5' 8" - Born: Maryland or Virginia - Trade: Tailor - Enlistment date: 12 May 1814 - Place: Dunhantown, Clermont County, OH - Period: War - BLW 14620-160-12 - Discharged at Zanesville, OH, on 10 Jun 1815.

McKnight, Joshua - Private - 12th Infantry - Company: Thomas Post - Age: 29 - Height: 5' 4" - Born: Somerset Cty, MD - Trade: Shipwright - Enlistment date: 19 Oct 1813 - Place: Wheeling, VA - Period: War - BLW 6989-160-12 - Discharged at Buffalo on 31 May 1815.

McKnight, Lewis - Seaman - US Sea Fencibles - Company: Simmones Bunbury - Enlistment date: 31 Jan 1814 - Period: 1 Yr - Discharged on 30 Jan 1815.

McLane, James - Private - 36th Infantry - Company: Samuel Raisin - Age: 25 - Height: 5' 8 1/4" - Born: Fairfax, VA - Trade: Sailor - Enlistment date: 21 Sep 1814 - Place: District of Columbia - Period: War - Died in hospital on 3 Jan 1815.

McLane, John - Private - 36th Infantry - Company: Samuel Raisin - Age: 18 -

Maryland Regulars in War of 1812

Height: 5' 5" - Born: Ireland - Trade: Farmer - Enlistment date: 24 Jun 1814 - Place: Fredericktown, MD - Period: War - Deserted on 14 Jan 1815.

McLaughlin, James - Musician - 17th Infantry - Age: 31 - Height: 5' 9 1/2" - Born: Maryland - Enlistment date: 18 Aug 1813.

McLaughlin, James - Musician - 36th Infantry - Company: Samuel Raisin - Age: 15 - Height: 4' - Born: Ireland - Trade: Bookbinder - Enlistment date: 17 Aug 1814 - Period: War - BLW 9816-160-12 - Discharged at Washington, DC, on 20 Mar 1815.

McLaughlin, James - Private - 28th Infantry - Company: George Stockton - Other regiment: 3rd Infantry (New) - Age: 29 - Height: 5' 10" - Born: St. Mary's Cty, MD - Trade: Hatter - Enlistment date: 1 Jun 1812 - Place: Elizabethtown, KY - Period: 5 Yrs - BLW 5730-160-12 - Wounded at Fort Meigs, OH; discharged at Detroit on 11 Nov 1815.

McLaughlin, James - Private - 36th Infantry - Age: 34 - Height: 5' 6" - Born: Ireland - Enlistment date: 8 Jun 1813 - Place: Baltimore.

McLaughlin, James - Private - 8th Infantry - Company: William Jones - Other regiment: 7th Infantry (New) - Age: 23 - Height: 5' 6" - Born: Prince George's Cty, MD - Trade: Farmer - Enlistment date: 17 Jun 1813 - Place: Georgia - Period: 5 Yrs - BLW 20113-160-12 - Discharged at Fort St. Marks, FL, on 16 Jun 1818.

McLaughlin, John - Private - 38th Infantry - Age: 26 - Born: Dublin, Ireland - Enlistment date: 20 Jul 1814 - Place: Baltimore - Period: War - Deserted.

McLaughlin, William - Private - 28th Infantry - Company: George Stockton - Age: 29 - Height: 5' 10" - Born: St. Mary's, MD - Trade: Millwright - Enlistment date: 1 Jun 1813 - Place: Elizabethtown, KY - Period: 5 Yrs - Discharged on 10 Nov 1815 on loss of arm during the siege of Fort Meigs.

McLean, James - Private - 36th Infantry - Pension: Land bounty to William McLean, father and heir at law of James McLean - BLW 9306-160-12.

McLean, James - Private - 36th Infantry - Age: 24 - Height: 5' 8" - Born: St. Mary's, MD - Enlistment date: 28 Aug 1813 - Place: Baltimore - Period: 1 Yr - Transferred to the U.S. Navy on 30 Oct 1813.

McLintick, Francis - Private - 36th Infantry - Age: 24 - Height: 5' 8" - Born: Pennsylvania - Enlistment date: 3 Jun 1813 - Place: Frederick, MD.

McMahon, Joseph - Private - 2nd Infantry - Age: 28 - Height: 5' 10 1/2" - Born: Armangh, Ireland - Trade: Sawyer - Enlistment date: 15 May 1807 - Place: Baltimore - Period: 5 Yrs.

McMullen, George - Private - 14th Infantry - Company: Samuel Lane - Other regiment: 4th Infantry (New) - Age: 24 - Height: 5' 8" - Born: Ireland - Enlistment date: 4 Jun 1812 - Place: Chambersburg, PA - Period: 5 Yrs - Prisoner of War (Halifax), sent to England; discharged on 11 Jun 1817.

McMullin, Thomas - Private - 14th Infantry - Company: David Cummings - Age: 32 - Height: 5' 8" - Born: Dublin, Ireland - Trade: Currier - Enlistment date: 5 Jul 1814 - Place: Alexandria, DC - Period: 5 Yrs - Deserted at Alexandria, DC, on 3 Aug 1814.

McNab, John - Private - 14th Infantry - Company: Reuben Gilder - Enlistment date: 3 May 1813 - Period: 5 Yrs - Pension: Land bounty to Isaac McNab, brother and other heirs at law of John McNab - BLW 25422-160-12 - Died at Sackets Harbor, NY, on 3 Nov 1813.

McNamar, John - Private - Artillery - Age: 22 - Height: 5' 9" - Born: Maryland - Trade: Blacksmith - Enlistment date: 24 Feb 1814 - Place: Plattsburgh, NY - Period: War.

McNamar, John - Fife Major - 14th Infantry - Age: 22 - Height: 5' 9" - Born: Montgomery Cty, MD - Trade: Blacksmith - Enlistment date: 27 Feb 1814 - Period: War - BLW 13947-160-12 - Discharged at Greenbush, NY, on 3 May 1815.

McNeir, George - Third Lieutenant - US Sea Fencibles - Company: William Addison - Commissioned as a 3rd lieutenant on 17 Ma 1813; resigned on 24 Nov 814.

McNiel, John - Private - 5th Infantry - Age: 27 - Height: 5' 9" - Born: Cecil Cty, MD - Trade: Basket maker - Enlistment date: 5 Nov 1814 - Place: Lancaster, PA - Period: 5 Yrs - Deserted at Harrisburgh, PA, on 13 Nov 1814.

McPherson, Abraham - Sergeant - 36th Infantry - Company: Joseph Hook - Age: 27 - Height: 5' 10" - Born: Baltimore - Trade: Sail maker - Enlistment date: 1 Mar 1814 - Place: Annapolis - Period: War - BLW 13422-160-12 - Discharged at Fort Covington, MD, on 30 Mar 1815.

McPherson, Abraham - Private - 36th Infantry - Age: 21 - Height: 5' 7" - Born: Baltimore - Enlistment date: 28 Apr 1813 - Place: Baltimore - Period: 1 Yr.

McPherson, Alexander - Private - 20th Infantry - Company: William Jett - Other regiment: 12th Infantry - Age: 22 - Height: 5' 3" - Born: Charles Cty, MD - Trade: Laborer - Enlistment date: 11 Apr 1813 - Period: 18 Mos - Transferred to 12th Infantry; discharged at Black Rock, NY, on 11 Oct 1814.

McPherson, Jacob - Corporal - 5th Infantry - Company: Richard Whartenby - Other regiment: 3rd Infantry (New) - Age: 27 - Height: 5' 8" - Born: Baltimore - Trade: Baker - Enlistment date: 26 Feb 1811 - Place: Baltimore - Period: 5 Yrs - Discharged on 26 Feb 1816.

McPherson, William - Sergeant - 38th Infantry - Company: Isaac Aldridge - Age: 21 - Height: 5' 6 1/2" - Born: Baltimore - Trade: Cooper - Enlistment date: 28 May 1813 - Period: War - Pension: Wife Hannah, SO-2127, SC-13264, WO-34011, WC-24372 - BLW 14192-160-12 - Discharged at Baltimore on 3 Apr 1815.

McQuay, Thomas - Private - 36th Infantry - Company: Joseph Merrick - Age: 37 - Height: 5' 5 1/2" - Born: Talbot Cty, MD - Enlistment date: 1 Jun 1813 - Place: Easton, MD - Period: 1 Yr.

McQuay, William - Private - 3rd Infantry - Company: James Dinkins - Age: 24 - Height: 5' 4 1/2" - Born: Maryland - Trade: Laborer - Enlistment date: 9 Sep 1808 - Place: Easton, MD - Period: 5 Yrs - Discharged at New Orleans on 9 Apr 1815.

McQuery, John - Private - 36th Infantry - Company: Samuel Raisin - Age: 29 - Height: 6' 2" - Born: Hanover Cty, VA - Trade: Carpenter - Enlistment date: 3 Oct 1814 - Place: Manchester, VA - Period: War - BLW 15328-160-12 - Discharged at Washington, DC, on 20 Mar 1815.

Maryland Regulars in War of 1812

McQuinn, Donald - Corporal - 14th Infantry - Company: Isaac Barnard - Enlistment date: 30 Nov 1812 - Period: 5 Yrs - Pension: Old War IF-25274 - BLW 10687-160-12 - Prisoner of War, exchanged on 15 Apr 1815; discharged at Washington, DC, on 5 Oct 1814, wounded on 24 Jun 1813.

McRae, Duncan F. - Third Lieutenant - 12th Infantry - Company: James Paxton - Other regiment: Rifles - Age: 26 - Height: 5' 11" - Born: Maryland - Trade: Laborer - Enlistment date: 19 May 1812 - Place: Morgantown, VA - Period: 5 Yrs - Promoted to ensign, US Rifles, from private, 12th Infantry, on 17 Mar 1814; promoted to 3rd lieutenant on 1 Oct 1814; discharged on 15 Jun 1815.

McRay, William - Private - 14th Infantry - Prisoner of War (Halifax), sent to England.

McRhea, Edward - Private - 14th Infantry - Prisoner of War (Halifax), discharged from Halifax on 3 Feb 1814.

Mead, Benjamin - Private - 16th Infantry - Company: William Smith - Age: 40 - Height: 5' 6" - Born: Charles Cty, MD - Trade: Laborer - Enlistment date: 9 Apr 1813 - Place: Wilmington, DE - Period: 5 Yrs - Discharged at Buffalo on 19 Jun 1815, scrotal hernia.

Meagle, Joseph - Private - 2nd Infantry - Company: Mathew Arbuckle - Other regiment: 1st Infantry (New) - Age: 21 - Height: 5' 6" - Born: Fredericktown, MD - Trade: Shoemaker - Enlistment date: 10 Feb 1807 - Place: Fredericktown, MD - Period: 5 Yrs - Re-enlisted on 15 Jul 1813 for five years; discharged on 27 May 1817.

Meagle, Valentine - Private - 26th Infantry - Company: William Bezeau - Age: 21 - Height: 5' 6" - Born: Frederick, MD - Trade: Shoemaker - Enlistment date: 5 Nov 1814 - Place: Baltimore - Period: War - Discharged at Philadelphia on 23 Mar 1815.

Mechane, Almond - Private - 14th Infantry - Prisoner of War (Halifax), discharged from Halifax on 3 Feb 1814.

Medcalf, John - Private - 42nd Infantry - Company: Edmund Duval - Age: 39 - Height: 5' 2" - Born: Annapolis - Trade: Waterman - Enlistment date: 25 Dec 1813 - Place: Bladensburg, MD - Period: 5 Yrs - Pension: Heirs received half pay for five years in lieu of land bounty - Died at New York on 18 Jun 1815.

Medcalf, Thomas - Sergeant - 14th Infantry - Company: David Cummings - Age: 37 - Height: 5' 10 1/2" - Born: Tawney, Frederick Cty, MD - Trade: Blacksmith - Enlistment date: 15 Oct 1814 - Place: Baltimore - Period: 5 Yrs.

Medcap, William - Private - 38th Infantry - Company: John Mowton - BLW 12138-160-12.

Medley, Basil - Private - 12th Infantry - Company: Andrew Madison - Other regiment: 8th Infantry (New) - Age: 23 - Height: 5' 10 1/2" - Born: Montgomery Cty, MD - Trade: Laborer - Enlistment date: 2 Oct 1812 - Place: Harpers Ferry, VA - Period: 5 Yrs - BLW 16911-160-12 - Discharged at Fort Osage, MO, on 2 Oct 1817.

Medtart, Joshua - First Lieutenant - 38th Infantry - Company: John Rothrock

By War of 1812 Soc. in MD

- Pension: Land bounty to Ann Mary Medtart, widow of Joshua Medtart - BLW 27113-160-50 - Commissioned as a 2nd lieutenant on 20 May 1813; promoted to 1st lieutenant on 15 Aug 1813; discharged on 15 Jun 1815.

Meek, Samuel - Private - 17th Infantry - Company: William Bradford - Other regiment: 3rd Infantry (New) - Age: 41 - Height: 5' 7" - Born: Maryland - Trade: Shoemaker - Enlistment date: 23 Mar 1813 - Place: Newtown (OH or KY) - Period: 5 Yrs - Discharged on 17 May 1815 by habeas corpus.

Meek, Thomas - Private - 14th Infantry - Company: Joseph Marshall - Age: 29 - Height: 5' 6" - Born: Baltimore - Trade: Cooper - Enlistment date: 24 Mar 1814 - Place: Baltimore - Period: War - BLW 408-160-12 - Discharged at Greenbush, NY, on 3 May 1815.

Meeks, James P. - Seaman - US Sea Fencibles - Company: Simmones Bunbury - Age: 21 - Height: 5' 7" - Born: Baltimore - Enlistment date: 31 Jan 1814 - Period: 1 Yr - Discharged at Fort McHenry on 2 Jan 1815.

Meguides, John - Private - Light Artillery - Age: 27 - Height: 6' 3" - Born: Baltimore - Enlistment date: 6 Jul 1814 - Place: Boston.

Melcher, Isaac - Private - US Volunteers - Company: Stephen Moore - Enlistment date: 8 Sep 1812 - Period: 1 Yr.

Mellon, George - Private - 12th Infantry - Company: Thomas Sangster - Other regiment: 8th Infantry (New) - Age: 35 - Height: 5' 3" - Born: Baltimore - Trade: Hatter - Enlistment date: 28 Apr 1812 - Place: Alexandria, DC - Period: 5 Yrs - Discharged on 28 Apr 1817.

Melville, William - Private - 14th Infantry - Company: David Cummings - Age: 26 - Height: 5' 5 1/2" - Born: Rhode Island - Trade: Shoemaker - Enlistment date: 30 May 1814 - Place: Baltimore - Period: War - BLW 9844-160-12 - Discharged at Baltimore on 16 Mar 1815.

Melvin, James - 38th Infantry - Company: George Porter - Pension: Wife Sarah S., SO-20736, SC-19746, WO-26495, WC-14397.

Melvin, Scarborough - Private - 14th Infantry - Company: Thomas Montgomery - Other regiment: 4th Infantry (New) - Age: 28 - Height: 5' 5" - Born: Virginia - Trade: Laborer - Enlistment date: 15 Sep 1812 - Period: 5 Yrs - Pension: Old War WF-16401 Rejected, SO-25647, SC-20223 - Discharged at Camp Montpelier, MS, on 16 Sep 1817.

Melvin, William - Private - 14th Infantry - Prisoner of War (Halifax), sent to England.

Mendenhall, Elijah B. - Private - 38th Infantry - Company: Isaac Aldridge - Age: 35 - Height: 6' 1" - Born: Chester, PA - Trade: Teacher - Enlistment date: 16 Jan 1815 - Place: Baltimore - Period: War - BLW 272-320-12 - Discharged at Baltimore on 6 Apr 1815.

Mercer, Jacob - Private - 2nd Artillery - Company: Stanton Sholes - Other regiment: Corps of Artillery - Age: 28 - Height: 5' 8" - Born: Washington Cty, MD - Trade: Carpenter - Enlistment date: 7 Apr 1813 - Place: Greensburg, Beaver Cty, PA - Period: 5 Yrs - Discharged at Fort Mackinaw, MI, on 16 Apr 1818.

Mercer, John - Private - 14th Infantry - Company: Thomas Montgomery - Age: 35 - Height: 5' 8" - Born: Cecil Cty, MD - Trade: Carpenter -

Maryland Regulars in War of 1812

Enlistment date: 14 Jun 1812 - Place: Wilmington, DE - Period: 5 Yrs - Discharged at Greenbush, NY, on 18 Mar 1815 for a rupture.

Mercer, Thomas - Sergeant - 1st Artillery - Company: Hopley Yeaton - Other regiment: Corps of Artillery - Age: 27 - Height: 6' - Born: Baltimore - Trade: Laborer - Enlistment date: 4 Jul 1812 - Place: Fort Nelson, VA - Period: 5 Yrs - Pension: Land bounty to Thomas Mercer, son and other heirs at law of Thomas Mercer - BLW 16956-160-12 - Died on 9 May 1816.

Mercer, William Newton - Surgeon - 38th Infantry - Other regiment: ex 22nd Infantry - Born: Virginia - Commissioned as a hospital surgeon's mate on 18 May 1813; promoted to surgeon, 22nd Infantry, on 7 Aug 1813; transferred to 38th Infantry on 11 Apr 1814; promoted to hospital surgeon on 22 Nov 1814; discharged on 15 Jun 1815.

Merchant, Joel - Private - 14th Infantry - Company: David Cummings - Age: 18 - Height: 5' 8" - Born: Berkley, VA - Trade: Farmer - Enlistment date: 12 Nov 1814 - Place: Snowden's Camp, VA - Period: 5 Yrs - Deserted at Fort McHenry on 15 Apr 1815.

Merchant, Richard - Private - 38th Infantry - Company: James Hook - Age: 24 - Height: 5' 3" - Born: Baltimore - Trade: Printer - Enlistment date: 23 Nov 1814 - Place: Baltimore - Period: War - BLW 25047-160-12 - Discharged on 31 Mar 1815.

Merchant, Richard - Private - US Volunteers - Company: Stephen Moore - Enlistment date: 8 Sep 1812 - Period: 1 Yr.

Meredith, Benjamin R. - Private - 36th Infantry - Company: Joseph Hook - Age: 20 - Height: 5' 11" - Born: Queen Anne, MD - Trade: Tailor - Enlistment date: 1 Mar 1814 - Place: Annapolis - Period: War - BLW 16535-160-12 - Discharged at Fort Covington, MD, on 30 Mar 1815.

Merrick, Joseph J. - Captain - 36th Infantry - Born: Maryland - BLW 1480-160-50 - Commissioned as a captain on 30 Apr 1813; discharged on 15 Jun 1815.

Merrick, William D. - First Lieutenant - 36th Infantry - Company: Joseph Merrick - Born: Maryland - BLW 18557-160-50 - Commissioned as a 3rd lieutenant on 30 Apr 1813; promoted to 2nd lieutenant on 15 Aug 1813; promoted to 1st lieutenant on 30 Sep 1814; discharged on 15 Jun 1815.

Merridue, William - Private - 36th Infantry - Company: Thomas Carbery - Age: 22 - Height: 5' 6 1/2" - Born: Brunigueore, England - Trade: Seaman - Enlistment date: 19 Feb 1814 - Place: Georgetown, DC - Period: War - BLW 11826-160-12 - Discharged at Baltimore on 31 Mar 1815.

Merriman, Archibald - Private - 22nd Infantry - Company: Thomas Lawrence - Other regiment: 2nd Infantry (New) - Age: 35 - Height: 5' 11 1/2" - Born: Baltimore - Trade: Weaver - Enlistment date: 12 May 1813 - Place: Somerset, PA - Period: 5 Yrs - Discharged on 3 Mar 1816 after furnishing a substitute.

Merritt, Thurston - Private - 14th Infantry - Company: Reuben Gilder - Age: 42 - Height: 5' 7" - Born: Massachusetts - Enlistment date: 18 Apr 1813 - Period: 18 Mos - Discharged on 26 Nov 1814.

Merry, Joseph - Private - Light Artillery - Company: John Bell - Age: 22 -

By War of 1812 Soc. in MD

Height: 5' 4" - Born: Maryland - Trade: Mariner - Enlistment date: 20 Jun 1814 - Place: Boston - Period: War - Discharged on 31 Mar 1815.

Merryman, John - Private - 14th Infantry - Company: Samuel Lane - Other regiment: 4th Infantry (New) - Age: 21 - Height: 5' 9" - Born: Montgomery Cty, MD - Trade: Laborer - Enlistment date: 25 May 1812 - Place: Allegany Cty, MD - Period: 5 Yrs - Discharged at Raleigh, NC, on 4 Jun 1817.

Messick, Nehemiah - Private - Light Artillery - Company: Charles Sarrabee - Age: 21 - Height: 5' 7" - Born: Princess Anne, MD - Trade: Seaman - Enlistment date: 13 Jan 1814 - Place: Boston - Period: 5 Yrs - Discharged on 13 Oct 1818.

Messings, John - Corporal - 17th Infantry - Company: Harris Hickman - Age: 40 - Height: 5' 10" - Born: Fredericktown, MD - Trade: Blacksmith - Enlistment date: 9 Apr 1814 - Place: Chillicothe, OH - Period: War - BLW 14023-160-12 - Discharged at Chillicothe, OH, on 9 Jun 1815.

Mestler, Coonrod - Seaman - US Sea Fencibles - Company: William Addison.

Metakern, Seth - Private - 14th Infantry - Prisoner of War (Halifax), discharged from Halifax on 3 Feb 1814.

Metcalf, Mordreai - Corporal - 22nd Infantry - Company: Willis Foulk - Other regiment: 2nd Infantry (New) - Age: 20 - Height: 5' 11" - Born: Fredericktown, MD - Trade: Miller - Enlistment date: 7 Jun 1814 - Period: 5 Yrs - Sent to state prison for five years on 31 Jan 1818.

Metcalf, Thomas - Private - 14th Infantry - Company: David Cummings - Pension: Wife Mary, WO-4821, WC-3096 - BLW 24426-160-50 - Also served in Captain Durbin's Company, MD Militia.

Mezzick, Covington - Private - 38th Infantry - Company: James Haslett - Age: 19 - Height: 6' 8" - Born: Dorchester Cty, MD - Trade: Farmer - Enlistment date: 19 Jul 1813 - Period: 1 Yr - BLW 12691-160-12 - Re-enlisted at Craney Island, VA, on 1 Mar 1814 for the war; discharged in Mar 1815.

Mezzick, Luke - Private - 38th Infantry - Company: John Rothrock - Age: 24 - Height: 5' 3 1/2" - Born: Somerset Cty, MD - Trade: Carpenter - Enlistment date: 10 Aug 1813 - Period: 1 Yr - BLW 11906-160-12 - Discharged at Norfolk, VA, on 15 Mar 1815.

Michaels, Andrew - Private - 38th Infantry - Other regiment: 4th Infantry (New) - Age: 22 - Height: 5' 8 3/4" - Born: Frederick, MD - Trade: Farmer - Enlistment date: 24 Oct 1814 - Place: Baltimore - Period: 5 Yrs - Discharged on 23 Oct 1819.

Michaels, John - Private - 18th Infantry - Company: Henry Taylor - Other regiment: 4th Infantry (New) - Age: 21 - Height: 5' 7 1/2" - Born: Baltimore - Trade: Shoemaker - Enlistment date: 21 Jun 1813 - Place: New Bern, NC - Period: 5 Yrs - Discharged at Pensacola, FL, on 30 Jun 1818.

Middleton, Hugh - Fifer - 38th Infantry - Pension: No pension claim - BLW 15944-160-50.

Mietellan, John - Private - 14th Infantry - Prisoner of War (Halifax), discharged from Halifax on 3 Feb 1814.

Miles, Charles - Private - 14th Infantry - Company: Reuben Gilder - Age: 23 - Height: 5' 6" - Born: Maryland - Trade: Chair maker - Enlistment date: 11

Jun 1813 - Period: 18 Mos - Discharged at Greenbush, NY, on 5 May 1815.

Miles, Israel - Private - 14th Infantry - Company: Samuel Lane - Other regiment: 4th Infantry (New) - Age: 22 - Height: 6' 1" - Born: Bartle, VA - Trade: Miller - Enlistment date: 12 Jun 1812 - Place: Williamsport, MD - Period: 5 Yrs - Discharged on 11 Jun 1817.

Miles, James - Private - 22nd Infantry - Company: Thomas Lawrence - Age: 23 - Height: 5' 8 1/2" - Born: Hagerstown, MD - Trade: Saddler - Enlistment date: 27 May 1813 - Place: Pittsburgh - Period: 5 Yrs - Discharged on 27 May 1818 and re-enlisted.

Miles, John - Seaman - US Sea Fencibles - Company: William Addison.

Miles, Joseph - Private - Light Dragoons - Age: 21 - Height: 5' 8" - Born: Delaware - Trade: Blacksmith - Enlistment date: 25 Feb 1813 - Place: Elkton, MD - Period: War.

Miles, Morris - Private - 14th Infantry - Company: Reuben Gilder - Other regiment: 4th Infantry (new) - Age: 25 - Height: 5' 6" - Born: Prince George's Cty, MD - Trade: Farmer - Enlistment date: 11 Feb 1813 - Period: 5 Yrs - BLW 19452-160-12 - Wounded at La Cole Mill, LC, on 30 Mar 1814; discharged at Fort Hawkins on 11 Feb 1818.

Miles, Stanislaw - Sergeant - 42nd Infantry - Company: Edmund Duval - Age: 23 - Height: 5' 10" - Born: Port Tobacco, MD - Trade: Joiner - Enlistment date: 26 Feb 1814 - Place: Washington - Period: War - Discharged at New York on 1 Jun 1815.

Millan, Abraham - Private - 12th Infantry - Company: James Charlton - Other regiment: 8th Infantry (New) - Age: 31 - Height: 6' - Born: Kent Cty, MD - Trade: Laborer - Enlistment date: 23 May 1812 - Place: Morgantown, VA - Period: 5 Yrs - Discharged on 23 May 1817.

Miller John - Private - 4th Rifles - Company: Matthew Magee - Age: 30 - Height: 5' 4" - Born: Baltimore - Trade: Hair dresser - Enlistment date: 14 Nov 1814 - Period: 5 Yrs.

Miller, Edward - Private - 26th Infantry - Company: William Bezeau - Age: 21 - Height: 5' 7" - Born: Tennessee - Trade: Farmer - Enlistment date: 21 Nov 1814 - Place: Baltimore - Period: War - BLW 493-160-12 - Discharged at Philadelphia on 23 Mar 1815.

Miller, Ephraim - Private - 28th Infantry - Age: 27 - Height: 5' 11" - Born: Maryland - Enlistment date: 22 May 1814.

Miller, Francis - Musician - 5th Infantry - Company: Richard Bell - Enlistment date: 29 Jul 1812 - Place: Baltimore - Period: 5 Yrs - Pension: Land bounty to Matilda Miller, mother and only heir at law of Francis Miller - BLW 25806-160-12 - Died at Plattsburgh, NY, on 10 Jul 1814.

Miller, Frederick - Private - 14th Infantry - Company: James McDonald - Age: 17 - Height: 5' 2" - Born: Baltimore - Trade: Laborer - Enlistment date: 27 Jan 1814 - Place: Baltimore - Period: 5 Yrs - Pension: Land bounty to Sarah Miller, mother and heir at law of Frederick Miller - BLW 9687-160-12 - Died on 2 Jun 1814.

Miller, Gattel - Private - 14th Infantry - Prisoner of War (Halifax); died at Halifax on 18 Oct 1813.

By War of 1812 Soc. in MD

Miller, George - Private - 14th Infantry - Company: Samuel Lane - Age: 45 - Height: 5' 10" - Born: Pennsylvania - Enlistment date: 27 Jun 1812 - Period: 18 Mos - Died at Sackets Harbor, NY, on 14 Jan 1814.

Miller, Henry - Private - 36th Infantry - Age: 28 - Height: 5' 11" - Born: Chester, MD - Enlistment date: 24 May 1813 - Place: Baltimore.

Miller, Henry - Private - 36th Infantry - Company: Samuel Raisin - Age: 36 - Height: 5' 8" - Born: Berkley, VA - Trade: Shoemaker - Enlistment date: 2 Aug 1814 - Place: Richmond, VA - Period: 5 Yrs.

Miller, Henry - Private - 38th Infantry - Company: Shepperd Leakin - Age: 35 - Height: 5' 6" - Born: Vinn, Germany - Trade: Miller - Enlistment date: 16 Aug 1814 - Place: Baltimore - Period: War - BLW 186-160-12 - Discharged at Fort McHenry on 28 Mar 1815.

Miller, Isaac - Private - 38th Infantry - Company: John Brookes - Age: 35 - Height: 5' 8" - Born: Somerset Cty, MD - Trade: Blacksmith - Enlistment date: 1 Oct 1813 - Period: 1 Yr - Pension: Land bounty to Lovey Gullett, sister and other heirs at law of Isaac Miller - BLW 26454-160-12 - Re-enlisted at Craney Island, VA, on 23 Feb 1814 for the war; discharged on 15 Mar 1815.

Miller, Jacob - Private - 36th Infantry - Company: Charles Randolph - Age: 27 - Height: 5' 5" - Born: Frederick, MD - Enlistment date: 18 May 1813 - Place: Frederick, MD - Period: 1 Yr - Pension: SO-24629, SC-20197.

Miller, James - Private - 14th Infantry - Company: David Cummings - Enlistment date: 18 Aug 1813 - Period: 5 Yrs - BLW 7048-160-12 - Prisoner of War (Halifax), sent to England; discharged at Fort McHenry on 30 Apr 1815 for old age and rheumatism.

Miller, James - Private - 19th Infantry - Company: Stephen Lee - Age: 29 - Height: 5' 7 1/2" - Born: Baltimore Cty - Trade: Carpenter - Enlistment date: 27 Aug 1813 - Period: War - Prisoner of War (Quebec), captured at Fort Niagara, NY, 19 Dec 1813; exchanged 11 May 1814; discharged at Greenbush, NY, on 8 May 1815.

Miller, James - Private - 12th Infantry - Company: Thomas Sangster - Age: 23 - Height: 5' 8" - Born: Hagerstown, MD - Trade: Farrier - Enlistment date: 12 Jun 1812 - Place: Chambersburg, PA - Period: 18 Mos - Discharged on 12 Dec 1813.

Miller, James - Private - 14th Infantry - Company: Reuben Gilder - Other regiment: 4th Infantry (New) - Age: 18 - Height: 5' 2" - Born: Maryland - Trade: Laborer - Enlistment date: 12 May 1813 - Place: Summerset - Period: 5 Yrs - Deserted on 15 Jul 1815.

Miller, John - Captain - 2nd Infantry - Commissioned as an ensign on 8 Feb 1802; promoted to 2nd lieutenant on 30 Sep 1803; promoted to 1st lieutenant on 20 Jun 1806; promoted to captain on 12 Mar 1812; discharged on 15 Jun 1815.

Miller, John - Private - 36th Infantry - Age: 27 - Height: 5' 9" - Born: Sweden - Enlistment date: 6 May 1813 - Place: Baltimore.

Miller, John - Sergeant - 36th Infantry - Age: 20 - Height: 5' 8" - Born: Maryland - Trade: Butcher - Enlistment date: 10 Dec 1814 - Period: War -

Maryland Regulars in War of 1812

BLW 420-320-12 - Discharged at Georgetown, DC, on 13 Mar 1815.

Miller, John - Private - 14th Infantry - Company: David Cummings - Age: 31 - Height: 5' 6 1/2" - Born: Centre Cty, PA - Trade: Laborer - Enlistment date: 7 Dec 1814 - Place: Williamsport, MD - Period: War - Discharged on 18 Mar 1815.

Miller, John - Private - Artillery - Other regiment: Corps of Artillery - Age: 35 - Height: 5' 5" - Born: Maryland - Trade: Butcher - Enlistment date: 12 Jun 1813 - Place: Lancaster - Period: 5 Yrs - Deserted on 16 Jun 1816.

Miller, Joseph - Musician - 5th Infantry - Company: Richard Whartenby - Other regiment: 3rd Infantry (New) - Age: 25 - Height: 5' 6 1/2" - Born: Baltimore - Trade: Shoemaker - Enlistment date: 4 Apr 1812 - Place: Baltimore - Period: 5 Yrs - Discharged on 3 Apr 1817.

Miller, Michael S. - Private - 14th Infantry - Company: Thomas Montgomery - Other regiment: 4th Infantry (New) - Age: 26 - Height: 5' 8" - Born: Philadelphia - Trade: Tailor - Enlistment date: 25 Dec 1812 - Place: Hagerstown, MD - Period: 5 Yrs - Discharged on 25 Dec 1817.

Miller, Peter - Private - 2nd Artillery - Company: Frederick Evans - Age: 27 - Height: 5' 10" - Born: Norway - Trade: Seaman - Enlistment date: Sep 1813 - Place: Baltimore - Period: War - Discharged on 24 Mar 1815.

Miller, Peter F. - Drummer - 1st Infantry - Company: Horatio Stark - Age: 22 - Height: 5' 6" - Born: Amden, Prussia - Trade: Saddler - Enlistment date: 1 May 1812 - Place: Cincinnati - Period: 5 Yrs.

Miller, Samuel - Private - 14th Infantry - Company: Samuel Lane - Age: 35 - Height: 5' 4" - Born: Cecil Cty, MD - Enlistment date: 18 Aug 1812 - Period: 5 Yrs - Discharged at Plattsburgh, NY, on 11 Jul 1814 for infirmities.

Miller, William - Private - 16th Infantry - Company: William Davenport - Age: 35 - Height: 5' 6" - Born: Baltimore - Trade: Shoemaker - Enlistment date: 1 Mar 1813 - Period: 5 Yrs - Deserted at Burlington, VT, on 8 Mar 1814.

Miller, William - Private - 36th Infantry - Company: Mortimer Hall - Age: 18 - Height: 5' 7" - Born: Baltimore - Enlistment date: 26 Sep 1813 - Place: Baltimore - Period: 1 Yr - Discharged on 19 Dec 1813.

Miller, William - Sergeant - 38th Infantry - Company: Shepperd Leakin - Age: 25 - Height: 5' 8" - Born: Baltimore - Trade: Blacksmith - Enlistment date: 14 Mar 1814 - Period: War - BLW 490-160-12 - Discharged at Fort McHenry on 28 Mar 1815.

Milligan, James - Sergeant - 3rd Rifles - Company: William Parker - Other regiment: Rifles - Age: 23 - Height: 6' 1 3/4" - Born: Washington, MD - Trade: Laborer - Enlistment date: 23 Oct 1814 - Place: West Liberty - Period: 5 Yrs - Discharged on 30 Oct 1819.

Mills, David - Private - 36th Infantry - Age: 34 - Height: 6' 2" - Born: Pennsylvania - Enlistment date: 25 May 1813 - Place: Fredericktown, MD - Period: 1 Yr.

Mills, David - Private - 36th Infantry - Company: Joseph Merrick - Age: 19 - Height: 5' 4" - Born: Baltimore - Enlistment date: 17 May 1813 - Place: Baltimore - Period: 1 Yr - Transferred to the U.S. Navy on 26 Oct 1813.

By War of 1812 Soc. in MD

Mills, George - Private - 2nd Artillery - Company: Nathan Towson - Pension: Land bounty to Jesse Mills, Jr., brother and other heirs at law of George Mills - BLW 25910-160-12.

Mills, James - Sergeant - 38th Infantry - Company: Isaac Aldridge - Age: 38 - Height: 5' 5 1/4" - Born: Maryland - Trade: Baker - Enlistment date: 19 Nov 1814 - Place: Baltimore - Period: War - Discharged at Baltimore on 6 Apr 1815.

Mills, John - Private - 38th Infantry - Company: Isaac Aldridge - Age: 18 - Height: 5' 1" - Born: Baltimore City - Trade: Laborer - Enlistment date: 11 Jan 1815 - Period: War - Discharged at Baltimore on 6 Apr 1815.

Mills, John - Private - 36th Infantry - Age: 21 - Height: 5' 3 1/2" - Born: Ireland - Trade: Laborer - Enlistment date: 10 Jan 1815 - Period: War - Discharged on 13 Mar 1815.

Mills, Samuel - Private - 5th Infantry - Other regiment: 3rd Infantry (New) - Age: 30 - Height: 5' 6" - Born: Baltimore - Trade: House joiner - Enlistment date: 17 May 1813 - Place: Baltimore - Period: 5 Yrs.

Mills, Stephen - Private - 14th Infantry - Company: David Cummings - Enlistment date: 6 Mar 1813 - Period: 18 Mos - Prisoner of War, sent to England.

Mills, Truman - Private - 20th Infantry - Company: Matthew Payne - Other regiment: 4th Infantry (New) - Age: 27 - Height: 5' 9" - Born: Prince George's Cty, MD - Trade: Farmer - Enlistment date: 30 Sep 1814 - Place: Alexandria, DC - Period: 5 Yrs - Discharged at Montpelier, VT, on 30 Sep 1819 and re-enlisted.

Mills, William - Private - 36th Infantry - Company: Samuel Raisin - Age: 22 - Height: 5' 9" - Born: Fairfax, VA - Trade: Farmer - Enlistment date: 13 Aug 1814 - Place: Georgetown, DC - Period: War - BLW 10422-160-12 - Discharged at Washington, DC, on 20 Mar 1815.

Mills, William G. - First Lieutenant - 14th Infantry - Commissioned as an ensign on 30 Apr 1812; promoted to 2nd lieutenant on 13 Mar 1813; promoted to 1st lieutenant on 14 Nov 1813; discharged on 15 Jun 1814.

Miltenberger, Anthony - Captain - 38th Infantry - BLW 11587-160-50 - Commissioned as a captain on 20 May 1813; resigned on 1 May 1814.

Milton, Thomas - Sergeant - 36th Infantry - Company: Thomas Carbery - Age: 33 - Height: 6' - Born: St. Mary's, MD - Trade: Farmer - Enlistment date: 13 Apr 1814 - Place: Leonardstown, MD - Period: War - BLW 16825-160-12 - Discharged at Baltimore on 31 Mar 1815.

Minches, Joseph - Private - 36th Infantry - Company: Joseph Hook - Age: 21 - Height: 5' 6" - Born: Baltimore - Enlistment date: 8 May 1813 - Place: Baltimore - Period: 1 Yr.

Minifie, Christopher - Private - 36th Infantry - Age: 45 - Height: 5' 9" - Born: Exeter, NH - Enlistment date: 1 Mar 1814 - Place: Annapolis.

Minter, Joseph - Private - 38th Infantry - Company: James Smith - BLW 8857-160-12.

Missings, John - Private - 19th Infantry - Age: 40 - Height: 5' 10" - Born: Montgomery Cty, MD - Enlistment date: 9 Apr 1814 - Place: Chillicothe,

Maryland Regulars in War of 1812

OH - Period: War.

Mitchel, James - Private - 36th Infantry - Company: Joseph Hook - Age: 28 - Height: 5' 7 1/2" - Born: Ireland - Enlistment date: 15 May 1815 - Place: Baltimore - Period: 1 Yr - Deserted on 7 Jun 1815.

Mitchell, Charles - Private - 14th Infantry - Company: David Cummings - Age: 20 - Height: 5' 6 1/2" - Born: England - Enlistment date: 18 Jan 1815 - Place: Salisbury, MD - Period: War - BLW 807-320-12 - Discharged at Baltimore City on 16 Mar 1815.

Mitchell, Charles - Private - 14th Infantry - Company: David Cummings - Enlistment date: 13 Mar 1812 - Period: 5 Yrs - Prisoner of War, sent to England; deserted at Baltimore on 20 Apr 1815.

Mitchell, George E. - Lieutenant Colonel - 3rd Artillery - Born: Maryland - Commissioned as a major, 3rd Artillery, on 1 May 1812; promoted to lieutenant colonel on 3 Mar 1813; transferred to Corps of Artillery on 12 May 1814; transferred to 3rd Artillery on 1 Jun 1821; resigned on 1 Jun 1821; died on 28 Jun 1832.

Mitchell, James - Private - 38th Infantry - Company: Charles Stansbury - BLW 11840-160-12.

Mitchell, Joseph - Private - 38th Infantry - Company: Charles Stansbury - Age: 30 - Height: 5' 10 1/2" - Born: County Galway, Ireland - Trade: Weaver - Enlistment date: 25 Oct 1814 - Place: Baltimore - Period: War - Discharged at Baltimore on 6 Apr 1815.

Mitchell, Reuben - Private - 5th Infantry - Age: 27 - Height: 5' 7 3/4" - Born: Dorchester Cty, MD - Trade: Seaman - Enlistment date: 21 Apr 1813 - Place: Maryland - Period: 5 Yrs.

Mitchell, Samuel - Private - 36th Infantry - Company: Thomas Carbery - Age: 18 - Height: 5' 4 1/2" - Born: Calvert, MD - Trade: Planter - Enlistment date: 19 Feb 1814 - Place: Georgetown, DC - Period: 5 Yrs.

Mitchell, Thomas - Private - 14th Infantry - Company: Reuben Gilder - Other regiment: 4th Infantry (New) - Age: 16 - Height: 5' 1" - Born: District of Columbia - Trade: Laborer - Enlistment date: 26 Apr 1813 - Place: Baltimore - Period: 5 Yrs - Discharged on 25 Apr 1818.

Mitchell, William - Private - 14th Infantry - Age: 46 - Prisoner of War (Quebec), died at Quebec on 4 Aug 1813.

Mixture, Ezra - Sergeant - 26th Infantry - Company: William Bezeau - Age: 27 - Height: 5' 6 1/2" - Born: Stafford, Tolland, CT - Trade: Shoemaker - Enlistment date: 29 Sep 1814 - Place: Baltimore - Period: War - BLW 9437-160-12 - Discharged at Philadelphia on 23 Mar 1815.

Moffit, Charles - Private - 36th Infantry - Age: 23 - Height: 5' 7 1/2" - Born: New Windsor, Orange Cty, NY - Enlistment date: 27 May 1813 - Place: Baltimore - Period: 1 Yr - Deserted on 9 Jun 1813.

Monahan, John - Private - 17th Infantry - Company: Benjamin Sanders - Other regiment: 3rd Infantry (New) - Age: 27 - Height: 5' 10" - Born: Maryland - Trade: Shoemaker - Enlistment date: 25 Feb 1814 - Place: Kentucky - Period: 5 Yrs - Discharged on 25 Feb 1819.

Monero, Peter - Corporal - 38th Infantry - Company: Isaac Aldridge - Age: 26

By War of 1812 Soc. in MD

- Height: 5' 3/4" - Born: Hanover, PA - Trade: Farmer - Enlistment date: 11 Nov 1814 - Place: Baltimore - Period: War - Discharged at Baltimore on 6 Apr 1815.
- Money, Thomas M. - Private - 20th Infantry - Age: 36 - Height: 5' 8" - Born: Charles Cty, MD - Trade: Cooper - Enlistment date: 28 Apr 1812 - Period: 5 Yrs - Pension: Wife Margaret, Old War Widow File 10530 - Died on 4 Feb 1814.
- Monjoy, George - Private - 2nd Artillery - Company: Nathan Towson - Pension: Land bounty to Jane Mills, sister and other heirs at law of George Monjoy - BLW 16531-160-12.
- Monroe, John S. - Private - 38th Infantry - Company: John Rothrock - BLW 19569-160-12.
- Montcrief, Robert - Private - 38th Infantry - Company: Charles Stansbury - Age: 23 - Height: 5' 7" - Born: County Longford, Ireland - Trade: Farmer - Enlistment date: 17 Oct 1814 - Place: Baltimore - Period: War - BLW 12864-160-12 - Discharged at Baltimore on 6 Apr 1815.
- Montgomery, Archibald - Seaman - US Sea Fencibles - Company: Simmones Bunbury - Enlistment date: 18 Jan 1814 - Period: 1 Yr - Discharged at Fort McHenry on 17 Jan 1815.
- Montgomery, Thomas - Major - 14th Infantry - Born: Maryland - Commissioned as a captain on 12 Mar 1812; promoted to major on 21 Dec 1814; discharged on 15 Jun 1815.
- Moon, Joseph - Private - 14th Infantry - Prisoner of War (Halifax), sent to England.
- Mooney, William S. - Private - 36th Infantry - Company: Joseph Hook - Age: 18 - Height: 5' 3" - Born: Alexandria, DC - Enlistment date: 12 May 1813 - Place: Baltimore - Period: 1 Yr.
- Moore, Benjamin - Private - 3rd Rifles - Company: John Blount - Other regiment: Rifles - Age: 25 - Height: 5' 9 1/2" - Born: Amelia Cty, VA - Trade: Farmer - Enlistment date: 23 Feb 1814 - Place: Salisbury, NC - Period: War - BLW 19137-160-12 - Enlisted in Light Dragoons and transferred to 3rd Infantry; discharged at Washington on 6 Apr 1815.
- Moore, David - Private - 38th Infantry - Company: James Smith - BLW 4394-160-12.
- Moore, Ezekiel - Private - 38th Infantry - Age: 21 - Height: 5' 8 1/2" - Born: Cecil Cty, MD - Trade: Rope maker - Enlistment date: 1 Nov 1814 - Place: Baltimore - Period: War - BLW 309-160-12 - Discharged at Baltimore on 6 Apr 1815.
- Moore, Henry - Corporal - 2nd Artillery - Company: Nathan Towson - Pension: Land bounty to Margaret Moore, daughter and only heir at law of Henry Moore - BLW 26231-160-12.
- Moore, James - Private - 14th Infantry - Company: Joseph Marshall - Age: 22 - Height: 5' 5" - Born: Delaware - Trade: Tailor - Enlistment date: 13 Aug 1814 - Place: Elkton, MD - Period: 5 Yrs - BLW 19995-160-12 - Discharged at Greenbush, NY, on 4 May 1815 on Surgeon's Certificate, rupture.
- Moore, James - Private - 20th Infantry - Company: Byrd Willis - Age: 25 -

Maryland Regulars in War of 1812

Height: 5' 6" - Born: St. Mary's, MD - Trade: Tailor - Enlistment date: 9 Jun 1813 - Place: Stevensburg, VA - Period: 18 Mos - Discharged at Burlington, VT, on 24 Mar 1814 on Surgeon's Certificate, fractured arm.

Moore, John - Private - 19th Infantry - Age: 21 - Height: 5' 10" - Born: Maryland - Enlistment date: 8 Aug 1812 - Period: 18 Mos.

Moore, Lyman - Private - 14th Infantry - Company: Reuben Gilder - BLW 17716-160-12.

Moore, Samuel - Private - 14th Infantry - Company: William McIlvane - Age: 24 - Height: 5' 10" - Born: Montgomery Cty, MD - Trade: Carpenter - Enlistment date: 5 Mar 1813 - Period: War - BLW 88-160-12 - Discharged at Greenbush, NY, on 5 May 1815.

Moore, Samuel - Private - 5th Infantry - Company: Leroy Opie - Other regiment: 3rd Infantry (New) - Age: 21 - Height: 5' 10" - Born: Shephersburg, Cumberland Cty, PA - Trade: Blacksmith - Enlistment date: 4 Aug 1812 - Place: Baltimore - Period: 5 Yrs - Discharged on 2 Oct 1817.

Moore, Samuel C. - Private - 38th Infantry - Company: John Rothrock - Age: 23 - Height: 5' 10" - Born: Somerset Cty, MD - Trade: Carpenter - Enlistment date: 9 Sep 1813 - Period: 1 Yr - Pension: Land bounty to John B. Moore and other heirs at law of Samuel C. Moore - BLW 15909-160-12 - Died in 1815.

Moore, Simeon - Private - 14th Infantry.

Moore, Stephen H. - Captain - US Volunteers - Enlistment date: 8 Sep 1812 - Period: 1 Yr - Pension: Land bounty to Jane Moore, widow of Stephen H. Moore - BLW 9756-160-50 - Wounded at York, UC, on 27 Apr 1813.

Moore, Symond - Corporal - 14th Infantry - Company: Reuben Gilder - Age: 26 - Height: 5' 5 3/4" - Born: Minnack, Duchess Cty, NY - Trade: Distiller - Enlistment date: 12 May 1812 - Place: Baltimore - Period: 5 Yrs - Prisoner of War (Halifax); discharged at Greenbush, NY, on 1 Jun 1815 for debility.

Moore, William - Private - 42nd Infantry - Age: 31 - Height: 5' 4 1/2" - Born: Baltimore City - Enlistment date: 28 Feb 1814 - Deserted at New Castle, DE, on 18 Mar 1814.

Moore, William - Private - 5th Infantry - Company: James Dorman - Age: 29 - Height: 5' 10 1/2" - Born: Harford Cty, MD - Trade: Tailor - Enlistment date: 27 Apr 1812 - Place: Alexandria, DC - Period: 5 Yrs - Discharged at Pittsfield on 13 Jun 1815 on Surgeon's Certificate, wounded at Black Rock, NY.

Moorland, Alfred - Private - 1st Rifles - Company: Daniel Appling - Other regiment: Rifles - Age: 21 - Height: 5' 8" - Born: Maryland - Trade: Saddler - Enlistment date: 6 Jun 1813 - Place: Kentucky - Period: War - Discharged at Fort Gibson, NY, on 11 Jun 1815.

Moorland, Thomas H. - Private - 28th Infantry - Company: Joseph Belt - Age: 40 - Height: 5' 10" - Born: Maryland - Trade: Carpenter - Enlistment date: 20 Jan 1814 - Place: Winchester - Period: War - BLW 1978-160-12 - Discharged at Olympian Springs, KY, on 31 Mar 1815.

Moran, William - Private - 2nd Infantry - Company: Hezekiah Bradley - Age: 31 - Height: 5' 9" - Born: Montgomery Cty, MD - Trade: Soldier - Enlistment

By War of 1812 Soc. in MD

date: 1 Mar 1814 - Place: Washington, MS - Period: 5 Yrs - Died at New Orleans on 25 Dec 1815.

Moran, William - Sergeant - 23rd Infantry - Company: Lizur Canfield - Age: 34 - Height: 5' 7 3/4" - Born: Baltimore - Trade: Hatter - Enlistment date: 28 Aug 1812 - Period: 5 Yrs - BLW 17997-160-12 - Discharged at Sackets Harbor, NY, for disability.

Morand, Thomas - Private - 36th Infantry - Company: Joseph Hook - Age: 30 - Height: 5' 7 3/4' - Born: Maryland - Trade: Cooper - Enlistment date: 29 Nov 1813 - Period: 1 Yr - BLW 13543-160-12 - Re-enlisted at Baltimore on 5 Apr 1814 for the war; discharged at Fort Covington, MD, on 30 Mar 1815.

Moraro, Peter - Private - 14th Infantry - Company: Joseph Marshall - Age: 24 - Height: 5' 9" - Born: Bordeaux, France - Trade: Seaman - Enlistment date: 13 Jun 1814 - Place: Baltimore - Period: War - BLW 92-160-12 - Discharged at Greenbush, NY, on 3 May 1815.

Moreland, Elisha - Private - 38th Infantry - Company: Isaac Aldridge - Age: 18 - Height: 5' 3 1/2" - Born: New Jersey - Trade: Brick layer - Enlistment date: 9 Jan 1815 - Place: Baltimore - Period: War - BLW 126-320-12 - Discharged at Baltimore on 6 Apr 1815.

Moreland, Theodore - Private - 36th Infantry - Company: Samuel Raisin - Age: 21 - Height: 5' 5" - Born: Charles Cty, MD - Trade: Laborer - Enlistment date: 21 Aug 1814 - Period: War - Pension: Land bounty to Mathew Moreland, brother and other heirs at law of Theodore Moreland - BLW 25774-160-12 - Died on 27 Jan 1815.

Moreland, Thomas - Private - 2nd Artillery - Company: John Peyton - Age: 39 - Height: 5' 7 1/2" - Born: Anne Arundel Cty, MD - Trade: Shoemaker - Enlistment date: 6 Jun 1814 - Place: Winchester - Period: War - Discharged at Norfolk, VA, on 31 Mar 1815.

Moreland, Thomas S. - Private - 1st Infantry - Company: Simon Owens - Age: 34 - Height: 5' 10" - Born: Anne Arundel, MD - Trade: Shoemaker - Enlistment date: 1 Apr 1809 - Place: Winchester - Period: 5 Yrs - Discharged at St. Louis on 31 Mar 1814.

Morford, John - Private - 14th Infantry - Company: Joseph Marshall - Age: 24 - Height: 6' 2" - Born: Harford Cty, MD - Trade: Laborer - Enlistment date: 5 Aug 1814 - Place: Baltimore - Period: War - BLW 3240-160-12 - Discharged at Greenbush, NY, on 3 May 1815.

Morgan, Charles - Private - 36th Infantry - Company: Thomas Carbery - Pension: No pension claim - BLW 30-160-50.

Morgan, James - Private - 14th Infantry - Company: Reuben Gilder.

Morgan, John - Private - 36th Infantry - Company: Samuel Raisin - Age: 26 - Height: 5' 9" - Born: Ireland - Trade: Currier - Enlistment date: 12 Nov 1814 - Place: Fredericktown, MD - Period: War - BLW 13535-160-12 - Discharged at Washington, DC, on 20 Mar 1815.

Morgan, John - Private - 1st Rifles - Company: William Smyth - Age: 20 - Height: 5' 5" - Born: Havre de Grace, MD - Trade: Farmer - Enlistment date: 28 Feb 1814 - Place: Plattsburgh, PA - Period: War - Discharged at Plattsburgh, NY, on 25 Aug 1815.

Maryland Regulars in War of 1812

Morgan, John - Seaman - US Sea Fencibles - Company: John Gill - Age: 21 - Height: 6' 1/2" - Born: Baltimore City - Enlistment date: 12 Jan 1814 - Period: 1 Yr - Deserted.

Morgan, Lodowick - Captain - 1st Rifles - Pension: Heirs received half pay for five years in lieu of land bounty - Commissioned as a 2nd lieutenant, Rifles, on 3 May 1808; promoted to 1st lieutenant on 21 May 1809; promoted to captain on 1 Jul 1811; promoted to major on 24 Jan 1814; killed at Fort Erie, UC, on 12 Aug 1814.

Morgan, Thomas - Sergeant - 2nd Artillery - Company: Frederick Evans - Age: 23 - Height: 5' 6" - Born: Maryland - Enlistment date: 31 May 1813 - Place: Annapolis - Period: 5 Yrs - Discharged at Fort McHenry on 31 May 1818 as a Quartermaster Sergeant.

Morrand, Thomas - Private - 36th Infantry - Company: Joseph Hook - BLW 13643-160-12.

Morris, Fisher - Private - 14th Infantry - Company: David Cummings - Age: 19 - Height: 5' 5" - Born: Calvert Cty, MD - Trade: Laborer - Enlistment date: 2 Nov 1814 - Place: Baltimore - Period: War - Discharged on 18 Mar 1815.

Morris, George - Seaman - US Sea Fencibles - Company: Simmones Bunbury - Enlistment date: 3 Nov 1813 - Period: 1 Yr - Discharged on 13 Jan 1815.

Morris, George - Seaman - US Sea Fencibles - Company: Simmones Bunbury - Enlistment date: 20 Jan 1815 - Period: 1 Yr - Discharged at Fort McHenry on 24 Mar 1815.

Morris, Isaac - Private - 14th Infantry - Age: 17 - Height: 5' 5" - Born: Worcester Cty, MD - Trade: Farmer - Enlistment date: 18 Aug 1814 - Period: 5 Yrs.

Morris, Isaiah - Private - 3rd Infantry - Company: James Woodruff - Age: 30 - Height: 5' 8" - Born: Harford Cty, MD - Trade: Shoemaker - Enlistment date: 14 Nov 1813 - Place: Washington, MT - Period: War - BLW 3723-160-12 - Discharged at New Orleans on 9 Apr 1815.

Morris, Jesse - Private - 14th Infantry - Age: 32 - Height: 5' 9" - Born: Worcester Cty, MD - Enlistment date: 23 Apr 1814 - Place: Salisbury, MD.

Morris, John - Private - 38th Infantry - Pension: Wife May, Old War WF-11565.

Morris, John - Private - 38th Infantry - Company: Charles Stansbury - Other regiment: 14th Infantry then 4th Infantry (New) - Age: 19 - Height: 5' 11 1/2" - Born: Baltimore - Trade: Farmer - Enlistment date: 4 Dec 1814 - Period: 5 Yrs - Pension: Land bounty to David Morris, brother and only heir at law of John Morris - BLW 22948-160-12 - Transferred to 38th Infantry on 30 Jul 1815; discharged at Fort Crawford on 3 Dec 1819.

Morris, John P. - Private - 6th Infantry - Company: Gad. Humphreys - Age: 28 - Height: 5' 10" - Born: Worcester Cty, MD - Trade: Laborer - Place: Philadelphia - Period: 5 Yrs - Dishonorable discharged on 18 Apr 1815, deserted.

Morris, Justus - Private - 14th Infantry - Company: Joseph Marshall - Age: 35 - Height: 5' 9" - Born: Worcester Cty, MD - Trade: Farmer - Enlistment date:

By War of 1812 Soc. in MD

25 Apr 1814 - Place: Salisbury, MD - Period: War - BLW 5681-160-12 - Discharged at Greenbush, NY, on 3 May 1815.

Morris, Morris - Private - 14th Infantry - Company: Richard Arett - Age: 40 - Height: 5' 8" - Born: Montgomery Cty, PA - Trade: Farmer - Enlistment date: 13 Aug 1812 - Place: Baltimore - Period: 5 Yrs - BLW 7054-160-12 - Prisoner of War (Halifax); discharged at Greenbush, NY, on 19 Jun 1815 for old age and infirmity.

Morris, Moses - Private - 14th Infantry - Company: David Cummings - Age: 19 - Height: 5' 5" - Born: Loudoun Cty, VA - Trade: Brick maker - Enlistment date: 6 Nov 1814 - Place: Snowden's Camp, VA - Period: 5 Yrs - Deserted at Fort McHenry on 15 Apr 1815.

Morris, Nathan - Private - 38th Infantry - Company: Charles Stansbury - Age: 26 - Height: 5' 9 3/4" - Born: Washington, MD - Trade: Tailor - Enlistment date: 12 Mar 1814 - Place: Georgetown, DC - Period: War - BLW 2872-160-12 - Discharged at Baltimore on 31 May 1815.

Morris, Thomas - Private - 34th Infantry - Company: Benjamin Poland - Age: 22 - Height: 5' 7 1/2" - Born: Harford Cty, MD - Trade: Cooper - Enlistment date: 27 Mar 1814 - Place: Belfast, MD - Period: War - BLW 9220-160-12 - Discharged at Portland, ME, on 21 Mar 1815.

Morris, William - Private - 38th Infantry - Company: John Mowton - Age: 34 - Height: 5' 4" - Born: Maryland - Trade: Shoemaker - Enlistment date: 26 Jul 1814 - Place: Norfolk, VA - Period: War - BLW 584-160-12 - Discharged at Craney Island, VA, on 15 Mar 1815.

Morrison, George - Private - 36th Infantry - Age: 25 - Height: 5' 6" - Born: New Hampshire - Enlistment date: 5 Sep 1813 - Place: York, PA - Period: 1 Yr.

Morrison, John - Private - 2nd Artillery - Company: Nathan Towson - Pension: Old War IF-4238.

Morrow, John - Private - 36th Infantry - Company: Thomas Carbery - Age: 39 - Height: 5' 10" - Born: Cumberland Cty, PA - Trade: Laborer - Enlistment date: 5 Aug 1814 - Place: Georgetown, DC - Period: War - BLW 3513-160-12 - Discharged at Baltimore on 31 Mar 1815.

Morrow, Robert - Corporal - 36th Infantry - Company: Thomas Carbery - Enlistment date: 29 Oct 1813 - Period: 1 Yr - Discharged on 28 Oct 1814.

Morrow, William - Seaman - US Sea Fencibles - Company: Simmones Bunbury - Age: 27 - Height: 5' - Born: Kent Cty, MD.

Morton, John - Private - 20th Infantry - Company: William Jett - Age: 36 - Height: 5' 9 1/4" - Born: Baltimore - Trade: Planter - Enlistment date: 11 Aug 1814 - Period: War.

Morton, William - Private - 26th Infantry - Company: William Bezeau - Age: 21 - Height: 5' 8" - Born: Baltimore - Enlistment date: 9 Nov 1814 - Place: Baltimore.

Mosbury, Abraham - Private - 2nd Light Dragoons - Company: William Littlejohn - Age: 21 - Height: 5' 11" - Born: Frederick, MD - Trade: Miller - Enlistment date: 14 Sep 1813 - Place: Leesburg, VA - Period: 5 Yrs.

Moseley, Peter - Waiter - 38th Infantry - Company: John Brookes.

Maryland Regulars in War of 1812

Moses, Charles - Drummer - 14th Infantry - Prisoner of War (Halifax), discharged from Halifax on 3 Feb 1814.

Moses, Philip - Sergeant - 14th Infantry - Other regiment: 4th Rifles - Enlistment date: 28 Apr 1813 - Period: War - Commissioned as an ensign in 4th Rifles.

Moss, James - Private - 38th Infantry - Company: James Smith - BLW 13923-160-12.

Moss, John - Sergeant - 7th Infantry - Company: Carey Nicholas - Age: 27 - Height: 5' 7" - Born: Maryland - Trade: Shoemaker - Enlistment date: 6 Jan 1814 - Place: Tennessee - Period: War - Discharged at New Orleans on 30 Apr 1815.

Motzenbaucher, Jacob - Private - 5th Infantry - Company: James Dorman - Other regiment: 3rd Infantry (New) - Age: 30 - Height: 5' 7" - Born: Manchim, Lancaster Cty, PA - Trade: Stone mason - Enlistment date: 18 Mar 1812 - Place: Hagerstown, MD - Period: 5 Yrs - On board the fleet on Lake Champlain; deserted at Buffalo, NY, on 10 Apr 1815.

Mowbray, Aaron - Private - 14th Infantry - Company: Clement Sullivan - Enlistment date: 25 Apr 1812 - Period: 5 Yrs - Pension: Land bounty to William Mowbray, brother and other heirs at law of Aaron Mowbray - BLW 23329-160-12 - Died on or about 19 Sep 1812.

Mowton, John - Captain - 38th Infantry - BLW 21948-160-50 - Commissioned as a 1st lieutenant on 20 May 1813; promoted to captain on 1 May 1814; discharged on 15 Jun 1815; died on 10 May 1865.

Mowton, Martin - Private - 3rd Artillery - Company: James House - Age: 40 - Height: 5' 9 1/2" - Born: St. Mary's, MD - Trade: Laborer - Enlistment date: 24 Oct 1806 - Place: Fort Wolcott, RI - Period: 5 yrs - Re-enlisted at Fort Washington on 10 Nov 1811 for five years; discharged on 22 Apr 1815 for old age.

Moxley, Nehemiah - Private - 36th Infantry - Company: Joseph Hook - Race: 28 - Age: 28 - Height: 5' 8" - Born: Montgomery Cty, MD - Trade: Laborer - Enlistment date: 15 Apr 1814 - Place: Annapolis - Period: War - BLW 199-160-12 - Discharged at Fort Covington, MD, on 30 Mar 1815.

Mudd, Aloysius - Sergeant - 5th Infantry - Company: William Henshaw - Other regiment: 3rd Infantry (New) - Age: 21 - Height: 6' - Born: Baltimore - Trade: Farmer - Enlistment date: 22 Jun 1813 - Place: Baltimore - Period: 5 Yrs - Discharged on 22 Jun 1818.

Mudd, Mason - Second Lieutenant - 14th Infantry - Company: Samuel Lane - Age: 24 - Height: 6' 2" - Born: Charles Cty, MD - Enlistment date: 4 Jun 1812 - Place: Fredericktown, MD - Period: 1 Yr - Enlisted as a sergeant, 14th Infantry, on 4 Jun 1812; commissioned as an ensign on 13 Mar 1813; promoted to 3rd lieutenant on 12 May 1813; promoted to 2nd lieutenant on 29 Jun 1814; discharged on 15 Jun 1815.

Mulholland, Patrick - Private - 14th Infantry - Other regiment: 5th Infantry - Enlistment date: 1 Mar 1812 - Transferred to 5th Infantry, Captain Henshaw's Company, on 30 Apr 1814.

Mulhorn, Edwin - Private - 12th Infantry - Company: Andrew Madison - Age:

27 - Height: 5' 6 1/2" - Born: Anne Arundel, MD - Trade: Chair maker - Enlistment date: 29 Mar 1814 - Place: New Market, VA - Period: War - Discharged at Buffalo on 31 May 1815.

Mulligan, Patrick - Private - 14th Infantry - Company: Reuben Gilder - Age: 45 - Height: 5' 8" - Born: Ireland - Enlistment date: 27 Jan 1813 - Period: 18 Mos - Discharged at Plattsburgh, NY, on 27 Jul 1814.

Mullin, Charles - Private - 17th Infantry - Company: Caleb Holder - Other regiment: 3rd Infantry (New) - Age: 40 - Height: 5' 5" - Born: Cecil Cty, MD - Trade: Farmer - Enlistment date: 16 Dec 1812 - Period: 5 Yrs - Discharged on 8 Jun 1816 on Surgeon's Certificate.

Mullin, William - Sergeant - 36th Infantry - Company: Joseph Hook - Age: 26 - Height: 5' 5 3/4" - Born: England - Trade: Fisherman - Enlistment date: 4 May 1813 - Place: Baltimore - Period: 1 Yr - BLW 2837-160-12 - Re-enlisted at Fort Washington, MD, on 16 May 1814 for the war; discharged at Fort Covington, MD, on 30 Mar 1815.

Mullins, Owen - Private - 14th Infantry - Prisoner of War (Halifax), discharged from Halifax on 3 Feb 1814.

Mulvy, Michael - Private - 22nd Infantry - Company: Daniel McFarland - Age: 19 - Height: 5' 4" - Born: Maryland - Trade: Stone cutter - Enlistment date: 22 Jul 1812 - Place: Washington, PA - Period: 5 Yrs - Died at Franklin, NY, on 1 Jan 1814.

Mumford, Jesse - Private - 36th Infantry - Company: Joseph Hook - Age: 4 - Height: 5' 6 1/2" - Born: Maryland - Enlistment date: 24 May 1813 - Place: Annapolis - Period: 1 Yr.

Munjar, William - Private - 36th Infantry - Company: Joseph Hook - Age: 22 - Height: 5' 7" - Born: Chestertown, MD - Trade: Farmer - Enlistment date: 19 Jun 1813 - Period: 1 Yr - Pension: Wife Mary, SO-27282, SC-19236, WO-44153, WC-34492 - BLW 12513-160-12 - Re-enlisted at Annapolis on 1 Mar 1814 for the war; discharged at Fort Covington, MD, on 30 Mar 1815.

Munson, Chauncey - Private - 36th Infantry - Company: Thomas Carbery - Enlistment date: 28 Jun 1813 - Period: 1 Yr - Transferred to the U.S. Navy on 26 Oct 1813.

Munson, Freeman - Private - 14th Infantry - Company: James McDonald - Age: 25 - Height: 5' 11" - Born: Connecticut - Enlistment date: 24 Jan 1814 - Period: 5 Yrs - Deserted at Albany, NY, on 20 May 1814.

Murdock, George - First Lieutenant - 14th Infantry - Born: Maryland - Commissioned as a 2nd lieutenant on 15 Apr 1812; promoted to 1st lieutenant on 14 Nov 1813; discharged on 15 Jun 1815.

Murnett, William - Private - 14th Infantry - Company: Joseph Marshall - Other regiment: 4th Infantry (New) - Age: 21 - Height: 5' 7" - Born: Calvert Cty, MD - Trade: Sailor - Enlistment date: 11 Jun 1814 - Place: Salisbury, MD - Period: 5 Yrs - Deserted at Fort Severn, MD, on 14 Jul 1815.

Murphey, Samuel - Private - 2nd Infantry - Company: Perrin Willis - Age: 21 - Height: 6' - Born: Montgomery Cty, MD - Trade: House joiner - Enlistment date: 21 Nov 1814 - Place: Washington, DC - Period: 5 Yrs - Discharged at Carlisle Barracks, PA, on 7 Jul 1815 because of a rupture.

Maryland Regulars in War of 1812

Murphey, Stephen - Private - 36th Infantry - Company: Joseph Hook - Age: 20 - Height: 5' 3" - Born: Queen Anne's Cty, MD - Trade: Shoemaker - Enlistment date: 21 Mar 1814 - Place: Sandy Point, VA - Period: War - BLW 9633-160-12 - Discharged at Fort Covington, MD, on 30 Mar 1815.

Murphy Jr., John - Private - 2nd Infantry - Company: Reuben Chamberlain - Age: 21 - Height: 6' 1" - Born: Montgomery Cty, MD - Trade: Farmer - Enlistment date: 24 Jan 1808 - Place: Fredericktown, MD - Period: 5 Yrs - Died in 1814.

Murphy, Benjamin - Private - 14th Infantry - Company: Joseph Marshall - Age: 24 - Height: 6' - Born: Dauphin Cty, PA - Trade: Blacksmith - Enlistment date: 1 Sep 1814 - Place: Baltimore - Period: War - BLW 14372-160-12 - Discharged at Greenbush, NY, on 2 May 1815.

Murphy, Daniel - Private - 14th Infantry - Company: David Cummings - Age: 32 - Height: 5' 6" - Born: Derry, Ireland - Trade: Laborer - Enlistment date: 4 Oct 1814 - Place: Baltimore - Period: War - BLW 3082-160-12 - Discharged at Baltimore on 16 Mar 1815.

Murphy, Daniel - Private - 36th Infantry - Age: 42 - Height: 5' 5" - Born: Londonderry, Ireland - Enlistment date: 27 Aug 1812 - Place: Havre de Grace, MD.

Murphy, Henry - Private - 36th Infantry - Company: Thomas Carbery - Age: 30 - Height: 5' 11" - Born: Maryland - Trade: Cooper - Enlistment date: 23 Apr 1814 - Place: Leonardstown, MD - Period: War - BLW 13537-160-12.

Murphy, James - Private - 14th Infantry - Company: Reuben Gilder - Age: 25 - Height: 5' 5" - Born: Clonmel, Ireland - Trade: Soap boiler - Enlistment date: 2 Mar 1813 - Period: 5 Yrs - Pension: Placed on the pension roll on 9 Aug 1815 - BLW 14283-160-12 - Discharged at Greenbush, NY, on 4 May 1815 on Surgeon's Certificate because of wounds received in leg and wrist; died on 24 Jun 1818.

Murphy, James - Private - 38th Infantry - Age: 22 - Height: 5' 7" - Born: Ireland - Trade: Farmer - Enlistment date: 12 Oct 1814 - Place: Baltimore - Period: War - Deserted in Oct 1814.

Murphy, John - Private - 38th Infantry - Company: Shepperd Leakin - Age: 19 - Height: 5' 7" - Born: North Carolina - Trade: Joiner - Enlistment date: 1 Sep 1814 - Place: Baltimore - Period: War - Discharged on 27 Mar 1815.

Murphy, John - Corporal - 14th Infantry - Company: Samuel Lane - Other regiment: 4th Infantry (New) - Age: 21 - Height: 5' 8 1/4" - Born: Fort Cumberland, MD - Trade: Carpenter - Enlistment date: 2 Jun 1812 - Place: Fort Cumberland, MD - Period: 5 Yrs - BLW 14010-160-12 - Prisoner of War (Halifax), discharged from Halifax on 3 Feb 1814, exchanged on 15 Apr 1814; discharged at Raleigh, NC, on 4 Jun 1817.

Murphy, Josiah - Private - 2nd Artillery - Company: Isaac Roach - Other regiment: Corps of Artillery - Age: 19 - Height: 5' 6" - Born: Maryland - Trade: Laborer - Enlistment date: 13 Jun 1811 - Place: Fort Washington, MD - Period: 5 Yrs.

Murphy, Michael - Private - 2nd Infantry - Company: William Boots - Age: 29 - Height: 5' 6" - Born: Dublin, Ireland - Trade: Farmer - Enlistment date:

By War of 1812 Soc. in MD

30 Jan 1807 - Place: Baltimore - Period: 5 Yrs.

Murphy, Thomas - Private - 16th Infantry - Company: Thomas Mahon - Age: 19 - Height: 5' 6 1/2" - Born: Frederick, MD - Enlistment date: 7 Jan 1814 - Place: Lancaster, PA - Period: 5 Yrs - Deserted at Buffalo on 1 Jun 1815.

Murphy, Thomas - Private - 36th Infantry - Company: Joseph Merrick - Age: 18 - Height: 5' 4" - Born: Prince George's Cty, MD - Enlistment date: 1 Jun 1813 - Place: Baltimore - Period: 1 Yr - Discharged on 19 Dec 1813.

Murphy, William - Private - 22nd Infantry - Company: David Millikin - Other regiment: 2nd Infantry (New) - Age: 21 - Height: 5' 10" - Born: Frederick, MD - Trade: Forgeman - Enlistment date: 8 Jun 1812 - Place: Connellsville, PA - Period: 5 Yrs.

Murray, Amon - Private - 14th Infantry - Company: Clement Sullivan - Enlistment date: 20 May 1812 - Period: 5 Yrs - Pension: Land bounty to Ann Lanham and other heirs at law of Amon Murray - BLW 7090-160-12 - Died on 3 Dec 1812.

Murray, Edward - Private - 5th Infantry - Company: James Dorman - Other regiment: 3rd Infantry (New) - Age: 28 - Height: 5' 9" - Born: Ireland - Trade: Tailor - Enlistment date: 10 Aug 1812 - Place: Baltimore - Period: 5 Yrs - Discharged at Greenbush, NY, for incapacity.

Murray, George - Private - 38th Infantry - Company: Shepperd Leakin - Pension: Land bounty to Hannah Kees, daughter and other heirs at law of George Murray - BLW 26078-160-12.

Murray, John - Private - 36th Infantry - Company: Thomas Carbery - Age: 25 - Height: 5' 7" - Born: Ireland - Trade: Blacksmith - Enlistment date: 26 May 1814 - Place: Leonardstown, MD - Period: War - BLW 10907-160-12 - Discharged at Baltimore on 31 Mar 1815.

Murray, Robert - Private - 5th Infantry - Company: Colin Buckner - Age: 22 - Height: 5' 4" - Born: York Cty, PA - Trade: Seaman - Enlistment date: 10 May 1813 - Place: Baltimore - Period: 5 Yrs - On board the fleet with Commodore Chauncey.

Murray, Thomas - Private - 14th Infantry - Company: Joseph Marshall - Age: 38 - Height: 5' 7 1/2" - Born: Down, Ireland - Trade: Fisherman - Enlistment date: 24 Sep 1814 - Place: Baltimore - Period: War - BLW 437-160-12 - Discharged at Greenbush, NY, on 3 May 1815.

Murray, Thomas - Private - 38th Infantry - Company: Isaac Aldridge - Age: 22 - Height: 5' 4 1/2" - Born: Wales - Trade: Seaman - Enlistment date: 12 Nov 1814 - Place: Baltimore - Period: War - BLW 18780-160-12 - Discharged at Baltimore on 6 Apr 1815.

Murray, Thomas G. - Captain - 1st Artillery - Other regiment: Corps of Artillery - Born: Maryland - Commissioned as a 2nd lieutenant, Artillery, on 17 Jan 1805; promoted to 1st lieutenant on 11 Apr 1807; promoted to captain on 10 Feb 1813; transferred to Corps of Artillery on 12 May 1814; died on 28 Sep 1817.

Murrow, Isaac - Private - 12th Infantry - Company: Thomas Sangster - Other regiment: 8th Infantry (New) - Age: 25 - Height: 5' 4" - Born: Worcester Cty, MD - Trade: Mariner - Enlistment date: 21 May 1812 - Place:

Alexandria, DC - Period: 5 Yrs - Discharged on 21 May 1817.

Murry, Alexander - Private - 36th Infantry - Company: Thomas Carbery - Age: 14 - Height: 4' 8 3/4" - Born: St. Mary's, MD - Trade: Tailor - Enlistment date: 29 Mar 1814 - Place: Leonardstown, MD - Period: 5 Yrs.

Murry, John - Private - 38th Infantry - Company: John Buck - BLW 17602-160-12.

Murry, John - Private - 36th Infantry - Company: Thomas Carbery - Age: 23 - Height: 5' 6" - Born: Ireland - Enlistment date: 30 Mar 1814 - Place: Georgetown, DC - Period: War.

Myerheifer, John - Corporal - 38th Infantry - Company: John Rothrock - Age: 30 - Height: 5' 9" - Born: Fredericktown, MD - Trade: Saddler - Enlistment date: 1 Mar 1814 - Place: Craney Island, VA - Period: War - BLW 9894-160-12 - Discharged at Craney Island, VA, on 15 Mar 1815.

Myers, Abel - Private - 35th Infantry - Company: Isaac Preston - Age: 21 - Height: 5' 6" - Born: Maryland - Trade: Seaman - Enlistment date: 27 Aug 1814 - Place: Norfolk, VA - Period: War - Discharged on 11 Nov 1814 by civil authority, being improperly enlisted.

Myers, Frederick - Private - 5th Infantry - Company: William Henshaw - Age: 30 - Height: 5' 6" - Born: Hanover, Germany - Trade: Miller - Enlistment date: 1 Oct 1812 - Place: Hagerstown, MD - Period: 5 Yrs.

Myers, George - Private - 38th Infantry - Company: Isaac Aldridge - Age: 18 - Height: 5' 2" - Born: Baltimore - Trade: Baker - Enlistment date: 4 Feb 1815 - Period: War - BLW 409-320-12 - Discharged at Baltimore on 6 Apr 1815.

Myers, John - Private - US Volunteers - Company: Stephen Moore - Enlistment date: 8 Sep 1812 - Period: 1 Yr.

Myers, John - Private - 14th Infantry - Enlistment date: 26 Feb 1813 - Period: 18 Mos - Prisoner of War, sent to England; discharged at Baltimore on 31 Mar 1815.

Myers, John - Private - 14th Infantry - Company: Thomas Montgomery - Age: 41 - Height: 5' 4" - Born: England - Trade: Sailor - Enlistment date: 18 Apr 1813 - Place: Frederick, MD - Period: 5 Yrs - On board the fleet on Lake Champlain; deserted in Mississippi on 1 Jun 1816.

Myers, Michael - Private - 14th Infantry - Prisoner of War (Halifax), sent to England.

Myers, Michael - Private - 14th Infantry - Company: Samuel Lane - Age: 18 - Height: 5' 7" - Enlistment date: 13 Jun 1812 - Place: Lewistown - Period: 5 Yrs - Wounded at Black Rock, NY, on 28 Nov 1813; deserted at Fort McHenry on 20 Apr 1815.

Myers, Michael - Private - 20th Infantry - Company: John Duval - Age: 34 - Height: 5' 6" - Born: Frederick, MD - Trade: Shoemaker - Enlistment date: 19 Jul 1814 - Period: War - BLW 11907-160-12 - Discharged at Norfolk, VA, on 20 Mar 1815.

N

Nabb, Charles - Private - 16th Infantry - Company: Robert Gray - Age: 35 -

Height: 5' 9" - Born: Queen Anne's Cty, MD - Trade: Farmer - Enlistment date: 30 Sep 1813 - Place: Philadelphia - Period: 1 Yr - BLW 9177-160-12 - Re-enlisted on 1 Apr 1814 for five years; discharged at Sackets Harbor, NY, on 23 Feb 1816.

Nabb, Edward - Private - 5th Infantry - Company: William Bird - Age: 22 - Height: 5' 8" - Born: Cecil Cty, MD - Trade: Cabinet maker - Enlistment date: 22 Feb 1814 - Place: Harrisburg, PA - Period: 5 Yrs.

Nagle, Henry - Private - 38th Infantry - Company: John Rothrock - BLW 25778-160-12.

Nagle, Joseph - Private - 36th Infantry - Company: Joseph Hook - Age: 25 - Height: 5' 8 3/4" - Born: Lancaster, PA - Enlistment date: 19 Apr 1813 - Place: Westminster, MD - Period: 1 Yr.

Nailor, Benjamin - Private - 14th Infantry - Company: Reuben Gilder - Age: 22 - Height: 5' 7" - Born: Maryland - Trade: Laborer - Enlistment date: 28 Feb 1814 - Period: War.

Nailor, John - Private - 20th Infantry - Company: Charles Gee - Age: 25 - Height: 5' 5" - Born: Frederick, MD - Trade: Saddle maker - Enlistment date: 5 Jan 1814 - Period: War - Discharged at Norfolk, VA, on 20 Mar 1814.

Nangle, Anthony - Private - 38th Infantry - Company: Shepperd Leakin - BLW 10908-160-12.

Napear, George - Private - 14th Infantry - Company: Thomas Montgomery - Pension: Land bounty to Jesse W. Napear, brother and other heirs at law of George Napear - BLW 23777-160-12.

Naperney, John - Private - 14th Infantry - Prisoner of War (Halifax), sent to England.

Nary, Michael - Seaman - US Sea Fencibles - Company: John Gill - Age: 28 - Height: 5' 3" - Born: Baltimore City - Enlistment date: 10 Jan 1814 - Period: 1 Yr.

Nash, Archibald - Private - 14th Infantry - Company: Samuel Lane - Age: 24 - Height: 5' 2" - Born: Maryland - Enlistment date: 21 May 1812 - Place: Westminster, MD - Period: 1 Yr - Killed at Beaver Dams, UC, on 24 Jun 1813.

Neal, Charles - Private - 1st Rifles - Other regiment: 4th Infantry then 3rd Infantry (New) - Age: 28 - Height: 6' - Born: Baltimore - Trade: Cordwainer - Enlistment date: 9 Oct 1812 - Place: Shepherdstown - Period: 5 Yrs - Transferred to 4th Infantry on 9 Aug 1814; discharged at Fort Wayne, IN, on 9 Oct 1817.

Neal, James - Private - 5th Infantry - Company: Richard Whartenby - Age: 24 - Height: 5' 6" - Born: Baltimore - Trade: Blacksmith - Enlistment date: 25 Feb 1814 - Place: Plattsburgh, NY - Period: War - Discharged on 29 Mar 1815.

Neal, John - Private - Light Artillery - Company: Arthur Thornton - Age: 32 - Height: 5' 6" - Born: Maryland - Trade: Hatter - Enlistment date: 22 May 1812 - Place: District of Columbia - Period: 5 Yrs - Discharged on 23 May 1817.

Neal, Thomas - Private - 5th Infantry - Company: Colin Buckner - Age: 19 -

Height: 5' 4 1/2" - Born: Baltimore - Trade: Tinner - Enlistment date: 23 Feb 1813 - Period: 5 Yrs.

Neale, Francis J. - First Lieutenant - 36th Infantry - Born: Maryland - Commissioned as a 1st lieutenant on 30 Apr 1813; discharged on 15 Jun 1815.

Neale, Henry C. - Captain - 36th Infantry - Born: Maryland - Commissioned as a 1st lieutenant, 9th Infantry, on 8 Jan 1799; discharged on 15 Jun 1800; commissioned as a captain, 36th Infantry, on 30 Apr 1813; discharged on 15 Jun 1815.

Neale, James - First Lieutenant - 36th Infantry - Born: Maryland - Commissioned as a 1st lieutenant on 30 Apr 1813; dropped on 30 Sep 1814.

Nealle, John - Private - 36th Infantry - Company: Joseph Nelson - Age: 23 - Height: 5' 8" - Born: Greenbrier, VA - Enlistment date: 19 Sep 1814 - Place: Norfolk, VA - Period: War - Deserted at Norfolk, VA, in 1814.

Nebro, Godfry - Corporal - 14th Infantry - Enlistment date: 2 Jan 1813 - Period: 18 Mos - Prisoner of War, sent to England; discharged on 31 Mar 1815.

Neil, Jeremiah - Private - 14th Infantry - Company: Joseph Marshall - Age: 28 - Height: 5' 10 1/2" - Born: Baltimore - Enlistment date: 18 Jun 1814 - Place: Baltimore - Period: War - Pension: Heirs received half pay for five years in lieu of land bounty - Died at Utica, NY, on 12 Jan 1815.

Neilson, William W. - Captain - 1st Light Dragoons - Born: Maryland - Commissioned as a 2nd lieutenant, 1st Light Dragoons, on 1 May 1812; promoted to 1st lieutenant on 7 Jun 1813; transferred to 4th Infantry on 17 May 1815; promoted to captain on 1 Dec 1816; resigned on 31 Jul 1817.

Nelly, George - Private - 36th Infantry - Company: Joseph Hook - Age: 37 - Height: 5' 8 1/2" - Born: Montgomery Cty, MD - Trade: Carpenter - Enlistment date: 1 Aug 1814 - Place: Georgetown, DC - Period: War - BLW 569-160-12 - Discharged at Fort Covington, MD, on 30 Mar 1815.

Nelmes, Jonathan - Private - 38th Infantry - Company: John Mowton - Pension: Land bounty to Sarah Nelmes and other heirs at law of Jonathan Nelmes - BLW 11144-160-12.

Nelson, George - Private - 1st Rifles - Company: William Smyth - Other regiment: Corps of Artillery - Age: 26 - Height: 5' 4" - Born: Baltimore - Trade: Shoemaker - Enlistment date: 12 Aug 1812 - Place: Shepherdstown - Period: 5 Yrs - Transferred to Corps of Artillery on 14 Jun 1817; discharged at New Orleans on 11 Aug 1817.

Nelson, Joseph S. - Captain - 14th Infantry - Other regiment: 36th Infantry - Born: Maryland - Pension: Land bounty to Hannah Nelson, widow of Joseph S. Nelson - BLW 18041-160-50 - Commissioned as a 1st lieutenant, 14th Infantry, on 12 Mar 1812; promoted to captain, 36th Infantry, on 30 Apr 1813; discharged on 15 Jun 1815; died on 27 Feb 1843.

Nelson, William - Private - 5th Infantry - Company: James Dorman - Other regiment: 3rd Infantry (New) - Age: 20 - Height: 5' 8 3/4" - Born: Charles Cty, MD - Trade: Farmer - Enlistment date: 4 May 1812 - Place: Alexandria, DC - Period: 5 Yrs - Discharged on 3 May 1817.

By War of 1812 Soc. in MD

Nelson, Zephaniah - Sergeant - 36th Infantry - Company: Thomas Carbery - Age: 25 - Height: 6' - Born: St. Mary's Cty, MD - Trade: Farmer - Enlistment date: 19 Mar 1814 - Place: Leonardstown, MD - Period: War - BLW 1960-160-12 - Discharged at Baltimore on 31 Mar 1815.

Nesbitt, Shadrack - Private - 14th Infantry - Company: Reuben Gilder - Other regiment: 4th Infantry (New) - Age: 21 - Height: 5' 6" - Born: Dorchester Cty, MD - Trade: Farmer - Enlistment date: 19 May 1812 - Place: Maryland - Period: 5 Yrs - Prisoner of War (Halifax), captured at Stoney Creek, UC, on 12 Jun 1813; discharged at Montpelier, MS, on 18 May 1817.

Netheway, Samuel - Sergeant - 36th Infantry - Age: 29 - Height: 5' 11" - Born: Shenandoah Cty, VA - Trade: Boatman - Enlistment date: 3 Jul 1814 - Place: Lynchburg, VA - Period: 5 Yrs.

Nett, Henry - Private - 5th Infantry - Company: William Henshaw - Other regiment: 3rd Infantry (New) - Age: 36 - Height: 5' 11" - Born: Denmark - Trade: Sailor - Enlistment date: 10 Oct 1813 - Place: Baltimore - Period: 5 Yrs - BLW 20064-160-12 - Discharged at Fort Mackinaw, MI, on 18 Oct 1818.

Neville, Peter T. - Corporal - 36th Infantry - Company: Samuel Raisin - Age: 22 - Height: 5' 8 1/4" - Born: Alexandria, DC - Trade: Butcher - Enlistment date: 26 Sep 1814 - Place: Georgetown, DC - Period: War - BLW 14833-160-12 - Discharged at Washington, DC, on 20 Mar 1815.

Nevitt, Thomas - Private - 10th Infantry - Company: Joseph Clay - Age: 35 - Height: 5' 10 1/2" - Born: Charles Cty, MD - Trade: Carpenter - Enlistment date: 25 Jan 1814 - Place: Washington, DC - Period: War - BLW 84-160-12 - Discharged at Buffalo on 31 May 1815.

Newburgh, John V. - Private - US Volunteers - Company: Stephen Moore - Enlistment date: 8 Sep 1812 - Period: 1 Yr - Deserted on 5 Dec 1812.

Newcomb, George - Corporal - 14th Infantry - Company: Robert Kent - Age: 30 - Height: 5' 6" - Born: Maryland - Trade: Coach maker - Enlistment date: 15 Apr 1813 - Place: Maryland - Period: 5 Yrs - Discharged on 17 Mar 1818.

Newcommer, Jacob - Private - 38th Infantry - Company: Shepperd Leakin - Age: 28 - Height: 5' 7 1/4" - Born: Taney, MD - Trade: Farmer - Enlistment date: 14 Aug 1813 - Period: 1 Yr - Re-enlisted on 17 Jun 1814 for the war; deserted in Nov 1814.

Newell, William - Private - 26th Infantry - Company: William Bezeau - Age: 21 - Height: 5' 10" - Born: Brandywine, DE - Enlistment date: 5 Nov 1814 - Place: Baltimore - Discharged at Baltimore.

Newit, Edward - Seaman - US Sea Fencibles - Company: William Addison.

Newland, Thomas - Private - 14th Infantry - Prisoner of War (Halifax), sent to England.

Newman, Francis - Captain - 1st Artillery - Other regiment: Corps of Artillery - Commissioned as a 2nd lieutenant, Artillery, on 25 Jan 1803; promoted to 1st Lieutenant on 13 Mar 1805; promoted to captain on 1 Oct 1809; transferred to 1st Artillery on 12 Mar 1812; transferred to Corps of Artillery on 12 May 1814; discharged on 15 Jun 1815.

Newman, Stokely - Private - 14th Infantry - Company: Reuben Gilder - Age:

26 - Height: 5' 11 1/2" - Born: Lewistown, DE - Trade: Blacksmith - Enlistment date: 12 Aug 1812 - Place: Chester - Period: 5 Yrs - Pension: Old War IF-25327 - BLW 3990-160-12 - Wounded at Cook's Mills, UC, on 19 Oct 1814; Prisoner of War (Halifax); discharged at Greenbush, NY, on 28 Mar 1815, wounds.

Newman, Thomas - Private - 38th Infantry - Company: John Rothrock - BLW 8420-160-12.

Newman, William - Musician - 12th Infantry - Company: Thomas Sangster - Other regiment: 8th Infantry (New) - Age: 16 - Height: 5' 5" - Born: Baltimore - Trade: Baker - Enlistment date: 13 Jul 1812 - Place: Camp Buffalo, VA - Period: 5 Yrs - Discharged on 12 Jul 1817.

Newton, John - Private - 14th Infantry - Company: Kenneth McKenzie - Enlistment date: 12 Dec 1812 - Period: 18 Mos - Discharged at Greenbush, NY, on 23 Jun 1814.

Newton, William - Private - 36th Infantry - Company: Charles Randolph - Age: 21 - Height: 5' 8" - Born: District of Columbia - Enlistment date: 28 May 1813 - Place: Frederick, MD - Period: 1 Yr.

Nice, David - Private - 17th Infantry - Company: John Chunn - Other regiment: 3rd Infantry (New) - Age: 23 - Height: 5' 9 1/2" - Born: Baltimore Cty - Trade: Shoemaker - Enlistment date: 15 May 1813 - Period: 5 Yrs - Discharged at Fort Harrison, IN, on 15 May 1818.

Nicholas, Cary - Major - 14th Infantry - Born: Virginia - Place: Kentucky - Commissioned as a 1st lieutenant, 7th Infantry, on 3 May 1808; promoted to captain on 1 Mar 1811; promoted to major, 14th Infantry, on 15 May 1814; discharged on 15 Jun 1815.

Nicholas, Francis - Private - 36th Infantry - Company: Samuel Raisin - BLW 2740-160-12.

Nicholas, George - Surgeon's Mate - 14th Infantry - Commissioned as a surgeon's mate, 14th Infantry, on 14 Oct 1812; died on 17 Mar 1813.

Nicholls, William - Private - 14th Infantry - Company: William McIlvane - Age: 40 - Height: 5' 6 3/4" - Born: Delaware - Trade: Seaman - Enlistment date: 13 Apr 1813 - Period: 5 Yrs - BLW 14245-160-12 - On board the fleet with Commodore Chauncey; discharged at Greenbush, NY, on 4 May 1815, for infirmities.

Nichols, Elisha - Private - 38th Infantry - Company: John Mowton - BLW 12136-160-12.

Nichols, Isaac - Private - 14th Infantry - Company: Joseph Marshall - Age: 21 - Height: 5' 7 1/2" - Born: Worcester, MD - Trade: Shoemaker - Enlistment date: 4 Jul 1814 - Place: Boonesboro, MD - Period: War - Pension: Wife Mary Ann, WO-26989, WC-30748 - BLW 558-160-12 - Discharged at Greenbush, NY, on 3 May 1815.

Nichols, James - Private - 36th Infantry - Company: Samuel Raisin - Age: 22 - Height: 5' 7 1/2" - Born: Dorchester Cty, MD - Enlistment date: 17 May 1813 - Place: Cambridge, MD - Period: 1 Yr.

Nichols, James - Private - 2nd Light Dragoons - Company: William Littlejohn - Other regiment: Light Dragoons - Age: 22 - Height: 5' 7" - Born: Talbot

By War of 1812 Soc. in MD

Cty, MD - Trade: Carpenter - Enlistment date: 25 May 1814 - Place: Baltimore - Period: War - BLW 10160-160-12 - Discharged at Baltimore on 10 May 1815.

Nichols, John - Private - 10th Infantry - Company: Josiah Woods - Age: 43 - Height: 5' 1/2" - Born: Maryland - Trade: Laborer - Enlistment date: 1 Jun 1813 - Period: 5 Yrs - Discharged on 28 Feb 1814 on Surgeon's Certificate.

Nichols, Mills - Private - 38th Infantry - Company: John Mowton - BLW 12137-160-12.

Nichols, Pimbleton - Private - 5th Infantry - Company: Colin Buckner - Age: 40 - Height: 5' 8 1/2" - Born: Baltimore - Trade: Fisherman - Enlistment date: 11 Apr 1813 - Place: Baltimore - Period: 5 Yrs - Discharged at Sackets Harbor, NY, on 17 Mar 1816, rheumatism.

Nicholson, Benjamin - Captain - 14th Infantry - Born: Maryland - Commissioned as a 1st lieutenant on 12 Mar 1812; promoted to captain on 3 Mar 1813; died on 13 May 1813 from wounds received at York, UC, on 27 Apr 1813.

Nicholson, John - Private - 24th Infantry - Company: John Rodgers - Other regiment: 7th Infantry (New) - Age: 28 - Height: 5' 5" - Born: Maryland - Trade: Shoemaker - Enlistment date: 10 Mar 1813 - Period: 5 Yrs - Prisoner of War (Quebec), captured at Fort Niagara, NY, on 19 Dec 1813 and exchanged on 11 May 1814; discharged on 21 Dec 1815 at Fort Jackson, AL.

Nicum, John - Private - 17th Infantry - Company: Harris Hickman - Age: 40 - Height: 5' 10" - Born: Maryland - Enlistment date: 23 Apr 1814 - Place: Franklin Cty, OH - Pension: Land bounty to John Nicum & other heirs at law of John Nicum - BLW 23025-160-12 - Died on 2 Nov 1814.

Nigh, John - Private - Light Dragoons - Age: 26 - Height: 5' 6" - Born: Maryland - Trade: Hatter - Enlistment date: 10 Feb 1813 - Place: Fredericktown, MD - Period: 5 Yrs.

Nihart, Jacob - Private - 38th Infantry - Company: James Hook - Age: 40 - Height: 5' 8 1/2" - Born: Frederick, MD - Enlistment date: 29 Dec 1813 - Place: Cumberland - Period: 1 Yr - Discharged on 29 Dec 1814.

Nines, George - Private - 2nd Artillery - Company: Nathan Towson - Enlistment date: 30 Jun 1812 - Period: 5 Yrs - Deserted on 21 Aug 1812.

Nixon, William - Private - 14th Infantry - Company: Samuel Lane - Age: 21 - Height: 5' 4" - Born: Burlington, NJ - Trade: Tailor - Enlistment date: 2 May 1812 - Place: Shepherdstown - Period: 5 Yrs - Prisoner of War (Halifax), discharged from Halifax on 3 Feb 1814, exchanged on 15 Apr 1814; discharged on 2 May 1817.

Noble, John - Private - 26th Infantry - Company: William Bezeau - Age: 23 - Height: 5' 5" - Born: Frederick, MD - Trade: Shoemaker - Enlistment date: 18 Nov 1814 - Place: Baltimore - Period: War - BLW 17994-160-12 - Discharged at Philadelphia on 23 Mar 1815.

Noggle, Anthony - Private - 38th Infantry - Company: Shepperd Leakin - Age: 39 - Height: 5' 6" - Born: Pennsylvania - Enlistment date: 4 Jul 1814 - Place: Baltimore - Period: War - Discharged at Fort McHenry on 28 Mar 1815.

Maryland Regulars in War of 1812

Noland, Elias - Private - 5th Infantry - Company: James Dorman - Other regiment: 3rd Infantry (New) - Age: 30 - Height: 5' 7" - Born: Maryland - Trade: Laborer - Enlistment date: 9 May 1812 - Place: Baltimore - Period: 5 Yrs.

Nolen, Barnabas - Private - 26th Infantry (OH) - Other regiment: 25th Infantry - Age: 19 - Height: 5' 8" - Born: Frederick Cty, MD - Trade: Laborer - Enlistment date: 12 Feb 1813 - Place: Ohio - Period: War - Discharged at Sackets Harbor, NY, on 17 May 1815.

Noles, Thomas - Private - 16th Infantry - Company: Miles Greenwood - Age: 21 - Height: 5' 9" - Born: Maryland - Trade: Carter - Enlistment date: 14 Aug 1812 - Place: Philadelphia - Period: 5 Yrs - Discharged at Sackets Harbor, NY, on 23 Mar 1816, lame arm.

Norman, John - Private - 14th Infantry - Company: James McDonald - Age: 25 - Height: 5' 7 1/2" - Born: West Indies - Enlistment date: 1 Mar 1814 - Place: Baltimore - Period: War - Deserted to the enemy at Chippewa, UC, on 17 Oct 1814.

Norman, Stokely - Private - 14th Infantry - Pension: Placed on the pension rolls on 18 Apr 1816.

Norris, Abraham - Private - 2nd Infantry - Company: Reuben Chamberlain - Other regiment: 1st Infantry (New) - Age: 22 - Height: 5' 11" - Born: Maryland - Trade: Cooper - Enlistment date: 2 Jun 1812 - Place: Natchez, MS - Period: 5 Yrs - Died at New Orleans on 28 Mar 1816.

Norris, Isaac - Private - 5th Infantry - Company: Colin Buckner - Age: 28 - Height: 5' 6" - Born: Harford Cty, MD - Trade: Cooper - Enlistment date: 5 May 1813 - Place: Maryland - Period: 5 Yrs.

Norris, John B. - Private - 36th Infantry - Company: Thomas Carbery - Age: 35 - Height: 5' 10" - Born: St. Mary's Cty, MD - Trade: Farmer - Enlistment date: 14 Feb 1814 - Place: Leonardstown, MD - Period: 5 Yrs - BLW 2339-160-12 - Discharged at Baltimore on 31 Mar 1815.

Norris, Patrick - Private - 14th Infantry - Company: Joseph Marshall - Age: 33 - Height: 5' 9" - Born: Boonsbury, MD - Trade: Cooper - Enlistment date: 5 Aug 1814 - Place: Boonesboro, MD - Period: 5 Yrs - Pension: Placed on the pension rolls on 25 Dec 1819 - BLW 23543-160-12 - Discharged at Greenbush, NY, on 4 Apr 1815, wounded in left shoulder.

Norris, William - Private - 14th Infantry - Company: Joseph Marshall - Age: 38 - Height: 5' 10" - Born: Maryland - Trade: Cooper - Enlistment date: 15 Aug 1814 - Period: War - Pension: Land bounty to George Norris, son and other heirs at law of William Norris - BLW 26899-160-12 - Died at Greenbush, NY, on 22 Jan 1815 from diarrhea.

North, Henry - Private - 38th Infantry - Company: George Keyser - Age: 19 - Height: 5' 4 1/2" - Born: Dorchester Cty, MD - Trade: Waterman - Enlistment date: 18 Nov 1814 - Place: Baltimore - Period: War - Pension: Land bounty to Mary North, sister and other heir at law of Henry North - BLW 27004-160-12 - Died on 24 Jan 1815.

North, James - Private - 14th Infantry - Company: Clement Sullivan - Enlistment date: 15 Jul 1812 - Period: 5 Yrs.

By War of 1812 Soc. in MD

North, James - Private - 14th Infantry - Company: Thomas Kearney - Enlistment date: 15 May 1812 - Period: 5 Yrs - Pension: Land bounty to James North, son and other heirs at law of James North - BLW 26979-160-12 - Died on 27 Dec 1812.

Northrup, John - Private - 38th Infantry - Company: Shepperd Leakin - Age: 23 - Height: 5' 3 1/2" - Born: South Kingston, RI - Trade: Seaman - Enlistment date: 2 Sep 1814 - Place: Baltimore - Period: War - BLW 21187-160-12 - Discharged at Fort McHenry on 28 Mar 1815.

Northum, Miles - Private - 38th Infantry - Company: John Rothrock - BLW 20835-160-12.

Northup, John - Private - 38th Infantry - Company: Shepperd Leakin - BLW 21187-160-12.

Norton, John - Private - 36th Infantry - Age: 19 - Height: 5' 11" - Born: New York - Trade: Mariner - Enlistment date: 15 Nov 1814 - Period: War.

Norton, Theophilus J. - Sergeant - 14th Infantry - Company: Kenneth McKenzie - Age: 23 - Height: 5' 6" - Born: Philadelphia - Trade: Comb maker - Enlistment date: 16 Jan 1813 - Place: Baltimore - Period: 18 Mos - Pension: Old War IF-25332 - BLW 2602-160-12 - Discharged at Baltimore on 28 Nov 1814 on Surgeon's Certificate, lost a leg at Fort George, UC, on 27 May 1813.

Norwich, Henry - Private - 36th Infantry - Company: Joseph Hook - Age: 40 - Height: 5' 4" - Born: Lancaster, PA - Enlistment date: 6 Jul 1813 - Place: Frederick, MD - Period: 1 Yr - Deserted on 9 Jul 1813.

Notter, Benjamin - Private - 14th Infantry - Company: Reuben Gilder - Age: 48 - Height: 5' 8" - Born: Ireland - Trade: Shoemaker - Enlistment date: 5 Sep 1812 - Place: Baltimore - Period: 5 Yrs - BLW 9795-160-12 - Discharged at Greenbush, NY, on 1 Jun 1815 for age and infirmities.

Nowland, John - Private - 38th Infantry - Age: 20 - Height: 5' 2 1/2" - Born: Philadelphia - Trade: Cordwainer - Enlistment date: 9 Dec 1814 - Place: Fredericktown, MD - Period: War - BLW 5506-160-12 - Discharged at Baltimore on 6 Apr 1815.

Nowland, Patrick - Private - 36th Infantry - Company: Samuel Raisin - Age: 25 - Height: 5' 6" - Born: Ireland - Trade: Soldier - Enlistment date: 9 Oct 1814 - Place: Richmond, VA - Period: War.

Nowlton, Ebenezer - Private - 14th Infantry - Prisoner of War (Halifax), discharged from Halifax on 3 Feb 1814.

Noyer, John - Private - 14th Infantry - Company: Reuben Gilder - Born: Maryland - Enlistment date: 12 Nov 1812 - Period: 18 Mos - Discharged at Plattsburgh, NY, on 25 May 1814.

Nugent, Patrick - Private - 36th Infantry - Age: 18 - Height: 5' 4" - Born: Philadelphia - Trade: Laborer - Enlistment date: 27 Sep 1814 - Period: War - BLW 423-160-12 - Discharged at Georgetown, DC, on 13 Mar 1815.

Nunamacker, Michael - Private - 5th Infantry - Company: Alexander McIlhenney - Age: 33 - Height: 5' 6" - Born: Maryland - Trade: Laborer - Enlistment date: 10 Apr 1813 - Place: Lancaster, PA - Period: 5 Yrs - Died on 12 Jan 1814.

Maryland Regulars in War of 1812

Nurser, John - Private - 14th Infantry - Company: Robert Kent - Enlistment date: 12 Nov 1812 - Period: 18 Mos.

Nusser, John - Private - 36th Infantry - Company: Joseph Hook - Age: 33 - Height: 5' 7 1/2" - Born: Fredericktown, MD - Trade: Carpenter - Enlistment date: 23 Jul 1814 - Place: Baltimore - Period: War - BLW 10063-160-12 - Discharged at Fort Covington, MD, on 30 Mar 1815.

Nybro, Godfred - Private - 14th Infantry - Prisoner of War (Halifax), sent to England, captured at Beaver Dams on 24 Jun 1813.

Nye, John - Private - 14th Infantry - Company: Thomas Montgomery - Age: 27 - Height: 5' 6" - Born: Somerset Cty, PA - Trade: Hatter - Enlistment date: 10 Feb 1813 - Period: 5 Yrs - Pension: Old War IF-26748 - BLW 2645-160-12 - Discharged at Washington, DC, on 31 May 1815 due to loss of right arm, wounded in battle in Jun 1813.

O

Oakley, John - Private - 2nd Light Dragoons - Company: William Littlejohn - Other regiment: Light Dragoons - Age: 23 - Height: 6' - Born: Harford Cty, MD - Trade: Cooper - Enlistment date: 6 May 1814 - Place: Baltimore - Period: War - BLW 7047-160-12 - Discharged at Carlisle Barracks, PA, on 4 May 1815.

Oaks, James - Private - 38th Infantry - Company: James Smith - Pension: Land bounty to Polly J. Norman, daughter and other heirs at law of James Oaks - BLW 26844-160-12.

Oates, John - Private - 14th Infantry - Company: William McIlvane - Other regiment: 4th Infantry (New) - Age: 37 - Height: 5' 4 1/2" - Born: Accomack Cty, VA - Trade: Barber - Enlistment date: 15 Jun 1812 - Period: 5 Yrs - Prisoner of War (Halifax), captured at Beaver Dams on 24 Jun 1813, discharged from Halifax on 3 Feb 1813, exchanged on 15 Apr 1814; discharged at Fort Hawkins on 15 Jun 1817.

Oble, John - Private - 7th Infantry - Company: Walter Overton - Age: 23 - Height: 5' 10" - Born: Maryland - Trade: Farmer - Enlistment date: 15 Aug 1808 - Period: 5 Yrs - BLW 6263-160-12 - Re-enlisted at Fort Claiborne on 1 Mar 1813 for the war; discharged at Fort Claiborne on 30 Apr 1815.

O'Brian, Christopher - Private - 36th Infantry - Company: Samuel Raisin - Age: 26 - Height: 5' 5 1/2" - Born: Ireland - Trade: Weaver - Enlistment date: 9 Oct 1814 - Place: Manchester - Period: War - BLW 1063-160-12 - Discharged at Washington, DC, on 20 Mar 1815.

O'Brian, Edmund - Private - 38th Infantry - Company: Thomas Sangster - Other regiment: 14th Infantry then 4th Infantry (New) - Age: 24 - Height: 5' 5" - Born: Ireland - Trade: Sawyer - Enlistment date: 4 Oct 1814 - Place: Baltimore - Period: 5 Yrs - Discharged at Montpelier, AL, on 3 Oct 1819.

O'Brian, Mesback - Sergeant - 17th Infantry - Company: Martin Hawkins - Age: 40 - Height: 6' - Born: Frederick Cty, MD - Trade: Blacksmith - Enlistment date: 1 Oct 1813 - Period: War - BLW 8710-160-12 - Discharged at Chillicothe, OH, on 9 Jun 1815.

O'Brian, Thomas - Private - 36th Infantry - Company: Joseph Hook - Age: 19

By War of 1812 Soc. in MD

- Height: 5' 5 1/2" - Born: New York City - Trade: Tailor - Enlistment date: 7 Aug 1814 - Place: Georgetown, DC - Period: War - BLW 15107-160-12 - Discharged at Fort Covington, MD, on 30 Mar 1815.

O'Brien, James - Private - 14th Infantry - Company: David Cummings - Age: 18 - Height: 5' 4 1/2" - Born: Ireland - Trade: Glasseitte - Enlistment date: 18 Nov 1813 - Place: Salisbury, MD - Period: War - BLW 17255-160-12 - Discharged on 18 Mar 1815.

O'Brien, Thomas - Private - 14th Infantry - Company: David Cummings - Age: 26 - Height: 6' 1" - Born: Baltimore - Trade: Hatter - Enlistment date: Oct 1814 - Place: Baltimore - Period: War - BLW 15443-160-12 - Discharged at Baltimore on 16 Mar 1815.

O'Bryan, Shadrack - Sergeant - 22nd Infantry - Company: John Foster - Other regiment: 2nd Infantry (New) - Age: 32 - Height: 5' 7 1/2" - Born: Montgomery Cty, MD - Trade: Cutter - Enlistment date: 6 Mar 1813 - Place: Connelstown, PA - Period: 5 Yrs - Discharged at Sackets Harbor, NY, on 2 Apr 1816.

O'Conner, Dennis - Private - 36th Infantry - Company: Thomas Carbery - Age: 36 - Height: 5' 9 7/8" - Born: Anne Arundel Cty, MD - Trade: Tanner - Enlistment date: 2 Mar 1814 - Place: Leonardstown, MD - Period: War - BLW 10062-160-12 - Discharged at Baltimore on 31 Mar 1815.

O'Conner, Jeremiah - Private - 38th Infantry - Company: Charles Stansbury - Age: 35 - Height: 5' 4" - Born: Ireland - Trade: Cooper - Enlistment date: 9 Dec 1814 - Place: Cumberland - Period: War - Pension: Land bounty to Thomas O'Conner, brother and only heir at law of Jeremiah O'Conner - BLW 18443-160-12 - Died on 6 Feb 1815.

Oden, Benjamin - Private - 14th Infantry - Company: David Cummings - Age: 31 - Height: 6' - Born: Halifax, VA - Trade: Farmer - Enlistment date: 4 Nov 1814 - Place: Baltimore - Period: War - BLW 1646-160-12 - Discharged at Baltimore on 16 Mar 1815.

Oden, Francis - Private - 36th Infantry - Company: Joseph Hook - Age: 31 - Height: 5' 10 1/2" - Born: Charles Cty, MD - Trade: Carpenter - Enlistment date: 4 Apr 1814 - Place: Linchburgh - Period: War.

Odenbaugh, Charles - Private - 22nd Infantry - Company: Daniel McFarland - Age: 41 - Height: 5' 7 3/4" - Born: Baltimore - Trade: Cabinet maker - Enlistment date: 29 May 1812 - Place: Washington, DC - Period: 5 Yrs - Died at Fort Niagara, NY, on 22 Apr 1813 from dropsy.

O'Donald, John - Private - 5th Infantry - Company: Richard Whartenby - Other regiment: 3rd Infantry (New) - Age: 40 - Height: 5' 7" - Born: County Tumanagh, Ireland - Trade: Shoemaker - Enlistment date: 1 Jul 1812 - Place: Baltimore - Period: 5 Yrs - BLW 12207-160-12 - Discharged on 1 Jul 1817.

O'Donald, Peter - Private - 14th Infantry - Company: Thomas Montgomery - Other regiment: 4th Infantry (New) - Age: 43 - Height: 5' 2 1/2" - Born: Ireland - Trade: Seaman - Enlistment date: 11 Apr 1813 - Place: Baltimore - Period: 5 Yrs - Discharged at Fort Scott, GA, on 10 Apr 1818.

O'Donnel, John - Private - 14th Infantry - Company: David Cummings - Age: 26 - Height: 5' 6" - Born: Calcutta - Trade: Carver - Enlistment date: 13 Jan

Maryland Regulars in War of 1812

1815 - Place: Baltimore - Period: War - Discharged on 16 Mar 1815.

Ogle, James - Private - 2nd Light Dragoons - Company: George Haig - Other regiment: Light Dragoons - Age: 23 - Height: 5' 8 1/2" - Born: Baltimore - Trade: Tailor - Enlistment date: 11 Jul 1812 - Place: Leesburg - Period: 5 Yrs - BLW 12429-160-12 - Discharged on 21 Jul 1815 on Surgeon's Certificate.

O'Hara, John - Private - 36th Infantry - Company: Joseph Hook - Age: 26 - Height: 5' 7" - Born: Ireland - Enlistment date: 24 Jun 1813 - Place: Baltimore - Period: 1 Yr - Transferred to the U.S. Navy on 26 Oct 1813.

O'Harra, Arthur - Private - 14th Infantry - Company: Samuel Lane - Other regiment: 4th Infantry (New) - Age: 22 - Height: 5' 5 1/2" - Born: Washington, MD - Enlistment date: 6 Jun 1812 - Place: Hagerstown, MD - Period: 5 Yrs - Discharged on 6 Jun 1817.

Oiler, George - Private - 14th Infantry - Company: Samuel Lane - Age: 30 - Height: 5' 7" - Born: Frederick, MD - Enlistment date: 16 May 1812 - Period: 5 Yrs - Pension: Wife Nancy, Old War Widow File 10839 - Prisoner of War (Halifax); died at Halifax on 19 Sep 1813.

Oldham, Richard - Private - 36th Infantry - Company: Charles Randolph - Age: 24 - Height: 5' 8" - Born: Virginia - Enlistment date: 30 Aug 1813 - Place: Waynesburgh, PA - Period: 1 Yr.

Oliver, Harmon - Private - 38th Infantry - Company: John Brookes - BLW 3610-160-12.

Oliver, Moses - Private - 2nd Infantry - Company: John Brahan - Other regiment: 1st Infantry (New) - Age: 25 - Height: 5' 6" - Born: Clark Cty, MD - Trade: Laborer - Enlistment date: 17 Aug 1807 - Place: Fredericksburg, VA - Period: 5 Yrs - BLW 13771-160-12 - Discharged at Baton Rouge, LA, on 2 Nov 1817.

Oliver, William - Private - 14th Infantry - Company: Reuben Gilder - Enlistment date: 1 May 1813 - Period: 5 Yrs - Pension: Land bounty to Joshua Oliver, brother and other heirs at law of William Oliver - BLW 19853-160-12 - Died on 14 Oct 1813.

Oliver, Zephaniah - Private - 2nd Infantry - Company: John Brahan - Other regiment: 1st Infantry (New) - Age: 24 - Height: 5' 8 3/4" - Born: St. Mary's Cty, MD - Trade: Laborer - Enlistment date: 14 Sep 1807 - Place: Fredericksburg, VA - Period: 5 Yrs - Re-enlisted at Fort Stoddert on 14 Jun 1812 for five years; discharged on 14 Jun 1817.

O'Neal, Con. - Private - 14th Infantry - Enlistment date: 30 Oct 1812 - Period: 18 Mos - Prisoner of War (Halifax), sent to England; discharged on 31 Mar 1815.

O'Neal, Patrick - Private - 5th Infantry - Other regiment: 3rd Infantry (New) - Age: 28 - Height: 5' 7 1/2" - Born: Ireland - Trade: Artificer - Enlistment date: 31 Mar 1813 - Place: Baltimore - Period: 5 Yrs - Died on 29 May 1816.

O'Neal, William - Private - 36th Infantry - Company: Joseph Hook - Age: 23 - Height: 5' 6" - Born: Ireland - Trade: Tailor - Enlistment date: 8 Jul 1814 - Place: Baltimore - Period: War.

O'Neale, Hugh - Corporal - 36th Infantry - Company: Samuel Raisin - Age:

By War of 1812 Soc. in MD

28 - Height: 5' 8" - Born: Ireland - Trade: Weaver - Enlistment date: 11 Nov 1814 - Period: War - BLW 5041-160-12 - Discharged at Washington, DC, on 20 Mar 1815.

O'Neil, Henry - Captain - 38th Infantry Commissioned as a captain on 20 May 1813; discharged on 15 Jun 1815.

Oram, Benjamin - Musician - 5th Infantry - Company: William Henshaw - Other regiment: 3rd Infantry (New) - Age: 14 - Height: 4' 6" - Born: Baltimore Cty - Trade: Comb maker - Enlistment date: 29 Apr 1819 - Place: Baltimore - Period: 5 Yrs - Re-enlisted in 1817.

Oram, Isaiah - Seaman - US Sea Fencibles - Company: Simmones Bunbury - Age: 21 - Height: 5' 6 1/2" - Born: Hookstown, MD - Enlistment date: 3 Jan 1814 - Period: 1 Yr.

Oram, John - Seaman - US Sea Fencibles - Company: Simmones Bunbury - Age: 22 - Height: 5' 10" - Born: Hookstown, MD - Enlistment date: 18 Nov 1813 - Period: 1 Yr - Discharged on 17 Nov 1814.

Oram, William - Private - 36th Infantry - Company: Samuel Raisin - Age: 18 - Height: 5' 5" - Born: London, England - Trade: Mariner - Enlistment date: 21 Oct 1814 - Place: Georgetown, DC - Period: War - BLW 26166-160-12 - Discharged on 30 Mar 1815.

Ord, James - First Lieutenant - 36th Infantry - Company: Joseph Nelson - BLW 11990-160-50 - Commissioned as a 1st lieutenant on 30 Apr 1813; resigned on 14 Feb 1815.

Orison, Jacob - Private - 38th Infantry - Company: Charles Stansbury - Age: 25 - Height: 5' 7" - Born: Montgomery Cty, MD - Trade: Weaver - Enlistment date: 8 May 1814 - Period: War - BLW 21903-160-12 - Discharged at Baltimore on 6 Apr 1815.

Orn, Benjamin - Private - 14th Infantry - Company: Reuben Gilder - Age: 19 - Height: 5' 7" - Born: Pennsylvania - Enlistment date: 11 Oct 1813 - Period: 5 Yrs - Pension: Land bounty to Joseph Orn, brother and heir at law of Benjamin Orn - BLW 26305-160-12 - Died at Black Rock, NY, on 29 Dec 1814 from sickness.

Orr, George - Private - 14th Infantry - Company: Reuben Gilder - Other regiment: 4th Infantry (New) - Age: 17 - Height: 5' 6" - Born: Montgomery Cty, MD - Trade: Laborer - Enlistment date: 3 Sep 1813 - Place: Maryland - Period: 5 Yrs - Discharged on 2 Sep 1818.

Orr, John - Private - 36th Infantry - Company: Joseph Merrick - Age: 23 - Height: 5' 7" - Born: New Hampshire - Enlistment date: 13 May 1813 - Place: Baltimore - Period: 1 Yr - Transferred to the U.S. Navy on 30 Oct 1813.

Orr, William - Private - 20th Infantry - Company: John Stanard - Age: 26 - Height: 6' - Born: Maryland - Trade: Shoemaker - Enlistment date: 11 Aug 1812 - Place: Leesburg, VA - Period: 5 Yrs - Discharged at Pittsburgh on 17 Aug 1815 because of a ulcerated leg.

O'Sary, Patrick - Private - 36th Infantry - Company: Thomas Carbery - Age: 38 - Height: 5' 5" - Born: Ireland - Trade: Laborer - Enlistment date: 24 Jul 1814 - Place: Washington, DC - Period: War - Discharged at Baltimore on

Maryland Regulars in War of 1812

31 Mar 1815.

Osbaldestel, Thomas - Private - 36th Infantry - Company: Joseph Hook - Age: 20 - Height: 5' 1" - Born: England - Trade: Mariner - Enlistment date: 15 Mar 1814 - Place: Fort Washington, MD - Period: War - BLW 3014-160-12 - Discharged at Fort Covington, MD, on 30 Mar 1815.

Osborne, Horatio - Private - 17th Infantry - Company: Martin Hawkins - Age: 24 - Height: 5' 8" - Born: Maryland - Trade: Tanner - Enlistment date: 21 Dec 1813 - Period: 18 Mos.

Osborne, Thomas - Sergeant - 14th Infantry - Company: Kenneth McKenzie - Enlistment date: 30 Nov 1814 - Period: 18 Mos - Prisoner of War (Halifax), discharged from Halifax on 3 Feb 1814, exchanged on 15 Apr 1814; discharged at Greenbush, NY, on 23 Jun 1814.

Osburn, Moses - Private - 14th Infantry - Company: Samuel Lane - Enlistment date: 10 Jun 1812 - Period: 5 Yrs - Pension: Land bounty to Jane Skeen, daughter and other heirs at law of Moses Osburn - BLW 26468-160-12 - Killed at Carlisle Barracks, PA, on 20 Aug 1812.

Ostatar, Gabriel - Private - Light Dragoons - Age: 31 - Height: 5' 10" - Born: Pennsylvania - Trade: Blacksmith - Enlistment date: 11 Feb 1813 - Place: Baltimore - Period: 5 Yrs.

Ostender, Gabriel - Private - 14th Infantry - Prisoner of War (Halifax), discharged from Halifax on 3 Feb 1814, exchanged on 15 Apr 1814.

Otlinger, Meschesba - Sergeant - 17th Infantry - Age: 40 - Height: 6' - Born: Maryland - Trade: Blacksmith - Enlistment date: 1 Oct 1813 - Period: War.

Ourster, Alexander - Private - 39th Infantry - Company: Alfred Douglass - Age: 31 - Height: 5' 10" - Born: Baltimore - Enlistment date: 5 Nov 1813 - Place: Gallatin, TN - Period: 1 Yr.

Ouselburgh, Philip - Private - 14th Infantry - Prisoner of War (Halifax), discharged from Halifax on 3 Feb 1814.

Outerbridge, John Collin - Private - 38th Infantry - Company: John Brookes - Age: 39 - Height: 5' 11" - Born: Somerset Cty, MD - Trade: Mariner - Enlistment date: 5 May 1814 - Place: Craney Island, VA - Period: War - BLW 17498-160-12 - Discharged at Washington, DC, on 15 Mar 1815.

Owen, John - Musician - 38th Infantry - Company: John Rothrock - BLW 10496-160-12.

Owen, William - Private - 14th Infantry - Company: Joseph Marshall - Discharged on 15 Mar 1815.

Owens, Denwood - Private - 14th Infantry - Company: David Cummings - Born: Worcester, MD - Period: 5 Yrs - Deserted at Salisbury, MD, on 1 Sep 1814.

Owens, John - Private - 14th Infantry - Company: William McIlvane - Other regiment: 4th Infantry (New) then Corps of Artillery - Age: 35 - Height: 5' 5 1/2" - Born: Wales - Trade: Laborer - Enlistment date: 21 Jul 1813 - Place: Baltimore - Period: 5 Yrs - Pension: Land bounty to Hampton Owens, Alsey Owens, Marvell Owens, James Owens, William Owens and Richard Owens, brothers and sisters and only heirs at law of John Owens - BLW 28085-160-12 - Transferred to Corps of Artillery; died on 27 Sep 1819.

By War of 1812 Soc. in MD

Owens, John - Private - 14th Infantry - Prisoner of War, exchanged on 15 Apr 1815.

Owens, Peregrine - Private - 12th Infantry - Company: Thomas Post - Age: 23 - Height: 5' 10" - Born: Baltimore - Trade: Farmer - Enlistment date: 19 Oct 1813 - Place: Hagerstown, MD - Period: 5 Yrs - Discharged at Washington, DC.

Owens, Thomas - Private - 17th Infantry - Company: John Chunn - Other regiment: 3rd Infantry (New) - Age: 38 - Height: 5' 7" - Born: Montgomery Cty, MD - Trade: Farmer - Enlistment date: 25 Oct 1813 - Place: Kentucky - Period: 18 Mos - Discharged on 2 May 1815.

Owings, Peregrine - Private - 26th Infantry - Company: William Bezeau - Age: 23 - Height: 5' 11 1/2" - Born: Baltimore - Enlistment date: 16 May 1814 - Place: Baltimore - Period: War.

Oxford, Anthony - Sergeant Major - 36th Infantry - Age: 16 - Height: 4' 10 1/2" - Born: Prince George's Cty, MD - Trade: Butcher - Enlistment date: 18 Jan 1815 - Period: War - Discharged at Georgetown, DC, on 13 Mar 1815.

Oxford, Francis - Private - 36th Infantry - Age: 22 - Height: 5' 6" - Born: Prince George's Cty, MD - Enlistment date: 29 May 1813 - Place: Baltimore - Period: 1 Yr - Transferred to the U.S. Navy on 30 Oct 1813.

P

Paddy, James - Private - Light Artillery - Company: Benjamin Branch - Age: 20 - Height: 5' 7 3/4" - Born: Charles Cty, MD - Trade: Miller - Enlistment date: 10 Mar 1812 - Place: Washington, DC - Period: 5 Yrs - BLW 9436-160-12 - Discharged at Fort Trumbull, CT, on 9 Mar 1817.

Paddy, William - Musician - Light Artillery - Company: Benjamin Branch - Age: 18 - Height: 4' 9/12" - Born: Charles Cty, MD - Trade: Farmer - Enlistment date: 7 Mar 1812 - Place: Washington, DC - Period: 5 Yrs - Discharged on 6 Mar 1817.

Padgett, John - Corporal - 36th Infantry - Company: Joseph Hook - Age: 21 - Height: 5' 9" - Born: Charles Cty, MD - Enlistment date: 27 Jun 1813 - Place: Baltimore - Period: 1 Yr - Died on 7 Feb 1814.

Page, Jenkins - Seaman - US Sea Fencibles - Company: Simmones Bunbury - Age: 27 - Height: 5' 7 1/4" - Born: Dorchester Cty, MD - Enlistment date: 6 Dec 1813 - Period: 1 Yr - Discharged on 5 Dec 1814.

Page, Meshach R. - Private - 36th Infantry - Company: Thomas Carbery - Enlistment date: 1 Mar 1813 - Place: Bladensburg, MD - Period: 1 Yr.

Paine, James - Private - 5th Infantry - Company: Leroy Opie - Age: 33 - Height: 5' 11" - Born: Frederick, MD - Trade: Farmer - Enlistment date: 5 Jun 1813 - Place: Baltimore - Period: 5 Yrs.

Paine, John - Private - 14th Infantry - Company: David Cummings - Age: 23 - Height: 5' 5" - Born: Loudoun Cty, VA - Trade: Laborer - Enlistment date: 22 Oct 1814 - Place: Boonesboro, MD - Period: 5 Yrs - Deserted at Fort McHenry on 19 Apr 1815.

Paine, William G. - Musician - 14th Infantry - Company: Thomas Montgomery - Age: 25 - Height: 5' 7" - Born: Connecticut - Enlistment date:

12 Jun 1812 - Period: 5 Yrs - Wounded at Williamsburg, NY; died at Burlington, VT, on 23 Feb 1814.

Pale, James - Private - 38th Infantry - Company: Charles Stansbury - BLW 14775-160-12.

Palmer, Adam - Private - 14th Infantry - Company: Samuel Lane - Age: 38 - Height: 5' 7 1/2" - Born: Pennsylvania - Enlistment date: 29 Jun 1812 - Place: Chambersburg, PA - Period: 18 Mos - Deserted on 21 Sep 1812.

Palmer, William - Private - 1st Artillery - Company: James Reed - Age: 23 - Height: 5' 8" - Born: Prince George's Cty, MD - Trade: Farmer - Enlistment date: 21 May 1814 - Place: Alexandria, DC - Period: War - Discharged at Fort Washington, MD, on 7 Mar 1815.

Panham, Andrew - Private - 36th Infantry - Company: Joseph Hook - Age: 33 - Height: 5' 11" - Born: Washington, NY - Enlistment date: 2 Apr 1814 - Place: Fort Washington, MD - Period: War - Deserted on 5 May 1814.

Pannell, Edward - Private - 38th Infantry - Company: James Hook - Age: 29 - Height: 5' 2 1/2" - Born: Baltimore - Trade: Mason - Enlistment date: 24 May 1814 - Place: Baltimore - Period: War - Discharged at Fort Covington, MD, on 31 Mar 1815.

Paradise, John - Private - 14th Infantry - Company: Richard Arell - Other regiment: 4th Infantry (New) - Age: 22 - Height: 5' 6 1/2" - Born: Worcester, MD - Trade: Farmer - Enlistment date: 3 Sep 1812 - Place: Snow Hill, MD - Period: 5 Yrs - Prisoner of War (Halifax), discharged from Halifax on 3 Feb 1814, exchanged on 15 Apr 1814; discharged on 2 Sep 1817.

Park, David - Private - 4th Rifles - Company: Joseph Kean - Age: 28 - Height: 5' 9 1/2" - Born: Maryland - Trade: Laborer - Enlistment date: 29 May 1814 - Place: Uniontown, PA - Period: 5 Yrs.

Parkans, Archibald - Private - 14th Infantry - Company: Reuben Gilder.

Parker, Aaron - Private - 38th Infantry - Company: James Smith - BLW 26423-160-12.

Parker, Archibald - Private - 38th Infantry - Company: James Hook - Age: 26 - Height: 5' 11 1/2" - Born: Harford Cty, MD - Trade: Cabinet maker - Enlistment date: 26 Feb 1814 - Place: Baltimore - Period: War - Pension: Land bounty to John Parker, brother and other heirs at law of Archibald Parker - BLW 27156-160-42 - Discharged on 31 Mar 1815.

Parker, Caleb - Surgeon's Mate - 36th Infantry - Commissioned as a surgeon's mate, 36th Infantry, on 26 May 1814; discharged on 15 Jun 1815.

Parker, Elisha - Private - 38th Infantry - Company: John Mowton - BLW 11279-160-12.

Parker, George - Private - 36th Infantry - Company: Mortimer Hall - Age: 26 - Height: 5' 8" - Born: Pennsylvania - Enlistment date: 2 Aug 1813 - Place: Baltimore - Period: 1 Yr.

Parker, Henry - Third Lieutenant - 14th Infantry - Commissioned as an ensign on 12 May 1813; promoted to 3rd lieutenant on 14 Nov 1813; died from wounds received at La Cole Mill, LC, on 30 Mar 1814.

Parker, Jonathan R. - Private - 20th Infantry - Company: Richard Pollard - Age: 33 - Height: 5' 5 1/2" - Born: Annapolis - Trade: Baker - Enlistment date: 20 Nov 1812 - Period: 5 Yrs - Discharged on 1 May 1816 for inability.

Parker, Peter - Sergeant - 38th Infantry - Company: John Buck - Age: 30 - Height: 5' 11 1/2" - Born: Harford Cty, MD - Trade: Shoemaker - Enlistment date: 6 Mar 1814 - Place: Baltimore - Period: War - BLW 10563-160-12 - Discharged at Baltimore on 30 Mar 1815.

Parker, Price - Sergeant - 38th Infantry - Company: Isaac Aldridge - Age: 32 - Height: 5' 11 1/4" - Born: Pennsylvania - Trade: Laborer - Enlistment date: 17 Dec 1814 - Place: Baltimore - Period: War - BLW 238-320-12 - Discharged at Baltimore on 6 Apr 1815.

Parkinson, William - Corporal - 5th Infantry - Company: Leroy Opie - Age: 28 - Height: 5' 10 1/2" - Born: Cumberland, MD - Trade: Carpenter - Enlistment date: 31 May 1813 - Place: Baltimore - Period: 5 Yrs - Discharged at Fort Erie, UC, on 21 Jun 1815 for inability.

Parks, Edward - Private - 2nd Artillery - Company: James Barker - Age: 27 - Height: 5' 9 1/2" - Born: Baltimore - Trade: Shoemaker - Enlistment date: 8 Feb 1814 - Place: Philadelphia - Period: War - Discharged at Philadelphia on 6 May 1815.

Parks, Francis - Private - 38th Infantry - Company: James Smith - BLW 4271-160-12.

Parks, Selby - Private - 14th Infantry - Company: Richard Arell - Other regiment: 4th Infantry (New) - Age: 20 - Height: 5' 9 3/4" - Born: Somerset Cty, MD - Trade: Sailor - Enlistment date: 13 Apr 1812 - Place: Baltimore - Period: 5 Yrs - Prisoner of War (Halifax), captured at Beaver Dams on 24 Jun 1813, discharged from Halifax on 3 Feb 1813, exchanged on 15 Apr 1814; deserted at Fort Severns on 19 Jul 1815.

Parks, Selvey - Private - 23rd Infantry - Age: 19 - Height: 5' 8 1/2" - Born: Baltimore - Enlistment date: 30 May 1814.

Parks, William - Private - 36th Infantry - Age: 30 - Height: 5' 5 1/2" - Born: Ireland - Enlistment date: 16 Sep 1813 - Place: York, PA - Period: 1 Yr.

Parlate, Zacheus - Private - 38th Infantry - Company: James Smith - Age: 21 - Height: 5' 9 1/2" - Born: Baltimore - Trade: Blacksmith - Enlistment date: 18 Feb 1814 - Place: Craney Island, VA - Period: War - Discharged at Craney Island, VA, on 15 Mar 1815.

Parley, Thomas - Private - 38th Infantry - Company: John Mowton - BLW 2710-160-12.

Parmeley, John - Private - 14th Infantry - Company: Thomas Kearney - Enlistment date: Jan 1812 - Period: 5 Yrs.

Parmillion, Thomas - Private - 38th Infantry - Company: John Rothrock - Age: 22 - Height: 5' 10" - Born: Elk Ridge, MD - Trade: Farmer - Enlistment date: 12 Mar 1814 - Place: Craney Island, VA - Period: War - BLW 17208-160-12 - Discharged at Craney Island, VA, on 15 Mar 1815.

Parr, Henry - Private - 36th Infantry - Company: Joseph Hook - Age: 20 - Height: 5' 6 1/2" - Born: Alleghany Cty, PA - Trade: Weaver - Enlistment date: 18 Sep 1813 - Period: 1 Yr - BLW 17092-160-12 - Re-enlisted at Fredericktown, MD, on 28 Feb 1814 for the war.

Parsons, Jesse - Private - 5th Infantry - Company: Benjamin Wallace - Age:

Maryland Regulars in War of 1812

23 - Height: 5' 9 1/2" - Born: Maryland - Trade: Laborer - Enlistment date: 29 Nov 1814 - Place: Harrisburg, PA - Period: 5 Yrs.

Parsons, William - Drummer - 14th Infantry - Company: Richard Arell - Other regiment: 4th Infantry (New) - Age: 22 - Height: 5' 5 1/2" - Born: Anne Arundel Cty, MD - Trade: Barber - Enlistment date: 13 Oct 1812 - Place: Baltimore - Period: 5 Yrs - Deserted at Charleston, SC, on 20 May 1816.

Partridge, Asa - Third Lieutenant - 14th Infantry - Born: Vermont - BLW 24513-160-12 - Sergeant and quartermaster sergeant in 14th Infantry between 3 May and Sep 1813; commissioned as an ensign on 24 Sep 1813; promoted to 3rd lieutenant on 14 Nov 1813; resigned on 30 May 1814.

Pasley, David - Private - 14th Infantry - Company: Thomas Montgomery - Enlistment date: 13 Apr 1812 - Period: 5 Yrs - Died on 21 Dec 1812.

Passwater, Jeremiah - Private - 14th Infantry - Company: Reuben Gilder - Age: 24 - Height: 5' 8" - Born: Delaware - Enlistment date: 12 Dec 1812 - Period: 18 Mos - Discharged on 12 Aug 1814.

Pasture, Charles - Private - 5th Infantry - Company: Colin Buckner - Age: 30 - Height: 5' 3" - Born: Norfolk, VA - Trade: Seaman - Enlistment date: 4 Feb 1813 - Place: Baltimore - Period: 5 Yrs - On board the fleet with Commodore Chauncey; discharged on 21 Jan 1814.

Pate, James - Private - 38th Infantry - Company: Charles Stansbury - Age: 27 - Height: 5' 6 1/2" - Born: Dorchester, VA - Trade: Pilot - Enlistment date: 24 Oct 1814 - Place: Baltimore - Period: War - BLW 14775-160-12 - Discharged in Mar or Apr 1815.

Paterson, Aaron - Private - Light Dragoons - Age: 22 - Height: 5' 8" - Born: Leicester Cty, PA - Trade: Tanner - Enlistment date: 3 May 1814 - Place: Baltimore - Period: War - Deserted at Baltimore.

Patterson, Adam - Private - 2nd Artillery - Company: Nathan Towson - Other regiment: Corps of Artillery - Age: 26 - Height: 5' 5" - Born: Ireland - Trade: Stone cutter - Enlistment date: 15 Jun 1812 - Place: Baltimore - Period: 5 Yrs - BLW 17339-160-12 - Discharged at Sackets Harbor, NY, on 14 Jun 1817.

Patterson, Archibald - Private - 14th Infantry - Company: David Cummings - Enlistment date: 16 Apr 1812 - Period: 5 Yrs - Prisoner of War (Halifax), sent to England.

Patterson, Charles G. - Private - 14th Infantry - Company: David Cummings - Age: 22 - Height: 5' 6" - Born: Rotterdam - Trade: Sail maker - Enlistment date: 18 Jun 1814 - Place: Baltimore - Period: War - Deserted on 24 Jul 1814.

Patterson, James - Private - 37th Infantry - Company: Samuel Northrop - Age: 29 - Height: 5' 7" - Born: Baltimore - Trade: Currier - Enlistment date: 17 Feb 1815 - Period: 5 Yrs.

Patterson, James D. - Private - 2nd Light Dragoons - Company: William Littlejohn - Age: 26 - Height: 6' - Born: Calvert Cty, MD - Trade: Clerk - Enlistment date: 1 Jan 1814 - Place: Alexandria, DC - Period: War - Discharged at Fort Washington, MD, on 7 Mar 1815.

Patterson, John - Private - US Volunteers - Company: Stephen Moore - Enlistment date: 8 Sep 1812 - Period: 1 Yr.

By War of 1812 Soc. in MD

Patterson, Marks - Private - 26th Infantry - Company: William Bezeau - Age: 21 - Height: 5' 10" - Born: New Jersey - Trade: Farmer - Enlistment date: 31 Oct 1814 - Place: Baltimore - Discharged at Philadelphia on 23 Mar 1815.

Patterson, Thomas - Private - 14th Infantry - Company: Richard Arell - Other regiment: 4th Infantry (New) - Age: 25 - Height: 5' 7 1/2" - Born: Stafford, VA - Trade: Carpenter - Enlistment date: 3 Jul 1812 - Place: Baltimore - Period: 5 Yrs - Prisoner of War (Halifax), discharged from Halifax on 3 Feb 1814, exchanged on 15 Apr 1814; deserted at Fort Severn on 15 Jul 1815.

Patterson, Thomas G. - Seaman - US Sea Fencibles - Company: Simmones Bunbury - Age: 24 - Height: 5' 7" - Born: Pennsylvania - Enlistment date: 1 Jul 1814 - Period: 1 Yr.

Patterson, William - Seaman - US Sea Fencibles - Company: John Gill - Age: 17 - Height: 5' 4" - Born: West Point, MD - Enlistment date: 10 Feb 1814 - Period: 1 Yr - Discharged on 9 Feb 1815.

Patton, David - Corporal - 14th Infantry - Enlistment date: 11 Nov 1812 - Period: 18 Mos - Prisoner of War (Halifax), sent to England; discharged on 31 Mar 1815.

Patton, John - Private - 1st Infantry - Company: Horatio Stark - Other regiment: 3rd Infantry (New) - Age: 21 - Height: 6' - Born: Maryland - Trade: Laborer - Enlistment date: 26 Jun 1812 - Place: Lebanon, Warren Cty, OH - Period: 5 Yrs - BLW 12037-160-12 - Discharged at Fort Dearborn, IL, on 26 Jun 1817.

Paul, John - Private - 2nd Rifles - Company: John O'Fallon - Age: 35 - Height: 5' 8" - Born: Maryland - Trade: Farmer - Enlistment date: 18 Aug 1814 - Place: Ohio or Newport, KY - Period: 5 Yrs - Discharged at Detroit on 23 Sep 1815, Surgeon's Certificate.

Paul, Michael - Private - Light Artillery - Company: George Morris - Other regiment: Corps of Artillery - Age: 29 - Height: 5' 6" - Born: Fredericktown, MD - Trade: Tailor - Enlistment date: 10 Mar 1812 - Place: District of Columbia - Period: 5 Yrs - BLW 9480-160-12 - Discharged at Fort Independence, MA, on 13 Mar 1817.

Paulding, Jonathan - Private - 5th Infantry - Company: Richard Whartenby - Age: 33 - Height: 5' 9" - Born: Philadelphia - Trade: Brick layer - Enlistment date: 2 Apr 1814 - Place: Baltimore Cty - Period: 5 Yrs - Discharged at Buffalo on 21 Jun 1815 for weakness.

Paxton, William - Private - 26th Infantry - Company: William Bezeau - Age: 21 - Height: 5' 8 1/2" - Born: Bedford, PA - Trade: Farmer - Enlistment date: 7 Oct 1814 - Place: Baltimore - Period: War - BLW 17993-160-12 - Discharged at Philadelphia on 23 Mar 1815.

Payne, Benjamin D. - Private - 38th Infantry - Company: John Mowton - BLW 20875-160-12.

Payne, Richard - Private - 14th Infantry - Prisoner of War (Halifax), discharged from Halifax on 3 Feb 1814.

Payne, Samuel L. - Private - 38th Infantry - Company: John Mowton - Age: 22 - Height: 5' 8 1/2" - Born: Harford Cty, MD - Trade: Weaver - Enlistment date: 20 Aug 1813 - Period: 1 Yr - Pension: Old War IF-3628 - Re-enlisted at Craney Island, VA, on 21 Mar 1814 for the war; discharged at Craney Island, VA, on 15 Mar 1815.

Peach, Isaac - Private - 14th Infantry - Company: Reuben Gilder - Age: 38 - Height: 5' 6" - Born: Maryland - Trade: Farmer - Enlistment date: 8 Oct 1813 - Period: 5 Yrs - Died at Cheektowaga Hospital, NY, on 8 Feb 1815.

Peach, John - Private - 1st Light Dragoons - Company: Arthur Hayne - Other regiment: Corps of Artillery - Age: 31 - Height: 5' 8 3/4" - Born: Prince George's Cty, MD - Trade: Tailor - Enlistment date: 16 Jun 1812 - Place: Baltimore - Period: 5 Yrs - Transferred to the Corps of Artillery; deserted from Fort Lewis on 26 Jul 1816.

Peach, William B. - Private - 36th Infantry - Age: 19 - Height: 5' 3 1/2" - Born: Marblehead, MA - Enlistment date: 30 Apr 1813 - Place: Baltimore - Period: 1 Yr - Transferred to the U.S. Navy in Oct 1813.

Pearce, Maurice - Private - 5th Infantry - Company: James Dorman - Other regiment: 3rd Infantry (New) - Age: 22 - Height: 5' 6" - Born: Prince George's Cty, MD - Trade: Shoemaker - Enlistment date: 5 May 1812 - Place: Alexandria, DC - Period: 5 Yrs - Discharged at Buffalo on 21 Jun 1815, hernia.

Pearce, Philip - Private - 5th Infantry - Company: William Bird - Age: 42 - Height: 5' 9" - Born: New Jersey - Trade: Farmer - Enlistment date: 23 Mar 1813 - Place: Baltimore - Period: 5 Yrs.

Pearce, Thomas - Private - 36th Infantry - Company: Joseph Hook - Age: 28 - Height: 5' 6 1/2" - Born: Philadelphia - Enlistment date: 4 May 1813 - Place: Westminster, MD - Period: 1 Yr.

Peardon, Samuel - Private - 18th Infantry - Company: William Taylor - Other regiment: 4th Infantry (New) - Age: 40 - Height: 5' 9 1/2" - Born: Harford Cty, MD - Trade: Joiner - Enlistment date: 15 Sep 1814 - Place: Chester, SC - Period: 5 Yrs - Discharged at Sullivan Island on 5 Apr 1816 on Surgeon's Certificate.

Peasely, David - Private - 14th Infantry - Pension: Heirs received half pay for five years in lieu of land bounty - Died on 21 Nov 1814.

Peasley, Ithream - Private - 38th Infantry - Company: Anthony Miltenberger - Age: 32 - Height: 5' 7" - Born: Baltimore - Trade: Cooper - Enlistment date: 17 Jun 1813 - Period: 1 Yr - BLW 197-160-12 - Re-enlisted at Craney Island, VA, on 14 Apr 1814 for the war; discharged at Craney Island, VA, on 15 Mar 1815.

Pebase, John - Private - 14th Infantry - Company: Henry Fleming - Prisoner of War, received from Chazy on 11 May 1814.

Pechrand, Joseph - Private - 36th Infantry - Deserted in 1815 as a recruit.

Peck, John - Private - 2nd Infantry - Company: John Miller - Age: 20 - Height: 5' 10 1/4" - Born: Pennsylvania - Enlistment date: 17 Feb 1813 - Place: Hagerstown, MD - Period: 5 Yrs.

Peck, John - Private - 26th Infantry - Company: William Bezeau - Age: 26 - Height: 5' 7" - Born: Prince William's Cty, VA - Trade: Pilot - Enlistment date: 21 Oct 1814 - Place: Baltimore - Period: War - BLW 9291-160-12 - Discharged at Philadelphia on 21 Mar 1815.

By War of 1812 Soc. in MD

Peddicord, Leakin D. - Private - 7th Infantry - Company: Walter Overton - Other regiment: 6th Infantry (New) - Age: 44 - Height: 5' 10" - Born: Maryland - Trade: School master - Enlistment date: 24 Sep 1813 - Place: Fort Claiborne, AL - Period: War.

Peeples, Alexander - Private - 19th Infantry - Company: Richard Talbott - Age: 24 - Height: 5' 7" - Born: Maryland - Trade: Farmer - Enlistment date: 11 Jun 1814 - Place: Urbana, OH - Period: War - BLW 3503-160-12 - Discharged at Chillicothe, OH, on 5 Jun 1815.

Peirce, John - Private - 36th Infantry - Company: Mortimer Hall - Age: 29 - Height: 5' 6" - Born: France - Enlistment date: 20 Sep 1813 - Place: Baltimore - Period: 1 Yr - Deserted at Baltimore on 21 Sep 1813.

Pemberton, Nicholas - Private - 5th Infantry - Company: Richard Bell - Age: 40 - Height: 5' 8 1/2" - Born: Baltimore - Trade: Farmer - Enlistment date: 11 Apr 1813 - Place: Baltimore - Period: 5 Yrs.

Pendel, Thomas - Sergeant - 2nd Artillery - Company: Nathan Towson - Age: 25 - Height: 5' 11" - Trade: Tailor - Enlistment date: 27 Jun 1812 - Period: 5 Yrs - Wounded at Black Rock, NY; discharged at Boston on 12 Jun 1815 on Surgeon's Certificate.

Pendel, Thomas - Sergeant Major - 2nd Artillery - Pension: Old War IF-20123.

Pender, Isaac - Private - 36th Infantry - Company: Joseph Hook - Age: 20 - Height: 5' 7" - Born: New Jersey - Trade: Shoemaker - Enlistment date: 11 May 1814 - Place: Annapolis - Period: War - Discharged at Fort Covington, MD, on 30 Mar 1815.

Penman, John - Private - US Volunteers - Company: Stephen Moore - Enlistment date: 8 Sep 1812 - Period: 1 Yr.

Pennell, Edward - Private - 38th Infantry - Company: James Hook - BLW 5650-160-12.

Pennell, Edward - Private - 36th Infantry - Company: Mortimer Hall - Age: 28 - Height: 5' 2" - Born: Baltimore - Enlistment date: 5 May 1813 - Place: Havre de Grace, MD - Period: 1 Yr.

Pennington, Edmund - Private - 2nd Infantry - Company: John Miller - Age: 23 - Height: 5' 6" - Born: Sunbury, MD / PA - Trade: Nailor - Enlistment date: 19 Mar 1810 - Place: Sunbury - Period: 5 Yrs - Discharged at New Orleans on 19 Mar 1815.

Pennington, Ephraim - Private - 12th Infantry - Company: Thomas Sangster - Other regiment: 8th Infantry (New) - Age: 21 - Height: 5' 8" - Born: Cecil Cty, MD - Trade: Cooper - Enlistment date: 16 Jun 1812 - Place: Elkton, MD - Period: 5 Yrs - BLW 15313-160-12 - Discharged at Fort Clark, IL, on 16 Jun 1817.

Penny, John - Sergeant - 1st Artillery - Company: Lewis Howard - Age: 35 - Height: 5' 8 1/2" - Born: Queen Anne's Cty, MD - Trade: Laborer - Enlistment date: 13 Jan 1809 - Place: Fort Mackinaw, MI - Period: 5 Yrs - Prisoner of War paroled at Fort Fayette; discharged on 31 Jan 1814.

Pennywell, Patty - Private - 16th Infantry - Company: John Baldy - Age: 20 - Height: 5' 1" - Born: Worcester, MD - Trade: Laborer - Enlistment date: 9

Mar 1814 - Place: Philadelphia - Period: 5 Yrs - Discharged at Buffalo on 17 Jun 1815, mental incapacity.

Pentz, Jacob - Private - 26th Infantry - Company: William Bezeau - Age: 37 - Height: 5' 7 1/4" - Born: Pennsylvania - Trade: Butcher - Enlistment date: 8 Oct 1814 - Place: Baltimore - Period: War - Discharged at Philadelphia on 26 Mar 1815.

Perch, Thomas - Private - 36th Infantry - Company: Samuel Raisin - Age: 25 - Height: 5' 7" - Born: Stafford, VA - Trade: Baker - Enlistment date: 1 Jul 1814 - Place: Manchester - Period: War - Deserted.

Peregoy, William - Gunner - US Sea Fencibles - Company: John Gill - Enlistment date: 20 Dec 1813 - Period: 1 Yr.

Peregoy, William - Private - US Volunteers - Company: Stephen Moore - Enlistment date: 8 Sep 1812 - Period: 1 Yr.

Perkins, Archibald - Private - 36th Infantry - Age: 43 - Height: 5' 7 1/2" - Born: Castine, NH - Enlistment date: 5 May 1813 - Place: Baltimore - Period: 1 Yr - Deserted on 9 May 1813.

Perkins, Archibald - Private - 14th Infantry - Company: Reuben Gilder - Age: 42 - Height: 5' 9" - Born: Casteen Cty, ME - Trade: Sailor - Enlistment date: 1 Mar 1814 - Place: Plattsburgh, NY - Period: War - BLW 1673-160-12 - Discharged at Greenbush, NY, on 14 May 1815.

Perkins, William - Private - 38th Infantry - Company: John Brookes - Pension: Wife Dorothy, SO-25936, SC-16226, WO-43868, WC-34230 - BLW 27454-160-42.

Perrigue, James - Corporal - 14th Infantry - Company: William McIlvane - Enlistment date: 29 Jan 1813 - Period: 18 Mos - Prisoner of War (Halifax), discharged from Halifax on 3 Feb 1814, exchanged on 15 Apr 1814.

Perry, James - Private - Light Artillery - Company: John Bell - Race: 18 - Age: 18 - Height: 5' 6" - Born: Baltimore or Queen Anne's County, MD - Trade: Mariner - Enlistment date: 3 Nov 1814 - Place: Concord or Boston, MA - Period: 5 Yrs - Discharged on 31 Mar 1815 on account of being a Negro and not a fit companion for the American soldier.

Perry, James - Private - 5th Infantry - Company: James Dorman - Other regiment: 3rd Infantry (New) - Age: 24 - Height: 5' 8" - Born: Baltimore - Trade: Brick maker - Enlistment date: 4 May 1812 - Place: Alexandria, DC - Period: 5 Yrs - Pension: Placed on the pension roll on 11 Dec 1815 - Discharged at Pittsburgh on 16 Aug 1815, wounded in battle of Lyon's Creek.

Perry, James - Private - Light Artillery - Company: John Bell - Race: 18 - Age: 18 - Height: 5' 6" - Born: Maryland - Trade: Mariner - Enlistment date: 3 Nov 1814 - Place: Concord, MA - Period: 5 Yrs - Discharged on 31 Mar 1815 for being a Negro, and not a fit companion for the American soldier.

Perry, Peter - Private - 22nd Infantry - Company: John Foster - Age: 44 - Height: 5' 9" - Born: Maryland - Trade: Shoemaker - Enlistment date: 14 May 1813 - Place: Uniontown, PA - Period: 5 Yrs - Discharged at Sackets Harbor, NY, on 11 Aug 1815, lameness.

Perry, Peter - Private - 2nd Infantry - Company: George Salmon - Other regiment: 1st Infantry (New) - Age: 26 - Height: 5' 6" - Born: Maryland - Trade: Miller - Enlistment date: 11 Jan 1803 - Period: 5 Yrs - Re-enlisted on 12 Oct 1807 and 10 Jul 1812; discharged on 9 Jul 1817.

Perry, Richard - Private - 14th Infantry - Company: Kenneth McKenzie - Enlistment date: 24 Nov 1812 - Period: 18 Mos - Prisoner of War (Halifax); re-enlisted on 1 May 1814.

Perry, William - Private - 14th Infantry - Company: William McIlvane - Age: 28 - Height: 5' 6" - Born: Harford Cty, MD - Trade: Cooper - Enlistment date: 31 Dec 1812 - Period: 5 Yrs - Pension: Old War IF-3671 - BLW 6977-160-12 - Wounded at Plattsburg, NY, by bayonet; discharged at Greenbush, NY, on 1 May 1815 by Surgeon's Certificate.

Persons, James - Private - 14th Infantry - Died at Sackets Harbor, NY, on 29 Oct 1813.

Peters, Conrad - Private - 2nd Artillery - Company: Frederick Evans - Age: 41 - Height: 5' 8" - Born: Pennsylvania - Trade: Brick layer - Enlistment date: 23 Nov 1812 - Place: Annapolis - Period: 5 Yrs - Discharged on 25 Apr 1815.

Peters, William - Seaman - US Sea Fencibles - Company: William Addison.

Peters, William - Private - 2nd Artillery - Company: Nathan Towson - Age: 30 - Died from wounds received during the Battle of Stoney Creek.

Peters, William - Private - US Volunteers - Company: Stephen Moore - Enlistment date: 8 Sep 1812 - Period: 1 Yr.

Petitt, Jonas - Private - 19th Infantry - Company: Carey Trimble - Age: 26 - Height: 5' 5 1/2" - Born: Maryland - Trade: Farmer - Enlistment date: 15 Feb 1814 - Place: Chillicothe, OH - Period: War - BLW 7824-160-12 - Discharged at Chillicothe, OH, on 6 Jun 1815.

Petry, John - Private - 14th Infantry - Company: Henry Fleming - Age: 49 - Height: 5' 9 1/2" - Born: Philadelphia - Trade: Saddler - Enlistment date: 6 Mar 1813 - Period: 5 Yrs - Discharged at Plattsburgh, NY, on 17 May 1814 on Surgeon's Certificate, hernia.

Pettit, John - Private - 36th Infantry - Company: Charles Randolph - Pension: No pension claim - BLW 138-160-50.

Pew, Jesse - Private - 14th Infantry - Company: Thomas Kearney - Enlistment date: 16 May 1812 - Period: 18 Mos.

Philips, Edwin - Private - 18th Infantry - Company: Henry Taylor - Age: 27 - Height: 5' 5 3/4" - Born: Maryland - Trade: Sailor - Enlistment date: 4 Apr 1814 - Place: Fort Mechanic, Charleston, SC - Period: War - Deserted in 1815.

Phillips, Abraham - Private - 14th Infantry - Company: William McIlvane - Age: 53 - Height: 5' 11" - Born: Ireland - Trade: Weaver - Enlistment date: 23 Apr 1813 - Period: 5 Yrs - BLW 9501-160-12 - Discharged at Greenbush, NY, on 1 May 1815 for old age.

Phillips, Daniel - Private - 36th Infantry - Company: Thomas Carbery - Enlistment date: 13 Dec 1814 - Period: War.

Phillips, Elijah - Private - 14th Infantry - Company: Thomas Kearney - BLW 19103-160-12.

Maryland Regulars in War of 1812

Phillips, Evan - Private - 14th Infantry - Company: Thomas Montgomery - Enlistment date: 8 Jun 1812 - Period: 5 Yrs - BLW 7595-160-12 - Discharged at Plattsburgh, NY, on 30 Mar 1814 for lame shoulder.

Phillips, Henry - Private - 32nd Infantry - Company: James Clark - Age: 23 - Height: 5' 11" - Born: Worcester, MD - Trade: Laborer - Enlistment date: 13 Jun 1813 - Place: Lewistown, DE - Period: 1 Yr - Discharged at Sandy Hook, NJ, on 30 Jun 1814.

Phillips, John - Private - 1st Rifles - Company: William Smith - Age: 41 - Height: 5' 6" - Born: Charles Cty, MD - Trade: Farmer - Enlistment date: 4 Sep 1812 - Place: Elizabeth - Period: 5 Yrs - Discharged on 4 Sep 1817.

Phillips, John - Private - 38th Infantry - Age: 22 - Height: 5' 5" - Born: Delaware - Enlistment date: 7 Jan 1814 - Place: Baltimore - Period: 1 Yr.

Phillips, John C. - Sergeant - 14th Infantry - Company: Isaac Barnard - Age: 24 - Height: 6' 2" - Born: New Jersey - Trade: Carpenter - Enlistment date: 26 May 1812 - Period: 5 Yrs - BLW 9646-160-12 - Discharged at Washington, DC, on 23 Mar 1814.

Phillips, Samuel - Private - 14th Infantry - Company: Richard Arell - Other regiment: 4th Infantry (New) - Age: 23 - Height: 5' 8" - Born: Cooperstown, NY - Trade: Farmer - Enlistment date: 15 Sep 1812 - Place: Pennsylvania - Period: 5 Yrs - Prisoner of War (Halifax), captured at Beaver Dams on 24 Jun 1813, discharged from Halifax on 3 Feb 1813, exchanged on 15 Apr 1815; discharged on 15 Sep 1817.

Phillips, William - Private - 19th Infantry - Company: George Kesling - Age: 23 - Height: 6' - Born: Talbot Cty, MD - Enlistment date: 30 Mar 1814 - Place: Adelphi, Ross Cty, OH - Period: War.

Phobus, William - Corporal - 41st Infantry - Company: Charles Humphrey - Age: 25 - Height: 5' 7 1/2" - Born: Princess Ann's Cty, MD - Trade: Shoemaker - Enlistment date: 29 Nov 1813 - Place: New York - Period: War - Discharged on 31 May 1815.

Pickett, John - Private - 12th Infantry - Company: Thomas Post - Other regiment: 8th Infantry (New) - Age: 25 - Height: 5' 9" - Born: Washington Cty, MD - Trade: Laborer - Enlistment date: 4 Oct 1813 - Place: Hagerstown, MD - Period: 5 Yrs - Pension: Land bounty to John Pickett and other heirs at law of John Pickett - BLW 24797-160-12 - Died at Fort Clark, IL, on 20 Jul 1816.

Pickett, John H. - Private - 2nd Light Dragoons - Company: William Littlejohn - Age: 24 - Height: 5' 11" - Born: Baltimore - Trade: Carpenter - Enlistment date: 22 Sep 1813 - Place: Baltimore - Period: 5 Yrs - Deserted at Baltimore on 23 Feb 1815.

Picquet, William - Private - 36th Infantry - Age: 38 - Height: 5' 7" - Born: Mathews Cty, VA - Enlistment date: 10 Jul 1813 - Place: Cambridge, MD.

Piercal, Henry - Private - US Volunteers - Company: Stephen Moore - Enlistment date: 8 Sep 1812 - Period: 1 Yr.

Pierce, Jesse - Private - 5th Infantry - Age: 34 - Height: 5' 11 1/2" - Born: Harford Cty, MD - Trade: Laborer - Enlistment date: 18 Jul 1814 - Place: York, PA - Period: 5 Yrs.

By War of 1812 Soc. in MD

Pierce, Thomas - Private - 36th Infantry - Company: Joseph Merrick - Age: 19 - Height: 5' 6" - Born: Baltimore - Enlistment date: 18 May 1813 - Place: Baltimore - Period: 1 Yr - Discharged on 19 Dec 1813.

Pierce, William - Sergeant - 14th Infantry - Company: Thomas Montgomery - Other regiment: 4th Infantry (New) - Age: 19 - Height: 5' 5 1/2" - Born: Barnsborough, NJ - Trade: Carpenter - Enlistment date: 6 May 1812 - Place: Baltimore - Period: 5 Yrs - Deserted on 3 Oct 1815.

Piercy, Henry - Private - 5th Infantry - Company: William Bird - Age: 22 - Height: 5' 11" - Born: Natchez, MS - Trade: Potter - Enlistment date: 28 Mar 1813 - Place: Baltimore - Period: 5 Yrs - Discharged on 16 Apr 1814 for a rupture.

Piercy, Jacob - Private - 5th Infantry - Company: Colin Buckner - Age: 44 - Height: 5' 8" - Born: Philadelphia - Trade: Potter - Enlistment date: 26 Feb 1813 - Place: Baltimore - Period: 5 Yrs - BLW 5206-160-12 - Discharged at Greenbush, NY, on 12 Jun 1814 because of a rupture.

Pierson Jr., Daniel - Private - 38th Infantry - Company: Shepperd Leakin - Age: 21 - Height: 5' 8 1/2" - Born: Cumberland, Ireland - Trade: Seaman - Enlistment date: 30 Aug 1814 - Place: Annapolis - Period: War - BLW 5505-160-12 - Discharged on 27 Mar 1815.

Pierson Sr., Daniel - Private - 38th Infantry - Company: Shepperd Leakin - Age: 40 - Height: 5' 8 1/2" - Born: Essex Cty, NY - Trade: Shoemaker - Enlistment date: 9 Jul 1814 - Place: Baltimore - Period: War - BLW 5505-160-12 - Discharged at Fort McHenry on 28 Mar 1815.

Pike, David - Private - 2nd Artillery - Company: Nathan Towson - Enlistment date: 22 May 1812 - Period: 5 Yrs - Died in hospital during 1813.

Pike, John H. - Private - US Volunteers - Company: Stephen Moore - Enlistment date: 8 Sep 1812 - Period: 1 Yr.

Pilcher, Edward - Private - 1st Artillery - Company: Heman Fay - Age: 23 - Height: 5' 9" - Born: Stafford Cty, VA - Trade: Sailor - Enlistment date: 21 Apr 1814 - Place: Annapolis - Period: War - Discharged at Washington, DC, on 9 Mar 1815.

Piles, Henderson - Private - Light Dragoons - Company: James Reed - Age: 22 - Height: 5' 10 1/2" - Born: Prince George's Cty, MD - Trade: Farmer - Enlistment date: 16 Apr 1814 - Place: Alexandria, DC - Period: War - Discharged at Fort Washington, MD, on 7 Mar 1815.

Pinchon, John - Private - 38th Infantry - Company: John Buck - Age: 33 - Height: 5' 10" - Born: York, PA - Trade: Laborer - Enlistment date: 11 Mar 1814 - Place: Hagerstown, MD - Period: War - Pension: Old War WF-17106 Rejected - BLW 9864-160-12 - Discharged at Baltimore on 30 Mar 1815.

Pinder, Isaac - Private - 36th Infantry - Company: Joseph Hook - BLW 17046-160-12.

Pine, George - Private - 36th Infantry - Company: Joseph Hook - Age: 20 - Height: 5' 7" - Born: New Jersey - Trade: Laborer - Enlistment date: 15 Mar 1814 - Place: Fort Washington, MD - Period: War - BLW 20863-160-12 - Discharged on 2 Apr 1815.

Pine, George - Private - 36th Infantry - Company: Joseph Hook - Age: 23 -

Height: 5' 6" - Born: Pennsylvania - Enlistment date: 8 May 1813 - Place: Baltimore - Period: 1 Yr.

Pinkerton, Thomas - Garrison Surgeon's Mate - Medical Department - Commissioned as a surgeon's mate on 28 Apr 1809; resigned on 15 Apr 1813.

Pinkney, Ninian - Captain - 1st Infantry - Other regiment: 5th Infantry, 22nd Infantry, 6th Infantry (New) - Born: Maryland - Commissioned as a 1st lieutenant, 9th Infantry, on 8 Jan 1799; discharged on 15 Jun 1800; commissioned as a 1st lieutenant, 1st Infantry, on 16 Feb 1801; promoted to captain on 9 Dec 1807; promoted to major, 5th Infantry, on 20 Jan 1813; promoted to lieutenant colonel, 22nd Infantry on 15 Apr 1814; promoted to colonel, 6th Infantry on 13 May 1820; died on 16 Dec 1825.

Pitcher, Jacob - Private - 14th Infantry - Prisoner of War (Halifax), discharged from Halifax on 3 Feb 1814.

Pitman, Isaac - Private - 38th Infantry - Company: John Rothrock - BLW 7375-160-12.

Pitt, Robert - Sergeant - 14th Infantry - Company: David Cummings - Age: 30 - Height: 5' 8" - Born: Dorchester Cty, MD - Trade: Sailor - Enlistment date: 11 Mar 1814 - Place: Baltimore - Period: War - BLW 7568-160-12 - Discharged at Baltimore on 16 Mar 1815.

Pitts, James - Private - 3rd Infantry - Company: Hays White - Age: 26 - Height: 5' 6" - Born: Maryland - Trade: Laborer - Period: War - Discharged on 9 Apr 1815.

Pitts, Thomas - Private - 25th Infantry - Company: Daniel Ketchum - Age: 29 - Height: 5' 7" - Born: Anne Arundel Cty, MD - Trade: Hatter - Enlistment date: 28 Jan 1815 - Place: Hartford, CT - Period: War - BLW 26-320-12 - Discharged at Hartford, CT, on 24 Mar 1815.

Place, James - Private - 5th Infantry - Company: James Dorman - Other regiment: 3rd Infantry (New) - Age: 21 - Height: 5' 8 1/2" - Born: Providence, RI - Trade: Coach maker - Enlistment date: 7 Feb 1812 - Place: Baltimore - Period: 5 Yrs - On board the fleet on Lake Champlain since 9 Jun 1814.

Plain, George - Private - 26th Infantry - Company: William Bezeau - Age: 21 - Height: 5' 10" - Born: Philadelphia - Enlistment date: 29 Oct 1814 - Place: Baltimore - Period: War - Discharged at Philadelphia.

Plante, Edward T. - Private - 2nd Artillery - Company: Nathan Towson - Age: 19 - Prisoner of War (Quebec), captured during the Battle of Stoney Creek and released on 31 Oct 1813.

Plummer, Nathan - Private - 14th Infantry - Company: David Cummings - Age: 23 - Height: 5' 6 1/2" - Born: Sussex Cty, DE - Trade: Laborer - Enlistment date: 30 Dec 1814 - Place: Baltimore - Period: War - Discharged on 18 Mar 1815.

Plunkett, Peter - Private - 36th Infantry - Company: Joseph Hook - Age: 38 - Height: 5' 8 1/4" - Born: Pennsylvania - Enlistment date: 24 May 1813 - Place: Annapolis - Period: 1 Yr - Pension: Wife Lethe, Old War WF-12631.

Pocock, James - Private - 5th Infantry - Company: Leroy Opie - Age: 24 -

By War of 1812 Soc. in MD

Height: 5' 11 1/2" - Born: Baltimore - Trade: Cooper - Enlistment date: 24 Feb 1814 - Place: Lancaster, PA - Period: 5 Yrs.

Pocock, Thomas - Private - US Volunteers - Company: Stephen Moore - Enlistment date: 8 Sep 1812 - Period: 1 Yr.

Poff, Henry - Private - 38th Infantry - Company: John Mowton - Pension: Land bounty to John Poff, Polly Poff, Cather Crawford and Elizabeth Vernum, children and only heirs at law of Henry Poff - BLW 27157-160-42.

Poiter, Alexander - Private - 22nd Infantry - Company: Jacob Carmack - Age: 19 - Height: 5' 8" - Born: Baltimore - Trade: Joiner - Enlistment date: 11 Aug 1814 - Period: War - Discharged at Fort Fayette, PA, on 24 Mar 1815.

Poland, Aaron - Corporal - 38th Infantry - Company: James Hook - Pension: SO-7482, SC-17889.

Poling, William - Private - 36th Infantry - Company: Thomas Carbery - Age: 39 - Height: 5' 8" - Born: New Jersey - Trade: Farmer - Enlistment date: 9 Apr 1814 - Place: Georgetown, DC - Period: War - BLW 16168-160-12 - Discharged at Baltimore on 31 Mar 1815.

Pollard, Martin - Private - 14th Infantry - Enlistment date: 16 Mar 1813 - Period: War - Prisoner of War, sent to England; discharged on 31 Mar 1815.

Pollard, Robert - Private - 38th Infantry - Company: James Smith - BLW 8856-160-12.

Pollock, George W. - Private - 2nd Light Dragoons - Company: William Littlejohn - Other regiment: Light Dragoons - Age: 33 - Height: 5' 8 1/2" - Born: Snow Hill, MD - Trade: Hatter - Enlistment date: 7 Nov 1813 - Place: Alexandria, DC - Period: War - Pension: Land bounty to Elizabeth Pollock, widow and other heirs at law of George W. Pollock - BLW 10260-160-12 - Discharged at Carlisle Barracks, PA, on 4 May 1815.

Polson, Philip - Private - 1st Rifles - Company: Thomas Ramsey - Age: 21 - Height: 5' 7" - Born: Maryland - Trade: Cooper - Enlistment date: 6 Dec 1813 - Place: Cincinnati - Period: War - Deserted at Camp Springfield on 5 Aug 1814.

Pongue, William - Private - 44th Infantry - Company: Joseph Miles - Age: 23 - Height: 5' 7" - Born: Baltimore - Enlistment date: 2 Nov 1813 - Period: War.

Pool, George - Private - 1st Rifles - Company: Joshua Hamilton - Other regiment: Rifles - Age: 23 - Height: 5' 4" - Born: Maryland - Trade: Tailor - Enlistment date: 5 May 1813 - Place: Knoxville, TN - Period: War - Discharged at Fort Gibson, NY, on 11 Jun 1815.

Pool, John - Private - 30th Infantry - Age: 37 - Height: 5' 9" - Born: Fredericktown, MD - Enlistment date: 8 Apr 1814 - Place: Burlington, VT - Period: War.

Pool, John - Private - 16th Infantry - Company: John Machesney - Age: 36 - Height: 5' 9" - Born: Fredericktown, MD - Trade: Comb maker - Enlistment date: 25 Jul 1812 - Place: Philadelphia - Period: 5 Yrs.

Pool, Philemon - Private - 14th Infantry - Age: 27 - Height: 5' 8 1/2" - Born: Maryland - Trade: Blacksmith - Enlistment date: 25 Aug 1814 - Period: War.

Maryland Regulars in War of 1812

Poole, Benjamin F. - Private - 14th Infantry - Company: Reuben Gilder - Other regiment: 4th Infantry (New) - Age: 22 - Height: 5' 9" - Born: Boston - Trade: Hatter - Enlistment date: 14 Mar 1813 - Place: Baltimore - Period: 5 Yrs - BLW 19699-160-12 - Discharged at Fort Gadsden, FL, on 13 Mar 1818.

Pope, George - Artificer - 2nd Artillery - Company: Frederick Evans - Other regiment: Corps of Artillery - Age: 40 - Height: 5' 11" - Born: Baltimore Cty - Trade: Blacksmith - Enlistment date: 4 Aug 1812 - Place: Fort Severn, MD - Period: 5 Yrs - BLW 10462-160-12 - Discharged at Fort McHenry on 4 Aug 1817.

Pope, Henry - Private - 38th Infantry - Company: Charles Stansbury - Age: 24 - Height: 5' 2 1/2" - Born: Hanover, Germany - Trade: Tobacconist - Enlistment date: 10 Oct 1814 - Place: Baltimore - Period: War - BLW 241-160-12 - Discharged at Baltimore on 6 Apr 1815.

Pope, John - Sergeant - 23rd Infantry - Company: Frederick Brown - Age: 29 - Height: 5' 10" - Born: Baltimore - Trade: Cordwainer - Enlistment date: 8 Feb 1814 - Place: Albany, NY - Period: War - BLW 1931-160-12 - Discharged at Sackets Harbor, NY, in 1815.

Pope, Moses - Private - 5th Infantry - Company: James Dorman - Age: 21 - Height: 5' 8" - Born: Baltimore - Trade: Farmer - Enlistment date: 30 May 1812 - Place: Baltimore - Period: 5 Yrs - Killed during the Battle of Lyon's Creek on 19 Oct 1814.

Popkins, George - Private - 38th Infantry - Company: James Smith - BLW 14249-160-12.

Porter Ranville - Private - 2nd Artillery - Company: Nathan Towson - Enlistment date: 10 Mar 1813 - Period: 5 Yrs - Deserted on 26 Sep 1813.

Porter, George W. - Third Lieutenant - 38th Infantry - BLW 23557-160-50 - Commissioned as a 3rd lieutenant on 1 May 1814; discharged on 15 Jun 1815; died on 7 Nov 1836.

Porter, Gustavus H. - Private - 36th Infantry - Company: Thomas Carbery - Age: 17 - Height: 5' 1" - Born: Queen Anne's Cty, MD - Trade: Printer - Enlistment date: 27 Sep 1814 - Period: War - BLW 18097-160-12 - Discharged on 13 Mar 1815.

Porter, Isaac - Private - 36th Infantry - Company: Joseph Hook - Age: 23 - Height: 5' 9" - Born: Anne Arundel Cty, MD - Trade: Blacksmith - Enlistment date: 6 Aug 1814 - Place: Georgetown, DC - Period: War - Discharged at Fort Covington, MD, on 30 Mar 1815.

Porter, James - Private - 1st Rifles - Company: Thomas Ramsey - Age: 38 - Height: 5' 11" - Born: Baltimore - Trade: Farmer - Enlistment date: 17 May 1814 - Place: Cincinnati - Period: War - Deserted at Cincinnati on 20 May 1814.

Porter, John - Private - 36th Infantry - Company: Samuel Raisin - Age: 33 - Height: 5' 10" - Born: Delaware - Trade: Laborer - Enlistment date: 25 Jul 1814 - Place: Georgetown, DC - Period: War - Pension: Land bounty to Mary Porter, daughter and other heirs at law of John Porter - BLW 24575-160-12 - Discharged on 20 Mar 1815.

By War of 1812 Soc. in MD

Porter, William - Private - US Volunteers - Company: Stephen Moore - Enlistment date: 8 Sep 1812 - Period: 1 Yr - Wounded at York, UC, on 27 Apr 1813.

Porter, William C. - Corporal - 36th Infantry - Company: Samuel Raisin - Age: 20 - Height: 5' 8" - Born: Caroline Cty, MD - Enlistment date: 10 Aug 1813 - Place: Greensboro, MD - Period: 1 Yr - Pension: Wife Harriet, SO-13983, SC-8799, WO-20506, WC-15764.

Post, Thomas - Captain - 12th Infantry - Pension: Land bounty to Sarah Post, widow of Thomas Post - BLW 1892-163-50 - Commissioned as a 1st lieutenant on 25 Apr 1812; promoted to captain on 29 Mar 1813; discharged on 15 Jun 1815.

Potter, George - Private - 38th Infantry - Company: Shepperd Leakin - Age: 28 - Height: 5' 8" - Born: Montgomery Cty, MD - Trade: Plasterer - Enlistment date: 7 Sep 1814 - Place: Baltimore - Period: War - Discharged on 27 Mar 1815.

Potter, John - Private - 36th Infantry - Company: Joseph Hook - Age: 39 - Height: 5' 11" - Born: Anne Arundel Cty, MD - Trade: Cooper - Enlistment date: 24 May 1813 - Place: Baltimore - Period: 1 Yr - Prisoner of War (Halifax), captured at Washington, DC, on 24 Aug 1814; discharged on 9 May 1815.

Potter, Thomas - Servant - US Sea Fencibles - Company: William Addison.

Potts, John - Private - 5th Infantry - Company: Leroy Opie - Age: 21 - Height: 5' 8" - Born: Berkley, VA - Trade: Distiller - Enlistment date: 20 Dec 1812 - Place: Williamsport, MD - Period: 5 Yrs.

Potts, William - Private - 14th Infantry - Company: Reuben Gilder - Other regiment: 4th Infantry (New) - Age: 28 - Height: 6' 3 1/2" - Born: Montgomery Cty, PA - Trade: Laborer - Enlistment date: 26 Sep 1813 - Place: Baltimore - Period: 5 Yrs - Discharged on 27 Sep 1818.

Poulton, John - Private - 14th Infantry - Company: Reuben Gilder - Age: 24 - Height: 6' - Born: Leesburg, VA - Trade: Farmer - Enlistment date: 31 Dec 1812 - Place: Maryland - Period: 5 Yrs.

Powell, Handy - Private - 14th Infantry - Company: Thomas Montgomery - Other regiment: 4th Infantry (New) - Age: 36 - Height: 5' 7" - Born: Snow Hill, MD - Trade: Shoemaker - Enlistment date: 3 Feb 1813 - Place: Baltimore - Period: 5 Yrs - Deserted on 18 Aug 1815.

Powell, John - Private - 12th Infantry - Company: Thomas Sangster - Age: 43 - Height: 5' 8 1/2" - Born: Maryland - Trade: Planter - Enlistment date: 14 May 1812 - Place: Alexandria, DC - Period: 5 Yrs - Discharged at Burlington, VT, on 28 Mar 1814 for rheumatism and old age.

Powell, Richard - Private - 2nd Light Dragoons - Company: William Littlejohn - Other regiment: Corps of Artillery - Age: 25 - Height: 5' 5 1/2" - Born: North Carolina - Trade: Accountant - Enlistment date: 15 Jan 1814 - Place: Baltimore - Period: 5 Yrs - BLW 21164-160-12 - Transferred to Corps of Artillery in Jul 1815; discharged at Fort McHenry on 15 Jan 1819.

Powell, William - Private - 36th Infantry - Age: 21 - Height: 5' 7" - Born: Norfolk, VA - Enlistment date: 10 Jul 1814 - Place: Norfolk, VA.

Maryland Regulars in War of 1812

Powell, William - Private - 36th Infantry - Age: 17 - Height: 5' 3" - Born: Fairfax, VA - Enlistment date: 1 Mar 1814 - Place: Annapolis.

Powers, Joseph - Musician - 14th Infantry - Company: David Cummings - Enlistment date: 18 May 1812 - Period: 5 Yrs - Prisoner of War (Halifax), sent to England, captured at Beaver Dams on 24 Jun 1813.

Powers, Joseph - Fifer - 12th Infantry - Company: Thomas Sangster - Other regiment: 8th Infantry (New) - Age: 23 - Height: 5' 10" - Born: Baltimore - Trade: Carpenter - Enlistment date: 1 May 1815 - Place: Baltimore - Period: 5 Yrs.

Powers, William - Private - 20th Infantry - Company: Walter Hayes - Age: 25 - Height: 5' 10 1/2" - Born: Maryland - Trade: Cabinet maker - Enlistment date: 10 Nov 1814 - Period: 5 Yrs - Deserted at Craney Island, VA, on 5 May 1815.

Powley, George - Private - 14th Infantry - Company: William McIlvane - Age: 38 - Height: 5' 6" - Born: Baltimore - Trade: Shoemaker - Enlistment date: 17 Mar 1814 - Place: Towsontown, MD - Period: War - BLW 24295-160-12 - Discharged at Greenbush, NY, on 5 May 1815.

Prather, Charles - Sergeant - 17th Infantry - Company: Caleb Holder - Age: 30 - Height: 5' 10" - Born: Allegany Cty, MD - Trade: Accountant - Enlistment date: 23 Feb 1814 - Period: War - BLW 20232-160-12 - Discharged at Detroit on 30 Jul 1815.

Prejohn, John - Private - 14th Infantry - Company: Reuben Gilder - Age: 27 - Height: 5' 8 1/2" - Born: France - Enlistment date: 16 Apr 1813 - Period: 18 Mos - Discharged at Black Rock, NY, on 16 Oct 1814.

Prestman, Stephen Wilson - First Lieutenant - 5th Infantry - Born: Maryland - Commissioned as an ensign on 14 Apr 1812; promoted to 2nd lieutenant on 6 Jul 1812; promoted to 1st lieutenant on 1 May 1814; discharged on 15 Jun 1815.

Preston, Armond - Sergeant - 14th Infantry - Company: Richard Arell - Other regiment: 4th Infantry (New) - Age: 26 - Height: 5' 9 1/2" - Born: Poultary, Rutland Cty, VT - Trade: Carpenter - Enlistment date: 21 May 1821 - Place: Baltimore - Period: 5 Yrs - Prisoner of War (Halifax), discharged from Halifax on 3 Feb 1814, exchanged on 15 Apr 1814; deserted at Fort Severn on 14 Jul 1815.

Preston, Benjamin - Private - 5th Infantry - Company: William Bird - Age: 39 - Height: 5' 5" - Born: Harford Cty, MD - Trade: Farmer - Enlistment date: 19 May 1814 - Place: Baltimore - Period: 5 Yrs - Discharged at Buffalo on 21 Jun 1814 because of hard drinking.

Pretty, Wilson - Private - 36th Infantry - Company: Thomas Carbery - Age: 25 - Height: 5' 10" - Born: Albemarle Cty, VA - Trade: Carpenter - Enlistment date: 24 Feb 1814 - Place: Leonardstown, MD - Period: War - BLW 9793-160-12 - Discharged at Baltimore on 31 Mar 1815.

Pretzman, George - Private - 38th Infantry - Company: John Buck - Age: 23 - Height: 5' 10 1/2" - Born: Frederick, MD - Trade: Carpenter - Enlistment date: 9 Jun 1814 - Period: War - Pension: Land bounty to Catherine Shuff and Elizabeth Sigman, only heirs at law of George Pretzman - BLW 27403-

By War of 1812 Soc. in MD

160-42 - Died at Baltimore on 12 Dec 1814.

Pretzman, John - Private - 5th Infantry - Company: James Dorman - Other regiment: 3rd Infantry (New) - Age: 27 - Height: 6' 1 3/4" - Born: Maryland - Trade: Carpenter - Enlistment date: 8 Jun 1812 - Place: Emmetsburg MD - Period: 5 Yrs - Discharged on 8 Jun 1817.

Price, David - Private - 38th Infantry - Company: John Rothrock - BLW 7890-160-12.

Price, Evan T. - Corporal - 3rd Rifles - Company: Edward Carrington - Age: 23 - Height: 5' 10" - Born: Talbot Cty, MD - Trade: Brick layer - Enlistment date: 19 Aug 1814 - Period: 5 Yrs.

Price, Johnson - Private - 14th Infantry - Company: Thomas Montgomery - Other regiment: 4th Infantry (New) - Age: 18 - Height: 5' 6" - Born: Dorchester Cty, MD - Trade: Saddler - Enlistment date: 7 May 1812 - Place: Cambridge, MD - Period: 5 Yrs - Prisoner of War (Montreal); discharged on 7 May 1817.

Price, Joseph - Private - 14th Infantry - Company: William McIlvane.

Price, Joseph - Private - Light Dragoons - Age: 18 - Height: 5' 2" - Born: Cecil Cty, MD - Trade: Carpenter - Enlistment date: 1 Mar 1813 - Place: Baltimore - Period: War.

Price, Joseph - Private - 14th Infantry - Company: David Cummings - Age: 23 - Height: 5' 11 1/4" - Born: Baltimore - Trade: Cooper - Enlistment date: 14 Feb 1814 - Place: Baltimore - Period: War - Discharged on 18 Mar 1815.

Price, Joseph - Private - 14th Infantry - Company: Thomas Kearney - Age: 19 - Height: 5' 5 1/2" - Born: Cecil Cty, MD - Trade: Carpenter - Enlistment date: 28 Feb 1813 - Period: War - Discharged at Greenbush, NY, on 5 May 1815.

Price, Merritt - Private - 14th Infantry - Company: Clement Sullivan - Other regiment: 4th Infantry (New) - Age: 18 - Height: 5' 8" - Born: St. Mary's, MD - Trade: Farmer - Enlistment date: 8 Jun 1812 - Place: Leonardstown, MD - Period: 5 Yrs - Pension: SO-28669, SC-20741 - Prisoner of War (Halifax), captured at Beaver Dams on 24 Jun 1813, discharged from Halifax on 3 Feb 1813, exchanged on 15 Apr 1814; discharged at Montpelier, VT, on 8 Jun 1817.

Price, Robert - Private - 39th Infantry - Company: John Jones - Age: 45 - Height: 5' 8" - Born: Cecil Cty, MD - Enlistment date: 29 Sep 1813 - Place: Gallatin, TN - Period: 1 Yr.

Price, Samuel C. - Private - 5th Infantry - Company: Richard Whartenby - Other regiment: 3rd Infantry (New) - Age: 34 - Height: 5' 8 1/2" - Born: Philadelphia - Trade: Bookbinder - Enlistment date: 30 Dec 1811 - Place: Baltimore - Period: 5 Yrs - Discharged at Fort Mackinaw, MI, on 31 Dec 1816.

Price, Thomas - Sergeant - 36th Infantry - Company: Henry Neale - Age: 28 - Height: 5' 11" - Born: Annapolis - Trade: Harness maker - Enlistment date: 1 Jan 1814 - Period: 1 Yr - Discharged at Washington, DC, on 22 Mar 1815.

Price, Thomas - Private - US Volunteers - Company: Stephen Moore - Enlistment date: 8 Sep 1812 - Period: 1 Yr.

Maryland Regulars in War of 1812

Price, William - Private - 38th Infantry - Company: Shepperd Leakin - Age: 37 - Height: 5' 6 1/2" - Born: Marblehead, MA - Trade: Seaman - Enlistment date: 15 Aug 1814 - Place: Baltimore - Period: War - BLW 13805-160-12 - Discharged at Fort McHenry on 28 Mar 1815.

Price, William - Private - 36th Infantry - Company: Shepperd Leakin - Age: 28 - Height: 5' 8" - Born: Ireland - Trade: Carpenter - Enlistment date: 16 Nov 1814 - Period: War - BLW 13805-160-12 - Discharged on 13 Mar 1815.

Primer, Jacob - Private - 36th Infantry - Company: Joseph Merrick - Age: 29 - Height: 5' 8 3/4" - Born: Lebanon, PA - Enlistment date: 30 Apr 1813 - Place: Baltimore - Transferred to the U.S. Navy on 26 Oct 1813.

Pringle, William - Private - 5th Infantry - Company: Colin Buckner - Age: 25 - Height: 5' 6 1/4" - Born: Philadelphia - Trade: Mariner - Enlistment date: 5 Feb 1813 - Place: Baltimore - Period: 5 Yrs - Discharged on 31 May 1814 for injuries received in service.

Pritchard, Carvel - Private - 1st Rifles - Company: Thomas Ramsey - Age: 40 - Height: 5' 9" - Born: Cecil Cty, MD - Trade: Hatter - Enlistment date: 29 Jan 1814 - Place: Cincinnati - Period: 5 Yrs - Died, probably at Newport, KY, on 12 Jul 1814.

Pritchard, Harmon - Private - 28th Infantry - Company: George Stockton - Age: 21 - Height: 5' 11" - Born: Maryland - Trade: Farmer - Enlistment date: 27 Mar 1814 - Place: Kentucky - Period: War - BLW 22815-160-12 - Discharged on 30 Jun 1815.

Pritchard, James - Private - 19th Infantry - Company: Carey Trimble - Age: 36 - Height: 5' 9" - Born: Maryland - Trade: Laborer - Enlistment date: 14 Mar 1814 - Place: Newark, OH - Period: War - BLW 13637-160-12 - Discharged at Chillicothe, OH, on 6 Jun 1815.

Pritchard, John - Sergeant - 2nd Light Dragoons - Company: Samuel Harris - Other regiment: Corps of Artillery - Age: 20 - Height: 5' 8 1/2" - Born: Talbot Cty, MD - Trade: Farmer - Enlistment date: 11 Aug 1813 - Place: Easton, MD - Period: 5 Yrs - BLW 18590-160-12 - Discharged at Fort Washington, MD, on 11 Aug 1818.

Pritchard, Parvall - Private - 24th Infantry - Company: Robert Desha - Age: 40 - Height: 5' 9" - Born: Cecil Cty, MD - Trade: Hatter - Enlistment date: 28 Jul 1812 - Period: 18 Mos - Re-enlisted on 29 Jul 1814 for five years in the Rifles, Captain Thomas Ramsey's Company; died on 12 Jul 1814.

Pritchard, William - Private - 2nd Artillery - Company: Nathan Towson - Pension: Old War WF-22247 Rejected.

Probart, Robert - Private - 24th Infantry - Company: Benjamin Jones - Age: 39 - Height: 5' 6" - Born: Maryland - Trade: Carpenter - Enlistment date: 4 Aug 1814 - Period: War - BLW 2858-160-12 - Discharged at Camp Mandeville, AL, on 22 Mar 1815.

Probus, Henry - Private - 1st Infantry - Company: John Symmes - Age: 21 - Height: 5' 11" - Born: Maryland - Trade: Farmer - Enlistment date: 4 Apr 1814 - Place: Cincinnati - Period: War - Wounded at Chippewa, UC, on 25 Jul 1814.

Pry, Richard - Private - 1st Infantry - Company: Daniel Bissell - Other regiment: 2nd Infantry then 1st Infantry (New) - Age: 26 - Height: 5' 6" - Born: Washington Cty, MD - Trade: Wheelwright - Enlistment date: 25 Oct 1804 - Period: 5 Yrs - Re-enlisted on 25 Oct 1809 for five years; transferred to 2nd Infantry; re=enlisted on 1 Nov 1814 for five years; re-enlisted in 1819.

Pryor, John - Private - 14th Infantry - Company: Reuben Gilder - Died at Cheektowaga Hospital, NY, on 22 Nov 1814 from dysentery.

Pryse, Thomas - Sergeant - 36th Infantry - Pension: Land bounty to Charles Pryse, minor son and only heir at law of Thomas Pryse - BLW 24826-160-12; BLW 11306-160-12 Cancelled.

Purdue, James - Private - 19th Infantry - Age: 21 - Height: 5' 6" - Born: Cumberland, MD - Enlistment date: 11 Apr 1814 - Place: Portsmouth, OH.

Purnell, John - Private - 14th Infantry - Company: Reuben Gilder - Enlistment date: 12 Aug 1812 - Period: 5 Yrs - Pension: Old War IF-3748 - BLW 24222-160-12 - Discharged at Burlington, VT, on 30 Mar 1814, emaciation of right thigh and leg.

Purse, William - Private - 14th Infantry - Prisoner of War (Halifax), discharged from Halifax on 3 Feb 1814.

Putt, Conrad - Private - 36th Infantry - Age: 41 - Height: 5' 4" - Born: Germany - Enlistment date: 25 Jun 1813 - Place: Baltimore - Period: 1 Yr.

Pye, John - Private - 22nd Infantry - Company: John Pentland - Age: 27 - Height: 5' 11" - Born: Maryland - Trade: Laborer - Enlistment date: 17 Oct 1812 - Place: Shippensburg, PA - Period: 5 Yrs.

Q

Queen, Charles J. - First Lieutenant - 36th Infantry - Born: Maryland - Pension: Land bounty to Maria A. Queen, widow and heir at law of Charles J. Queen - BLW 423-160-50 - Commissioned as a 2nd lieutenant on 30 Apr 1813; promoted to 1st lieutenant on 30 Sep 1814; resigned on 31 Jan 1815.

Quillan, Thomas - Private - 14th Infantry - Company: James McDonald - Age: 26 - Height: 5' 5" - Born: Kent Cty, DE - Trade: Laborer - Enlistment date: 22 Aug 1813 - Period: 18 Mos - BLW 21123-160-12 - Re-enlisted on 31 Mar 1814 for the war; discharged on 3 May 1815.

Quinn, Edward - Private - 38th Infantry - Company: Shepperd Leakin - Age: 28 - Height: 5' 7" - Born: Ireland - Trade: Carpenter - Enlistment date: 7 Sep 1814 - Place: Baltimore - Period: War - Discharged at Fort McHenry on 27 Mar 1815.

Quinn, John - Private - 36th Infantry - Company: Charles Randolph - Age: 42 - Height: 5' 8" - Born: Pennsylvania - Enlistment date: 19 May 1813 - Place: Frederick, MD - Period: 1 Yr.

R

Radley, Robert - Private - 36th Infantry - Age: 42 - Height: 5' 5 1/4" - Born: Charleston, MD - Enlistment date: 3 May 1813 - Place: Baltimore - Period: 1 Yr - Died on 15 Aug 1813.

Ragan, William - Private - 19th Infantry - Company: Carey Trimble - Age: 41

- Height: 5' 8" - Born: Maryland - Trade: Farmer - Enlistment date: 1 Jan 1814 - Period: War - BLW 17043-160-12 - Discharged at Chillicothe, OH, on 7 Jun 1815.

Ragar, George - Corporal - 38th Infantry - Company: John Rothrock - BLW 8509-160-12.

Rage, James - Private - 14th Infantry - Company: Samuel Lane - Enlistment date: 19 Jun 1812 - Period: 5 Yrs - Killed at Carlisle Barracks, PA, on 25 Jul 1812.

Rainey, Joseph - Private - 36th Infantry - Company: Samuel Raisin - Enlistment date: 21 Jun 1814 - Period: War - Died in hospital on 21 Jan 1815.

Raisin, Samuel - Captain - 36th Infantry - Pension: Wife Mary Sidney, WO-1810; second husband Thomas Wilcoxon - Commissioned as a captain on 30 Apr 1813; discharged on 15 Jun 1815.

Raker, John F. - Private - 36th Infantry - Age: 22 - Height: 5' 7 3/4" - Born: Queen Anne's Cty, MD - Enlistment date: 27 May 1813 - Place: Centreville, MD - Rejected.

Ramsey, David - Sergeant - 38th Infantry - BLW 11044-160-12.

Ramsey, Hanson - Private - 14th Infantry - Company: Thomas Montgomery - Other regiment: 4th Infantry (New) - Age: 34 - Height: 5' 4" - Born: Talbot Cty, MD - Trade: Farmer - Enlistment date: 7 Apr 1813 - Place: Middleton, MD - Period: 5 Yrs - Discharged at Fort St. Marks, FL, on 6 Apr 1818.

Ramsey, John - Private - 27th Infantry (OH) - Age: 26 - Height: 5' 8" - Born: Maryland - Enlistment date: 25 Apr 1814 - Place: Steubenville, OH.

Ramsons, Henry - Private - 5th Infantry - Company: James Dorman - Other regiment: 3rd Infantry (New) - Age: 22 - Height: 5' 6 1/2" - Born: Frederick, MD - Trade: Farmer - Enlistment date: 26 May 1812 - Place: Fredericktown, MD - Period: 5 Yrs - Discharged at Greenbush, NY, on 1 May 1815 for incapacity.

Ranch, William - Private - 14th Infantry - Company: Joseph Marshall - BLW 5202-160-12.

Randall, Samuel G. - Private - 36th Infantry - Company: Joseph Hook - Age: 24 - Height: 5' 11" - Born: Rockingham, VA - Trade: Laborer - Enlistment date: 2 May 1814 - Place: Georgetown, DC - Period: War - BLW 25325-160-12 - Discharged at Fort Covington, MD, on 30 Mar 1815.

Randall, Thomas - Captain - 14th Infantry - Other regiment: Corps of Artillery - Born: Maryland - Pension: Old War IF-20160 - BLW 17610-160-50 - Commissioned as a 2nd lieutenant on 12 Mar 1812; promoted to 1st lieutenant on 13 Mar 1813; promoted to captain on 1 Dec 1815; transferred to Corps of Artillery on 17 May 1815; resigned on 15 Jun 1817.

Randall, William - Private - US Volunteers - Company: Stephen Moore - Enlistment date: 8 Sep 1812 - Period: 1 Yr - Pension: Placed on the pension rolls on 7 Sep 1820.

Randle, William - Private - 14th Infantry - Company: Joseph Marshall - Age: 19 - Height: 5' 9" - Born: Alexandria, DC - Trade: Ship's carpenter - Enlistment date: 18 Jul 1814 - Place: Alexandria, DC - Period: War -

By War of 1812 Soc. in MD

Discharged at Greenbush, NY, on 3 May 1815.

Randolph, Charles C. - Captain - 36th Infantry - Born: Virginia - BLW 89132-160-55 - Commissioned as a captain on 30 Apr 1813; discharged on 15 Jun 1815.

Randolph, David - Private - 36th Infantry - Company: Mortimer Hall - Age: 30 - Height: 5' 7" - Born: Maryland - Enlistment date: 31 May 1813 - Place: Havre de Grace, MD - Period: 1 Yr - Deserted at Little York, PA.

Randolph, Richard - Private - 36th Infantry - Company: Samuel Raisin - Age: 26 - Height: 6' 2" - Born: Henrico Cty, VA - Trade: Carpenter - Enlistment date: 3 Aug 1814 - Place: Petersburg, VA - Period: War - BLW 7783-160-12 - Discharged at Washington, DC, on 20 Mar 1815.

Raney, John - Private - 5th Infantry - Company: James Dorman - Age: 19 - Height: 5' 9" - Born: Ireland - Trade: Baker - Enlistment date: 14 Jul 1812 - Place: Baltimore - Period: 5 Yrs.

Rappin, Thomas - Private - 36th Infantry - Company: Charles Randolph - Age: 25 - Height: 5' 11" - Born: Frederick, MD - Enlistment date: 17 May 1813 - Place: Frederick, MD - Period: 1 Yr.

Ratcliff, Richard - Private - 36th Infantry - Company: Thomas Carbery - Age: 25 - Height: 5' 6 1/2" - Born: Charles Cty, MD - Trade: Farmer - Enlistment date: 5 Mar 1814 - Place: Georgetown, DC - Period: War - Discharged at Baltimore on 31 Mar 1815.

Rattle, John - Private - US Volunteers - Company: Stephen Moore - Enlistment date: 8 Sep 1812 - Period: 1 Yr.

Rattle, Robert - Private - 38th Infantry - Company: Thomas Sangster - Age: 21 - Height: 5' 4" - Born: Maryland - Trade: Cordwainer - Enlistment date: 4 Oct 1814 - Place: Baltimore - Period: 5 Yrs.

Raval, Joseph - Private - 36th Infantry - Age: 29 - Height: 5' 9 1/2" - Born: England - Trade: Distiller - Enlistment date: 12 Dec 1814 - Period: War - Discharged at Georgetown, DC, on 11 Mar 1815.

Rave, John - Private - 36th Infantry - Company: Thomas Carbery.

Rawlings, John C. - 36th Infantry - Company: William Scott - Pension: Wife Mary Ann, WO-5492, WC-3515.

Rawlings, Robert - Private - 36th Infantry - Company: Samuel Raisin - Age: 19 - Height: 5' 8" - Born: Fairfax, VA - Trade: Farmer - Enlistment date: 17 Aug 1814 - Place: Georgetown, DC - Period: War - Pension: Wife Susanna, SO-4387, SC-117, WO-13916, WC-7149 - BLW 13011-160-12 - Discharged at Washington, DC, on 20 Mar 1815.

Rawson, William - private - 36th Infantry - Company: Joseph Hook - Age: 34 - Height: 5' 6" - Born: Baltimore - Enlistment date: 16 Jun 1813 - Place: Baltimore - Period: 1 Yr - Deserted on 27 Jun 1813.

Ray, Abednego, J. - Private - 36th Infantry - Company: Thomas Carbery - Age: 27 - Height: 5' 10 1/4" - Born: St. Mary's, MD - Trade: Joiner - Enlistment date: 24 Feb 1814 - Place: Leonardstown, MD - Period: War - BLW 20373-160-12 - Discharged at Baltimore on 31 Mar 1815.

Ray, David - Private - 12th Infantry - Company: James Paston - Age: 44 - Height: 5' 8" - Born: Maryland - Trade: Farmer - Enlistment date: 15 May

1812 - Place: Alexandria, DC - Period: 18 Mos.

Ray, John - Private - 14th Infantry - Company: David Cummings - Enlistment date: 1 Jun 1812 - Period: 5 Yrs - Prisoner of War, sent to England; deserted at Fort McHenry on 28 Apr 1815.

Ray, Keller S. - Private - 2nd Light Dragoons - Company: George Haig - Other regiment: Corps of Artillery - Age: 26 - Height: 5' 7" - Born: Rockville, MD - Trade: Farmer - Enlistment date: 31 Oct 1812 - Place: Frederick, MD - Period: 5 Yrs - Transferred to Corps of Artillery; discharged on 31 Oct 1817.

Ray, Philip - Private - 1st Rifles - Company: Thomas Ramsey - Other regiment: Rifles - Age: 28 - Height: 5' 9" - Born: Maryland - Trade: Farmer - Enlistment date: 13 Aug 1813 - Place: Knoxville, TN - Period: War - Discharged at Buffalo on 6 Aug 1815.

Read, Benjamin - Private - 5th Infantry - Company: William Henshaw - Age: 27 - Height: 6' 1" - Born: Baltimore - Trade: Carpenter - Enlistment date: 19 Mar 1813 - Place: Baltimore - Period: 5 Yrs - Died at Buffalo on 18 Dec 1814.

Read, Joseph - Private - 14th Infantry - Company: Reuben Gilder - BLW 5146-160-12.

Read, Thomas - Private - 20th Infantry - Company: Walter Hayes - Age: 35 - Height: 5' 11 1/2" - Born: Charles Cty, MD - Trade: Farmer - Enlistment date: 13 Dec 1814 - Period: 5 Yrs - Deserted on 14 Mar 1815.

Reader, Joseph - Private - 14th Infantry - Company: Joseph Marshall - Other regiment: 4th Infantry (New) - Age: 31 - Height: 6' 1 3/4" - Born: Washington, MD - Trade: Cooper - Enlistment date: 9 Aug 1814 - Place: Boonesboro, MD - Period: 5 Yrs - Deserted at Fort Severn, MD, on 17 Jul 1815.

Readon, Patrick - Private - 26th Infantry - Company: William Bezeau - Age: 22 - Height: 5' 9" - Born: Baltimore - Trade: Farmer - Enlistment date: 5 Nov 1814 - Place: Baltimore - Period: War - BLW 7205-160-12 - Discharged at Philadelphia on 23 Mar 1815.

Ream, George - Private - 1st Artillery - Company: Henry Craig - Other regiment: Corps of Artillery - Age: 21 - Height: 5' 7" - Born: Frederick, MD - Trade: Shoemaker - Enlistment date: 5 Jul 1814 - Place: Youngstown, PA - Period: War - Discharged at Buffalo on 3 Jun 1815.

Reamer, Godfrey - Private - 38th Infantry - Company: John Buck - Age: 26 - Height: 5' 9 1/2" - Born: Philadelphia - Trade: Blacksmith - Enlistment date: 18 Apr 1814 - Place: Baltimore - Period: War - BLW 11823-160-12 - Discharged at Baltimore on 30 Mar 1815.

Reams, John - Private - 28th Infantry - Company: George Stockton - Age: 21 - Height: 5' 10" - Born: Frederick Cty, MD - Trade: Shoemaker - Enlistment date: 3 May 1814 - Place: Kentucky - Period: War - Discharged at Detroit on 30 Jun 1815.

Reardon, John W. - Corporal - 19th Infantry - Company: Thomas Read - Other regiment: 25th Infantry - Age: 21 - Height: 5' 8" - Born: Maryland - Enlistment date: 28 Aug 1812 - Place: Ohio - Period: 18 Mos - Discharged on 25 Oct 1814 because of wounds.

By War of 1812 Soc. in MD

Reason, Reuben - Private - 36th Infantry - Company: Joseph Hook - Age: 23 - Height: 5' 10" - Born: Culpeper, VA - Enlistment date: 18 Apr 1814 - Place: Fort Washington, MD - Period: War - Deserted on 30 Apr 1814.

Record, William - Private - 14th Infantry - Pension: Old War WF-17324 Rejected.

Redburn, James T. - Private - 17th Infantry - Company: David Holt - Age: 39 - Height: 5' 8" - Born: Maryland - Trade: Laborer - Enlistment date: 11 May 1814 - Place: Zanesville, OH - Period: War - Pension: Wife Jemima, WO-39501, WC-30636 - BLW 9423-160-12 - Discharged at Detroit on 30 Jul 1815.

Redd, John - Private - 38th Infantry - Company: John Mowton - BLW 8053-160-12.

Rederoop, Samuel - Private - 14th Infantry - Age: 25 - Height: 5' 11 1/2" - Born: Washington, MD - Trade: Hatter - Enlistment date: 6 Oct 1814 - Period: War.

Redgraves, Samuel - Private - 2nd Light Dragoons - Company: William Littlejohn - Other regiment: Light Dragoons - Age: 31 - Height: 5' 9" - Born: Kent Cty, MD - Trade: Mariner - Enlistment date: 7 Oct 1813 - Place: Elkton, MD - Period: 5 Yrs - Discharged on 7 Oct 1818.

Redman, Benjamin - Private - 36th Infantry - Company: Charles Randolph - Age: 42 - Height: 5' 5" - Born: St. Mary's, MD - Enlistment date: 17 May 1813 - Place: Frederick, MD - Period: 1 Yr.

Redman, Henry - Second Lieutenant - 36th Infantry - Commissioned as a 3rd lieutenant on 30 Apr 1813; promoted to 2nd lieutenant on 1 May 1814; discharged on 15 Jun 1815; died on 29 Oct 1823.

Redman, James - Seaman - US Sea Fencibles - Company: John Gill - Age: 18 - Height: 5' 1" - Born: St. Mary's, MD - Enlistment date: 6 Jan 1814 - Period: 1 Yr.

Redman, Joshua - Seaman - US Sea Fencibles - Company: John Gill - Age: 19 - Height: 5' 4" - Born: Somerset Cty, MD - Enlistment date: 8 Jan 1814 - Period: 1 Yr.

Redman, Simon M. - Private - 14th Infantry - Age: 26 - Height: 5' 6" - Born: Ireland - Enlistment date: 7 May 1814 - Place: Baltimore.

Redman, Thomas - Private - 38th Infantry - Company: Charles Stansbury - Age: 32 - Height: 5' 6 3/4" - Born: Dublin, Ireland - Trade: Laborer - Enlistment date: 20 Nov 1814 - Place: Fredericktown, MD - Period: War - Discharged at Baltimore on 6 Apr 1815.

Redmond, Martin - Private - 14th Infantry - Company: Richard Arell - Prisoner of War (Halifax), captured at Beaver Dams on 24 Jun 1813, discharged from Halifax on 3 Feb 1813.

Redmond, Thomas - Private - 38th Infantry - Company: Charles Stansbury - BLW 5500-160-12.

Reed, Henry - Private - 36th Infantry - Company: Joseph Hook - Enlistment date: 4 Jul 1813 - Period: 1 Yr - Transferred to the U.S. Navy on 8 Oct 1813.

Reed, James - Private - 5th Infantry - Company: Richard Bell - Age: 20 - Height: 5' 10" - Born: Baltimore - Trade: Carpenter - Enlistment date: 1 Mar

Maryland Regulars in War of 1812

1814 - Place: Harrisburg, PA - Period: 5 Yrs.

Reed, William - Private - 26th Infantry - Company: William Bezeau - Age: 21 - Height: 5' 6" - Born: Fort Penn, DE - Trade: Laborer - Enlistment date: 4 Nov 1814 - Place: Baltimore - Period: War - Discharged at Philadelphia on 23 Mar 1815.

Reeder, Robert D. - Private - 36th Infantry - Company: Henry Neale - Age: 30 - Height: 5' 6" - Born: St. Mary's Cty, MD - Trade: Laborer - Enlistment date: 4 Aug 1814 - Period: War - Pension: Land bounty to Helen Ann Reeder, daughter and only heir at law of Robert D. Reeder - BLW 18136-160-12 - Discharged on 24 Nov 1814, dysentery.

Reese, George - Private - 20th Infantry - Company: Walter Hayes - Age: 23 - Height: 6' 1" - Born: Baltimore - Trade: Tanner - Enlistment date: 20 Nov 1814 - Period: War - Discharged at Norfolk, VA, on 15 Mar 1815.

Reeves, Asa - Private - 7th Infantry - Company: Horatio Stark - Age: 21 - Height: 6' - Born: Maryland - Enlistment date: 22 Apr 1812 - Period: 5 Yrs - Died at Sackets Harbor, NY, on 27 Dec 1814.

Register, Thomas - Private - 14th Infantry - Company: Thomas Montgomery - Enlistment date: 22 Aug 1812 - Period: 18 Mos - Pension: SO-6489, SC-4040 - Discharged on 22 Feb 1814.

Reider, Elijah - Private - 12th Infantry - Company: Andrew Madison - Other regiment: 8th Infantry (New) - Age: 21 - Height: 5' 6" - Born: Charles Cty, MD - Trade: Blacksmith - Enlistment date: 21 Sep 1812 - Place: Harrisburg, PA - Period: 5 Yrs - Discharged on 21 Sep 1817.

Reinecker, Frederic - Private - 2nd Artillery - Company: Nathan Towson - Other regiment: Corps of Artillery - Age: 27 - Height: 5' 6" - Born: Hanover, Germany - Trade: Paper maker - Enlistment date: 20 Apr 1812 - Place: Baltimore - Period: 5 Yrs - Discharged at Fort Niagara, NY, on 20 Apr 1817.

Remmey, Ada - Private - 14th Infantry - Company: James McDonald - Age: 25 - Height: 5' 5" - Born: Baltimore - Trade: Miller - Enlistment date: 16 Sep 1813 - Period: War - Discharged at Baltimore on 31 May 1815.

Remmey, William - Private - 2nd Artillery - Company: Nathan Towson - Age: 21 - Height: 5' 5 3/4" - Born: Baltimore - Trade: Laborer - Enlistment date: 20 mar 1812 - Place: Baltimore - Period: 5 Yrs - Discharged on 20 May 1817.

Renark, Thomas - Private - 2nd Artillery - Company: Nathan Towson - Prisoner of War, captured during the Battle of Stoney Creek.

Renner, George - Private - 14th Infantry - Company: David Cummings - Age: 18 - Height: 5' 8" - Born: Orange, VA - Trade: Farmer - Enlistment date: 22 Oct 1814 - Place: Snowden's Camp, VA - Period: 5 Yrs.

Rennier, Andrew - Private - 14th Infantry - BLW 18964-160-12.

Renolds, Price - Servant - US Sea Fencibles - Company: Simmones Bunbury - Servant to Lieutenant Foy.

Renshaw, James - Private - 5th Infantry - Age: 22 - Height: 5' 6" - Born: Baltimore - Trade: Seaman - Enlistment date: 13 May 1813 - Place: Maryland - Period: 5 Yrs - On board the fleet with Commodore Chauncey, Sackets Harbor, 15 Jul 1813.

By War of 1812 Soc. in MD

Renshaw, James - Private - Rifles - Age: 33 - Height: 6' - Born: Harford Cty, MD - Enlistment date: 15 Apr 1814 - Place: Morgantown, VA.

Repiley, Alexander - Private - 19th Infantry - Company: George Kesling - Age: 25 - Height: 5' 4" - Born: Maryland - Enlistment date: 11 Jul 1814 - Place: Urbana, OH - Discharged at Zanesville, OH, on 27 Mar 1815.

Retallock, Simon - Private - 2nd Artillery - Company: Lloyd Beall - Other regiment: Corps of Artillery - Age: 37 - Height: 5' 6 3/4" - Born: Anne Arundel Cty, MD - Trade: Blacksmith - Enlistment date: 26 May 1812 - Place: Greenleaf Point, DC - Period: 5 Yrs - Discharged on 28 May 1817.

Reuben, Ball - Private - 38th Infantry - Company: John Brookes - Pension: Land bounty to William Ball, brother and other heirs at law of Reuben Ball - BLW 20092-160-12.

Rewark, Henry - Private - 20th Infantry - Company: Bernard Peyton - Age: 23 - Height: 5' 4 1/4" - Born: Kent Cty, MD - Trade: Planter - Enlistment date: 17 Mar 1814 - Place: Danville - Period: War - Discharged at Norfolk, VA, on 15 Mar 1815.

Reyling, William - Private - 38th Infantry - Company: John Buck - Age: 39 - Height: 5' 8" - Born: Cecil Cty, MD - Trade: Blacksmith - Enlistment date: 25 Feb 1814 - Period: War - Deserted at Ellicott's Mills, MD, on 23 Dec 1814.

Reynolds, Daniel - Private - 14th Infantry - Company: Reuben Gilder - Enlistment date: 10 Nov 1812 - Period: 5 Yrs - Pension: Land bounty to Agnes Reynolds, daughter and other heirs at law of Daniel Reynolds - BLW 21599-160-12 - Died at Sackets Harbor, NY, on 23 Jan 1814.

Reynolds, Gilbert - Private - 5th Infantry - Company: Richard Bell - Other regiment: Corps of Artillery - Age: 21 - Height: 5' 7 1/2" - Born: Gloucester, NJ - Trade: Baker - Enlistment date: 14 Jan 1812 - Place: Baltimore - Period: 5 Yrs - Prisoner of War (Halifax), captured at Stoney Creek, UC, on 12 Jun 1813; transferred to Corps of Artillery; discharged on 14 Jan 1817.

Rhea, John - Private - 14th Infantry - Prisoner of War (Halifax), captured at Beaver Dams on 24 Jun 1813, discharged from Halifax on 9 Nov 1813.

Rhoads, Samuel - Private - 26th Infantry - Company: William Bezeau - Age: 28 - Height: 5' 7 1/4" - Born: Salem, MA - Trade: Seaman - Enlistment date: 22 Oct 1814 - Place: Baltimore - Period: War.

Rhodes, John - Private - 4th Rifles - Company: Joseph Kean - Other regiment: Rifles - Age: 23 - Height: 5' 7 3/4" - Born: Harford Cty, MD - Trade: Laborer - Enlistment date: 2 Jun 1814 - Place: Uniontown, PA - Period: 5 Yrs - BLW 22384-160-12 - Discharged at Belle Fontaine, MO, on 2 Jun 1819 and re-enlisted.

Rhodes, John - Private - 1st Artillery - Company: George Armistead - Age: 38 - Height: 5' 9" - Born: Germany - Trade: Barber - Enlistment date: 29 Aug 1811 - Place: Fort McHenry - Period: 5 Yrs - Discharged on 6 Oct 1815 for disability.

Rhody, Henry - Private - 5th Infantry - Company: Colin Buckner - Age: 39 - Height: 5' 5" - Born: Germany - Trade: Laborer - Enlistment date: 12 Feb 1813 - Place: Baltimore - Period: 5 Yrs.

Maryland Regulars in War of 1812

Ricaud, John - First Lieutenant - 36th Infantry - Commissioned as a 1st lieutenant on 30 Apr 1813; dropped on 30 Sep 1814.

Rice, Jacob - Private - 14th Infantry - Company: William McIlvane - Age: 22 - Height: 6' 1" - Born: Washington, MD - Trade: Laborer - Enlistment date: 4 Oct 1813 - Period: 18 Mos - BLW 16590-160-12 - Re-enlisted at Baltimore on 30 May 1814 for the war; discharged at Greenbush, NY, on 3 May 1815.

Rice, John - Private - 14th Infantry - Company: Thomas Montgomery - Enlistment date: 22 Jul 1812 - Period: 5 Yrs - Prisoner of War, exchanged on 11 May 1814; deserted at Plattsburg, NY, in Jun 1814.

Rice, John - Private - 12th Infantry - Company: Thomas Moore - Age: 18 - Height: 5' 4 1/2" - Born: Allegany Cty, MD - Trade: Laborer - Enlistment date: 13 Jul 1813 - Place: Virginia - Period: War - Discharged at Buffalo on 31 May 1815.

Rice, Philip - Private - 38th Infantry - Company: Shepperd Leakin - Height: 5' 8" - Born: Frederick, MD - Trade: Wheelwright - Enlistment date: 2 Oct 1813 - Period: 1 Yr - Re-enlisted on 11 Mar 1814 for the war.

Rice, Samuel - Private - 26th Infantry - Company: William Bezeau - Age: 23 - Height: 5' 7" - Born: Lebanon Cty, PA - Trade: Blacksmith - Enlistment date: 22 Oct 1814 - Place: Baltimore - Period: War - Discharged at Philadelphia on 23 Mar 1815.

Rice, William - Private - 2nd Artillery - Company: Nathan Towson - Pension: Old War IF-2015 - BLW 7755-160-12.

Rich, Peter - First Lieutenant - 14th Infantry - Company: James McDonald - Commissioned as a 1st lieutenant on 12 Mar 1812; resigned on 29 Jun 1814.

Richards, George - Private - 36th Infantry - Age: 35 - Height: 5' 7" - Born: Baltimore - Enlistment date: 22 Aug 1813 - Place: Baltimore.

Richards, John - Private - 14th Infantry - Company: Thomas Kearney - Other regiment: 4th Infantry (New) - Age: 17 - Height: 5' 7" - Born: Winchester, VA - Trade: Laborer - Enlistment date: 22 Jun 1812 - Place: Western Post, MD - Period: 5 Yrs - Discharged on 22 Jun 1817.

Richards, John - Private - 38th Infantry - Company: Charles Stansbury - Age: 30 - Height: 5' 4 1/2" - Born: Maryland - Trade: Brick layer - Enlistment date: 5 Nov 1814 - Place: Baltimore - Period: War - Pension: Heirs received half pay for five years in lieu of land bounty - Died on 14 Mar 1815.

Richards, Noah - Private - 2nd Light Dragoons - Company: John Burd - Age: 21 - Height: 5' 5 1/4" - Born: Kent Cty, MD - Trade: Shoemaker - Enlistment date: 14 May 1812 - Place: Bedford, PA - Period: 5 Yrs - Discharged on 14 May 1817.

Richards, William L. - Private - 38th Infantry - Company: John Mowton - BLW 21616-160-12.

Richardson, Amos - Private - 38th Infantry - Company: Charles Stansbury - Age: 33 - Height: 5' 9" - Born: Dorchester Cty, MD - Trade: Cooper - Enlistment date: 24 Aug 1814 - Place: Annapolis - Period: War - Pension: Wife Eleanor, Old War Minor 480; heirs received half pay for five years in lieu of land bounty - Died on 12 Jan 1815.

Richardson, John - Private - 14th Infantry - Company: David Cummings -

By War of 1812 Soc. in MD

Age: 24 - Height: 5' 10" - Born: Worcester, MD - Trade: Farmer - Enlistment date: 11 Jul 1814 - Place: Salisbury, MD - Period: 5 Yrs.

Richardson, Sharon - Private - 36th Infantry - Company: Joseph Merrick - Age: 20 - Height: 5' 7" - Born: Medford, MA - Enlistment date: 28 Apr 1813 - Place: Baltimore - Period: 1 Yr.

Richardson, Solomon - Private - 36th Infantry - Company: Joseph Merrick - Age: 35 - Height: 5' 6" - Born: Queen Anne's Cty, MD - Enlistment date: 26 May 1813 - Place: Chestertown, MD - Period: 1 Yr.

Richardson, Thomas - Private - 14th Infantry - Company: David Cummings - Age: 18 - Height: 5' 7" - Born: Sussex Cty, DE - Trade: Plasterer - Enlistment date: 16 Oct 1814 - Place: Baltimore - Period: War - BLW 27034-160-12 - Discharged on 18 Mar 1815.

Richardson, William - Seaman - US Sea Fencibles - Company: Simmones Bunbury - Age: 28 - Height: 5' 6" - Born: Somerset Cty, MD - Enlistment date: 25 Dec 1813 - Period: 1 Yr - Discharged on 24 Dec 1814.

Richer, Robert - Private - 38th Infantry - Company: Shepperd Leakin - BLW 23655-160-12.

Richie, Allen - Private - 14th Infantry - Company: Samuel Lane - Age: 21 - Height: 5' 11 1/2" - Born: Ireland - Enlistment date: 4 Sep 1812 - Period: 5 Yrs - Prisoner of War, sent to England.

Rick, John - Seaman - US Sea Fencibles - Company: John Gill - Age: 25 - Height: 5' 8 1/2" - Born: Baltimore - Enlistment date: 7 Jan 1814 - Period: 1 Yr.

Ricker, John - Private - 14th Infantry - Company: Isaac Barnard - Age: 26 - Height: 5' 3" - Born: Baltimore - Trade: Chair maker - Enlistment date: 13 Aug 1814 - Period: War - Pension: Land bounty to John Ricker, son and other heirs at law of John Ricker, Sr. - BLW 25321-160-12 - Discharged on 5 Dec 1814 on Surgeon's Certificate, epileptic fits.

Ricketts, Benjamin - Third Lieutenant - 14th Infantry - Commissioned as a 3rd lieutenant on 12 May 1813; died in Nov 1813.

Ricketts, Charles - Private - 5th Infantry - Company: William Henshaw - Other regiment: 3rd Infantry (New) - Age: 26 - Height: 5' 7" - Born: Annapolis - Trade: Wheelwright - Enlistment date: 29 May 1813 - Place: Baltimore - Period: 5 Yrs - Discharged on 29 May 1818.

Ricketts, Rezin - Musician - 38th Infantry - Company: John Mowton - Age: 18 - Height: 5' 7 1/2" - Born: Baltimore - Trade: Painter - Enlistment date: 10 Jul 1813 - Period: 1 Yr - BLW 13301-160-12 - Re-enlisted at Craney Island, VA, on 21 Feb 1814 for the war; discharged at Craney Island, VA, on 15 Mar 1815.

Ricketts, William - Private - 14th Infantry - Company: Samuel Lane - Age: 25 - Height: 5' 11" - Born: Pennsylvania - Enlistment date: 21 Jul 1812 - Place: Bellefonte.

Ricks, James - Private - 5th Infantry - Company: Richard Whartenby - Other regiment: 3rd Infantry (New) - Age: 22 - Height: 5' 7" - Born: Northumberland Cty, VA - Trade: Baker - Enlistment date: 6 Jan 1812 - Place: Baltimore - Period: 5 Yrs - Discharged at Fort Mackinaw, MI, on 6

Maryland Regulars in War of 1812

Jan 1817.

Ricks, Joseph - Private - 14th Infantry - Company: Reuben Gilder - Age: 30 - Height: 5' 5" - Born: St. Mary's, MD - Trade: Plasterer - Enlistment date: 3 May 1813 - Place: Annapolis - Period: War - BLW 5146-160-12 - Discharged at Greenbush, NY, on 4 May 1815.

Rictor, Christian - Private - 14th Infantry - Company: Reuben Gilder - Age: 27 - Height: 5' 7" - Born: Pennsylvania - Enlistment date: 21 Jan 1814 - Period: War - Pension: Heirs received half pay for five years in lieu of land bounty - Wounded at Lyon's Creek, UC, on 19 Oct 1814; died at Williamsville, NY, on 18 Nov 1814.

Riddle, James W. - Second Lieutenant - 14th Infantry - Pension: SO-5755, SC-2810 - BLW 5819-156-50 - Commissioned as an ensign on 5 Apr 1814; promoted to 3rd lieutenant on 14 Nov 1814; promoted to 2nd lieutenant on 14 Nov 1814; resigned on 4 Mar 1815.

Riddle, Robert - Private - 14th Infantry - Prisoner of War (Halifax), discharged from Halifax on 3 Feb 1814.

Ridenbaugh, George - Private - 1st Infantry - Company: Daniel Bissell - Other regiment: 2nd Infantry - Height: 5' 9 1/2" - Born: Albemarle Cty, VA - Trade: Laborer - Enlistment date: 27 Jul 1805 - Period: 5 Yrs - Discharged on 28 Jul 1810; re-enlisted on 28 Sep 1810 in the 2nd Infantry for five years; died on 8 Apr 1815.

Ridenour, Conrad - Private - 14th Infantry - Company: Henry Fleming - Age: 30 - Height: 5' 9" - Born: Maryland - Trade: Comb maker - Enlistment date: 1 Feb 1813 - Period: 18 Mos - Prisoner of War (Halifax), captured at Beaver Dams on 24 Jun 1813; discharged on 20 Jun 1815.

Rider, Peter - Private - 7th Infantry - Company: Carey Nicholas - Other regiment: 1st Infantry (New) - Age: 40 - Height: 5' 7" - Born: Maryland - Trade: Farmer - Enlistment date: 1 Jan 1813 - Place: New Orleans - Period: 5 Yrs - Deserted on 4 Feb 1816.

Ridgeby, Absalom - Private - 38th Infantry - Age: 31 - Height: 6' - Born: Annapolis - Trade: Mariner - Enlistment date: 10 Oct 1814 - Place: Annapolis - Period: War.

Ridgeway, Fielder - Captain - 1st Rifles - Born: Maryland - Commissioned as a 1st lieutenant, Rifles, on 3 May 1808; promoted to captain on 31 Jul 1810; struck off on 11 May 1814.

Ridgway, James - Private - 38th Infantry - Company: James Smith - BLW 27704-160-42.

Rigby, Alexander - Private - 38th Infantry - Company: John Mowton - BLW 18210-160-12.

Rigby, Alexander - Private - 38th Infantry - Company: John Mowton - Age: 19 - Height: 5' 10" - Born: Maryland - Trade: Carpenter - Enlistment date: 22 Feb 1814 - Place: Craney Island, VA - Period: War - BLW 18210-160-12 - Discharged at Craney Island, VA, on 15 Mar 1815.

Riggin, Hamilton - Private - 38th Infantry - Company: John Brookes - Pension: Wife Nancy, WO-38617, WC-28626.

Riggin, Laban - Corporal - 14th Infantry - Company: David Cummings - Age:

By War of 1812 Soc. in MD

39 - Height: 5' 9" - Born: Wester, NC - Trade: Seaman - Enlistment date: 10 Jun 1812 - Period: 5 Yrs - Prisoner of War (Halifax), sent to England, captured at Beaver Dams on 24 Jun 1813; discharged on 10 Jun 1817.

Riggs, Clement A. - Musician - 42nd Infantry - Company: Edward Mendenhall - Other regiment: Corps of Artillery - Age: 18 - Height: 5' 3" - Born: Maryland - Enlistment date: 13 May 1814 - Place: Columbia, MD - Period: 5 Yrs - Transferred to Corps of Artillery on 30 Jun 1815; deserted on 18 Dec 1817.

Riley, Bennett - Captain - 1st Rifles - Other regiment: 5th Infantry (New) - Born: Maryland - BLW 22731-181-50 - Commissioned as an ensign, Rifles, on 19 Jan 1813; promoted to 3rd lieutenant on 12 Mar 1813; promoted to 2nd lieutenant on 15 Apr 1814; promoted to 1st lieutenant on 31 Mar 1817; promoted to captain on 6 Aug 1818; continued to served in the army, promoted to colonel on 2 Jun 1840; died on 9 Jun 1853.

Riley, Edward - Corporal - 2nd Infantry - Company: John Bowyer - Age: 20 - Height: 5' 10" - Born: Allegheny Cty, PA - Trade: Farmer - Enlistment date: 13 Sep 1807 - Place: Fredericktown, MD - Period: 5 Yrs - Re-enlisted; deserted at Mt. Vernon on 27 Jul 1812.

Riley, James - Sergeant - 38th Infantry - Company: Charles Stansbury - BLW 17936-160-12.

Riley, James - Private - 38th Infantry - Company: Charles Stansbury - Age: 27 - Height: 5' 10" - Born: Philadelphia - Trade: Laborer - Enlistment date: 9 Nov 1814 - Place: Cumberland - Period: War - BLW 17936-160-12 - Discharged at Baltimore on 6 Apr 1815.

Riley, Peter - Private - 5th Infantry - Company: Richard Bell - Other regiment: 3rd Infantry (New) - Age: 30 - Height: 5' 5" - Born: Baltimore - Trade: Cordwainer - Enlistment date: 22 Apr 1813 - Place: Baltimore - Period: 5 Yrs - Discharged on 22 Apr 1818.

Riley, William - Private - 5th Infantry - Company: Colin Buckner - Age: 42 - Height: 5' 9" - Born: Ireland - Trade: Ship's carpenter - Enlistment date: 11 Apr 1813 - Place: Baltimore - Period: 5 Yrs.

Riley, William - Private - 36th Infantry - Company: Samuel Raisin - Age: 25 - Height: 5' 6" - Born: England - Trade: Farmer - Enlistment date: 6 Dec 1814 - Place: F. Town - Period: War - Discharged at Washington, DC, on 20 Mar 1815.

Rimmer, John - Private - 14th Infantry - Enlistment date: 17 Apr 1812 - Period: 5 Yrs - Pension: Land bounty to Mary Ann Parker, sister and other heirs at law of John Rimmer - BLW 13137-160-12 - Died on 22 May 1812.

Rimmey, Adam - Sergeant - 14th Infantry - Company: Thomas Kearney - BLW 9559-160-12.

Rind, William A. - First Lieutenant - 36th Infantry - Company: Samuel Raisin - BLW 2112-160-50 - Commissioned as a 3rd lieutenant on 30 Apr 1813; promoted to 2nd lieutenant on 15 Aug 1813; promoted to 1st lieutenant on 30 Sep 1814; discharged on 15 Jun 1815.

Rinehart, John - Private - 38th Infantry - Company: Shepperd Leakin - Age: 38 - Height: 5' 9" - Born: Baltimore - Trade: Laborer - Enlistment date: 19

Maryland Regulars in War of 1812

Apr 1814 - Place: Baltimore - Period: 5 Yrs.

Rinnear, William - Private - 14th Infantry - Company: Joseph Marshall - Discharged on 3 Dec 1814 due to a rupture, served for four months.

Rison, John - Private - 38th Infantry - Company: John Buck - Age: 27 - Height: 5' 7 3/4" - Born: Charles Cty, MD - Trade: Sail maker - Enlistment date: 17 Apr 1814 - Place: Annapolis - Period: War - BLW 500-160-12 - Discharged at Baltimore on 30 Mar 1815.

Ristin, Bazil E. - Private - Light Artillery - Company: Arthur Thornton - Other regiment: Corps of Artillery - Age: 18 - Height: 5' 9 1/2" - Born: Prince George's Cty, MD - Trade: Farmer - Enlistment date: 13 Mar 1812 - Period: 5 Yrs - Discharged on 13 May 1817.

Riswick, Edward - Private - 36th Infantry - Company: Joseph Hook - Age: 22 - Height: 5' 2" - Born: Charles Cty, MD - Trade: Hatter - Enlistment date: 23 Jul 1814 - Place: Baltimore - Period: War - Pension: Wife Larriet E., WO-40016, WC-34463 - BLW 25775-160-12; BLW 290-160-12 Cancelled - Discharged at Fort Covington, MD, on 30 Mar 1815.

Ritchie, Isaac - Private - 36th Infantry - Age: 22 - Height: 5' 11" - Born: Virginia - Trade: Blacksmith - Enlistment date: 22 Jan 1814 - Period: War - BLW 879-160-12 - Discharged at Georgetown, DC, on 13 Mar 1815.

Ritchie, Robert - Private - 38th Infantry - Company: Shepperd Leakin - Age: 25 - Height: 5' 8" - Born: Montgomery Cty, MD - Trade: Farmer - Enlistment date: 1 May 1814 - Period: War - Discharged at Fort McHenry on 28 Mar 1815.

Ritchie, Thomas - First Lieutenant - 36th Infantry - Born: Maryland - BLW 6990-160-50 - Commissioned as a 1st lieutenant on 30 Apr 1813; discharged on 15 Jun 1815.

Ritchie, Thomas - Private - 36th Infantry - Company: Charles Randolph - Place: Frederick, MD - Pension: Wife Mary Ann, WO-21726, WC-27321.

Roach, James - Private - 5th Infantry - Company: Leroy Opie - Other regiment: 3rd Infantry (New) - Age: 32 - Height: 5' 6 3/4" - Born: Ireland - Trade: Laborer - Enlistment date: 19 Jun 1812 - Place: Baltimore - Period: 5 Yrs - Discharged on 19 Jun 1817.

Roach, Stephen - Private - 14th Infantry - Company: Thomas Montgomery - Enlistment date: 26 May 1812 - Period: 5 Yrs - Pension: Land bounty to James Roach and other heirs at law of Stephen Roach - BLW 17625-160-12 - Killed during the Battle of La Cole Mill on 30 Mar 1814.

Roan, Samuel - Private - 2nd Artillery - Company: Richard Zantzinger - Other regiment: Corps of Artillery - Age: 19 - Height: 5' 6" - Born: Maryland - Trade: Painter - Enlistment date: 4 Oct 1814 - Place: Lancaster, PA - Period: 5 Yrs - Discharged at Buffalo on 4 Oct 1819.

Roarke, Benjamin - Private - 2nd Artillery - Company: Sander Donoho - Age: 19 - Height: 5' 6" - Born: Worcester Cty, MD - Trade: Farmer - Enlistment date: 13 Aug 1812 - Place: Warrenton, GA - Period: 5 Yrs - Discharged on 1 Sep 1815.

Robb, John - Ensign - 7th Infantry - Company: Samuel Vail - Age: 26 - Height: 5' - Born: Maryland - Enlistment date: 3 Dec 1813 - Period: War - Served as a sergeant and quartermaster sergeant in the 7th Infantry; commissioned as an ensign on 22 Dec 1814; discharged on 15 Jun 1815.

Robbins, John - Private - 32nd Infantry - Company: Horatio Davis - Age: 23 - Height: 5' 8 1/2" - Born: Snow Hill, MD - Trade: Laborer - Enlistment date: 31 May 1813 - Period: 1 Yr - Discharged on 30 May 1814.

Roberts, Arthur - Private - 36th Infantry - Company: Thomas Carbery - Age: 32 - Height: 5' 2" - Born: Boston - Trade: Trader - Enlistment date: 19 Feb 1814 - Place: Georgetown, DC - Period: 5 Yrs.

Roberts, Benjamin - Private - 12th Infantry - Company: Thomas Moore - Age: 21 - Height: 5' 5" - Born: Maryland - Trade: Farmer - Enlistment date: 13 Jul 1813 - Place: Parkersburg, VA - Period: War - Discharged at Buffalo on 31 May 1815.

Roberts, Edward - Private - 14th Infantry - Company: Samuel Lane - Age: 26 - Height: 5' 9" - Born: New York - Enlistment date: 29 Jun 1812 - Place: Hagerstown, MD - Period: 5 Yrs - Deserted at Buffalo on 10 Oct 1812.

Roberts, Jacob - Private - 14th Infantry - Company: David Cummings - Age: 24 - Height: 5' 8" - Born: Baltimore - Trade: Laborer - Enlistment date: 22 May 1814 - Period: War - Discharged at Baltimore on 18 Mar 1815.

Roberts, James S. - Private - 2nd Rifles - Company: John O'Fallon - Age: 29 - Height: 5' 9" - Born: Baltimore - Trade: Carpenter - Enlistment date: 23 Jun 1814 - Place: Louisville, KY - Period: War - BLW 6178-160-12 - Discharged at Detroit on 30 Jun 1815.

Roberts, John - Private - 7th Infantry - Company: Jacob Miller - Age: 20 - Height: 5' 7 1/2" - Born: Maryland - Trade: Tobacconist - Enlistment date: 13 Jul 1814 - Place: Eddyville, KY - Period: War - BLW 4048-160-12 - Discharged at Hopkinsville, KY, on 14 Apr 1815.

Roberts, Robert - Private - 36th Infantry - Company: Joseph Hook - Age: 40 - Height: 5' 5" - Born: Queen Anne's Cty, MD - Trade: Farmer - Enlistment date: 1 Mar 1814 - Place: Annapolis - Period: War - BLW 3217-160-12 - Discharged at Fort Covington, MD, on 30 Mar 1815.

Roberts, William - Private - 14th Infantry - Prisoner of War (Halifax); died at Halifax on 28 Sep 1813.

Roberts, Zachariah - Private - 2nd Artillery - Company: Nathan Towson - Pension: Old War IF-4145.

Roberts, Zachariah - Private - 2nd Artillery - Company: Nathan Towson - Age: 34 - Height: 5' 7 1/2" - Born: Baltimore - Trade: Drummer - Enlistment date: 25 Jan 1812 - Place: Baltimore - Period: 5 Yrs - Pension: Placed on the pension rolls on 2 May 1815 - Wounded at Chippewa, UC, on 5 Jul 1814; discharged at Baltimore on 1 May 1815.

Robertson, John - Private - 36th Infantry - Age: 21 - Height: 5' 9 1/2" - Born: New York City - Enlistment date: 8 Dec 1814 - Place: Richmond, VA - Period: War - Deserted.

Robertson, Thomas - Seaman - US Sea Fencibles - Company: Simmones Bunbury - Enlistment date: 25 Jan 1814 - Period: 1 Yr - Discharged on 26 Jan 1815.

Robertson, William - Private - 36th Infantry - Company: Samuel Raisin - Age:

27 - Height: 5' 8" - Born: Amelia Cty, VA - Trade: Wheelwright - Enlistment date: 6 Aug 1814 - Place: Manchester - Period: War - Discharged at Washington, DC, on 20 Mar 1815.

Robey, Isaac - Private - 14th Infantry - Company: Thomas Montgomery - Pension: Land bounty to Elizabeth Robey and other heirs at Law of Isaac Robey - BLW 6616-160-12.

Robinett, Richard - Private - 42nd Infantry - Company: George Barker - Age: 25 - Height: 5' 6 1/2' - Born: Frederick, MD - Trade: Cooper - Enlistment date: 28 Oct 1814 - Place: New Castle, DE - Period: 5 Yrs - Deserted at Fort McHenry on 24 Aug 1815.

Robins, Daniel - Corporal - 38th Infantry - Company: John Buck - Age: 23 - Height: 5' 6" - Born: Baltimore - Trade: Laborer - Enlistment date: 30 Aug 1813 - Period: 1 Yr - Re-enlisted at Baltimore on 12 Dec 1814 for the war.

Robinson, Benjamin - Private - 14th Infantry - Company: James Britton - Enlistment date: 19 Apr 1813 - Period: 5 Yrs - Pension: Land bounty to Lydia Williams, daughter and other heirs at law of Benjamin Robinson - BLW 26603-160-12 - Died.

Robinson, Benjamin - Private - 38th Infantry - Company: Shepperd Leakin - Age: 18 - Height: 5' - Born: Boston - Trade: Mariner - Enlistment date: 28 Feb 1814 - Place: Baltimore - Period: War.

Robinson, Caleb R. - Second Lieutenant - US Sea Fencibles - Company: William Addison - Commissioned as a 2nd Lieutenant on 17 Mar 1814; died on 28 Jan 1815.

Robinson, Daniel - Private - 14th Infantry - Prisoner of War (Halifax), discharged from Halifax on 3 Feb 1814.

Robinson, Francis S. - Sergeant - 2nd Rifles - Company: Batteal Harrison - Age: 24 - Height: 5' 10 3/4" - Born: Maryland - Trade: House joiner - Enlistment date: 8 Aug 1814 - Place: Cincinnati - Period: War - Pension: SO-17685, SC-8464 - BLW 27124-160-42 - Discharged on 20 Jun 1815.

Robinson, Henry - Private - 38th Infantry - Company: John Brookes - Pension: Land bounty to James Robinson, bother and other heirs at law of Henry Robinson - BLW 14064-160-12.

Robinson, Hugh - Private - 14th Infantry - Company: Thomas Montgomery - Pension: Old War IF-4156 - BLW 7006-160-12.

Robinson, John - Private - 36th Infantry - Company: Thomas Carbery - Age: 24 - Height: 5' 10" - Born: Maryland - Enlistment date: 25 Mar 1814 - Place: Washington, DC - Period: War - Pension: Wife Elizabeth, SO-7281, WC-3485 - BLW 9987-160-12 - Discharged at Baltimore on 31 Mar 1815.

Robinson, John - Private - 2nd Light Dragoons - Company: Samuel Hopkins - Age: 22 - Height: 5' 9" - Born: Caroline Cty, MD - Trade: Printer - Enlistment date: 7 May 1814 - Place: Philadelphia - Period: 5 Yrs - Discharged at Washington, DC, on 20 Jan 1819.

Robinson, John - Private - 26th Infantry - Company: William Bezeau - Age: 25 - Height: 5' 10" - Born: Baltimore - Enlistment date: 27 Oct 1814 - Place: Baltimore - Deserted on 29 Oct 1814.

Robinson, John - Private - 38th Infantry - Company: Shepperd Leakin - Age:

By War of 1812 Soc. in MD

24 - Height: 5' 6 1/2" - Born: Hull, England - Trade: Farmer - Enlistment date: 14 Aug 1814 - Place: Annapolis - Period: War.

Robinson, John - Private - 3rd Artillery - Company: Alexander Fanning - Age: 22 - Height: 5' 5 1/2" - Born: Baltimore - Trade: Carpenter - Enlistment date: 22 Dec 1814 - Place: Buffalo - Period: 5 Yrs.

Robinson, Joseph - Private - 14th Infantry - Company: David Cummings - Age: 35 - Height: 5' 9" - Born: Harford Cty, MD - Trade: Carpenter - Enlistment date: 22 Oct 1814 - Place: Baltimore - Period: War - BLW 10355-160-12 - Discharged at Baltimore on 16 Mar 1815.

Robinson, Joseph - Private - 1st Light Dragoons - Company: Lemuel Gustine - Age: 34 - Height: 5' 6" - Born: Harford Cty, MD - Trade: Carpenter - Enlistment date: 22 May 1814 - Place: Baltimore - Period: War.

Robinson, Joseph - Private - 36th Infantry - Company: Joseph Hook - Age: 38 - Height: 5' 8" - Born: Harford Cty, MD - Enlistment date: 6 Jun 1813 - Place: Baltimore - Period: 1 Yr - Deserted on 17 Mar 1814.

Robinson, Joseph - Private - 36th Infantry - Company: Thomas Carbery - Age: 35 - Height: 5' 10" - Born: Cork, Ireland - Enlistment date: 27 Apr 1814 - Place: Georgetown, DC - Period: War - Died at Georgetown, DC, on 4 Oct 1814.

Robinson, Nicholas - Private - 14th Infantry - Prisoner of War (Halifax), discharged from Halifax on 3 Feb 1814.

Robinson, Nicholas N. - Second Lieutenant - 14th Infantry - Company: Henry Grindage - Pension: SO-7512, SC-3919 - BLW 11600-160-60 - Commissioned as an ensign on 2 Jul 1812; promoted to 3rd lieutenant on 13 May 1813; promoted to 2nd lieutenant on 12 May 1813; discharged on 15 Jun 1815.

Robinson, Thomas - Seaman - US Sea Fencibles - Company: Simmones Bunbury - Pension: Heirs received half pay for five years in lieu of land bounty - Died on 25 Apr 1827.

Robinson, Tobias - Private - 28th Infantry - Company: George Stockton - Age: 20 - Height: 5' 9" - Born: Maryland - Trade: Farmer - Enlistment date: 23 Mar 1814 - Place: Detroit - Period: War - Discharged at Detroit on 8 Jul 1815.

Robinson, William - Private - 36th Infantry - Company: Samuel Raisin - BLW 20506-160-12.

Robinson, William - Private - 38th Infantry - Company: Shepperd Leakin - Height: 5' 6" - Born: Dorchester Cty, MD - Trade: Mariner - Enlistment date: 26 Nov 1813 - Period: 1 Yr - Re-enlisted on 28 Feb 1814 for the war; discharged on 31 Mar 1815.

Robinson, William T. - Private - 14th Infantry - Age: 28 - Height: 5' 10 1/2" - Born: Virginia - Enlistment date: 8 Jul 1814 - Place: Alexandria, DC - Deserted.

Robson, Robert - Private - 38th Infantry - Company: Shepperd Leakin - Age: 25 - Height: 5' 6" - Born: Northumberland, England - Trade: Farmer - Enlistment date: 3 Sep 1814 - Place: Baltimore - Period: 5 Yrs.

Roby, Leonard - Private - 36th Infantry - Age: 27 - Height: 5' 7" - Born:

Maryland Regulars in War of 1812

Maryland - Trade: Plasterer - Enlistment date: 20 Jan 1815 - Period: War - Pension: Land bounty to Elizabeth Roby and other heirs at law of Leonard Roby - BLW 290-320-12 - Discharged at Georgetown, DC, on 13 Mar 1815.

Roby, Peter H. - Private - 45th Infantry - Company: Edward Tatnall - Age: 27 - Height: 5' 6" - Born: Charles Cty, MD - Trade: Farmer - Enlistment date: 28 May 1814 - Place: Washington, GA - Period: 5 Yrs - BLW 9320-160-12 - Discharged at Fort Moultrie, SC, on 16 Sep 1815, rupture.

Rodgers, Charles W. - Musician - 14th Infantry - Company: Reuben Gilder - Other regiment: Corps of Artillery - Age: 14 - Height: 4' 4" - Born: Philadelphia - Trade: Drummer - Enlistment date: 4 Jun 1812 - Place: West Chester, PA - Period: 5 Yrs - Transferred to Corps of Artillery on 4 Feb 1815; discharged at Fort McHenry on 4 Jun 1817.

Rodgers, George - Private - 14th Infantry - Company: Richard Arell - Other regiment: 4th Infantry (New) - Age: 23 - Height: 5' 5" - Born: Baltimore - Trade: Cooper - Enlistment date: 2 Jun 1812 - Place: Baltimore - Period: 5 Yrs - Prisoner of War, exchanged on 15 Apr 1814; deserted at Fort Severn on 16 Jul 1815.

Rodgers, Harrison G. - Ensign - 14th Infantry - Born: Maryland - Enlistment date: 9 Jul 1812 - Period: 18 Mos - Served as a quartermaster sergeant; commissioned as an ensign on 15 Apr 1814; Prisoner of War (Halifax), captured at Beaver Dams on 24 Jun 1813, discharged from Halifax on 3 Feb 1813, exchanged on 15 Apr 1814; struck off on 5 May 1814.

Rodgers, John - Private - 38th Infantry - Company: Shepperd Leakin - Age: 22 - Height: 5' 9" - Born: Cumberland, PA - Trade: Farmer - Enlistment date: 3 Sep 1814 - Place: Baltimore - Period: War - BLW 17314-160-12 - Discharged at Fort McHenry on 28 Mar 1815.

Rodgers, William - Private - 1st Artillery - Company: George Armistead - Age: 18 - Height: 5' 7" - Born: St. Mary's Cty, MD - Trade: Cabinet maker - Enlistment date: 2 Feb 1810 - Place: Fort McHenry - Period: 5 Yrs - Discharged on 1 Jun 1815.

Roe, John - Private - 12th Infantry - Company: Thomas Sangster - Other regiment: 8th Infantry (New) - Age: 15 - Height: 5' - Born: Alexandria, DC - Trade: Baker - Enlistment date: 17 Jul 1812 - Place: Baltimore - Period: 5 Yrs - Discharged on 17 Jul 1817.

Roe, Michael - Private - 2nd Artillery - Company: Nathan Towson - Age: 30 - Prisoner of War, captured during the Battle of Stoney Creek.

Rogers, Joseph - Seaman - US Sea Fencibles - Company: Simmones Bunbury - Enlistment date: 22 Mar 1814 - Period: 1 Yr - Discharged at Fort McHenry on 22 Mar 1815.

Rogers, William L. - First Lieutenant - 36th Infantry - Company: Thomas Carbery - Pension: Wife Jane C., WO-31900, WC-15415 - BLW 380-160-50 - Commissioned as a 2nd lieutenant on 30 Apr 1813; promoted to 1st lieutenant on 1 May 1814; discharged on 15 Jun 1815.

Roles, Reason - Private - 14th Infantry - Company: Henry Grindage - Other regiment: 27th Infantry - Age: 31 - Height: 5' 5 1/2" - Trade: Cordwainer - Enlistment date: 9 Aug 1812 - Period: 1 Yr - Prisoner of War (Halifax), captured at Stoney Creek, UC, on 12 Jun 1813; discharged on 7 Aug 1814; re-enlisted on 9 Aug 1814 for the war in the 27th Infantry, Captain J. Porter's Company; discharged at New York on 17 Jun 1815.

Rollins, George - Private - 36th Infantry - Company: Thomas Carbery - Age: 18 - Height: 5' 3" - Born: Charles Cty, MD - Trade: Farmer - Enlistment date: 26 Jun 1813 - Place: Washington, DC - Period: War - BLW 1502-160-12 - Discharged at Baltimore on 31 May 1815.

Rollins, John - Private - 36th Infantry - Company: Thomas Corcoran - Enlistment date: 6 May 1813 - Period: 1 Yr - Discharged at Annapolis on 6 May 1814.

Rolph, Stephen - Private - 39th Infantry - Age: 32 - Height: 5' 6" - Born: Caroline Cty, MD - Enlistment date: 18 Aug 1814 - Place: Knoxville, TN - Period: 5 Yrs.

Rolston, John - Private - 36th Infantry - Company: Joseph Hook - Age: 25 - Height: 5' 9" - Born: Ireland - Enlistment date: 16 May 1813 - Place: Baltimore - Period: 1 Yr - Deserted on 21 Nov 1813.

Roney, Patrick - Artificer - 2nd Artillery - Company: Frederick Evans - Age: 26 - Height: 5' 6" - Born: Ireland - Trade: Blacksmith - Enlistment date: 7 Dec 1810 - Place: Fort Washington, MD - Period: 5 Yrs - Discharged on 9 Dec 1815.

Rook, James - Private - 17th Infantry - Company: David Holt - Age: 23 - Height: 5' 8" - Born: Maryland - Trade: Laborer - Enlistment date: 14 Jun 1812 - Place: Zanesville, OH - Period: War - BLW 2659-160-12 - Discharged at Chillicothe, OH, on 4 Jun 1814.

Rook, John - Seaman - US Sea Fencibles - Company: William Addison.

Rook, William - Private - 36th Infantry - Company: Joseph Merrick - Age: 30 - Height: 5' 8 1/2" - Born: Queen Anne's Cty, MD - Enlistment date: 7 May 1813 - Place: Baltimore - Period: 1 Yr - Discharged by sentence of a court-martial.

Rooke, John A. - Gunner - US Sea Fencibles - Company: John Gill - Age: 40 - Height: 5' 7 1/2" - Born: Somerset Cty, MD.

Root, John - Private - 14th Infantry - Age: 28 - Born: New York - Prisoner of War (Quebec), died at Quebec on 22 Aug 1813.

Rose, James - Private - 38th Infantry - Company: John Brookes - BLW 10525-160-12.

Rose, Jonathan - Private - 14th Infantry - Company: Samuel Lane - Age: 27 - Height: 5' 11" - Enlistment date: 28 Jun 1812 - Period: 18 Mos - Prisoner of War (Halifax), discharged from Halifax on 3 Feb 1814; discharged at Plattsburgh, NY, on 5 Jul 1814.

Rose, Nathaniel - Private - 14th Infantry - Company: William McIlvane - Enlistment date: 7 Feb 1813 - Period: 18 Mos - Prisoner of War (Halifax), captured at Beaver Dams on 24 Jun 1813, discharged from Halifax on 3 Feb 1813, exchanged on 15 Apr 1814; discharged at Plattsburg, NY, on 6 Aug 1814.

Roseburg, John - Private - 36th Infantry - Company: Joseph Hook - Age: 24 - Height: 5' 2" - Born: Baltimore - Enlistment date: 3 May 1813 - Place:

Maryland Regulars in War of 1812

Baltimore - Period: 1 Yr.

Rosepaugh, Cornelius - Private - 2nd Artillery - Company: Nathan Towson - Age: 38 - Prisoner of War, captured during the Battle of Stoney Creek.

Rosepaugh, Cornelius - Private - 2nd Artillery - Company: Nathan Towson - Other regiment: Corps of Artillery - Age: 32 - Height: 5' 6" - Born: New Jersey - Trade: Laborer - Enlistment date: 2 May 1812 - Place: Baltimore - Period: 5 Yrs - Discharged at Fort Niagara, NY, on 1 May 1817.

Rosinguest, Oliver - Private - 36th Infantry - Company: Thomas Carbery - Age: 39 - Height: 5' 8" - Born: Prussia - Enlistment date: 18 May 1814 - Place: Washington, DC - Period: War - BLW 3348-160-12 - Discharged at Baltimore on 31 Mar 1815.

Ross, David - Private - 22nd Infantry - Pension: Heirs received half pay for five years in lieu of land bounty - Died on 25 Jul 1814.

Ross, David - Private - 14th Infantry - Prisoner of War (Halifax), died at Halifax on 17 Oct 1813.

Ross, James - Private - 14th Infantry - Prisoner of War, exchanged on 15 Apr 1814.

Ross, James - Private - 38th Infantry - Company: Sheppard Leakin - Age: 23 - Height: 5' 6 3/4" - Born: Dorchester Cty, MD - Trade: Laborer - Enlistment date: 15 Apr 1814 - Period: War.

Ross, Joseph - Private - 14th Infantry - Prisoner of War (Halifax), discharged from Halifax on 3 Feb 1814.

Ross, Joseph N. - Sergeant - 14th Infantry - Company: Richard Arell - Age: 25 - Height: 5' 7 1/2" - Born: Allegany Cty, MD - Trade: Farmer - Enlistment date: 8 Jun 1812 - Place: Allegany Cty, MD - Period: 5 Yrs - Pension: Wife Elizabeth, Old War IF-4130, Old War CF-627, WO-1178, WC-8192 - BLW 5709-160-12 - Discharged at Greenbush, NY, on 29 May 1815 for disability.

Ross, Peter - Private - 1st Artillery - Company: James Reed - Age: 39 - Height: 5' 6 1/4" - Born: Harford Cty, MD - Trade: Tailor - Enlistment date: 14 Apr 1814 - Place: Annapolis - Period: 5 Yrs - Discharged on 14 Apr 1819.

Ross, Samuel S. - Seaman - US Sea Fencibles - Company: Simmones Bunbury - Enlistment date: 26 Jan 1814 - Period: 1 Yr - Discharged on 25 Jan 1815.

Ross, Solomon - Private - 38th Infantry - Company: Isaac Aldridge - Age: 18 - Height: 6' 1/2" - Born: Somerset Cty, MD - Trade: Painter - Enlistment date: 4 Feb 1815 - Period: War - BLW 236-320-12 - Discharged at Baltimore on 6 Apr 1815.

Rotes, Peter - Private - 38th Infantry - Company: John Mowton - BLW 10255-160-12.

Rothrock, John - Captain - 38th Infantry - BLW 11338-160-50 - Commissioned as a captain on 20 May 1813; discharged on 15 Jun 1815.

Rothrock, Joseph - Private - 14th Infantry - Company: Samuel Lane - Age: 28 - Height: 5' 4 1/2" - Born: Maryland - Enlistment date: 2 Jun 1812 - Period: 18 Mos.

Roundsville, John - Private - 38th Infantry - Company: Shepperd Leakin - Age: 30 - Height: 5' 4 1/2" - Born: New Jersey - Trade: Farmer - Enlistment date: 20 Jul 1814 - Place: Hagerstown, MD - Period: War - BLW 10624-

By War of 1812 Soc. in MD

160-12 - Discharged at Fort McHenry on 28 Mar 1815.

Rourk, Hollingsworth - Private - 38th Infantry - Age: 35 - Height: 5' 6" - Born: Dorchester Cty, MD - Trade: Seaman - Enlistment date: 7 Oct 1814 - Place: Baltimore - Period: War.

Rouse, John - Private - 36th Infantry - Company: Samuel Raisin - Age: 22 - Height: 5' 5 1/4" - Born: Kent Cty, DE - Enlistment date: 8 Jul 1813 - Place: Kent Cty, MD - Period: 1 Yr.

Row, Michael - Private - 38th Infantry - Company: Charles Stansbury - Age: 19 - Height: 5' 3 1/2" - Born: Frederick, MD - Trade: Chair maker - Enlistment date: 11 Jan 1815 - Place: Fredericktown, MD - Period: War - BLW 12442-160-12 Cancelled - Discharged at Baltimore on 6 Apr 1815.

Rowe, John - Private - 36th Infantry - Company: Thomas Carbery - Age: 29 - Height: 5' 8" - Born: Georgia - Trade: Accountant - Enlistment date: 10 Apr 1814 - Place: Leonardstown, MD - Period: War - BLW 5461-160-12 - Discharged at Washington, DC, on 15 Mar 1815.

Rowe, Noah - Private - 38th Infantry - Company: Charles Stansbury - Age: 27 - Height: 5' 5" - Born: Baldwin, MA - Trade: Farmer - Enlistment date: 12 Aug 1814 - Place: Annapolis - Period: War - Died on 8 Feb 1815.

Rowland, George - Fife Major - 36th Infantry - Age: 32 - Height: 5' 11 1/2" - Born: New London, CT - Trade: Blacksmith - Enlistment date: 9 May 1814 - Place: Georgetown, DC - Period: War - BLW 7-160-12 - Discharged at Baltimore on 31 Mar 1815.

Rowland, John - Private - 5th Infantry - Company: Richard Whartenby - Age: 21 - Height: 5' 8" - Born: York, PA - Trade: Farmer - Enlistment date: 8 Mar 1811 - Place: Baltimore - Period: 5 Yrs.

Rozen, Henry - Private - 2nd Artillery - Company: Frederick Evans - Other regiment: Corps of Artillery - Age: 38 - Height: 5' - Born: Pennsylvania - Trade: Tailor - Enlistment date: 1 Jun 1814 - Place: Fort McHenry - Period: 5 Yrs - Deserted at Fort McHenry on 16 Oct 1815.

Ruddy, Samuel - Private - 14th Infantry - Company: David Cummings - Age: 25 - Height: 5' 11 1/2" - Born: Washington, MD - Trade: Hatter - Enlistment date: 6 Oct 1814 - Period: War - Discharged on 18 Mar 1815.

Rudolph, Titan K. - Private - 17th Infantry - Company: David Holt - Age: 21 - Height: 5' 10" - Born: Maryland - Trade: Shoemaker - Enlistment date: 26 May 1814 - Place: Warren, Trumbull Cty, OH - Period: War - BLW 7230-160-12 - Discharged at Chillicothe, OH, on 4 Jun 1814.

Rudy, Samuel - Private - 14th Infantry - BLW 16628-160-12.

Rumblay, William - Private - 38th Infantry - Company: John Rothrock - Age: 22 - Height: 5' 9 1/4" - Born: Somerset Cty, MD - Trade: Sailor - Enlistment date: 9 Oct 1813 - Place: Craney Island, VA - Period: War - BLW 27819-160-42 - Discharged at Craney Island, VA, on 15 Mar 1815.

Rummey, John - Private - 36th Infantry - Company: Joseph Hook - Age: 42 - Height: 5' 4" - Born: Chester, MD - Trade: Carpenter - Enlistment date: 14 Apr 1812 - Place: Georgetown, DC - Period: 5 Yrs.

Rumsey, Joseph - Private - 14th Infantry - Company: Isaac Barnard - Enlistment date: 5 Mar 1813 - Period: 5 Yrs - Pension: Wife Jeanette, SO-

30121, SC-21658, WO-24790, WC-26410 - Prisoner of War (Halifax), sent to England.

Ruppert, Joseph S. - 14th Infantry - Company: James Britton - Pension: Wife Amelia, WO-25577, WC-28420.

Rush, Samuel M. - Private - 36th Infantry - Company: Thomas Carbery - Age: 28 - Height: 6' - Born: Essex Cty, VA - Trade: Seaman - Enlistment date: 20 May 1814 - Place: Washington, DC - Period: War - Discharged at Baltimore on 31 Mar 1815.

Rusia, Peter F. - Musician - 36th Infantry - Company: Joseph Hook - Age: 30 - Height: 5' 7 1/2" - Born: England - Trade: Drummer - Enlistment date: 1 Mar 1814 - Place: Annapolis - Period: War - BLW 12789-160-12 - Discharged at Fort Covington, MD, on 30 Mar 1815.

Rusler, John - Private - 38th Infantry - Company: Shepperd Leakin - Age: 20 - Height: 5' 4" - Born: Frederick, MD - Trade: Farmer - Enlistment date: 12 Aug 1814 - Place: Baltimore - Period: War - BLW 3720-160-12 - Discharged at Fort McHenry on 28 Mar 1815.

Russell, Dewey - Private - 2nd Artillery - Company: Frederick Evans - Other regiment: Corps of Artillery - Age: 32 - Height: 5' 7" - Born: Hartford, CT - Trade: Laborer - Enlistment date: 2 Jan 1812 - Place: Baltimore - Period: 5 Yrs - Discharged on 2 Jan 1817.

Russell, James - Private - 14th Infantry - Company: David Cummings - Age: 25 - Height: 5' 8" - Born: Ireland - Trade: Laborer - Enlistment date: 9 Jul 1814 - Place: Elkton, MD - Period: 5 Yrs - Deserted at Baltimore on 18 Sep 1814.

Russell, John - Private - 38th Infantry - Company: Charles Stansbury - BLW 5501-160-12.

Russell, Robert - Private - 36th Infantry - Company: Thomas Carbery - Age: 18 - Height: 5' 5" - Born: Maryland - Trade: Baker - Enlistment date: 3 May 1814 - Place: Georgetown, DC - Period: War - BLW 27052-160-12 - Discharged at Baltimore on 31 Mar 1815.

Russell, Robert - Private - 36th Infantry - Company: Joseph Hook - Age: 29 - Height: 5' 6" - Born: Anne Arundel Cty, MD - Trade: Laborer - Enlistment date: 19 Jul 1814 - Place: Georgetown, DC - Period: War - BLW 1730-160-12 - Discharged at Fort Covington, MD, on 30 Mar 1815.

Russell, Thomas - Private - 14th Infantry - Company: Reuben Gilder - Enlistment date: 17 Oct 1812 - Period: 5 Yrs - Deserted at Plattsburgh, NY, on 14 Apr 1814.

Russum, Thomas - Private - 2nd Light Dragoons - Company: Samuel Harris - Age: 22 - Height: 5' 6" - Born: Talbot Cty, MD - Trade: Laborer - Enlistment date: 1 May 1813 - Place: Centreville, MD - Period: 5 Yrs - Deserted on 23 Aug 1815.

Russumo, Benjamin - Private - 2nd Light Dragoons - Company: Jonas Holland - Other regiment: Corps of Artillery - Age: 19 - Height: 5' 7" - Born: Hillsborough, MD - Trade: Tanner - Enlistment date: 24 Jul 1813 - Place: Centreville, MD - Period: 5 Yrs - Transferred to Corps of Artillery on 31 Aug 1815; deserted on 23 Aug 1815.

By War of 1812 Soc. in MD

Rust, James M. - Private - 36th Infantry - Company: Thomas Carbery - BLW 16750-160-12.

Ruth, Thomas C. - Private - 26th Infantry - Company: William Bezeau - Age: 16 - Height: 4' 9" - Born: Baltimore - Trade: Shoemaker - Enlistment date: 26 Oct 1814 - Place: Baltimore - Period: War - BLW 14777-160-12 - Discharged at Baltimore.

Rutherford, James - Private - 14th Infantry - Company: Samuel Lane - Age: 26 - Height: 5' 8" - Born: Virginia - Enlistment date: 8 Jun 1812 - Place: Shepherdstown - Period: 5 Yrs - Pension: Land bounty to Peter W. Kenny and other heirs at law of James Rutherford - BLW 27895-160-42 - Died at Black Rock, NY, on 8 Dec 1812.

Rutherford, William - Musician - 36th Infantry - Company: Thomas Carbery - Age: 15 - Height: 5' 1" - Born: Washington, DC - Trade: Butcher - Enlistment date: 19 Feb 1814 - Place: Georgetown, DC - Period: War - BLW 5578-160-12 - Discharged at Georgetown, DC, on 13 Mar 1815.

Rutter, John - Third Lieutenant - 14th Infantry - Born: Maryland - Commissioned as an ensign on 22 Nov 1814; promoted to 3rd lieutenant on 1 Jan 1815; resigned on 20 Mar 1815.

Rutter, Richard - Private - 38th Infantry - Company: Shepperd Leakin - Age: 26 - Height: 5' 10 1/2" - Born: Harford Cty, MD - Trade: Sailor - Enlistment date: 15 Feb 1814 - Period: War.

Rutter, Thomas B. - Military Store Keeper - Military Store Keeper - Born: Maryland - Commissioned as a military store keeper in March 1813; discharged on 2 Jun 1817.

Ryan, John - Private - 19th Infantry - Other regiment: 25th Infantry - Age: 30 - Height: 5' 10" - Born: Maryland - Trade: Farmer - Enlistment date: 24 Mar 1813 - Place: Ohio - Period: 5 Yrs - Transferred to Captain Benjamin Watson's Company, 25th Infantry; discharged on 20 Mar 1815 at Sackets Harbor, NY, because of scrotal hernia.

Ryan, Zachariah - Private - 5th Infantry - Company: Richard Bell - Age: 26 - Height: 5' 10" - Born: Anne Arundel Cty, MD - Trade: Carpenter - Enlistment date: 24 Feb 1814 - Place: Plattsburgh, NY - Period: War - Discharged at Greenbush, NY, on 4 Mar 1815.

Rye, Thomas - Private - 24th Infantry - Other regiment: 7th Infantry (New) - Age: 30 - Height: 5' 8 1/2" - Born: Maryland - Trade: Farmer - Enlistment date: 12 Jun 1813 - Period: 5 Yrs - BLW 5513-160-12 - Prisoner of War (Montreal), exchanged on 28 Apr 1814; discharged on 1 Dec 1815 on Surgeon's Certificate.

Ryland, William - Private - 36th Infantry - Company: Samuel Raisin - Age: 20 - Height: 5' 3" - Born: Kent Cty, MD - Enlistment date: 5 Oct 1813 - Period: 1 Yr - Pension: Land bounty to Henry Ryland, brother & other heirs at law of William Ryland - BLW 26563-160-12 - Died in hospital on 29 Jan 1815.

Ryne, William - Private - US Volunteers - Company: Stephen Moore - Enlistment date: 8 Sep 1812 - Period: 1 Yr.

Maryland Regulars in War of 1812

S

Saddler, Samuel - Private - Light Artillery - Company: John McIntosh - Age: 25 - Height: 5' 11" - Born: Maryland - Trade: Mariner - Enlistment date: 1 Jun 1814 - Place: Providence, RI - Period: War.

Saddler, William - Private - 44th Infantry - Company: Joseph Miles - Age: 29 - Height: 5' 10 1/2" - Born: Baltimore - Trade: Mariner - Enlistment date: 11 Jan 1814 - Place: New Orleans - Period: War - Discharged on 8 Apr 1815.

Sadler Jr., Joseph - Private - US Volunteers - Company: Stephen Moore - Enlistment date: 8 Sep 1812 - Period: 1 Yr.

Sadler, Augustus - Seaman - US Sea Fencibles - Company: William Addison.

Sadler, Benjamin - Private - 36th Infantry - Company: Samuel Raisin - Age: 29 - Height: 6' 1" - Born: Prince Edwards, VA - Trade: Brick layer - Enlistment date: 9 Jul 1814 - Place: Manchester - Period: War - BLW 20963-160-12 - Discharged at Washington, DC, on 20 Mar 1815.

Sadler, Joseph - Private - 36th Infantry - Company: Thomas Carbery - Age: 19 - Height: 5' 6 1/4" - Born: Maryland - Trade: Farmer - Enlistment date: 25 Feb 1814 - Place: Leonardstown, MD - Period: War - BLW 6472-160-12 - Discharged at Baltimore on 21 Mar 1815.

Sadler, Mitchell - Private - 38th Infantry - Company: Isaac Aldridge - Age: 20 - Height: 5' 3 1/2" - Born: Mathews Cty, VA - Trade: Waterman - Enlistment date: 14 Dec 1814 - Place: Baltimore - Period: War - BLW 5497-160-12.

Sadler, Samuel - Private - 36th Infantry - Enlistment date: 12 Aug 1814 - Period: War - BLW 16679-160-12 - Discharged at Georgetown, DC, on 23 Mar 1815.

Saffell, Charles - Private - 14th Infantry - Company: Joseph Marshall - Age: 32 - Height: 5' 10" - Born: Anne Arundel Cty, MD - Trade: Laborer - Enlistment date: 31 Jul 1814 - Place: Baltimore - Period: War - BLW 3224-160-12 - Discharged at Greenbush, NY, on 3 May 1815.

Saffell, James Grimes - Private - 39th Infantry - Company: William Walker - Age: 34 - Height: 5' 6" - Born: Maryland - Enlistment date: 3 Nov 1813 - Place: Knoxville, TN - Period: 1 Yr - BLW 41651-80-55.

Sales, John B. - Seaman - US Sea Fencibles - Company: John Gill - Age: 21 - Height: 5' 8" - Born: Baltimore - Enlistment date: 14 Jan 1814 - Period: 1 Yr - BLW 405-120-55.

Salkeld, David - Private - Light Dragoons - Age: 27 - Height: 5' 6" - Born: Frederick Cty, MD - Trade: Wheelwright - Enlistment date: 28 Jan 1813 - Place: Fredericktown, MD - Period: 5 Yrs.

Salter, Daniel - Private - 5th Infantry - Company: James Dorman - Other regiment: 3rd Infantry (New) - Age: 25 - Height: 5' 9" - Born: Shepherdstown, VA - Trade: Tailor - Enlistment date: 15 Feb 1813 - Place: Baltimore - Period: 5 yrs - Discharged at Pittsburgh on 16 Aug 1815 on Surgeon's Certificate.

Saltskiver, Andrew - Private - 7th Infantry - Company: Narcissus Broutin - Age: 27 - Height: 5' 6" - Born: Maryland - Trade: Farmer - Enlistment date: 22 Apr 1813 - Period: War - Discharged at Fort St. Philip, LA, on 30 Apr 1815.

By War of 1812 Soc. in MD

Sanders, Samuel - Private - 14th Infantry - Company: David Cummings - Age: 22 - Height: 5' 7 1/2" - Born: Harford Cty, MD - Trade: Cooper - Enlistment date: 28 Nov 1814 - Place: Baltimore - Period: War - Discharged on 18 Mar 1815.

Sanders, William G. - Second Lieutenant - 14th Infantry - Age: 21 - Pension: Wife Caroline E., WO-42217, WC-32824 - BLW 22922-160-50 - Commissioned as a ensign on 12 Mar 1812; promoted to 3rd lieutenant on 13 Mar 1813; promoted to 2nd lieutenant on 13 May 1813; Prisoner of War (Dartmouth), released on 31 May 1814; resigned on 5 Aug 1814; died on 1 Sep 1845.

Sanderson, Charles - Private - 36th Infantry - Company: Samuel Raisin - Age: 19 - Height: 5' 10" - Born: St. Mary's Cty, MD - Trade: Hatter - Enlistment date: 24 Sep 1814 - Period: War - BLW 10179-160-12 - Discharged at Washington, DC, on 20 Mar 1815.

Sands, Richard Martin - Second Lieutenant - 38th Infantry - Other regiment: 4th Infantry (New) - Born: Maryland - Pension: Land bounty to Adel Sands, widow of Richard M. Sands - BLW 18225-167-50 - Commissioned as an ensign on 20 May 1813; promoted to 3rd lieutenant on 22 Apr 1814; promoted to 2nd lieutenant on 9 Jul 1814; discharged on 15 Jun 1815; died on 13 Sep 1836.

Sansbury, John - Private - 20th Infantry - Company: William Jett - Age: 40 - Height: 5' 6 1/2" - Born: Prince George's Cty, MD - Trade: Farmer - Enlistment date: 20 Apr 1812 - Period: War - BLW 10014-160-12 - Discharged at Plattsburgh, NY, on Surgeon's Certificate.

Sappington, John - Private - 42nd Infantry - Company: Jonathan Robinson - Age: 31 - Height: 5' 7 3/4" - Born: Anne Arundel Cty, MD - Trade: Carpenter - Enlistment date: 13 Apr 1814 - Period: War - Discharged at Philadelphia on 19 May 1815.

Sargant, James - Private - 5th Infantry - Age: 43 - Height: 5' 8" - Born: Baltimore - Trade: Baker - Enlistment date: 6 May 1813 - Place: Baltimore - Period: 5 Yrs - Discharged on 1 Jun 1814 for disability.

Sargeant, John - Private - 36th Infantry - Age: 23 - Height: 5' 10 1/2" - Born: Maryland - Enlistment date: 4 Jul 1813 - Place: Anne Arundel Cty, MD.

Sargent, Moses - Private - 14th Infantry - Company: William McIlvane - Other regiment: 4th Infantry (New) - Enlistment date: 21 Apr 1812 - Period: 5 Yrs - Discharged on 21 Apr 1817.

Saunders, Frederick - Musician - Light Dragoons - Other regiment: Artillery - Age: 38 - Height: 5' 3" - Born: Baltimore - Trade: Trunk maker - Discharged at Camp Flourney, GA, on 20 Sep 1814; re-enlisted at Camp Huger, GA, on 22 Jan 1815 for five year in Lt John Erwings' Company, artillery; discharged in 1818.

Saunders, John - Private - 14th Infantry - Deserter, apprehended in Myerstown.

Saunders, John - Private - 5th Infantry - Age: 34 - Height: 5' 8" - Born: Annapolis - Trade: Seaman - Enlistment date: 11 Feb 1813 - Place: Baltimore - Period: 5 Yrs.

Maryland Regulars in War of 1812

Savery, John - Second Lieutenant - 38th Infantry - Commissioned as an ensign on 20 May 1813; promoted to 3rd lieutenant on 7 Jan 1814; promoted to 2nd lieutenant on 20 May 1814; discharged on 15 Jun 1815.

Sawils, John - Private - 36th Infantry - Company: Thomas Carbery - Enlistment date: 4 May 1814 - Period: 5 Yrs - Deserted on 5 May 1814.

Saxon, Meshack - Private - 1st Rifles - Company: Daniel Appling - Age: 35 - Height: 5' 10 1/2" - Born: Maryland - Trade: Shoemaker - Enlistment date: 8 Dec 1812 - Place: Lincoln Cty, KY - Period: 18 Mos - Discharged at Sackets Harbor, NY, on 2 Jun 1814.

Saxton, Edward - Private - 26th Infantry - Company: William Bezeau - Age: 20 - Height: 5' 4" - Born: Maryland - Trade: Saddler - Enlistment date: 24 Nov 1814 - Place: Philadelphia - Period: War.

Saylor, Daniel - Private - 16th Infantry - Company: George Steele - Other regiment: 2nd Infantry (New) - Age: 25 - Height: 5' 6 1/2" - Born: Maryland - Trade: Laborer - Enlistment date: 30 Jun 1814 - Place: Halifax, NC - Period: 5 Yrs - Discharged on 30 Jun 1819.

Scaggs, Richard D. - Private - 38th Infantry - Company: John Mowton - Age: 39 - Height: 5' 10" - Born: Frederick, MD - Trade: Tailor - Enlistment date: 29 Jun 1814 - Place: Norfolk, VA - Period: War - BLW 25935-160-12 - Discharged on 15 Mar 1815.

Scandling, Jacob - Private - 38th Infantry - Company: John Rothrock - BLW 16718-160-12.

Scansling, John - Drummer - 5th Infantry - Company: William Henshaw - Other regiment: 3rd Infantry (New) - Age: 18 - Height: 5' - Born: Baltimore - Trade: Coppersmith - Enlistment date: 4 May 1813 - Place: Baltimore - Period: 5 Yrs - Discharged on 4 May 1818.

Schamer, John - Private - 36th Infantry - Age: 19 - Height: 5' 8" - Born: Kent, VA - Enlistment date: 17 Apr 1813 - Place: Westminster, MD.

Schee, Augustus M. - Second Lieutenant - 14th Infantry - Company: Henry Grindage - Pension: Land bounty to Frances L. McDowell, formerly widow of Augustus M. Schee - BLW 94521-160-55 - Commissioned as a 2nd lieutenant on 12 Mar 1812; resigned on 10 Jan 1813.

Scheiltz, John - Private - 14th Infantry - Prisoner of War (Halifax), discharged from Halifax on 3 Feb 1814.

Schmuck, Jacob - Captain - 2nd Artillery - Other regiment: Corps of Artillery - Born: Pennsylvania - BLW 11738-160-12 - Served as a sergeant in 2nd Artillery; commissioned as a 3rd lieutenant on 10 Feb 1814; promoted to 2nd lieutenant on 1 May 1814; transferred to Corps of Artillery on 12 May 1814; promoted to 1st lieutenant on 20 Apr 1818; transferred to 4th Artillery on 1 Jun 1821; promoted to captain on 11 Apr 1825; died on 10 Apr 1835.

Schroyer, Conrad - Private - 14th Infantry - Company: Reuben Gilder - Age: 21 - Height: 5' 8" - Born: Maryland - Enlistment date: 21 Mar 1813 - Period: 18 Mos - Discharged at Black Rock, NY, on 21 Sep 1814.

Scott, Edward H. - Second Lieutenant - 36th Infantry - Commissioned as a 3rd lieutenant on 30 Apr 1813; promoted to 2nd lieutenant on 10 Aug 1814; resigned on 13 Feb 1815.

By War of 1812 Soc. in MD

Scott, James - Private - 14th Infantry - Company: Kenneth McKenzie - Age: 43 - Height: 5' 7" - Born: Connecticut - Trade: School master - Enlistment date: 6 Jan 1813 - Place: Baltimore - Period: 18 Mos - Prisoner of War (Halifax), sent to England; leg shot off by cannon ball in the attack on Fort George; discharged at Washington, DC, on 13 Jun 1815 on Surgeon's Certificate.

Scott, John - Private - 14th Infantry - Company: Robert Kent - Other regiment: Ordnance Department - Age: 32 - Height: 5' 10 1/2" - Born: Baltimore - Trade: Wheelwright - Enlistment date: 24 Jul 1812 - Place: Baltimore - Period: 5 Yrs - BLW 10166-160-12 - Prisoner of War (Halifax), sent to England, captured at Beaver Dams on 24 Jun 1813; transferred to Ordnance Department as an artificer; discharged at Richmond, VA, on 24 Jul 1817.

Scott, John - Private - 38th Infantry - Company: Charles Stansbury - BLW 24384-160-12.

Scott, John - Private - 36th Infantry - Company: Mortimer Hall - Age: 36 - Height: 5' 7" - Born: Salkers, Scotland - Enlistment date: 10 May 1813 - Place: Havre de Grace, MD - Period: 1 Yr.

Scott, John - Private - 38th Infantry - Company: Charles Stansbury - Age: 22 - Height: 5' 10" - Born: Chester, PA - Trade: Farmer - Enlistment date: 22 Nov 1814 - Place: Fredericktown, MD - Period: War - Discharged at Baltimore on 6 Apr 1815.

Scott, Joseph - Seaman - US Sea Fencibles - Company: William Addison.

Scott, Matthew - Private - 14th Infantry - Company: Thomas Montgomery - Other regiment: 4th Infantry (New) - Age: 35 - Height: 5' 6" - Born: Glasgow, Scotland - Trade: Joiner - Enlistment date: 14 May 1813 - Place: Baltimore - Period: 5 Yrs - Discharged on 14 May 1818.

Scott, Richard - Seaman - US Sea Fencibles - Company: John Gill - Age: 27 - Height: 5' 9" - Born: Baltimore - Enlistment date: 4 Jan 1814 - Period: 1 Yr.

Scott, Robert J. - Second Lieutenant - Corps of Artillery - Born: Virginia - Graduated from the US Military Academy on 2 Mar 1815; commissioned as a 3rd lieutenant on 2 Mar 1815; promoted to 2nd lieutenant on 15 Jun 1817; resigned on 4 Nov 1818; died in May 1834.

Scott, William - Private - 14th Infantry - Company: Thomas Kearney - Enlistment date: 25 Jul 1812 - Period: 5 Yrs - Pension: Old War IF-3861 - Discharged on 11 Jun 1814 for wounds received in service.

Scott, William - Lieutenant Colonel - 36th Infantry - Place: Maryland - Served as a officer in the U.S. Army between 1 May 1795 and 1 Jun 1802; commissioned as a lieutenant colonel, 36th Infantry, on 25 Mar 1813; discharged on 15 Jun 1815.

Scott, William - Private - 38th Infantry - Company: Thomas Sangster - Age: 19 - Height: 5' 6" - Born: Pennsylvania - Trade: Cooper - Enlistment date: 10 Nov 1814 - Place: Baltimore - Period: 5 Yrs.

Scott, William - Private - 14th Infantry - Company: David Cummings - Age: 22 - Height: 5' 8" - Born: England - Trade: Laborer - Enlistment date: 18 Nov 1814 - Place: Salisbury, MD - Period: War - Discharged on 18 Mar 1815.

Maryland Regulars in War of 1812

Scott, William - Private - 14th Infantry - Company: Reuben Gilder - Age: 26 - Height: 6' - Born: Ireland - Trade: Laborer - Enlistment date: 27 Feb 1813 - Period: 5 Yrs - Deserted at Plattsburgh, NY, on 18 Mar 1815.

Scott, William - Private - 14th Infantry - Company: Kenneth McKenzie - Enlistment date: 2 Jan 1813 - Period: 18 Mos - Prisoner of War (Halifax), discharged from Halifax on 3 Feb 1814, exchanged on 11 May 1814; discharged at Greenbush, NY, on 14 Jul 1814.

Scourci, John - Private - 36th Infantry - Age: 27 - Height: 5' 7" - Born: Boston - Enlistment date: 10 Jul 1814 - Place: Norfolk, VA.

Scracklin, Lewis - Seaman - US Sea Fencibles - Company: Simmones Bunbury.

Scriven, Richard - Private - 14th Infantry - Prisoner of War (Halifax), captured at Beaver Dams on 24 Jun 1813, discharged from Halifax on 9 Nov 1813.

Scrivener, Levi - Private - 44th Infantry - Company: Anatole Peychand - Age: 27 - Height: 5' 11 1/2" - Born: Montgomery Cty, MD - Trade: Millwright - Enlistment date: 16 May 1814 - Place: Powder Magazine Barracks, LA - Period: War - BLW 22260-160-12 - Discharged at New Orleans on 8 Apr 1815.

Seabreeze, William - Private - 14th Infantry - Company: Thomas Montgomery - Other regiment: 4th Infantry (New) - Age: 41 - Height: 5' 11" - Born: Somerset Cty, MD - Trade: Seaman - Enlistment date: 1 May 1812 - Place: Baltimore - Period: 5 Yrs - Discharged on 1 May 1817.

Seals, Francis - Private - 17th Infantry - Company: Harris Hickman - Other regiment: 3rd Infantry (New) - Age: 33 - Height: 5' 8 1/2" - Born: St. Mary's Cty, MD - Trade: Laborer - Enlistment date: 11 Oct 1813 - Place: Limestone, KY - Period: 5 Yrs - Discharged on 8 Mar 1816 after furnishing a substitute.

Sears, Abraham - Private - 14th Infantry - Prisoner of War (Halifax), captured at Beaver Dams on 24 Jun 1813, discharged from Halifax on 9 Nov 1813.

Sears, John - Private - 14th Infantry - Company: Samuel Lane - Age: 38 - Height: 5' 7" - Born: Frederick, MD - Enlistment date: 19 May 1812 - Place: Westminster, MD - Period: 18 Mos - Prisoner of War (Halifax), discharged from Halifax on 3 Feb 1814; discharged at Greenbush, NY, on 23 Jun 1814.

Sears, William - Private - 2nd Light Dragoons - Company: George Haig - Other regiment: Corps of Artillery - Age: 25 - Height: 5' 9" - Born: Cecil Cty, MD - Trade: Silversmith - Enlistment date: 9 May 1812 - Place: Baltimore - Period: 5 Yrs - Transferred to Corps of Artillery; discharged on 9 May 1817.

Sebraat, Hendrick - Sergeant - 38th Infantry - Company: James Hook - Age: 36 - Height: 5' 9" - Born: Rotterdam, Holland - Trade: Gardener - Enlistment date: 14 Apr 1814 - Place: Baltimore - Period: War - Discharged at Fort Covington, MD, on 31 Mar 1815.

Sefer, Joseph - Private - 36th Infantry - Age: 40 - Height: 5' 11" - Born: Frederick, MD - Enlistment date: 18 Apr 1813 - Place: Baltimore.

Segars, George - Private - 36th Infantry - Company: Joseph Hook - Age: 21 - Height: 5' 6" - Born: New York City - Enlistment date: 7 May 1813 - Place: Baltimore - Period: 1 Yr - Deserted on 10 Jul 1814.

By War of 1812 Soc. in MD

Segdick, John - Private - 7th Infantry - Company: James Doherty - Height: 5' 9" - Born: Maryland - Trade: Hatter - Enlistment date: 14 Sep 1808 - Period: 5 Yrs - Discharged at New Orleans on 8 Apr 1815.

Seidenstricker, Daniel - Private - 38th Infantry - Company: John Rothrock - Pension: Land bounty to Thomas Seidenstricker, son and only heir at law of Daniel Seidenstricker - BLW 27591-160-42.

Selbey, James - Private - 14th Infantry - Company: Samuel Lane - Age: 24 - Height: 6' 3/4" - Born: Maryland - Enlistment date: 20 Jun 1812 - Place: Westminster, MD - Period: 18 Mos.

Selby, Hezekiah - Private - 14th Infantry - Age: 31 - Height: 6' - Born: Maryland - Trade: Farmer - Enlistment date: 15 Apr 1812 - Period: 5 Yrs - BLW 6733-160-12 - Discharged at Washington, DC, on 14 Apr 1814, Surgeon's Certificate.

Selby, James - 14th Infantry - Company: Thomas Montgomery - Pension: Wife Ellse, WO-5789.

Selby, Micajah - Private - US Volunteers - Company: Stephen Moore - Enlistment date: 8 Sep 1812 - Period: 1 Yr.

Selman, James - Private - 35th Infantry - Company: Walter Cocke - Age: 28 - Height: 5' 5 1/4" - Born: Baltimore - Trade: Cooper - Enlistment date: 25 Feb 1814 - Place: Fort Nelson, VA - Period: War - Discharged at Norfolk, VA, on 20 Mar 1815.

Selrat, Henry Augusta - Sergeant - 38th Infantry - Company: James Hook - Pension: Land bounty to Henry Philip Selrat, son and only heir at law of Henry Augusta Selrat - BLW 25864-160-12.

Seltzer, Adam - Private - 2nd Light Dragoons - Company: John Burd - Age: 20 - Height: 5' 11" - Born: Hagerstown, MD - Trade: Shoemaker - Enlistment date: 4 Aug 1812 - Period: 18 Mos.

Senior, Jacob - Private - 2nd Rifles - Company: Batteal Harrison - Age: 21 - Height: 5' 6" - Born: Frederick, MD - Trade: Hatter - Enlistment date: 24 Aug 1814 - Place: Lancaster, OH - Period: War - Discharged at Detroit on 30 Jun 1815.

Senney, Solomon - Private - 5th Infantry - Company: John Taylor - Other regiment: 3rd Infantry (New) - Age: 29 - Height: 5' 10 1/2" - Born: Queenstown, MD - Trade: Blacksmith - Enlistment date: 22 Aug 1814 - Place: York, PA - Period: 5 Yrs - Re-enlisted in 1819.

Sentz, John - Private - 5th Infantry - Company: James Dorman - Other regiment: 3rd Infantry (New) - Age: 24 - Height: 5' 5" - Born: Dauphin Cty, PA - Trade: Laborer - Enlistment date: 8 Mar 1814 - Place: Baltimore - Period: 5 Yrs.

Sepechy, Michael - Corporal - 38th Infantry - Company: Shepperd Leakin - Age: 40 - Height: 5' 7" - Born: Hungary - Trade: Soldier - Enlistment date: 8 Apr 1814 - Place: Baltimore - Period: War - Discharged at Baltimore on 30 Mar 1815.

Serts, John P. - Private - 5th Infantry - Age: 22 - Height: 5' 5" - Born: France - Trade: Gardener - Enlistment date: 21 Apr 1813 - Place: Baltimore - Period: 5 Yrs.

Maryland Regulars in War of 1812

Sevier, Francis V. - Private - 14th Infantry - Company: Reuben Gilder - Other regiment: 4th Infantry (New) - Age: 22 - Height: 5' 7 3/4" - Born: Maryland - Trade: Tobacconist - Enlistment date: 7 Apr 1813 - Period: 5 Yrs - Discharged on 8 Apr 1818.

Sewall, Henry - Private - 36th Infantry - Company: Thomas Carbery - Enlistment date: 29 Nov 1813 - Period: 1 Yr - Discharged on 28 Nov 1814.

Sewell, Clement - First Lieutenant - 36th Infantry - Company: Thomas Carbery - Commissioned as a 2nd lieutenant on 30 Apr 1813; promoted to 1st lieutenant on 1 May 1814; dropped on 30 Sep 1814.

Sewell, James - Private - 14th Infantry - Company: Thomas Montgomery - Enlistment date: 20 May 1812 - Period: 5 Yrs.

Shackles, John A. - Private - 2nd Infantry - Company: John Brahan - Age: 23 - Height: 5' 7" - Born: Prince George's Cty, MD - Trade: Carpenter - Enlistment date: 12 Mar 1808 - Place: Charlottesville, VA - Period: 5 Yrs - Re-enlisted at Mount Vernon on 12 Dec 1812 for five years; discharged on 12 Dec 1817.

Shackly, Elijah - Musician - 38th Infantry - Company: John Brookes - BLW 18656-160-12.

Shade, William G. - Second Lieutenant - 14th Infantry - Born: Maryland - Commissioned as an ensign on 18 Mar 1813; promoted to 3rd lieutenant on 1 Oct 1813; promoted to 2nd lieutenant on 31 Aug 1814; discharged on 15 Jun 1815.

Shaeffer, Benjamin - Private - 12th Infantry - Company: Thomas Sangster - Age: 36 - Height: 5' 8" - Born: Baltimore - Trade: Blacksmith - Enlistment date: 20 Jul 1814 - Place: Morgantown, VA - Period: War - BLW 18048-160-12 - Discharged at Fort Covington, MD, on 30 May 1815.

Shaffer, Henry - Fifer - 14th Infantry - Company: Richard Arell - Other regiment: 4th Infantry (New) - Age: 24 - Height: 5' 8 1/2" - Born: Baltimore - Trade: Shoemaker - Enlistment date: 23 Apr 1812 - Place: Baltimore - Period: 5 Yrs - Deserted on 7 Aug 1815.

Shaffer, Henry - Private - 2nd Artillery - Company: Frederick Evans - Age: 14 - Height: 4' 7" - Born: Holland - Enlistment date: 7 May 1813 - Place: Annapolis - Period: 5 Yrs - Discharged on 6 May 1818.

Shaffer, Valentine - Private - 14th Infantry - Company: Richard Arell - Other regiment: 4th Infantry (New) - Age: 43 - Height: 5' 7" - Born: Germany - Trade: Carpenter - Enlistment date: 12 Dec 1812 - Place: Lancaster, PA - Period: 5 Yrs - Prisoner of War, exchanged on 15 Apr 1814; died in hospital on 15 Sep 1817.

Shamar, John - Private - 36th Infantry - Company: Joseph Hook - Pension: No pension claim - BLW 174-160-50.

Shane, Arthur - Sergeant - Light Artillery - Company: Luther Leonard - Age: 23 - Height: 5' 7" - Born: Harford Cty, MD - Trade: Boot maker - Enlistment date: 20 Dec 1811 - Place: Philadelphia - Period: 5 Yrs - Discharged at Fort McHenry on 30 Jun 1816 after furnishing a substitute.

Shane, George - Private - 12th Infantry - Company: James Paxton - Other regiment: 8th Infantry (New) - Age: 21 - Height: 5' 8" - Born: Funkstown,

By War of 1812 Soc. in MD

MD - Trade: Laborer - Enlistment date: 10 Oct 1813 - Place: Cumberland Cty, VA - Period: 5 Yrs - Deserted at Harmony, PA, on 11 Aug 1815.

Shannon, Samuel - Third Lieutenant - 14th Infantry - Company: Samuel Shannon - Other regiment: ex 36th Infantry - Age: 20 - Height: 5' 7" - Born: Pittsburgh - Trade: Shoemaker - Enlistment date: 19 Mar 1814 - Place: Washington, DC - Period: War - Pension: Wife Mary, Old War IF-12958 - BLW 6578-160-12 - Served as a sergeant in the 36th Infantry; commissioned as an ensign, 14th Infantry, on 27 Jul 1814; promoted to 3rd lieutenant on 5 Aug 1814; discharged on 15 Jun 1815; died on 4 Sep 1836.

Sharp, Francis H. - Private - 36th Infantry - Company: Samuel Raisin - Age: 34 - Height: 5' 11" - Born: Norfolk, VA - Trade: Ship's carpenter - Enlistment date: 18 Jun 1814 - Period: War - Pension: Land bounty to William Sharp, brother and other heirs at law of Francis H. Sharp - BLW 25185-160-12 - Died in hospital on 31 Jan 1815.

Sharp, William - Private - 5th Infantry - Other regiment: 3rd Infantry (New) - Age: 28 - Height: 5' 7 1/2" - Born: Baltimore - Trade: Carpenter - Enlistment date: 17 Apr 1813 - Period: 5 Yrs - BLW 17860-160-12.

Sharpe, John H. - Private - 24th Infantry - Company: William Allen - Age: 27 - Height: 6' - Born: Maryland - Trade: Blacksmith - Enlistment date: 22 Jul 1812 - Place: St. Genievere, MO - Period: 5 Yrs - BLW 13872-160-12 - Discharged at Newport, KY, on 21 Jul 1817.

Sharpless, William - Private - Ordnance Department - Company: Edwin Tyler - Age: 22 - Height: 5' 7" - Born: Talbot Cty, MD - Trade: Laborer - Enlistment date: 27 Dec 1813 - Place: Greenleaf Point, DC - Period: 5 Yrs - Discharged at Greenleaf Point, DC, on 27 Dec 1818.

Shartle, Henry - Seaman - US Sea Fencibles - Company: William Addison.

Shaver, George - Private - 14th Infantry - Age: 30 - Born: Palatine Cty, NY - Prisoner of War (Quebec), died at Quebec on 25 Jul 1813.

Shaw, John - Musician - 5th Infantry - Company: Richard Bell - Age: 14 - Height: 4' 5" - Born: Baltimore - Enlistment date: 11 Jul 1813 - Place: Baltimore - Period: 5 Yrs - Discharged on 11 Jul 1818.

Shaw, Luke L. - Private - 36th Infantry - Company: Samuel Raisin - Age: 19 - Height: 5' 8" - Born: Charles Cty, MD - Trade: Farmer - Enlistment date: 24 Oct 1814 - Period: War - Pension: Wife Lucinda Elizabeth, SO-890, SC-13107, WO-26303, WC-24519 - Discharged at Washington, DC, on 20 Mar 1815.

Shaw, Thomas - Private - 5th Infantry - Company: James Dorman - Other regiment: 3rd Infantry (New) - Age: 24 - Height: 6' 2 1/2" - Born: Baltimore - Trade: Wheelwright - Enlistment date: 23 Jun 1812 - Place: Baltimore - Period: 5 Yrs - Discharged on 25 Jun 1815 on Surgeon's Certificate for paralysis of the bladder.

Shay, Patrick - Private - 36th Infantry - Company: Samuel Raisin - Age: 25 - Height: 5' 7" - Born: Ireland - Trade: Laborer - Enlistment date: 12 Oct 1814 - Period: War.

Shayack, Samuel - Seaman - US Sea Fencibles - Company: John Gill - Age: 25 - Height: 6' 2 1/2" - Born: Baltimore - Enlistment date: 3 Jan 1814 -

Maryland Regulars in War of 1812

Period: 1 Yr.

Sheaves, Thomas - Private - 14th Infantry - Prisoner of War (Halifax), discharged from Halifax on 3 Feb 1814, exchanged on 15 Apr 1814.

Sheckles, Thomas C. - Private - 14th Infantry - Company: Thomas Montgomery - Other regiment: 4th Infantry (New) - Age: 35 - Height: 5' 7 1/2" - Born: Queen Anne's Cty, MD - Trade: Laborer - Enlistment date: 11 Apr 1812 - Place: Baltimore - Period: 5 Yrs - Deserted on 5 Aug 1815.

Sheffield, Henry - Private - 14th Infantry - Enlistment date: 23 Apr 1812 - Period: 5 Yrs - Prisoner of War (Halifax), exchanged on 15 Apr 1814.

Sheffield, Richard - Private - 42nd Infantry - Company: Edmund Duval - Age: 27 - Height: 5' 6" - Born: Baltimore - Trade: Mariner - Enlistment date: 21 Feb 1814 - Place: Bladensburg, MD - Period: War - Discharged on 20 Apr 1815.

Shehey, Michael - Seaman - US Sea Fencibles - Company: William Addison.

Shelby, Thomas - Private - 36th Infantry - Company: Joseph Hook - Age: 30 - Height: 5' 9 1/2" - Born: Prince George's Cty, MD - Trade: Farmer - Enlistment date: 1 Jun 1814 - Place: Georgetown, DC - Period: War - BLW 140-160-12 - Discharged at Fort Covington, MD, on 30 Mar 1815.

Shellinberg, George - Private - 2nd Artillery - Company: Frederick Evans - Age: 15 - Height: 4' 9" - Born: Baltimore - Trade: Yeoman - Enlistment date: 3 Apr 1814 - Place: Baltimore - Period: War - Discharged on 24 Mar 1815.

Shenabrough, Jacob - Private - 38th Infantry - Company: Shepperd Leakin - BLW 5093-160-12.

Shennebrook, Jacob - Private - 38th Infantry - Company: Shepperd Leakin - BLW 15148-160-12.

Shepard, Thomas - Private - 16th Infantry - Company: John Baldy - Age: 28 - Height: 5' 8" - Born: Maryland - Trade: Carpenter - Enlistment date: 20 May 1814 - Place: Philadelphia - Period: 5 Yrs - BLW 13499-160-12 - Discharged at Washington, DC, on 1 Jun 1815.

Shepherd, Charles - Private - 14th Infantry - Company: David Cummings - Age: 25 - Height: 5' 6 1/2" - Born: Anne Arundel Cty, MD - Trade: Shoemaker - Enlistment date: 22 Oct 1814 - Place: Baltimore - Period: War - BLW 11589-160-12 - Discharged on 18 Mar 1815.

Shepherd, John - Private - 36th Infantry - Company: Thomas Carbery - Enlistment date: 1 Dec 1813 - Period: 1 Yr - Discharged on 30 Nov 1814.

Sheppard, William H. - Sergeant - 16th Infantry - Company: Nathaniel McLaughlin - Age: 23 - Height: 5' 6" - Born: Baltimore - Trade: Printer - Enlistment date: 12 Oct 1813 - Place: Norristown, PA - Period: 5 Yrs.

Shepperd, James - Private - 14th Infantry - Company: Joseph Marshall - Age: 29 - Height: 5' 4" - Born: Anne Arundel Cty, MD - Enlistment date: 18 Jul 1814 - Place: Baltimore - Period: War - Died at Greenbush, NY, on 22 Jan 1815 from diarrhea.

Shepperd, John - Private - 14th Infantry - Company: David Cummings - Age: 32 - Height: 5' 6" - Born: Anne Arundel Cty, MD - Trade: Carpenter - Enlistment date: 9 Dec 1814 - Place: Baltimore - Period: War - BLW 2342-60-12 - Discharged at Baltimore on 16 Mar 1815.

By War of 1812 Soc. in MD

Sheridan, Abner - Private - 1st Light Dragoons - Company: Alexander Cummings - Age: 22 - Height: 5' 6" - Born: Baltimore - Trade: Farmer - Enlistment date: 23 Aug 1812 - Place: Lancaster, PA - Period: 5 Yrs - Discharged on 23 Aug 1817.

Sherington, Ezekiel - Private - 14th Infantry - Prisoner of War (Halifax), sent to England.

Sherman, Benjamin - Private - Light Dragoons - Age: 27 - Height: 5' 7" - Born: Massachusetts - Trade: Laborer - Enlistment date: 20 Nov 1812 - Place: Fredericktown, MD - Period: 5 Yrs.

Sherman, James - Private - 26th Infantry - Company: William Bezeau - Age: 21 - Height: 5' 8" - Born: Chester, MD - Enlistment date: 26 Oct 1814 - Place: Baltimore - Deserted on 30 Oct 1814.

Sherman, Lewis - Musician - US Volunteers - Company: Stephen Moore - Enlistment date: 8 Sep 1812 - Period: 1 Yr.

Sherman, Lewis J. - Seaman - US Sea Fencibles - Company: Simmones Bunbury - Age: 20 - Height: 5' 3" - Born: Baltimore - Enlistment date: 15 Dec 1813 - Period: 1 Yr - Discharged on 16 Jan 1815.

Shermentine, Thomas - Private - 36th Infantry - Company: Thomas Carbery - Age: 20 - Height: 5' 6" - Born: St. Mary's Cty, MD - Trade: Pilot - Enlistment date: 25 Feb 1814 - Place: Donaldtown - Period: War - BLW 3237-160-12 - Discharged at Baltimore on 31 Mar 1815.

Sherrard, Thomas - Private - 14th Infantry - Company: Reuben Gilder - Age: 55 - Height: 5' 7" - Born: Ireland - Enlistment date: 26 Jun 1813 - Period: 18 Mos - Discharged at Black Rock, NY, on 26 Dec 1814.

Shewmaker, Jacob - Private - 38th Infantry - Company: John Buck - BLW 10522-160-12.

Shields, William - Private - 14th Infantry - Company: David Cummings - Age: 24 - Height: 5' 10" - Born: Cecil Cty, MD - Trade: Farmer - Enlistment date: 2 Sep 1814 - Place: Baltimore - Period: 5 Yrs - Deserted on 10 Sep 1814.

Shilling, John - Private - 36th Infantry - Age: 33 - Height: 5' 7" - Born: Baltimore - Enlistment date: 24 Apr 1813 - Place: Baltimore.

Shilling, Nicholas - Private - 36th Infantry - Age: 25 - Height: 5' 7" - Born: Maryland - Enlistment date: 10 Jul 1814 - Place: Norfolk, VA.

Shilling, Philip - Private - 5th Infantry - Company: Richard Whartenby - Other regiment: 3rd Infantry (New) - Age: 33 - Height: 5' 7 1/2" - Born: Baltimore - Trade: Plasterer - Enlistment date: 30 Jul 1812 - Place: Baltimore - Period: 5 Yrs - Discharged on 25 Jan 1821.

Shillings, William - Private - 16th Infantry - Company: John Machesney - Age: 20 - Height: 5' 8" - Born: Baltimore - Enlistment date: 25 Nov 1814 - Place: Philadelphia - Period: 5 Yrs.

Shillingsbag, Christopher - Private - 36th Infantry - Age: 36 - Height: 5' 6" - Born: Baltimore - Enlistment date: 31 May 1813 - Place: Baltimore.

Shillingsburgh, George - Private - 36th Infantry - Age: 15 - Height: 5' - Born: Baltimore - Enlistment date: 6 Jan 1813 - Place: Baltimore - Period: 1 Yr.

Shinnick, Jacob - Quartermaster Sergeant - 14th Infantry - Pension: Placed on the pension rolls on 28 Mar 1834.

Maryland Regulars in War of 1812

Shipley, Elias - Private - 2nd Light Dragoons - Company: William Littlejohn - Age: 24 - Height: 5' 10" - Born: Maryland - Enlistment date: 25 Dec 1813 - Place: Frederick, MD - Period: 5 Yrs - Deserted at Baltimore.

Shipley, Peter - Private - 19th Infantry - Company: Carey Trimble - Age: 33 - Height: 5' 9" - Born: Maryland - Trade: Laborer - Enlistment date: 8 Jun 1814 - Place: Portsmouth, OH - Period: War - Pension: Old War IF-1369 - BLW 6559-160-12 - Discharged at Chillicothe, OH, on 6 Jun 1815.

Shipley, Samuel - Sergeant - 14th Infantry - Company: James McDonald - Other regiment: 4th Infantry (New) - Age: 21 - Height: 6' 1/2" - Born: Baltimore - Trade: Laborer - Enlistment date: 17 Aug 1813 - Period: 5 Yrs - Discharged on 17 Aug 1818.

Shipper, David - Seaman - US Sea Fencibles - Company: John Gill - Age: 20 - Height: 5' 6" - Born: Baltimore City - Enlistment date: 5 Jan 1814 - Period: 1 Yr.

Shivers, Thomas - Private - 14th Infantry - Company: Isaac Barnard - Enlistment date: 23 Oct 1812 - Period: 18 Mos - Prisoner of War (Halifax).

Shivers, Thomas - Private - Light Artillery - Company: Thomas Ketcham - Age: 21 - Height: 5' 5" - Born: Baltimore - Trade: Rope maker - Enlistment date: 2 May 1814 - Place: Boston - Period: War - Discharged at Plattsburgh, NY, on 6 Jun 1815.

Shoane, John - Private - 38th Infantry - Age: 22 - Height: 5' 9" - Born: Germany - Trade: Cooper - Enlistment date: 28 Dec 1814 - Place: Frederick, MD - Period: War - Discharged at Baltimore on 6 Apr 1815.

Shockley, David - Private - 36th Infantry - Company: Samuel Raisin - Age: 27 - Height: 5' 6" - Born: Dunenbury, VA - Trade: Carpenter - Enlistment date: 11 Jul 1814 - Period: 5 Yrs - Died in barracks on 24 Feb 1815.

Shockley, Elijah - Fifer - 38th Infantry - Company: John Brookes - Age: 18 - Height: 5' 5" - Born: Somerset Cty, MD - Trade: Miller - Enlistment date: 22 Jul 1813 - Period: 1 Yr - Re-enlisted at Norfolk, VA, on 2 Mar 1814 for the war; discharged at Craney Island, VA, on 15 Mar 1815.

Shockley, James - Private - 14th Infantry - Company: Clement Sullivan - Enlistment date: 26 Nov 1812 - Period: 5 Yrs - Pension: Land bounty to Sarah Coale, sister and other heirs at law of James Shockley - BLW 25389-160-12 - Died on 20 Nov 1812.

Shockley, Perkins - Private - 14th Infantry - Company: Clement Sullivan - Enlistment date: 25 May 1812 - Period: 5 Yrs - Pension: Land bounty to Sarah Coale, sister and other heirs at law of Perkins Shockley - BLW 25388-160-12 - Died on 25 Dec 1812.

Shoemaker, Jacob - Private - 38th Infantry - Company: John Buck - Age: 34 - Height: 5' 8" - Born: Delaware - Trade: Tailor - Enlistment date: 4 Jun 1814 - Place: Baltimore - Period: War - Discharged at Baltimore on 30 Mar 1815.

Shoke, Jacob - Private - 38th Infantry - Company: Charles Stansbury - Age: 35 - Height: 5' 7" - Born: Hamburg, Germany - Trade: Cooper - Enlistment date: 16 Dec 1814 - Place: Fredericktown, MD - Period: War - BLW 15217-160-12 - Discharged at Baltimore on 6 Apr 1815.

Shone, Abraham - Private - 38th Infantry - Company: James Hook - BLW

By War of 1812 Soc. in MD

24627-160-12.

Shoop, Martin - Private - 2nd Artillery - Company: Nathan Towson - Pension: Old War IF-15290 - BLW 8139-160-12.

Shorben, John - Seaman - US Sea Fencibles - Company: John Gill - Age: 26 - Height: 5' 10" - Born: Baltimore - Enlistment date: 3 Jan 1814 - Period: 1 Yr.

Shough, Jacob - Private - 32nd Infantry - Company: Samuel Borden - Age: 31 - Height: 5' 7" - Born: Washington Cty, MD - Trade: Gunsmith - Enlistment date: 4 Apr 1814 - Place: Philadelphia - Period: War - Discharged on 11 May 1815.

Shrader, John - Private - 17th Infantry - Company: Harris Hickman - Age: 40 - Height: 5' 9" - Born: Frederick Cty, MD - Trade: Farmer - Enlistment date: 27 Oct 1813 - Place: Hamilton, Butler Cty, OH - Period: War - BLW 3139-160-12 - Discharged at Chillicothe, OH, on 9 Jun 1815.

Shriver, James - Private - 19th Infantry - Age: 22 - Height: 5' 7" - Born: Georgetown, MD - Enlistment date: 25 Jul 1812 - Place: Zanesville, OH - Period: 5 Yrs.

Shroyer, Andrew - Private - 36th Infantry - Company: Samuel Raisin - Age: 23 - Height: 5' 10" - Born: Maryland - Trade: Blacksmith - Enlistment date: 29 Sep 1814 - Place: Manchester - Period: War - BLW 12461-160-12 - Discharged at Washington, Dc, on 20 Mar 1815.

Shutes, Thomas - Private - 14th Infantry - Company: Reuben Gilder - Age: 25 - Height: 5' 9" - Born: New York - Trade: Baker - Enlistment date: 15 Jan 1813 - Period: 5 Yrs.

Sickless, Michael - Private - 14th Infantry - Prisoner of War (Halifax), captured at Beaver Dams on 24 Jun 1813, discharged from Halifax on 3 Feb 1813.

Sidney, Edward - Sergeant - 29th Infantry - Company: Matthew Danvers - Age: 21 - Height: 5' 7" - Born: Fredericktown, MD - Trade: Sailor - Enlistment date: 17 Nov 1814 - Place: New York - Period: War - Discharged at Plattsburgh, NY, on 29 Jun 1815.

Silence, Nicholas - Private - 14th Infantry - Company: Isaac Barnard - Enlistment date: 5 Jan 1813 - Period: 5 Yrs - Prisoner of War (Halifax); deserted on 19 Jun 1814.

Silence, Nicholas - Private - Artillery - Age: 21 - Height: 5' 3 1/2" - Born: Baltimore - Enlistment date: 23 Apr 1814 - Place: Boston - Deserted.

Silence, Nicholas - Private - 14th Infantry - Other regiment: 4th Infantry (New) - Age: 18 - Height: 5' 2 1/2" - Born: Baltimore City - Trade: Bookbinder - Enlistment date: 5 Jun 1815 - Place: Baltimore - Period: 5 Yrs - Deserted at Fort Severn, MD, on 14 Jul 1815.

Siller Jr., Jacob - Private - 38th Infantry - Company: Shepperd Leakin - Age: 31 - Height: 5' 6" - Born: York, PA - Trade: Hatter - Enlistment date: 6 Jul 1814 - Place: Baltimore - Period: War - Discharged at Fort McHenry on 28 Mar 1815.

Simkins, William - Private - 36th Infantry - Age: 23 - Height: 5' 6 1/2" - Born: Cumberland, MD - Trade: Laborer - Enlistment date: 4 Jul 1812 - Place:

Maryland Regulars in War of 1812

Cumberland, MD - Re-enlisted at Annapolis on 21 Apr 1814 for the war.

Simmons, David - Private - 17th Infantry - Company: Caleb Holder - Other regiment: 3rd Infantry (New) - Age: 26 - Height: 5' 9" - Born: Maryland - Trade: Carpenter - Enlistment date: 18 Jul 1814 - Place: Maysville, KY - Period: 5 Yrs - Discharged on 18 Jul 1819.

Simmons, Ira - Private - 19th Infantry - Company: Carey Trimble - Other regiment: 17th Infantry - Age: 30 - Height: 5' 9" - Born: Maryland - Trade: Blacksmith - Enlistment date: 7 May 1814 - Place: Adelphi, Ross Cty, OH - Period: War - Pension: Wife Minta, WO-7504 - BLW 23562-160-12 - Discharged at Chillicothe, OH, on 6 Jun 1815.

Simmons, James - Private - 1st Artillery - Company: George Armistead - Pension: Wife Mary W. - BLW 55148-40-50, 55149-40-50, 5534-80-55.

Simmons, Levin B. - Corporal - 38th Infantry - Company: Isaac Aldridge - Age: 21 - Height: 5' 1" - Born: Dorchester Cty, MD - Trade: Clerk - Enlistment date: 23 Nov 1814 - Place: Baltimore - Period: War - BLW 9925-160-12 - Discharged at Baltimore on 6 Apr 1815.

Simmons, Simon - Private - 36th Infantry - Other regiment: ex 38th Infantry - Age: 36 - Height: 5' 5" - Born: Germany - Enlistment date: 19 Jun 1813 - Place: Baltimore - Period: 1 Yr - Transferred from Captain James Haslett's Company, 38th Infantry, to the U.S. Navy on 26 Oct 1813.

Simms, John - Private - 14th Infantry - Company: David Cummings - Age: 24 - Height: 5' 11" - Born: Frederick, MD - Trade: Farmer - Enlistment date: 29 Oct 1814 - Place: Baltimore - Period: War - Pension: No pension claim - BLW 2672-160-12 - Discharged on 18 Mar 1815.

Simms, Robert - Private - 38th Infantry - Company: James Smith - BLW 24480-160-12.

Simon, John - Private - 14th Infantry - Company: Samuel Lane - Deserted at Hagerstown on 22 May 1812.

Simonds, Daniel - Private - 17th Infantry - Company: David Holt - Age: 22 - Height: 5' 8 3/4" - Born: Montgomery Cty, MD - Trade: Laborer - Enlistment date: 9 Jun 1814 - Place: Limestone, KY - Period: War - Discharged at Chillicothe, OH, on 4 Jun 1814.

Simons, James - Seaman - US Sea Fencibles - Company: John Gill - Age: 28 - Height: 5' 5 1/4" - Born: Norwich, CT - Enlistment date: 12 Feb 1814 - Period: 1 Yr - BLW 25066-80-55.

Simpkins, Dickerson - Private - 12th Infantry - Company: Thomas Moore - Age: 17 - Height: 5' 3 1/2" - Born: Allegany Cty, MD - Trade: Shoemaker - Enlistment date: 24 Apr 1813 - Place: Morgantown, VA - Period: War - Discharged at Buffalo on 31 May 1815.

Simpson, Benjamin - Private - 38th Infantry - Company: John Brookes - Age: 21 - Height: 5' 5 1/2" - Born: Charles Cty, MD - Trade: Farmer - Enlistment date: 7 Apr 1814 - Place: Norfolk, VA - Period: War - BLW 1004-160-12 - Discharged at Chaney Island, VA, on 15 Mar 1815.

Simpson, George - Private - 14th Infantry - Company: Samuel Lane - Age: 40 - Height: 5' 4" - Born: Maryland - Enlistment date: 30 Aug 1812 - Period: 5 Yrs - Deserted at Black Rock, NY, on 6 Feb 1813.

By War of 1812 Soc. in MD

Simpson, Hanson - Private - 1st Light Dragoons - Company: Thomas Harrison - Age: 21 - Height: 5' 10" - Born: Maryland - Enlistment date: 19 Mar 1814.

Simpson, John - Private - 36th Infantry - Company: Hugh Deneale - Age: 21 - Height: 5' 9" - Born: Prince George's Cty, MD - Enlistment date: 26 Mar 1814 - Place: Annapolis - Period: War - Pension: Old War IF-3911 - Discharged at Washington, DC, on 27 Nov 1815.

Simpson, John - Private - 26th Infantry - Company: William Bezeau - Age: 35 - Height: 5' 11" - Born: England - Trade: Weaver - Enlistment date: 20 Jan 1815 - Place: Baltimore - Period: War - Deserted on 28 Jan 1815.

Simpson, Josiah. S. - Third Lieutenant - 14th Infantry - Commissioned as an ensign on 11 May 1814; promoted to 3rd lieutenant on 30 may 1814; struck off on 27 Jul 1814.

Simpson, Lewis - Private - 36th Infantry - Age: 20 - Height: 5' 7" - Born: Fredericksburg, VA - Trade: Tailor - Enlistment date: 17 Jan 1813 - Period: War - BLW 487-160-12 - Discharged at Georgetown, DC, on 18 Mar 1815.

Simpson, Mark - Sergeant Major - 14th Infantry - Enlistment date: 27 Jun 1812 - Period: 18 Mos - Prisoner of War (Halifax), captured at Beaver Dams on 24 Jun 1813, discharged from Halifax on 9 Nov 1813; discharged on 31 Mar 1815.

Simpson, Thomas - Private - 1st Artillery - Company: Hopley Yeaton - Other regiment: Corps of Artillery - Age: 26 - Height: 5' 6 1/2" - Born: Prince George's Cty, MD - Trade: Laborer - Enlistment date: 26 Apr 1814 - Place: Fort Nelson, VA - Period: 5 Yrs - BLW 1619-160-12 - Discharged on 31 May 1815.

Sims, Daniel - 38th Infantry - Company: John Rothrock - Pension: SO-23192, SC-19098 - BLW 25580-160-50.

Sims, William - Private - Light Dragoons - Company: James Reed - Age: 23 - Height: 5' 5 1/2" - Born: Prince George's Cty, MD - Trade: Seaman - Enlistment date: 6 Apr 1814 - Period: War - Discharged at Fort Washington, MD, on 7 Mar 1815.

Sinard, John - Private - US Volunteers - Company: Stephen Moore - Enlistment date: 8 Sep 1812 - Period: 1 Yr.

Sinclair, Robert - Private - 36th Infantry - Company: Thomas Carbery - Age: 25 - Height: 5' 8" - Born: Scotland - Trade: Weaver - Enlistment date: 4 Oct 1815 - Place: Georgetown, DC - Period: War - BLW 4130-160-12 - Discharged at Baltimore on 31 Mar 1815.

Singer, Thomas - Private - 19th Infantry - Age: 18 - Height: 5' 7" - Born: Maryland - Enlistment date: 13 Apr 1814 - Place: Adelphi, Ross Cty, OH.

Singery, Ingold - Private - 26th Infantry - Company: William Bezeau - Age: 32 - Height: 5' 7" - Born: Baltimore - Enlistment date: 11 Nov 1814 - Place: Baltimore.

Sinner, James - Private - 26th Infantry - Company: William Bezeau - Age: 23 - Height: 5' 6" - Born: Lancaster Cty, PA - Enlistment date: 28 Oct 1814 - Place: Baltimore - Died.

Sinton, Francis - Private - US Volunteers - Company: Stephen Moore - Enlistment date: 8 Sep 1812 - Period: 1 Yr.

Maryland Regulars in War of 1812

Sinton, Francis - Seaman - US Sea Fencibles - Company: John Gill - Age: 24 - Height: 5' 5" - Born: Baltimore - Enlistment date: 9 Jan 1814 - Period: 1 Yr.

Sipple, Richard - Private - 5th Infantry - Company: Colin Buckner - Age: 44 - Height: 5' 9" - Trade: Tailor - Enlistment date: 26 Mar 1813 - Place: Baltimore - Period: 5 Yrs - Pension: Land bounty to Samuel Sipple, son and other heirs at law of Richard Sipple - BLW 24674-160-12 - Discharged on 21 Jan 1814.

Sitler, Jacob - Private - 38th Infantry - Company: Shepperd Leakin - BLW 14214-160-12.

Sizemore, Hiram - Private - 38th Infantry - Company: James Smith - Pension: Land bounty to John Sizemore, son and other heirs at law of Hiram Sizemore - BLW 26762-160-12.

Skilley, Thomas - Private - 14th Infantry - Company: Joseph Marshall - Age: 24 - Height: 5' 6 1/2" - Born: Somerset Cty, MD - Trade: Sailor - Enlistment date: 8 Jul 1814 - Place: Salisbury, MD - Period: War.

Skinner, James A. - Sergeant - 8th Infantry - Company: Charles Crawford - Age: 25 - Height: 5' 10 3/4" - Born: Maryland - Trade: Farmer - Enlistment date: 29 Jun 1813 - Place: Lexington, GA - Period: 1 Yr - Died on 30 Oct 1814.

Skinner, William - Private - 38th Infantry - Company: John Mowton - BLW 12134-160-12.

Slaid, James - Corporal - 7th Infantry - Company: Jacob Miller - Other regiment: 1st Infantry (New) - Age: 28 - Height: 5' 7" - Born: Maryland - Trade: Farmer - Enlistment date: 29 Jul 1814 - Place: Hopkinsville, KY - Period: 5 Yrs - Discharged on 28 Jul 1819.

Slaine, William - Private - 14th Infantry - Prisoner of War (Halifax), sent to England.

Slaker, Zacheus - Gunner - US Sea Fencibles - Company: John Gill - Enlistment date: 22 Dec 1813 - Period: 1 Yr.

Slater, John R. - Private - 36th Infantry - Company: Thomas Carbery - Age: 27 - Height: 5' 8 3/4" - Born: Charles Cty, MD - Trade: Shoemaker - Enlistment date: 24 Feb 1814 - Place: Leonardstown, MD - Period: War - BLW 96-160-12 - Discharged at Baltimore on 31 Mar 1815.

Slaughter, Lee - Sergeant - 14th Infantry - Company: James McDonald - Other regiment: 4th Infantry (New) - Age: 19 - Height: 5' 5" - Born: Maryland - Trade: Farmer - Enlistment date: 6 Sep 1813 - Place: Denton, MD - Period: 5 Yrs - Promoted to 2nd lieutenant on 18 Apr 1818.

Sloan, William - Private - 14th Infantry - Company: David Cummings - Enlistment date: 1 Feb 1813 - Period: 5 Yrs - Pension: Old War IF-4674 - BLW 11596-160-12 - Prisoner of War, sent to England; discharged on 20 Apr 1815 on Surgeon's Certificate.

Sloan, William - Private - Light Dragoons - Age: 40 - Height: 5' 7" - Born: Ireland - Trade: Weaver - Enlistment date: 8 Feb 1813 - Period: 5 Yrs.

Sly, Thomas - Drum Major - 14th Infantry - Enlistment date: 10 Oct 1812 - Period: 18 Mos - Pension: Wife Margaret, no pension claim - BLW 68783-

By War of 1812 Soc. in MD

160-55 - Prisoner of War (Halifax), discharged from Halifax on 3 Feb 1814, exchanged on 15 Apr 1814; discharged on 2 Jun 1814.

Slye, George - Second Lieutenant - 36th Infantry - Company: Thomas Carbery - Pension: Land bounty to Mary A. Slye, widow of George Slye - BLW 12920-178-50 - Commissioned as a 3rd lieutenant on 30 Apr 1813; promoted to 2nd lieutenant on 21 Feb 1814; discharged on 15 Jun 1815.

Small, John D. - Private - US Volunteers - Company: Stephen Moore - Enlistment date: 8 Sep 1812 - Period: 1 Yr.

Small, John J. - Private - US Volunteers - Company: Stephen Moore - Enlistment date: 8 Sep 1812 - Period: 1 Yr.

Smallwood, Calvert - Musician - 36th Infantry - Company: Samuel Raisin - Age: 17 - Height: 5' 2" - Born: Charles Cty, MD - Trade: Shoemaker - Enlistment date: 21 Oct 1814 - Period: War - Discharged at Washington, DC, on 20 Mar 1815.

Smallwood, James - Private - 1st Rifles - Company: William Smyth - Age: 35 - Height: 6' 1" - Born: Port Tobacco, MD - Trade: Farmer - Enlistment date: 26 Feb 1814 - Place: Plattsburgh, NY - Period: War - Discharged at Plattsburgh, NY, on 24 Aug 1815.

Smiley, John - Private - 14th Infantry - Company: Samuel Lane - Age: 36 - Height: 6' - Born: Ireland - Enlistment date: 14 May 1812 - Place: Gettysburg, PA - Period: 5 Yrs - Prisoner of War (Halifax), sent to England; drown in Alabama on 8 May 1817.

Smith, Abijah - Private - 14th Infantry - Company: Richard Arell - Age: 19 - Height: 5' 7 1/2" - Born: Cumberland, NY - Trade: Cabinet maker - Enlistment date: 26 Feb 1813 - Place: Elkton, MD - Period: War - Pension: SO-22287, SC-11008 - BLW 3343-160-12 - Prisoner of War (Halifax), discharged from Halifax on 3 Feb 1814, exchanged on 15 Apr 1814; discharged at Greenbush, NY, on 3 May 1815.

Smith, Abijah - Private - Light Dragoons - Age: 19 - Height: 5' 7" - Born: New Jersey - Trade: Cabinet maker - Enlistment date: 1 Mar 1813 - Place: Baltimore - Period: 5 Yrs.

Smith, Alexander - Seaman - US Sea Fencibles - Company: Simmones Bunbury - Age: 18 - Height: 5' 7" - Born: Baltimore - Enlistment date: 15 Jan 1814 - Period: 1 Yr - Discharged at Fort McHenry on 14 Jan 1815.

Smith, Andrew - Private - 38th Infantry - Company: Charles Stansbury - BLW 5499-160-12.

Smith, Andrew - Private - 38th Infantry - Company: Charles Stansbury - Age: 25 - Height: 5' 8" - Born: Belfast, Ireland - Trade: Weaver - Enlistment date: 22 Nov 1814 - Place: Fredericktown, MD - Period: War - BLW 5499-160-12 - Discharged at Baltimore on 6 Apr 1815.

Smith, Bazil - Private - 12th Infantry - Company: Thomas Sangster - Other regiment: 8th Infantry (New) - Age: 18 - Height: 5' 2" - Born: Baltimore - Trade: Shoemaker - Enlistment date: 15 May 1812 - Place: Alexandria, DC - Period: 5 Yrs - Discharged on 15 May 1817.

Smith, Charles - Private - 14th Infantry - Age: 33 - Height: 5' 10 1/2" - Born: Germany - Enlistment date: 16 Mar 1814 - Place: Baltimore - Period: War -

Deserted from the 10th Infantry.

Smith, Charles - Private - 5th Infantry - Age: 27 - Height: 5' 8" - Born: Germany - Trade: Seaman - Enlistment date: 16 Apr 1813 - Place: Baltimore - Period: 5 Yrs.

Smith, Charles - Private - 15th Infantry - Company: Henry Van Dalsem - Age: 25 - Height: 5' 5" - Born: Maryland - Trade: Seaman - Enlistment date: 12 Jun 1812 - Place: Philadelphia - Period: 5 Yrs.

Smith, Charles - Private - 2nd Artillery - Company: Frederick Evans - Age: 22 - Height: 5' 9" - Born: Maryland - Trade: Carpenter - Enlistment date: 2 Apr 1814 - Place: Fort McHenry - Period: War - Discharged on 24 Mar 1815.

Smith, Christian - Private - 3rd Rifles - Company: Edward Carrington - Age: 22 - Height: 5' 10" - Born: Maryland - Trade: Farmer - Enlistment date: 3 Jun 1814 - Place: Clarksburg, VA - Period: War - Discharged at Washington, DC, on 31 Mar 1815.

Smith, Christian P. - Drum Major - 27th Infantry (OH) - Company: Alexander Hill - Other regiment: 19th Infantry - Age: 27 - Height: 5' 8" - Born: Summitburg, Frederick County, MD - Trade: Shoemaker - Enlistment date: 28 Jun 1813 - BLW 11360-160-12 - Enlisted on 18 Apr 1814 for War by Captain Spencer at Newark, OH; Captain George Kesling's Company, discharged on 27 Mar 1815 at Zanesville, OH.

Smith, David - Private - 14th Infantry - Company: Thomas Montgomery - Enlistment date: 5 Jan 1813 - Period: 5 Yrs - Pension: Land bounty to John Smith, brother and other heirs at law of David Smith - BLW 25228-160-12 - Died at Plattsburgh, NY, in Mar 1814 from sickness.

Smith, Edward M. - Quartermaster Sergeant - 38th Infantry - BLW 14855-160-12.

Smith, Elijah - Private - 38th Infantry - Company: Shepperd Leakin - Age: 39 - Height: 5' 9" - Born: Baltimore - Trade: Ship's carpenter - Enlistment date: 8 Mar 1814 - Period: War - BLW 252-160-12 - Discharged at Baltimore on 6 Apr 1815.

Smith, Ferdinand - Private - 14th Infantry - Company: David Cummings - Age: 22 - Height: 5' 11" - Born: Germany - Trade: Barber - Enlistment date: 7 Jul 1814 - Place: Baltimore - Period: War - Discharged at Baltimore on 14 Mar 1815.

Smith, Henry - Private - 38th Infantry - Company: James Hook - Age: 21 - Height: 5' 10" - Born: Hagerstown, MD - Trade: Joiner - Enlistment date: 15 Dec 1812 - Period: 1 Yr - BLW 2667-160-12 - Re-enlisted on 14 Mar 1814 for the war; discharged at Fort Covington, MD, on 31 Mar 1815.

Smith, Henry - Private - 14th Infantry - Company: Richard Arell - Other regiment: 4th Infantry (New) - Age: 19 - Height: 5' 6" - Born: Baltimore - Trade: Laborer - Enlistment date: 19 Nov 1812 - Place: Baltimore - Period: 5 Yrs - Prisoner of War, exchanged on 15 Apr 1814; deserted at Fort Severn on 15 Jul 1815.

Smith, Henry - Private - 5th Infantry - Company: James Dorman - Age: 22 - Height: 5' 7" - Born: Cecil Cty, MD - Trade: Laborer - Enlistment date: 11 Dec 1812 - Period: 5 Yrs.

By War of 1812 Soc. in MD

Smith, Henry - Private - Artillery - Age: 22 - Height: 5' 5" - Born: Baltimore - Enlistment date: 23 Apr 1814 - Place: Boston - Deserted.

Smith, Hugh - Private - 2nd Rifles - Company: Benjamin Desha - Age: 21 - Height: 6' - Born: Somerset Cty, MD - Trade: Farmer - Enlistment date: 20 Jun 1814 - Place: Russellville, KY - Period: War - Discharged at Detroit on 30 Jun 1815.

Smith, Jacob - Private - 14th Infantry - Company: Joseph Marshall - Age: 31 - Height: 5' 9" - Born: Essex or Sussex Cty, NJ or DE - Trade: Laborer - Enlistment date: 11 Aug 1814 - Place: Baltimore - Period: War - BLW 22339-160-12 - Discharged at Greenbush, NY, on 3 May 1815.

Smith, Jacob - Private - 14th Infantry - Company: Samuel Lane - Age: 44 - Height: 5' 4 1/2" - Born: Germany - Enlistment date: 31 Aug 1812 - Period: 18 Mos - Prisoner of War (Halifax); died at Halifax on 14 Sep 1813.

Smith, Jacob - Corporal - 18th Infantry - Company: William Taylor - Other regiment: 4th Infantry (New) - Born: Frederick Cty, MD - Trade: Tailor - Enlistment date: 15 Aug 1814 - Place: Charleston, SC - Period: 5 Yrs - Discharged on 14 Aug 1819.

Smith, James - Captain - 38th Infantry - Commissioned as a 1st lieutenant on 20 May 1813; promoted to captain on 20 May 1814; discharged on 15 Jun 1815.

Smith, James - Private - Light Dragoons - Age: 27 - Height: 5' 6" - Born: County Tyrone, Ireland - Trade: Cabinet maker - Enlistment date: 9 May 1814 - Place: Baltimore - Period: War.

Smith, James - Private - 15th Infantry - Company: Jacob Howell - Age: 18 - Height: 5' 4" - Born: Baltimore - Trade: Laborer - Enlistment date: 12 Sep 1814 - Place: Philadelphia - Period: War.

Smith, James - Private - 36th Infantry - Company: Charles Randolph - Age: 34 - Height: 5' 8" - Born: Lancaster, PA - Enlistment date: 4 Sep 1813 - Place: Waynesburg, PA - Period: 1 Yr.

Smith, James - Private - 27th Infantry - Age: 26 - Height: 5' 6 1/2" - Born: Baltimore - Trade: Seaman - Enlistment date: 17 Aug 1814 - Period: 5 Yrs.

Smith, James - Corporal - 26th Infantry - Company: William Bezeau - Age: 27 - Height: 5' 9 1/2" - Born: Lancaster Cty, PA - Trade: Shoemaker - Enlistment date: 10 Oct 1814 - Place: Baltimore - Period: War.

Smith, James H. - Corporal - 38th Infantry - Company: Shepperd Leakin - Age: 25 - Height: 5' 8" - Born: Dublin, Ireland - Trade: Printer - Enlistment date: 9 Jul 1814 - Place: Baltimore - Period: War - BLW 9317-160-12 - Discharged at Fort McHenry on 28 Mar 1815.

Smith, James S. - Private - 36th Infantry - Company: Samuel Raisin - Age: 24 - Height: 6' - Born: Goochland Cty, VA - Trade: Shoemaker - Enlistment date: 9 Oct 1814 - Place: Manchester - Period: War - BLW 23361-160-12 - Discharged at Washington, DC, on 20 Mar 1815.

Smith, James W. - 36th Infantry - Company: Samuel Raisin - Pension: Wife Elizabeth A., SO-25459, SC-20143, WO-27960, WC-20634 - Also served in Captain John Pollock's Company, VA Militia.

Smith, Jeremiah - Private - 5th Infantry - Company: Richard Whartenby - Age:

Maryland Regulars in War of 1812

23 - Height: 5' 9" - Born: Berks Cty, PA - Trade: Tailor - Enlistment date: 29 Jan 1812 - Place: Baltimore - Period: 5 Yrs.

Smith, John - Private - 14th Infantry - Prisoner of War (Halifax), discharged from Halifax on 3 Feb 1814.

Smith, John - Private - 19th Infantry - Company: Henry Crittenden - Other regiment: 17th Infantry - Age: 19 - Height: 5' 9" - Born: Maryland - Trade: Shoemaker - Enlistment date: 3 Apr 1813 - Place: Mount Vernon, OH - Period: War - BLW 9887-160-12 - Discharged at Erie, PA, on 6 Apr 1815.

Smith, John - Private - 38th Infantry - Company: John Mowton - BLW 27122-160-42.

Smith, John - Private - 36th Infantry - Company: Thomas Carbery - Enlistment date: 10 May 1813 - Period: 1 Yr - Pension: Wife Linnie, WO-39291, WC-29236.

Smith, John - Private - 14th Infantry - Company: David Cummings - Age: 40 - Height: 5' 6" - Born: Fairfax Cty, VA - Trade: Farmer - Enlistment date: 6 Nov 1814 - Place: Snowden's Camp, VA - Period: War - BLW 10198-160-12 - Discharged on 18 Mar 1815.

Smith, John - Private - 5th Infantry - Age: 36 - Height: 5' 6" - Born: Maryland - Trade: Laborer - Enlistment date: 14 Feb 1814 - Place: Greencastle, PA - Period: 5 Yrs.

Smith, John - Private - 1st Artillery - Company: George Armistead - Age: 35 - Height: 5' 7" - Born: Maryland - Trade: Cooper - Enlistment date: 29 Jan 1810 - Place: Fort McHenry - Period: 5 Yrs - Discharged on 28 Jan 1815.

Smith, John - Private - Artillery - Age: 22 - Height: 5' 7 1/4" - Born: Baltimore - Enlistment date: 23 Sep 1814 - Place: New York - Period: War - Deserted.

Smith, John - Private - 4th Rifles - Company: Matthew Magee - Age: 33 - Height: 5' 4" - Born: Maryland - Trade: Shoemaker - Enlistment date: 8 Dec 1814 - Period: 5 Yrs.

Smith, John - Private - 6th Infantry - Company: John Walworth - Age: 26 - Height: 5' 5 1/4" - Born: Maryland - Trade: Sail maker - Enlistment date: 12 Feb 1814 - Place: New York - Period: War - Dishonorable discharged at Champlain, NY, on 24 Jul 1815.

Smith, John - Private - 14th Infantry - Company: Samuel Lane - Age: 29 - Height: 5' 6 1/2" - Born: Ireland - Enlistment date: 28 Jul 1812 - Period: 18 Mos - Prisoner of War (Halifax), sent to England; discharged on 31 Mar 1815.

Smith, John - Private - 5th Infantry - Company: Richard Whartenby - Age: 24 - Height: 5' 7" - Born: Holland - Trade: Shoemaker - Enlistment date: 1 May 1812 - Place: Baltimore - Period: 5 Yrs - Discharged on 1 May 1817.

Smith, John - Private - 14th Infantry - Company: Isaac Barnard - Enlistment date: 4 Jan 1813 - Period: 18 Mos - Discharged at Greenbush, NY, on 8 Jul 1814.

Smith, John - Private - 36th Infantry - Company: Thomas Carbery - Period: War - Deserted on 15 Aug 1814.

Smith, John - Private - 26th Infantry - Company: William Bezeau - Age: 36 - Height: 5' 10 1/12" - Born: Hagerstown, MD - Trade: Plasterer - Enlistment date: 29 Oct 1814 - Place: Baltimore - Period: War - Discharged at Philadelphia on 23 Mar 1815.

Smith, John 2nd - Private - 19th Infantry - Company: Joel Collins - Age: 24 - Height: 5' 9" - Born: Maryland - Trade: Blacksmith - Enlistment date: 30 Apr 1814 - Place: Detroit - Period: War - BLW 3811-160-12 - Discharged at Detroit on 20 Jul 1815.

Smith, John D. - Private - 38th Infantry - Pension: Land bounty to William Smith, son and other heirs at law of John D. Smith - BLW 27067-160-12.

Smith, John J. - Private - 38th Infantry - Company: John Buck - BLW 17980-160-12.

Smith, John P. - Corporal - 1st Artillery - Company: George Armistead - Pension: Old War WF-17933 Rejected.

Smith, John S. - Private - 38th Infantry - Company: Shepperd Leakin - Age: 38 - Height: 5' 11" - Born: Massachusetts - Trade: Carpenter - Enlistment date: 26 Feb 1814 - Place: Baltimore - Period: War - BLW 23361-160-12 - Discharged at Baltimore on 30 Mar 1815.

Smith, John W. - First Lieutenant - 14th Infantry - Born: Pennsylvania - Commissioned as a 1st lieutenant on 12 Mar 1812; resigned on 20 Jun 1812.

Smith, Joseph - Private - 2nd Light Dragoons - Company: William Littlejohn - Age: 20 - Height: 5' 7 1/2" - Born: Calvert Cty, MD - Trade: Sailor - Enlistment date: 19 Feb 1814 - Place: Alexandria, DC - Period: War - Deserted in May 1814.

Smith, Joseph D. - Private - 19th Infantry - Company: William Gill - Age: 40 - Height: 5' 10 1/2" - Born: Harford Cty, MD - Trade: Shoemaker - Enlistment date: 17 Mar 1814 - Period: War - BLW 13651-160-12 - Discharged at Chillicothe, OH, on 5 Jun 1815.

Smith, Livingston - Private - 14th Infantry - Age: 44 - Height: 5' 6 1/2" - Born: New Castle - Enlistment date: 2 Jun 1814 - Place: Alexandria, DC - Period: 5 Yrs.

Smith, Michael - Private - 2nd Infantry - Age: 56 - Height: 5' 5" - Born: Fredericktown, MD - Trade: Baker - Enlistment date: 1 Feb 1813 - Discharged at Sackets Harbor, NY, on Surgeon's Certificate, old age.

Smith, Nathan - Private - 10th Infantry - Company: Robert Mitchell - Age: 21 - Height: 5' 7" - Born: Maryland - Trade: Farmer.

Smith, Nathan C. - Private - 14th Infantry - Company: Reuben Gilder - Age: 33 - Height: 5' 8 3/4" - Born: Kent Cty, MD - Trade: Tailor - Enlistment date: 17 Mar 1813 - Period: War - BLW 10097-160-12 - Discharged at Greenbush, NY, on 4 May 1815.

Smith, Nathaniel - Private - 2nd Light Dragoons - Company: William Littlejohn - Age: 24 - Height: 5' 10" - Born: Prince George's Cty, MD - Trade: Laborer - Enlistment date: 16 Feb 1814 - Place: Alexandria, DC - Period: War - Deserted at Baltimore on 12 Dec 1814.

Smith, Nathaniel - Private - 2nd Light Dragoons - Company: William Littlejohn - Other regiment: Light Dragoons - Age: 25 - Height: 5' 9" - Born: Franklin Cty, PA - Trade: Shoemaker - Enlistment date: 27 Apr 1814 - Place: Baltimore - Period: War - BLW 6479-160-12 - Prisoner of War, captured at

Bladensburg, paroled on 25 Aug 1814; discharged at Carlisle Barracks, PA, on 4 May 1815.

Smith, Nicholas - Private - 4th Rifles - Company: Matthew Magee - Other regiment: Rifles - Age: 30 - Height: 5' 7" - Born: Baltimore - Trade: Carpenter - Enlistment date: 18 Aug 1814 - Place: Mercer - Period: 5 Yrs - BLW 22974-160-12 - Discharged at Fort Crawford, IL, on 18 Aug 1819.

Smith, Peter - Seaman - US Sea Fencibles - Company: Simmones Bunbury - Enlistment date: 29 Jan 1814 - Period: 1 Yr - Pension: Navy Widow File 1413 Rejected - Died at Baltimore on 27 Feb 1814.

Smith, Peter - Private - 5th Infantry - Age: 27 - Height: 5' 3 1/2" - Born: Ireland - Trade: Seaman - Enlistment date: 8 Feb 1813 - Place: Baltimore - Period: 5 Yrs.

Smith, Randolph - Private - 2nd Artillery - Company: John Peyton - Age: 28 - Height: 5' 9 1/2" - Born: Frederick, MD - Trade: Sailor - Enlistment date: 7 Jun 1814 - Place: Winchester - Period: War.

Smith, Randolph - Private - 12th Infantry - Company: James Paxton - Age: 26 - Height: 5' 9 1/2" - Born: Charles Cty, MD - Trade: Sailor - Enlistment date: 29 Aug 1812 - Place: Virginia - Period: 18 Mos - Discharged at Burlington, VT, on 1 Mar 1814.

Smith, Richard - Private - 17th Infantry - Company: William Bradford - Other regiment: 3rd Infantry (New) - Age: 21 - Height: 5' 5" - Born: Queen Anne's Cty, MD - Trade: Hatter - Enlistment date: 18 Jul 1812 - Place: St. Clairsville, OH - Period: 5 Yrs - BLW 15591-160-12 - Prisoner of War (Halifax); captured at Sackets Harbor, NY, on 12 Jun 1813; exchanged on 31 Mar 1814; discharged at Fort Howard, WI, on 17 Jul 1817.

Smith, Richard - Private - 1st Artillery - Company: George Armistead - Other regiment: Corps of Artillery - Age: 22 - Height: 5' 8" - Born: England - Trade: Carpenter - Enlistment date: 20 Mar 1810 - Place: Fort McHenry - Period: 5 Yrs - Discharged on 19 Mar 1815.

Smith, Richard - Private - 35th Infantry - Company: Francis Walker - Age: 24 - Height: 5' 5 1/2" - Born: Baltimore - Trade: Shoemaker - Enlistment date: 22 Feb 1814 - Place: Craney Island, VA - Period: War - Discharged at Norfolk, VA, on 15 Mar 1815.

Smith, Robert - Private - 38th Infantry - Pension: Land bounty to Margaret Smith, widow and heir at law of Robert Smith - BLW 17276-160-12.

Smith, Robert S. - Private - 36th Infantry - Company: Samuel Raisin - BLW 23362-160-12.

Smith, Samuel - Private - 14th Infantry - Company: Reuben Gilder - Age: 27 - Height: 5' 5 1/2" - Born: Philadelphia - Trade: Seaman - Enlistment date: 11 Mar 1814 - Place: Plattsburgh, NY - Period: War.

Smith, Thomas - Private - 38th Infantry - Company: James Smith - BLW 18821-160-12.

Smith, Thomas - Private - 38th Infantry - Company: Charles Stansbury - Age: 35 - Height: 5' 8 1/2" - Born: Maryland - Enlistment date: 25 Oct 1814 - Period: War - BLW 2728-160-12.

Smith, Thomas - Private - 20th Infantry - Company: Byrd Willis - Other

By War of 1812 Soc. in MD

regiment: 8th Infantry (New) - Age: 28 - Height: 5' 11" - Born: Harford Cty, MD - Trade: Shoemaker - Enlistment date: 4 May 1812 - Place: Jerusalem, VA - Period: 5 Yrs - Discharged on 4 May 1818.

Smith, Thomas - Private - 38th Infantry - Company: Charles Stansbury - Age: 23 - Height: 5' 8" - Born: Middleboro, Zealand - Trade: Laborer - Enlistment date: 26 Sep 1814 - Place: Annapolis - Period: War - BLW 2728-160-12 - Discharged at Baltimore on 6 Apr 1815.

Smith, Thomas - Private - 13th Infantry - Company: John Fink - Age: 24 - Height: 5' 6" - Born: Anne Arundel Cty, MD - Trade: Nailor - Enlistment date: 8 Oct 1814 - Place: Albany, NY - Period: War - Discharged at Sackets Harbor, NY, on 30 Jun 1815.

Smith, Thomas S. - Private - 14th Infantry - Company: David Cummings - Age: 35 - Height: 5' 4 1/2" - Born: Worcester, MD - Trade: Shoemaker - Enlistment date: 6 Jul 1814 - Place: Salisbury, MD - Period: War - Discharged on 18 Mar 1815.

Smith, William - Private - 38th Infantry - Company: John Brookes - BLW 6197-160-12.

Smith, William - Private - 14th Infantry - Company: Thomas Kearney - Enlistment date: 5 Jun 1812 - Period: 5 Yrs - Pension: Land bounty to Elizabeth Smith, daughter and other heirs at law of William Smith - BLW 18234-160-12 - Died on or about 15 Aug 1813.

Smith, William - Private - 22nd Infantry - Company: Daniel McFarland - Age: 25 - Height: 5' 7 1/2" - Born: Frederick, MD - Trade: Sailor - Enlistment date: 13 Aug 1812 - Place: Fayette, PA - Period: 5 Yrs - Discharged at Burlington, VT, on 19 Apr 1814 on Surgeon's Certificate for asthma.

Smith, William - Private - 1st Rifles - Company: William Smyth - Age: 23 - Height: 6' 1" - Born: Frederick, MD - Trade: Hatter - Enlistment date: 2 Mar 1814 - Place: Plattsburgh, NY - Period: War - Deserted at Champlain, NY, on 4 Oct 1814.

Smith, William - Private - 2nd Artillery - Company: Frederick Evans - Other regiment: Corps of Artillery - Age: 25 - Height: 5' 8" - Born: Maryland - Trade: Laborer - Enlistment date: 26 Jan 1812 - Place: Annapolis - Period: 5 Yrs - BLW 10024-160-12 - Discharged at Washington, DC, on 30 Jun 1817.

Smith, William - Private - 14th Infantry - Company: Samuel Lane - Enlistment date: 19 May 1812 - Period: 5 Yrs - Deserted at Hagerstown on 23 Jun 1812.

Smith, William - Private - 14th Infantry - Company: Thomas Montgomery - Age: 21 - Height: 5' 11" - Born: Frederick, MD - Enlistment date: 30 Mar 1813 - Place: Fredericktown, MD - Period: 18 Mos - Discharged at Plattsburgh, NY, in Mar 1814 and re-enlisted.

Smith, William - Seaman - US Sea Fencibles - Company: John Gill - Age: 20 - Height: 5' 8" - Born: Baltimore - Enlistment date: 16 Jan 1814 - Period: 1 Yr.

Smith, William - Private - 14th Infantry - Company: David Cummings - Age: 21 - Height: 5' 8" - Born: Frederick, MD - Trade: Laborer - Enlistment date: 6 Jan 1815 - Place: Baltimore - Period: War - Discharged on 18 Mar 1815.

Maryland Regulars in War of 1812

Smith, William - Private - 14th Infantry - Age: 36 - Prisoner of War (Quebec), died at Quebec on 16 Jul 1813.

Smith, William - Private - 35th Infantry - Company: Benjamin Hardaway - Age: 28 - Height: 5' 6 1/2" - Born: Maryland - Trade: Saddler - Enlistment date: 27 Apr 1814 - Place: Windsor, NC - Period: War - BLW 8279-160-12 - Discharged at Norfolk, VA, on 15 Mar 1815.

Smith, William - Private - 36th Infantry - Company: Samuel Raisin - Age: 17 - Height: 5' - Born: Campbell, VA - Trade: Laborer - Enlistment date: 19 Apr 1813 - Place: Baltimore - Period: 1 Yr - Pension: Wife Mary, WO-9108, WC-15854 - BLW 27725-160-42 - Re-enlisted on 8 Jun 1814 for the war; discharged on 20 Mar 1815.

Smith, William - Fifer - 36th Infantry - Company: Samuel Raisin - Age: 29 - Height: 5' 9" - Born: Ireland - Enlistment date: 30 Nov 1814 - Place: Fredericktown, MD - Period: War.

Smith, William - Private - 20th Infantry - Company: Bernard Peyton - Age: 22 - Height: 5' 3 1/2" - Born: Baltimore - Trade: Sailor - Enlistment date: 5 Jan 1813 - Place: Norfolk, VA - Period: War - Discharged at Norfolk, VA, on 15 Mar 1815.

Smithers, John - Private - 14th Infantry - Company: David Cummings - Age: 35 - Height: 5' 6 1/4" - Born: Prince William's Cty, VA - Trade: Baker - Enlistment date: 17 Jun 1814 - Place: Baltimore - Period: War - Discharged on 18 Mar 1815.

Smithers, John A. - Private - 2nd Artillery - Company: Nathan Towson - Enlistment date: 24 Apr 1812 - Period: 5 Yrs - Discharged on 25 Sep 1812 for rheumatism.

Smithson, Luther - Seaman - US Sea Fencibles - Company: John Gill - Age: 22 - Height: 6' - Born: Harford Cty, MD - Enlistment date: 6 Jan 1814 - Period: 1 Yr.

Smithson, Shelea - Private - 38th Infantry - Company: James Smith - Age: 34 - Height: 5' 8 1/4" - Born: Charles Cty, MD - Enlistment date: 10 Jan 1814 - Place: Drumfires, MD - Period: 1 Yr - Discharged on 10 Jan 1815.

Smoot, John W. - First Lieutenant - 5th Infantry - Other regiment: 4th Rifles - Born: Maryland - BLW 22684-160-50 - Commissioned as a 2nd lieutenant on 3 Jan 1812; promoted to 1st lieutenant on 21 Feb 1814; transferred to 4th Rifles on 28 Apr 1814; discharged on 15 Jun 1185.

Smother, Aderick - Servant - US Sea Fencibles - Company: Simmones Bunbury - Servant to Captain Bunbury.

Smothers, John - Servant - US Sea Fencibles - Company: Simmones Bunbury - Discharged at Fort McHenry on 24 Mar 1815.

Smyth, James - Private - 28th Infantry - Age: 39 - Height: 5' 8" - Born: Baltimore - Trade: Farmer - Enlistment date: 18 Mar 1814 - Place: Detroit - Period: War.

Sneed, George - Private - 14th Infantry - Company: Joseph Marshall - Age: 18 - Height: 5' 5" - Born: Worcester Cty, MD - Enlistment date: 11 Jun 1814 - Place: Snow Hill, MD - Period: War - Deserted at Huntington, NJ, on 14 Oct 1814.

By War of 1812 Soc. in MD

Sneed, Tully - Private - 14th Infantry - Company: Joseph Marshall - Age: 19 - Height: 5' 7" - Born: Worcester, MD - Trade: Shoemaker - Enlistment date: 13 Jun 1814 - Place: Snow Hill, MD - Period: War - BLW 6553-160-12 - Discharged at Greenbush, NY, on 3 May 1815.

Sneeds, Abraham C. - Private - 14th Infantry - Other regiment: 4th Infantry (New) - Age: 30 - Height: 5' 8" - Born: Orange Cty, NY - Trade: Currier - Enlistment date: 23 May 1812 - Place: Bladensburg, MD - Period: 5 Yrs - Prisoner of War (Halifax), captured at Beaver Dams on 24 Jun 1813, discharged from Halifax on 3 Feb 1813; deserted at Fort Sevens on 16 Jul 1815.

Snider, Adam - Private - 19th Infantry - Company: George Kesling - Age: 19 - Height: 5' 10" - Born: Maryland - Trade: Laborer - Enlistment date: 3 Jan 1815 - Place: Franklinton, Franklin Cty, OH - Period: War - BLW 465-320-14 - Discharged at Zanesville, OH, on 27 Mar 1815.

Snider, George - Private - Light Dragoons - Age: 18 - Height: 5' 6" - Born: Hagerstown, MD - Trade: Laborer - Enlistment date: 1 Mar 1813 - Place: Baltimore - Period: 5 Yrs.

Snider, Thomas - Private - 26th Infantry - Company: William Bezeau - Age: 24 - Height: 5' 9" - Born: Maryland - Enlistment date: 16 Nov 1814 - Place: Carlisle, PA - Deserted on 18 Nov 1814.

Snow, Samuel - Private - 2nd Light Dragoons - Company: William Littlejohn - Age: 21 - Height: 5' 5" - Born: Colbert, MD - Trade: Butcher - Enlistment date: 3 Oct 1812 - Place: Fredericktown, MD - Period: 5 Yrs - Deserted at Greenbush, NY, on 11 Mar 1815.

Snyder, George - Private - 14th Infantry - Company: Richard Arell - Enlistment date: 1 Mar 1813 - Period: 5 Yrs - Deserted at Buffalo on 1 Jan 1815.

Snyder, George - Private - 14th Infantry - Enlistment date: 6 Dec 1812 - Period: 5 Yrs - Prisoner of War, exchanged on 15 Apr 1814.

Snyder, George D. - First Lieutenant - 5th Infantry - Born: Pennsylvania - Commissioned as an ensign on 30 Apr 1812; promoted to 2nd lieutenant on 1 Sep 1812; promoted to 1st lieutenant on 25 Jun 1814; discharged on 15 Jun 1815.

Snyder, John - Private - 14th Infantry - Company: David Cummings - Age: 28 - Height: 5' 7 1/2" - Born: Maryland - Trade: Laborer - Enlistment date: 7 Sep 1814 - Place: Williamsport, MD - Period: War - Deserted on 1 Jan 1815.

Snyder, John - Private - 36th Infantry - Other regiment: Ordnance Department - Age: 19 - Height: 5' 10" - Born: Hagerstown, MD - Trade: Hatter - Enlistment date: 13 Feb 1815 - Place: Richmond, VA - Period: 5 Yrs - Transferred to Ordnance Department on 21 Jun 1815; discharged at Washington, DC, on 14 Feb 1820.

Solomon, Joseph - Private - 38th Infantry - Company: John Rothrock - BLW 5745-160-12.

Soper, James - Private - 2nd Artillery - Company: James Barker - Age: 21 - Height: 5' 8 3/4" - Born: Monmouth, MD - Trade: Farmer - Enlistment date: 14 Oct 1814 - Place: Philadelphia - Period: War - Discharged at Philadelphia on 16 May 1815.

Soran, Thomas - Private - 14th Infantry - Company: Reuben Gilder - Other regiment: 4th Infantry (New) - Age: 42 - Height: 5' 8 1/4" - Born: Dublin, Ireland - Trade: Carpenter - Enlistment date: 26 Jan 1813 - Place: Baltimore - Period: 5 Yrs - Prisoner of War (Halifax), captured at Stoney Creek, UC, on 12 Jun 1813.

Southern, William - Private - 38th Infantry - Company: John Mowton - Age: 20 - Height: 5' 9 1/4" - Born: Bladensburg, MD - Trade: Farmer - Enlistment date: 4 Sep 1814 - Place: Norfolk, VA - Period: War - BLW 1125-160-12 - Discharged at Craney Island, VA, on 15 Mar 1815.

Sowders, Jacob - Private - 14th Infantry - Company: David Cummings - Age: 24 - Height: 5' 10" - Born: Germany - Trade: Laborer - Enlistment date: 22 Oct 1814 - Place: Baltimore - Period: War - BLW 7283-160-12 - Discharged on 18 Mar 1815.

Sowders, Valentine - Musician - 22nd Infantry - Company: Daniel McFarland - Age: 56 - Height: 5' 8 1/4" - Born: Prince George's Cty, MD - Trade: Laborer - Enlistment date: 1 Jul 1812 - Place: Allegany Cty, MD - Period: 5 Yrs - Discharged on 19 Apr 1814 on Surgeon's Certificate for old age.

Sowers, Jacob - Private - 2nd Light Dragoons - Company: John Burd - Age: 35 - Height: 6' - Born: Frederick, MD - Trade: Farmer - Enlistment date: 26 Mar 1814 - Period: 5 Yrs.

Spalding, Richard T. - Private - 36th Infantry - Company: Thomas Carbery - Enlistment date: 4 Oct 1813 - Period: 1 Yr - Discharged on 3 Oct 1814.

Spalding, William - Private - 14th Infantry - Enlistment date: 22 May 1812 - Period: 18 Mos - Prisoner of War (Halifax), sent to England, captured at Beaver Dams on 24 Jun 1813; discharged on 31 Mar 1815.

Sparkes, John B. - Second Lieutenant - 14th Infantry - Commissioned as a 2nd lieutenant on 12 Mar 1812; died on 15 Jul 1813.

Sparks, Jacob - Private - 18th Infantry - Company: William Taylor - Other regiment: 4th Infantry (New) - Age: 35 - Height: 5' 9 1/2" - Born: Baltimore - Trade: Farmer - Enlistment date: 23 Sep 1814 - Place: Spartanburg, SC - Period: 5 Yrs - Deserted in 1816.

Sparks, John K. - Sergeant - 36th Infantry - Company: Joseph Merrick - Age: 28 - Height: 5' 9 1/4" - Born: Queen Anne's Cty, MD - Enlistment date: 20 May 1813 - Place: Centreville, MD.

Sparks, William - Seaman - US Sea Fencibles - Company: Simmones Bunbury - Enlistment date: 21 Feb 1814 - Period: 1 Yr - Pension: Wife Jane, SO-2037, SC-10334, WO-24846, WC-15462 - Discharged on 20 Feb 1815.

Sparrow, Cephors - Private - 36th Infantry - Age: 19 - Height: 5' 11 1/2" - Born: Maryland - Enlistment date: 25 Feb 1814 - Place: Leonardstown, MD - Period: War.

Sparrow, David - Private - 38th Infantry - Company: John Rothrock - BLW 26375-160-12.

Sparrow, John - Private - 42nd Infantry - Company: Thomas Barker - Age: 35 - Height: 5' 6" - Born: Maryland - Trade: Seaman - Enlistment date: 8 Feb 1814 - Place: New York - Period: War - Discharged on 12 May 1815.

By War of 1812 Soc. in MD

Spaulding, James - Private - 36th Infantry - Age: 18 - Height: 5' 3" - Born: St. Mary's, MD - Trade: Carpenter - Enlistment date: 26 Sep 1814 - Period: War - Discharged on 13 Mar 1815.

Spaulding, John - Private - 36th Infantry - Age: 22 - Height: 5' 8" - Born: Maryland - Trade: Laborer - Enlistment date: 18 Jan 1815 - Period: War - Pension: Land bounty to Ignatius King, heir at law of John Spaulding - BLW 74-320-12 - Discharged on 13 Mar 1815.

Speak, John - Private - 36th Infantry - Company: Samuel Raisin - Age: 34 - Height: 5' 7 1/4" - Born: Fairfax, VA - Trade: Farmer - Enlistment date: 24 Sep 1814 - Period: War - BLW 26675-160-12.

Speake, Edward - Private - Light Dragoons - Other regiment: Artillery - Age: 38 - Height: 5' 10 1/2" - Born: Prince George's Cty, MD - Trade: Farmer - Enlistment date: 15 Mar 1814 - Place: Alexandria, DC - Period: War - Served with Captain James Reed's artillery company; discharged at Fort Washington, MD, on 7 Mar 1815.

Speakes, Edward L. - Private - US Volunteers - Company: Stephen Moore - Enlistment date: 8 Sep 1812 - Period: 1 Yr.

Speaks, Henry - Private - 8th Infantry - Company: Charles Crawford - Other regiment: 7th Infantry (New) - Age: 21 - Height: 5' 6" - Born: Charles Cty, MD - Trade: Farmer - Enlistment date: 6 Jun 1814 - Place: Abbeville, SC - Period: 5 Yrs - Discharged at Fort Scott, GA, on 31 May 1819.

Spears, Samuel - Private - 2nd Infantry - Company: Perrin Willis - Other regiment: 1st Infantry (New) - Age: 16 - Height: 5' 7 1/2" - Born: Baltimore - Trade: Shoemaker - Enlistment date: 1 Jan 1814 - Place: Washington, DC - Period: 5 Yrs - Died at Baton Rouge, LA, on 26 Jun 1819.

Speck, Adam - Private - 36th Infantry - Company: Samuel Raisin - Age: 23 - Height: 5' 4" - Born: Pennsylvania - Trade: Hatter - Enlistment date: 30 Nov 1814 - Period: War - BLW 12572-160-12 - Discharged at Washington, DC, on 20 Mar 1815.

Speck, John - Private - 36th Infantry - Company: Joseph Hook - Age: 28 - Height: 5' 8 3/4" - Born: York, PA - Trade: Millwright - Enlistment date: 17 Apr 1813 - Place: Westminster, MD - Period: 1 Yr - BLW 12783-160-12 - Re-enlisted on 14 Mar 1814 for the war; discharged at Washington, DC, on 20 Mar 1815.

Spence, William - Private - 12th Infantry - Company: Thomas Sangster - Other regiment: 8th Infantry (New) - Age: 22 - Height: 5' 7 1/2" - Born: Hutchins, MD - Trade: Cooper - Enlistment date: 15 May 1812 - Place: Maryland - Period: 5 Yrs - Discharged on 18 May 1817.

Spencer, James - Private - 26th Infantry - Company: William Bezeau - Age: 34 - Height: 5' 9" - Born: Baltimore - Enlistment date: 16 Oct 1814 - Place: Baltimore - Period: War.

Spencer, William - Private - 14th Infantry - Company: William McIlvane - Age: 43 - Height: 5' 7" - Born: Baltimore - Trade: Shoemaker - Enlistment date: 20 Jun 1813 - Period: War - BLW 24333-160-12 - Discharged at Greenbush, NY, on 3 May 1815.

Sperling, Thomas - Private - 36th Infantry - Company: Joseph Hook - Age: 29

- Height: 5' 10" - Born: Fairfax Cty, VA - Trade: Farmer - Enlistment date: 1 Aug 1814 - Place: Georgetown, DC - Period: War - BLW 2016-160-12 - Discharged at Fort Covington, MD, on 30 Mar 1815.

Spicer, Peregrine - Private - 5th Infantry - Company: Richard Whartenby - Other regiment: 3rd Infantry (New) - Age: 25 - Height: 5' 10" - Born: Baltimore - Trade: Cooper - Enlistment date: 6 Jan 1812 - Place: Baltimore - Period: 5 Yrs - Discharged on 6 Jan 1817.

Spicer, Stephen - Private - 36th Infantry - Age: 20 - Height: 5' 4" - Born: Halifax, VA - Trade: Brick layer - Enlistment date: 5 Feb 1815 - Period: War - Discharged on 12 Mar 1815.

Spicer, Valentine - Sergeant - 38th Infantry - Company: Shepperd Leakin - Age: 23 - Height: 5' 9" - Born: Baltimore - Trade: Farmer - Enlistment date: 5 Dec 1813 - Period: 1 Yr - BLW 15896-160-12 - Discharged at Baltimore on 6 Apr 1815.

Spicknall, John - Second Lieutenant - 38th Infantry - Company: Shepperd Leakin - BLW 15550-160-50 - Commissioned as an ensign on 20 May 1813; promoted to 2nd lieutenant on 1 Oct 1814; discharged on 15 Jun 1815.

Spigler, Jacob - Private - 5th Infantry - Company: Richard Whartenby - Other regiment: 3rd Infantry (New) - Age: 32 - Height: 5' 9" - Born: Lancaster, PA - Trade: Blacksmith - Enlistment date: 22 Jan 1812 - Place: Baltimore - Period: 5 Yrs - BLW 11592-160-12 - Discharged at Fort Mackinaw, MI, on 22 Jan 1817.

Spiknell, Richard - Private - 36th Infantry - Company: Thomas Carbery - Enlistment date: 4 Aug 1814 - Period: War - BLW 13485-160-12 - Discharged at Baltimore on 31 Mar 1815.

Sprott, Samuel - Corporal - 38th Infantry - Company: James Smith - Age: 22 - Height: 5' 6 1/2" - Born: Queen Anne, MD - Trade: Carpenter - Enlistment date: 19 Feb 1814 - Place: Craney Island, VA - Period: War - Discharged at Craney Island, VA, on 15 Mar 1815.

Spry, Samuel - Corporal - 38th Infantry - Company: James Smith - BLW 17245-160-12.

Spunie, Lot - Private - 14th Infantry - Prisoner of War (Halifax), captured at Beaver Dams on 24 Jun 1813, discharged from Halifax on 3 Feb 1813.

Spunogle, Samuel - Private - 36th Infantry - Company: Samuel Raisin - Age: 18 - Height: 5' 4" - Born: District of Columbia - Trade: Shoemaker - Enlistment date: 25 Oct 1814 - Period: War - BLW 276-160-12 - Discharged at Washington, DC, on 20 Mar 1815.

Spurrier, Lancelot - Private - 14th Infantry - Company: Samuel Lane - Other regiment: Light Artillery - Age: 21 - Height: 5' 11" - Born: Frederick, MD - Enlistment date: 21 May 1812 - Place: Westminster, MD - Period: 18 Mos - Prisoner of War, exchanged on 15 Apr 1814; re-enlisted at Boston on 30 Apr 1814 for the war in the Light Artillery; discharged on 6 Jan 1815.

Squire, Charles C. - Private - 38th Infantry - Company: John Rothrock - BLW 15176-160-12.

St. Clair, John - Private - 5th Infantry - Company: Colin Buckner - Age: 19 - Height: 5' 10" - Born: Germany - Trade: Baker - Enlistment date: 10 Mar

By War of 1812 Soc. in MD

1813 - Place: Baltimore - Period: 5 Yrs - Discharged on 10 Mar 1818.

Stackpole, Conrad - Private - 19th Infantry - Age: 18 - Height: 5' 5" - Born: Allegany Cty, MD - Enlistment date: 13 Jul 1812 - Period: 5 Yrs.

Stacks, Moses R. - Private - 38th Infantry - Company: John Rothrock - Pension: Wife Sarah J., WO-38532, WC-29872.

Stacks, Thomas - Private - 36th Infantry - Company: Samuel Raisin - Age: 19 - Height: 5' 8 1/2" - Born: Baltimore - Trade: Laborer - Enlistment date: 17 Aug 1814 - Period: War.

Staines, Moses - Private - 1st Artillery - Company: Henry Craig - Age: 30 - Height: 6' 3/4" - Born: Talbot Cty, MD - Trade: Carpenter - Enlistment date: 14 May 1814 - Place: Pittsburgh - Period: War - Discharged at Buffalo on 8 Jun 1815.

Stake, Andrew - Sergeant - 1st Infantry - Company: Simon Owens - Age: 23 - Height: 5' 7 1/2" - Born: Washington Cty, MD - Trade: Nailor - Enlistment date: 8 Jan 1810 - Place: Newport, KY - Period: 5 Yrs - Died at Belle Fontaine, MO, on 20 Oct 1813.

Staley, John - Private - 14th Infantry - Company: Samuel Lane - Age: 25 - Height: 5' 11" - Born: Jefferson Cty, VA - Enlistment date: 14 May 1812 - Place: Shepherdstown - Period: 5 Yrs - Prisoner of War (Halifax), sent to England.

Stall, John - Musician - 19th Infantry - Company: Richard Talbott - Age: 30 - Height: 5' 10" - Born: Maryland - Trade: Farmer - Enlistment date: 17 Apr 1814 - Place: Cincinnati - Period: War.

Stallica, Joseph - Private - 38th Infantry - Company: John Mowton - BLW 12135-160-12.

Stallings, Richard - Private - 38th Infantry - Company: Isaac Aldridge - Age: 38 - Height: 6' 1" - Born: Calvert Cty, MD - Trade: Farmer - Enlistment date: 16 Jan 1815 - Place: Baltimore - Period: War - BLW 316-320-12 - Discharged at Baltimore on 6 Apr 1815.

Stamm, Joseph - Sergeant - 24th Infantry - Company: Samuel Brady - Other regiment: 8th Infantry (New) - Age: 25 - Height: 5' 4 1/2" - Born: Hagerstown, MD - Trade: Stone mason - Enlistment date: 5 Apr 1814 - Place: Kaskaskia, IL - Period: 5 Yrs - Discharged at Fort Covington, MD, on 5 Apr 1819.

Stanheart, James - Private - 26th Infantry - Company: William Bezeau - Age: 22 - Height: 6' - Born: Holland - Trade: Baker - Enlistment date: 3 Oct 1814 - Place: Baltimore - Period: War - BLW 9865-160-12 - Discharged at Baltimore.

Stanley, James - Private - 36th Infantry - Company: Samuel Raisin - Age: 25 - Height: 5' 5" - Born: Fish Kills, NY - Enlistment date: 29 Sep 1813 - Place: Baltimore - Period: 1 Yr.

Stansburg, Charles - Captain - 38th Infantry - Born: Maryland - Commissioned as a captain on 20 May 1813; discharged on 15 Jun 1815.

Stansbury, Dixon - Captain - 1st Infantry - Born: Maryland - Pension: Land bounty to Sarah Sansbury, widow and heir at law of Dixon Stansbury - BLW 2167-160-50 - Commissioned as an ensign on 13 Jun 1808; promoted to 2nd lieutenant on 1 Apr 1810; promoted to 1st lieutenant on 20 Jan 1813; promoted to captain on 30 Jun 1814; resigned on 31 Jan 1815.

Stansbury, Samuel - Private - 2nd Artillery - Company: Nathan Towson - Age: 25 - Pension: Land bounty to John Stansbury, brother and other heirs at law of Samuel Stansbury - BLW 24624-160-12 - Prisoner of War (Quebec), captured during the Battle of Stoney Creek and released on 31 Oct 1813.

Stansbury, William - Private - 14th Infantry - Company: Joseph Marshall - Age: 16 - Height: 5' 4" - Born: Annapolis - Trade: Laborer - Enlistment date: 25 Jun 1812 - Place: Annapolis - Period: 5 Yrs - BLW 4486-160-12 - Discharged at Greenbush, NY, on 4 May 1816 for inability.

Staples, John - Quartermaster Sergeant - 36th Infantry - Age: 39 - Height: 5' 7" - Born: Cork, Ireland - Trade: Chair maker - Enlistment date: 24 Feb 1814 - Place: Georgetown, DC - Period: War - BLW 540-160-12 - Discharged at Baltimore on 31 Mar 1815.

Stapleton, John - Private - 26th Infantry - Company: William Bezeau - Age: 36 - Height: 5' 9" - Born: Harford Cty, MD - Trade: Tailor - Enlistment date: 18 Dec 1814 - Period: War - BLW 462-320-12 - Discharged at close of war.

Star, Edward - Private - 36th Infantry - Age: 36 - Height: 5' 5" - Born: Ireland - Enlistment date: 7 May 1813 - Place: Baltimore.

Starkey, Nathan - Private - 14th Infantry - Company: Reuben Gilder - Enlistment date: 28 Dec 1812 - Period: 18 Mos - Discharged at Chazy, NY, on 12 Aug 1814.

Starks, Ebenezer - Private - 22nd Infantry - Company: John Pentland - Other regiment: 2nd Infantry (New) - Age: 35 - Height: 5' 7 1/4" - Born: Maryland - Trade: Saddler - Enlistment date: 12 Oct 1812 - Place: Lewiston - Period: 5 Yrs - Discharged at Greenbush, NY, on 12 Dec 1815.

Starky, Joseph - Private - 43rd Infantry - Company: Robert Love - Age: 43 - Height: 5' 5" - Born: Baltimore - Trade: Farmer - Enlistment date: 21 Dec 1813 - Place: Asheville, NC - Period: 5 Yrs - Discharged on 5 Mar 1814, over age.

Statham, John D. - Private - 38th Infantry - Company: John Mowton - BLW 3561-160-12.

Staunton, John (Stanton) - Private - 17th Infantry - Company: Benjamin Sanders - Other regiment: 3rd Infantry (New) - Age: 14 - Height: 4' 10" - Born: Manchester, Washington Cty, MD - Trade: Shoemaker - Enlistment date: 1 Jan 1814 - Place: Lexington or Maysville, KY - Period: 5 Yrs.

Stayton, Henry - Musician - 14th Infantry - Company: David Cummings - Age: 23 - Height: 5' 10" - Born: Baltimore - Trade: Laborer - Enlistment date: 13 Apr 1814 - Place: Baltimore - Period: War - BLW 11827-160-12 - Discharged at Baltimore on 16 Mar 1815.

Stedman, Thomas - Musician - 38th Infantry - Company: Isaac Aldridge - Pension: No pension claim - BLW 941-160-12.

Steeds, John - Sergeant - 2nd Artillery - Company: Nathan Towson - Enlistment date: 12 Apr 1812 - Period: 5 Yrs - Killed at Chippewa, UC, on 15 Oct 1814.

Steel, Samuel H. - Private - 14th Infantry - Company: Henry Fleming - Age:

By War of 1812 Soc. in MD

28 - Height: 5' 8" - Born: Cheshire Cty, NH - Trade: Laborer - Enlistment date: 26 Jan 1813 - Period: 5 Yrs - Pension: Old War IF-26415 - BLW 2853-160-12 - Discharged at Washington, DC, on 23 Sep 1814 on Surgeon's Certificate, gunshot wound received on 24 Jun 1813.

Steele, Samuel - Private - 36th Infantry - Company: Samuel Raisin - Age: 34 - Height: 6' 1 1/2" - Born: Rockbridge, VA - Trade: Farmer - Enlistment date: 26 Sep 1814 - Place: Manchester - Period: 5 Yrs.

Steigelman, Henry - Private - 2nd Artillery - Company: Nathan Towson - Pension: Land bounty to John Steigelman, brother and other heirs at law of Henry Steigelman - BLW 26268-160-12.

Steinback, Rudolph - Corporal - 6th Infantry - Company: Thomas Davis - Height: 5' 7" - Born: Philadelphia - Enlistment date: 4 Apr 1807 - Place: Baltimore - Period: 5 Yrs.

Steinnety, Francis - Private - 5th Infantry - Company: James Dorman - Other regiment: 3rd Infantry (New) - Age: 23 - Height: 5' 7 3/4" - Born: Pennsylvania - Trade: Shoemaker - Enlistment date: 23 Apr 1812 - Place: Baltimore - Period: 5 Yrs - Discharged on 22 Apr 1817.

Steller, William - Private - 14th Infantry - Company: Joseph Marshall - Age: 37 - Height: 5' 6 1/2" - Born: Germany - Trade: Barber - Enlistment date: 20 May 1814 - Place: Baltimore - Period: War - BLW 4601-160-12 - Discharged at Greenbush, NY, on 16 May 1815.

Stephens, Edward - Private - 20th Infantry - Company: Matthew Payne - Age: 25 - Height: 5' 5 1/2" - Born: Prince George's Cty, MD - Trade: Sailor - Enlistment date: 14 Nov 1814 - Place: Alexandria, DC - Period: 5 Yrs.

Stephens, James - Private - 22nd Infantry - Company: John Foster - Age: 38 - Height: 5' 6" - Born: Prince George's Cty, MD - Trade: Gunsmith - Enlistment date: 18 Nov 1812 - Period: 5 Yrs - Died at French Mills, NY, on 15 Jan 1814 from dysentery.

Stephens, James L. - Gunner - US Sea Fencibles - Company: William Addison - Pension: Old War IF-25852.

Stephens, John - Private - 14th Infantry - Prisoner of War (Halifax), discharged from Halifax on 3 Feb 1814.

Stephens, John - Private - 14th Infantry - Company: Clement Sullivan - Enlistment date: 11 May 1812 - Place: Annapolis - Period: 5 Yrs - Pension: Land bounty to James M. Stephen, son and other heirs at law of John Stephen - BLW 26390-160-12 - Died at Black Rock, NY, on 2 Dec 1812.

Stephens, John - Private - 5th Infantry - Company: James Dorman - Age: 42 - Height: 5' 5" - Born: Wexford, Ireland - Trade: Merchant - Enlistment date: 21 May 1812 - Place: Frederick, MD - Period: 5 Yrs - Prisoner of War (Halifax), captured at Stoney Creek, UC, on 7 Jun 1813; discharged at Buffalo on 2 Jun 1815 on Surgeon's Certificate.

Stephens, Levin - Private - 39th Infantry - Company: George Hallam - Age: 20 - Height: 5' 8" - Born: Dawson, MD - Enlistment date: 22 Oct 1813 - Place: Lebanon, TN - Period: 1 Yr.

Stephens, Michael - Private - 24th Infantry - Company: William Allen - Age: 22 - Height: 5' 7" - Born: Maryland - Trade: Shoemaker - Enlistment date: 7

Mar 1813 - Place: St. Genevieve, MO - Period: 5 Yrs - Pension: Old War IF-25850 - BLW 2538-160-12 - Wounded at Mackinaw Island, MI, on 4 Aug 1814; discharged on 5 Mar 1815.

Stephens, Samuel - Corporal - 17th Infantry - Company: Martin Hawkins - Age: 35 - Height: 5' 9" - Born: Maryland - Trade: Shoemaker - Enlistment date: 26 Dec 1813 - Period: 18 Mos - On board the fleet on Lake Huron since 16 Aug 1814.

Stephens, Solomon - Private - 1st Light Dragoons - Company: George Birch - Age: 36 - Height: 5' 10" - Born: Talbot Cty, MD - Trade: Saddler - Enlistment date: 9 Jan 1813 - Place: Centreville, MD - Period: 5 Yrs - Discharged at Greenbush, NY, on 30 May 1815.

Stephens, Timothy - Seaman - US Sea Fencibles - Company: Simmones Bunbury - Age: 22 - Height: 5' 7 1/2" - Born: New Bedford, MA - Enlistment date: 25 Jul 1814 - Period: 1 Yr - Deserted on 17 Mar 1815.

Stephens, William - Private - 5th Infantry - Company: Richard Whartenby - Other regiment: 3rd Infantry (New) - Age: 26 - Height: 5' 3" - Born: Baltimore - Trade: Shoemaker - Enlistment date: 16 Feb 1814 - Place: Baltimore - Period: 5 Yrs - Discharged on 16 Feb 1816.

Stephens, William - Private - 14th Infantry - Company: James McDonald - Age: 24 - Height: 5' 5" - Born: Caroline Cty, MD - Trade: Farmer - Enlistment date: 10 Feb 1814 - Period: 5 Yrs - Died at Burlington, VT, on 29 Sep 1814 from wounds received at Crab Island.

Stephenson, William - Private - 14th Infantry - Company: Joseph Marshall - Age: 24 - Height: 5' 8" - Born: Delaware - Trade: Brick layer - Enlistment date: 5 Mar 1814 - Place: Baltimore - Period: War - Pension: Land bounty to James Cochrane, nephew and other heirs at law of William Stephenson - BLW 14287-160-12 - Re-enlisted; died at Utica, NY, on 17 Jan 1815 from diarrhea.

Sterling, Jeremiah - Private - 26th Infantry - Company: William Bezeau - Age: 22 - Height: 5' 6" - Born: Caroline Cty, MD - Trade: Farmer - Enlistment date: 12 Feb 1815 - Period: War - BLW 735-320-12 - Discharged at Baltimore.

Sterling, Jeremiah N. - Ensign - 14th Infantry - Company: Clement Sullivan - BLW 22332-160-12 Cancelled - Served as a sergeant, 14th Infantry; commissioned as an ensign on 13 Mar 1813; resigned on 31 May 1813.

Sterling, Levi - Private - 14th Infantry - Company: David Cummings - Age: 21 - Height: 5' 7" - Born: Annapolis - Trade: Laborer - Enlistment date: 4 Feb 1815 - Place: Baltimore - Period: War - Discharged at Baltimore on 18 Mar 1815.

Sterling, Richard - Private - 36th Infantry - Company: Joseph Hook - Age: 42 - Height: 6' 2" - Born: Calvert, MD - Enlistment date: 28 May 1813 - Place: Baltimore - Period: 1 Yr.

Sterling, Zachariah - Private - 36th Infantry - Company: Joseph Hook - Age: 19 - Height: 5' 5" - Born: Calvert, MD - Enlistment date: 11 Jun 1813 - Period: 1 Yr.

Sterrett, Robert - Seaman - US Sea Fencibles - Company: Simmones Bunbury

By War of 1812 Soc. in MD

- Age: 34 - Height: 5' 10" - Born: Norfolk, VA - Enlistment date: 6 Jan 1814 - Period: 1 Yr - Discharged on 5 Jan 1815.

Steuart, William - Lieutenant Colonel - 38th Infantry - Born: Maryland - Place: Maryland - Commissioned as a lieutenant colonel on 19 May 1813; discharged on 15 Jun 1815.

Stevens, George - Private - 14th Infantry - Company: Henry Grindage - Enlistment date: 24 Jul 1812 - Period: 5 Yrs - Pension: Land bounty to Ann Rossell, sister and only heir at law of George Stevens - BLW 24486-160-12 - Died on 30 Dec 1812.

Stevens, James L. - Gunner - US Sea Fencibles - Company: John Gill - Age: 31 - Height: 5' 7" - Born: Charleston, SC - Enlistment date: 29 Dec 1813 - Period: 1 Yr.

Stevens, Joseph - Private - 36th Infantry - Company: Joseph Hook - Age: 33 - Height: 5' 7" - Born: Baltimore - Enlistment date: 7 May 1813 - Place: Baltimore.

Stevens, Samuel - Private - 21st Infantry - Company: Benjamin Ropes - Other regiment: 5th Infantry (New) - Age: 29 - Height: 5' 9 1/4" - Born: Baltimore - Trade: Carpenter - Enlistment date: 20 Jul 1813 - Place: Boston - Period: 5 Yrs - Deserted at Detroit on 19 May 1816.

Stevens, William - Private - 26th Infantry - Company: William Bezeau - Age: 21 - Height: 5' 4" - Born: Kent Cty, MD - Trade: Shoemaker - Enlistment date: 22 Oct 1814 - Place: Baltimore - Period: War - BLW 14473-160-12 - Discharged at Philadelphia on 23 Mar 1815.

Stevenson, John - Private - 5th Infantry - Company: James Dorman - Other regiment: 3rd Infantry (New) - Age: 20 - Height: 5' 9" - Born: Harford Cty, MD - Trade: Laborer - Enlistment date: 14 May 1812 - Place: Baltimore - Period: 5 Yrs - Discharged at Buffalo on 21 Jun 1815, disability.

Stevenson, Richard - Corporal - Corps of Artillery - Company: Evans Humphrey - Age: 32 - Height: 5' 8 1/2" - Born: Maryland - Trade: Farmer - Enlistment date: 2 Sep 1814 - Place: New Orleans - Period: War - Discharged on 9 Apr 1815.

Steward, Charles - Private - 2nd Light Dragoons - Company: William Littlejohn - Other regiment: Light Dragoons - Age: 21 - Height: 5' 6 3/4" - Born: Charles Cty, MD - Trade: Farmer - Enlistment date: 28 Dec 1813 - Place: Alexandria, DC - Period: War - BLW 12574-160-12 - Discharged at Carlisle Barracks, PA, on 4 May 1815.

Steward, Edward (Stewart) - Private - 17th Infantry - Company: David Holt - Age: 31 - Height: 6' 1/2" - Born: Charles Cty, MD - Trade: Cooper - Enlistment date: 18 May 1814 - Place: Limestone, KY - Period: War - BLW 4496-160-12 - Discharged at Chillicothe, OH, on 4 Jun 1814.

Steward, John - Private - 2nd Artillery - Company: Nathan Towson - Killed during the Battle of Chippewa.

Steward, Joseph - Sergeant - 2nd Artillery - Company: John Peyton - Other regiment: Corps of Artillery - Age: 24 - Height: 5' 10 1/2" - Born: Anne Arundel Cty, MD - Trade: Tailor - Enlistment date: 5 Apr 1814 - Place: Winchester - Period: War - BLW 13139-160-12 - Discharged at Norfolk,

Maryland Regulars in War of 1812

VA, on 8 May 1815.

Steward, Joseph - Private - 2nd Artillery - Company: Nathan Towson - Enlistment date: 16 Apr 1812 - Period: 5 Yrs.

Steward, William - Private - 14th Infantry - Company: David Cummings - Age: 27 - Height: 5' 6 1/2" - Born: Ireland - Trade: Laborer - Enlistment date: 20 Oct 1814 - Place: Baltimore - Period: War - Deserted on 16 Nov 1814.

Stewart, Charles - Private - 14th Infantry - Company: David Cummings - Age: 31 - Height: 5' 5" - Born: Ireland - Trade: Laborer - Enlistment date: 28 Nov 1814 - Place: Baltimore - Period: War - Deserted on 16 Dec 1814.

Stewart, David - Private - 36th Infantry - Company: Hugh Deneale - Pension: Old War IF-18150 Rejected - BLW 11923-160-50.

Stewart, Edward - Private - 14th Infantry - Company: Reuben Gilder - Other regiment: 4th Infantry (New) - Age: 19 - Height: 5' 6" - Born: Baltimore - Trade: Dyer - Enlistment date: 27 Feb 1814 - Place: Plattsburgh, NY - Period: 5 Yrs.

Stewart, James - Corporal - 14th Infantry - Company: Reuben Gilder - Other regiment: 4th Infantry (New) - Age: 21 - Height: 5' 7" - Born: Maryland - Trade: Laborer - Enlistment date: 23 Mar 1813 - Period: 5 Yrs - Discharged on 23 Mar 1818.

Stewart, John - Private - 12th Infantry - Company: Thomas Sangster - Age: 43 - Height: 5' 8" - Born: Maryland - Trade: Farmer - Enlistment date: 26 Jun 1812 - Place: Baltimore - Period: 5 Yrs - BLW 17710-160-12 - Discharged on 1 Sep 1813. incapable to perform the duties of a soldier.

Stewart, John C. - Private - 38th Infantry - Company: Thomas Sangster - Age: 18 - Height: 4' 9" - Born: Baltimore City - Trade: Cooper - Enlistment date: 31 Dec 1814 - Place: Baltimore - Period: 5 Yrs.

Stewart, Joseph - Private - 36th Infantry - Company: Hugh Deneale - Age: 30 - Height: 5' 10" - Born: Queen Anne's Cty, MD - Enlistment date: 28 Jul 1813 - Place: Waynesburg, PA - Pension: Wife Elizabeth Ann, WO-13103, WC-12152.

Stewart, Robert - Private - 38th Infantry - Company: John Brookes - Age: 21 - Height: 5' 10" - Born: Somerset Cty, MD - Trade: Farmer - Enlistment date: 24 Jul 1813 - Period: 1 Yr - BLW 3576-160-12 - Re-enlisted at Craney Island, VA, on 17 War 1814 for the war; discharged at Craney Island, VA, on 15 Mar 1815.

Stewart, Robert - Private - 36th Infantry - Company: Joseph Hook - Age: 33 - Height: 5' 8" - Born: Anne Arundel, MD - Enlistment date: 4 May 1813 - Place: Baltimore.

Stewart, Thomas - Sergeant - 38th Infantry - Company: Shepperd Leakin - Age: 22 - Height: 5' 9" - Born: Baltimore City - Trade: Shoemaker - Enlistment date: 1 Jul 1814 - Place: Baltimore - Period: 5 Yrs.

Stewart, William - Private - 38th Infantry - Company: John Rothrock - Age: 33 - Height: 5' 8" - Born: Talbot Cty, MD - Trade: Miller - Enlistment date: 22 Mar 1814 - Place: Richmond, VA - Period: War - BLW 5864-160-12 - Discharged at Craney Island, VA, on 15 Mar 1815.

By War of 1812 Soc. in MD

Stewart, William L. - Sergeant - 38th Infantry - Company: Charles Stansbury - Age: 21 - Height: 5' 11" - Born: Dorchester Cty, MD - Trade: Cordwainer - Enlistment date: 14 Jul 1814 - Place: Baltimore - Period: War - BLW 4710-160-12 - Discharged at Baltimore on 6 Apr 1815.

Sticher, John - First Lieutenant - 38th Infantry - Commissioned as a 2nd lieutenant on 20 May 1813; promoted to 1st lieutenant on 9 Jul 1814; discharged on 15 Jun 1815.

Sticker, Jacob - Private - 19th Infantry - Company: Joel Collins - Age: 25 - Height: 5' 6" - Born: Maryland - Trade: Laborer - Enlistment date: 7 Aug 1813 - Period: 1 Yr - BLW 7447-160-12 - Re-enlisted at Fort Malden, UC, on 6 Aug 1814 for the war; discharged at Detroit on 20 Jul 1815.

Stickling, Francis - Private - 12th Infantry - Company: Thomas Post - Age: 33 - Height: 5' 6 1/2" - Born: Prince George's Cty, MD - Trade: Laborer - Enlistment date: 22 Mar 1814 - Period: War.

Stikes, James S. - Private - 5th Infantry - Age: 21 - Height: 5' 8" - Born: Baltimore - Trade: Baker - Enlistment date: 8 Feb 1813 - Place: Baltimore - Period: 5 Yrs.

Stilwell, John - Private - 12th Infantry - Company: Thomas Sangster - Age: 22 - Height: 5' 10" - Born: Ireland - Trade: Tailor - Enlistment date: 22 Jun 1812 - Place: Baltimore - Period: 18 Mos - Discharged on 22 Dec 1813.

Stincicum, Norman - Private - 38th Infantry - Age: 29 - Height: 5' 8 3/4" - Born: Baltimore - Trade: Carpenter - Enlistment date: 11 Feb 1815 - Period: War.

Stinson, Stephen - Seaman - US Sea Fencibles - Company: John Gill - Age: 28 - Height: 5' 10" - Born: Berwick, MA - Enlistment date: 12 Jan 1814 - Period: 1 Yr.

Stirling, John - Private - 36th Infantry - Company: Joseph Hook - Age: 27 - Height: 5' 7" - Born: Calvert, MD - Enlistment date: 2 Jun 1813 - Place: Baltimore - Period: 1 Yr.

Stiver, John - Private - 5th Infantry - Company: Richard Bell - Age: 42 - Height: 5' 4" - Born: Germany - Trade: Blacksmith - Enlistment date: 9 Mar 1813 - Place: Baltimore - Period: 5 Yrs - Discharged on 9 Mar 1818.

Stivers, Elihu - Sergeant - 36th Infantry - Company: Samuel Raisin - Age: 24 - Height: 6' - Born: Newark, NJ - Trade: Millwright - Enlistment date: 24 Jun 1814 - Place: Georgetown, DC - Period: War - Discharged at Washington, DC, on 20 Mar 1815.

Stocker, Jesse L. - Sergeant - 14th Infantry - Company: Isaac Barnard - Age: 27 - Height: 5' 10 1/2" - Born: Talbot Cty, MD - Trade: Cooper - Enlistment date: 12 Jan 1813 - Period: War - BLW 406-160-12 - Prisoner of War (Halifax), captured at Beaver Dams on 24 Jun 1813; discharged at Baltimore on 30 Apr 1815.

Stoden, Henry - Private - 14th Infantry - Enlistment date: 26 Nov 1812 - Period: 5 Yrs - Prisoner of War (Halifax), discharged from Halifax on 3 Feb 1814, exchanged on 15 Apr 1814; deserted at Schenectady, NY, on 19 Jan 1819.

Stoke, Zacheous - Gunner - US Sea Fencibles - Company: William Addison.

Stokeley, Newman - Private - 14th Infantry - Company: Reuben Gilder - BLW 3990-160-12.

Stokes, David - Private - 5th Infantry - Company: Colin Buckner - Age: 21 - Height: 5' 11" - Born: Baltimore - Trade: Cordwainer - Enlistment date: 18 Feb 1813 - Place: Baltimore - Period: 5 Yrs.

Stokes, John - Private - 36th Infantry - Company: Charles Randolph - Age: 30 - Height: 5' 4" - Born: Delaware - Trade: Farmer - Enlistment date: 21 Aug 1812 - Period: 1 Yr - BLW 25572-160-12 - Re-enlisted at Annapolis on 8 Mar 1814 for the war; Prisoner of War (Halifax), captured at Washington, DC, on 24 Aug 1814; discharged at Washington, DC, on 12 May 1815.

Stoky, John - Private - 36th Infantry - Company: Hugh Deneale - BLW 25572-160-12.

Stone, Samuel - Private - 14th Infantry - Company: Clement Sullivan - Enlistment date: 29 May 1812 - Period: 5 Yrs - Pension: Land bounty to Francis Stone and other heirs at law of Samuel Stone - BLW 10020-160-12 - Died on 29 Nov 1812.

Stonebracker, Peter - Private - 35th Infantry - Company: Isaac Preston - Age: 30 - Height: 5' 6" - Born: Washington Cty, MD - Trade: Miller - Enlistment date: 29 Apr 1814 - Place: Winchester, VA - Period: War - Discharged at Norfolk, VA, on 18 Mar 1815.

Stonebraker, Peter - Private - 12th Infantry - Age: 29 - Height: 5' 7" - Born: Washington, MD - Trade: Farmer - Enlistment date: 1 May 1813 - Place: New Market, VA - Period: 18 Mos.

Stoner, John - Musician - 13th Infantry - Company: Hugh Martin - Age: 44 - Height: 5' 11" - Born: Maryland - Trade: Laborer - Enlistment date: 30 Jun 1812 - Place: Johnstown - Period: 5 Yrs - Died on 28 Feb 1815.

Stoner, John - Private - 1st Infantry - Company: John Whistler - Other regiment: 3rd Infantry (New) - Age: 21 - Height: 5' 10" - Born: Baltimore - Trade: Joiner - Enlistment date: 5 Jun 1807 - Place: Lancaster, PA - Period: 5 Yrs - Re-enlisted at Detroit on 5 Mar 1812 for five years; Prisoner of War on parole; discharged at Detroit on 9 Jun 1816 on Surgeon's Certificate, wounded in foot.

Stott, Jacob - Artificer - 5th Infantry - Company: Edward Upham - Enlistment date: 3 Feb 1814 - Period: 5 Yrs.

Stoub, Adam - Private - 36th Infantry - Company: Hugh Deneale - Age: 43 - Height: 5' 8" - Born: Frederick, MD - Enlistment date: 16 Jun 1813 - Place: Frederick, MD - Period: 1 Yr.

Stouffer, John - Musician - 36th Infantry - Company: Charles Randolph - Age: 22 - Height: 5' 4" - Born: Frederick, MD - Enlistment date: 25 May 1813 - Place: Frederick, MD - Period: 1 Yr.

Stout, John - Seaman - US Sea Fencibles - Company: John Gill - Age: 21 - Height: 5' 6 1/2" - Born: Philadelphia - Enlistment date: 29 Dec 1812 - Period: 1 Yr.

Stout, Peter - Private - 14th Infantry - Company: William McIlvane - Age: 40 - Height: 4' 8" - Born: Northampton Cty, PA - Trade: Laborer - Enlistment date: 2 Jun 1812 - Period: 5 Yrs - Pension: Old War IF-3982 - BLW 24997-

By War of 1812 Soc. in MD

160-12 - Prisoner of War, captured at Stoney Creek, wounded in shoulder blade; also wounded at La Cole Mill, Canada, wound to left knee; discharged at Burlington, VT, 18 Jan 1815 on Surgeon's Certificate.

Stoutsberger, Andrew - Private - US Volunteers - Company: Stephen Moore - Enlistment date: 8 Sep 1812 - Period: 1 Yr.

Stover, Charles - Private - 36th Infantry - Age: 27 - Height: 5' 10 3/4" - Born: Woodbridge, NJ - Enlistment date: 20 Apr 1813 - Place: Baltimore - Period: 1 Yr.

Straham, John - Private - 14th Infantry - Company: Reuben Gilder - Age: 34 - Height: 5' 7" - Born: Colerain, PA - Trade: Shoemaker - Enlistment date: 20 Jun 1813 - Place: Baltimore - Period: War - BLW 17211-160-12 - Discharged at Greenbush, NY, on 4 May 1815.

Stranghn, William - Corporal - 2nd Artillery - Company: Isaac Roach - Other regiment: Corps of Artillery - Age: 24 - Height: 5' 7 1/2" - Born: Queen Anne's Cty, MD - Trade: Brick layer - Enlistment date: 30 Jul 1814 - Place: Annapolis - Period: 5 Yrs - Discharged on 29 Jul 1819.

Straughn, William - Sergeant - 36th Infantry - Company: Joseph Merrick - Age: 22 - Height: 5' 7 3/4" - Born: Queen Anne's Cty, MD - Enlistment date: 24 May 1812 - Place: Centreville, MD - Period: 1 Yr.

Street, Jacob G. - Private - 5th Infantry - Company: Richard Whartenby - Other regiment: 3rd Infantry (New) - Age: 22 - Height: 5' 8 1/2" - Born: Montgomery Cty, PA - Trade: Hatter - Enlistment date: 11 Apr 1811 - Place: Baltimore - Period: 5 Yrs - Discharged on 11 Apr 1816.

Street, William - Private - 26th Infantry - Company: William Bezeau - Age: 21 - Height: 5' 10 1/2" - Born: Washington, DC - Trade: Farmer - Enlistment date: 6 Oct 1814 - Place: Baltimore - Period: War - BLW 7206-160-12 - Discharged at Philadelphia on 23 Mar 1815.

Streets, Ishmael - Private - 14th Infantry - Company: Reuben Gilder - Other regiment: 4th Infantry (New) - Age: 22 - Height: 6' - Born: Maryland - Trade: Laborer - Enlistment date: 18 May 1812 - Place: Maryland - Period: 5 Yrs - Prisoner of War (Halifax); discharged on 17 May 1817.

Streets, James - Corporal - 36th Infantry - Company: Joseph Hook - Age: 18 - Height: 5' 7 1/2" - Born: Talbot Cty, MD - Trade: Farmer - Enlistment date: 12 Jul 1814 - Place: Baltimore - Period: War - BLW 9702-160-12 - Discharged at Fort Covington, MD, on 30 Mar 1815.

Stretch, William - Drummer - 36th Infantry - Company: Hugh Deneale - Age: 20 - Height: 5' 5 1/2" - Born: New Jersey - Enlistment date: 9 Aug 1813 - Place: Bellaire, MD - Period: 1 Yr - Discharged on 17 Dec 1813.

Stricker, Terick - Private - 14th Infantry - Company: Thomas Montgomery - Age: 30 - Height: 5' 10 1/2" - Born: Northampton Cty, PA - Trade: Blacksmith - Enlistment date: 17 Jul 1812 - Place: Chester, PA - Period: 5 Yrs - BLW 15068-160-12 - Discharged on 23 Apr 1815 for inability.

Strickland, George - Private - 2nd Artillery - Company: Nathan Towson - Enlistment date: 2 Jul 1812 - Period: 5 Yrs - Deserted on 2 Aug 1812.

Strickland, William - Private - 19th Infantry - Company: Richard Talbott - Age: 35 - Height: 5' 6" - Born: Maryland - Trade: Tailor - Enlistment date:

Maryland Regulars in War of 1812

4 Feb 1814 - Place: Xenia, OH - Period: War.

Stromatt, John E. - Private - Light Dragoons - Company: James Reed - Age: 21 - Height: 6' - Born: Charles Cty, MD - Trade: Blacksmith - Enlistment date: 12 Mar 1814 - Place: Alexandria, DC - Period: War - Discharged at Fort Washington, MD, on 7 May 1815.

Strong, Abraham - Private - 38th Infantry - Company: John Hook - Age: 40 - Height: 5' 8 1/2" - Born: Washington, MD - Trade: Joiner - Enlistment date: 3 Aug 1814 - Place: Baltimore - Period: War - Discharged at Fort Covington, MD, on 31 Mar 1815.

Strong, Benjamin - Private - 17th Infantry - Company: Caleb Holder - Age: 26 - Height: 5' 9" - Born: Kent Cty, MD - Trade: Tailor - Enlistment date: 10 Jul 1814 - Period: War - BLW 4099-160-12 - Discharged at Detroit on 30 Jul 1815.

Strong, Jesse - Private - 17th Infantry - Company: William Adair - Age: 18 - Height: 5' 9 1/2" - Born: Maryland - Trade: Carpenter - Enlistment date: 26 May 1814 - Place: Russellville, KY - Period: War - BLW 11271-160-12 - Discharged at Chillicothe, OH, on 4 Jun 1814.

Stroud, John - Private - 14th Infantry - Company: Samuel Lane - Other regiment: 4th Infantry (New) - Age: 32 - Height: 5' 8" - Born: Allegheny Cty, PA - Enlistment date: 10 Jun 1812 - Place: Waynesburg, PA - Period: 5 Yrs - Prisoner of War (Halifax), captured at Beaver Dams on 24 Jun 1813, discharged from Halifax on 9 Nov 1813; discharged on 10 Jun 1817.

Strowd, John - Private - 12th Infantry - Company: Charles Page - Age: 30 - Height: 5' 7" - Born: Harford Cty, MD - Trade: Cooper - Enlistment date: 30 Jun 1812 - Period: War - BLW 9409-160-12 - Discharged at Pittsburgh on 19 Aug 1815 on Surgeon's Certificate, injury of the side.

Struckler, Robert - Private - 38th Infantry - Company: James Smith - Age: 22 - Height: 5' 5" - Born: Prince George's Cty, MD - Trade: Laborer - Enlistment date: 30 Mar 1814 - Place: Norfolk, VA - Period: War.

Stuart, Alexander - Major - 36th Infantry - Other regiment: 5th Infantry - Born: Maryland - Place: Maryland - Commissioned as a major, 36th Infantry, on 25 Mar 1813; transferred to the 5th Infantry on 10 Oct 1814; discharged on 15 Jun 1815.

Stuart, David - Private - 14th Infantry - Prisoner of War (Halifax), discharged from Halifax on 3 Feb 1814.

Stuart, James - Private - 36th Infantry - Company: Joseph Hook - Age: 18 - Height: 5' - Born: Virginia - Enlistment date: 5 Jul 1813 - Place: Baltimore - Period: 1 Yr - Discharged on 15 Sep 1813.

Stuart, Peter - Private - 36th Infantry - Company: Joseph Hook - Age: 19 - Height: 5' 6" - Born: Caroline Cty, MD - Trade: Farmer - Enlistment date: 29 Apr 1813 - Place: Cambridge, MD - Period: 1 Yr - BLW 16948-160-12 - Re-enlisted on 1 Mar 1814 for the war; discharged at Fort Covington, MD, on 30 Mar 1815.

Stuart, Thomas - Private - 36th Infantry - Company: Samuel Raisin - Age: 21 - Height: 5' 5" - Born: Kent Cty, DE - Enlistment date: 26 May 1812 - Place: Chester.

By War of 1812 Soc. in MD

Stuart, Thomas - Private - 14th Infantry - Company: David Cummings - Enlistment date: 14 May 1812 - Period: 5 Yrs - Prisoner of War (Halifax), sent to England.

Stubblefield, George - Private - 26th Infantry - Company: William Bezeau - Age: 21 - Height: 5' 4" - Born: Virginia - Trade: Laborer - Enlistment date: 8 Nov 1814 - Place: Baltimore - Period: War - BLW 5319-160-12 - Discharged at Philadelphia on 23 Mar 1815.

Stubbs, John W. - Private - 36th Infantry - Company: Samuel Raisin - Age: 21 - Height: 5' 9" - Born: Pennsylvania - Trade: Soldier - Enlistment date: 7 Jul 1814 - Place: Manchester - Period: War - Discharged by civil authority on 19 Jan 1815.

Studley, John - Private - 5th Infantry - Company: William Henshaw - Age: 23 - Height: 5' 10" - Born: Boston - Trade: Seaman - Enlistment date: 3 Apr 1813 - Place: Baltimore - Period: 5 Yrs.

Sturdevant, Thomas - Private - 14th Infantry - Prisoner of War (Halifax), discharged from Halifax on 3 Feb 1814.

Stutsman, Daniel - Musician - 19th Infantry - Company: Richard Talbott - Age: 22 - Height: 5' 4 1/2" - Born: Washington Cty, MD - Trade: Sickle maker - Enlistment date: 11 Apr 1814 - Place: Franklin Cty, OH - Period: War - BLW 15444-160-12 - Discharged at Chillicothe, OH, on 5 Jun 1815.

Styers, Cornelius - Private - 5th Infantry - Company: Leroy Opie - Other regiment: 3rd Infantry (New) - Age: 47 - Height: 5' 8" - Born: Frederick, MD - Trade: Shoemaker - Enlistment date: 22 Mar 1812 - Place: Virginia - Period: 5 Yrs - Discharged on 22 Mar 1817.

Suit, Nathaniel - Sergeant - 14th Infantry - Company: Thomas Montgomery - Enlistment date: 28 May 1812 - Period: 5 Yrs - Pension: Land bounty to John Smith Suit, brother and other heirs at law of Nathaniel Suit - BLW 26697-160-12 - Killed during the Battle of La Cole Mill on 30 Mar 1814.

Suit, William - Private - 36th Infantry - Company: Thomas Carbery - Age: 21 - Height: 5' 8" - Born: St. Mary's, MD - Trade: Farmer - Enlistment date: 13 Mar 1814 - Place: Leonardstown, MD - Period: War - BLW 11908-160-12 - Discharged at Baltimore on 31 Mar 1815.

Suiter, George - Sergeant - 36th Infantry - Company: Joseph Hook - Age: 33 - Height: 5' 9" - Born: Baltimore - Trade: Hatter - Enlistment date: 27 Mar 1814 - Place: Baltimore - Period: War - Discharged at Fort Covington, MD, on 30 Mar 1815.

Suitor, Arthur - Private - 36th Infantry - Company: Samuel Raisin - Other regiment: Ordnance Department - Age: 14 - Height: 4' 8" - Born: Southampton, VA - Trade: Laborer - Enlistment date: 31 May 1814 - Place: Jerusalem - Period: 5 Yrs - Transferred to the U.S. Ordnance Department, Lieutenant Baden's Detachment on 23 Mar 1815; discharged on 30 May 1819 and re-enlisted.

Sullivan, Clement - Captain - 14th Infantry - Commissioned as a captain on 28 Mar 1812; died on 14 Dec 1812.

Suman, Isaac - Private - 12th Infantry - Company: Andrew Madison - Age: 19 - Height: 5' 7" - Born: Frederick, MD - Trade: Farmer - Enlistment date: 5

Feb 1814 - Place: Hagerstown, MD - Period: 5 Yrs - Discharged at Buffalo on 6 Jun 1815 for injury of his right thigh received before enlistment.

Summers, Anthony - Private - 14th Infantry - Company: Samuel Lane - Other regiment: 4th Infantry (New) - Age: 26 - Height: 5' 8" - Born: Montgomery Cty, PA - Trade: Hatter - Enlistment date: 4 May 1812 - Place: Carlisle, PA - Period: 5 Yrs - Died at Creek Agency on 22 Aug 1816.

Summers, Frederick - Private - 17th Infantry - Company: David Holt - Age: 21 - Height: 5' 9" - Born: Washington Cty, MD - Trade: Laborer - Enlistment date: 23 May 1814 - Place: Limestone, KY - Period: War - Discharged at Chillicothe, OH, on 4 Jun 1814.

Summers, Lazarus - Private - 38th Infantry - Company: Charles Stansbury - Age: 19 - Height: 5' 5" - Born: Somerset Cty, MD - Trade: Farmer - Enlistment date: 16 Nov 1814 - Place: Baltimore - Period: War - BLW 5498-160-12 - Discharged at Baltimore on 6 Apr 1815.

Summers, Noah - Private - 38th Infantry - Company: John Brookes - Age: 25 - Height: 5' 11" - Born: Somerset Cty, MD - Trade: Ship's carpenter - Enlistment date: 10 Mar 1814 - Place: Craney Island, VA - Period: War - BLW 302-160-12 - Discharged at Craney Island, VA, on 15 Mar 1815.

Summers, Samuel - Private - 20th Infantry - Company: William Jett - Age: 21 - Height: 5' 6 3/4" - Born: Aninessix, MD - Trade: Seaman - Enlistment date: 12 Aug 1813 - Period: 5 Yrs - Deserted at Craney Island, VA, on 8 May 1815.

Summers, William - Private - 2nd Infantry - Company: William Bootes - Other regiment: 1st Infantry (New) - Age: 31 - Height: 5' 9" - Born: Fredericktown, MD - Trade: Coppersmith - Enlistment date: 8 Apr 1808 - Period: 5 Yrs - Re-enlisted on 1 Jan 1813 for five years; re-enlisted on 12 Dec 1817.

Sunderland, George - Private - 20th Infantry - Company: John Stanard - Other regiment: 8th Infantry (New) - Age: 25 - Height: 5' 10 1/2" - Born: Maryland - Trade: Cooper - Enlistment date: 15 May 1812 - Place: Chesterfield Cty, VA - Period: 5 Yrs - Discharged at New Orleans on 15 May 1817.

Surges, Nathaniel - Private - 27th Infantry - Company: Allen Reynolds - Age: 25 - Height: 5' 9" - Born: Dover, MD - Trade: Paper hanger - Enlistment date: 15 Aug 1814 - Place: New York - Period: War - BLW 15114-160-12 - Discharged on 20 Jun 1815.

Sutherley, Joseph - Private - 14th Infantry - Prisoner of War (Halifax), discharged from Halifax on 3 Feb 1814.

Swails, James - Sergeant - 5th Infantry - Company: William Henshaw - Age: 22 - Height: 5' 10 3/4" - Born: St. Mary's Cty, MD - Trade: Farmer - Enlistment date: 6 Apr 1812 - Place: Hagerstown, MD - Period: 5 Yrs.

Swain, Paul - Private - 14th Infantry - Company: Samuel Lane - Enlistment date: 18 May 1812 - Period: 5 Yrs - Pension: Heirs received half pay for five years in lieu of land bounty - Killed at Black Rock, NY, on 28 Nov 1812.

Swaizey, Jarrett - Private - 14th Infantry - Company: Thomas Montgomery - Other regiment: 4th Infantry (New) - Age: 17 - Height: 5' 5 1/2" - Born: Nantucket, MA - Trade: Seaman - Enlistment date: 8 May 1813 - Place: Alexandria, DC - Period: 5 Yrs - Deserted on 6 Aug 1815.

By War of 1812 Soc. in MD

Swallow, William - Private - 14th Infantry - Company: Joseph Marshall - Other regiment: 4th Infantry (New) - Age: 40 - Height: 5' 6" - Born: Columbia, VA - Trade: Farmer - Enlistment date: 7 Aug 1814 - Place: Alexandria, DC - Period: 5 Yrs - Deserted at Annapolis on 14 Oct 1815.

Swanistead, James - Private - 36th Infantry - Company: Joseph Hook - Age: 36 - Height: 5' 8 3/4" - Born: Calvert Cty, MD - Enlistment date: 7 Jun 1813 - Place: Baltimore - Period: 1 Yr.

Swann, Samuel - Private - 14th Infantry - Company: Richard Arell - Other regiment: 4th Infantry (New) - Age: 39 - Height: 5' 9" - Born: Cumberland, PA - Trade: Farmer - Enlistment date: 12 Jun 1812 - Place: Bladensburg, MD - Period: 5 Yrs - Prisoner of War (Halifax), captured at Beaver Dams on 24 Jun 1813, discharged from Halifax on 3 Feb 1813, exchanged on 15 Apr 1814; discharged at Annapolis on 18 Oct 1815 for inability.

Swarthwood, Joseph - Private - 14th Infantry - Company: Thomas Montgomery - Age: 18 - Height: 5' 6" - Born: New Jersey - Trade: Farmer - Enlistment date: 13 Jun 1812 - Place: Lewistown, PA - Period: 5 Yrs - Pension: Old War IF-25869 - BLW 26584-160-12 - Discharged at Greenbush, NY, on 24 Aug 1815, wounded in right leg by a musket ball.

Swartwood, Jacob - Private - 14th Infantry - Company: Samuel Lane - Age: 23 - Height: 5' 11" - Born: Maryland - Enlistment date: 10 Jun 1812 - Place: Waynesburg, PA.

Swasey, James H. - Private - 14th Infantry - Company: Thomas Montgomery - Enlistment date: 18 May 1812 - On board the fleet on Lake Champlain.

Sweaney, William - Private - 36th Infantry - Company: Charles Randolph - Age: 21 - Height: 5' 9" - Born: Maryland - Enlistment date: 20 Jul 1813 - Place: Waynesburg, PA - Period: 1 Yr - Pension: Wife Polly, SO-10538, SC-14285, WO-20587, WC-35364.

Sweeney, Daniel - Private - 2nd Artillery - Company: John Goodall - Age: 31 - Height: 5' 10" - Born: Baltimore - Trade: Sailor - Enlistment date: 26 Jun 1813 - Place: Petersburg, VA - Period: 5 Yrs - Discharged at Fort Nelson, VA, on 26 Jun 1818.

Sweeney, James - Private - 38th Infantry - Company: John Mowton - Age: 24 - Height: 5' 8" - Born: Maryland - Trade: Fallow chandler - Enlistment date: 27 Feb 1814 - Place: Alexandria, DC - Period: War.

Sweeney, James - Private - 12th Infantry - Company: James Charlton - Other regiment: 8th Infantry (New) - Age: 34 - Height: 5' 7 1/4" - Born: Ireland - Trade: Weaver - Enlistment date: 11 May 1812 - Place: Maryland - Period: 5 Yrs - Pension: Land bounty to John Sweeney, son & other heirs at law of James Sweeney - BLW 23993-160-12 - Died on 21 Aug 1815.

Sweeney, John - Private - 2nd Artillery - Company: Frederick Evans - Age: 36 - Height: 5' 8 7/8" - Born: Pennsylvania - Trade: Hatter - Enlistment date: 2 Feb 1814 - Place: Fort McHenry - Period: 5 Yrs.

Sweet, Rowland - Private - Corps of Artillery - Company: Richard Zantzinger - Other regiment: Corps of Artillery - Age: 31 - Height: 5' 7 1/2" - Born: Rhode Island - Trade: Farmer - Enlistment date: 29 Jun 1812 - Place: Baltimore - Period: 5 Yrs - Deserted at Fort Mifflin, PA, on 15 Sep 1815.

Sweeting, Benjamin B. - First Lieutenant - 38th Infantry - Pension: Land bounty to Mary Sweeting, widow of Benjamin Sweeting - BLW 5975-162-50 - Commissioned as a 2nd lieutenant on 20 May 1813; promoted to 1st lieutenant on 22 Apr 1813; resigned on 8 May 1814.

Sweeting, James - Private - 14th Infantry - Company: Henry Grindage - Enlistment date: 22 Jun 1812 - Period: 5 Yrs - Pension: Land bounty to Ann Sweeting, sister and other heirs at law of James Sweeting - BLW 15860-160-12 - Died in Oct 1814.

Swetsel, John - Private - 38th Infantry - Company: Shepperd Leakin - Age: 22 - Height: 5' 9" - Born: Fayette, VA - Trade: Farmer - Enlistment date: 5 Aug 1814 - Place: Baltimore - Period: War - BLW 14360-160-12 - Discharged at Fort McHenry on 28 Mar 1815.

Swift, Francis - Private - 26th Infantry - Company: William Bezeau - Age: 22 - Height: 5' 7" - Born: Falmouth, MA - Trade: Seaman - Enlistment date: 28 Oct 1814 - Place: Baltimore - Period: War - BLW 17349-160-12 - Discharged at Philadelphia on 22 Mar 1815.

Swift, Jarvis - Private - 36th Infantry - Company: Joseph Merrick - Age: 24 - Height: 5' 6 1/2" - Born: Falmouth, MA - Enlistment date: 10 May 1813 - Place: Baltimore - Transferred to the U.S. Navy on 26 Oct 1813.

Swift, John - Private - 38th Infantry - Company: James Smith - BLW 3158-160-12.

Swift, John - Seaman - US Sea Fencibles - Company: John Gill - Age: 21 - Height: 5' 11 1/2" - Born: Philadelphia - Enlistment date: 15 Jan 1814 - Period: 1 Yr.

Swift, Thomas - Private - 38th Infantry - Age: 36 - Height: 5' 8 1/2" - Born: Somerset Cty, MD - Trade: Carpenter - Enlistment date: 20 Oct 1814 - Place: Newtown, MD - Period: War.

Swigget, Ralph - Private - 14th Infantry - Company: David Cummings - Enlistment date: 9 Nov 1814 - Period: War - Pension: Land bounty to Peter C. Swigget, brother and other heirs at law of Ralph Swigget - BLW 23972-160-12 - Died on 15 Feb 1815.

Swiler, John - Sergeant - 19th Infantry - Company: George Kesling - Age: 22 - Height: 5' 8" - Born: Maryland - Trade: Potter - Enlistment date: 9 May 1814 - Period: War - BLW 19703-160-12 - Discharged at Zanesville, OH, on 27 Mar 1815.

Swilers, James - Private - 14th Infantry - Company: Reuben Gilder - Other regiment: 4th Infantry (New) - Age: 21 - Height: 5' 6" - Born: District of Columbia - Trade: Barber - Enlistment date: 3 Apr 1813 - Period: 5 Yrs - On board the fleet on Lake Champlain; deserted at Fort Severns on 18 Jul 1815.

Swindle, Samuel - Private - 38th Infantry - Company: Shepperd Leakin - Age: 39 - Height: 5' 3" - Born: Connecticut - Trade: Painter - Enlistment date: 25 Aug 1814 - Place: Baltimore - Period: War - BLW 11410-160-12 - Discharged at Fort McHenry on 28 Mar 1815.

Switzer, Henry - Private - 38th Infantry - Company: Shepperd Leakin - Age: 24 - Height: 5' 5" - Born: Germany - Trade: Baker - Enlistment date: 1 Jan 1814 - Place: Baltimore - Period: 1 Yr - Re-enlisted on 10 Feb 1814 for the war; deserted on 5 Feb 1815.

Switzer, John - Drum Major - 2nd Artillery - Company: Nathan Towson - Pension: Old War IF-18319 Rejected.

Switzer, John - Musician - 14th Infantry - Company: James McDonald - Age: 42 - Height: 5' 6" - Born: Lancaster, PA - Trade: Blacksmith - Enlistment date: 28 Feb 1814 - Period: 5 Yrs - BLW 2671-160-12 - Discharged at Greenbush, NY, on 1 May 1815 on Surgeon's Certificate.

Switzer, John - Private - 5th Infantry - Age: 25 - Height: 5' 7" - Born: Germany - Trade: Baker - Enlistment date: 17 May 1813 - Place: Baltimore - Period: 5 Yrs.

Sykes, Nathaniel - Private - 14th Infantry - Company: Joseph Marshall - Age: 25 - Height: 5' 6" - Born: New York - Trade: Saddler - Enlistment date: 17 Jun 1814 - Place: Baltimore - Period: War - BLW 19046-160-12 - Discharged at Greenbush, NY, on 3 May 1815.

Sylvester, Joseph - Private - 3rd Infantry - Company: John McClelland - Age: 27 - Height: 5' 6" - Born: Maryland - Trade: Painter - Enlistment date: 12 May 1813 - Place: Washington, MS - Period: War - BLW 14921-160-12 - Discharged at New Orleans on 9 Apr 1815.

Symington, John - Colonel - Ordnance Department - Born: Delaware - Graduated from the US Military Academy on 2 Mar 1815; commissioned as a 3rd lieutenant on 2 Mar 1815; promoted to 2nd lieutenant on 8 Apr 1818; promoted to 1st lieutenant on 17 May 1820; transferred to the 1st Artillery on 1 Jun 1821; promoted to captain, Ordnance, on 30 May 1832; promoted to major on 27 Mar 1842; promoted to colonel on 3 Aug 1861; retired on 1 Jun 1863; died on 4 Apr 1864.

Symonds, Daniel - Private - 17th Infantry - Age: 22 - Height: 5' 8 3/4" - Born: Montgomery Cty, MD - Trade: Laborer - Enlistment date: 9 Jun 1814 - Place: Limestone, KY - Period: War - BLW 22269-160-12.

Syms, John - Private - 36th Infantry - Company: Joseph Hook - Age: 14 - Height: 4' 7 1/2" - Born: Frederick Cty, MD - Trade: Laborer - Enlistment date: 2 Jun 1814 - Place: Frederick, MD - Period: War - Pension: Wife Margaret, WO-26379, WC-15321 - BLW 10613-160-12 - Discharged at Fort Covington, MD, on 30 Mar 1815.

T

Tabbs, Edward B. - Private - 36th Infantry - Company: Joseph Hook - Age: 27 - Height: 5' 7" - Born: Princess Anne, VA - Trade: Seaman - Enlistment date: 29 Mar 1814 - Place: Fort Washington, MD - Period: War - Deserted at Baltimore on 26 Jan 1815.

Taggart, Archibald - Private - 14th Infantry - Company: Samuel Lane - Age: 36 - Height: 5' 11" - Born: Maryland - Enlistment date: 17 May 1812 - Place: Shepherdstown - Period: 5 Yrs - Prisoner of War, sent to England.

Taggert, Samuel - Private - 36th Infantry - Company: Joseph Hook - Age: 33 - Height: 5' 7 1/2" - Born: Hampshire Cty, VA - Trade: Hatter - Enlistment date: 1 Mar 1814 - Place: Annapolis - Period: War - BLW 20417-160-12 - Discharged at Fort Covington, MD, on 30 Mar 1815.

Maryland Regulars in War of 1812

Taggert, Thomas - Private - 14th Infantry - Enlistment date: 6 Jul 1812 - Period: 18 Mos - Prisoner of War, sent to England; discharged on 31 Mar 1815.

Talbert, Elisha - Private - 4th Rifles - Company: Thomas Beall - Other regiment: Rifles - Age: 21 - Height: 5' 10" - Born: Baltimore - Trade: Farmer - Enlistment date: 29 Jul 1814 - Period: War - BLW 24279-160-12 - Discharged at Greenbush, NY, on 21 Jun 1815.

Talbot, Benjamin - Private - 14th Infantry - Company: Isaac Barnard - Enlistment date: 9 Nov 1812 - Period: 18 Mos - Prisoner of War (Halifax), discharged from Halifax on 3 Feb 1814, exchanged on 15 Apr 1814; discharged at Greenbush, NY, on 23 Jun 1814.

Talbot, Elijah - Private - 14th Infantry - Company: William McIlvane - Age: 21 - Height: 5' 10" - Born: Maryland - Enlistment date: 21 Jan 1813 - Period: 18 Mos - Prisoner of War (Halifax), captured at Beaver Dams on 24 Jun 1813, discharged from Halifax on 3 Feb 1813, exchanged on 15 Apr 1814; discharged at Plattsburgh, NY, on 20 Jul 1814.

Talbot, Joseph - Seaman - US Sea Fencibles - Company: Simmones Bunbury - Age: 33 - Height: 5' 5" - Born: Massachusetts - Enlistment date: 2 Nov 1814 - Period: 1 Yr - Discharged at Fort McHenry on 24 Mar 1815.

Talbot, William - Private - 36th Infantry - Company: Charles Randolph - Pension: SO-7955, SC-12262.

Talbott, David - Sergeant - Light Artillery - Company: James Gibson - Age: 24 - Height: 5' 8" - Born: Frederick, MD - Trade: Wheelwright - Enlistment date: 4 Mar 1812 - Place: Lancaster, PA - Period: 5 yrs - Prisoner of War, captured at Queenston Heights, UC, and paroled on 25 Mar 1813; discharged on 5 Mar 1817.

Talbott, John H. - Sergeant - 14th Infantry - Company: Thomas Montgomery - Other regiment: 4th Infantry (New) - Age: 31 - Height: 5' 10 1/2" - Born: Baltimore - Trade: Tailor - Enlistment date: 31 Mar 1813 - Place: District of Columbia - Period: 5 Yrs - Discharged at Fort Severn, MD, on 18 Oct 1815 on Surgeon's Certificate.

Talbott, John H. - Private - 14th Infantry - Company: William McIlvane - Enlistment date: 3 Feb 1813 - Period: 18 Mos - Prisoner of War (Halifax), captured at Beaver Dams on 24 Jun 1813, discharged from Halifax on 3 Feb 1813, exchanged on 15 Apr 1814; discharged at Plattsburgh, NY, on 2 Aug 1814.

Talbott, John H. - Private - 14th Infantry - Company: Clement Sullivan - Age: 30 - Height: 6' 2 1/2" - Born: Baltimore - Trade: Farmer - Enlistment date: 3 Oct 1814 - Place: Baltimore - Period: War - BLW 13460-160-12 - Discharged on 18 Mar 1815.

Tallhammer, Mathias - Private - 38th Infantry - Company: Shepperd Leakin - Age: 36 - Height: 5' 6" - Born: Frederick, MD - Trade: Farmer - Enlistment date: 18 Aug 1814 - Place: Baltimore - Period: War - Deserted on 6 Feb 1815.

Tanner, Jacob - Sergeant - 38th Infantry - Company: Shepperd Leakin - Age: 28 - Height: 5' 7" - Born: Emmetsburg, MD - Trade: Tailor - Enlistment date:

By War of 1812 Soc. in MD

29 Dec 1813 - Period: 1 Yr - Pension: No pension claim - BLW 24781-160-12 - Re-enlisted on 21 Jun 1814 for the war; discharged at Fort McHenry on 28 Mar 1815.

Tanner, Jesse - Private - 36th Infantry - Company: Thomas Carbery - Age: 18 - Height: 5' 7" - Born: Washington, DC - Enlistment date: 14 Apr 1814 - Place: Washington, DC - Period: War - BLW 163-160-12 - Discharged at Baltimore on 31 Mar 1815.

Tap, William - Private - 14th Infantry - Company: Clement Sullivan - Enlistment date: 6 May 1812 - Period: 5 Yrs - Died on 1 Dec 1812.

Tapnian, Obadiah - Private - 5th Infantry - Company: Leroy Opie - Age: 22 - Height: 5' 7 1/2" - Born: Allegany Cty, MD - Trade: Cordwainer - Enlistment date: 7 Jun 1813 - Place: Baltimore - Period: 5 Yrs.

Tarman, Joshua - Private - 14th Infantry - Company: Samuel Lane - Age: 20 - Height: 5' 9 1/2" - Born: Maryland - Enlistment date: 11 Jun 1812 - Place: Shepherdstown - Period: 5 Yrs - Prisoner of War, sent to England.

Tarr, Daniel - Private - 14th Infantry - Company: Joseph Marshall - Enlistment date: 27 Jul 1814 - Period: War - Died at Greenbush, NY, on 19 Jan 1815.

Tarr, William - Private - 14th Infantry - Company: Joseph Marshall - Age: 20 - Height: 5' 7" - Born: Worcester, MD - Trade: Farmer - Enlistment date: 14 Jul 1814 - Place: Snow Hill, MD - Period: War - BLW 6243-160-12 - Discharged at Greenbush, NY, on 16 Mar 1815.

Taylor Jr., John - Private - 26th Infantry - Company: William Bezeau - Age: 19 - Height: 5' 7" - Born: Maryland - Trade: Farmer - Enlistment date: 21 Nov 1814 - Place: Philadelphia - Period: War.

Taylor, Abraham S. - Private - 1st Light Dragoons - Company: Arthur Hayne - Age: 25 - Height: 5' 9" - Born: Baltimore - Trade: Boot maker - Enlistment date: 4 Jan 1812 - Place: Carlisle, PA - Period: 5 Yrs - Discharged on 21 Jul 1815, invalid.

Taylor, Andrew - Private - 22nd Infantry - Company: Willis Foulk - Age: 24 - Height: 5' 10 1/2" - Born: Maryland - Enlistment date: 24 May 1814 - Period: War - Pension: Land bounty to Andrew Taylor, father & other heirs at law of Andrew Taylor - BLW 26381-160-12 - Died at Fort Erie, UC, on 1 Sep 1814 from camp disease.

Taylor, Frederick - Private - 22nd Infantry - Company: Daniel McFarland - Age: 24 - Height: 5' 10" - Born: Frederick, MD - Trade: Shoemaker - Enlistment date: 12 Jun 1812 - Place: Uniontown, PA - Period: 5 Yrs - Deserted at French Mills, NY, on 24 Jan 1814.

Taylor, George - Private - 36th Infantry - Company: Thomas Carbery - Age: 16 - Height: 4' 11" - Born: St. Mary's, MD - Trade: Laborer - Enlistment date: 23 Jan 1815 - Period: War - BLW 491-320-12 - Discharged at Georgetown, DC, on 12 Mar 1815.

Taylor, Jesse - Private - 38th Infantry - Company: John Brookes - BLW 19927-160-12.

Taylor, John - Private - 14th Infantry - Company: Joseph Marshall - Age: 43 - Height: 5' 9" - Born: Jersey - Enlistment date: 4 Aug 1812 - Period: 5 Yrs - Pension: Wife Deborah, Old War CF-1068; received half pay for five years in lieu of bounty land - BLW 26779-160-12 - Died at Black Rock, NY, on 21 Feb 1813.

Taylor, John - Private - 4th Infantry - Company: Abraham Hawkins - Age: 22 - Height: 5' 5 1/2" - Born: Annapolis - Trade: Farmer - Enlistment date: 2 May 1814 - Period: War - Died at Plattsburgh, NY, in 1814.

Taylor, John - Private - 14th Infantry - Company: William McIlvane - Enlistment date: 1 Jan 1813 - Period: 18 Mos - Prisoner of War, exchanged on 15 Apr 1814.

Taylor, John - Private - 14th Infantry - Company: James McDonald - Age: 35 - Height: 5' 6" - Born: Frederick, MD - Trade: Weaver - Enlistment date: 1 Apr 1814 - Place: Williamsport, MD - Period: 5 Yrs - Discharged at Greenbush, NY, on 1 Jun 1815, disability.

Taylor, John - Private - 36th Infantry - Company: Joseph Hook - Age: 41 - Height: 5' 10" - Born: Pennsylvania - Enlistment date: 7 Jun 1813 - Place: Baltimore - Period: 1 Yr - Deserted on 17 Jun 1813.

Taylor, John N. - Private - 32nd Infantry - Company: George Goodman - Age: 31 - Height: 5' 8" - Born: Maryland - Trade: Miller - Enlistment date: 20 Jan 1814 - Place: Shippensburg, PA - Period: 1 Yr - Pension: Wife Deborah, Old war Minor File 1170; heirs received half pay for five years in lieu of land bounty - Re-enlisted on 16 Feb 1814 for five years; died on 14 Feb 1815.

Taylor, Joseph - Private - 17th Infantry - Company: Caleb Holder - Other regiment: 3rd Infantry (New) - Age: 41 - Height: 5' 8" - Born: Washington Cty, MD - Trade: Farmer - Enlistment date: 23 Sep 1814 - Place: Dayton, OH - Period: 5 Yrs - Discharged on 9 Jun 1816 on Surgeon's Certificate.

Taylor, Joseph - Private - 22nd Infantry - Company: Daniel McFarland - Age: 19 - Height: 5' 8" - Born: Maryland - Trade: Tailor - Enlistment date: 15 Jun 1812 - Place: Washington, PA - Period: 5 Yrs - Deserted on 3 Oct 1813.

Taylor, Joseph - Private - 36th Infantry - Company: Samuel Raisin - Age: 26 - Height: 5' 5" - Born: Worcester, MD - Trade: Sailor - Enlistment date: 13 Aug 1813 - Place: Snow Hill, MD - Period: 1 Yr - Re-enlisted at Sandy Point, VA, on 19 Mar 1814 for the war.

Taylor, Moses - Private - 36th Infantry - Age: 43 - Height: 5' 7" - Born: St. Mary's Cty, MD - Trade: Sailor - Enlistment date: 22 Jan 1815 - Period: War - BLW 498-320-12 - Discharged at Georgetown, DC, on 13 Mar 1815.

Taylor, Richard - Private - 14th Infantry - Company: Reuben Gilder - Other regiment: 4th Infantry (New) - Age: 26 - Height: 5' 8" - Born: Juniata Cty, PA - Trade: Farmer - Enlistment date: 1 May 1812 - Place: Havre de Grace, MD - Period: 5 Yrs - Pension: Wife Ann M., SO-2946, SC-5817, WO-20589, WC-27867 - Prisoner of War, exchanged on 11 May 1814; deserted on 1 Jun 1816.

Taylor, Samuel - Private - 14th Infantry - Age: 40 - Height: 5' 9" - Born: Accomack Cty, VA - Enlistment date: 14 Jul 1814 - Place: Salisbury, MD.

Taylor, Samuel W. - Private - 38th Infantry - Company: Shepperd Leakin - Age: 37 - Height: 6' - Born: Maryland - Trade: Farmer - Enlistment date: 3 Apr 1814 - Period: War - BLW 3558-160-12 - Discharged at Baltimore on 30 Mar 1815.

By War of 1812 Soc. in MD

Taylor, Thomas - Private - 14th Infantry - Company: William McIlvane - Age: 35 - Height: 5' 8" - Born: Ireland - Trade: Distiller - Enlistment date: 23 Mar 1813 - Period: 5 Yrs - Pension: Old War IF-25875 - BLW 14090-160-12 - Discharged at Greenbush, NY, on 4 May 1815, ruptured.

Taylor, Thomas H. - Private - 14th Infantry - Company: Reuben Gilder - Age: 28 - Height: 5' 6" - Born: Bleturbet, Caran Cty, Ireland - Trade: Paper maker - Enlistment date: 5 Feb 1813 - Place: Baltimore - Period: 5 Yrs - Pension: Wife Elizabeth, SO-23364, SC-18552, WO-27115, WC-28257 - Prisoner of War, exchanged on 11 May 1814.

Taylor, William - Private - 14th Infantry - Company: Reuben Gilder - Age: 40 - Height: 5' 11" - Born: Eastern Shore, MD - Trade: Farmer - Enlistment date: 12 Aug 1812 - Period: 5 Yrs - Pension: Land bounty to Esther Denston, daughter and only heir at law of William Taylor - BLW 26010-160-12 - Died at Sackets Harbor, NY, in 1813.

Tayman, Joseph - Private - 38th Infantry - Company: John Brookes - Age: 24 - Height: 6' 1" - Born: Prince George's Cty, MD - Trade: Farmer - Enlistment date: 24 Jul 1813 - Period: 1 Yr - Pension: Wife Elizabeth Jane, WO-27114, WC-17692 - Re-enlisted at Norfolk, VA, on 1 Mar 1814 for the war; discharged at Craney Island, VA, on 15 Mar 1815.

Teaggs, Richard - Private - 38th Infantry - Company: John Mowton - BLW 25935-160-12.

Teal, Nicholas - Private - 2nd Artillery - Company: Daniel Cushing - Age: 49 - Height: 6' - Born: Baltimore Cty - Enlistment date: 19 Aug 1812 - Period: 18 Mos.

Teel, Nicholas - Private - 27th Infantry (OH) - Company: Alexander Hill - Other regiment: 19th Infantry - Age: 40 - Height: 6' - Born: Maryland - Trade: Millwright - Enlistment date: 22 Apr 1814 - Place: Steubenville, OH - Period: War - BLW 6990-160-12 - Discharged at Chillicothe, OH, on 6 Jun 1815.

Teenis, Robert - Private - 5th Infantry - Company: Colin Buckner - Other regiment: 3rd Infantry (New) - Age: 26 - Height: 5' 2 3/4" - Born: Queen Anne, MD - Trade: Farmer - Enlistment date: 11 Apr 1813 - Period: 5 Yrs - Discharged on 11 Apr 1818.

Tennison, Samuel - 36th Infantry - Company: Thomas Carbery - Pension: WO-22512, WC-10173.

Terney, John - Private - 14th Infantry - Prisoner of War (Halifax), sent to England.

Tesh, Peter - Private - 19th Infantry - Company: William Gill - Age: 18 - Height: 5' 5 1/2" - Born: Maryland - Trade: Farmer - Enlistment date: 17 Aug 1814 - Period: War - Pension: SO-8122, SC-15357 - BLW 5891-160-12 - Discharged at Chillicothe, OH, on 5 Jun 1815.

Tevis, Peter - Corporal - 17th Infantry - Age: 34 - Height: 5' 10" - Born: Maryland - Trade: Laborer - Enlistment date: 28 Feb 1814 - Period: War - Pension: Heirs received half pay for five years in lieu of land bounty - Died on 10 Jan 1815.

Thistlewood, Charles - Private - 14th Infantry - Company: Henry Fleming -

Maryland Regulars in War of 1812

Age: 25 - Height: 5' - Born: Kent Cty, DE - Trade: Tailor - Enlistment date: 24 Dec 1812 - Period: 18 Mos - Prisoner of War (Halifax), sent to England; discharged on 31 Mar 1815.

Thomas, Allen - Musician - 2nd Artillery - Company: Nathan Towson - BLW 10595-160-12.

Thomas, Benjamin - Private - 2nd Artillery - Company: James Barker - Age: 22 - Height: 5' 7" - Born: Maryland - Trade: Cooper - Enlistment date: 7 May 1812 - Place: Wilmington, DE - Period: 18 Mos.

Thomas, David - Private - 36th Infantry - Company: Charles Randolph - Age: 44 - Height: 5' 6" - Born: Frederick, MD - Enlistment date: 11 Aug 1813 - Place: Waynesburg, PA - Period: 1 Yr - Discharged on 25 Sep 1813 after furnishing a substitute.

Thomas, Edward - Corporal - 2nd Artillery - Company: John Goodall - Age: 30 - Height: 5' 10 3/4" - Born: Dorchester Cty, MD - Enlistment date: 2 May 1814 - Place: Norfolk, VA - Period: War.

Thomas, John - Private - 37th Infantry - Company: Stephen Tilden - Age: 23 - Height: 5' 7" - Born: Baltimore - Trade: Seaman - Enlistment date: 9 Dec 1813 - Place: Hartford, CT - Period: 1 Yr - BLW 18341-160-12 - Re-enlisted at Fort Trumbull on 30 Mar 1814 for the war; discharged at New London, CT, on 10 May 1815.

Thomas, John H. - Corporal - 5th Infantry - Company: Colin Buckner - Other regiment: 3rd Infantry (New) - Age: 35 - Height: 5' 6 1/2" - Born: Elkton, MD - Trade: Miller - Enlistment date: 24 Feb 1813 - Place: Baltimore - Period: 5 Yrs - Discharged on 24 Feb 1818.

Thomas, Josiah - Private - 38th Infantry - Company: John Mowton - Age: 25 - Height: 5' 7" - Born: St. Mary's, MD - Trade: Pilot - Enlistment date: 23 Apr 1814 - Place: Richmond, VA - Period: War - BLW 4107-160-12 - Discharged on 15 Mar 1815.

Thomas, Peter - Private - 5th Infantry - Company: James Dorman - Other regiment: 3rd Infantry (New) - Age: 27 - Height: 5' 10" - Born: Maryland - Trade: Farmer - Enlistment date: 21 May 1812 - Place: Baltimore - Period: 5 Yrs - Discharged on 20 May 1817.

Thomas, William - Private - 14th Infantry - Company: Thomas Montgomery - Age: 42 - Height: 5' 5" - Born: Warwick, England - Trade: Carpenter - Enlistment date: 10 Dec 1812 - Place: Hagerstown, MD - Period: 5 Yrs - Pension: Old War WF-18442 Rejected - BLW 9765-160-12 - Discharged at Greenbush, NY, on 4 May 1815 on Surgeon's Certificate.

Thomas, William - Private - 38th Infantry - Company: Shepperd Leakin - Age: 20 - Height: 5' 5" - Born: Dorchester Cty, MD - Trade: Shoemaker - Enlistment date: 6 Sep 1814 - Place: Baltimore - Period: War - BLW 15613-160-12.

Thomas, William - Private - 14th Infantry - Company: William McIlvane - Age: 22 - Height: 5' 5" - Born: Pennsylvania - Enlistment date: 30 Jan 1813 - Period: 18 Mos - Prisoner of War (Halifax), discharged from Halifax on 3 Feb 1814, exchanged on 15 Apr 1814; discharged at Plattsburgh, NY, on 28 Jul 1814.

By War of 1812 Soc. in MD

Thomason, William - Private - 5th Infantry - Company: Leroy Opie - Other regiment: 3rd Infantry (New) - Age: 21 - Height: 5' 8 1/2" - Born: Baltimore - Trade: Laborer - Enlistment date: 26 May 1812 - Place: Baltimore - Period: 5 Yrs - Discharged on 26 May 1817.

Thompson, Amos - Private - 14th Infantry - Company: Reuben Gilder - Enlistment date: 7 Oct 1812 - Period: 5 Yrs - Pension: Land bounty to William Thompson, brother and other heirs at law of Amos Thompson - BLW 25557-160-12 - Died on 29 Dec 1813.

Thompson, Barney - Private - 36th Infantry - Company: Joseph Hook - Age: 22 - Height: 5' 10" - Born: Perth Amboy, NJ - Enlistment date: 5 May 1813 - Place: Baltimore - Period: 1 Yr.

Thompson, Daniel D. - Private - 38th Infantry - Company: Shepperd Leakin - Age: 37 - Height: 5' 7" - Born: Harford Cty, MD - Trade: Shoemaker - Enlistment date: 17 Mar 1814 - Period: 1 Yr - Re-enlisted on 10 Mar 1814 for the war; discharged on 20 Mar 1815.

Thompson, Elisha - Seaman - US Sea Fencibles - Company: Simmones Bunbury - Enlistment date: 13 Feb 1815 - Period: 1 Yr - Discharged at Fort McHenry on 24 Mar 1815.

Thompson, George - Private - 12th Infantry - Company: Andrew Madison - Other regiment: 8th Infantry (New) - Age: 19 - Height: 5' 10 1/2" - Born: Frederick Cty, MD - Trade: Miller - Enlistment date: 10 Oct 1812 - Place: Harpers Ferry, VA - Period: 5 Yrs - Discharged on 10 Oct 1817.

Thompson, Gilbert - Private - 38th Infantry - Company: John Mowton - Age: 27 - Height: 5' 9" - Born: Harford Cty, MD - Trade: Farmer - Enlistment date: 21 Feb 1814 - Place: Craney Island, VA - Period: War - BLW 14452-160-12 - Discharged at Craney Island, VA, on 15 Mar 1815.

Thompson, James - Private - 38th Infantry - Company: John Rothrock - Pension: Land bounty to Ann B. Lacoste, daughter and only heir at law of James Thompson - BLW 26918-160-12.

Thompson, James - Private - 37th Infantry - Company: Chauncey Ives - Age: 42 - Height: 5' 9 1/2" - Born: Baltimore - Trade: Farmer - Enlistment date: 12 Jun 1813 - Period: 1 Yr - Discharged on 1 Aug 1813.

Thompson, John - Private - 38th Infantry - Company: John Mowton - BLW 10290-160-12.

Thompson, John - Private - 14th Infantry - Company: Reuben Gilder - Other regiment: 4th Infantry (New) - Enlistment date: 17 Sep 1812 - Period: 5 Yrs - Discharged on 23 Jul 1815, inability.

Thompson, John - Private - 40th Infantry - Company: Seth Nye - Age: 23 - Height: 5' 9" - Born: Baltimore - Trade: Mariner - Enlistment date: 1 Oct 1813 - Place: Boston - Period: War - Discharged on 12 Oct 1813, previously enlisted in the U.S. Navy, serving on the U.S. Sloop Siren.

Thompson, John - Private - 26th Infantry - Company: William Bezeau - Age: 25 - Height: 5' 10" - Born: Cecil Cty, MD - Enlistment date: 1 Nov 1814 - Place: Baltimore - Period: War - Deserted in 1814.

Thompson, John - Private - 1st Artillery - Company: Heman Fay - Age: 27 - Height: 5' 9" - Born: Maryland - Enlistment date: 7 Sep 1814 - Place:

Maryland Regulars in War of 1812

Annapolis - Period: 5 Yrs.

Thompson, John - Private - Corps of Artillery - Company: Evans Humphrey - Age: 40 - Height: 5' 8" - Born: Baltimore Cty - Enlistment date: 27 Jul 1814 - Place: New Orleans - Period: War - Discharged on 9 Apr 1815.

Thompson, John - Private - 27th Infantry - Company: Aaron Craine - Age: 23 - Height: 5' 7" - Born: Maryland - Trade: Wheelwright - Enlistment date: 7 Jun 1814 - Period: 5 Yrs - Deserted at Sandy Hook, NJ, on 5 Dec 1814.

Thompson, John W. - First Lieutenant - 14th Infantry - Born: Maryland - Commissioned as a 2nd lieutenant on 12 Mar 1812; promoted to 1st lieutenant on 26 Jun 1813; discharged on 15 Jun 1815.

Thompson, Jonathan - Private - 14th Infantry - Company: Richard Arell - Other regiment: 4th Infantry (New) - Age: 40 - Height: 5' 8" - Born: Lancaster, PA - Trade: Miller - Enlistment date: 29 Apr 1812 - Place: Hagerstown, MD - Period: 5 Yrs - Prisoner of War (Halifax), discharged from Halifax on 3 Feb 1814, exchanged on 15 Apr 1814; discharged on 29 Apr 1817.

Thompson, Richard - Seaman - US Sea Fencibles - Company: Simmones Bunbury - Age: 27 - Height: 5' 9 1/2" - Born: Ireland - Enlistment date: 16 Mar 1812 - Period: 1 Yr - Discharged on 16 Mar 1815.

Thompson, Richard - Private - 2nd Light Dragoons - Company: John Burd - Age: 21 - Height: 5' 9 3/4" - Born: Frederick Cty, MD - Trade: Laborer - Enlistment date: 14 May 1812 - Period: 5 Yrs - Proved to be a minor.

Thompson, Richard - Private - 2nd Artillery - Company: Philemon Hawkins - Other regiment: Corps of Artillery - Age: 39 - Height: 5' 9" - Born: Queen Anne, MD - Trade: Farmer - Enlistment date: 1 Mar 1814 - Period: 5 Yrs - BLW 14945-160-12 - Discharged at Castle Pinckney, SC, on 8 Jan 1815; disability.

Thompson, Samuel A. - Private - 2nd Infantry - Company: Mathew Arbuckle - Other regiment: 1st Infantry (New) - Age: 28 - Height: 5' 10" - Born: Queenstown, MD - Trade: Physician - Enlistment date: 6 Oct 1806 - Place: Winchester, VA - Period: 5 Yrs - Re-enlisted at Fort Stoddert on 7 Oct 1811 for five years; discharged on 6 Oct 1816.

Thompson, Stan L. - Sergeant - 44th Infantry - Company: William Butler - Other regiment: 1st Infantry (New) - Age: 23 - Height: 5' 10" - Born: St. Mary's, MD - Enlistment date: 11 Jul 1814 - Place: New Orleans - Period: 5 Yrs - Discharged on 4 Apr 1816.

Thompson, Thomas - Private - 2nd Light Dragoons - Company: William Littlejohn - Other regiment: Light Dragoons - Age: 31 - Height: 6' - Born: Charles Cty, MD - Trade: Farmer - Enlistment date: 27 Dec 1813 - Place: Alexandria, DC - Period: War - BLW 214-160-12 - Discharged at Carlisle Barracks, PA, on 4 May 1815.

Thompson, Thomas - Private - Light Artillery - Company: Benjamin Branch - Other regiment: Corps of Artillery - Age: 24 - Height: 5' 10" - Born: Frederick Cty, MD - Trade: Harness maker - Enlistment date: 19 Apr 1812 - Place: Washington, DC - Period: 5 Yrs - Discharged on 18 Apr 1817.

Thompson, Valentine - Private - 14th Infantry - Company: Thomas

By War of 1812 Soc. in MD

Montgomery - Age: 47 - Height: 5' 4 1/2" - Born: Boonebury, MD - Trade: Laborer - Enlistment date: 21 Jun 1812 - Place: Hagerstown, MD - Period: 5 Yrs - BLW 21055-160-12 - Discharged at Greenbush, NY, on 4 May 1815, old age.

Thompson, William - Second Lieutenant - 14th Infantry - Born: Maryland - Enlistment date: 10 May 1812 - Period: 5 Yrs - BLW 24274-160-12 - Served as a sergeant in 14th Infantry between 10 May 1812 and Mar 1813; commissioned as an ensign on 13 Mar 1813; promoted to 3rd lieutenant on 12 May 1813; promoted to 2nd lieutenant on 5 Aug 1814; discharged on 15 Jun 1814; re-instated as a 2nd lieutenant in Rifle Regiment on 27 Oct 1820; discharged on 1 Jun 1821; died on 8 Aug 1844.

Thompson, William - Private - 25th Infantry - Company: Benjamin Watson - Age: 37 - Height: 5' 8" - Born: St. Mary's, MD - Trade: Farmer - Enlistment date: 11 Jan 1815 - Place: Norwich, VT - Period: War - BLW 109-320-12 - Discharged at Hartford, CT, On 24 Mar 1815.

Thompson, William - Private - 35th Infantry - Company: Meriwether Taliaferros - Age: 28 - Height: 5' 2" - Born: Georgetown, MD - Trade: Sailor - Enlistment date: 18 Feb 1814 - Place: Craney Island, VA - Period: War - Discharged at Norfolk, VA, on 31 Mar 1815.

Thomson, Thomas - Private - 14th Infantry - Company: James McDonald - Age: 25 - Height: 5' 11" - Born: Connecticut - Enlistment date: 24 Jan 1814 - Period: 5 Yrs - Discharged on 19 May 1814.

Thorn, Ephraim - Private - 1st Artillery - Company: Hopley Yeaton - Age: 26 - Height: 6' 1 1/2" - Born: Prince George's Cty, MD - Trade: Shoemaker - Enlistment date: 8 Dec 1810 - Place: Washington, DC - Period: 5 Yrs - Died at Fort Nelson, VA, on 22 Mar 1815.

Thorne, John - Private - 6th Infantry - Company: Gad. Humphreys - Age: 30 - Height: 5' 3" - Born: Baltimore - Trade: Seaman - Enlistment date: 30 Dec 1813 - Period: War - Deserted at Sandy Hill on 21 Jun 1814.

Thornell, William - Private - 14th Infantry - Company: Reuben Gilder - Age: 23 - Height: 5' 7" - Born: Baltimore - Trade: Brick layer - Enlistment date: 22 Jul 1813 - Place: Baltimore - Period: War - BLW 6549-160-12 - Discharged at Greenbush, NY, on 4 May 1815 because of wounds.

Thornley, William F. - Private - 6th Infantry - Age: 18 - Height: 4' 6" - Born: Maryland - Enlistment date: 23 May 1814 - Place: Philadelphia - Period: 5 Yrs.

Thornton, Jesse - Private - 12th Infantry - Company: Thomas Sangster - Age: 24 - Height: 5' 10" - Born: Cecil Cty, MD - Trade: Blacksmith - Enlistment date: 3 Jun 1812 - Place: Cecil Cty, MD - Period: 5 Yrs - BLW 3894-160-12 - Discharged at Buffalo on 6 Jun 1815, loss of left eye.

Thorpe, John - Private - 10th Infantry - Company: Emanuel Leigh - Age: 35 - Height: 5' 10 3/4" - Born: Baltimore - Trade: Mason - Enlistment date: 21 Mar 1814 - Place: Warren - Period: 5 Yrs - Deserted on 21 Jun 1815.

Thrush, George - Private - 38th Infantry - Company: Shepperd Leakin - Age: 40 - Height: 5' 10" - Born: Baltimore - Trade: Laborer - Period: War - Discharged on 11 Jan 1815 because of old age.

Maryland Regulars in War of 1812

Tibbetts, Abraham - 14th Infantry - Company: Peter Chadwick - Pension: SO-11484, SC-6887.

Tibbins, Francis - Private - 14th Infantry - Company: Reuben Gilder.

Tichenea, William - Sergeant - 12th Infantry - Age: 22 - Height: 5' 11" - Born: Allegany Cty, MD - Trade: Farmer - Enlistment date: 2 Apr 1814 - Place: Clarksburg, VA - Period: War - Discharged at Winchester, VA, on 18 Mar 1815.

Tilghman, Matthew - Private - 1st Infantry - Company: Hezekiah Johnson - Age: 38 - Height: 5' 9" - Born: Baltimore - Trade: Seaman - Enlistment date: 15 Jan 1815 - Place: New Brunswick, NJ - Period: War - Discharged on 15 Mar 1815.

Tilkins, John - Private - 14th Infantry - Prisoner of War (Halifax), discharged from Halifax on 3 Feb 1814.

Tillett, James - Drummer - 38th Infantry - Pension: Wife Minerva, Old War WF-18499 Rejected.

Tilly, George - Private - 38th Infantry - Company: John Brookes - BLW 15121-160-12.

Timms, James R. - Sergeant - 20th Infantry - Company: John Stanard - Age: 26 - Height: 5' 8" - Born: Charles Cty, MD - Trade: Farmer - Enlistment date: 26 Jul 1812 - Period: 5 Yrs - Pension: Land bounty to Joseph Timmes, father & only heir at law of James R. Timms - BLW 25190-160-12 - Died at Williamsville, NY, on 3 Mar 1813.

Tinges, John - Private - 26th Infantry - Company: William Bezeau - Age: 18 - Height: 5' 11" - Born: Baltimore - Trade: Shoemaker - Enlistment date: 9 Jan 1815 - Place: Baltimore - Period: War.

Tinnus, John - Private - 1st Infantry - Company: John Whistler - Age: 34 - Height: 5' 7 1/2" - Born: Charles Cty, MD - Trade: Turner - Enlistment date: 7 Mar 1810 - Place: Detroit - Period: 5 Yrs - Prisoner of War on parole; died at Greenbush, NY, on 17 May 1815.

Tipson, Bryant - Private - 1st Rifles - Company: Joseph Calhoun - Age: 32 - Height: 5' 8" - Born: Baltimore - Trade: Carpenter - Enlistment date: 27 Aug 1814 - Place: Knoxville, TN - Period: 5 Yrs - Deserted on 26 Aug 1819.

Tipton, Bryant - Private - 1st Rifles - Company: Joseph Calhoun - Other regiment: Rifles - Age: 32 - Height: 5' 8" - Born: Baltimore - Trade: Carpenter - Enlistment date: 27 Aug 1814 - Place: Knoxville, TN - Period: 5 Yrs - Discharged on 26 Aug 1819.

Tivis, Peter - Corporal - 17th Infantry - Company: Henry Crittenden - Age: 34 - Height: 5' 10" - Born: Maryland - Trade: Farmer - Enlistment date: 24 Feb 1814 - Period: War - Died on 10 Jan 1815.

Tobias, Samuel - Private - 14th Infantry - Company: Clement Sullivan - Age: 28 - Height: 5' 9 1/2" - Born: Duchess Cty, NY - Trade: Stone mason - Enlistment date: 29 Sep 1814 - Place: Baltimore - Period: War - BLW 14789-160-12 - Discharged at Baltimore on 16 Mar 1815.

Tobin, John - Sergeant - 38th Infantry - Company: Charles Stansbury - Age: 39 - Height: 5' 1 1/2" - Born: Ireland - Trade: Mariner - Enlistment date: 18 Oct 1814 - Place: Baltimore - Period: War.

By War of 1812 Soc. in MD

Todd, George - Seaman - US Sea Fencibles - Company: Simmones Bunbury - Age: 35 - Height: 5' 5" - Born: Ireland - Enlistment date: 2 Sep 1814 - Period: 1 Yr - Discharged at Fort McHenry on 24 Mar 1815.

Todd, John - Private - 14th Infantry - Company: David Cummings - Enlistment date: 14 May 1812 - Period: 5 Yrs - BLW 12969-160-12 - Prisoner of War (Halifax), sent to England; discharged on 30 Apr 1815 on Surgeon's Certificate.

Tolson, James - Private - 1st Light Dragoons - Company: Arthur Hayne - Age: 25 - Height: 5' 7 1/2" - Born: Centreville, MD - Trade: Farmer - Enlistment date: 7 May 1813 - Place: Centreville, MD - Period: 5 Yrs - Deserted at Fort Wood, NY, on 28 May 1817.

Tom, Henry - Private - 5th Infantry - Company: William Henshaw - Other regiment: 3rd Infantry (New) - Age: 23 - Height: 6' - Born: Hagerstown, MD - Trade: Farmer - Enlistment date: 6 Oct 1812 - Place: Pennsylvania - Period: 5 Yrs - Discharged on 5 Oct 1817.

Tomson, Henry T. - Private - 18th Infantry - Company: John Blount - Age: 22 - Height: 5' 8" - Born: Baltimore - Trade: Mariner - Enlistment date: 25 Sep 1813 - Period: War - Discharged on 28 Mar 1815.

Tonlinson, John - Private - 14th Infantry - Prisoner of War (Halifax), sent to England, captured at Beaver Dams on 24 Jun 1813.

Tonsong, Francis T. - Sergeant - 36th Infantry - Company: Thomas Carbery - BLW 4487-160-12.

Topham, Samuel - Private - 36th Infantry - Age: 28 - Height: 5' 9 1/2" - Born: Montgomery Cty, MD - Enlistment date: 9 May 1814 - Place: Georgetown, DC.

Tousong, Francis T. - Sergeant - 38th Infantry - Company: Charles Stansbury - Age: 20 - Height: 5' 3 3/4" - Born: Jacquenel, St. Domingo (Haiti) - Trade: Shoemaker - Enlistment date: 8 Aug 1814 - Place: Baltimore - Period: War - BLW 4487-160-12 - Discharged at Baltimore on 6 Apr 1815.

Towner, Greshone - Private - 36th Infantry - Company: Hugh Deneale - Age: 25 - Height: 5' 6" - Born: New York - Enlistment date: 19 Sep 1813 - Place: York, PA - Period: 1 Yr - Deserted on 20 Dec 1813.

Townes, George - Private - 36th Infantry - Company: Mortimer Hall - Age: 25 - Height: 5' 3" - Born: New Haven, CT - Trade: Sailor - Enlistment date: 29 Aug 1813 - Place: Baltimore - Period: 1 Yr - Re-enlisted on 1 Mar 1814 for the war.

Townsend, Marshal - Private - 7th Infantry - Company: Walter Overton - Other regiment: 1st Infantry (New) - Age: 20 - Height: 5' 9" - Born: Maryland - Trade: Saddler - Enlistment date: 11 Oct 1808 - Period: 5 Yrs - Re-enlisted at Fort Claiborne on 26 Oct 1813 for five years; died in New Orleans on 24 Jul 1816.

Townsend, Samuel - Corporal - 1st Artillery - Company: George Armistead - Age: 26 - Height: 5' 6 1/4" - Born: Maryland - Trade: Blacksmith - Enlistment date: 22 Feb 1812 - Place: Fort McHenry - Period: 5 Yrs - Discharged at New York City on 21 Jul 1814, invalid.

Towson, Benjamin - Private - 17th Infantry - Company: William Adair - Age:

44 - Height: 5' 10" - Born: Maryland - Enlistment date: 8 Feb 1814 - Period: War - Pension: Heirs received half pay for five years in lieu of land bounty - Died on 13 Dec 1814.

Towson, Charles B. - Private - 12th Infantry - Company: Thomas Post - Other regiment: 8th Infantry (New) - Age: 47 - Height: 5' 8 1/2" - Born: Baltimore - Trade: Shoemaker - Enlistment date: 1 Feb 1814 - Place: Staunton, VA - Period: 5 Yrs - Discharged at Fort Covington, MD, on 1 Feb 1819.

Towson, Nathan - Captain - 2nd Artillery - Other regiment: Corps of Artillery - Born: Maryland - BLW 18665-160-50 - Commissioned as a captain, 2nd Artillery, on 12 Mar 1812; transferred to Corps of Artillery on 12 May 1814; transferred to Light Artillery on 17 May 1815; promoted to colonel, 2nd Artillery, on 1 Jun 1821; died on 20 Jul 1854.

Towson, William - Musician - 5th Infantry - Company: William Henshaw - Age: 21 - Height: 5' 8 1/4" - Born: Towsontown, MD - Trade: Tanner - Enlistment date: 19 Apr 1813 - Place: Baltimore - Period: 5 Yrs - Discharged at Buffalo on 21 Jun 1815 on Surgeon's Certificate, diseased lungs.

Tracey, Richard - Private - 38th Infantry - Company: John Rothrock - Age: 30 - Height: 5' 7" - Born: Baltimore - Trade: Blacksmith - Enlistment date: 13 Jan 1814 - Period: War - Pension: Land bounty to Amelia Tracey, sister & other heirs at law of Richard Tracey - BLW 23228-160-12 - Died on 2 Oct 1814.

Tracey, William - Private - 1st Infantry - Company: Simon Owens - Other regiment: 3rd Infantry (New) - Age: 25 - Height: 5' 11" - Born: Maryland - Trade: Laborer - Enlistment date: 13 Dec 1811 - Place: Fort Belle Fontaine, MO - Period: 5 Yrs - Discharged on 13 Dec 1816.

Tracey, William - Private - Light Artillery - Company: Arthur Thornton - Age: 31 - Height: 5' 8" - Born: Montgomery Cty, MD - Trade: Potter - Enlistment date: 18 Mar 1812 - Period: 5 Yrs.

Tracey, William - Private - US Volunteers - Company: Stephen Moore - Enlistment date: 8 Sep 1812 - Period: 1 Yr - Discharged at Baltimore on 24 Sep 1812, unfit for service.

Tracy, Basil - Musician - 36th Infantry - Company: Joseph Hook - Age: 19 - Height: 5' 10" - Born: Baltimore - Enlistment date: 24 Apr 1813 - Place: Westminster, MD - Period: 1 Yr.

Tracy, Garrard - Private - 29th Infantry - Company: John Rochester - Other regiment: 6th Infantry (New) - Age: 27 - Height: 5' 7 1/2" - Born: Baltimore - Trade: Miller - Enlistment date: 8 Mar 1814 - Place: Sackets Harbor, NY - Period: 5 Yrs - Discharged at Plattsburgh, NY, on 20 Oct 1818, disability.

Tracy, Jonathan - Private - 14th Infantry - Company: Clement Sullivan - Age: 22 - Height: 5' 6 1/2" - Born: Talbot Cty, MD - Trade: Stone cutter - Enlistment date: 12 Mar 1814 - Place: Baltimore - Period: 5 Yrs - Deserted at Baltimore on 8 Apr 1814.

Tracy, Joseph - Private - 38th Infantry - Company: Sheppard Leakin - Age: 35 - Height: 5' 7" - Born: Baltimore - Trade: Miller - Enlistment date: 19 Aug 1814 - Place: Baltimore - Period: War - BLW 845-160-12 - Discharged at Fort McHenry on 28 Mar 1815.

By War of 1812 Soc. in MD

Tracy, Moses - Private - 14th Infantry - Company: Clement Sullivan - Age: 28 - Height: 5' 4" - Born: Baltimore - Trade: Laborer - Enlistment date: 8 May 1814 - Place: Baltimore - Period: War - BLW 847-160-12 - Discharged at Baltimore on 16 Mar 1815.

Tracy, Richard - Private - 36th Infantry - Company: Joseph Hook - Age: 31 - Height: 5' 9 1/2" - Born: Baltimore - Enlistment date: 8 May 1813 - Place: Westminster, MD - Period: 1 Yr - Deserted on 24 Aug 1813.

Tracy, William - Private - 14th Infantry - Company: William McIlvane - Enlistment date: 21 Dec 1812 - Period: 5 Yrs - Pension: Land bounty to Caroline Tracy, daughter and only heir at law of William Tracy - BLW 26905-160-12 - Prisoner of War (Halifax), captured at Beaver Dams on 24 Jun 1813, discharged from Halifax on 3 Feb 1813; died at Whitehall in 1814.

Trader, James - Private - 22nd Infantry - Company: George Baker - Age: 18 - Height: 5' 8 3/4" - Born: Worcester, MD - Trade: Tailor - Enlistment date: 15 Jan 1813 - Place: Cumberland, MD - Period: War - Discharged at Sackets Harbor, NY, on 24 May 1815.

Trafford, Bradford - Private - 36th Infantry - Company: Joseph Hook - Age: 25 - Height: 5' 6" - Born: Maryland - Enlistment date: 23 Jul 1813 - Place: Baltimore - Period: 1 Yr.

Tragoe, Henry - Drummer - 38th Infantry - Company: James Hook - BLW 10391-160-12.

Tragoe, John - Musician - 38th Infantry - Company: Isaac Aldridge - Age: 23 - Height: 5' 5" - Born: Dorchester Cty, MD - Trade: Waterman - Enlistment date: 17 Nov 1814 - Place: Baltimore - Period: War - BLW 10365-160-12 - Discharged at Baltimore on 6 Apr 1815.

Tragor, Levin - Private - 38th Infantry - Company: Isaac Aldridge - Age: 18 - Height: 5' 11 1/2" - Born: Dorchester Cty, MD - Trade: Waterman - Enlistment date: 28 Nov 1813 - Place: Baltimore - Period: War - BLW 3163-160-12 - Discharged at Baltimore on 6 Apr 1815.

Trainor, James - Private - 2nd Artillery - Company: Nathan Towson - Age: 18 - Height: 5' 5" - Born: Baltimore - Trade: Blacksmith - Enlistment date: 12 Apr 1812 - Place: Baltimore - Period: 5 Yrs - Prisoner of War, captured during the Battle of Stoney Creek; discharged at Fort Niagara, NY, on 12 Apr 1817.

Tranor, John - Private - 26th Infantry - Company: William Bezeau - Age: 28 - Height: 5' 10" - Born: Sopus, NY - Trade: Laborer - Enlistment date: 8 Nov 1814 - Place: Baltimore - Period: War - Discharged at Philadelphia on 23 Mar 1815.

Tranquille, Lewis - Seaman - US Sea Fencibles - Company: Simmones Bunbury - Age: 24 - Height: 5' 7" - Born: St. Domingo (Haiti) - Enlistment date: 3 Aug 1814 - Period: 1 Yr - Discharged at Fort McHenry on 24 Mar 1815.

Trapin, Joseph - Private - 5th Infantry - Company: Richard Bell - Age: 32 - Height: 5' 10" - Born: Baltimore - Trade: Mason - Enlistment date: 4 Mar 1814 - Place: Harrisburg, PA - Period: 5 Yrs - Died at Buffalo on 13 Feb 1815.

Traveller, William - Musician - 38th Infantry - Company: James Smith - BLW 25791-160-12.

Travelles, John - Seaman - US Sea Fencibles - Company: Simmones Bunbury - Age: 29 - Height: 5' 7 1/2" - Born: Hookstown, MD - Enlistment date: 3 Jan 1814 - Period: 1 Yr - Discharged on 2 Jan 1814.

Travis, Thomas - Private - 36th Infantry - Company: Samuel Raisin - Age: 24 - Height: 5' 9" - Born: Dorchester Cty, MD - Trade: Sailor - Enlistment date: 5 Aug 1814 - Place: Annapolis - Period: War.

Travlot, John - Seaman - US Sea Fencibles - Company: Simmones Bunbury.

Trenary, Thomas - Private - 14th Infantry - Company: Samuel Lane - Height: 5' 9" - Born: Maryland - Enlistment date: 13 Jan 1812 - Place: Cumberland, MD - Period: 5 Yrs - Deserted at Fort Niagara, NY, on 31 Oct 1814.

Trice, Thomas - Private - 35th Infantry - Company: Isaac Preston - Age: 23 - Height: 5' 8" - Born: Baltimore - Trade: Printer - Enlistment date: 16 Jul 1813 - Place: Winchester, VA - Period: 1 Yr - Re-enlisted at Craney Island, VA, on 22 Feb 1814 for the war; discharged on 15 Mar 1815.

Trigo, Henry - Musician - 38th Infantry - Company: John Buck - Age: 17 - Height: 5' 4 1/2" - Born: Dorchester Cty, MD - Trade: Seaman - Enlistment date: 22 Nov 1813 - Period: 1 Yr - Re-enlisted at Baltimore on 3 Nov 1814 for the war; discharged at Fort Covington, MD, on 31 Mar 1815.

Trillish, George - Sergeant - 38th Infantry - Company: John Buck - Other regiment: 4th Infantry (New) - Age: 28 - Height: 5' 5 1/4" - Born: Germany - Trade: Painter - Enlistment date: 15 Jan 1814 - Place: Baltimore - Period: 1 Yr - BLW 11642-160-12 - Re-enlisted at Baltimore on 28 Feb 1814 for the war; died on 21 Jun 1818.

Trimble, James - Private - 5th Infantry - Company: James Dorman - Other regiment: 3rd Infantry (New) - Age: 29 - Height: 5' 7 3/4" - Born: County Donegal, Ireland - Trade: Farmer - Enlistment date: 14 Apr 1812 - Place: Hagerstown, MD - Period: 5 Yrs - Discharged on 3 Apr 1817.

Trimble, John - Quarter Gunner - US Sea Fencibles - Company: John Gill - Age: 33 - Height: 5' 6 1/2" - Born: Baltimore - Enlistment date: 1 Jan 1814 - Period: 1 Yr.

Trott, Henry - Private - 14th Infantry - Age: 25 - Height: 5 '6 1/4" - Born: Maryland - Enlistment date: 6 May 1812 - Place: Newport - Period: 5 Yrs - BLW 26657-160-12 - Discharged at Greenbush, NY, on 22 Oct 1813 on Surgeon's Certificate.

Trottenburger, William - Private - 38th Infantry - Company: James Hook - Age: 37 - Height: 5' 5 1/2" - Born: Dusseldorf, Germany - Trade: Rope maker - Enlistment date: 24 May 1814 - Place: Baltimore - Period: War - BLW 300-160-12 - Discharged at Fort Covington, MD, on 31 Mar 1815.

Troy, John - Private - 38th Infantry - Company: John Buck - Age: 38 - Height: 5' 11" - Born: Cecil Cty, MD - Trade: Carpenter - Enlistment date: 19 Feb 1814 - Place: Harford Cty, MD - Period: War - BLW 7496-160-12 - Discharged at Baltimore on 30 Mar 1815.

Trueblood, Alfred - Private - 38th Infantry - Company: John Brookes - BLW 21952-160-12.

By War of 1812 Soc. in MD

Truebridge, J. - Private - 14th Infantry - Prisoner of War (Halifax), discharged from Halifax on 3 Feb 1814.

Trumbo, John - Artificer - 1st Artillery - Company: George Armistead - Age: 31 - Height: 5' 7 3/4" - Born: Baltimore - Trade: Carpenter - Enlistment date: 19 Aug 1811 - Place: Fort McHenry - Period: 5 Yrs - Discharged on 19 Aug 1816.

Tucker, Benjamin - Private - 36th Infantry - Age: 35 - Height: 5' 8" - Born: Maryland - Trade: Laborer - Enlistment date: 22 Dec 1814 - Period: War - Pension: Land bounty to Ellen Tucker, Thomas Tucker, William Tucker, Mary Tucker, Elizabeth Tucker and Henry Tucker, heirs at law of Benjamin Tucker - BLW 460-320-12 - Discharged at Georgetown, DC, on 12 Mar 1815.

Tucker, Benjamin - Private - 14th Infantry - Company: William McIlvane - Other regiment: 4th Infantry (New) - Age: 38 - Height: 5' 7" - Born: Calvert Cty, MD - Trade: Sail maker - Enlistment date: 25 Aug 1812 - Period: 5 Yrs - BLW 17682-160-12 - Prisoner of War (Halifax), captured at Beaver Dams on 24 Jun 1813, discharged from Halifax on 3 Feb 1813, exchanged on 15 Apr 1814' discharged at Fort Crawford, GA, on 17 Nov 1816 on Surgeon's Certificate.

Tucker, Joseph - Private - 14th Infantry - Company: William McIlvane - Enlistment date: 8 Jan 1813 - Period: 5 Yrs - Prisoner of War, exchanged on 15 Apr 1814; deserted at Greenbush, NY, on 17 Jun 1814.

Tucker, Joshua - Seaman - US Sea Fencibles - Company: John Gill - Age: 20 - Height: 5' 2" - Born: Baltimore - Enlistment date: 4 Jan 1814 - Period: 1 Yr.

Tucker, Joshua - Private - 26th Infantry - Company: William Bezeau - Age: 21 - Height: 5' 1" - Born: St. Mary's, MD - Trade: Tailor - Enlistment date: 26 Sep 1814 - Place: Baltimore - Period: War - BLW 13346-160-12 - Discharged at Philadelphia on 23 Mar 1815.

Tucker, Levi - Private - 12th Infantry - Company: James Paxton - Age: 21 - Height: 5' 9" - Born: Maryland - Trade: Laborer - Enlistment date: 2 Jun 1812 - Place: Morgantown, VA - Period: 18 Mos - Discharged at French Mills on 2 Dec 1813.

Tucker, Richard - Private - Light Dragoons - Company: James Reed - Other regiment: Corps of Artillery - Age: 38 - Height: 5' 9" - Born: Anne Arundel Cty, MD - Trade: Carpenter - Enlistment date: 2 Apr 1814 - Place: Alexandria, DC - Period: War - BLW 518-160-12 - Discharged at Fort Washington, MD, on 7 Mar 1815.

Tucker, Stephen - Private - 3rd Infantry - Company: James Denking - Age: 25 - Height: 5' 6" - Born: Maryland - Trade: Blacksmith - Enlistment date: 28 Jul 1808 - Place: Virginia - Period: 5 Yrs - Discharged on 21 Jul 1813.

Tucker, William L. - Private - 36th Infantry - Company: Joseph Hook - Age: 27 - Height: 5' 5 1/2" - Born: Talbot Cty, MD - Trade: Wheelwright - Enlistment date: 18 Aug 1813 - Place: Snow Hill, MD - Period: 1 Yr - BLW 6190-160-12 - Re-enlisted at Sandy Point, VA, on 12 Mar 1814 for the war; discharged at Fort Covington, MD, on 30 mar 1815.

Maryland Regulars in War of 1812

Tufts, Samuel - Private - US Volunteers - Company: Stephen Moore - Enlistment date: 8 Sep 1812 - Period: 1 Yr.

Tuling, Jeremiah - Private - 14th Infantry - Age: 29 - Height: 5' 9" - Born: Philadelphia - Trade: Baker - Enlistment date: 9 Aug 1814 - Period: War.

Tull, James - Sergeant - 36th Infantry - Company: Samuel Raisin - Age: 19 - Height: 5' 8 1/2" - Born: Caroline Cty, MD - Enlistment date: 16 Aug 1813 - Place: Greensboro, MD - Period: 1 Yr.

Tully, Charles - Private - 36th Infantry - Company: Joseph Hook - Age: 22 - Height: 5' 5 1/2" - Born: Montgomery Cty, MD - Trade: Blacksmith - Enlistment date: 11 May 1813 - Place: Sandy Point, VA - Period: War - BLW 4970-160-12 - Discharged at Fort Covington, MD, on 30 Mar 1815.

Turnbaugh, Philip - Private - 36th Infantry - Company: Joseph Hook - Age: 44 - Height: 5' 10" - Born: Baltimore - Enlistment date: 18 May 1813 - Place: Westminster, MD - Period: 1 Yr.

Turnee, Daniel C. - Private - 16th Infantry - Company: Nathaniel McLaughlin - Age: 23 - Height: 6' 1 1/4" - Born: Maryland - Trade: Chair maker - Enlistment date: 2 Apr 1814 - Place: Philadelphia - Period: 5 Yrs - Discharged on 1 Jun 1815.

Turnelly, Patrick - Private - 14th Infantry - Company: Clement Sullivan - Age: 26 - Height: 5' 2 1/2" - Born: Ireland - Trade: Weaver - Enlistment date: 22 Oct 1814 - Place: Hagerstown, MD - Period: 5 Yrs - Deserted at Baltimore on 27 Nov 1814.

Turner, Edward - Private - 2nd Infantry - Company: Thomas Swaine - Other regiment: 1st Infantry (New) - Age: 35 - Height: 5' 6" - Born: Dublin, Ireland - Trade: Stocking weaver - Enlistment date: 6 Aug 1807 - Place: Frederick, MD - Period: 5 Yrs - Re-enlisted at Fort Stoddert on 8 May 1812 for five years; discharged on 9 May 1817.

Turner, Jesse - Private - 12th Infantry - Company: Zackeville Morgan - Other regiment: 8th Infantry (New) - Age: 26 - Height: 5' 7" - Born: Prince George's Cty, MD - Trade: Laborer - Enlistment date: 15 May 1812 - Place: Winchester - Period: 5 Yrs - Discharged on 15 May 1817.

Turner, John - Corporal - 36th Infantry - Company: Samuel Raisin - Age: 22 - Height: 5' 6 1/2" - Born: Queen Anne, MD - Enlistment date: 8 Jul 1813 - Place: Kent Cty, MD - Period: 1 Yr.

Turner, John - Private - 36th Infantry - Company: Thomas Carbery - Age: 20 - Height: 5' 1 1/2" - Born: Virginia - Trade: Barber - Enlistment date: 24 Feb 1814 - Place: Leonardstown, MD - Period: War.

Turner, Joshua - Corporal - 5th Infantry - Company: Richard Bell - Age: 19 - Height: 5' 5" - Born: Baltimore Cty - Trade: Farmer - Enlistment date: 11 May 1814 - Place: Plattsburgh, NY - Period: War - BLW 3241-160-12 - Discharged at Buffalo on 31 Mar 1815.

Turner, Walter - Private - 14th Infantry - Company: Richard Arell - Other regiment: 4th Infantry (New) - Age: 25 - Height: 5' 10 1/2" - Born: Worcester, MD - Trade: Shoemaker - Enlistment date: 23 May 1812 - Place: Salisbury, MD - Period: 5 Yrs - Prisoner of War (Halifax), discharged from Halifax on 3 Feb 1814, exchanged on 15 Apr 1814; deserted at Fort Severn

By War of 1812 Soc. in MD

on 15 Jul 1815.

Turner, William - Surgeon - 8th Military District - Company: District Staff - Born: Maryland - Pension: WC-6795 - Garrison surgeon's mate from 9 Jul 1810; Prisoner of War, captured at Detroit; surgeon in 17th Infantry from 7 Apr 1813; discharged on 31 Jan 1815; served as a surgeon's mate in the 1st Artillery.

Turner, William - Private - 36th Infantry - Company: Joseph Hook - Age: 23 - Height: 5' 11" - Born: Henrico Cty, VA - Trade: Carpenter - Enlistment date: 24 Apr 1814 - Place: Fort Washington, MD - Period: War - Discharged at Fort Covington, MD, on 30 Mar 1815.

Tuttle, Thomas - Private - 36th Infantry - Company: Joseph Hook - Age: 21 - Height: 5' 6 1/4" - Born: Georgetown, DC - Trade: Carpenter - Enlistment date: 21 Jun 1814 - Place: Georgetown, DC - Period: War - BLW 550-160-12 - Discharged at Fort Covington, MD, on 31 Mar 1815.

Tutweiter, David - Third Lieutenant - 14th Infantry - Commissioned as an ensign on 22 Nov 1814; promoted to 3rd lieutenant on 1 Jan 1815; discharged on 15 Jul 1815.

Tvel, William - Private - 14th Infantry - Company: Thomas Kearney - Pension: Old War WF-18583 Rejected.

Twaits, Benjamin - Private - 14th Infantry - Company: Samuel Lane - Age: 36 - Height: 5' 10" - Born: England - Enlistment date: 28 May 1812 - Place: Carlisle, PA - Period: 5 Yrs - Deserted to the British and was reported shot by the British for being a British deserter.

Twigg, Thomas - Private - 38th Infantry - Company: Shepperd Leakin - Age: 26 - Height: 5' 9" - Born: Somerset Cty, MD - Trade: Shoemaker - Enlistment date: 30 Aug 1814 - Place: Baltimore - Period: War - Pension: Wife Mahala, WO-4971, WC-2516 - Discharged at Fort McHenry on 28 Mar 1815.

Twilley, Elijah - Private - 2nd Artillery - Company: Sanders Donoho - Other regiment: Corps of Artillery - Age: 18 - Height: 5' 7 1/4" - Born: Maryland - Trade: Farmer - Enlistment date: 12 Aug 1812 - Place: Warrenson, GA - Period: 5 Yrs - Discharged on 12 Aug 1817.

Tydd, Samuel - Private - 14th Infantry - Company: Reuben Gilder - Age: 36 - Height: 5' 8" - Born: Baltimore - Trade: Shoemaker - Enlistment date: 13 Mar 1814 - Place: Towsontown, MD - Period: War - BLW 3225-160-12 - Discharged at Greenbush, NY, on 4 May 1815.

Tyler, John - Seaman - US Sea Fencibles - Company: Simmones Bunbury - Age: 28 - Height: 5' 10" - Born: Dorchester Cty, MD - Enlistment date: 6 Jan 1814 - Period: 1 Yr - Discharged at Fort McHenry on 24 Mar 1815.

U

Ug, Augustus (or Viz) - Private - 38th Infantry - Company: Charles Stansbury - BLW 6545-160-12.

Underwood, James - Private - US Volunteers - Company: Stephen Moore - Enlistment date: 8 Sep 1812 - Period: 1 Yr.

Maryland Regulars in War of 1812

Upcraft, Thomas - Private - 38th Infantry - Company: Shepperd Leakin - Age: 28 - Height: 5' 10 1/2" - Born: Frederick, MD - Trade: Farmer - Enlistment date: 9 Apr 1814 - Period: War - Pension: Land bounty to Margaret Cochran, sister & other heirs at law of Thomas Upcraft - BLW 23500-160-12 - Died in regimental hospital on 27 Jan 1815.

Updegraff, David - Private - 14th Infantry - Company: Joseph Marshall - Age: 30 - Height: 5' 10 1/2" - Born: Frederick, MD - Enlistment date: 30 May 1814 - Place: Sharpsburg, MD - Period: War - Died at Manlins, NY, on 7 Nov 1814 from flux.

Updegraft, Jesse - Corporal - 14th Infantry - Company: Clement Sullivan - Enlistment date: 23 Jul 1812 - Period: 18 Mos - Prisoner of War, exchanged on 15 Apr 1815.

Updegrove, George - Private - 26th Infantry - Company: William Bezeau - Age: 21 - Height: 5' 8" - Born: Berks Cty, PA - Trade: Farmer - Enlistment date: 8 Nov 1814 - Place: Baltimore - Period: War.

Urie, Zacharias - Private - 2nd Artillery - Company: Nathan Towson - Other regiment: Corps of Artillery - Age: 43 - Height: 6' - Born: Pennsylvania - Trade: Carpenter - Enlistment date: 12 Apr 1812 - Period: 5 Yrs - Discharged at Buffalo on 2 Aug 1815, wounded at Black Rock, NY, on 17 Mar 1814.

Usher, Edward - Private - 26th Infantry - Company: William Bezeau - Age: 21 - Height: 5' 5" - Born: Baltimore City - Trade: Carpenter - Enlistment date: 18 Nov 1814 - Place: Baltimore - Period: War - BLW 7208-160-12 - Discharged at Philadelphia on 22 Mar 1815.

Usher, Edward - Private - 14th Infantry - Company: Isaac Barnard - Enlistment date: 13 Jan 1813 - Period: 18 Mos - Prisoner of War (Halifax), captured at Beaver Dams on 24 Jun 1813, discharged from Halifax on 3 Feb 1813, exchanged on 15 Apr 1814; discharged at Greenbush, NY, on 14 Jul 1815.

Usher, James - Private - 38th Infantry - Age: 18 - Height: 5' 4" - Born: Baltimore - Trade: Tobacconist - Enlistment date: 6 Mar 1814 - Place: Norfolk, VA - Period: War - BLW 5429-160-12 - Discharged at Craney Island, VA, on 15 Mar 1815.

Usher, John - Private - 27th Infantry (OH) - Other regiment: 19th Infantry - Age: 29 - Height: 5' 5" - Born: Maryland - Trade: Shoemaker - Enlistment date: 16 Apr 1814 - Place: Zanesville, OH - Period: War - BLW 9669-160-12 - Deserted on 28 Sep 1814.

V

Vaherting, William - Private - 38th Infantry - Company: John Brookes - BLW 2748-160-12.

Valentine, Peter - Private - 3rd Infantry - Company: Robert Moore - Age: 24 - Height: 5' 6 3/4" - Born: Havre de Grace, MD - Trade: Sailor - Enlistment date: 28 Aug 1808 - Place: Bellaire, MD - Period: 5 Yrs - Re-enlisted at English Turn, LA, on 9 Jun 1813 for the war; discharged at New Orleans on 9 Mar 1815.

Valiant, James - Private - 36th Infantry - Company: Samuel Raisin - Age: 19 - Height: 5' 6" - Born: Dorchester Cty, MD - Enlistment date: 21 Jan 1813 -

Place: Cambridge, MD - Period: 1 Yr.

Valiant, John - Gunner - US Sea Fencibles - Company: Simmones Bunbury - Age: 30 - Height: 5' 4 1/2" - Born: Talbot Cty, MD - Enlistment date: 18 Nov 1813 - Period: 1 Yr - Discharged on 17 Nov 1814.

Valiant, William - Private - 14th Infantry - Company: Thomas Kearney - Enlistment date: 19 May 1812 - Period: 5 Yrs - Pension: Land bounty to Margaret Perry, mother and heir at law of William Valiant - BLW 24594-160-12 - Prisoner of War (Halifax); died at Halifax on 26 Sep 1813.

Van Burgen, William D. - Private - US Volunteers - Company: Stephen Moore - Enlistment date: 8 Sep 1812 - Period: 1 Yr.

Vanbibber, Isaac - Private - 14th Infantry - Company: Samuel Lane - Age: 32 - Height: 5' 7" - Born: Baltimore - Trade: Farmer - Enlistment date: 8 Oct 1812 - Place: Baltimore - Period: 5 Yrs - Pension: Land bounty to Isaac Vanbibber and other heirs at law of Isaac Vanbibber - BLW 27813-160-42 - Prisoner of War, exchanged on 15 Apr 1814; discharged at Greenbush, NY, on 17 Mar 1815.

Vance, Thomas - Private - 14th Infantry - Company: Joseph Marshall - Age: 25 - Height: 5' 9 1/2" - Born: Somerset Cty, MD - Trade: Shoemaker - Enlistment date: 20 Apr 1814 - Place: Salisbury, MD - Period: War - BLW 9557-160-12 - Discharged at Greenbush, NY, on 3 May 1815.

Vanderhacken, Francis - Private - 2nd Artillery - Company: Thomas Biddle - Age: 21 - Height: 5' 4" - Born: Baltimore City - Trade: Mariner - Enlistment date: 16 Feb 1814 - Place: Philadelphia - Period: War - Discharged at Buffalo on 3 Jun 1815.

Vanosdall, Cortland - Private - 36th Infantry - Company: Joseph Hook - Age: 30 - Height: 5' 8 1/2" - Born: Essex Cty, NJ - Trade: Painter - Enlistment date: 13 May 1814 - Place: Georgetown, DC - Period: War - BLW 21735-160-12 - Discharged at Fort Covington, MD, on 30 Mar 1815.

Vanscyoe, William - Private - 14th Infantry - Company: Samuel Lane - Age: 37 - Height: 5' 7 1/2" - Born: Pennsylvania - Enlistment date: 22 Jun 1812 - Period: 18 Mos.

Varnes, John - Private - 14th Infantry - Company: William McIlvane - Age: 35 - Height: 5' 6" - Born: Loudoun Cty, VA - Trade: Laborer - Enlistment date: 11 Mar 1814 - Period: War - BLW 9705-160-12 - Discharged at Greenbush, NY, on 4 May 1815.

Varney, Edmund - Private - 36th Infantry - Company: Joseph Hook - Age: 22 - Height: 5' 6" - Born: Baltimore - Enlistment date: 12 Jul 1813 - Place: Baltimore - Period: 1 Yr.

Vashall, Joseph - Corporal - 3rd Infantry - Company: James Dinkins - Age: 30 - Height: 5' 8" - Born: Baltimore - Trade: Distiller - Enlistment date: 27 Aug 1808 - Place: Harrisburg, PA - Period: 5 Yrs - Re-enlisted at Washington, MS, on 27 Aug 1813 for the war; discharged at New Orleans on 9 Apr 1815.

Vaugh, Caleb - Sergeant - 38th Infantry - Company: Shepperd Leakin - BLW 9842-160-12.

Vaughan, Caleb - Sergeant - 14th Infantry - Company: Thomas Montgomery

Maryland Regulars in War of 1812

- Enlistment date: 26 Dec 1812 - Period: 18 Mos - Discharged at Plattsburgh, NY, on 26 Jun 1814.

Vaughan, David - Private - 36th Infantry - Company: Joseph Hook - Age: 21 - Height: 5' 5 1/4" - Born: Baltimore - Trade: Carpenter - Enlistment date: 17 Apr 1813 - Place: Westminster, MD - Period: 1 Yr - Re-enlisted at Fort Washington, MD, on 15 Mar 1814 for the war; deserted on 15 Apr 1815.

Vaughan, Thomas - Private - 5th Infantry - Company: James Dorman - Other regiment: Corps of Artillery - Age: 40 - Height: 5' 10 1/2" - Born: Wiley, Hampford County, England - Trade: Skinner - Enlistment date: 25 Jan 1812 - Place: Hagerstown, MD - Period: 5 Yrs - Discharged at Fort Mifflin, PA, on 30 Jan 1817.

Vaughn, Caleb - Private - 38th Infantry - Company: Shepperd Leakin - Age: 24 - Height: 5' 6 1/2" - Born: Baltimore City - Trade: Cordwainer - Enlistment date: 10 Jul 1814 - Place: Baltimore - Period: War - BLW 9842-160-12 - Discharged at Fort McHenry on 28 Mar 1815.

Vaughn, James - Private - 14th Infantry - Company: William McIlvane - Other regiment: 4th Infantry (New) - Age: 31 - Height: 5' 7" - Born: Lancaster, PA - Trade: Laborer - Enlistment date: 25 Feb 1813 - Period: 5 Yrs - BLW 2993-160-12 - Prisoner of War (Halifax), discharged from Halifax on 3 Feb 1814, exchanged on 15 Apr 1814; discharged at Washington, DC, on 28 Jul 1815 for inability.

Veltz, Philip - Private - 36th Infantry - Age: 43 - Height: 5' 6 1/2" - Born: Lancaster, PA - Enlistment date: 9 Jun 1812 - Place: Havre de Grace, MD.

Venables, John - Private - 14th Infantry - Company: Joseph Marshall - Age: 32 - Height: 5' 10" - Born: Somerset Cty, MD - Enlistment date: 20 Mar 1814 - Place: Salisbury, MD - Period: War - Pension: Land bounty to William Venables, Nancy White and Polly Gravener, children and only heirs at law of John Venables - BLW 27378-160-42.

Ventaugh, Peter - Private - 38th Infantry - Company: Charles Stansbury - Age: 24 - Height: 5' 2" - Born: Venice, Italy - Trade: Laborer - Enlistment date: 5 Dec 1814 - Place: Fredericktown, MD - Period: War - Discharged at Baltimore on 6 Apr 1815.

Vermillion, Benjamin - Private - 5th Infantry - Company: James Dorman - Other regiment: 3rd Infantry (New) - Age: 32 - Height: 5' 11 1/2" - Born: Prince George's Cty, MD - Trade: Farmer - Enlistment date: 23 May 1812 - Place: Alexandria, DC - Period: 5 Yrs - BLW 9983-160-12 - Discharged at Fort Mackinaw, MI, on 22 May 1817.

Vermillion, John - Private - 2nd Infantry - Company: Perrin Willis - Age: 15 - Height: 4' 5 1/2" - Born: Montgomery Cty, MD - Trade: Laborer - Enlistment date: 14 Jan 1815 - Place: Washington, DC - Period: 5 Yrs - Deserted on 16 Feb 1815.

Vermillion, Levi - Private - 2nd Light Dragoons - Company: William Littlejohn - Age: 20 - Height: 5' 6 1/2" - Born: Maryland - Trade: Laborer - Enlistment date: 29 Aug 1813 - Period: 5 Yrs - Deserted.

Veverlan, John - Private - 14th Infantry - Prisoner of War (Halifax), died at Halifax on 10 Sep 1183.

By War of 1812 Soc. in MD

Vidal, Antonia - Private - 5th Infantry - Company: Colin Buckner - Age: 40 - Height: 5' 10" - Born: France - Trade: Laborer - Enlistment date: 9 Mar 1813 - Place: Baltimore - Period: 5 Yrs - Deserted on 9 Jan 1814.

Viers, George - Private - 6th Infantry - Company: John Chapman - Age: 26 - Height: 5' 8" - Born: Montgomery Cty, MD - Trade: Hatter - Enlistment date: 22 Feb 1811 - Place: New York - Period: 5 Yrs - Discharged at Fort Niagara, NY, on 23 Feb 1816.

Vincent, Nehemiah - Private - 2nd Light Dragoons - Company: Samuel Harris - Age: 28 - Height: 5' 6" - Born: Centreville, MD - Trade: Farmer - Enlistment date: 28 Apr 1813 - Place: Centreville, MD - Period: 5 Yrs - Deserted at Fort Lewis, NY, on 2 Jun 1816.

Vincent, Stephen S. - Private - 14th Infantry - Enlistment date: 11 May 1813 - Period: 5 Yrs - Discharged on Habeas Corpus, 23 Aug 1813, being a minor.

Vintree, Jacob - Private - 5th Infantry - Company: Richard Whartenby - Other regiment: 3rd Infantry (New) - Age: 28 - Height: 5' 6 1/2" - Born: Baltimore - Trade: Printer - Enlistment date: 14 Apr 1812 - Place: Baltimore - Period: 5 Yrs - BLW 13135-160-12 - Discharged at Fort Mackinaw, MI, on 13 Apr 1817.

Vinyard, Francis - Private - 14th Infantry - Company: Joseph Marshall - Age: 25 - Height: 5' 7" - Born: Accomack Cty, VA - Trade: Stage driver - Enlistment date: 1 Sep 1813 - Place: Baltimore - Period: War - BLW 451-160-12 - Discharged at Greenbush, NY, on 3 May 1815.

Vinyard, James - Seaman - US Sea Fencibles - Company: William Addison - Enlistment date: 7 Feb 1814 - Period: 1 Yr - Discharged at Baltimore on 7 Jan 1815.

Vlach, Thomas - Private - 35th Infantry - Company: James Belsche - Age: 29 - Height: 5' 10" - Born: Prince George's Cty, MD - Trade: Plasterer - Enlistment date: 20 Nov 1814 - Place: Norfolk, VA - Period: War - Died at Camp Defiance, VA, on 4 Feb 1815.

Voice, John - Private - 14th Infantry - Company: William McIlvane - Pension: Old War IF-26483.

Voorhees, Isaac - Private - 1st Infantry - Company: Horatio Stark - Other regiment: 3rd Infantry (New) - Age: 29 - Height: 5' 9" - Born: Kent Cty, MD - Trade: Saddler - Enlistment date: 24 Mar 1812 - Place: Chillicothe, OH - Period: 5 Yrs - Discharged on 24 Mar 1817.

Voss, Ebenezer - Private - 38th Infantry - Company: James Hook - Age: 30 - Height: 5' 5 1/4" - Born: Dorchester Cty, MD - Trade: Mariner - Enlistment date: 6 Feb 1814 - Place: Baltimore - Period: War.

W

Waddle, Leonard - Private - 14th Infantry - Company: Richard Arell - Age: 22 - Height: 5' 3" - Born: Baltimore - Trade: Cooper - Enlistment date: 1 Jul 1812 - Place: Baltimore - Period: 5 Yrs - Prisoner of War (Halifax), discharged from Halifax on 3 Feb 1814, exchanged on 15 Apr 1814; discharged on 23 Mar 1821.

Waddle, Robert - Private - 10th Infantry - Company: James Clay - Age: 24 -

Height: 5' 9 1/4" - Born: Maryland - Trade: Hatter - Enlistment date: 13 May 1813 - Place: Guilford Cty, NC - Period: 5 Yrs.

Wade, John - Private - 14th Infantry - Company: James McDonald - Age: 25 - Height: 5' 6" - Born: Norfolk, VA - Trade: Sailor - Enlistment date: 19 Apr 1813 - Period: 18 Mos - BLW 10-160-12 - Re-enlisted at Baltimore on 15 Mar 1814 for the war; discharged at Greenbush, NY, on 9 Apr 1815.

Wade, William - Private - 26th Infantry - Company: William Bezeau - Age: 21 - Height: 5' 8 1/2" - Born: Dauphin Cty, PA - Enlistment date: 16 Oct 1814 - Place: Baltimore - Period: War.

Wadkins, John - Private - 2nd Infantry - Company: Turbey Thomas - Age: 31 - Height: 5' 8" - Born: Frederick, MD - Trade: Farmer - Enlistment date: 3 Apr 1814 - Place: Washington, MS - Period: War - BLW 9393-160-12 - Discharged at New Orleans on 9 Apr 1815.

Waggoner, Jacob - Private - 26th Infantry - Company: William Bezeau - Age: 21 - Height: 5' 4" - Born: Holland - Trade: Weaver - Enlistment date: 27 Oct 1814 - Place: Baltimore - Period: War - Discharged at Philadelphia on 23 Mar 1815.

Waggoner, Jeremiah - Private - 14th Infantry - Period: 18 Mos - Prisoner of War (Halifax), sent to England; discharged on 31 Mar 1815.

Waggoner, Lewis - Private - 14th Infantry - Company: Samuel Lane - Other regiment: 4th Infantry (New) - Age: 23 - Height: 5' 5" - Born: Lancaster Cty, PA - Trade: Saddler - Enlistment date: 17 Aug 1812 - Place: Nottingham, MD - Period: 5 Yrs - Discharged on 16 Aug 1817.

Wagoner, John - Private - 36th Infantry - Age: 21 - Height: 5' 9" - Born: Maryland - Enlistment date: 9 Apr 1814 - Place: Georgetown, DC.

Wailino, Timothy - Servant - US Sea Fencibles - Company: Simmones Bunbury.

Waldron, Richard - Corporal - 5th Infantry - Company: Richard Whartenby - Other regiment: Corps of Artillery - Born: Prince George's Cty, MD - Enlistment date: 20 Apr 1813 - Place: Baltimore - Period: 5 Yrs - Transferred to Corps of Artillery at Plattsburgh, NY, on 1 Aug 1814; discharged at Plattsburgh, NY, on 9 Nov 1814 on Surgeon's Certificate, injury in the line of duty.

Walker, Andrew - Private - 14th Infantry - Company: Reuben Gilder - Other regiment: 4th Infantry (New) - Age: 30 - Height: 5' 11 3/4" - Born: County Donegal, Ireland - Trade: Weaver - Enlistment date: 23 Mar 1813 - Place: Baltimore - Period: 5 Yrs - BLW 850-160-12 - Discharged at Annapolis on 16 Oct 1816 for varicose veins and rheumatism.

Walker, Chechester - Private - 38th Infantry - Company: James Smith - BLW 5739-160-12.

Walker, Edward - Private - 38th Infantry - Company: James Smith - Pension: No pension claim - BLW 425-160-12.

Walker, Fielder - Private - Ordnance Department - Company: Samuel Perkins - Age: 21 - Height: 5' 6 1/2" - Born: Prince George's Cty, MD - Trade: Farmer - Enlistment date: 27 Dec 1809 - Period: 5 Yrs - BLW 147-160-50 - Re-enlisted at Greenleaf Point, DC, on 14 Dec 1814 for the war; discharged

By War of 1812 Soc. in MD

at Washington, DC, on 7 Mar 1815.

Walker, John - Private - 14th Infantry - Company: Thomas Montgomery - Other regiment: 4th Infantry (New) - Age: 32 - Height: 5' 10" - Born: Harrisburgh, PA - Trade: Laborer - Enlistment date: 14 May 1812 - Place: Carlisle, PA - Period: 5 Yrs - Discharged on 14 May 1817.

Walker, John - Private - 26th Infantry - Company: William Bezeau - Age: 21 - Height: 5' 2" - Born: Philadelphia - Trade: Seaman - Enlistment date: 19 Oct 1814 - Place: Baltimore - Period: War - BLW 9866-160-12 - Discharged at Baltimore.

Walker, John - Private - 5th Infantry - Company: James Dorman - Other regiment: 3rd Infantry (New) - Age: 22 - Height: 5' 9" - Born: Lancaster, PA - Trade: Wagoner - Enlistment date: 4 Feb 1812 - Place: Hagerstown, MD - Period: 5 Yrs - Discharged on 3 Feb 1817.

Walker, Richard - Private - 2nd Artillery - Company: James Barker - Age: 44 - Height: 5' 6" - Born: Maryland - Trade: Blacksmith - Enlistment date: 14 Jan 1814 - Place: Reading, PA - Period: 5 Yrs - Deserted on 24 Sep 1814.

Walker, William - Corporal - 2nd Artillery - Company: Nathan Towson - Pension: Wife Edith, SO-76, SC-474, WO-17045, WC-16511.

Walker, William - Private - 12th Infantry - Company: Thomas Sangster - Other regiment: 8th Infantry (New) - Age: 21 - Height: 5' 8" - Born: St. Mary's, MD - Trade: Planter - Enlistment date: 23 Apr 1812 - Place: Rockville, MD - Period: 5 Yrs - Discharged on 22 Apr 1817.

Walker, William T. - Musician - 5th Infantry - Company: James Dorman - Other regiment: 3rd Infantry (New) - Age: 20 - Height: 6' 1/2" - Born: Virginia - Trade: Carpenter - Enlistment date: 1 Sep 1812 - Place: Maryland - Period: 5 Yrs - Discharged on 31 Aug 1817.

Walker, Zachariah - Private - 2nd Light Dragoons - Company: William Littlejohn - Other regiment: Light Dragoons - Age: 19 - Height: 5' 7" - Born: Anne Arundel Cty, MD - Trade: Sailor - Enlistment date: 25 Feb 1814 - Place: Anne Arundel Cty, MD - Period: War - BLW 417-160-12 - Discharged at Carlisle Barracks, PA, on 4 May 1815.

Wallace, James - Seaman - US Sea Fencibles - Company: William Addison.

Wallace, Joseph - Private - 38th Infantry - Company: Shepperd Leakin - Age: 27 - Height: 5' 7" - Born: Buckingham, PA - Enlistment date: 20 Aug 1814 - Place: Baltimore - Period: War - Deserted on 25 Sep 1814.

Wallace, Richard - Private - 12th Infantry - Company: Thomas Sangster - Age: 32 - Height: 5' 6" - Born: Maryland - Trade: Farmer - Enlistment date: 21 May 1812 - Place: Alexandria, DC - Period: 18 Mos - Discharged at Albany, NY, on 21 Nov 1813.

Wallace, Samuel H. - Private - 38th Infantry - Company: James Hook - Age: 30 - Height: 5' 11" - Born: Philadelphia Cty, PA - Trade: Wagoner - Enlistment date: 15 Dec 1814 - Place: Baltimore - Period: War - BLW 51-320-12 - Discharged at Fort Covington, MD, on 31 Mar 1815.

Wallace, Thomas - Private - 2nd Artillery - Company: Nathan Towson - Enlistment date: 21 May 1812 - Period: 5 Yrs.

Wallace, William - Corporal - 5th Infantry - Company: James Dorman - Other regiment: 3rd Infantry (New) - Age: 22 - Height: 5' 11" - Born: Ireland - Trade: Soap boiler - Enlistment date: 9 Jun 1813 - Place: Baltimore - Period: 5 Yrs - Discharged on 8 Jun 1815.

Wallagn, Thomas - Private - 2nd Artillery - Company: Nathan Towson - Other regiment: Corps of Artillery - Age: 34 - Height: 5' 3 1/2" - Born: New Orleans - Trade: Ship's carpenter - Enlistment date: 1 Jun 1812 - Place: Baltimore - Period: 5 Yrs - Discharged at Fort Niagara, NY, on 1 Jun 1817.

Wallerman, John - Private - 5th Infantry - Company: Richard Whartenby - Age: 30 - Height: 5' 6" - Born: Germany - Trade: Whitesmith - Enlistment date: 13 Feb 1813 - Place: Baltimore - Period: 5 Yrs.

Walling, Joseph - Private - 38th Infantry - Company: John Rothrock - BLW 9127-160-12.

Walsh, Moses - Seaman - US Sea Fencibles - Company: William Addison.

Walters, Henry - Musician - 19th Infantry - Company: Wilson Elliott - Other regiment: 25th Infantry - Age: 24 - Height: 5' 8 1/2" - Born: Maryland - Trade: Farmer - Enlistment date: 30 Jul 1812 - Place: Warren, Trumbull Cty, OH - Period: 5 Yrs - Pension: Old War IF-18788 Rejected - Wounded at Mississinewa, IN, on 18 Dec 1812; discharged on 27 Jul 1817.

Walters, James - Private - 14th Infantry - Company: Reuben Gilder - Other regiment: 4th Infantry (New) - Age: 43 - Height: 6' 6" - Born: St. Mary's, MD - Trade: Blacksmith - Enlistment date: 1 Jan 1812 - Place: Alexandria, DC - Period: 5 Yrs - Discharged on 31 Dec 1817.

Walton, Benjamin S. - Private - 26th Infantry - Company: William Bezeau - Age: 23 - Height: 5' 8 1/2" - Born: York, York Cty, PA - Trade: Tailor - Enlistment date: 21 Oct 1814 - Place: Baltimore - Period: War - BLW 10775-160-12 - Discharged at Philadelphia on 23 Mar 1815.

Walton, Edward - Private - 14th Infantry - Company: Henry Fleming - Age: 22 - Height: 5' 2 1/4" - Trade: Cordwainer - Enlistment date: 3 Feb 1813 - Period: 18 Mos - Prisoner of War (Quebec), captured during the Battle of Stoney Creek and released on 31 Oct 1813; discharged on 7 Aug 1814.

Wann, James - Private - 14th Infantry - Company: William McIlvane - Enlistment date: 25 Feb 1813 - Period: 5 Yrs.

Waples, Burton - Private - 14th Infantry - Company: Henry Fleming - Enlistment date: 7 Nov 1812 - Period: 18 Mos - Pension: SO-4769, SC-10428 - Prisoner of War (Halifax), captured at Beaver Dams on 24 Jun 1813, discharged from Halifax on 3 Feb 1813, exchanged on 15 Apr 1814; discharged at Greenbush, NY, on 23 Jun 1814.

Ward, Ebenezer - Private - 1st Artillery - Company: Michael Walsh - Other regiment: Corps of Artillery - Height: 5' 10" - Born: Maryland - Trade: Drummer - Enlistment date: 13 Sep 1809 - Period: 5 Yrs - BLW 19371-160-12 - Re-enlisted on 18 Nov 1813 for five years; discharged at New Orleans on 20 June 1818 and re-enlisted.

Ward, Edward - Private - 38th Infantry - Company: Shepperd Leakin - Age: 22 - Height: 5' 11" - Born: Ireland - Trade: Cooper - Enlistment date: 5 Sep 1814 - Place: Baltimore - Period: War - Deserted at Baltimore on 1 Jan 1815.

Ward, Joseph - Private - 38th Infantry - Company: James Hook - Age: 22 -

By War of 1812 Soc. in MD

Height: 5' 7 1/2" - Born: Baltimore City - Trade: Painter - Enlistment date: 20 Dec 1814 - Place: Baltimore - Period: War - Discharged on 31 Mar 1815.

Ward, Thomas - Private - 38th Infantry - Company: Shepperd Leakin - Age: 36 - Height: 5' 9 1/2" - Born: Somerset Cty, MD - Trade: Cordwainer - Enlistment date: 19 Jul 1814 - Place: Baltimore - Period: War - BLW 16650-160-12 - Discharged at Fort McHenry on 28 Mar 1815.

Wardle, John G. - Private - 2nd Light Dragoons - Company: William Littlejohn - Age: 16 - Height: 5' 6" - Born: Cecil Cty, MD - Enlistment date: 29 Dec 1813 - Place: Baltimore - Period: 5 Yrs.

Ware, James - Private - 26th Infantry - Company: William Bezeau - Age: 24 - Height: 5' 6" - Born: Allegany Cty, MD - Trade: Shoemaker - Enlistment date: 3 Oct 1814 - Place: Baltimore - Period: War - Discharged at Baltimore.

Ware, William - Private - 1st Rifles - Company: William Smyth - Other regiment: Rifles - Age: 43 - Height: 5' 11 1/4" - Born: Port Tobacco, MD - Trade: Laborer - Enlistment date: 26 Feb 1812 - Place: Charles Cty, MD - Period: 5 Yrs - Discharged on 30 Mar 1817.

Warfield, George - Seaman - US Sea Fencibles - Company: Simmones Bunbury - Age: 22 - Height: 5' 10 1/4" - Born: Anne Arundel Cty, MD - Enlistment date: 8 Jan 1814 - Period: 1 Yr - Discharged on 7 Jan 1815.

Waring, John - Captain - 14th Infantry - Pension: Land bounty to Jane B. Waring, widow of John Waring - BLW 4949-161-50 - Commissioned as a 2nd lieutenant on 12 Mar 1812; promoted to 1st lieutenant on 13 Mar 1813; promoted to captain on 14 Nov 1814; discharged on 15 Jun 1815.

Warman, Solomon - Private - 1st Rifles - Company: Joshua Hamilton - Other regiment: Rifles - Age: 42 - Height: 6' 1 3/4" - Born: Maryland - Trade: Millwright - Enlistment date: 8 May 1813 - Place: Rutledge, TN - Period: 5 Yrs - BLW 22300-160-12 - Discharged at Belle Fontaine, MO, on 4 Dec 1818.

Warman, Solomon - Private - Rifles - Company: George Gray - Age: 42 - Height: 6' 1 3/4" - Born: Maryland - Trade: Millwright - Enlistment date: 8 May 1813 - Place: Rutledge, TN - Period: 5 Yrs - BLW 22300-160-12 - Discharged at Fort Belle Fontaine, MO, on 4 Dec 1818.

Warner, Thomas - Ensign - US Volunteers - Company: Stephen Moore - Enlistment date: 8 Sep 1812 - Period: 1 Yr - Pension: Land bounty to Mary Ann Warner, widow of Thomas Warner - BLW 7691-160-50.

Warnick, William - Sergeant - 36th Infantry - Company: Thomas Carbery - Age: 22 - Height: 5' 7 1/4" - Born: Belfast, Ireland - Enlistment date: 26 Apr 1814 - Place: Georgetown, DC - Period: War.

Warren Benney W. - Private - 27th Infantry - Age: 23 - Height: 5' 9 1/2" - Born: Maryland - Enlistment date: 25 Apr 1814 - Place: Newark, NJ.

Warren, Benny W. - Private - 27th Infantry (OH) - Age: 23 - Height: 5' 9 1/2" - Born: Maryland - Enlistment date: 25 Apr 1814 - Place: Newark, OH.

Warren, Edward - Private - 18th Infantry - Company: William Blount - Age: 22 - Height: 5' 8" - Born: Dorchester Cty, MD - Trade: Hatter - Enlistment date: 18 Sep 1814 - Period: War - Discharged on 28 Mar 1815.

Warrick, John - Seaman - US Sea Fencibles - Company: John Gill - Age: 20 -

Height: 5' 7 1/2" - Born: Harford Cty, MD - Enlistment date: 9 Jan 1814 - Period: 1 Yr.

Washington, Lloyd - Private - 5th Infantry - Company: Colin Buckner - Age: 23 - Height: 6' 2" - Born: Anne Arundel, MD - Enlistment date: 5 Mar 1813 - Period: 5 Yrs.

Washington, William - Second Lieutenant - 36th Infantry - Other regiment: ex 25th Infantry - Born: Virginia - Commissioned as an ensign, 25th Infantry, on 23 Apr 1814; transferred to the 36th Infantry on 23 May 1814; promoted to 2nd lieutenant on 1 Jan 1815; discharged on 15 Jun 1815; died on 6 Mar 1830.

Waterbury, Thaddeus - Private - 6th Infantry - Company: John Walworth - Age: 35 - Height: 5' 4" - Born: Prince George's Cty, MD - Trade: Shoemaker - Enlistment date: 30 Sep 1813 - Place: New York - Period: War - Discharged at Plattsburgh, NY, on 24 Jul 1815.

Waterfield, William - Private - 14th Infantry - Company: Joseph Marshall - Age: 25 - Height: 5' 7" - Born: Accomack Cty, VA - Enlistment date: 4 Jul 1814 - Place: Snow Hill, MD - Period: War - Pension: Land bounty to John Waterfield, brother and other heirs at law of William Waterfield - BLW 15974-160-12 - Died at Buffalo on 20 Dec 1814.

Waters, John - Private - 38th Infantry - Company: Isaac Aldridge - Age: 18 - Height: 5' 2 1/4" - Born: Somerset Cty, MD - Trade: Waterman - Enlistment date: 12 Dec 1814 - Place: Baltimore - Period: War - BLW 237-320-12 - Discharged at Baltimore on 6 Apr 1815.

Waters, Levi - Sergeant - 36th Infantry - Age: 29 - Height: 5' 10" - Born: Charles Cty, MD - Trade: Carpenter - Enlistment date: 3 Aug 1814.

Wather, Laurence - Private - 36th Infantry - Company: Thomas Carbery - Age: 19 - Height: 5' 6 1/4" - Born: Maryland - Trade: Plasterer - Enlistment date: 14 Mar 1814 - Place: Leonardstown, MD - Period: War - Discharged at Baltimore on 31 Mar 1815.

Wathers, George - Private - 5th Infantry - Age: 22 - Height: 5' 6" - Born: Virginia - Trade: Carpenter - Enlistment date: 4 May 1813 - Place: Baltimore - Period: 5 Yrs.

Watkins, Gassaway - Second Lieutenant - 38th Infantry - Commissioned as an ensign, 38th Infantry, on 20 May 1813; promoted to 3rd lieutenant on 31 Dec 1813; promoted to 2nd lieutenant on 20 May 1814; died on 8 Sep 1814.

Watkins, James - Private - 44th Infantry - Company: Isaac Baker - Age: 33 - Height: 6' - Born: Caroline Cty, MD - Trade: Blacksmith - Enlistment date: 18 Apr 1814 - Place: Columbia, TN - Period: War - BLW 7466-160-12 - Discharged at Fort St. John, LA, on 18 Apr 1815.

Watkins, James - Sergeant - 12th Infantry - Company: Thomas Moore - Other regiment: 8th Infantry (New) - Age: 21 - Height: 5' 8 1/2" - Born: Maryland - Trade: Cabinet Maker - Enlistment date: 1 May 1812 - Place: Martinsburg, VA - Period: 5 Yrs - Discharged on 31 May 1817.

Watkins, John - Private - 2nd Infantry - Company: Reuben Chamberlain - Age: 27 - Height: 5' 10" - Born: Frederick Cty, MD - Trade: Farmer - Enlistment date: 16 Mar 1808 - Place: Fredericktown, MD - Period: 5 Yrs.

By War of 1812 Soc. in MD

Watkins, Tobias - Surgeon - 38th Infantry - Born: Maryland - BLW 22788-160-50 - Commissioned as a surgeon, 38th Infantry, on 20 May 1813; promoted to hospital surgeon on 30 Mar 1814; promoted to assistant surgeon on 18 Apr 1818; discharged on 1 Jun 1821; died on 14 Nov 1855.

Watkins, William - Private - 14th Infantry - Company: David Cummings - Age: 22 - Height: 5' 7 1/2" - Born: Baltimore - Enlistment date: 11 Jul 1814 - Place: Baltimore - Period: War - Discharged on 18 Mar 1815.

Watkins, William - Private - US Volunteers - Company: Stephen Moore - Enlistment date: 8 Sep 1812 - Period: 1 Yr.

Wats, Richard - Private - Light Dragoons - Age: 36 - Height: 5' 6" - Born: Baltimore - Trade: Collier - Enlistment date: 23 Mar 1813 - Place: Red Hook, NY - Period: 5 Yrs.

Watson, Archibald - Private - 19th Infantry - Company: George Kesling - Other regiment: 3rd Infantry (New) - Age: 19 - Height: 5' 6" - Born: Maryland - Trade: Farmer - Enlistment date: 4 Feb 1815 - Place: Zanesville, OH - Period: 5 Yrs - Pension: Wife Cassandra, Old War IF-18829 Rejected, WO-11477, WC-27889, SO-18339, SC-21692; also served in Capt. Thomas. Mountjoy's Company, OH Militia; lived in Indiana - Discharged on 4 Feb 1820.

Watson, James - Private - 26th Infantry - Company: William Bezeau - Age: 23 - Height: 5' 8" - Born: Ireland - Enlistment date: 10 Sep 1814 - Place: Baltimore - Period: War - Deserted on 10 Sep 1814.

Watson, Jesse - Private - 1st Infantry - Company: Horatio Stark - Other regiment: 3rd Infantry (New) - Age: 38 - Height: 5' 8" - Born: Maryland - Trade: Soldier - Enlistment date: 24 Dec 1811 - Place: Fort Madison, IA - Period: 5 Yrs - Discharged at Sackets Harbor, NY, on 5 Jul 1815 for disability.

Watson, John - Private - 38th Infantry - Age: 21 - Height: 5' 9 1/2" - Born: Scotland - Trade: Cordwainer - Enlistment date: 4 Oct 1814 - Place: Baltimore - Period: War - Deserted on 5 Oct 1814.

Watson, Major - Private - 14th Infantry - Prisoner of War (Halifax), sent to England.

Watson, Mathew - Corporal - 2nd Artillery - Company: Nathan Towson - Enlistment date: 16 Oct 1812 - Period: 18 Mos - Died on 27 Aug 1813.

Watson, William - Private - 14th Infantry - Period: 18 Mos - Prisoner of War (Halifax), sent to England.

Watson, William - Private - 26th Infantry - Company: William Bezeau - Age: 25 - Height: 5' 7" - Born: Harford Cty, MD - Trade: Seaman - Enlistment date: 14 Dec 1814 - Period: War - BLW 9875-160-12 - Discharged at Baltimore.

Watt, Archibald - Private - 4th Rifles - Company: Matthew Magee - Age: 29 - Height: 5' 8" - Born: Maryland - Trade: Farmer - Enlistment date: 3 May 1814 - Period: 5 Yrs - Deserted at Pittsburgh on 4 May 1814.

Watts, John - Private - 36th Infantry - Company: Thomas Carbery - Age: 33 - Height: 5' 10" - Born: Mecklenburg, VA - Trade: Clerk - Enlistment date: 4 May 1814 - Place: Georgetown, DC - Period: War - BLW 8791-160-12 -

Discharged at Baltimore on 31 Mar 1815.

Watts, Richard K. - First Lieutenant - 36th Infantry - Commissioned as a 2nd lieutenant on 30 Apr 1813; promoted to 1st lieutenant on 1 May 1814; discharged on 15 Jun 1815.

Watts, Robert - Private - 38th Infantry - Company: John Rothrock - Pension: Old War MC-878, heirs received half pay for five years in lieu of land bounty - Died on 5 Jan 1815.

Wayne, John - Private - 14th Infantry - Age: 38 - Born: Virginia - Prisoner of War (Quebec), died at Quebec on 6 Oct 1813.

Wayne, Joseph - Private - 14th Infantry - Company: Henry Fleming - Enlistment date: 23 Jan1813 - Period: War - Pension: Land bounty to John L. Wirt, sole heir at law of Joseph Wayne - BLW 26945-160-12 - Prisoner of War, died about 1 Aug 1813.

Wayne, Michael - Private - 14th Infantry - Prisoner of War (Halifax), sent to England.

Weaver, Cornelius - Private - 14th Infantry - Age: 24 - Height: 5' 8" - Born: Pennsylvania - Enlistment date: 28 Mar 1814 - Place: Baltimore - Period: War.

Weaver, Jacob - Private - 36th Infantry - Company: Joseph Hook - Age: 36 - Height: 5' 6" - Born: Frederick, MD - Enlistment date: 2 May 1813 - Place: Westminster, MD - Period: 1 Yr.

Weaver, John - Private - 36th Infantry - Company: Thomas Carbery - Enlistment date: 4 Oct 1813 - Period: 1 Yr - Discharged on 3 Oct 1814.

Weaver, Lewis - Sergeant - 5th Infantry - Company: Richard Whartenby - Other regiment: 3rd Infantry (New) - Age: 22 - Height: 5' 10" - Born: Philadelphia - Trade: Baker - Enlistment date: 1 Apr 1811 - Place: Baltimore - Period: 5 Yrs - Discharged on 1 Apr 1816.

Weaver, William - Private - 14th Infantry - Company: Thomas Montgomery - Enlistment date: 24 Dec 1812 - Period: 18 Mos - Discharged at Sackets Harbor, NY, on 19 Sep 1814.

Weaverling, John - Private - 2nd Light Dragoons - Company: John Burd - Age: 29 - Height: 5' 9" - Born: Washington, MD - Trade: Laborer - Enlistment date: 25 Aug 1812 - Period: 5 Yrs - Pension: Land bounty to Jacob Weaverling, brother & other heirs at law of John Weaverling - BLW 26895-160-12 - Prisoner of War (Halifax), captured at Beaver Dams on 24 Jun 1813, died at Halifax.

Webb, Elisha - Private - 14th Infantry - Company: Reuben Gilder - Age: 33 - Height: 6' 1" - Born: Snow Hill, MD - Trade: Farmer - Enlistment date: 15 Aug 1812 - Period: 5 Yrs - BLW 4425-160-12 - Discharged at Greenbush, NY, on 24 May 1815 on Surgeon's Certificate, rheumatism.

Webb, John - Private - 14th Infantry - Company: David Cummings - Age: 30 - Height: 5' 10 1/2" - Born: Cecil Cty, MD - Trade: Weaver - Enlistment date: 30 Sep 1814 - Place: Baltimore - Period: War - BLW 27610-160-42 - Discharged on 16 Mar 1815.

Webb, John - Private - 16th Infantry - Company: George Steele - Age: 22 - Height: 5' 10" - Born: Maryland - Trade: Sailor - Enlistment date: 10 Aug

By War of 1812 Soc. in MD

1814 - Place: Philadelphia - Period: 5 Yrs.

Webb, Lambert - Corporal - 38th Infantry - Company: John Buck - Age: 19 - Height: 5' 8" - Born: Harford Cty, MD - Trade: Farmer - Enlistment date: 1 Jul 1813 - Period: 1 Yr - Pension: Land bounty to William Webb, brother and other heirs at law of Lambert Webb - BLW 16720-160-12 - Re-enlisted at Baltimore on 28 Feb 1814 for the war; died in regimental hospital on 23 Feb 1815.

Webb, Thomas - Private - 38th Infantry - Company: James Hook - Age: 21 - Height: 5' 7" - Born: Annapolis - Trade: Farmer - Enlistment date: 12 Apr 1814 - Place: Annapolis - Period: War - BLW 1680-160-12 - Discharged at Fort Covington, MD, on 31 Mar 1815.

Webb, Thomas H. - Private - 36th Infantry - Company: Charles Randolph - Age: 26 - Height: 5' 7" - Born: Prince George's Cty, MD - Enlistment date: 27 May 1813 - Place: Frederick, MD - Period: 1 Yr.

Webb, William - Private - US Volunteers - Company: Stephen Moore - Enlistment date: 8 Sep 1812 - Discharged on 31 May 1813, inability.

Webb, William - Private - 35th Infantry - Company: Isaac Preston - Age: 34 - Height: 6' - Born: Baltimore - Trade: Tailor - Enlistment date: 20 Aug 1814 - Place: Norfolk, VA - Period: War - BLW 17807-160-12 - Discharged at Norfolk, VA, on 20 Mar 1815.

Webster, George - Private - 12th Infantry - Company: Thomas Sangster - Age: 24 - Height: 5' 8" - Born: Prince George's Cty, MD - Trade: Painter - Enlistment date: 20 May 1812 - Place: Alexandria, DC - Period: 18 Mos - Discharged on 20 Nov 1813.

Webster, George T. - Private - Light Dragoons - Company: James Reed - Age: 26 - Height: 5' 8 1/2" - Born: Prince George's Cty, MD - Trade: Painter - Enlistment date: 23 Apr 1814 - Place: Alexandria, DC - Period: War - Discharged at Fort Washington, MD, on 7 Mar 1815.

Webster, Thomas - Private - 38th Infantry - Company: Shepperd Leakin - Age: 31 - Height: 5' 9 1/4" - Born: Hertfordshire, England - Trade: Farmer - Enlistment date: 25 Aug 1814 - Place: Annapolis - Period: War - Deserted on 8 Feb 1815.

Webster, William - Private - 2nd Artillery - Company: Samuel Archer - Age: 38 - Height: 5' 7 1/2" - Born: Baltimore - Trade: Carpenter - Enlistment date: 30 Jun 1812 - Place: Norfolk, VA - Period: 5 Yrs - Discharged at Philadelphia on 26 Jun 1815 for inability.

Webster, William - Private - 25th Infantry - Company: Benjamin Watson - Age: 27 - Height: 5' 7" - Born: Baltimore City - Trade: Seaman - Enlistment date: 30 Dec 1814 - Place: New York - Period: 5 Yrs - Deserted on 3 Jan 1815.

Wedding, John - Private - 14th Infantry - Company: Thomas Kearney - Enlistment date: 9 Jun 1812 - Period: 5 Yrs - Pension: Land bounty to John Frederick Wedding, son and other heirs at law of John Wedding - BLW 23830-160-12 - Died on 7 Dec 1812.

Weddings, Horatio - Private - 14th Infantry - Company: Samuel Lane - Other regiment: 4th Infantry (New) - Age: 26 - Height: 6' 1" - Born: Charles Cty,

MD - Trade: Farmer - Enlistment date: 19 May 1812 - Place: Hagerstown, MD - Period: 5 Yrs - Discharged on 19 May 1817.

Weedon, William - Private - 2nd Artillery - Company: John Goodall - Age: 35 - Height: 6' 1/2" - Born: Prince George's Cty, MD - Trade: Stone mason - Enlistment date: 7 May 1814 - Place: Petersburg, VA - Period: War - Discharged at Norfolk, VA, on 22 May 1815.

Weekly, William - Private - 36th Infantry - Company: Joseph Hook - Age: 19 - Height: 5' 8 1/2" - Born: Montgomery Cty, MD - Enlistment date: 29 Apr 1813 - Place: Westminster, MD - Period: 1 Yr.

Weems, Daniel - Private - 14th Infantry - Company: Reuben Gilder - Enlistment date: 12 Dec 1812 - Period: 5 Yrs - Killed during the Battle of Lyon's Creek on 19 Oct 1814.

Weightman, William - Private - 14th Infantry - Company: Joseph Marshall - Age: 22 - Height: 5' 11" - Born: Alexandria, DC - Trade: Cooper - Enlistment date: 7 May 1814 - Place: Baltimore - Period: War - Pension: Wife Ann Eliza, WO-18323, WC-7962 - Discharged at Greenbush, NY, on 3 May 1815.

Weimer, Thomas - Private - 36th Infantry - Company: Thomas Carbery - Age: 22 - Height: 5' 4" - Born: Georgetown, DC - Trade: Butcher - Enlistment date: 21 Apr 1814 - Place: Georgetown, DC - Period: War - BLW 6022-160-12 - Discharged at Baltimore on 31 Mar 1815.

Weir, Bezelial - Private - 38th Infantry - Company: Thomas Sangster - Age: 29 - Height: 6' 2" - Born: Pittsylvania, VA - Trade: Carpenter - Enlistment date: 11 Aug 1814 - Place: Annapolis - Period: 5 Yrs.

Weiser, John - Private - 20th Infantry - Pension: Heirs received half pay for five years in lieu of land bounty - Died on 5 Jan 1815.

Welch, Andrew - Corporal - Corps of Artillery - Company: John Peyton - Age: 38 - Height: 5' 10" - Born: Cecil Cty, MD - Trade: Shoemaker - Enlistment date: 15 Sep 1814 - Place: Norfolk, VA - Period: War.

Welch, Henry - Private - 2nd Light Dragoons - Company: John Burd - Age: 22 - Height: 5' 11" - Born: Maryland - Trade: Shoemaker - Enlistment date: 1 Jun 1812 - Place: Bedford, PA - Period: 18 Mos - Prisoner of War (Halifax), sent to England, captured at Beaver Dams on 24 Jun 1813; discharged on 31 Mar 1815.

Welch, Moses - Seaman - US Sea Fencibles - Company: John Gill - Enlistment date: 20 Dec 1813 - Period: 1 Yr.

Welch, Moses - Private - US Volunteers - Company: Stephen Moore - Enlistment date: 8 Sep 1812 - Period: 1 Yr.

Welch, William - Private - 26th Infantry - Company: William Bezeau - Age: 26 - Height: 5' 6" - Born: Ireland - Enlistment date: 27 Sep 1814 - Place: Baltimore - Period: War - BLW 14825-160-12 - Discharged at Philadelphia on 23 Mar 1815.

Welcoping, Washington (Coxen) - Private - 36th Infantry - Company: Thomas Carbery - Age: 26 - Height: 5' 9 1/2" - Born: Prince George's Cty, MD - Trade: Farmer - Enlistment date: 5 Apr 1814 - Place: Georgetown, DC - Period: War - Discharged on 16 Mar 1815.

By War of 1812 Soc. in MD

Weldon, Samuel - Private - 2nd Artillery - Company: John Peyton - Age: 28 - Height: 5' 2" - Born: Cecil Cty, MD - Trade: Gunsmith - Enlistment date: 10 Apr 1814 - Place: Winchester - Period: War.

Weller, Benjamin - Private - 14th Infantry - Company: Thomas Montgomery - Other regiment: 4th Infantry (New) - Age: 38 - Height: 5' 3" - Born: Connecticut - Trade: Laborer - Enlistment date: 13 Apr 1812 - Place: Baltimore - Period: 5 Yrs - Discharged on 13 Apr 1817.

Weller, John - Corporal - 38th Infantry - Company: John Buck - Age: 20 - Height: 5' 4" - Born: Baltimore - Trade: Laborer - Enlistment date: 24 May 1814 - Place: Baltimore - Period: 5 Yrs.

Wells, Alexander - Private - 6th Infantry - Company: Alexander Thompson - Other regiment: 2nd Infantry (New) - Age: 29 - Height: 5' 8" - Born: Baltimore - Trade: Silversmith - Enlistment date: 9 Jun 1812 - Place: Baltimore - Period: 5 Yrs - BLW 15950-160-12 - Discharged at Fort Niagara, NY, on 9 Jun 1817.

Wells, Benjamin - Private - 36th Infantry - Age: 38 - Height: 5' 7 1/2" - Born: Sussex, VA - Trade: Farmer - Enlistment date: 20 Jul 1814 - Place: Lynchburg, VA - Period: War - Deserted.

Wells, James - Private - 19th Infantry - Age: 19 - Height: 5' 3" - Born: Allegany Cty, MD - Enlistment date: 26 Jul 1812 - Period: 5 Yrs.

Wells, James - Private - 16th Infantry - Company: George Steele - Other regiment: 2nd Infantry (New) - Age: 27 - Height: 5' 10 1/2" - Born: Harford Cty, MD - Trade: Barber - Enlistment date: 14 Mar 1814 - Place: Lancaster, PA - Period: 5 Yrs - Pension: Heirs received half pay for five years in lieu of land bounty - Discharged at Sackets Harbor, NY, on 4 Apr 1816 for rheumatism.

Wells, Thomas - Corporal - 12th Infantry - Company: Thomas Moore - Other regiment: 8th Infantry (New) - Age: 20 - Height: 5' 6" - Born: Kent Cty, MD - Trade: Laborer - Enlistment date: 22 Jun 1812 - Place: Morgantown, VA - Period: 5 Yrs - Discharged on 26 Jun 1817.

Welsh, Daniel - Corporal - 14th Infantry - Company: Samuel Lane - Age: 26 - Height: 5' 7" - Born: Maryland - Enlistment date: 19 Sep 1812 - Period: 5 Yrs - Pension: Land bounty to Robert A. Welsh, brother and other heirs at law of Daniel Welsh - BLW 13143-160-12 - Died at French Mills, NY, in Jan 1814.

Welsh, John - Seaman - US Sea Fencibles - Company: Simmones Bunbury - Age: 19 - Height: 5' 7 1/2" - Born: Baltimore City - Enlistment date: 3 Jan 1814 - Period: 1 Yr - Died on 24 Nov 1814.

Welsh, John - Private - 14th Infantry - Company: Thomas Montgomery - Enlistment date: 11 Jun 1812 - Period: 5 Yrs - Pension: Land bounty to Henry Welsh, brother and other heirs at law of John Welsh - BLW 13323-160-12 - Died on 24 Jun 1813.

Welsh, John - Private - 36th Infantry - Company: Joseph Hook - Age: 22 - Height: 5' 6" - Born: Philadelphia - Enlistment date: 4 May 1813 - Place: Baltimore - Period: 1 Yr.

Welsh, Peter - Private - 16th Infantry - Company: William Smith - Age: 39 -

Height: 5' 4" - Born: Baltimore - Trade: Saddle tree maker - Enlistment date: 6 Nov 1812 - Place: New Jersey - Period: 5 Yrs.

Welsh, Pierce - Seaman - US Sea Fencibles - Company: Simmones Bunbury - Age: 25 - Height: 5' 6" - Born: Baltimore - Enlistment date: 18 Jan 1814 - Period: 1 Yr - Discharged on 24 Mar 1815.

Welsh, Samuel - Private - 1st Rifles - Company: Henry Van Swearingen - Age: 38 - Height: 6' 2" - Born: Anne Arundel Cty, MD - Trade: Farmer - Enlistment date: 10 Dec 1814 - Place: Shepherdstown, VA - Period: 5 Yrs.

Welsh, Walter - Private - 36th Infantry - Company: Joseph Hook - Age: 28 - Height: 5' 7 3/4" - Born: Duchess Cty, NY - Trade: Carpenter - Enlistment date: 8 Jul 1814 - Place: Georgetown, DC - Period: War - BLW 11631-160-12 - Discharged at Fort Covington, MD, on 30 mar 1815.

Weltenburg, William - Private - 38th Infantry - Company: John Mowton - BLW 10075-160-12.

Wendell, Oliver - Sergeant - 38th Infantry - Company: John Brookes - BLW 11926-160-12.

Wentz, Frederick - Private - 2nd Artillery - Company: Nathan Towson - Pension: Old War IF-3518 - BLW 2899-160-12.

Wertz, David - Private - 36th Infantry - Pension: Land bounty to Samuel Wertz, brother and other heirs at law of David Wertz - BLW 26885-160-12.

Wertz, Francis - Private - 14th Infantry - Company: Samuel Lane - Age: 24 - Height: 5' 7" - Born: Virginia - Enlistment date: 7 Nov 1814 - Place: Steubenville, OH - Period: War - Deserted from both the 17th and 22nd Infantries.

Wertz, Samuel - Private - 38th Infantry - Company: John Buck - Age: 22 - Height: 5' 6 1/2" - Born: Bedford, PA - Trade: Farmer - Enlistment date: 6 Mar 1814 - Place: Hagerstown, MD - Period: War - Pension: Old War IF-3515 - BLW 877-160-12 - Discharged at Baltimore on 8 Feb 1815, accidental hand wound.

West, Charles - Sergeant - 14th Infantry - Company: David Cummings - Other regiment: 4th Infantry (New) - Age: 21 - Height: 5' 7 1/2" - Born: Dorchester Cty, MD - Trade: Coppersmith - Enlistment date: 20 May 1812 - Period: 5 Yrs - BLW 9833-160-12 - Prisoner of War (Halifax), captured at Beaver Dams on 24 Jun 1813; discharged at Raleigh, NC, on 19 May 1817.

West, David - Private - 19th Infantry - Company: William Gill - Other regiment: 17th Infantry then 3rd Infantry (New) - Age: 17 - Height: 5' 4" - Born: Harford Cty, MD - Trade: Laborer - Enlistment date: 19 Oct 1814 - Period: 5 Yrs - Pension: Wife Lydia, WO-40099, WC-31394, SO-20403, SC-9876 - Discharged at Fort Howard, WI, on 19 Oct 1819.

West, Hugh S. - Private - 2nd Artillery - Company: Nathan Towson - Age: 33 - Prisoner of War (Quebec), captured during the Battle of Stoney Creek and released on 31 Oct 1813.

West, Hugh S. - Private - 2nd Artillery - Company: Nathan Towson - Age: 30 - Height: 5' 9" - Born: Virginia - Trade: Shoemaker - Enlistment date: 20 Apr 1812 - Place: Baltimore - Period: 5 Yrs - Prisoner of War (Halifax), sent to England, captured at Beaver Dams on 24 Jun 1813; discharged on 28 Sep

By War of 1812 Soc. in MD

1815.

West, Ignatius - Private - 1st Rifles - Company: Daniel Appling - Age: 34 - Height: 5' 8" - Born: Maryland - Trade: Blacksmith - Enlistment date: 16 Jul 1813 - Place: Kentucky - Period: War - Discharged at Conjocta Creek, NY, on 11 Jun 1815 ???.

West, Isaac - Private - 38th Infantry - Company: John Buck - Age: 22 - Height: 5' 7" - Born: Frederick Cty, MD - Trade: Cooper - Enlistment date: 9 Apr 1814 - Place: Libertytown, MD - Period: War - BLW 4155-160-12 - Discharged at Baltimore on 30 Mar 1815.

West, James - Private - 5th Infantry - Company: Leroy Opie - Other regiment: 3rd Infantry (New) - Age: 21 - Height: 5' 11" - Born: Anne Arundel, MD - Trade: Farmer - Enlistment date: 8 Apr 1813 - Place: Baltimore - Period: 5 Yrs - BLW 22375-160-12 - Discharged at Fort Howard, WI, on 8 Apr 1818.

West, Robert D. - Sergeant - 28th Infantry - Company: Joseph Belt - Age: 22 - Height: 5' 7" - Born: Maryland - Trade: Hatter - Enlistment date: 25 Aug 1814 - Place: Winchester, KY ??? - Period: War - BLW 19217-160-12 - Discharged at Olympian Springs, KY, on 31 Mar 1815.

West, Thomas - Private - 38th Infantry - Company: John Brookes - BLW 14727-160-12.

West, Thomas - Private - 22nd Infantry - Company: David Espy - Age: 35 - Height: 5' 8" - Born: Maryland - Trade: Blacksmith - Enlistment date: 27 Oct 1813 - Place: Gettysburg, PA - Period: 5 Yrs - Died at Sackets Harbor, NY, on 5 Jan 1815.

Westbrook, Nicholas - Private - 14th Infantry - Company: Samuel Lane - Wounded at Black Rock, NY, on 28 Nov 1812 and died from wounds on 29 Nov 1812.

Westcott, Amasa - Private - 14th Infantry - Company: James Gale - Pension: Wife Anna, SO-2072, SC-9483, WO-12881, WC-24518 - BLW 4624-160-12.

Weston, John - Corporal - 36th Infantry - Company: Joseph Merrick - Age: 24 - Height: 5' 9" - Born: Plymouth, MA - Enlistment date: 1 May 1813 - Place: Baltimore - Period: 1 Yr.

Westwood, Thomas - Seaman - US Sea Fencibles - Company: John Gill - Age: 21 - Height: 5' 1" - Born: Baltimore - Enlistment date: 3 Jan 1814 - Period: 1 Yr.

Weylie, Ephraim - Private - 14th Infantry - Company: Thomas Montgomery - BLW 213-160-50.

Whalen, William - Private - 1st Rifles - Company: Thomas Ramsey - Age: 17 - Height: 5' 4" - Born: Maryland - Trade: Farmer - Enlistment date: 10 Feb 1814 - Place: Louisville, KY - Period: 5 Yrs - Discharged at Detroit on 28 Sep 1815 for disability.

Whaler, Caleb - Private - 14th Infantry - Died at French Mills, NY, on 18 Jan 1814 from sickness.

Whalin, William - Private - 1st Rifles - Company: Thomas Ramsey - Age: 17 - Height: 5' 4" - Born: Maryland - Trade: Farmer - Enlistment date: 10 Feb 1814 - Place: Lewisville, KY - Period: 5 Yrs - BLW 7112-160-12 -

Maryland Regulars in War of 1812

Discharged at Detroit on 23 Sep 1815, Surgeon's Certificate.

Wheatley, Caleb - Corporal - 36th Infantry - Company: Samuel Raisin - Age: 21 - Height: 5' 5 5/8" - Born: Greensborough, MD - Enlistment date: 11 Aug 1813 - Place: Greenborough, MD - Period: 1 Yr - Pension: Wife Margaret, WO-30497, WC-27576 - Discharged at Piscataway on 4 Aug 1814.

Wheatley, John - Private - 38th Infantry - Company: James Hook - Age: 20 - Height: 5' 4 3/4" - Born: Dorchester Cty, MD - Trade: Cooper - Enlistment date: 14 Mar 1814 - Period: War - Deserted at Fort McHenry on 20 Nov 1814.

Wheatley, John - Private - 36th Infantry - Company: Thomas Carbery - Age: 18 - Height: 5' 8" - Born: St. Mary's Cty, MD - Trade: Farmer - Enlistment date: 25 Mar 1814 - Place: Leonardstown, MD - Period: War - BLW 7351-160-12 - Discharged at Baltimore on 31 Mar 1815.

Wheatley, Matthew - Private - 36th Infantry - Company: Thomas Carbery - Age: 22 - Height: 5' 10 1/2" - Born: St. Mary's Cty, MD - Trade: Farmer - Enlistment date: 5 May 1814 - Place: Leonardstown, MD - Period: War - BLW 7352-160-12 - Discharged at Baltimore on 31 Mar 1815.

Wheatley, William - Private - 16th Infantry - Company: William Smith - Age: 29 - Height: 5' 4" - Born: Maryland - Trade: Sailor - Enlistment date: 6 May 1813 - Period: 5 Yrs - Died on 11 Dec 1813.

Wheatly, James - Private - 14th Infantry - Company: Clement Sullivan - Enlistment date: 30 May 1812 - Period: 5 Yrs - Pension: Land bounty to Joseph Wheatly, brother and other heirs at law of James Wheatly - BLW 13552-160-12 - Died on 7 Dec 1812.

Wheatly, Noah - Corporal - 7th Infantry - Company: Enos Cutler - Other regiment: 1st Artillery then Corps of Artillery - Age: 30 - Height: 5' 8 1/2" - Born: Maryland - Trade: Laborer - Enlistment date: 4 Feb 1812 - Place: New York - Period: 5 Yrs - Transferred to 1st Artillery, Captain George Armistead's Company, on 30 Sep 1812; discharged on 18 Feb 1817.

Wheeler, Aloysius - Private - 36th Infantry - Company: Thomas Carbery - Age: 19 - Height: 5' 6" - Born: Washington, DC - Trade: Gunsmith - Enlistment date: 14 Apr 1814 - Place: Washington, DC - Period: War - BLW 220-160-12 - Discharged at Baltimore on 31 Mar 1815.

Wheeler, Anthony - Private - 14th Infantry - Company: David Cummings - Other regiment: Ordnance Department - Age: 21 - Height: 5' 8 1/2" - Born: Dorchester Cty, MD - Trade: Ship joiner - Enlistment date: 17 Jun 1812 - Period: 5 Yrs - Prisoner of War (Halifax), sent to England, captured at Beaver Dams on 24 Jun 1813; transferred to the Ordnance Department on 21 Jun 1815; discharged at Baltimore on 14 Jun 1817.

Wheeler, Anthony - Private - 2nd Artillery - Company: James Barker - Other regiment: Corps of Artillery - Age: 26 - Height: 5' 8 3/4" - Born: Cambridge, MD - Trade: Farm laborer - Enlistment date: 9 Apr 1814 - Place: Philadelphia - Period: 5 Yrs - Discharged on 9 Apr 1919.

Wheeler, Benedict - Private - 2nd Light Dragoons - Company: William Littlejohn - Other regiment: Light Dragoons - Age: 22 - Height: 5' 5" - Born: Charles Cty, MD - Trade: Brick layer - Enlistment date: 11 Feb 1814 - Place:

Alexandria, DC - Period: War - BLW 228-160-12 - Discharged at Carlisle Barracks, PA, on 4 May 1815.

Wheeler, Doyles - Private - 10th Infantry - Company: Robert Mitchell - Age: 16 - Height: 5' 2" - Born: Maryland - Trade: Laborer - Enlistment date: 1 May 1814 - Place: Washington, DC - Period: War - Discharged at Buffalo on 31 May 1815.

Wheeler, Francis J. - Second Lieutenant - 36th Infantry - Company: Mortimer Hall - Commissioned as a 3rd lieutenant on 30 Apr 1813; promoted to 2nd lieutenant on 1 May 1814; discharged on 1 Jun 1814.

Wheeler, Garrett - Private - 16th Infantry - Company: Thomas Horrell - Age: 22 - Height: 5' 7" - Born: Baltimore - Enlistment date: 24 Mar 1814 - Place: Carlisle, PA - Period: 5 Yrs.

Wheeler, Greenbury - Private - 17th Infantry - Company: Benjamin Sanders - Age: 21 - Height: 5' 9" - Born: Maryland - Enlistment date: 28 Apr 1814 - Period: 5 Yrs - Pension: Land bounty to William Wheeler, brother & other heirs at law of Greenbury Wheeler - BLW 25731-160-12 - Died at Buffalo on 22 Nov 1814 from fever.

Wheeler, Jared - Private - 16th Infantry - Company: Robert Gray - Age: 23 - Height: 5' 7" - Born: Baltimore - Trade: Laborer - Enlistment date: 24 Mar 1814 - Place: Harrisburg, PA - Period: 5 Yrs - Discharged at Sackets Harbor, NY, on 23 Dec 1818 and re-enlisted.

Wheeler, John - Corporal - 18th Infantry - Company: John Blount - Other regiment: 4th Infantry (New) - Age: 36 - Height: 6' 3/4" - Born: Baltimore - Trade: Gunsmith - Enlistment date: 1 Nov 1813 - Place: Washington, NC - Period: 5 Yrs - Died in Aug 1818.

Wheeler, Joseph - Private - 14th Infantry - Company: Samuel Lane - Age: 20 - Height: 5' 8" - Born: Delaware - Enlistment date: 6 Aug 1812 - Period: 18 Mos - Deserted on 12 Jul 1813.

Wheeler, Lucius - Private - 36th Infantry - Age: 19 - Height: 5' 6" - Born: Washington, DC - Enlistment date: 14 Apr 1814 - Place: Washington, DC.

Wheeler, Richard J. - Sergeant - 20th Infantry - Company: John Thornton - Other regiment: 4th Infantry (New) - Age: 29 - Height: 5' 8 1/2" - Born: Harford Cty, MD - Trade: Farmer - Enlistment date: 18 Jul 1812 - Period: 5 Yrs - BLW 1512-160-12 - Discharged at Annapolis on 16 Oct 1815 for atrophy.

Wheeler, William - Artificer - 2nd Artillery - Company: Nathan Towson - Age: 25 - Prisoner of War (Quebec), captured during the Battle of Stoney Creek and released on 31 Oct 1813.

Wheeler, William - Private - 2nd Artillery - Company: Nathan Towson - Other regiment: Corps of Artillery - Age: 24 - Height: 5' 7 1/2" - Born: Baltimore - Trade: Wheelwright - Enlistment date: 25 Jun 1812 - Place: Baltimore - Period: 5 Yrs - BLW 12089-160-12 - Prisoner of War (Halifax); discharged at Fort Niagara, NY, on 24 Jun 1817.

Wheelock, James - Private - 14th Infantry - Other regiment: 4th Infantry (New) - Age: 48 - Height: 5' 5 1/2" - Born: Dublin, Ireland - Trade: Laborer - Enlistment date: 23 Sep 1812 - Place: Baltimore - Period: 5 Yrs - Prisoner of War (Halifax), captured at Stoney Creek, UC, on 12 Jun 1813; deserted on 16 Jul 1815.

Whelan, Martin - Private - 5th Infantry - Company: Leroy Opie - Age: 30 - Height: 5' 10 1/2" - Born: Ireland - Trade: Farmer - Enlistment date: 3 Jun 1813 - Place: Baltimore - Period: 5 Yrs.

Whelen, Martin - Musician - 5th Infantry - Company: James Dorman - Other regiment: 3rd Infantry (New) - Age: 24 - Height: 6' - Born: Montgomery Cty, MD - Trade: Millwright - Enlistment date: 5 Apr 1813 - Place: Baltimore - Period: 5 Yrs - Discharged on 5 Apr 1818 and re-enlisted.

Whetman, Stephen - Private - 14th Infantry - Prisoner of War (Halifax), sent to England.

Whetson, George - Seaman - US Sea Fencibles - Company: William Addison - Discharged at Baltimore on 7 Jan 1815.

Whetton, John - Private - 14th Infantry - Prisoner of War (Halifax), sent to England.

Whiffrey, Andrew - Private - 36th Infantry - Age: 27 - Height: 5' 9 1/2" - Born: New Market, Frederick Cty, MD - Enlistment date: 23 Nov 1814 - Place: Norfolk, VA - Deserted.

While, Jacob - Private - 14th Infantry - Company: Henry Grindage - Pension: Land bounty to Joseph While, father and only heir at law of Jacob While - BLW 24888-160-12.

Whistler Jr., John - First Lieutenant - 19th Infantry - Company: Harris Hickman - Born: Maryland - Commissioned as an ensign on 12 Mar 1812; promoted to 2nd lieutenant on 30 Mar 1813; promoted to 1st lieutenant on 20 Nov 1813; Prisoner of War on Parole, captured at Detroit on 16 Aug 1812; wounded at Brownstown, MI, on 9 Aug 1812; died in Dec 1813.

Whistler, John - Captain - 1st Infantry - Born: England - Served during the Revolutionary War; commissioned as a lieutenant in the U.S. Levies in 1791; commissioned as an ensign, 1st Infantry, on 11 Apr 1792; transferred to 1st Sub-Legion on 4 Sep 1792; promoted to lieutenant on 27 Nov 1792; transferred to 1st Infantry on 1 Nov 1796; promoted to captain on 1 Jul 1797; discharged on 15 Jun 1815; died on 3 Sep 1829.

Whistler, William - Captain - 1st Infantry - Other regiment: 3rd Infantry (New) - Born: Maryland - Pension: Wife Julia, Old War WF-27680 - BLW 4956-160-50 - Commissioned as a 2nd lieutenant, 1st Infantry, on 8 Jun 1801; promoted to 1st lieutenant on 4 Mar 1807; promoted to captain on 31 Dec 1812; transferred to 3rd Infantry (New) on 17 May 1815; retired on 9 Oct 1861 with the rank of colonel; died on 4 Dec 1863.

Whitcomb, Robert - Musician - 36th Infantry - Company: Samuel Raisin - Age: 16 - Height: 4' 3" - Born: England - Trade: Seaman - Enlistment date: 23 Jul 1814 - Place: Norfolk, VA - Period: War - BLW 12210-160-12 - Discharged at Washington, DC, on 20 Mar 1815.

White, Alfred - Private - 6th Infantry - Company: Robert Patterson - Age: 35 - Height: 5' 7" - Born: Maryland - Enlistment date: 17 Aug 1814 - Place: Philadelphia - Period: 5 Yrs - Deserted on 20 Mar 1815.

White, Benedict - Corporal - 2nd Infantry - Company: Perrin Willis - Age: 23

By War of 1812 Soc. in MD

- Height: 5' 5" - Born: Prince George's Cty, MD - Trade: Hatter - Enlistment date: 6 Feb 1815 - Place: Washington, DC - Period: 5 Yrs.

White, Benedict - Third Lieutenant - 36th Infantry - Company: Thomas Carbery - Born: Maryland - Commissioned as a 3rd lieutenant on 30 Apr 1813; resigned in Jul 1813.

White, Benjamin - Quarter Gunner - US Sea Fencibles - Company: Simmones Bunbury - Age: 32 - Height: 5' 8" - Born: Talbot Cty, MD - Enlistment date: 2 Dec 1813 - Period: 1 Yr - Discharged on 28 Jul 1814.

White, Burgess B. - First Lieutenant - 1st Artillery - Company: James Many - Other regiment: Corps of Artillery - Born: Maryland - BLW 3740-156-50 - Commissioned as a 2nd lieutenant, 1st Artillery, on 15 Apr 1812; transferred to Corps of Artillery on 12 May 1814; promoted to 1st lieutenant on 29 Jul 1814; resigned on 30 Sep 1815.

White, Charles - Musician - 1st Artillery - Company: John Walbach - Age: 22 - Height: 5' 6 1/2" - Born: Baltimore - Trade: Laborer - Enlistment date: 13 Nov 1812 - Place: Fort Constitution, MA - Period: 5 Yrs - Discharged on 23 Aug 1817.

White, Charles - Private - 17th Infantry - Company: Caleb Holder - Other regiment: 3rd Infantry (New) - Age: 29 - Height: 5' 6 1/2" - Born: Anne Arundel Cty, MD - Trade: Hatter - Enlistment date: 9 May 1812 - Period: 5 Yrs - BLW 7643-160-12 - Discharged on 15 Jul 1816 after furnishing a substitute.

White, Charles - Seaman - US Sea Fencibles - Company: Simmones Bunbury - Enlistment date: 19 Jan 1814 - Period: 1 Yr - Discharged on 18 Jan 1815.

White, David - Private - 38th Infantry - Company: John Buck - Age: 29 - Height: 5' 3" - Born: Connecticut - Enlistment date: 3 Jan 1814 - Place: Cumberland - Period: 1 Yr - Died at Baltimore, on 15 Jan 1815.

White, George - Private - 2nd Artillery - Company: Stanton Sholes - Age: 24 - Height: 5' 6" - Born: Maryland - Trade: Farmer - Enlistment date: 16 Jan 1813 - Place: Beaver Cty, PA - Period: 5 Yrs - Pension: Old War IF-3538 - Discharged on 16 Oct 1815 because of blindness of the left eye, and loss of the left great toe.

White, Hays G. - Major - 20th Infantry - Other regiment: 36th Infantry then 44th Infantry - Born: North Carolina - Place: North Carolina - Commissioned as a 1st Lieutenant, 3rd Infantry, on 1 Jul 1808; promoted to Captain on 1 Jul 1808; promoted to major, 20th Infantry, on 31 May 1814; transferred to 36th Infantry on 10 Oct 1814; transferred to 44th Infantry on 18 Nov 1814; discharged on 15 Jun 1815.

White, Hiram M. - Private - 14th Infantry - Company: Charles Baisler - Pension: Wife Mary Ann, SO-33457, WO-53070 - Also served in Captain Fuller's Company, 4th Infantry.

White, Jacob - Private - 14th Infantry - Company: Henry Grindage - Enlistment date: 22 Aug 1812 - Period: 5 Yrs - Pension: Land bounty to Joseph W. White, father & only heir at law of Jacob White - BLW 24888-160-12 - Died on 9 Apr 1813.

White, James - Corporal - 38th Infantry - Company: James Hook - BLW

Maryland Regulars in War of 1812

20523-160-12.

White, James - Private - 36th Infantry - Age: 27 - Height: 6' 1 1/2" - Born: Poughkeepsie, NY - Enlistment date: 21 Nov 1814 - Place: Norfolk, VA.

White, James - Corporal - 36th Infantry - Company: Thomas Carbery - Age: 25 - Height: 5' 4" - Born: England - Trade: Shoemaker - Enlistment date: 27 Feb 1814 - Place: Leonardstown, MD - Period: 5 Yrs.

White, John - Musician - 36th Infantry - Age: 16 - Height: 4' 10" - Born: Washington, DC - Trade: Laborer - Enlistment date: 18 Jan 1815 - Period: War - BLW 386-320-12 - Discharged at Georgetown, DC, on 13 Mar 1815.

White, John - Private - 25th Infantry - Company: George Howard - Age: 25 - Height: 5' 6" - Born: Baltimore - Trade: Shoemaker - Enlistment date: 12 Aug 1813 - Place: Greenwich - Period: 5 Yrs - Discharged at Sackets Harbor, NY, on 21 Mar 1816 on Surgeon's Certificate.

White, John - Private - 14th Infantry - Company: David Cummings - Enlistment date: 21 Jun 1812 - Period: 5 Yrs - Prisoner of War (Halifax), sent to England, captured at Beaver Dams on 24 Jun 1813.

White, John - Surgeon - 32nd Infantry - Born: Maryland - Place: Maryland - Commissioned as a surgeon, 32nd Infantry, on 17 May 1813; resigned on 9 Nov 1813.

White, John - Private - 16th Infantry - Company: James McElroy - Other regiment: 2nd Infantry (New) - Age: 20 - Height: 5' 4 1/2" - Born: Maryland - Trade: Mariner - Enlistment date: 20 Oct 1812 - Place: Philadelphia - Period: 5 Yrs - BLW 9892-160-12 - Discharged at Sackets Harbor, NY, on 15 Aug 1815.

White, John B. - Private - 1st Infantry - Company: Hezekiah Johnson - Age: 32 - Height: 5' 7 1/2" - Born: Dorchester Cty, MD - Trade: Carpenter - Enlistment date: 3 Oct 1813 - Place: Philadelphia - Period: 5 Yrs.

White, Jonas - Private - 36th Infantry - Age: 27 - Height: 6' 1/2" - Deserted.

White, Joseph P. - Gunner - US Sea Fencibles - Company: Simmones Bunbury - Age: 45 - Height: 5' 6" - Born: Saybrook, CT - Enlistment date: 8 Nov 1813 - Period: 1 Yr - Discharged on 8 Nov 1814.

White, Joseph S. - Private - 36th Infantry - Company: Joseph Hook - Age: 32 - Height: 5' 8" - Born: Charles Cty, MD - Trade: Plasterer - Enlistment date: 21 Jul 1814 - Place: Georgetown, DC - Period: War - BLW 195-160-12 - Discharged at Washington, DC, on 10 Apr 1815.

White, Mark - Private - 36th Infantry - Age: 18 - Height: 5' 7" - Born: Orange, NC - Enlistment date: 19 Mar 1814 - Place: Fort Washington, MD - Period: War.

White, Robert - Private - 38th Infantry - Company: John Brookes - BLW 2704-160-12.

White, William - Recruit - 26th Infantry - Company: William Bezeau - Race: 17 - Age: 17 - Height: 4' 8" - Born: Maryland - Trade: Laborer - Enlistment date: 11 Oct 1814 - Place: Philadelphia - Period: War - Listed as "Colored" and as "Col'd" in his service record.

White, William - Private - 23rd Infantry - Company: Frederick Brown - Age: 34 - Height: 5' 7 3/4" - Born: Baltimore - Trade: Seaman - Enlistment date:

By War of 1812 Soc. in MD

13 Feb 1814 - Period: War - On board the fleet on Lake Erie.

White, William N. - Seaman - US Sea Fencibles - Company: John Gill - Age: 18 - Height: 6' - Born: Rockingham, VT - Enlistment date: 10 Feb 1814 - Period: 1 Yr.

Whitehead, Thomas - Private - 36th Infantry - Company: Samuel Raisin - Age: 19 - Height: 5' 8" - Born: Ireland - Trade: Stone mason - Enlistment date: 11 Dec 1814 - Period: War - Discharged at Washington, DC, in Apr 1815.

Whitehouse, John - Private - 1st Rifles - Company: Lodowick Morgan - Age: 19 - Height: 5' 9" - Born: Montgomery Cty, MD - Enlistment date: 19 Jan 1813 - Place: Shepherdstown, VA.

Whitehouse, John - Private - 4th Infantry - Company: John Smith - Other regiment: 5th Infantry (New) - Age: 18 - Height: 5' 11" - Born: Montgomery Cty, MD - Trade: Cordwainer - Enlistment date: 28 Feb 1814 - Place: Plattsburgh, NY - Period: 5 Yrs - Deserted at Detroit on 19 Sep 1815.

Whiteley, David R. - Private - 2nd Artillery - Company: Nathan Towson - Other regiment: Corps of Artillery - Age: 34 - Height: 5' 10" - Born: Baltimore - Trade: Laborer - Enlistment date: 11 Apr 1812 - Place: Baltimore - Period: 5 Yrs - BLW 12395-160-12 - Prisoner of War (Halifax); discharged at Fort Niagara, NY, on 2 Jun 1817.

Whiteley, David R. - Private - 2nd Artillery - Company: Nathan Towson - Age: 26 - Prisoner of War (Quebec), captured during the Battle of Stoney Creek and released on 31 Oct 1813.

Whitis, John - Private - 38th Infantry - Company: John Brookes - BLW 16344-160-12.

Whitlock, Ephraim L. - Lieutenant Colonel - 14th Infantry - Place: New Jersey - Served as a captain during the Revolutionary War; commissioned as a major, 15th Infantry, 1 May 1812; promoted to lieutenant colonel, 14th Infantry, on 14 Nov 1813; discharged on 15 Jun 1815.

Whitmore, Humphrey - Private - 14th Infantry - Age: 40 - Height: 5' 8" - Born: Prince George's Cty, MD - Enlistment date: 1 Jun 1814 - Place: Alexandria, DC - Period: War - Pension: Land bounty to Humphrey Whitmore, son and other heirs at law of Humphrey Whitmore - BLW 18667-160-12 - Died on 31 Jul 1814.

Whitney, David - Private - 38th Infantry - Company: John Brookes - BLW 17274-160-12.

Whitney, William - Private - 2nd Light Dragoons - Company: John Burd - Other regiment: Light Dragoons - Age: 23 - Height: 5' 8" - Born: Fairfield, CT - Trade: Shoemaker - Enlistment date: 20 Sep 1814 - Place: Baltimore - Period: War - BLW 15886-160-12 - Discharged at Carlisle Barracks, PA, on 10 May 1815.

Whittaker, Jesse W. - Sergeant - 2nd Light Dragoons - Company: Henry Hall - Age: 24 - Height: 6' 1 1/2" - Born: Cecil Cty, MD - Trade: Laborer - Enlistment date: 25 May 1812 - Place: Havre de Grace, MD - Period: 5 Yrs.

Whittington, John - Private - 35th Infantry - Company: Walter Cocke - Age: 38 - Height: 5' 10 1/2" - Born: Calvert Cty, MD - Trade: Farmer - Enlistment date: 22 Apr 1814 - Place: Middleton, VA - Period: War - Discharged at

Norfolk, VA, on 20 Mar 1815.

Whittle, Charles - Private - 5th Infantry - Company: Colin Buckner - Other regiment: 3rd Infantry (New) - Age: 28 - Height: 5' 6" - Born: Loudoun Cty, VA - Trade: Cooper - Enlistment date: 22 Apr 1813 - Place: Baltimore - Period: 5 Yrs - Discharged on 22 Apr 1818.

Whitton, William - Private - 14th Infantry - Company: William McIlvane - Age: 47 - Height: 5' 8" - Born: Maryland - Trade: Cooper - Enlistment date: 7 Sep 1812 - Period: 5 Yrs - BLW 19272-160-12 - Discharged at Greenbush, NY, on 1 May 1815, hernia.

Whitty, Arthur - Corporal - 14th Infantry - Company: David Cummings - Other regiment: 4th Infantry (New) - Age: 24 - Height: 5' 8" - Born: Maryland - Trade: Laborer - Enlistment date: 22 Oct 1814 - Place: Hagerstown, MD - Period: 5 Yrs - Died at Fort St. Carlos de Barrancas, FL, on 7 Nov 1818.

Wiant, Henry - Private - 36th Infantry - Company: Charles Randolph - Age: 24 - Height: 5' 11" - Born: Washington, MD - Enlistment date: 11 Jun 1813 - Place: Frederick, MD - Period: 1 Yr - Discharged on 25 Mar 1814.

Wickham, Andrew - Private - 38th Infantry - Company: Charles Stansbury - Age: 33 - Height: 5' 6 1/2" - Born: Frederick Cty, MD - Trade: Distiller - Enlistment date: 11 Dec 1814 - Place: Fredericktown, MD - Period: War - Discharged at Baltimore on 6 Apr 1815.

Wickham, David - Private - 36th Infantry - Company: Joseph Nelson - Enlistment date: 20 Jun 1814 - Period: War - Pension: Land bounty to Levi A. Wickham, brother and other heirs at law of David Wickham - BLW 22881-160-12 - Died on 17 Nov 1814.

Wicks, John - Private - 22nd Infantry - Company: James Pentland - Age: 25 - Height: 5' 9" - Born: Hagerstown, MD - Trade: Farmer - Enlistment date: 3 Oct 1812 - Place: Shirlysburg - Period: 18 Mos - Discharged at Burlington, VT, on 10 Apr 1814.

Wickware, John - Private - 44th Infantry - Company: Isaac Baker - Age: 23 - Height: 6' 1 1/2" - Born: Caroline Cty, MD - Trade: Farmer - Enlistment date: 2 Jun 1814 - Place: Gallatin, TN - Period: War - Died on 12 Mar 1815.

Wiegley, Jacob - Corporal - 38th Infantry - Company: John Rothrock - BLW 8510-160-12.

Wigfield, Robert - Private - 36th Infantry - Company: Joseph Hook - Age: 35 - Height: 5' 7 1/2" - Born: Prince George's Cty, MD - Trade: Carpenter - Enlistment date: 3 Jun 1814 - Place: Georgetown, DC - Period: War - Pension: Land bounty to widow and other heirs at law of Robert Wigfield - BLW 2316-160-12 - Discharged at Fort Covington, MD, on 30 Mar 1815.

Wigley, Jacob - Corporal - 38th Infantry - Company: John Rothrock - Age: 25 - Height: 6' 2 1/2" - Born: Washington, MD - Trade: Brick layer - Enlistment date: 20 Feb 1814 - Place: Craney Island, VA - Period: War - Discharged at Craney Island, VA, on 15 Mar 1815.

Wilbin, Walter - Sergeant - 2nd Infantry - Company: Perrin Willis - Age: 25 - Height: 5' 9" - Born: Prince George's Cty, MD - Trade: Farmer - Enlistment date: 26 Oct 1814 - Place: Washington, DC - Period: 5 Yrs.

By War of 1812 Soc. in MD

Wildt, Jacob - Private - 14th Infantry - Company: Reuben Gilder.

Wilett, Jacob - Private - 14th Infantry - Company: Reuben Gilder - Age: 24 - Height: 5' 6" - Born: Germany - Trade: Baker - Enlistment date: 7 Apr 1813 - Place: Frederick, MD - Period: War - Discharged at Greenbush, NY, on 4 May 1815.

Wiley, Ephraim - Private - 36th Infantry - Company: Samuel Raisin - Age: 25 - Height: 5' 4" - Born: Maryland - Enlistment date: 19 Jul 1813 - Place: Waynesburg, PA - Period: 1 Yr - Pension: Wife Elizabeth, WO-11218, WC-6617.

Wiley, Ephraim - Private - 14th Infantry - Company: Thomas Montgomery - Enlistment date: 24 Dec 1812 - Period: 18 Mos - Discharged at Plattsburgh, NY, on 24 Jun 1814.

Wilgiss, William - Private - 38th Infantry - Company: Shepperd Leakin - Pension: Old War IF-25492 - BLW 13909-160-12.

Wilhelm, Charles - Private - 14th Infantry - Company: Richard Arell - Other regiment: 4th Infantry (New) - Age: 33 - Height: 5' 5 1/4" - Born: Germany - Trade: Mason - Enlistment date: 21 Dec 1812 - Place: Lancaster, PA - Period: 5 Yrs - Prisoner of War, exchanged on 15 Apr 1814; discharged on 31 Dec 1815 after furnishing a substitute.

Wilkes, Thomas - Corporal - 14th Infantry - Company: Joseph Marshall - Age: 35 - Height: 5' 9" - Born: Baltimore - Trade: Laborer - Enlistment date: 2 Oct 1812 - Period: 18 Mos - BLW 2670-160-12 - Re-enlisted at Baltimore on 9 Jul 1814 for the war; discharged at Greenbush, NY, on 3 May 1815.

Wilkey, Frederick - Corporal - 38th Infantry - Company: John Rothrock - Age: 22 - Height: 5' 8" - Born: Philadelphia - Trade: Butcher - Enlistment date: 6 Jun 1813 - Place: Craney Island, VA - Period: War.

Wilkins, Elisha - Private - 14th Infantry - Company: Henry Grindage - Enlistment date: 22 Aug 1812 - Period: 5 Yrs - Pension: Land bounty to Sarah Wilkins, sister and only heir at law of Elisha Wilkins - BLW 27558-160-42 - Died on 8 Apr 1813.

Wilkins, James - Private - 14th Infantry - Company: Henry Grindage - Enlistment date: 15 Aug 1812 - Period: 5 Yrs - Pension: Land bounty to Sarah Wilkins, sister and only heir at law of James Wilkins - BLW 27559-160-42 - Prisoner of War (Quebec), died on 1 Nov 1813.

Wilkins, John - Private - 14th Infantry - Prisoner of War (Halifax), discharged from Halifax on 3 Feb 1814.

Wilkins, John - Private - 16th Infantry - Company: Robert Gray - Other regiment: 2nd Infantry (New) - Age: 38 - Height: 5' 3 1/2" - Born: Oldtown, MD - Trade: Cordwainer - Enlistment date: 12 Apr 1813 - Place: Philadelphia - Period: 5 Yrs - BLW 16877-160-12 - Discharged on 11 Apr 1818.

Wilkins, Robert - Private - 5th Infantry - Company: Richard Whartenby - Age: 35 - Height: 5' 10 1/2" - Born: Ireland - Trade: Saddler - Enlistment date: 12 Feb 1812 - Place: Baltimore - Period: 5 Yrs.

Wilkins, William - Private - 14th Infantry - Company: Thomas Montgomery - Other regiment: 4th Infantry (New) - Age: 24 - Height: 5' 8" - Born: Snow

Maryland Regulars in War of 1812

Hill, MD - Trade: Farmer - Enlistment date: 15 Aug 1812 - Place: Snow Hill, MD - Period: 5 Yrs - Prisoner of War (Halifax), captured at Stoney Creek, UC, on 12 Jun 1813.

Wilkinson, James - Major General - General Staff - Born: Maryland - Served during the Revolutionary War; commissioned as a lieutenant colonel commandant, 2nd Infantry, on 22 Oct 1791; promoted to brigadier general on 5 Mar 1792; promoted to major general on 2 Mar 1813; discharged on 15 Jun 1815; died on 28 Dec 1825.

Wilkinson, James - Private - 5th Infantry - Other regiment: 4th Infantry (New) - Age: 23 - Height: 5' 6" - Born: Baltimore Cty - Trade: Laborer - Enlistment date: 23 Feb 1813 - Period: 5 Yrs - BLW 17114-160-12 - Discharged on 22 Feb 1818.

Wilkinson, James - Private - 14th Infantry - Company: William McIlvane - Enlistment date: 1 Mar 1813 - Period: 5 Yrs - Prisoner of War (Halifax), captured at Beaver Dams on 24 Jun 1813, discharged from Halifax on 3 Feb 1813, exchanged on 15 Apr 1814.

Wilkinson, James Biddle - Captain - 2nd Infantry - Born: Maryland - Commissioned as a 2nd lieutenant, 4th Infantry, on 16 Feb 1801; transferred to 2nd Infantry on 1 Apr 1802; promoted to 1st lieutenant on 30 Sep 1803; promoted to captain on 8 Oct 1808; died on 7 Sep 1813.

Wilkinson, Philip - Private - 5th Infantry - Company: Richard Whartenby - Other regiment: 3rd Infantry (New) - Age: 27 - Height: 5' 6" - Born: Calvert, MD - Trade: Tailor - Enlistment date: 27 Dec 1811 - Place: Baltimore - Period: 5 Yrs - Discharged on 27 Dec 1816.

Wilkinson, Richard - Private - 14th Infantry - Company: David Cummings - Age: 34 - Height: 5' 9 1/2" - Born: Baltimore - Trade: Laborer - Enlistment date: 5 Oct 1814 - Place: Baltimore - Period: War - BLW 12238-160-12 - Discharged at Baltimore on 16 Mar 1815.

Wilkinson, Walter - Captain - 24th Infantry - Born: Maryland - Commissioned as a 1st lieutenant on 2 Jul 1812; promoted to captain on 15 Aug 1813; discharged on 15 Jun 1815.

Willelmi, Charles - Private - 14th Infantry - Prisoner of War (Halifax), captured at Beaver Dams on 24 Jun 1813, discharged from Halifax on 3 Feb 1813.

Willen, Littleton - Private - 38th Infantry - Company: John Brookes - Height: 5' 9" - Born: Maryland - Enlistment date: 27 Jul 1812 - Period: 1 Yr - Re-enlisted at Craney Island, VA, on 17 Apr 1814 for the war; died on 9 May 1814.

Willey, Henry - Private - 36th Infantry - Company: Joseph Hook - Age: 19 - Height: 5' 7" - Born: Pendleton, VA - Trade: Tinner - Enlistment date: 6 Mar 1814 - Place: Fredericktown, MD - Period: War - BLW 3037-160-12 - Discharged at Fort Covington, MD, on 30 Mar 1815.

William, Thomas - Private - 38th Infantry - Company: Shepperd Leakin - BLW 15613-160-12.

Williams, Andrew - Private - 14th Infantry - Prisoner of War (Halifax), captured at Beaver Dams on 24 Jun 1813, discharged from Halifax on 9 Nov

1813.

Williams, Andrew - Private - 38th Infantry - Company: James Hook - Age: 38 - Height: 5' 4 1/2" - Born: Baltimore Cty - Trade: Tailor - Enlistment date: 13 Feb 1814 - Period: War - BLW 6922-160-12 - Discharged at Fort Covington, MD, on 3 Mar 1815.

Williams, Barton - Private - 1st Light Dragoons - Company: Archibald Sneed - Age: 45 - Height: 5' 9" - Born: Prince George's Cty, MD - Trade: Farmer - Enlistment date: 11 Aug 1812 - Place: Virginia - Period: 5 Yrs - Discharged at Fort Hawkins, GA, on 9 Aug 1815 on general debility.

Williams, Bazil - Private - 7th Infantry - Company: Walter Overton - Age: 32 - Height: 5' 10 1/4" - Born: Maryland - Trade: Farmer - Enlistment date: 3 Sep 1808 - Period: 5 Yrs - Discharged on 2 Sep 1813.

Williams, Edmond - Private - 36th Infantry - Age: 21 - Height: 5' 9 1/2" - Born: Monroe, VA - Trade: Farmer - Enlistment date: 13 Dec 1814 - Place: Norfolk, VA - Period: War - Deserted on 12 Feb 1815.

Williams, Eli - Private - 16th Infantry - Company: William Smith - Age: 18 - Height: 5' 4" - Born: Baltimore - Trade: Shoemaker - Enlistment date: 24 Apr 1813 - Period: 5 Yrs.

Williams, Eli - Private - 14th Infantry - Company: Reuben Gilder - Other regiment: 4th Infantry (New) - Age: 18 - Height: 5' 5" - Born: Alexandria, DC - Trade: Carpenter - Enlistment date: 1 Jun 1812 - Place: Shepherdstown, VA - Period: 5 Yrs - Pension: SO-21115 - Also served in Captain John Berry's Company, OH Militia; discharged on 1 Jun 1817.

Williams, George - Private - 36th Infantry - Company: Thomas Carbery - Enlistment date: 4 Oct 1813 - Period: 1 Yr - Discharged on 3 Oct 1814.

Williams, George - Private - 26th Infantry - Company: William Bezeau - Age: 21 - Height: 5' 5" - Born: Lancaster Cty, PA - Trade: Farrier - Enlistment date: 20 Oct 1814 - Place: Baltimore - Period: War - BLW 17027-160-12 - Discharged at Philadelphia on 23 Mar 1815.

Williams, George - Private - Corps of Artillery - Company: Frederick Evans - Age: 18 - Height: 5' 8" - Born: Pennsylvania - Trade: Cabinet maker - Enlistment date: 4 Apr 1814 - Place: Fort McHenry - Period: 5 Yrs - Discharged on 22 Apr 1815 for deafness.

Williams, George H. - Private - 5th Infantry - Age: 33 - Height: 6' 2" - Born: Baltimore - Trade: Wheelwright - Enlistment date: 15 Jan 1814 - Place: Chambersburg, PA - Period: 5 Yrs - Deserted.

Williams, Jacob - Private - 36th Infantry - Company: Joseph Hook - Age: 35 - Height: 5' 10" - Born: Harford Cty, MD - Trade: Shoemaker - Enlistment date: 8 Aug 1814 - Place: Frederick, MD - Period: War - Deserted on 27 Mar 1815.

Williams, James - Seaman - US Sea Fencibles - Company: Simmones Bunbury - Age: 22 - Height: 5' 4 1/2" - Born: Stonington, CT - Enlistment date: 12 Feb 1814 - Period: 1 Yr.

Williams, James - Seaman - US Sea Fencibles - Company: Simmones Bunbury - Enlistment date: 11 Mar 1814 - Period: 1 Yr - Discharged on 10 Mar 1815.

Maryland Regulars in War of 1812

Williams, Jeremiah - Private - 36th Infantry - Company: Joseph Hook - Age: 27 - Height: 5' 6" - Born: Virginia - Enlistment date: 5 May 1813 - Place: Baltimore - Died on 14 Jul 1813.

Williams, Job - Servant - US Sea Fencibles - Company: William Addison.

Williams, John - Private - 39th Infantry - Age: 33 - Height: 5' 11 1/2" - Born: Baltimore - Enlistment date: 19 Jan 1815.

Williams, John - Private - US Volunteers - Company: Stephen Moore - Enlistment date: 8 Sep 1812 - Period: 1 Yr.

Williams, John - Private - 36th Infantry - Company: Joseph Merrick - Age: 24 - Height: 5' 10 1/2" - Born: Rehoboth, MA - Enlistment date: 26 Apr 1813 - Place: Baltimore - Period: 1 Yr.

Williams, John - Private - 38th Infantry - Age: 20 - Height: 5' 5 1/2" - Born: Harford Cty, MD - Trade: Seaman - Enlistment date: 14 Apr 1814 - Period: War.

Williams, John - Private - 14th Infantry - Company: Joseph Marshall - Age: 15 - Height: 5' 4" - Born: Balltown, NY - Trade: Farmer - Enlistment date: 2 Jan 1815 - Period: 5 Yrs - Discharged by *habeas corpus*.

Williams, John - Private - 36th Infantry - Other regiment: Ordnance Department - Age: 25 - Height: 5' 6" - Born: Mathews Cty, VA - Trade: Farmer - Enlistment date: 5 Jul 1814 - Place: Manchester, VA - Period: 5 Yrs - Transferred to Ordnance Department, Lt. N. Braden's detachment; deserted at Greenleaf Point, DC, on 1 Apr 1815.

Williams, John - Sergeant - 5th Infantry - Company: George Brooks - Age: 22 - Height: 5' 6" - Born: Boston - Trade: Laborer - Enlistment date: 18 Apr 1813 - Place: Maryland - Period: 5 Yrs - Deserted on 15 Sep 1813.

Williams, John R. - Private - 36th Infantry - Age: 20 - Height: 5' 8" - Born: Orange, VA - Enlistment date: 24 Sep 1814 - Place: Norfolk, VA - Period: War - Deserted.

Williams, Joseph - Private - 7th Infantry - Company: Walter Overton - Other regiment: 1st Infantry (New) - Age: 40 - Height: 5' 9" - Born: Maryland - Trade: Farmer - Enlistment date: 23 Feb 1814 - Place: Fort Claiborne, AL - Period: War - Discharged on 16 Nov 1816.

Williams, Laban - Private - 10th Infantry - Company: George Strothers - Age: 37 - Height: 5' 7 1/2" - Born: Prince George's Cty, MD - Trade: Farmer - Enlistment date: 25 Jun 1814 - Place: Mt. Pleasant, NC - Period: War - BLW 3246-160-12 - Discharged at Washington, DC, on 10 Apr 1815.

Williams, Lancelot - Private - 2nd Artillery - Company: Frederick Evans - Age: 44 - Height: 5' 8 1/2" - Born: Maryland - Trade: Distiller - Enlistment date: 12 Feb 1814 - Period: 5 Yrs - Discharged on 10 Aug 1815, unfit for service.

Williams, Lot - Private - Light Dragoons - Pension: Heirs received half pay for five years in lieu of land bounty - Died on 24 Nov 1814.

Williams, Luther - Sergeant - 22nd Infantry - Company: Willis Foulk - Age: 22 - Height: 5' 10 1/2" - Born: Talbot Cty, MD - Trade: Physician - Enlistment date: 24 Apr 1814 - Place: Pittsburgh - Period: War - Discharged at Sackets Harbor, NY, on 24 May 1815.

By War of 1812 Soc. in MD

Williams, Richard - Seaman - US Sea Fencibles - Company: William Addison.

Williams, Richard - Private - 2nd Artillery - Company: Nathan Towson - Killed during the Battle of Stoney Creek.

Williams, Robert - Private - 36th Infantry - Company: Joseph Merrick - Age: 18 - Height: 5' 8" - Born: Baltimore - Enlistment date: 18 May 1813 - Place: Baltimore - Period: 1 Yr - Discharged on 19 Dec 1813.

Williams, Spencer - Private - 14th Infantry - Company: Reuben Gilder - Age: 43 - Height: 5' 7" - Born: Maryland - Trade: Shoemaker - Enlistment date: 30 Jun 1813 - Period: War.

Williams, Thomas - Private - 14th Infantry - Company: William McIlvane - Pension: Wife Ellen, WO-37248 - Prisoner of War (Halifax), discharged from Halifax on 3 Feb 1814.

Williams, Thomas - Private - 16th Infantry - Age: 40 - Height: 5' 6" - Born: Baltimore - Enlistment date: 6 Mar 1814 - Place: Philadelphia.

Williams, Thomas G. - Private - 20th Infantry - Company: William Jett - Other regiment: 35th Infantry - Age: 50 - Height: 5' 11" - Born: St. Mary's Cty, MD - Trade: Farmer - Enlistment date: 25 Jul 1813 - Place: Alexandria, DC - Period: 5 Yrs - BLW 485-160-12 - Transferred to 35th Infantry on 20 Mar 1815; discharged at Norfolk, VA, on 26 Jul 1815 on Surgeon's Certificate for old age and rheumatism.

Williams, Thomas L. - Private - US Volunteers - Company: Stephen Moore - Enlistment date: 8 Sep 1812 - Period: 1 Yr.

Williams, William - Seaman - US Sea Fencibles - Company: William Addison.

Williams, William - Corporal - 38th Infantry - Company: Shepperd Leakin - Race: 22 - Age: 22 - Height: 5' 8" - Born: Baltimore - Trade: Laborer - Enlistment date: 5 Apr 1814 - Period: War - A Mulatto who also used the name "Frederick Hall".

Williams, William - Private - 38th Infantry - Company: John Buck - Age: 25 - Height: 5' 10 1/2" - Born: Anne Arundel, MD - Trade: Shoemaker - Enlistment date: 14 Apr 1814 - Place: Baltimore - Period: War - BLW 13457-160-12 - Discharged at Baltimore on 30 Mar 1815.

Williams, William - Private - 36th Infantry - Company: Samuel Raisin - Age: 24 - Height: 5' 8" - Born: Mathews Cty, VA - Trade: Cooper - Enlistment date: 24 Jun 1814 - Period: 5 Yrs - BLW 13322-160-12 - Discharged at Washington, DC, on 20 Mar 1815.

Williams, William - Private - 36th Infantry - Age: 18 - Height: 5' 6 1/2" - Born: St. Mary's Cty, MD - Trade: Laborer - Enlistment date: 3 May 1814 - Place: Leonardstown, MD - Period: 1 Yr - BLW 1613-160-12 - Re-enlisted on 22 Nov 1814 for the war; discharged at Georgetown, DC, on 13 Mar 1815.

Williams, William - Private - 36th Infantry - Company: Charles Randolph - Enlistment date: 25 Aug 1813 - Period: 1 Yr - Deserted on 13 Sep 1813.

Williams, William - Private - 5th Infantry - Age: 20 - Height: 5' 8" - Born: Baltimore - Trade: Cooper - Enlistment date: 31 Mar 1813 - Place: Baltimore - Period: 5 Yrs - BLW 10773-160-12.

Williams, William - Private - 14th Infantry - Died at Sackets Harbor, NY, on

16 Jan 1814.

Williams, William J. - Sergeant - 38th Infantry - Company: John Buck - Pension: Placed on the pension rolls on 15 Dec 1814 - BLW 13998-160-12 - Died in 1825.

Williams, Zadock - Private - 36th Infantry - Age: 36 - Height: 5' 10" - Born: Maryland - Trade: Laborer - Enlistment date: 3 Feb 1815 - Place: Georgetown, DC - Period: War - BLW 14-320-12 - Discharged at Georgetown, DC, on 12 Mar 1815.

Williamson, George - Private - 36th Infantry - Company: Thomas Carbery - Age: 27 - Height: 5' 6 1/4" - Born: Virginia - Enlistment date: 25 Feb 1814 - Place: Leonardstown, MD - Period: War - Discharged on 31 Mar 1815.

Williamson, Hoyt - Private - 14th Infantry - Company: Thomas Montgomery - Other regiment: 4th Infantry (New) - Age: 25 - Height: 5' 10" - Born: Fairfield, CT - Trade: Shoemaker - Enlistment date: 21 May 1812 - Place: Baltimore - Period: 5 Yrs - Discharged on 21 May 1817.

Williamson, Thomas - Sergeant - 3rd Rifles - Company: Edward Carrington - Other regiment: Rifles - Age: 26 - Height: 5' 11" - Born: Baltimore - Trade: Fuller - Enlistment date: 5 May 1814 - Place: West Liberty - Period: War - BLW 23800-160-12.

Williamson, Thomas - Surgeon's Mate - 36th Infantry - Born: Maryland - Commissioned as a surgeon's mate, 36th Infantry, on 30 Apr 1813; resigned in Jun 1813.

Willimea, John N. - Private - 14th Infantry - Company: William McIlvane - Enlistment date: 6 Oct 1812 - Period: 5 Yrs - Prisoner of War, exchanged on 15 Apr 1814; discharged at Greenbush, NY, on 29 Mar 1815.

Willin, Loudor - Private - 14th Infantry - Company: Thomas Montgomery - Enlistment date: 10 May 1813 - Period: 5 Yrs - Pension: Land bounty to Nancy Westcoat, sister and other heirs at law of Loudor Willin - BLW 27031-160-12 - Died on 29 Dec 1813.

Williner, John - Private - 14th Infantry - Prisoner of War (Halifax), discharged from Halifax on 3 Feb 1814.

Willing, Samuel - Private - 36th Infantry - Company: Joseph Hook - Age: 21 - Height: 5' 10" - Born: Eastern Shore, MD - Enlistment date: 23 May 1813 - Place: Baltimore - Period: 1 Yr - Died on 6 Apr 1814.

Willingham, William - Private - 3rd Rifles - Company: Edward Carrington - Age: 32 - Height: 5' 9 1/2" - Born: West Liberty, MD - Trade: Laborer - Enlistment date: 14 Jun 1814 - Place: Truerty Town - Period: War - Pension: Land bounty to James Willingham, only heir at law of William Willingham - BLW 27132-160-42 - Died on 12 Feb 1815.

Willis, John - Sergeant - 14th Infantry - Company: William McIlvane - Age: 51 - Height: 5' 9" - Born: Baltimore - Trade: Hatter - Enlistment date: 3 Jun 1812 - Period: 5 Yrs - Prisoner of War, exchanged on 15 Apr 1814; discharged at Greenbush, NY, on 24 May 1815, wounded at Beaver Dams with gunshot wound in his arm.

Willis, Richard - Private - 28th Infantry - Company: Joseph Belt - Other regiment: 3rd Infantry (New) - Age: 21 - Height: 5' 5" - Born: Maryland -

By War of 1812 Soc. in MD

- Trade: Farmer - Enlistment date: 23 Aug 1814 - Place: Augusta, MD - Period: 5 Yrs - BLW 23275-160-12 - Discharged at Fort Dearborn, IL, on 28 Aug 1819.
- Willis, Stephen - Private - 14th Infantry - Prisoner of War (Halifax), sent to England, captured at Beaver Dams on 24 Jun 1813.
- Willison, Jeremiah - Private - Rifles - Age: 29 - Height: 6' 2" - Born: Allegany Cty, MD - Trade: Laborer - Enlistment date: 11 Feb 1815 - Place: Morgantown, VA - Period: War - Discharged at Washington, DC, on 29 Mar 1815.
- Willoughby, George - Corporal - 5th Infantry - Company: Richard Whartenby - Age: 24 - Height: 5' 9" - Born: Caroline Cty, MD - Trade: Miller - Enlistment date: 17 Jul 1811 - Place: Eastown - Period: 5 Yrs - Prisoner of War (Halifax), captured at Stoney Creek, UC, on 12 Jun 1813.
- Wills, John - Private - 2nd Light Dragoons - Company: William Littlejohn - Other regiment: Light Dragoons - Age: 20 - Height: 5' 8 1/4" - Born: Prince George's Cty, MD - Trade: Shoemaker - Enlistment date: 3 Feb 1814 - Place: Alexandria, DC - Period: War - BLW 1553-160-12 - Discharged at Carlisle Barracks, PA, on 4 May 1815.
- Wills, Miles - Private - 36th Infantry - Company: Samuel Raisin - Age: 23 - Height: 5' 8" - Born: Lunenburg Cty, VA - Trade: Farmer - Enlistment date: 21 Jun 1814 - Period: War - BLW 20028-160-12 - Discharged at Washington, DC, on 20 Mar 1815.
- Willson, John - Private - 36th Infantry - Age: 28 - Height: 5' 6 3/4" - Born: Portsmouth, NH - Enlistment date: 7 May 1813 - Place: Baltimore.
- Willson, John - Private - 15th Infantry - Age: 22 - Height: 5' 6" - Born: Baltimore - Enlistment date: 24 Sep 1814 - Place: Elizabethtown, NJ - Deserted.
- Wilmore, John - Private - 26th Infantry - Company: William Bezeau - Age: 25 - Height: 5' 7 1/2" - Born: Maryland - Trade: Shoemaker - Enlistment date: 17 Oct 1814 - Place: Philadelphia - Period: War - BLW 19599-160-12 - Discharged on 20 Mar 1815.
- Wilson, Alexander - Private - 14th Infantry - Company: Henry Grindage - Age: 26 - Prisoner of War, captured during the Battle of Stoney Creek.
- Wilson, Alexander - Sergeant - 14th Infantry - Company: Richard Arell - Other regiment: 4th Infantry (New) - Age: 27 - Height: 5' 4" - Born: Hunterdon Cty, NJ - Trade: Shoemaker - Enlistment date: 6 Oct 1812 - Place: Baltimore - Period: 5 Yrs - Prisoner of War, exchanged on 15 Apr 1814; discharged on 6 Oct 1817.
- Wilson, Alexander - Private - 2nd Artillery - Company: Stanton Sholes - Other regiment: Corps of Artillery - Age: 20 - Height: 6' - Born: Cecil Cty, MD - Trade: Miller - Enlistment date: 8 Feb 1813 - Place: Washington Cty, PA - Period: 5 Yrs - Discharged on 15 Jan 1816 after furnishing a substitute.
- Wilson, Amos - Private - 36th Infantry - Age: 24 - Height: 5' 11" - Born: Pennsylvania - Enlistment date: 9 Dec 1813 - Place: Baltimore.
- Wilson, Azariah - Private - 14th Infantry - Enlistment date: 20 Oct 1812 - Period: 5 Yrs - Pension: Land bounty to Samuel Wilson, heir at law of

Maryland Regulars in War of 1812

Azariah Wilson - BLW 3231-160-12 - Died on 7 Jan 1814.

Wilson, Benjamin E. - Private - 22nd Infantry - Company: John Foster - Age: 43 - Height: 5' 9 1/2" - Born: Maryland - Trade: Shoemaker - Enlistment date: 13 Mar 1813 - Period: 5 Yrs - Deserted at French Mills, NY, on 23 Jan 1814.

Wilson, Charles - Seaman - US Sea Fencibles - Company: William Addison.

Wilson, Charles - Private - 36th Infantry - Age: 33 - Height: 5' 9" - Born: North Carolina - Enlistment date: 9 Sep 1813 - Place: Baltimore.

Wilson, Daniel - Private - 14th Infantry - Company: Samuel Lane - Age: 36 - Height: 5' 8" - Born: Marlborough, MD - Trade: Carpenter - Enlistment date: 1 Aug 1812 - Period: 5 Yrs - Pension: Old War IF-25952 - BLW 7932-160-12 - Prisoner of War (Halifax), captured at Beaver Dams on 24 Jun 1813, discharged from Halifax on 3 Feb 1813, severely wounded in the leg at Beaver Dams, exchanged on 15 Apr 1814; discharged at Greenbush, NY, on 20 Jun 1814 on Surgeon's Certificate.

Wilson, Edward - First Lieutenant - 14th Infantry - Company: Thomas Montgomery - BLW 2269-160-50 - Commissioned as a 2nd lieutenant on 12 Mar 1812; promoted to 1st lieutenant on 12 May 1813; discharged on 15 Jun 1815.

Wilson, George - Artificer - 2nd Artillery - Company: Nathan Towson - Age: 25 - Height: 5' 7" - Born: Pennsylvania - Trade: Cabinet maker - Enlistment date: 16 Mar 1813 - Place: Black Rock, NY - Period: 5 Yrs.

Wilson, George - Private - 5th Infantry - Company: James Dorman - Other regiment: 3rd Infantry (New) - Age: 24 - Height: 5' 8" - Born: Lancaster, PA - Trade: Shoemaker - Enlistment date: 12 Feb 1812 - Place: Hagerstown, MD - Period: 5 Yrs - Deserted at Pittsburgh on 23 Aug 1815.

Wilson, George - Private - 12th Infantry - Company: Thomas Sangster - Age: 27 - Height: 5' 7 3/4" - Born: Ireland - Trade: Weaver - Enlistment date: 14 may 1812 - Place: Havre de Grace, MD - Period: 5 Yrs - Possible deserter from the 5th Infantry.

Wilson, George C. - Gunner - US Sea Fencibles - Company: Simmones Bunbury - Age: 28 - Height: 5' 11 3/4" - Born: King & Queen Cty, VA - Enlistment date: 7 Dec 1813 - Period: 1 Yr - Pension: Old War IF-25953 - Re-enlisted on 7 Dec 1814 for one year; discharged at Fort McHenry on 24 Mar 1815.

Wilson, James - Private - 14th Infantry - Company: David Cummings - Age: 30 - Height: 5' 9 1/2" - Born: Little York, PA - Trade: Brick layer - Enlistment date: 23 Nov 1814 - Place: Baltimore - Period: War - BLW 10064-160-12 - Discharged at Baltimore on 16 Mar 1815.

Wilson, James - Private - 14th Infantry - Age: 28 - Height: 5' 6 1/2" - Born: England - Trade: Sailor - Enlistment date: 28 Dec 1814 - Period: War - Discharged at Baltimore on 16 Mar 1815.

Wilson, James - Private - 38th Infantry - Age: 25 - Height: 5' 10" - Born: Baltimore - Trade: Shoemaker - Enlistment date: 22 Feb 1814 - Period: War.

Wilson, James H. - Private - 38th Infantry - Age: 18 - Height: 5' 5" - Born: Bladen, NC - Enlistment date: 25 Aug 1814 - Place: Baltimore - Period: War

By War of 1812 Soc. in MD

- Claimed to be a minor and he was discharged.

Wilson, John - Private - 38th Infantry - Company: James Hook - BLW 22009-160-12.

Wilson, John - Private - 17th Infantry - Company: Henry Crittenden - Height: 5' 6 1/4" - Born: Baltimore - Trade: Cooper - Enlistment date: 22 Mar 1813 - Place: Washington, DC - Period: 5 Yrs - Deserted on 21 Apr 1815.

Wilson, John - Private - Rifles - Age: 30 - Height: 5' 8" - Born: Havre de Grace, MD - Enlistment date: 20 Aug 1814.

Wilson, John - Private - 22nd Infantry - Company: Willis Foulk - Age: 25 - Height: 5' 11" - Born: Baltimore - Trade: Cooper - Enlistment date: 2 Apr 1814 - Period: War - Deserted at Buffalo on 2 Oct 1814.

Wilson, John - Private - 5th Infantry - Company: Colin Buckner - Age: 32 - Height: 5' 7" - Born: Delaware - Trade: Miller - Enlistment date: 13 May 1813 - Place: Baltimore - Period: 5 Yrs.

Wilson, John - Private - 5th Infantry - Company: Leroy Opie - Age: 34 - Height: 6' - Born: Baltimore - Trade: Cooper - Enlistment date: 17 Apr 1814 - Place: Lancaster, PA - Period: 5 Yrs.

Wilson, Joseph C. - Private - 5th Infantry - Age: 29 - Height: 5' 10" - Born: Baltimore - Trade: Baker - Enlistment date: 10 May 1813 - Place: Baltimore - Period: 5 Yrs - BLW 10770-160-12.

Wilson, Joshua - Private - 36th Infantry - Company: Joseph Hook - Age: 18 - Height: 5' 6" - Born: Baltimore - Enlistment date: 24 May 1813 - Place: Westminster, MD - Period: 1 Yr.

Wilson, Lazarue - Private - 38th Infantry - Company: John Brookes - BLW 5414-160-12.

Wilson, Moses - Private - 5th Infantry - Company: Colin Buckner - Age: 23 - Height: 5' 8 1/2" - Born: New Jersey - Trade: Millwright - Enlistment date: 16 Apr 1813 - Place: Baltimore - Period: 5 Yrs.

Wilson, Robert M. - Private - 1st Infantry - Company: John Symmes - Age: 40 - Height: 5' 10" - Born: Maryland - Trade: Weaver - Enlistment date: 6 Mar 1813 - Place: Nashville, TN - Period: 5 Yrs - Discharged at Sackets Harbor, NY, on 10 Jun 1815.

Wilson, Robert M. - Private - 1st Infantry - Company: John Symmes - Age: 41 - Height: 5' 10" - Born: Maryland - Trade: Weaver - Enlistment date: 6 Mar 1813 - Place: Nashville, TN - Period: 5 Yrs - Wounded at Chippewa, UC, on 25 Jul 1814; discharged at Sackets Harbor, NY, on 10 Jun 1815.

Wilson, Spedden - Corporal - 2nd Artillery - Company: Frederick Evans - Other regiment: Corps of Artillery - Age: 23 - Height: 5' 9 1/2" - Born: Dorchester Cty, MD - Trade: Laborer - Enlistment date: 24 Apr 1812 - Place: Annapolis - Period: 5 Yrs.

Wilson, William - Major - 1st Artillery - Other regiment: Corps of Artillery - Born: Maryland - Commissioned as a lieutenant, 1st Artillery and Engineers, on 17 Jul 1794; resigned on 24 Sep 1799; commissioned as a lieutenant, 2nd Artillery and Engineers, on 16 Feb 1801; transferred to Artillery on 1 Apr 1802; promoted to captain on 3 May 1808; transferred to Corps of Artillery on 12 May 1814; transferred to 3rd Artillery on 1 Jun 1821; promoted to major on 8 May 1822; died on 15 Sep 1825.

Wilson, William - Gunner - US Sea Fencibles - Company: Simmones Bunbury - Enlistment date: 24 Jan 1814 - Period: 1 Yr - Discharged on 23 Jan 1815.

Wilson, William - Private - 1st Artillery - Company: Heman Fay - Age: 34 - Height: 5' 10" - Born: New York - Trade: Miller - Enlistment date: 6 Apr 1814 - Place: Annapolis - Period: War.

Wilson, William - Drummer - 36th Infantry - Company: Hugh Deneale - Age: 34 - Height: 5' 10 1/2" - Born: Pennsylvania - Enlistment date: 5 Sep 1813 - Place: York, PA - Period: 1 Yr.

Wilson, William W. - Private - 14th Infantry - Company: Thomas Montgomery - Enlistment date: 1 Dec 1812 - Period: 18 Mos - Died at Burlington, VT, in Apr 1814 from sickness.

Wilt, Jacob - Private - 14th Infantry - Company: Reuben Gilder - BLW 9389-160-12.

Winder, William Henry - Brigadier General - 14th Infantry - Other regiment: General Staff - Born: Maryland - Place: Maryland - Pension: Land bounty to Gertrude Winder, widow of William H. Winder - BLW 14081-160-50 - Commissioned as a lieutenant colonel on 16 Mar 1812; promoted to colonel, 14th Infantry, on 6 Jul 1812; promoted to brigadier general on 12 Mar 1813; Prisoner of War, captured at Stoney Creek, UC, on 12 Jun 1813; discharged on 15 Jun 1815; died on 24 May 1824.

Winders, Thomas - Private - 17th Infantry - Company: Henry Crittenden - Age: 38 - Height: 5' 10" - Born: Maryland - Trade: Laborer - Enlistment date: 14 Jun 1812 - Period: War - BLW 8616-160-12 - Discharged at Chillicothe, OH, on 9 Jun 1815.

Windsor, Richard - Private - 24th Infantry - Other regiment: 8th Infantry (New) - Age: 40 - Height: 5' 10" - Born: Maryland - Trade: Farmer - Enlistment date: 27 Feb 1815 - Place: Belle Fontaine, MO - Period: 5 Yrs.

Wine, Jacob - Private - 16th Infantry - Company: Thomas Horrell - Age: 18 - Height: 5' 8" - Born: Hagerstown, MD - Enlistment date: 23 Nov 1813 - Place: Little York, PA - Period: 5 Yrs - Discharged at Buffalo on 19 Jun 1815 for epilepsy.

Wine, John - Private - 16th Infantry - Company: Thomas Horrell - Age: 17 - Height: 4' 4" - Born: Hagerstown, MD - Enlistment date: 13 Feb 1814 - Place: Norristown, PA - Period: 5 Yrs - Discharged at Buffalo on 19 Jun 1815.

Winebrener, Peter - Private - 35th Infantry - Company: Francis Walker - Age: 20 - Born: Frederick, MD - Enlistment date: 10 Aug 1813 - Place: Richmond, VA - Period: 1 Yr.

Wineman, Mathias - Private - 2nd Artillery - Company: Nathan Towson - Other regiment: Corps of Artillery - Age: 34 - Height: 5' 5 1/2" - Born: Baltimore - Trade: Shoemaker - Enlistment date: 15 Apr 1812 - Place: Baltimore - Period: 5 Yrs - BLW 11901-160-12 - Prisoner of War; discharged at Fort Niagara, NY, on 15 Apr 1817.

Wingard, James C. - Sergeant - 17th Infantry - Company: Martin Hawkins - Age: 34 - Height: 5' 11" - Born: Maryland - Trade: Carpenter - Enlistment date: 11 May 1813 - Period: War - Pension: Wife Elizabeth Klink, Old War WF-12042 - BLW 21047-160-12 - Discharged at Newport, KY, on 10 May 1815.

Wingate, John - Private - 26th Infantry - Company: William Bezeau - Age: 21 - Height: 5' 7" - Born: Cecil Cty, MD - Trade: Laborer - Enlistment date: 4 Nov 1814 - Place: Baltimore - Period: War - Discharged at Philadelphia on 23 Mar 1815.

Wingendour, Peter - Private - 5th Infantry - Company: Richard Whartenby - Other regiment: 3rd Infantry (New) - Age: 35 - Height: 5' 10" - Born: Baltimore - Trade: Mariner - Enlistment date: 23 Mar 1812 - Place: Baltimore - Period: 5 Yrs - Deserted on 20 Aug 1815.

Winks, James R. - Private - 36th Infantry - Company: Thomas Carbery - Age: 23 - Height: 5' 10" - Born: Baltimore - Trade: Rope maker - Enlistment date: 27 Feb 1814 - Place: Georgetown, DC - Period: War - BLW 87-160-12 - Discharged at Baltimore on 31 Mar 1815.

Winn, Aaron - Private - 14th Infantry - Company: Thomas Montgomery - Pension: Land bounty to Moses Wnn, brother and other heirs at law of Aaron Winn - BLW 14936-160-12.

Winn, Clement - Private - 14th Infantry - Company: David Cummings - Enlistment date: 24 Jun 1812 - Period: 5 Yrs - Prisoner of War (Halifax), sent to England, captured at Beaver Dams on 24 Jun 1813; deserted at Fort McHenry on 19 Apr 1815.

Winn, Henry - Recruit - 26th Infantry - Company: William Bezeau - Race: 25 - Age: 25 - Height: 5' 5" - Born: Maryland - Trade: Farmer - Enlistment date: 9 Oct 1814 - Place: Philadelphia - Period: War - Listed as "Col'd" in his service record; discharged on 20 Mar 1815.

Winn, James - Recruit - 26th Infantry - Company: William Bezeau - Race: 22 - Age: 22 - Height: 5' 5" - Born: Maryland - Trade: Laborer - Enlistment date: 10 Oct 1814 - Place: Philadelphia - Period: 5 Yrs - BLW 10299-160-12 - Listed as "Col'd" in his service record; discharged at Philadelphia on 25 Mar 1815.

Winner, Michael - Private - 14th Infantry - Prisoner of War (Halifax), sent to England.

Winniscot, Charles - Private - 2nd Infantry - Company: William Boots - Age: 21 - Height: 5' 9" - Born: Cumberland Cty, VA - Trade: Farmer - Enlistment date: 2 Feb 1807 - Place: Hagerstown, MD - Period: 5 Yrs - Discharged at Fort Stoddard, AL, on 19 Sep 1812.

Winslow, Isaac - Private - 1st Light Dragoons - Company: Selleck Osborn - Age: 26 - Height: 5' 6 3/4" - Born: Queen Anne, MD - Trade: Carpenter - Enlistment date: 18 Apr 1810 - Place: New Bedford - Period: 5 Yrs - Discharged at Greenbush, NY, on 18 Apr 1815.

Winsor, Lazarus - Private - 38th Infantry - Company: John Brookes - Age: 27 - Height: 5' 7 1/2" - Born: Somerset Cty, MD - Trade: Farmer - Enlistment date: 11 Aug 1813 - Period: 1 Yr - Re-enlisted at Norfolk, VA, on 17 Mar 1814 for the war; discharged at Washington, DC, on 12 Apr 1815.

Wintz, George - Private - 38th Infantry - Company: John Mowton - Pension:

Maryland Regulars in War of 1812

SO-26532, SC-21063, WO-11177, WC-13098 - BLW 13093-160-12.

Wise, John - Private - 15th Infantry - Company: White Young - Other regiment: Corps of Artillery - Age: 22 - Height: 5' 6" - Born: Baltimore - Trade: Plasterer - Enlistment date: 6 Apr 1812 - Place: Philadelphia - Period: 5 Yrs - Discharged on 26 Apr 1817.

Wiseman, William - Private - 8th Infantry - Company: David Twiggs - Age: 22 - Height: 5' 7 1/2" - Born: Washington, MD - Trade: Carpenter - Enlistment date: 3 Sep 1813 - Place: Georgia - Period: 5 Yrs - Discharged on 21 Aug 1815 for debility.

Wittenberg, William - Private - 38th Infantry - Company: John Mowton - Age: 38 - Height: 5' 11" - Born: Baltimore - Trade: Butcher - Enlistment date: 22 Feb 1814 - Place: Craney Island, VA - Period: War - Discharged at Craney Island, VA, on 15 Mar 1815.

Wittlurger, George - Private - 2nd Artillery - Company: Nathan Towson - Enlistment date: 1 Jul 1812 - Period: 5 Yrs - Deserted on 19 Aug 1812.

Wolcott, William - Private - 27th Infantry (OH) - Age: 53 - Height: 6' 1 1/2" - Born: Maryland - Enlistment date: 27 Dec 1813 - Place: Newark, OH.

Wolfrom, William - Private - 14th Infantry - Prisoner of War (Halifax), discharged from Halifax on 3 Feb 1814.

Wolslager, Daniel - Private - 36th Infantry - Company: Joseph Hook - Age: 21 - Height: 5' 5" - Born: Hagerstown, MD - Enlistment date: 22 May 1813 - Place: Baltimore - Period: 1 Yr.

Wood, Aaron - Private - 36th Infantry - Company: Joseph Hook - Born: Frederick, MD - Enlistment date: 27 Jun 1813 - Place: Fredericktown, MD - Period: 1 Yr - Deserted on 12 Jul 1813.

Wood, George - Private - 22nd Infantry - Company: Thomas Lawrence - Other regiment: 2nd Infantry (New) - Age: 19 - Height: 5' 6" - Born: Hagerstown, MD - Trade: Tailor - Enlistment date: 12 Dec 1813 - Place: Franklin Cty, PA - Period: 5 Yrs - Discharged on 12 Dec 1817.

Wood, Joel P. - Private - 1st Artillery - Company: George Armistead - Pension: Wife Elizabeth, WO-41373, WC-42006.

Wood, John - Seaman - US Sea Fencibles - Company: Simmones Bunbury - Age: 26 - Height: 5' 9 3/4" - Born: Essex, MA - Enlistment date: 1 Nov 1813 - Period: 1 Yr - Re-enlisted on 25 Nov 1814 for one year; discharged on 24 Mar 1815.

Wood, John - Private - 27th Infantry - Company: Aaron Craine - Age: 28 - Height: 5' 6" - Born: Bladensburg, MD - Trade: Hair dresser - Enlistment date: 21 Sep 1814 - Place: New York - Period: War - BLW 3444-160-12 - Discharged at New York on 29 Jun 1815.

Wood, Moses - Private - 2nd Artillery - Company: Nathan Towson - Enlistment date: 3 Jun 1812 - Period: 5 Yrs - Pension: Land bounty to Elisha Wood and Mary Riley, children and only heirs at law of Moses Wood - BLW 27929-160-42 - Died in 1812.

Wood, Nicholas - Private - 36th Infantry - Company: Mortimer Hall - Age: 46 - Height: 5' 8" - Born: Holland - Enlistment date: 29 Aug 1813 - Place: Baltimore - Period: 1 Yr - Died on 10 Feb 1814.

By War of 1812 Soc. in MD

Wood, Pratt - Private - 14th Infantry - Prisoner of War (Halifax), discharged from Halifax on 3 Feb 1814.

Wood, Robert - Private - 38th Infantry - Company: James Hook - Age: 20 - Height: 5' 4" - Born: Baltimore Cty - Trade: Farmer - Enlistment date: 13 Mar 1814 - Place: Baltimore - Period: War - BLW 5729-160-12 - Discharged at Fort Covington, MD, on 31 Mar 1815.

Wood, Robert - Private - 38th Infantry - Company: James Hook - Age: 21 - Height: 5' 5" - Born: Anne Arundel, MD - Trade: Laborer - Enlistment date: 11 Dec 1813 - Period: 1 Yr - BLW 5729-160-12 - Re-enlisted on 6 Apr 1814 for the war.

Wood, William - Private - 36th Infantry - Company: Thomas Carbery - Age: 23 - Height: 5' 11 1/2" - Born: St. Mary's Cty, MD - Trade: Farmer - Enlistment date: 22 Sep 1814 - Place: Washington, DC - Period: War - BLW 3344-160-12 - Discharged on 31 Mar 1815.

Wood, William - Private - 1st Artillery - Company: James House - Age: 26 - Height: 5' 7 3/4" - Born: Baltimore - Trade: Seaman - Enlistment date: 16 Oct 1812 - Place: Fort Wolcott, RI - Period: 5 Yrs - Deserted at Fort Pike, NY, on 5 Oct 1815.

Wood, Zachariah - Private - 38th Infantry - Company: John Rothrock - Age: 30 - Height: 5' 11" - Born: Elk Ridge, MD - Trade: Ship joiner - Enlistment date: 24 Feb 1814 - Place: Craney Island, VA - Period: War - BLW 7533-160-12 - Discharged at Craney Island, VA, on 15 Mar 1815.

Woodard, Abraham - Private - 36th Infantry - Company: Mortimer Hall - Enlistment date: 9 Aug 1813 - Period: 1 Yr.

Woodard, James - Private - 12th Infantry - Company: Andrew Madison - Age: 47 - Height: 5' 9" - Born: Montgomery Cty, MD - Trade: Blacksmith - Enlistment date: 6 Mar 1813 - Place: Newton, VA - Period: War - Discharged at Buffalo on 31 May 1815.

Woodbridge, Cuttison Norton - Private - 5th Infantry - Other regiment: 8th Infantry (New) - Age: 40 - Height: 5' 8" - Born: Boston - Trade: Artificer - Enlistment date: 11 Feb 1813 - Place: Baltimore - Period: 5 Yrs - BLW 24360-160-12.

Woodland, Samuel - Private - 36th Infantry - Age: 43 - Height: 5' 9" - Born: Kent Cty, MD - Enlistment date: 19 Mar 1814 - Place: Sandy Point, VA - Died in 1814.

Woodland, Samuel - Private - 36th Infantry - Company: Samuel Raisin - Age: 43 - Height: 5' 7 3/4" - Born: Kent Cty, MD - Enlistment date: 16 Aug 1813 - Place: Chestertown, MD - Period: 1 Yr.

Woodland, William - Private - 14th Infantry - Company: William McIlvane - Age: 37 - Height: 5' 5" - Born: Maryland - Trade: Tailor - Enlistment date: 15 May 1812 - Period: 5 Yrs - BLW 12446-160-12 - Prisoner of War (Halifax), captured at Beaver Dams on 24 Jun 1813, discharged from Halifax on 3 Feb 1813, exchanged on 15 Apr 1814; discharged at Greenbush, NY, on 4 May 1814 on Surgeon's Certificate, rheumatism.

Woodland, Zebediah - Private - 14th Infantry - Company: Isaac Barnard - Age: 25 - Height: 5' 5" - Trade: Blacksmith - Enlistment date: 20 Nov 1812 -

Period: 18 Mos - Prisoner of War (Halifax), captured at Stoney Creek, UC, on 12 Jun 1813; discharged on 7 Aug 1814.

Woodland, Zebediah - Private - 38th Infantry - Company: Shepperd Leakin - Other regiment: Ordnance Department - Age: 26 - Height: 5' 5" - Born: Vienna, MD - Trade: Blacksmith - Enlistment date: 2 Sep 1814 - Place: Baltimore - Period: 5 Yrs - Transferred to Ordnance Department on 21 Jun 1815.

Woodruff, Ichabod - Private - 26th Infantry - Company: William Bezeau - Age: 35 - Height: 5' 8" - Born: Elizabeth, NJ - Trade: Nailor - Enlistment date: 26 Oct 1814 - Place: Baltimore - Period: War - BLW 13689-160-12 - Discharged at Philadelphia on 23 Mar 1815.

Woods, Lewis - Seaman - US Sea Fencibles - Company: Simmones Bunbury - Age: 20 - Height: 5' 7" - Born: Havre de Grace, MD - Enlistment date: 30 Dec 1813 - Period: 1 Yr.

Woodward, Amon - Third Lieutenant - 36th Infantry - Company: Thomas Carbery - BLW 4964-160-50 - Commissioned as an ensign on 10 Jun 1814; promoted to 3rd lieutenant, on 10 Aug 1814; discharged on 15 Jun 1815.

Woodward, Amos - First Lieutenant - 14th Infantry - Company: David Cummings - Born: Maryland - Pension: Wife Barbara, SO-9741, SC-12568, WO-18360, SC-9021 - Commissioned as a 2nd lieutenant on 4 Jun 1812; promoted to 1st lieutenant on 13 May 1813; discharged on 15 Jun 1815.

Woodward, L. - Private - 14th Infantry.

Woody, Abraham - Private - 36th Infantry - Age: 45 - Height: 5' 11" - Born: Prince George's Cty, MD - Enlistment date: 9 Aug 1813 - Place: Baltimore.

Wooker, Henry - Private - 2nd Infantry - Company: John Bowyer - Age: 35 - Height: 5' 6" - Born: Germany - Trade: Chair maker - Enlistment date: 23 Feb 1808 - Place: Fredericktown, MD - Period: 5 Yrs - Died in 1812.

Woolan, Joseph - Private - 36th Infantry - Company: Samuel Raisin - Age: 36 - Height: 5' 8" - Born: Dorchester Cty, MD - Trade: Cooper - Enlistment date: 20 Jul 1814 - Period: War - BLW 21080-160-12 - Discharged at Washington, DC, on 20 Mar 1815.

Woolfred, John - Private - 14th Infantry - Prisoner of War (Halifax), discharged from Halifax on 3 Feb 1814.

Woolslager, Daniel - Private - 38th Infantry - Age: 23 - Height: 5' 5 1/2" - Born: Hagerstown, MD - Trade: Shoemaker - Enlistment date: 19 Oct 1814 - Place: Baltimore - Period: War.

Wooster, John - Private - 14th Infantry - Company: David Cummings - Age: 23 - Height: 5' 11" - Born: Talbot Cty, MD - Trade: Farmer - Enlistment date: 10 Oct 1814 - Place: Baltimore - Period: War - BLW 18346-160-12 - Discharged at Baltimore on 16 Mar 1815.

Woosters, Shadrick - Private - 36th Infantry - Company: Samuel Raisin - Age: 36 - Height: 5' 6 1/2" - Born: Talbot Cty, MD - Enlistment date: 11 May 1813 - Place: Cambridge, MD - Period: 1 Yr.

Wooten Jr., Bennett - Private - 36th Infantry - Company: Thomas Carbery - Enlistment date: 29 Nov 1813 - Period: 1 Yr - Discharged on 28 Nov 1814.

Wooten Sr., Bennett - Private - 36th Infantry - Company: Thomas Carbery -

By War of 1812 Soc. in MD

Enlistment date: 4 Jan 1814 - Period: 1 Yr - Discharged on 3 Jan 1815.

Wooters, David - Private - 36th Infantry - Company: Samuel Raisin - Age: 19 - Height: 5' 6 1/2" - Born: Talbot Cty, MD - Enlistment date: 11 May 1813 - Place: Cambridge, MD - Period: 1 Yr - Died on 15 May 1813.

Wooters, John - Private - 36th Infantry - Company: Joseph Merrick - Age: 22 - Height: 5' 11" - Born: Talbot Cty, MD - Enlistment date: 11 May 1813 - Place: Cambridge, MD - Period: 1 Yr.

Worland, Henry - Private - 2nd Rifles - Company: Benjamin Desha - Age: 27 - Height: 5' 4 1/2" - Born: Washington, MD - Trade: Laborer - Enlistment date: 17 Oct 1814 - Period: War.

Worrels, Joseph - Private - 8th Infantry - Age: 18 - Height: 5' 3" - Born: Maryland - Enlistment date: 20 Jul 1814 - Place: Edgefield, SC.

Worseley, George - Private - 26th Infantry - Company: William Bezeau - Age: 37 - Height: 6' 1" - Born: Harford Cty, MD - Trade: Millwright - Enlistment date: 23 Nov 1814 - Period: War - Discharged at Baltimore on 15 Mar 1815.

Worth, Thomas - Private - 36th Infantry - Company: Mortimer Hall - Age: 32 - Height: 5' 7 1/2" - Born: Baltimore - Enlistment date: 15 Aug 1813 - Place: Baltimore - Period: 1 Yr.

Worthington, John - Corporal - 14th Infantry - Company: Joseph Marshall - Age: 22 - Height: 5' 2" - Born: Worcester, MD - Trade: Shoemaker - Enlistment date: 11 Jul 1814 - Place: Salisbury, MD - Period: War - Pension: Land bounty to Peggy Worthington, sister and heir at law of John Worthington - BLW 23428-160-12 - Discharged at Greenbush, NY, on 3 May 1815.

Worthington, John - Private - 14th Infantry - Company: Thomas Montgomery - Enlistment date: 26 Nov 1812 - Period: 18 Mos - Discharged on 26 May 1814.

Wother, Christian - Private - 2nd Light Dragoons - Company: William Littlejohn - Age: 27 - Height: 5' 8 1/2" - Born: Bornier, Prussia - Trade: Brass founder - Enlistment date: 15 Mar 1813 - Place: Baltimore - Period: 5 Yrs - Discharged on 18 Apr 1815.

Wright, Caleb - Private - 43rd Infantry - Company: George Dabney - Age: 40 - Height: 5' 4" - Born: Baltimore - Trade: Laborer - Enlistment date: 13 Apr 1814 - Place: Roxbrough - Period: War - BLW 2634-160-12 - Discharged at Fort Hampton on 1 Aug 1815.

Wright, Clinton - Cornet - 2nd Light Dragoons - Born: Maryland - Commissioned as a cornet, Light Dragoons, on 19 Jan 1813; transferred to 2nd Light Dragoons, on 29 Jan 1813; promoted to 3rd lieutenant on 29 Apr 1813; promoted to 2nd lieutenant on 19 Apr 1814; transferred to Light Dragoons on 12 May 1814; transferred to 4th Infantry on 17 May 1815; drowned on 24 Feb 1818.

Wright, George - Private - 36th Infantry - Company: Samuel Raisin - Age: 23 - Height: 5' 10 1/2" - Born: Loudoun Cty, VA - Trade: Blacksmith - Enlistment date: 26 Aug 1814 - Place: Manchester - Period: War - BLW 23244-160-12 - Discharged at Washington, DC, on 20 Mar 1815.

Wright, Ignatius - Private - 14th Infantry - Company: Reuben Gilder - Other regiment: 4th Infantry (New) - Age: 23 - Height: 5' 6 1/2" - Born: Virginia - Trade: Cooper - Enlistment date: 14 Apr 1813 - Place: Alexandria, DC - Period: 5 Yrs - Deserted on 17 Jul 1815.

Wright, Jesse T. - Sergeant - 5th Infantry - Company: William Smith - Other regiment: 3rd Infantry (New) - Age: 21 - Height: 5' 10 1/2" - Born: Liberty, VA - Trade: Clerk - Enlistment date: 3 Jul 1812 - Place: Baltimore - Period: 5 Yrs - Discharged on 2 Jul 1817.

Wright, John - Private - 14th Infantry - Prisoner of War (Halifax), discharged from Halifax on 3 Feb 1814, exchanged on 15 Apr 1814.

Wright, John - Private - 14th Infantry - Company: Samuel Lane - Other regiment: 4th Infantry (New) - Age: 22 - Height: 5' 5 1/2" - Born: Salem, NJ - Trade: Shoemaker - Enlistment date: 9 Jun 1812 - Place: Hagerstown, MD - Period: 5 Yrs - Discharged on 9 Jun 1817.

Wright, John - Private - 36th Infantry - Age: 30 - Height: 5' 9" - Born: Wales - Enlistment date: 29 Nov 1814 - Place: Petersburg, VA.

Wright, Lerick - Private - 14th Infantry - Prisoner of War (Halifax), discharged from Halifax on 3 Feb 1814.

Wright, Nathaniel - Private - 24th Infantry - Company: Walter Wilkinson - Age: 25 - Born: Baltimore Cty - Trade: Farmer - Enlistment date: 21 Jul 1814 - Period: War - BLW 19268-160-12 - Discharged at Belle Fontaine, MO, on 25 Jul 1815.

Wright, Reuben - Corporal - 17th Infantry - Company: Benjamin Sanders - Age: 24 - Height: 5' 11" - Born: Frederick Cty, MD - Trade: Wagon maker - Enlistment date: 25 Mar 1814 - Place: Kentucky - Period: War - BLW 5880-160-12 - Discharged at Chillicothe, OH, on 7 Jun 1815.

Wright, Samuel - Private - 1st Artillery - Company: George Armistead - Age: 25 - Height: 5' 8" - Born: Pennsylvania - Trade: Plasterer - Enlistment date: 27 Dec 1810 - Place: Fort McHenry - Period: 5 Yrs - Discharged at Fort Columbus, NY, on 13 Mar 1815, disability.

Wright, Thomas - Private - 26th Infantry - Company: William Bezeau - Age: 29 - Height: 5' 7" - Born: Harford Cty, MD - Trade: Tinman - Enlistment date: 30 Sep 1814 - Place: Baltimore - Period: War - BLW 13317-160-12 - Discharged at Philadelphia on 28 Mar 1815.

Wright, William - Private - 38th Infantry - Company: Shepperd Leakin - Age: 38 - Height: 5' 2" - Born: Dorchester Cty, MD - Trade: Farmer - Enlistment date: 25 Feb 1814 - Period: War - Deserted on 10 Mar 1814.

Wright, William - Private - 14th Infantry - Enlistment date: 24 Jul 1813 - Period: 18 Mos.

Wright, William - Ensign - 12th Infantry - Company: Thomas Sangster - Age: 20 - Height: 5' 8" - Born: Maryland - Trade: House carpenter - Enlistment date: 24 Jul 1812 - Served as a sergeant in 12th Infantry; commissioned as an ensign on 23 Oct 1813; promoted to 3rd lieutenant on 5 Apr 1814; discharged on 15 Jun 1815; died on 4 Sep 1816.

Wright, Zadock - Sergeant - 14th Infantry - Company: William McIlvane - Enlistment date: 23 May 1812 - Period: 5 Yrs - Pension: Land bounty to William Wright, brother & only heir at law of Zadock Wright - BLW 27167-

By War of 1812 Soc. in MD

160-42 - Killed during the Battle of Lyon's Creek on 19 Oct 1814.

Wright, Zenick - Private - 14th Infantry - Prisoner of War, exchanged on 15 Apr 1814.

Write, Israel - Private - 36th Infantry - Age: 39 - Height: 5' 9" - Enlistment date: 26 Apr 1814 - Place: Georgetown, DC.

Wyatt, Joseph S. - Private - 14th Infantry - Company: William McIlvane.

Wyatt, William - Sergeant - 36th Infantry - Company: Thomas Carbery - BLW 20179-160-12.

Wyatt, William - Private - 38th Infantry - Company: Charles Stansbury - Age: 22 - Height: 5' 8" - Born: Kent Cty, MD - Trade: Plasterer - Enlistment date: 28 Mar 1814 - Place: Georgetown, DC - Period: War - BLW 6217-160-12 - Discharged at Baltimore on 31 May 1815.

Wyatt, William - Private - 38th Infantry - Company: Charles Stansbury - Age: 40 - Height: 5' 6 1/2" - Born: Maryland - Trade: Laborer - Enlistment date: 24 Nov 1814 - Place: Cumberland - Period: War - BLW 6217-160-12 - Discharged at Baltimore on 6 Apr 1815.

Wyman, John - Private - 38th Infantry - Company: Thomas Sangster - Age: 29 - Height: 5' 5 1/2" - Born: Pennsylvania - Trade: Blacksmith - Enlistment date: 10 Nov 1814 - Place: Baltimore - Period: 5 Yrs - Discharged on 21 Jul 1815, invalid.

Wymes, David - Private - 40th Infantry - Company: John Leonard - Age: 33 - Height: 5' 7 1/2" - Born: Arundel, ME - Trade: Laborer - Enlistment date: 2 Dec 1813 - Place: Roxbury, MA - Period: War - Discharged at Portland, ME, on 9 Apr 1815.

Wymes, David - Private - 13th Infantry - Company: Myndert Cox - Age: 32 - Height: 5' 7 1/2" - Born: Annapolis - Trade: Farmer - Enlistment date: 2 Jun 1812 - Place: Canandaigua, NY - Prisoner of War, captured at Queenston Heights, UC, on 3 Oct 1812 and paroled; discharged on 11 Mar 1814.

Wynne, Aaron - Private - 14th Infantry - Company: Thomas Montgomery - Enlistment date: 20 Apr 1812 - Period: 5 Yrs - Killed on the Canadian shore on 27 Nov 1813.

Y

Yantz, John - 38th Infantry - Company: John Buck - Pension: SO-27907, SC-20445.

Yard, Charles - Private - 36th Infantry - Company: Joseph Merrick - Age: 22 - Height: 5' 8" - Born: Barbertown, NJ - Enlistment date: 30 Apr 1813 - Place: Baltimore - Period: 1 Yr.

Yard, Francis - Private - 36th Infantry - Company: Joseph Merrick - Transferred to the U.S. Navy on 5 Nov 1813.

Yates, David - Private - 14th Infantry - Company: David Cummings - Age: 18 - Height: 5' 4" - Born: Georgetown, DC - Trade: Sailor - Enlistment date: 6 Nov 1814 - Place: Snowden's Camp, VA - Period: 5 Yrs - Discharged on 12 May 1815.

Yates, Henry D. - Sergeant - 14th Infantry - Company: David Cummings - Age: 25 - Height: 5' 8" - Born: Charles Cty, MD - Trade: Saddler -

Enlistment date: 23 Jun 1812 - Period: 18 Mos - BLW 118-160-12 - Prisoner of War (Halifax), captured at Beaver Dams on 24 Jun 1813; discharged at Baltimore on 30 Apr 1815.

Yates, Joseph - Private - 38th Infantry - Company: John Mowton - BLW 6256-160-12.

Yohn, John - Private - 5th Infantry - Company: Colin Buckner - Age: 34 - Height: 5' 10" - Born: Baltimore - Trade: Farmer - Enlistment date: 9 Mar 1813 - Place: Baltimore - Period: 5 Yrs - Pension: Heirs received half pay for five years in lieu of land bounty - Died at Sackets Harbor, NY, on 10 Nov 1813.

York, Tobias - Recruit - 26th Infantry - Company: William Bezeau - Race: 30 - Age: 30 - Height: 5' 10" - Born: Maryland - Trade: Laborer - Enlistment date: 13 Oct 1814 - Place: Philadelphia - Period: War - Listed as "Col'd" in his service record; deserted on 11 Nov 1814.

Young, Benjamin - Private - 36th Infantry - Company: Joseph Hook - Age: 19 - Height: 5' 8" - Born: Philadelphia - Trade: Cooper - Enlistment date: 6 Aug 1814 - Place: Georgetown, DC - Period: War - BLW 3156-160-12 - Discharged at Fort Covington, MD, on 30 Mar 1815.

Young, Benjamin - Private - 14th Infantry - Company: Isaac Barnard - Enlistment date: 2 Oct 1812 - Period: 5 Yrs - Pension: Land bounty to William B. Young, son and only heir at law of Benjamin Young - BLW 27542-160-42 - Died on 30 Jun 1813.

Young, Benjamin - Private - 1st Artillery - Company: Heman Fay - Other regiment: Corps of Artillery - Age: 37 - Height: 5' 6 3/4" - Born: England - Trade: Laborer - Enlistment date: 19 Apr 1814 - Place: Annapolis - Period: War - BLW 9664-160-12 - Discharged at Washington, DC, on 9 May 1815.

Young, Caleb - Private - 5th Infantry - Company: Colin Buckner - Age: 21 - Height: 5' 6" - Born: Baltimore - Trade: Shoemaker - Enlistment date: 4 May 1813 - Place: Baltimore - Period: 5 Yrs.

Young, Henry - Private - 36th Infantry - Company: Joseph Merrick - Age: 23 - Height: 5' 9" - Born: Baltimore - Enlistment date: 22 Apr 1813 - Place: Baltimore - Period: 1 Yr.

Young, John - Private - 38th Infantry - Company: John Brookes - Pension: Wife Susannah, WO-24172, WC-25833 - BLW 25820-160-12.

Young, John - Surgeon - 14th Infantry - Born: Maryland - Commissioned as a surgeon, 14th Infantry, on 4 Jun 1812; discharged on 15 Jun 1815.

Young, John - Private - 2nd Artillery - Company: Nathan Towson - BLW 8123-160-12.

Young, John - Private - 2nd Light Dragoons - Company: Samuel Hopkins - Age: 33 - Height: 5' 6" - Born: Fredericktown, MD - Trade: Saddler - Enlistment date: 5 Sep 1812 - Place: Carlisle, PA - Period: 5 Yrs.

Young, John - Private - 36th Infantry - Company: Thomas Carbery - Age: 21 - Height: 5' 11" - Born: Leonardstown, MD - Trade: Laborer - Enlistment date: 23 Mar 1814 - Place: Washington, DC - Period: War - BLW 85-160-12 - Discharged at Baltimore on 31 Mar 1815.

Young, John - Private - 38th Infantry - Company: Shepperd Leakin - Age: 21

- Height: 6' - Born: Frederick, MD - Trade: Laborer - Enlistment date: 20 Feb 1814 - Place: Hagerstown, MD - Period: War - BLW 25820-160-12 - Discharged at Baltimore on 30 Mar 1815.

Young, Joseph - Private - 36th Infantry - Company: James Merrick - Age: 26 - Height: 5' 7" - Born: Newport, RI - Enlistment date: 27 May 1813 - Place: Baltimore - Period: 1 Yr.

Young, Matthias - Private - 19th Infantry - Company: Richard Talbott - Age: 23 - Height: 5' 9 1/2" - Born: Maryland - Trade: Tanner - Enlistment date: 28 Jan 1814 - Place: Franklinton, Franklin Cty, OH - Period: 1 Yr - Discharged on 23 Jan 1815.

Young, Peter - Quarter Gunner - US Sea Fencibles - Company: Simmones Bunbury - Age: 38 - Height: 5' 9 1/2" - Born: Lancaster, PA - Enlistment date: 30 Dec 1813 - Period: 1 Yr - Re-enlisted on 29 Nov 1814 for one year; discharged at Fort McHenry on 24 Mar 1815.

Young, Richard - Musician - 1st Artillery - Company: Hopley Yeaton - Age: 27 - Height: 5' 7 1/4" - Born: Prince George's Cty, MD - Trade: Boat maker - Enlistment date: 24 May 1811 - Place: Fort Nelson, VA - Period: 5 Yrs - Discharged on 28 May 1816 and re-enlisted.

Young, Thomas - Private - 36th Infantry - Company: Thomas Carbery - Age: 24 - Height: 6' 1" - Born: Maryland - Enlistment date: 16 Apr 1814 - Place: Washington, DC - Period: War - Pension: Land bounty to Eleanor Young and other heirs at law of Thomas Young - BLW 4118-160-12 - Died on 6 Dec 1814.

Young, William - Private - 26th Infantry - Company: William Bezeau - Age: 25 - Height: 5' 5" - Born: Fredericktown, MD - Trade: Laborer - Enlistment date: 29 Sep 1814 - Place: Baltimore - Period: War - BLW 4102-160-12 - Discharged at Philadelphia on 23 Mar 1815.

Young, William - Private - 5th Infantry - Company: William Oliver - Age: 26 - Height: 5' 6" - Born: Princess Anne, MD - Enlistment date: 26 Aug 1813 - Place: Philadelphia - Period: 5 Yrs - Discharged at Greenbush, NY, on 24 May 1815, rheumatism and palsy.

Z

Zelhart, Frederick - Artificer - 5th Infantry - Company: Richard Bell - Age: 27 - Height: 5' 7" - Born: Yankstown, MD - Trade: Blacksmith - Enlistment date: 26 Apr 1813 - Place: Hagerstown, MD - Period: 5 Yrs.

Zell, George - Private - 14th Infantry - Company: Reuben Gilder - Enlistment date: 7 Jan 1812 - Period: 18 Mos - Pension: Wife Jane, WO-36166, WC-27363 - Discharged at Plattsburgh, NY, on 6 Jul 1814.

Zimmer, Peter D. - Corporal - 14th Infantry - Company: Samuel Lane - Other regiment: 4th Infantry (New) - Age: 26 - Height: 5' 4" - Born: Hanover, PA - Trade: Saddler - Enlistment date: 4 Aug 1812 - Place: Baltimore - Period: 5 Yrs - Prisoner of War (Halifax).

Zimmerman, Frederick - Private - 17th Infantry - Company: Harris Hickman - Age: 30 - Height: 6' 2" - Born: Maryland - Trade: Laborer - Enlistment date: 14 Mar 1814 - Place: Dayton, OH - Period: War - BLW 8258-160-12

Maryland Regulars in War of 1812

- Discharged at Chillicothe, OH, on 9 Jun 1815.

By War of 1812 Soc. in MD

The Bibliography

Baker II, Harrison Scott, *American Prisoners of War Held at Halifax During the War of 1812*, (Heritage Books, Inc.: Westminster, MD, 2004), volumes 1 and 2.

Cullum, George W., Brevet Major General, *Biographical Register of the Officers and Graduates of the U.S. Military Academy at West Point, N.Y.*, (Houghton, Mifflin and Company: Riverside Press, Cambridge, 1891), volume 1, third edition.

Fredriksen, John C., *The United States Army in the War of 1812*, (McFarland & Company, Inc.: Jefferson, NC, 2009).

Graves, Donald E., *Red Coats & Grey Jackets: The Battle of Chippawa, 5 July 1814*, (Dundurn Press Limited, Toronto, Canada: 1994), Appendices, American Regulars, Militia and Native Warriors killed at Chippawa, page 173.

Graves, Donald E., *Where right and Glory Lead! The Battle of Lundy's Lane, 1814*, (Robin Brass Studio: Toronto, Canada, 1993).

Heitman, Francis B., *Historical Register and Dictionary of the United States Army From Its Organization, September 29, 1789, to March 2, 1903*, Volume I and II, (Genealogical Publishing Company, Baltimore, Maryland: 1994).

Johnson, Eric Eugene, *American Prisoners of War Held at Quebec During the War of 1812*, (Heritage Books, Inc.: Westminster, MD, 2011).

Malcomson, Robert, *Historical Dictionary of the War of 1812*, (Scarecrow Press, Inc.: Lanham, MD, 2006).

Powell, Colonel William H. (US Army), *List of Officers of the Army of the United States from 1779-1900*, (L. R. Hamersly & Co.: New York 1900).

Quimby, Robert S., *The U.S. Army in the War of 1812: An Operational and Command Study*, (Michigan State University Press; East Lansing, MI, 977), volumes 1 and 2.

Pensioners of the U.S., 1818, Invalids & Half Pay Pensioners, Widows & Orphans, (Washington, D.C.: E. De Krafft, 1818).

Maryland Regulars in War of 1812

Records Relating to War of 1812 Prisoners of War, 1812; (National Archives Microfilm Publication M2019); Records of the Adjutant General's Office, 1780's-1917; Record Group 94, compiled by Clair Prechtel-Kluskens; National Archives, Washington, D.C.

Register of Enlistments in the U.S. Army, 1798-1914; (National Archives Microfilm Publication M233, 81 rolls); Records of the Adjutant General's Office, 1780's-1917, Record Group 94; National Archives, Washington, D.C.

Report from the Secretary of War in Obedience to the Resolutions of the Senate of the 5th and 30th of June, 1834, and the 3rd of March, 1835, in Relation to the Pension Establishment of the United States, (1835 Pension Rolls), (Duff Green: Washington, D.C.,1835).

United States. Bureau of Land Management, General Land Office Records. Automated Records Project; *Federal Land Patents*, State Volumes. http://www.glorecords.blm.gov/. Springfield, Virginia: Bureau of Land Management, Eastern States, 2007.

War of 1812 Military Bounty Land Warrants, 1815-1858; National Archives Microfilm Publication M848; Records of the Bureau of Land Management, Record Group 49; National Archives, Washington, D.C.

War of 1812 Pension Applications; National Archives Microfilm Publication M313; Records of the Department of Veterans Affairs, Record Group Number 15; National Archives, Washington D.C.

The Citizen Soldiers at North Point and Fort McHenry, September 12 & 13, 1814, (James Young: Baltimore, 1889), pp. 65-67, Muster Roll of Captain William H. Addison's Sea Fencibles Company, September-October 1814, and Muster Roll of Captain M. Simmones Bunbury's Sea Fencibles Company, July-August 1814 and September-October 1814.

To my family - John, Guy, Joseph and Kayleigh – may this work inspire you to do great things. CC Jr.

Heritage Books by Christos Christou:

Abstracts of Kent County, Maryland Wills, Volume 1: 1777–1816
Christos Christou and John Anthony Barnhouser

Abstracts of Kent County, Maryland Wills. Volume 2: 1816–1867
Christos Christou and John Anthony Barnhouser

Colonial Families of the Eastern Shore of Maryland, Volume 4
Christos Christou and F. Edward Wright

Heritage Books by the Society of the War of 1812 in the State of Maryland:

Maryland Regulars in the War of 1812
Transcribed by Eric Eugene Johnson; Foreword by Christos Christou

Heritage Books by the Society of the War of 1812
in the State of Ohio:

Transcribed by Harrison Scott Baker

*American Prisoners of War Held at Bermuda,
Cape of Good Hope and Jamaica During the War of 1812*

*American Prisoners of War Held at Barbados,
Newfoundland and New Providence During the War of 1812*

*American Prisoners of War Held at Halifax
During the War of 1812, Volume I and II*

Transcribed by Eric Eugene Johnson

American Prisoners of War Held at Dartmoor During the War of 1812

*American Prisoners of War Held in Montreal
and Quebec During the War of 1812*

*American Prisoners of War Held at Quebec
During the War of 1812, 8 June 1813–11 December 1814*

*American Prisoners of War Paroled at Dartmouth,
Halifax, Jamaica and Odiham During the War of 1812*

Black Regulars in the War of 1812

Black Regulars and Militiamen in the War of 1812

Ohio and the War of 1812: A Collection of Lists, Musters and Essays

Ohio's Regulars in the War of 1812

www.ingramcontent.com/pod-product-compliance
Lightning Source LLC
Chambersburg PA
CBHW081343230426
43667CB00017B/2710